MODULATION, NOISE, AND SPECTRAL ANALYSIS

Applied to Information Transmission

PHILIP F. PANTER, Ph. D.

Director, Guidance and Control Laboratory
ITT Federal Laboratories

McGRAW-HILL BOOK COMPANY

New York   San Francisco   Toronto   London   Sydney

MODULATION, NOISE, AND SPECTRAL ANALYSIS

3456789 HD 987

48446

PREFACE

This book is an outgrowth from a set of lecture notes on modulation theory which were prepared for the evening education program at the ITT Federal Laboratories. The subject is also offered in the graduate school of Newark College of Engineering in a two-term course entitled Modulation Theory and Communication Systems Design.

This text is intended for graduate students and practicing engineers who are interested in obtaining a thorough understanding of the modern concepts of modulation theory and who wish to develop the ability to keep up with the state of the art of communication systems. Special attention is given to the order of presentation to effect a logical and understandable evolution of the material. The book covers the fundamental principles of modulation theory and modulation systems which are commonly used in the field of communication. The subject is treated on a unified basis consistent with modern statistical theory of communication. Basic concepts common to all modulation systems are stressed and quantitative criteria are developed by which the performance of these systems can be measured. The fundamental role of system bandwidth and noise in transmission of information is emphasized. In order to render the book self-sufficient, the essential prerequisite mathematical background is included in several introductory chapters. Some of the mathematical review topics are application of Fourier analysis to spectral problems, transmission of periodic and transient signals through linear systems, random signal theory and random processes, with special application to noise problems which are commonly encountered in modulation systems. The concept of generalized harmonic analysis is introduced to relate the statistical description of a random process to its spectrum. The introductory mathematical topics are specifically geared to the subject matter of "modulation" in such a manner as to provide the reader with a self-contained treatise without constantly referring him to other sources. This feature should be of great help to practicing engineers who require a ready reference to these mathematical tools.

In broad outline, following the introductory mathematical material, the topics covered are:

1. Linear modulation systems, including amplitude modulation, double- and single-sideband modulation, and vestigal sideband modulation. The concept of the analytic signal and its application to the single sideband is introduced.

2. Frequency modulation (FM). Among the topics discussed are spectral analysis, including combined amplitude and frequency modulation; distortion of frequency-modulated signals through linear systems, using both the Fourier and asymptotic method; transient response in FM, interference in FM reception, including common and adjacent-channel interference and multipath transmission; generation, detection, and signal-to-noise improvement in FM systems. The application of the principle of frequency-compressive feedback to FM systems is introduced and the advantages of its use in ranging systems are discussed. The subject of threshold extension, which has recently gained popularity in the design of communication system, is then introduced. Frequency-compressive feedback and phase-locked loop demodulation are presented in this connection.

3. Pulse modulation. Some of the topics covered are the sampling principle; spectral analysis of modulated pulses; time-division multiplex systems, with particular consideration being given to the problem of bandwidth; signal-to-noise ratio; and crosstalk. The basic concepts of information theory are then discussed, with special emphasis on the application to pulse-code modulation. This is followed by a comprehensive treatise on pulse-code modulation systems including quantization noise, fluctuation noise output, signal-to-noise improvement ratio, and the use of pulse regeneration in PCM systems. The rate of transmission over a noisy channel is also considered. The basic principles of various delta modulation systems are then discussed, including a comparison between delta modulation and pulse-code modulation from the point of view of information theory and system performance.

4. Digital data modulation systems. The performance of the various digital data transmission systems is analyzed from the aspect of error rate, system bandwidth, and threshold. Finally a comparison is made of the various modulation systems on the basis of their

figures of merit of performance criteria established for analog and digital modulation systems.

In the selection of material for this book the author has drawn freely from the vast literature on the various topics under discussion. He has endeavored to give credit to the original publications and contributors by referring to the list of references at the end of each chapter. No attempt has been made to provide a complete bibliography. The author is grateful to the publishers of the numerous books and magazines for permission to use selected reference material and reproduce pertinent graphs and diagrams. Any inadvertent omission of acknowledgment will be gladly corrected in future printings in response to notes from the contributors.

The author is indebted to many of his associates and to the management of the ITT Federal Laboratories for encouragement in the preparation of the manuscript. He is also gratefully indebted to Mrs. Gilda Edwards, who typed the rough draft in its many phases of evolution, as well as the final manuscript, while carrying out her normal secretarial duties.

Because of the size and scope of the subject, it has been impossible to cover all aspects of modulation theory. In fact, the author has been faced with many difficult decisions in the selection of the topics in order to offer a self-contained, extensive, and up-to-date coverage while keeping a reasonable limitation on size.

Philip F. Panter

CONTENTS

Preface   iii

1 METHODS OF MODULATION AND EVOLUTION OF MODULATION
  SYSTEMS   1

2 MATHEMATICAL BACKGROUND I: REVIEW OF FOURIER SERIES
  AND FOURIER TRANSFORM   8

3 SIGNAL TRANSMISSION THROUGH LINEAR SYSTEMS   63

4 MATHEMATICAL BACKGROUND II: RANDOM SIGNAL THEORY —
  APPLICATION TO NOISE PROBLEMS.   101

5 LINEAR MODULATION SYSTEMS.   171

6 LINEAR DEMODULATION OR DETECTION   206

7 EXPONENTIAL MODULATION: BASIC PRINCIPLES AND
  SPECTRAL DISTRIBUTION.   236

8 DISTORTION OF FREQUENCY-MODULATED SIGNALS THROUGH
  LINEAR SYSTEMS: FOURIER METHOD   273

9 DISTORTION OF FREQUENCY-MODULATED SIGNALS THROUGH
  LINEAR SYSTEMS: DYNAMIC OR ASYMPTOTIC METHOD   301

10 TRANSIENT RESPONSE IN FREQUENCY MODULATION   325

11 INTERFERENCE IN FREQUENCY-MODULATION RECEPTION:
   COMMON- AND ADJACENT-CHANNEL INTERFERENCE AND
   MULTIPATH TRANSMISSION   350

12 GENERATION OF FREQUENCY-MODULATED SIGNALS   381

13 DETECTION OF FREQUENCY-MODULATED SIGNALS   405

14 SIGNAL-TO-NOISE IMPROVEMENT IN FM SYSTEMS   427

15 APPLICATION OF NEGATIVE FEEDBACK TO FREQUENCY-
   MODULATED SYSTEMS   461

vii

16 THRESHOLD EXTENSION IN FREQUENCY-MODULATION RECEIVERS 478

17 THE SAMPLING PRINCIPLE AND INTRODUCTION TO PULSE MODULATION 505

18 TIME-DIVISION MULTIPLEX SYSTEMS (TDM) 548

19 INTRODUCTION TO INFORMATION THEORY — WITH SPECIAL APPLICATION TO PCM 590

20 PRINCIPLES OF PULSE-CODE MODULATION (PCM) 619

21 OUTPUT SIGNAL-TO-NOISE IMPROVEMENT RATIO IN PCM SYSTEMS 653

22 DELTA MODULATION (DM) 679

23 DIGITAL DATA MODULATION SYSTEMS 700

24 COMPARATIVE ANALYSIS OF MODULATION SYSTEMS 737

Index 751

1

METHODS OF MODULATION AND EVOLUTION OF MODULATION SYSTEMS[1-3]

All information-carrying signals must be transmitted over some medium separating the transmitter from the receiver. At the transmitting end, the message is rarely produced in a form suitable for direct transmission over the medium. Efficiency of transmission requires that this information be processed in some manner before being transmitted over the intervening medium. Modulation may be generally defined as the process whereby a message is transformed from its original form into a signal that is more suitable for transmission over the medium between the transmitter and receiver. The process of modulation may be realized by use of a high-frequency carrier, varying one of its parameters as a linear function of the instantaneous value of the message to be transmitted. Mathematically, modulation may be considered as a process of mapping from a message space to a signal space. At the receiver, this process is reversed by demodulation methods. It will be shown that the transformation from message space into signal space is equivalent to a translation of spectrum. In linear systems, like amplitude modulation, the spectral components which specify the message are merely translated without any change in their relative energy distribution, while in nonlinear systems, like frequency modulation,

1

such a process involves the generation of new frequencies and a different energy distribution of the message-carrying spectral components.

In this chapter, we shall outline some of the modulation methods and modulation systems which are being used to meet increasing communication requirements.

1.1 Processing of Message for Efficient Transmission

We have noted previously that at the transmitter end the message is normally in a form not suitable for direct transmission over the medium, and consequently the message must be processed before its propagation over the transmitting medium. This process may involve the following operations: (1) coding—to minimize the effects of disturbing noise sources which are inherent in the communication system, and (2) modulation—to translate the message spectrum into a frequency region which is more suited for efficient propagation and for meeting the overall transmission requirements of the system. The modulation process also permits multiplex transmission over a common medium, i.e., simultaneous transmission of several different messages having overlapping spectra over the same transmission medium.

Two common methods are used for multiplex transmission: these are frequency-division multiplexing and time-division multiplexing. In a frequency-division multiplex system, each message modulates a high-frequency sinusoidal carrier, and all carriers are transmitted simultaneously over the same medium. At the receiver, the carriers are separated by means of selective filters, and the messages are then recovered by the process of demodulation. In time-division multiplex systems, each message is used to pulse a carrier. However, each pulsed carrier is allocated a different time interval for its transmission, and thus at each instant of time only one carrier is being transmitted. In time-division multiplexing each channel utilizes the entire frequency spectrum which is occupied by the system, whereas in frequency division the spectrum of the system is the composite of the spectra of the individual modulation channels.

In some systems, especially in telemetry, multiple modulation operations are used, in which each message modulates corresponding carriers suitably separated in frequency. The complex spectrum of the so-called subcarriers is used to modulate the final carrier which is transmitted over the transmission medium.

1.2   Evolution of Various Modulation Systems

1. *Continuous-wave* *(CW)* *Modulation.*   The process of modulation
implies a change in some parameters of the carrier by the message
signal. We usually refer to the carrier whose parameter is being
varied linearly with the value of the message function as the modulated wave, while the modulating wave is defined as a signal which
carries the message specification.

Early experimenters were interested primarily in amplitude-
modulation (AM) and single-sideband (SSB) systems. The sinusoidal
carrier was found to be easy to generate, modulate, and transmit.
The allocation of AM broadcasting channels 10 kc apart, which were
rapidly used up commercially, gave an impetus to many experimenters in the field of communication to experiment with other modulation methods in the hope of narrowing the broadcasting channels.
Frequency modulation (FM) was considered as a possible scheme
for accommodating many channels about 100 cps wide within the
broadcast band. In 1922, J. R. Carson published his famous mathematical paper[4] in which he corrected the narrow-band fallacy by
proving that frequency modulation would not provide a narrower
band than amplitude modulation, but might actually require a wider
band. His analysis led him also to believe that FM inherently produced distortion in the signal. While Carson was correct in his conclusion about the narrow-band fallacy, his general statement about
noise-reducing schemes proposed subsequently by E. H. Armstrong
was too hasty and was in a great measure responsible for the tem-
porary setback in the development of FM systems. In a paper published in 1928,[5] in which he analyzed a method proposed by
Armstrong[6] for reduction of atmospheric disturbances, Carson
states that "In fact, as more and more schemes are analyzed and
tested and as the essential nature of the problem is more clearly
perceived, we are unavoidably forced to the conclusion that static,
like the poor, will always be with us." It appears, however, that
Carson was in error on both scores. In order to minimize the disturbances due to random noise, Armstrong conceived the idea of
wideband frequency modulation. He reasoned that the effect of random noise is primarily to amplitude-modulate the carrier without
producing consistently large frequency deviations. Thus a signal
having a large-frequency deviation should easily be separable from
random noise, provided that the receiver is designed to be insensitive to amplitude variations. This reasoning led him to the invention
of wideband FM in combination with an amplitude limiter-discriminator in the receiver. The substantial reduction of noise in a FM

receiver by the use of a limiter was indeed a startling discovery, contrary to the behavior of AM systems, because experience with such systems had shown that the noise contribution to the modulation of the carrier could not be eliminated without partial elimination of the message.

The principle of improving system performance and increasing output signal-to-noise ratio in the presence of noise by widening, rather than narrowing the bandwidth of the modulated carrier, gave subsequently birth to many other similar modulation schemes, as we shall see later on. However, inherent in all these modulation methods, where bandwidth is traded for signal-to-noise improvement, there exists a high threshold that the signal must exceed in order to have an operational system. The problem of lowering the threshold, which is usually referred to as threshold extension, is of great importance in long-range wideband communication systems, and several schemes are in operation for lowering the threshold, as will be discussed in Chap. 16.

2. *Pulse Modulation.* In pulse-modulation systems, the unmodulated carrier is usually a series of regularly recurrent pulses; information is conveyed by modulating some parameter of the transmitted pulses such as the amplitude, duration, time of occurrence, or shape of pulse. This type of modulation is based on the "sampling principle," which states that a continuous message waveform that has a spectrum of finite width could be recovered from a set of discrete samples whose rate is slightly higher than twice the highest signal frequency. In pulse-amplitude modulation (PAM), the series of periodically recurring pulses are modulated in amplitude by the corresponding instantaneous samples of the message function. In pulse-time modulation (PTM), the instantaneous samples of the message function are used to vary the time of occurrence of some parameter of the pulsed carrier. Pulse-duration and pulse-position modulation are particular forms of pulse-time modulation. In pulse-duration modulation (PDM), or as it is sometimes called pulse-length modulation, the time of occurrence of either the leading or trailing edge of each pulse, or both, is varied from its unmodulated position by the instantaneous samples of the modulating wave. In pulse-position modulation (PPM), the value of each instantaneous sample is used to vary the position in time of a pulse, relative to its unmodulated time of occurrence.

Pulse-amplitude modulation does not offer any significant advantages over AM or FM. On the other hand, pulse-time (PDM and PPM) systems exhibit the feature of trading bandwidth for signal-to-noise ratio characteristic of FM, provided the peak interference is less than the peak signal, which imposes a sharp threshold for acceptable performance, and represents a limitation on the signal-to-noise improvement which may be realized with a given amount of average transmitted power.

Pulse-code modulation (PCM) represents a major breakthrough in the art of communication. As in the other pulse systems, each modulating wave is sampled periodically at a rate in excess of twice its highest-frequency component. However, in PCM, the samples are quantized into discrete steps; i.e., within a specified range of expected sample values, only certain discrete levels are allowed. Each quantized sample is assigned a code pattern of a series of un-modulated pulses which are transmitted by the pulsed carrier. At the receiver end, each code pattern is identified, decoded, and used to produce a voltage proportional to the original quantized sample. In this system, if the noise peaks at the receiver do not exceed one-half of a code pulse, the demodulation process will not introduce any error. In PCM systems, we have two distinct sources of errors: one due to the process of quantization which introduces a maximum error of one-half a quantum step in the value of each transmitted sample, and the second due to noise peaks which may introduce wrong pulses or obliterate some of the pulses in the code pattern.

The quantizing errors, called quantization noise, are a function of the maximum number of steps used in the system and thus may be considered as controllable noise. PCM exhibits the outstanding property of repeated regeneration and transmission without significant accumulation of errors in the pulse-code pattern, for the code can be reconstructed in exactly the same form as it was first generated at the transmitter and retransmitted again for further processing by the next repeater station.

Although PCM is considered to be the most efficient among the existing communication systems, the circuitry used in modulation and demodulation is quite complex. A more recent pulse-modulation system which requires wider bandwidth than PCM but whose circuitry is much simpler is known as delta modulation. Delta modulation is a one-digit PCM system, and like PCM may be classified as a code-modulation system. As in PCM, the range of signal amplitudes is quantized, and binary pulses are produced at the sending end at regular intervals. However, in delta-modulation systems, instead of the absolute quantized signal amplitude being transmitted at each sampling, only the changes in signal amplitude from sampling to sampling instant are transmitted. At each sampling, the

presence or absence of only one transmitted pulse contains the intelligence. Thus in a delta-modulation system, the transmitted
pulses carry the information corresponding to the derivative of the
amplitude of the modulation signal, and at the receiver these pulses
are integrated to obtain the original waveform of the signal. A comparison of the two systems, delta and pulse-code modulation, will
show that delta modulation is more advantageous under certain conditions for voice communication, from the point of view of both
quality and simplicity. However, delta modulation requires much
larger bandwidth than PCM if the desired quality of voice transmission is moderate or high. In order to increase the efficiency of
delta modulation, higher orders of amplitude quantizing must be
used, which in turn will make the system less attractive because
the simplicity of the basic one-digit system may easily be lost.

3. *Compound-modulation Systems.* Combinations of various
modulation methods are in common use, especially in telemetry,
where several subcarriers are frequency-modulated by the message
functions, and then the sum of the resulting FM signals is made to
frequency-modulate the final carrier; this double-modulation technique is known as FM/FM. The number of subcarriers is dependent
upon the capacity requirements of the telemetry system and the
commonly accepted telemetry frequency standards known as the
IRIG (Inter-Range Instrumentation Group) subcarrier frequencies.

In many applications, the subcarriers are pulsed for time-division multiplexing and then used to frequency-modulate the carrier;
this may be PAM/FM, PDM/FM, PPM/FM, or PCM/FM.

In more complex modulation systems a triple-modulation process
may be used such as PAM/FM/FM, PDM/FM/FM, etc. However,
in our discussions we shall be mainly concerned with single-modu-
lation processes.

References

1. Baghdady, E. J. (ed.): ''Lectures on Communication System
   Theory,'' chap. 19, McGraw-Hill Book Company, New York,
   1961.
2. Black, H. S.: ''Modulation Theory,'' chap. 2, D. Van Nostrand
   Company, Inc., Princeton, N. J., 1953.
3. Schwartz, M.: ''Information Transmission, Modulation, and
   Noise,'' chap. 3, McGraw-Hill Book Company, New York, 1959.
4. Carson, J. R.: Notes on the Theory of Modulation, *Proc. IRE*,
   February, 1922.

5. Carson, J. R.: The Reduction of Atmospheric Disturbances, *Proc. IRE*, July, 1928.
6. Armstrong, E. H.: Methods of Reducing the Effect of Atmospheric Disturbances, *Proc. IRE*, January, 1928.

METHODS OF MODULATION

5. Carson, J.: The Reduction of Atmospheric Disturbances,
   Proc. IRE, July, 1928.
6. Armstrong, E.: A Method of Reducing the Effect of Atmospheric Disturbances, *Proc. IRE*, January, 1928.

2

MATHEMATICAL BACKGROUND I: REVIEW OF FOURIER SERIES AND FOURIER TRANSFORM

The study of modulation theory requires the use of rather spec-
ialized analytical tools. To this end, we shall briefly review some
basic concepts associated with the application of Fourier series and
Fourier transforms to the analysis of modulation theory. The
Laplace transform and its derived Z transform will not be reviewed because of their limited use in our future discussions. In
our analysis, these transforms, together with some basic concepts
derived from them, will be used in the conversion from the time
domain to the frequency domain of typical communication signals, a
transformation which is useful in the evaluation of system bandwidth
requirements. Only basic concepts will be presented in this chapter; for complete discussions of these techniques, standard text-
books (see references at end of chapter) should be consulted.

In communication problems, we shall encounter three distinct
types of signals and three distinct and different corresponding types
of spectra. In spite of this basic distinction, the term "spectrum"
is quite often used loosely, without sufficient attention being paid to
the three different types of spectra.

The first type of signal to be discussed is the periodic function
of time, which can be written in the form of Fourier series. A

second type of signal encountered in communication work may be loosely described as the transient. More specifically, any voltage whose square when integrated over all time is finite (such as the impulse response of a linear filter) may be dealt with by means of the Fourier transform. The third type of signal which is encountered in communication problems is the random time function such as amplifier output noise or speech. The first two types, namely, the periodic and the transient signals, will be discussed in this chapter; the third type, the random time function, will be discussed in Chap. 4.

2.1 The Vector Diagram and Spectrum Representation of Complex Signals[1-3]

Before proceeding with the analysis of complex signals in terms of sinusoids of a Fourier series, we shall consider a few aspects of the simple rotating vector diagram and its spectrum representation. This method of presentation of an unmodulated carrier signal will then be applied to more complex signals, containing a number of sinusoidal components. Also, modulated carriers, both amplitude and phase, will be shown to consist of a number of steady-state sinusoidal components which can be represented by the same methods.

1. *The Rotating Vector.* The presentation of a sinusoidal signal by a rotating vector is well known from elementary theory. A sinusoid can be represented by the projection of the rotating vector on the real or imaginary axis, as shown in Fig. 2-1. The vector of length I rotating at a constant angular frequency ω_0 rad/sec can be written as the vector sum of the projections, namely,

$$i(t) = Ie^{j\omega_0 t} = I(\cos \omega_0 t + j \sin \omega_0 t) \qquad (2-1)$$

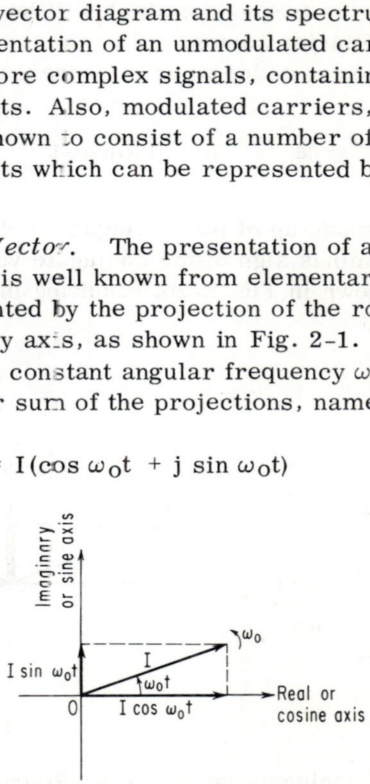

Fig. 2-1. The simple rotating vector.

The real part of the exponential representation of the rotating vector is called the inphase component of the vector, and the imaginary
part is the quadrature component. Thus, $\cos \omega_0 t$ can be written as
$\text{Re} [e^{j\omega_0 t}]$, meaning the real part of the bracketed quantity and
$\sin \omega_0 t$ as $\text{Im} [e^{j\omega_0 t}]$, the imaginary part of the quantity.

A simple alternative way of representing a sinusoid is by means
of conjugate vectors. A cosine wave may be considered as the resultant of two conjugate vectors of magnitude $I/2$, rotating in opposite directions with constant angular frequency $\pm \omega_0$, as shown in
Fig. 2-2a.

$$i(t) = \frac{I}{2} (e^{j\omega_0 t} + e^{-j\omega_0 t}) = I \cos \omega_0 t \qquad (2\text{-}2)$$

Since these conjugate vectors rotate in opposite directions with
angular frequency of $\pm \omega_0$ rad/sec, this representation gives rise to
the concept of negative frequency, which will be used in the exponential form of the Fourier series.

In a similar manner, a sine wave is represented by

$$i(t) = \frac{I}{2} (e^{j\omega_0 t} - e^{-j\omega_0 t}) = jI \sin \omega_0 t \qquad (2\text{-}3)$$

Here again, i(t) is made up of two conjugate vectors of magnitude
$I/2$. However, the minus sign of the conjugate vector indicates a
shift of 180°, as shown in Fig. 2-2b, resulting in the quadrature
component $I \sin \omega_0 t$.

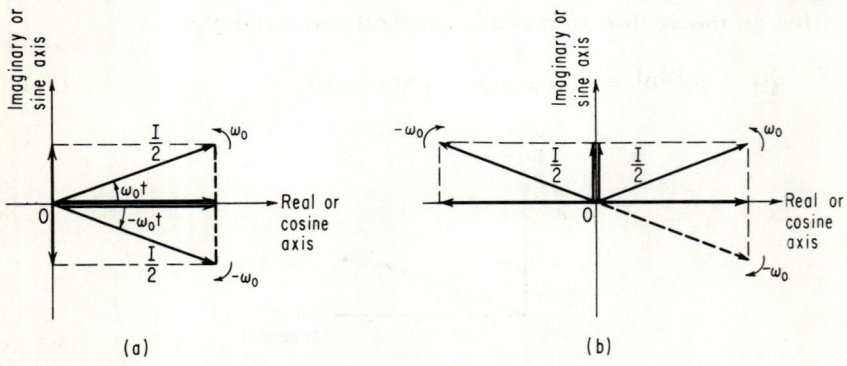

Fig. 2-2. A pair of conjugate vectors: (a) Representation of a cosine
wave; (b) representation of a sine wave.

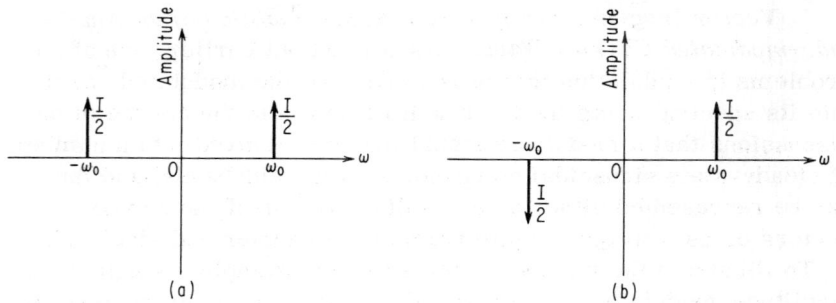

Fig. 2-3. Spectrum representation of a sinusoidal wave: (a) Cosine wave; (b) sine wave.

2. *Spectrum Representation of a Sinusoid.*   The rotating vector may also be represented by a frequency characteristic or spectrum, as a vertical line on a frequency axis. It follows therefore from Eqs. (2-2) and (2-3) that a sinusoid is represented by two spectral lines, as shown in Fig. 2-3. In Fig. 2-3a, we represent the spectrum of a cosine wave which consists of a symmetrical pair of spectral lines located at $\omega = \pm\omega_0$ and of equal amplitudes $I/2$ of the same sign. The spectrum of a sine wave, as shown in Fig. 2-3b, consists of a skew-symmetrical pair, in keeping with the presentation of Eq. (2-3).

Similarly, the spectrum of a complex wave consisting of a number of sinusoids is shown in Fig. 2-4, where the cosine spectrum is symmetrical and the sine spectrum is skew-symmetrical.

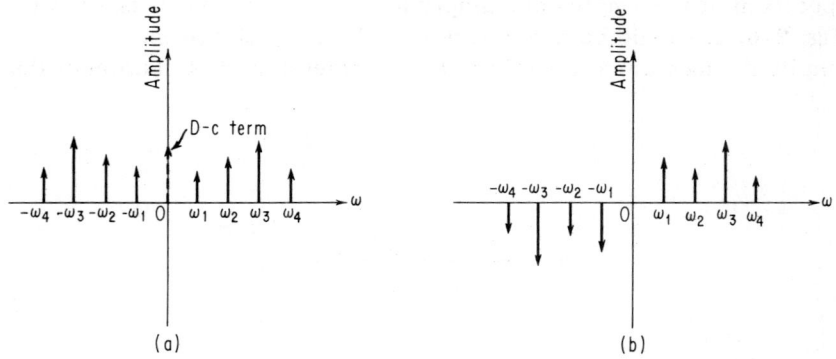

Fig. 2-4. Spectrum of a complex wave: (a) Cosine terms; (b) sine terms.

3. *Vector Diagram and Spectrum Representation of an Ampli-tude-modulated Carrier Wave.*   As pointed out earlier, one of the problems in modulation theory is to resolve the modulated carrier into its spectral components. It will be shown in the course of our discussions that a modulated signal may be resolved into a number of steady-state sinusoidal components called sidebands, and thus can be represented either as a resultant vector of the component vectors or as a frequency spectrum of the carrier and sidebands.

To illustrate the method, consider as an example a single-tone amplitude-modulated carrier signal (the application to more complex modulation forms will be demonstrated in subsequent chapters). This is given by the expression

$$i(t) = I(1 + m_a \cos pt) \cos \omega_c t \qquad (2-4)$$

where ω_c and p are the angular frequency of the modulated carrier and modulating signal, respectively, and m_a is the modulation index. This can be resolved into three steady-state components—the carrier and two sidebands—as follows:

$$i(t) = I \cos \omega_c t + \frac{m_a I}{2} \cos (\omega_c + p)t + \frac{m_a I}{2} \cos (\omega_c - p)t \qquad (2-5)$$

Each steady-state component can be represented by a pair of conjugate vectors. However, the picture is greatly simplified by reducing the whole angular speed of the vectors by the carrier angular frequency ω_c and showing the sideband vectors rotating at speeds $\pm p$ on this stationary vector diagram, as indicated in Fig. 2-5. The spectrum of the single-tone amplitude-modulated signal is shown in Fig. 2-6. It should be noted here that the essential property of amplitude-modulated signals is the symmetrical distribution of the

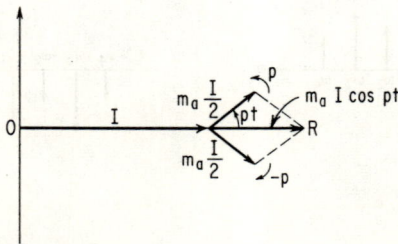

Fig. 2-5. Vector diagram of single-tone amplitude modulation.

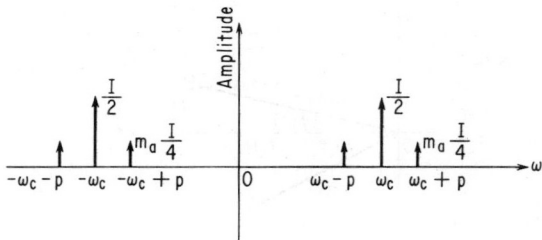

Fig. 2-6. Spectrum of single-tone amplitude-modulated signal.

spectral terms about the origin. This is contrasted with the spectrum of narrow-band phase modulation, which we shall discuss later on.

4. *Vector Representation of General RF Signal.* The concept of a rotating vector to represent a sinusoidal signal can be extended to a general RF signal modulated in both amplitude and phase. Let

$$e(t) = V(t) \cos [\omega_c t + \phi(t)] = \mathrm{Re} \left[V(t) e^{j[\omega_c t + \phi(t)]} \right] \qquad (2\text{-}6)$$

where $V(t)$ represents the envelope of the modulated carrier, and $\phi(t)$ is the modulated phase. This may be expressed in the following form:

$$e(t) = x(t) \cos \omega_c t - y(t) \sin \omega_c t = \mathrm{Re} [E(t) e^{j\omega_c t}] \qquad (2\text{-}7)$$

where $\qquad E(t) = V(t) e^{j\phi(t)} = x(t) + jy(t) \qquad (2\text{-}8)$

Hence $\qquad x(t) = V(t) \cos \phi(t) \qquad$ and $\qquad y(t) = V(t) \sin \phi(t)$

We may represent $e(t)$ as the projection of a rotating vector on a fixed reference axis, as shown in Fig. 2-7. The vector rotates with

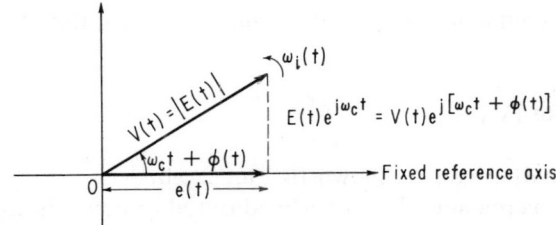

Fig. 2-7. Fixed-reference vector diagram.

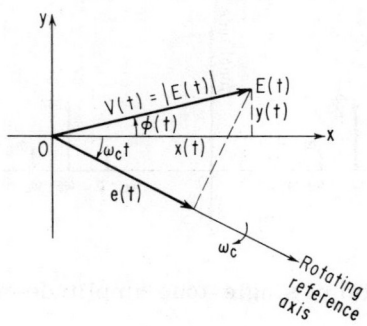

Fig. 2-8. Rotating-reference vector diagram.

an angular frequency $\omega_i(t)$ which is not constant but is given by

$$\omega_i(t) = \omega_c + \frac{d\phi(t)}{dt} \qquad (2-9)$$

As before, we can slow the vector down by subtracting the carrier angular frequency, as shown in the rotating-reference vector diagram of Fig. 2-8. In this presentation, $e(t)$ is given by the projection of the slowed-down vector on the rotating reference. This is often the more useful presentation where the slow variations of the envelope and phase of the modulated carrier signal are shown with respect to the carrier vector which rotates with an angular frequency ω_c rad/sec.

In amplitude modulation, only the amplitude changes and the general expression of Eq. (2-7) reduces to

$$e(t) = \text{Re} \left[V(t) e^{j\phi_0} \cdot e^{j\omega_c t} \right] \qquad (2-10)$$

The rotating reference vector diagram for the AM case is shown in Fig. 2-9a. In frequency or phase modulation, only the phase changes, and we obtain

$$e(t) = \text{Re} \left[V_0 e^{j\phi(t)} \cdot e^{j\omega_c t} \right] \qquad (2-11)$$

and the vector diagram is shown in Fig. 2-9b.

The vector representation of a modulated carrier in amplitude and phase employing the concept of a slowed-down vector on a rotating reference will be used throughout this book.

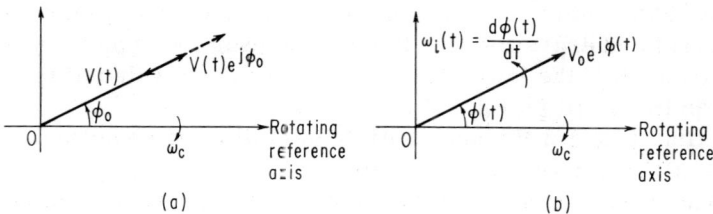

Fig. 2-9. Rotating-reference vector diagram of a modulated carrier:
(a) Amplitude modulation; (b) phase modulation.

2.2   Review of Fourier Series: The Discrete Spectrum[4-6]

The first type of signal to be discussed is the periodic function
of time. Let f(t) be a periodic function of time satisfying certain
conditions. Then f(t) may be represented in the frequency domain
by an infinite number of sinusoidal components which are harmonically related to one another. The magnitude and phase of these frequency components are given by a Fourier series expansion of f(t).

$$f(t) = \frac{a_0}{2} + \sum_{n=1}^{\infty} (a_n \cos n\omega_0 t + b_n \sin n\omega_0 t) \qquad (2\text{-}12)$$

$$\text{where } a_n = \frac{2}{T} \int_{-T/2}^{T/2} f(t) \cos n\omega_0 t \, dt \qquad (2\text{-}13)$$

$$b_n = \frac{2}{T} \int_{-T/2}^{T/2} f(t) \sin n\omega_0 t \, dt \qquad (2\text{-}14)$$

and $\omega_0 = 2\pi/T$, the fundamental frequency. It follows from Eq.
(2-13) that a_0 is given by

$$a_0 = \frac{2}{T} \int_{-T/2}^{T/2} f(t) \, dt \qquad (2\text{-}15)$$

and that $a_0/2$ represents the average value or the d-c term of f(t)
over one complete cycle.

It should be noted here that, in evaluating the coefficients above, the interval of integration should be taken over a complete cycle, and in general, if the lower limit has an arbitrary value of t_0, then the upper limit will be $(t_0 + T)$.

The time function $f(t)$ must satisfy the following conditions:
$f(t) = f(t + T)$; i.e., $f(t)$ is periodic.
$f(t)$ has a finite number of discontinuities (piecewise-continuous):

$$\int_{-T/2}^{T/2} |f(t)|\, dt < \infty$$

These conditions, known as Dirichlet's conditions, are practically always satisfied in physical problems.

Equation (2-12) may conveniently be expressed in the form

$$f(t) = \frac{a_0}{2} + \sum_{n=1}^{\infty} \sqrt{a_n^2 + b_n^2}\ \cos(n\omega_0 t - \theta_n) \qquad (2\text{-}16)$$

where $\theta_n = \tan^{-1}(b_n/a_n)$.

The series expansion as given by Eq. (2-12) reduces to a special form depending on whether $f(t)$ is even or odd. If $f(t)$ is even, namely, $f(-t) = f(t)$, the expression $f(t) \sin n\omega_0 t$ is odd, and b_n is zero for all n; the expansion reduces simply to a cosine series. Similarly, if $f(t)$ is odd, then $f(-t) = -f(t)$, and $f(t) \cos n\omega_0 t$ is odd, a_n is zero, and the result is a sine series. Expressing $f(t)$ as the sum of even and odd components, we have

$$f(t) = f_e(t) + f_o(t) \qquad (2\text{-}17)$$

$$f(-t) = f_e(t) - f_o(t) \qquad (2\text{-}18)$$

and $\quad f_e(t) = \tfrac{1}{2}\,[f(t) + f(-t)] \qquad (2\text{-}19)$

$$f_o(t) = \tfrac{1}{2}\,[f(t) - f(-t)] \qquad (2\text{-}20)$$

Generally, the Fourier series expansion of $f(t)$ will contain both sines and cosines.

The exponential form of the Fourier series is sometimes used, especially when the amplitude and phase characteristic of the discrete spectra are of interest. This alternative form of the series is given by

$$f(t) = \sum_{n=-\infty}^{\infty} C_n e^{jn\omega_o t} = C_o + 2 \sum_{n=1}^{\infty} |C_n| \cos (n\omega_o t - \theta_n) \qquad (2\text{-}21)$$

where C_n is defined as the complex quantity. $C_n = \frac{1}{2}(a_n - jb_n)$; $C_{-n} = \frac{1}{2}(a_n + jb_n) = C_n^*$, the complex conjugate of C_n, which follows from the relations that $a_{-n} = a_n$ and $b_{-n} = -b_n$; and

$$C_n = \frac{1}{T} \int_{-T/2}^{T/2} f(t) e^{-jn\omega_o t} \, dt, \qquad n = 0, \pm 1, \pm 2, \pm 3, \ldots \qquad (2\text{-}22)$$

Equation (2-22) is a representation of the periodic function f(t) in the frequency domain. The function C_n is in general complex and is known as the Fourier transform of f(t); it contains the amplitude and phase characteristics of the complex spectrum which consists of discrete lines. Similarly, Eq. (2-21) is a representation of the inverse transformation of C_n in its time domain.

It follows that

$$C_o = \frac{a_o}{2} \qquad \text{the average value of f(t) over the period T}$$

The amplitude spectrum of the component frequencies is given by

$$|C_n| = \frac{1}{2} \sqrt{a_n^2 + b_n^2} \qquad (2\text{-}23)$$

and the phase is given by

$$\theta_n = \tan^{-1} \frac{b_n}{a_n} \qquad (2\text{-}24)$$

The rms or effective value of the periodic function f(t) may be found as follows:

$$f(t)_{rms} = \sqrt{\frac{1}{T} \int_o^T f^2(t) \, dt} \qquad (2\text{-}25)$$

This expression can be evaluated by the use of Eq. (2-16), namely,

$$f(t)_{rms} = \sqrt{\frac{1}{T} \int_0^T \left[\frac{a_0}{2} + \sum_{n=1}^{\infty} \sqrt{a_n^2 + b_n^2} \, \cos\,(n\omega_0 t - \theta_n) \right]^2 dt}$$

$$(2-26)$$

However, the mathematical manipulation can be simplified by the use of the exponential form of the Fourier series as given by Eq. (2-21), as follows:

$$f(t)_{rms} = \sqrt{\frac{1}{T} \int_0^T f^2(t)\,dt} = \sqrt{\frac{1}{T} \int_0^T f(t) \left[\sum_{n=-\infty}^{\infty} C_n e^{jn\omega_0 t} \right] dt}$$

$$= \sqrt{\sum_{n=-\infty}^{\infty} C_n \frac{1}{T} \int_0^T f(t) e^{jn\omega_0 t}\,dt} = \sqrt{\sum_{n=-\infty}^{\infty} C_n C_{-n}}$$

$$= \sqrt{\sum_{n=-\infty}^{\infty} |C_n|^2} = \sqrt{\frac{a_0^2}{4} + \sum_{n=1}^{\infty} \frac{a_n^2 + b_n^2}{2}}$$

$$(2-27)$$

This relationship is called Parseval's theorem. Similarly, it can be shown that the average value of the product of two functions over a complete period T is given by

$$\frac{1}{T} \int_0^T f_1(t) f_2(t)\,dt = \sum_{n=-\infty}^{\infty} (C_n)_1 \cdot (C_{-n})_2$$

$$(2-28)$$

The physical significance of Parseval's relation is that the effective value of a sum of sinusoids of different frequencies is independent of their phase relationship and is determined only by the effective values of the harmonics. This means that the total average power in a periodic function is equal to the sum of the powers associated with the individual Fourier components.

$$P_{av} = \frac{1}{T} \int_o^T f^2(t) \, dt = C_o^2 + 2 \sum_{n=1}^{\infty} |C_n|^2 = \overline{f^2(t)} \qquad (2\text{-}29)$$

The following examples will serve to illustrate the application of the Fourier series formulas to the analysis of periodic waveforms.

1. *The Square Wave.* This is shown in Fig. 2-10, where the

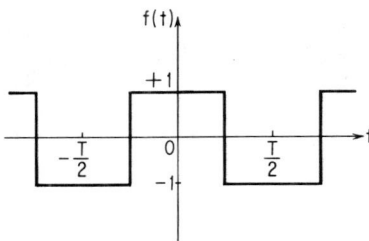

Fig. 2-10. The square-wave symmetric form.

origin has been chosen at the center of the wave to render f(t) even. From the figure, it is seen that

$$f(t) = -1, \qquad -\frac{T}{2} < t < -\frac{T}{4}$$

$$= +1, \qquad -\frac{T}{4} < t < \frac{T}{4}$$

$$= -1, \qquad \frac{T}{4} < t < \frac{T}{2}$$

and $f(t + T) = f(t)$

Since f(t) is even, $b_n = 0$; also since the average value of f(t) over one complete cycle is zero, therefore $a_o = 0$, and the series presentation of f(t) will be a cosine series only. Thus

$$a_n = \frac{2}{T} \int_{-T/2}^{T/2} f(t) \cos n\omega_o t \, dt$$

$$= \frac{2}{T} \int_{-T/2}^{-T/4} (-1) \cos n\omega_0 t \, dt + \frac{2}{T} \int_{-T/4}^{T/4} (+1) \cos n\omega_0 t \, dt$$

$$+ \frac{2}{T} \int_{T/4}^{T/2} (-1) \cos n\omega_0 t \, dt$$

$$= \frac{2/T}{n\omega_0} \left\{ -[\sin n\omega_0 t]_{-T/2}^{-T/4} + [\sin n\omega_0 t]_{-T/4}^{+T/4} - [\sin n\omega_0 t]_{T/4}^{T/2} \right\}$$

$$= \frac{4}{n\pi} \sin \frac{n\pi}{2} \tag{2-30}$$

Therefore $a_n = 0,$ n even

$$= + \frac{4}{n\pi}, \qquad n = 1, 5, 9, \cdots$$

$$= - \frac{4}{n\pi}, \qquad n = 3, 7, 11, \cdots$$

and $$f(t) = \frac{4}{\pi} (\cos \omega_0 t - \tfrac{1}{3} \cos 3\omega_0 t + \tfrac{1}{5} \cos 5\omega_0 t - \cdots) \tag{2-31}$$

2. *The Triangular Wave.*   The triangular wave is illustrated in Fig. 2-11.

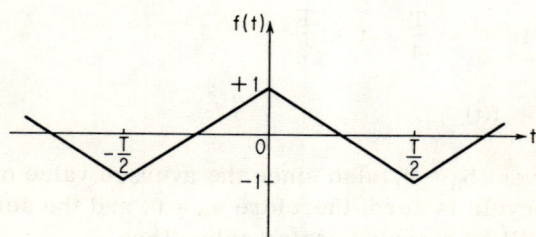

Fig. 2-11. The triangular wave

$$f(t) = 1 + \frac{4}{T} t, \qquad -\frac{T}{2} < t < 0$$

$$= 1 - \frac{4}{T} t, \qquad 0 < t < \frac{T}{2}$$

Since the function is even, $b_n = 0$; also since the average value is zero, $a_o = 0$.

$$a_n = \frac{2}{T} \int_{-T/2}^{T/2} f(t) \cos n\omega_0 t \, dt = \frac{2}{T} \int_{-T/2}^{0} (1 + \frac{4}{T})t \cos n\omega_0 t \, dt$$

$$+ \frac{2}{T \cdot} \int_{0}^{T/2} (1 - \frac{4}{T})t \cos n\omega_0 t \, dt = -\frac{16}{T^2} \int_{0}^{T/2} t \cos n\omega_0 t \, dt$$

$$= \frac{4}{\pi^2 n^2} (1 - \cos n\pi) \qquad\qquad (2\text{-}32)$$

or $\qquad a_n = \frac{8}{\pi^2 n^2}$, $\qquad$ n odd

$\qquad\qquad = 0$, $\qquad\qquad$ n even

and $\quad f(t) = \frac{8}{\pi^2} \left(\cos \omega_0 t + \frac{1}{3^2} \cos 3\omega_0 t + \frac{1}{5^2} \cos 5\omega_0 t + \cdots \right)$ $\quad(2\text{-}33)$

3. *The Sawtooth Wave.* The sawtooth wave is shown in Fig. 2-12.

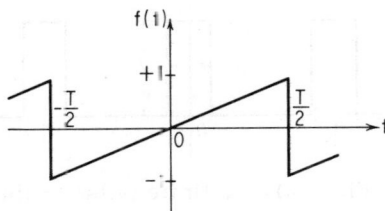

Fig. 2-12. The sawtooth wave.

$$f(t) = \frac{2}{T} t, \qquad -\frac{T}{2} < 0 < \frac{T}{2}$$

$\qquad f(t + T) = f(t)$

since $f(-t) = -f(t)$, the function is odd and $a_n = 0$.

$$b_n = \frac{2}{T} \int_{-T/2}^{T/2} f(t) \sin n\omega_0 t \, dt = \frac{4}{T^2} \int_{-T/2}^{T/2} t \sin n\omega_0 t \, dt$$

$$= \frac{4}{T^2} \left\{ \left[-\frac{1}{n\omega_0} t \cos n\omega_0 t \right]_{-T/2}^{T/2} + \frac{1}{n\omega_0} \int_{-T/2}^{T/2} \cos n\omega_0 t \, dt \right\}$$

$$= -\frac{1}{n\pi} [\cos n\pi + \cos (-n\pi)] = -\frac{2}{n\pi} \cos n\pi \qquad (2\text{-}34)$$

Therefore $b_n = +\dfrac{2}{n\pi}$,   n odd

$$= -\frac{2}{n\pi}, \qquad \text{n even}$$

and $f(t) = \dfrac{2}{\pi} (\sin \omega_0 t - \tfrac{1}{2} \sin 2\omega_0 t + \tfrac{1}{3} \sin 3\omega_0 t - \cdots)$ (2-35)

4. *The Infinite Pulse Train.*   The following examples will demonstrate the usefulness of the complex form of the Fourier series.
Consider an infinite periodic pulse train as shown in Fig. 2-13, of

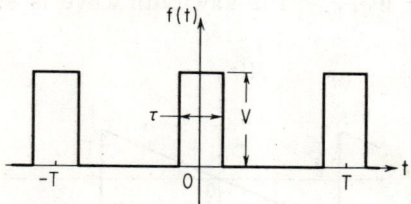

Fig. 2-13. Infinite pulse train.

width τ, period T, and constant voltage V; the origin has been
chosen to coincide with the center of the pulse. From Eq. (2-22),

$$C_n = \frac{1}{T} \int_{-\tau/2}^{\tau/2} V e^{-jn\omega_0 t} \, dt = \frac{1}{T} \left[-\frac{V}{jn\omega_0} e^{-jn\omega_0 t} \right]_{-\tau/2}^{\tau/2}$$

$$= \frac{V}{T} \frac{e^{jn\omega_0 \tau/2} - e^{-jn\omega_0 \tau/2}}{jn\omega_0} = \frac{2V}{n\omega_0 T} \sin n\omega_0 \frac{\tau}{2} \qquad (2\text{-}36)$$

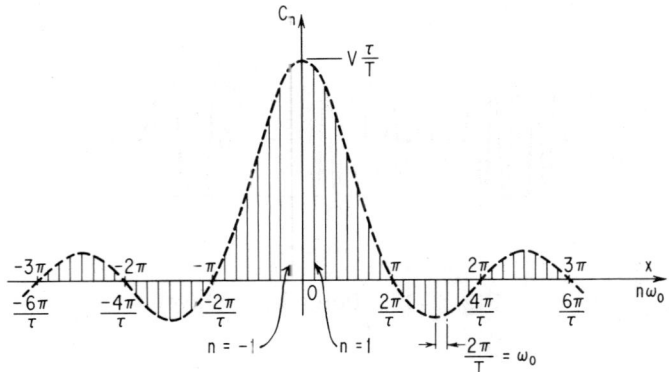

Fig. 2-14. Frequency spectrum of the periodic pulse train.

This may be written in the form

$$C_n = V \frac{\tau}{T} \frac{\sin (n\omega_0\tau/2)}{n\omega_0\tau/2} \qquad (2\text{-}37)$$

which is of the well-known form $(\sin x)/x$, where $x = n\omega_0\tau/2$. The envelope of the plot of C_n is shown in Fig. 2-14 where the spacing between successive lines is $\omega_0 = 2\pi/T$, the fundamental angular frequency, and the zeros are separated by $2\pi/\tau$.

From Eq. (2-21), the exponential form of the Fourier series for the periodic pulse train of Fig. 2-13 is given by

$$f(t) = V \frac{\tau}{T} \sum_{n=-\infty}^{\infty} \frac{\sin (n\omega_0\tau/2)}{n\omega_0\tau/2} e^{jn\omega_0 t}$$

$$= V \frac{\tau}{T} \left[1 + 2 \sum_{n=1}^{\infty} \frac{\sin (n\omega_0\tau/2)}{n\omega_0\tau/2} \cos n\omega_0 t \right] \qquad (2\text{-}38)$$

Only discrete angular frequencies are present, namely, $n\omega_0$.

Note that as the fundamental period T decreases (more pulses per second), the frequency lines move out farther. Alternatively, as T increases, the lines crowd in, and ultimately approach an almost smooth frequency spectrum.

5. *Pulsed RF Wave.* As a second example, consider a pulsed RF wave of angular frequency ω_c as shown in Fig. 2-15, where

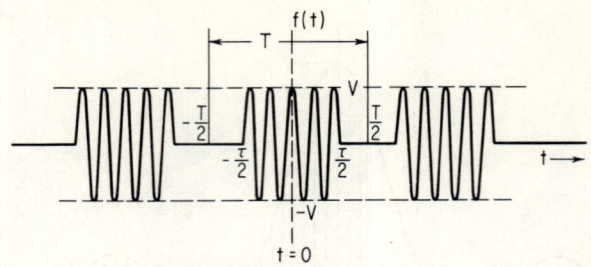

Fig. 2-15. Pulsed RF wave.

$$f(t) = 0, \qquad\qquad -\frac{T}{2} \le t < -\frac{\tau}{2}$$

$$f(t) = V \cos \omega_c t, \qquad -\frac{\tau}{2} \le t \le \frac{\tau}{2}$$

$$f(t) = 0, \qquad\qquad \frac{\tau}{2} < t \le \frac{T}{2}$$

and $\omega_0 = \dfrac{2\pi}{T}$

$$C_n = \frac{1}{T} \int_{-\tau/2}^{\tau/2} f(t) e^{-jn\omega_0 t}\, dt = \frac{V}{T} \int_{-\tau/2}^{\tau/2} \cos \omega_c t e^{-jn\omega_0 t}\, dt$$

$$= \frac{V}{2T} \left[\frac{e^{j(n\omega_0 - \omega_c)\tau/2} - e^{-j(n\omega_0 - \omega_c)\tau/2}}{j(n\omega_0 - \omega_c)} \right.$$

$$\left. + \frac{e^{j(n\omega_0 + \omega_c)\tau/2} - e^{-j(n\omega_0 + \omega_c)\tau/2}}{j(n\omega_0 + \omega_c)} \right]$$

$$= V \frac{\tau}{2T} \left[\frac{\sin (n\omega_0 - \omega_c)\tau/2}{(n\omega_0 - \omega_c)\tau/2} + \frac{\sin (n\omega_0 + \omega_c)\tau/2}{(n\omega_0 + \omega_c)\tau/2} \right] \qquad (2\text{-}39)$$

The frequency spectrum of the pulsed RF wave is shown in Fig. 2-16. Equation (2-39) is identical with Eq. (2-37) except for the symmetrical translation of the original frequency components $n\omega_0$ to both sides of the origin, namely, $(n\omega_0 - \omega_c)$ and $(n\omega_0 + \omega_c)$.

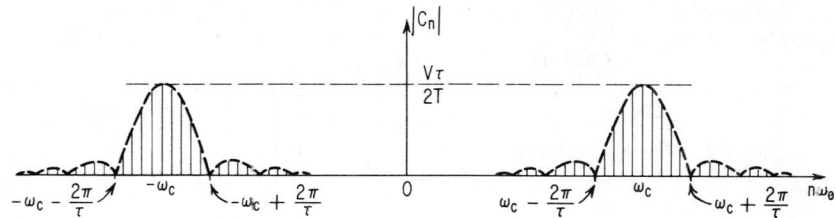

Fig. 2-16. Frequency spectrum of pulsed RF wave.

The last example may be generalized as follows. Let

$$f(t) = V(t) \cos \omega_c t \tag{2-40}$$

where $V(t)$ is the envelope of the RF cosine wave, and $f(t)$ is periodic with period T. Therefore

$$C_n = \frac{1}{T} \int_0^T V(t) \cos \omega_c t \, e^{-jn\omega_o t} \, dt$$

$$\tag{2-41}$$

$$= \frac{1}{2T} \int_0^T V(t) e^{-j(n\omega_o - \omega_c)t} \, dt + \frac{1}{2T} \int_0^T V(t) e^{-j(n\omega_o + \omega_c)t} \, dt$$

Both integrals represent a translation of the spectrum of $V(t)$ about $\pm \omega_c$.

In summary, a periodic function can be said to have a power spectrum which takes the form of a series of discrete lines at harmonically related frequencies. Thus, if a filter is used to measure the power spectrum of a periodic function and if this filter passband is placed about one of the harmonic frequencies, a certain amount of filter output power will be measured, regardless of how narrow the filter bandwidth is made. Thus for periodic functions we have finite amounts of power at discrete frequencies. The Fourier series may be used to represent nonperiodic functions over any selected time period. Of course, the Fourier series expression in such a case will represent a nonperiodic function only in the chosen interval of time and not elsewhere.

2.3   Correlation, Autocorrelation, and Cross-correlation of Periodic Functions [7]

Consider the periodic functions $f_1(t)$ and $f_2(t)$ of the same period T whose Fourier transforms are $(C_n)_1$ and $(C_n)_2$, respectively. Using Eq. (2-22), we obtain the following relations:

$$(C_n)_1 = \frac{1}{T} \int_{-T/2}^{T/2} f_1(t) e^{-jn\omega_0 t}\, dt \qquad (2\text{-}42)$$

$$\text{and} \quad (C_n)_2 = \frac{1}{T} \int_{-T/2}^{T/2} f_2(t) e^{-jn\omega_0 t}\, dt \qquad (2\text{-}43)$$

Form the expression

$$R_{12}(\tau) = \frac{1}{T} \int_{-T/2}^{T/2} f_1(t) f_2(t + \tau)\, dt \qquad (2\text{-}44)$$

where τ is a continuous time displacement in the range $(-\infty, \infty)$ independent of t. The expression $R_{12}(\tau)$ is of considerable interest in the general theory of harmonic analysis. We shall presently prove that its Fourier transform is $(C_n)_1^* \cdot (C_n)_2$.

$$R_{12}(\tau) = \frac{1}{T} \int_{-T/2}^{T/2} f_1(t) f_2(t + \tau)\, dt$$

$$= \frac{1}{T} \int_{-T/2}^{T/2} f_1(t) \left[\sum_{n=-\infty}^{\infty} (C_n)_2 e^{jn\omega_0(t + \tau)} \right] dt$$

$$= \sum_{n=-\infty}^{\infty} \left[(C_n)_2 e^{jn\omega_0 \tau} \cdot \frac{1}{T} \int_{-T/2}^{T/2} f_1(t) e^{jn\omega_0 t}\, dt \right]$$

$$= \sum_{n=-\infty}^{\infty} \left[(C_n)_1^* \cdot (C_n)_2 \right] e^{jn\omega_0 \tau} \tag{2-45}$$

Equation (2-45) is of the form

$$R_{12}(\tau) = \sum_{n=-\infty}^{\infty} C_n e^{jn\omega_0 \tau} \tag{2-46}$$

where $R_{12}(\tau) = \dfrac{1}{T} \displaystyle\int_{-T/2}^{T/2} f_1(t)_1 f_2(t + \tau)\, dt$

is a periodic function of period T with a complex spectrum $C_n = (C_n)_1^* \cdot (C_n)_2$. Therefore

$$C_n = (C_n)_1^* \cdot (C_n)_2 = \dfrac{1}{T} \int_{-T/2}^{T/2} R_{12}(\tau) e^{-jn\omega_0 \tau}\, d\tau \tag{2-47}$$

where $R_{12}(\tau)$ is evaluated first from Eq. (2-44). Thus we have shown that $R_{12}(\tau)$ and $C_n = (C_n)_1^* \cdot (C_n)_2$ form a Fourier transform pair. This relation is called the correlation theorem for periodic functions, and the expression of Eq. (2-44) is called the correlation function.

We shall see later on that the concept of correlation is of the utmost importance in the treatment of transient and random functions.

In Eq. (2-45), we consider the special case where $f_1(t) = f_2(t) = f(t)$, and we obtain

$$\dfrac{1}{T} \int_{-T/2}^{T/2} f(t) f(t + \tau)\, dt = \sum_{n=-\infty}^{\infty} |C_n|^2 e^{jn\omega_0 \tau} \tag{2-48}$$

If we further consider the special case for $\tau = 0$, then Eq. (2-48) reduces to

$$\dfrac{1}{T} \int_{-T/2}^{T/2} f^2(t)\, dt = \sum_{n=-\infty}^{\infty} |C_n|^2 \tag{2-49}$$

which is Parseval's relation of the last section. The left-hand side of Eq. (2-48) is called the autocorrelation function of f(t) and is denoted by $R(\tau)$. Thus

$$R(\tau) = \frac{1}{T} \int_{-T/2}^{T/2} f(t)f(t + \tau) \, dt \qquad (2\text{-}50)$$

We denote the power spectrum $|C_n|^2$ by $S(n\omega_0)$, and we obtain from Eq. (2-48) the relation

$$R(\tau) = \sum_{n=-\infty}^{\infty} S(n\omega_0)e^{jn\omega_0\tau} \qquad (2\text{-}51)$$

and inversely

$$S(n\omega_0) = \frac{1}{T} \int_{-T/2}^{T/2} R(\tau)e^{-jn\omega_0\tau} \, d\tau = |C_n|^2 \qquad (2\text{-}52)$$

Equations (2-51) and (2-52) are the mathematical representation of the autocorrelation theorem for periodic functions, which states that the autocorrelation function and the power spectrum of a periodic function are Fourier transforms of each other. We also note here that since the power spectrum of a periodic function is independent of the phase angles of its harmonics, it follows that periodic functions having the same harmonic amplitudes but differing in their phase angles have the same autocorrelation function.

From Eq. (2-48) we conclude that the autocorrelation function $R(\tau)$ is an even function of τ. Thus

$$R(-\tau) = R(\tau) \qquad (2\text{-}53)$$

$$R(\tau) = \sum_{n=-\infty}^{\infty} S(n\omega_0) \cos n\omega_0\tau$$

$$= S(0) + 2 \sum_{n=1}^{\infty} S(n\omega_0) \cos n\omega_0\tau \qquad (2\text{-}54)$$

and $S(n\omega_0) = \frac{1}{T} \int_{-T/2}^{T/2} R(\tau) \cos n\omega_0 \tau \, d\tau$

(2-55)

The cross-correlation between two periodic functions $f_1(t)$ and $f_2(t)$ of the same fundamental frequency is similarly defined as

$$R_{12}(\tau) = \frac{1}{T} \int_{-T/2}^{T/2} f_1(t) f_2(t + \tau) \, dt$$

(2-56)

and the cross-power spectrum of the function as

$$S_{12}(n\omega_0) = (C_n)_1^* \cdot (C_n)_2$$

(2-57)

which is generally a complex function. Similarly,

$$R_{21}(\tau) = \frac{1}{T} \int_{-T/2}^{T/2} f_2(t) f_1(t + \tau) \, dt$$

(2-58)

and $S_{21}(n\omega_0) = (C_n)_2^* \cdot (C_n)_1 = S_{12}^*(n\omega_0)$

(2-59)

It can also be shown that

$$R_{12}(-\tau) = R_{21}(\tau)$$

(2-60)

Since $R_{12}(\tau)$ is a periodic function of the same fundamental frequency as $f_1(t)$ and $f_2(t)$, we have

$$R_{12}(\tau) = \sum_{n=-\infty}^{\infty} S_{12}(n\omega_0) e^{jn\omega_0 \tau}$$

(2-61)

and $S_{12}(n\omega_0) = \frac{1}{T} \int_{-T/2}^{T/2} R_{12}(\tau) e^{-jn\omega_0 \tau} \, d\tau$

(2-62)

$$\text{Also} \qquad R_{21}(\tau) = \sum_{n=-\infty}^{\infty} S_{21}(n\omega_0)e^{jn\omega_0\tau} \qquad\qquad (2\text{-}63)$$

$$\text{and} \quad S_{21}(n\omega_0) = \frac{1}{T}\int_{-T/2}^{T/2} R_{21}(\tau)e^{-jn\omega_0\tau}\,d\tau \qquad\qquad (2\text{-}64)$$

Equations (2-61) and (2-62) form a Fourier transform pair for the cross-correlation of two periodic functions of the same fundamental frequency; similarly Eqs. (2-63) and (2-64).

The application of correlation function to several periodic functions is illustrated by the following examples.

1. Let $f(t) = A \cos(\omega_0 t + \theta)$.

The power spectrum $\qquad\qquad S(n\omega_0) = |C_n|^2 = \dfrac{A^2}{4}, \qquad n = \pm 1$

The autocorrelation function $\qquad R(\tau) = \displaystyle\sum_{n=-\infty}^{\infty} S(n\omega_0) \cos n\omega_0 \tau$

$$= \frac{A^2}{4}\left[\cos(-\omega_0\tau) + \cos \omega_0\tau\right]$$

$$= \frac{A^2}{2} \cos \omega_0\tau$$

Thus the autocorrelation function of a sinusoid is a cosine function of τ, independent of the phase angle θ, as expected.

2. Let $f(t)$ represent a square wave, as shown in Fig. 2-10. Using the equations derived earlier, we have

$$|a_n| = \frac{4}{n\pi}$$

Hence, $\quad S(n\omega_0) = |C_n|^2 = \tfrac{1}{4}\,|a_n|^2 = \dfrac{4}{\pi^2 n^2}$

The autocorrelation function is given by

$$R(\tau) = 2\sum_{n=1}^{\infty} S(n\omega_0) \cos n\omega_0\tau = \sum_{n=1}^{\infty} \frac{8}{\pi^2 n^2} \cos n\omega_0\tau$$

which represents a Fourier series expansion of a triangular wave, as shown in Fig. 2-11. Thus we have shown that the autocorrelation function of a symmetrical square-wave time function is given by a triangular-wave function of τ. For additional interesting examples, especially illustrating cross-correlation, the reader is referred to Lee.[7]

2.4   The Fourier Transform: The Continuous Spectrum[4,5,8]

Nonperiodic functions can best be analyzed by the use of the Fourier integral which is the extension of the Fourier series. It is a very powerful tool in the analysis of all types of linear systems and provides a clear picture of the physical situation in a practical problem. In the following we shall review some of the basic properties of the Fourier transform and illustrate its use.

Consider a nonperiodic function $f(t) = f_1(t) + jf_2(t)$ of a real variable t which satisfies the condition

$$\int_{-\infty}^{\infty} |f(t)| \ dt < \infty$$

Such a function may be transformed from the time domain to the frequency domain and vice versa by the following transform pair.

$$f(t) = \frac{1}{2\pi} \int_{-\infty}^{\infty} F(j\omega)e^{j\omega t} \ d\omega \qquad (2\text{-}65)$$

$$F(j\omega) = \int_{-\infty}^{\infty} f(t)e^{-j\omega t} \ dt \qquad (2\text{-}66)$$

The function $F(j\omega)$ is called the Fourier transform of $f(t)$, and $f(t)$ is called the inverse Fourier transform of $F(j\omega)$. The integral in Eq. (2-66) analyzes the time function into a continuous spectrum, whereas the integral in Eq. (2-65) synthesizes the spectrum back into a time function. The duality concept between frequency and time domains in the analysis of signals and linear networks is very useful and will be utilized extensively throughout the book.

In a physical system, $f(t)$ is a real function of time, and $F(j\omega)$ is, in general, a complex function of ω and may be written as

$$F(j\omega) = |F(j\omega)| \ e^{j\,\theta(\omega)} = R(\omega) + jX(\omega) = A(\omega) \ e^{j\theta(\omega)} \qquad (2\text{-}67)$$

where $A(\omega) = \sqrt{R^2(\omega) + X^2(\omega)}$,   amplitude spectrum

and $\theta(\omega) = \tan^{-1} \dfrac{X(\omega)}{R(\omega)}$,   phase spectrum

Equations (2-65) and (2-66) are sometimes written in a symmetrical form:

$$f(t) = \frac{1}{\sqrt{2\pi}} \int_{-\infty}^{\infty} F(j\omega) \, e^{j\omega t} \, d\omega \qquad (2\text{-}68)$$

$$F(j\omega) = \frac{1}{\sqrt{2\pi}} \int_{-\infty}^{\infty} f(t) \, e^{-j\omega t} \, dt \qquad (2\text{-}69)$$

The transform pair represented by Eqs. (2-65) and (2-66) may be rewritten in the following form:

$$f(t) = \frac{1}{2\pi} \int_{-\infty}^{\infty} [R(\omega) \cos \omega t - X(\omega) \sin \omega t] \, d\omega$$

$$+ \frac{1}{2\pi} \int_{-\infty}^{\infty} [R(\omega) \sin \omega t + X(\omega) \cos \omega t] \, d\omega \qquad (2\text{-}70)$$

$$\text{and } F(j\omega) = R(\omega) + jX(\omega) = \int_{-\infty}^{\infty} f(t) \, (\cos \omega t - j \sin \omega t) \, dt \qquad (2\text{-}71)$$

From these equations, it follows that if $f(t)$ is a real function of time, then

$$R(\omega) = \int_{-\infty}^{\infty} f(t) \cos \omega t \, dt \qquad (2\text{-}72)$$

$$\text{and } \quad X(\omega) = -\int_{-\infty}^{\infty} f(t) \sin \omega t \, dt \qquad (2\text{-}73)$$

It follows from the last two equations that

$$R(-\omega) = R(\omega)$$

$$X(-\omega) = -X(\omega)$$

and therefore $F(-j\omega) = F^*(j\omega)$,   conjugate of $F(j\omega)$

Thus: for $f(t)$ real, $R(\omega)$ is even, $X(\omega)$ is odd, and $F(-\omega)$ is the conjugate of $F(\omega)$. Equation (2-70) reduces to

$$f(t) = \frac{1}{2\pi} \int_{-\infty}^{\infty} [R(\omega) \cos \omega t - X(\omega) \sin \omega t] \, d\omega$$

$$= \frac{1}{\pi} \int_0^\infty [R(\omega) \cos \omega t - X(\omega) \sin \omega t] \, d\omega$$

$$= \frac{1}{\pi} \int_0^\infty A(\omega) \cos [\omega t + \theta(\omega)] \, d\omega \qquad (2\text{-}74)$$

Furthermore, if f(t) is even, namely, $f(-t) = f(t)$, then from Eqs. (2-72) to (2-74), we obtain

$$R(\omega) = 2 \int_0^\infty f_e(t) \cos \omega t \, dt, \qquad \text{even function of } \omega$$

$$X(\omega) = 0$$

$$f_e(t) = \frac{1}{\pi} \int_0^\infty R(\omega) \cos \omega t \, d\omega \qquad (2\text{-}75)$$

Thus the complex transform $F(j\omega)$ reduces to a real function of ω. If f(t) is odd, $f(-t) = -f(t)$, and

$$R(\omega) = 0$$

$$X(\omega) = -2 \int_0^\infty f_o(t) \sin \omega t \, dt, \qquad \text{odd function of } \omega$$

$$f_o(t) = -\frac{1}{\pi} \int_0^\infty X(\omega) \sin \omega t \, d\omega \qquad (2\text{-}76)$$

The transform $F(j\omega)$ becomes pure imaginary.
Since from Eqs. (2-19) and (2-20), we have

$$f_e(t) = \frac{f(t) + f(-t)}{2}$$

$$f_o(t) = \frac{f(t) - f(-t)}{2}$$

it follows that for the class of functions where $f(t) = 0$ for $t < 0$ we obtain

$$f(t) = 2f_e(t) = 2f_0(t), \qquad t > 0$$

and $\quad f(t) = \dfrac{2}{\pi} \displaystyle\int_0^\infty R(\omega) \cos \omega t \, d\omega = \dfrac{-2}{\pi} \displaystyle\int_0^\infty X(\omega) \sin \omega t \, d\omega \quad$ (2-77)

2.5 Some General Properties of the Fourier Transform[4-6,8]

In this section, some useful theorems are derived directly from Eqs. (2-65) and (2-66), and their physical significance is emphasized. These theorems will be found useful in the application of the Fourier transform to the problem of transmission of nonperiodic signals through linear networks.

1. *The Superposition Theorem.*   Let $F_1(j\omega)$ and $F_2(j\omega)$ be the Fourier transforms of $f_1(t)$ and $f_2(t)$, respectively. Then

$$f_1(t) + f_2(t) = \frac{1}{2\pi} \int_{-\infty}^\infty [F_1(j\omega) + F_2(j\omega)] \, e^{j\omega t} \, d\omega \qquad (2\text{-}78)$$

This follows directly from the definition of the Fourier transform. It follows therefore that the spectrum of the sum of two waveforms is equal to the sum of their individual spectra.

2. *Change of Sign of the Variable*

If $\qquad F(j\omega) = \displaystyle\int_{-\infty}^\infty f(t)e^{-j\omega t} \, dt$

then $\quad F(-j\omega) = \displaystyle\int_{-\infty}^\infty f(-t)e^{-j\omega t} \, dt \qquad\qquad\qquad$ (2-79)

This result is obtained directly from Eqs. (2-65) and (2-66) by replacing the variable ω with $-\omega$, t with $-t$, dt with $-dt$, and changing the signs of the infinite limits.

3. *Translation of the Variable: The Shift Theorem*

If $\qquad F(j\omega) = \int_{-\infty}^{\infty} f(t)e^{-j\omega t}\, dt$

then $\quad F(j\omega)e^{\pm j\omega t_0} = \int_{-\infty}^{\infty} f(t \pm t_0)e^{-j\omega t}\, dt$ $\qquad\qquad$ (2-80)

Proof:

$$\frac{1}{2\pi}\int_{-\infty}^{\infty} F(j\omega)e^{\pm j\omega t_0}e^{j\omega t}\, d\omega = \frac{1}{2\pi}\int_{-\infty}^{\infty} F(j\omega)e^{j\omega(t \pm t_0)}\, d\omega = f(t \pm t_0)$$

The physical significance of this expression is that if the spectrum undergoes modification by an admittance $e^{j\omega t_0}$, then the corresponding time function is undisturbed but delayed by the time t_0. Similarly, if

$$F(j\omega) = \int_{-\infty}^{\infty} f(t)e^{-j\omega t}\, dt$$

then $\quad F[j(\omega \mp \omega_0)] = \int_{-\infty}^{\infty} [f(t)e^{\pm j\omega_0 t}]e^{-j\omega t}\, dt$ $\qquad\qquad$ (2-81)

Proof:

$$\int_{-\infty}^{\infty} f(t)e^{\pm j\omega_0 t} \cdot e^{-j\omega t}\, dt = \int_{-\infty}^{\infty} f(t)e^{-jt(\omega \mp \omega_0)}\, dt = F[j(\omega \mp \omega_0)]$$

The significance of this expression is that if a time wave $f(t)$ is made the envelope of a carrier wave $e^{j\omega_0 t}$ (product modulation), the spectrum is placed symmetrically on either side of ω_0.

Thus the Fourier transform of $f(t) \cos \omega_0 t$ is

$$\frac{F[j(\omega - \omega_0)] + F[j(\omega + \omega_0)]}{2}$$

and of $f(t) \sin \omega_0 t$ is

$$\frac{F[j(\omega - \omega_0)] - F[j(\omega - \omega_0)]}{2j}$$

4. *Interchange of Functions. The Duality Theorem.*　　If

$$F(j\omega) = \int_{-\infty}^{\infty} f(t)e^{-j\omega t}\, dt$$

then　$f(\pm j\omega) = \dfrac{1}{2\pi} \int_{-\infty}^{\infty} F(\mp t)e^{-j\omega t}\, dt$　　　　　　　　　(2-82)

This follows directly from the definition of a Fourier transform.
For

$$F(-j\omega) = \int_{-\infty}^{\infty} f(t)e^{j\omega t}\, dt$$

Interchanging ω and t

$$F(-t) = \int_{-\infty}^{\infty} f(j\omega)e^{j\omega t}\, d\omega$$

so that $(1/2\pi)\, F(-t)$ and $f(j\omega)$ form a Fourier pair. Similarly, if

$$f(t) = \frac{1}{2\pi} \int_{-\infty}^{\infty} F(j\omega)e^{j\omega t}\, d\omega$$

$$f(-t) = \frac{1}{2\pi} \int_{-\infty}^{\infty} F(j\omega)e^{-j\omega t}\, d\omega$$

and　$f(-j\omega) = \dfrac{1}{2\pi} \int_{-\infty}^{\infty} F(t)e^{-j\omega t}\, dt$

Thus, if the spectrum of $f(t)$ is $F(j\omega)$, then the spectrum of $F(\pm t)$
is $2\pi f(\mp j\omega)$; if the signal of $F(j\omega)$ is $f(t)$, then the signal of $f(\pm j\omega)$
is $(1/2\pi)\, F(\mp t)$.

5. *Transform of a Product.*　　If

$$F(j\omega) = \int_{-\infty}^{\infty} f(t)e^{-j\omega t}\, dt$$

where $f(t) = f_1(t) \cdot f_2(t)$ and $F_1(j\omega)$, $F_2(j\omega)$ are the spectra of $f_1(t)$ and $f_2(t)$, respectively, then

$$F(j\omega) = \frac{1}{2\pi} \int_{-\infty}^{\infty} F_1(j\omega_1) \cdot F_2[j(\omega - \omega_1)] \, d\omega_1 \qquad (2\text{-}83)$$

Proof:

$$F(j\omega) = \int_{-\infty}^{\infty} f_1(t_1) \cdot f_2(t_1) e^{-j\omega t_1} \, dt_1$$

$$= \int_{-\infty}^{\infty} \left[\frac{1}{2\pi} \int_{-\infty}^{\infty} F_1(j\omega_1) e^{j\omega_1 t_1} \, d\omega_1 \right] f_2(t_1) e^{-j\omega t_1} \, dt_1$$

$$= \frac{1}{2\pi} \int_{-\infty}^{\infty} F_1(j\omega_1) \, d\omega_1 \left[\int_{-\infty}^{\infty} f_2(t_1) e^{-j(\omega - \omega_1)t_1} \, dt_1 \right]$$

$$= \frac{1}{2\pi} \int_{-\infty}^{\infty} F_1(j\omega_1) \cdot F_2[j(\omega - \omega_1)] \, d\omega_1$$

$$= \frac{1}{2\pi} \int_{-\infty}^{\infty} F_1[j(\omega - \omega_1)] \cdot F_2(j\omega_1) \, d\omega_1$$

The physical significance of this result is that the spectrum of the product of two time functions equals $(1/2\pi) \times$ the convolution of their corresponding spectra. Similarly, it can be shown that if

$$F(j\omega) = F_1(j\omega) F_2(j\omega)$$

and $\qquad f(t) = \frac{1}{2\pi} \int_{-\infty}^{\infty} F(j\omega) e^{j\omega t} \, d\omega$

then $\qquad f(t) = \int_{-\infty}^{\infty} f_1(t_1) f_2(t - t_1) \, dt_1$

$$= \int_{-\infty}^{\infty} f_1(t - t_1) f_2(t_1) \, dt_1 \qquad (2\text{-}84)$$

i.e., the spectrum of a convolution of two time functions equals the product of the spectra.

6. *Differentiation with Respect to the Variable.* We shall show that

$$(j\omega)^n F(j\omega) = \int_{-\infty}^{\infty} \frac{df^n(t)}{dt^n} e^{-j\omega t} dt \qquad (2\text{-}85)$$

and

$$(-t)^n f(t) = \frac{1}{2\pi} \int_{-\infty}^{\infty} \frac{d^n F(j\omega)}{d(j\omega)^n} e^{j\omega t} d\omega \qquad (2\text{-}86)$$

This can be shown by direct differentiation of Eqs. (2-65) and (2-66). Differentiating Eq. (2-65) n times with respect to t,

$$\frac{d^n f(t)}{dt^n} = \frac{1}{2\pi} \int_{-\infty}^{\infty} (j\omega)^n F(j\omega) e^{j\omega t} d\omega \qquad (2\text{-}87)$$

Hence $(j\omega)^n F(j\omega)$ is the transform of $d^n f(t)/dt^n$. Differentiating Eq. (2-66) n times with respect to $j\omega$,

$$\frac{d^n F(j\omega)}{d(j\omega)^n} = \int_{-\infty}^{\infty} (-t)^n f(t) e^{-j\omega t} dt \qquad (2\text{-}88)$$

Hence $(-t)^n f(t)$ is the inverse transform of $d^n F(j\omega)/d(j\omega)^n$.

7. *Integration with Respect to the Variable.* We shall show that

$$\frac{1}{j\omega} F(j\omega) = \int_{-\infty}^{\infty} \left[\int_{-\infty}^{t} f(t) \, dt \right] e^{-j\omega t} dt \qquad (2\text{-}89)$$

and

$$-\frac{1}{t} f(t) = \frac{1}{2\pi} \int_{-\infty}^{\infty} \left[j \int_{-\infty}^{\infty} F(j\omega) \right] e^{j\omega t} d\omega \qquad (2\text{-}90)$$

Let $\Phi(j\omega)$ denote the Fourier transform of $\phi(t)$ where $f(t) = d\phi(t)/dt$ or $\phi(t) = \int_{-\infty}^{t} f(t)\, dt$. Therefore, from Eq. (2-85), $j\omega\Phi(j\omega)$ is the Fourier transform of $d\phi(t)/dt = f(t)$, or

$$j\omega\,\Phi(j\omega) = F(j\omega)$$

$$\Phi(j\omega) = \frac{F(j\omega)}{j\omega}$$

We conclude therefore that $F(j\omega)/j\omega$ is the Fourier transform of $\phi(t) = \int_{-\infty}^{t} f(t)\, dt$. Similarly, Eq. (2-90) is derived.

8. *Sine and Cosine Fourier Integrals*

$$\tfrac{1}{2}\,[F(j\omega) + F(-j\omega)] = \int_{-\infty}^{\infty} f(t)\, \cos \omega t\, dt \qquad (2\text{-}91)$$

$$\tfrac{1}{2}\,[F(j\omega) - F(-j\omega)] = -j\int_{-\infty}^{\infty} f(t)\, \sin \omega t\, dt \qquad (2\text{-}92)$$

This follows directly from the definition of Fourier transform or from the shift theorem.

9. *The Multiplication Theorem*

$$\int_{-\infty}^{\infty} f_1(t) f_2(t)\, dt = \frac{1}{2\pi} \int_{-\infty}^{\infty} F_1^*(j\omega)\, F_2(j\omega)\, d\omega$$

$$= \frac{1}{2\pi} \int_{-\infty}^{\infty} F_1(j\omega)\, F_2^*(j\omega)\, d\omega \qquad (2\text{-}93)$$

Proof:

$$\int_{-\infty}^{\infty} f_1(t) f_2(t)\, dt = \frac{1}{2\pi} \int_{-\infty}^{\infty} f_1(t)\, dt \int_{-\infty}^{\infty} F_2(j\omega) e^{j\omega t}\, d\omega$$

$$= \frac{1}{2\pi} \int_{-\infty}^{\infty} F_2(j\omega)\, d\omega \int_{-\infty}^{\infty} f_1(t) e^{j\omega t}\, dt$$

$$= \frac{1}{2\pi} \int_{-\infty}^{\infty} F_1^*(j\omega)\, F_2(j\omega)\, d\omega$$

Similarly, it may be shown that

$$\int_{-\infty}^{\infty} f_1(t)f_2(t)\ dt\ =\ \frac{1}{2\pi}\int_{-\infty}^{\infty} F_1(j\omega)F_2^*(j\omega)\ d\omega$$

10. Energy Integrals: Parseval's Theorem

$$\int_{-\infty}^{\infty} [f(t)]^2\ dt\ =\ \frac{1}{2\pi}\int_{-\infty}^{\infty} |F(j\omega)|^2\ d\omega \qquad (2\text{-}94)$$

This is called Parseval's relation, which represents the integral of the energy due to the signal f(t). As in the case of the Fourier series, this is derived from the multiplication theorem by putting $f_1(t)\ =\ f_2(t)\ =\ f(t)$ in Eq. (2-93).

$$\int_{-\infty}^{\infty} [f(t)]^2\ dt\ =\ \frac{1}{2\pi}\int_{-\infty}^{\infty} F^*(j\omega)F(j\omega)\ d\omega\ =\ \frac{1}{2\pi}\int_{-\infty}^{\infty} |F(j\omega)|^2\ d\omega$$

$$=\ \frac{1}{2\pi}\int_{-\infty}^{\infty} S(\omega)\ d\omega$$

The term $S(\omega)\ =\ |F(j\omega)|^2$ is called the energy density function or the energy spectral density. Note the contrast with the previous case of periodic functions. In that instance the power was finite, but with aperiodic functions the total energy is finite and, consequently, the spectrum of f(t) is an energy density spectrum, which when integrated over all frequencies will yield a value equal to the total energy E of the time function f(t).

Thus $$E\ =\ \int_{-\infty}^{\infty} f^2(t)\ dt\ =\ \frac{1}{2\pi}\int_{-\infty}^{\infty} S(\omega)\ d\omega \qquad (2\text{-}95)$$

The following examples which are often encountered in modulation problems will serve to illustrate the application of the Fourier transforms to nonperiodic waveforms.

1. *The Rectangular Signal Pulse.* The rectangular pulse shown in Fig. 2-17a is defined as follows:

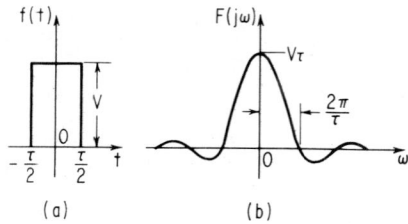

Fig. 2-17. Rectangular pulse and its spectrum.

$$f(t) = V, \qquad -\frac{\tau}{2} < t < +\frac{\tau}{2}$$

$$= 0, \qquad \text{elsewhere}$$

$$F(j\omega) = \int_{-\infty}^{\infty} f(t)e^{-j\omega t}\, dt = V \int_{-\tau/2}^{\tau/2} e^{-j\omega t}\, dt$$

$$= V\tau\, \frac{\sin(\omega\tau/2)}{\omega\tau/2} \qquad\qquad (2\text{-}93)$$

The Fourier spectrum $F(j\omega)$ is plotted in Fig. 2-21b; note that the zeros are separated by $2\pi/\tau$. The frequency spectrum of a single nonrecurrent pulse may be obtained by a limiting process from Eq. (2-37), by letting the period T approach infinity. As $T \rightarrow \infty$, the periodic pulse train reduces to a single pulse, and the spectral lines move together to form a continuous spectrum. Thus

$$F(j\omega) = \lim_{T \rightarrow \infty} C_n = V\tau\, \frac{\sin(\omega\tau/2)}{\omega\tau/2}$$

The total energy of the rectangular pulse (dissipated in 1 ohm resistance) is

$$E = \int_{-\infty}^{\infty} f^2(t)\, dt = V^2\tau$$

This can be checked with the expression given in Eq. (2-95) as follows:

$$E = \frac{1}{2\pi} \int_{-\infty}^{\infty} S(\omega) \, d\omega = \frac{1}{\pi} \int_{0}^{\infty} (V\tau)^2 \left[\frac{\sin (\omega\tau/2)}{\omega\tau/2} \right]^2 d\omega$$

$$= \frac{2V^2\tau}{\pi} \int_{0}^{\infty} \left(\frac{\sin x}{x} \right)^2 dx, \qquad x = \frac{\omega\tau}{2}$$

Since $\displaystyle\int_{0}^{\infty} \left(\frac{\sin x}{x} \right)^2 dx = \frac{\pi}{2}$, the total energy $E = V^2\tau$.

2. *The Rectangular Frequency Pulse.*   This is shown in Fig. 2-18a and is defined by

$$F(j\omega) = 1, \quad -\frac{\omega_c}{2} < \omega < +\frac{\omega_c}{2}$$

$$= 0, \qquad \text{elsewhere}$$

$$f(t) = \frac{1}{2\pi} \int_{-\infty}^{\infty} F(j\omega)e^{j\omega t} \, d\omega$$

$$= \frac{1}{2\pi} \int_{-\omega_c/2}^{\omega_c/2} e^{j\omega t} \, d\omega = \frac{1}{2\pi} \frac{1}{jt} (e^{j\omega_c t/2} - e^{-j\omega_c t/2})$$

$$= \frac{\omega_c}{2\pi} \frac{\sin (\omega_c/2) \, t}{(\omega_c/2) \, t} \tag{2-97}$$

which is plotted in Fig. 2-18b. This result can also be derived from

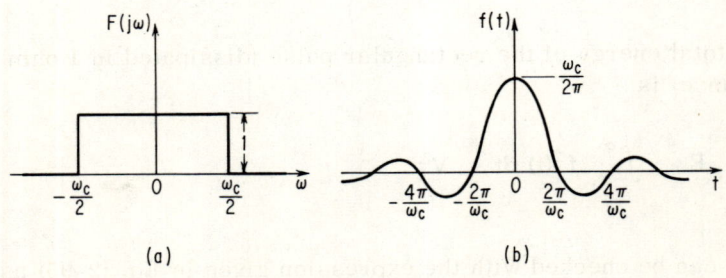

Fig. 2-18. Rectangular frequency pulse and its signal.

the last example by the use cf the duality theorem, which states that the signal of $f(\pm j\omega)$ is $(1/2\pi)$ $F(\mp t)$. Hence

$$f(t) = \frac{1}{2\pi} F(-t) = \frac{\omega_c}{2\pi} \frac{\sin(-t\omega_c/2)}{-t\omega_c/2} = \frac{\omega_c}{2\pi} \frac{\sin(\omega_c/2)t}{(\omega_c/2)t}$$

where $\tau/2$ is replaced by $\omega_c/2$.

3. *Rectangular Pulsed RF Signal.* This is shown in Fig. 2-19a where

$$f(t) = V \cos \omega_c t, \qquad -\frac{\tau}{2} < t < +\frac{\tau}{2}$$

$$f(t) = 0, \qquad\qquad \text{elsewhere}$$

$$F(j\omega) = \int_{-\infty}^{\infty} f(t)e^{-j\omega t} \, dt = V \int_{-\tau/2}^{\tau/2} \cos \omega_c t \, e^{-j\omega t} \, dt$$

Since $f(t)$ is even, we have

$$F(j\omega) = R(\omega) = 2 \int_{0}^{\tau/2} f(t) \cos \omega t \, dt = 2V \int_{0}^{\tau/2} \cos \omega t \cos \omega_c t \, dt$$

$$= V \int_{0}^{\tau/2} [\cos(\omega - \omega_c)t + \cos(\omega + \omega_c)t] \, dt$$

$$= V \left[\frac{\sin(\omega - \omega_c)t}{\omega - \omega_c} + \frac{\sin(\omega + \omega_c)t}{\omega + \omega_c} \right]_{0}^{\tau/2}$$

$$= \frac{V\tau}{2} \left\{ \frac{\sin[(\omega - \omega_c)\tau/2]}{(\omega - \omega_c)\tau/2} + \frac{\sin[(\omega + \omega_c)\tau/2]}{(\omega + \omega_c)\tau/2} \right\} \qquad (2\text{-}98)$$

The same results can be derived directly by the use of the shift theorem. The frequency spectrum $F(j\omega)$ is shown in Fig. 2-19b for $\omega_c \tau/2 \gg 1$. Note that the frequency spectrum is symmetrically shifted about $\omega = 0$.

In a similar manner, it can be shown that for

$$f(t) = V \sin \omega_c t, \qquad -\frac{\tau}{2} < t < \frac{\tau}{2}$$

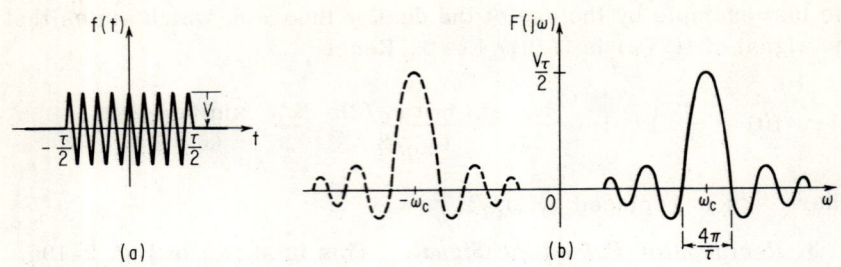

Fig. 2-19. Rectangular pulsed RF wave and its spectrum.

and $f(t) = 0,$ elsewhere

$$F(j\omega) = \frac{-jV\tau}{2}\left\{\frac{\sin\,[(\omega - \omega_C)\,\tau/2]}{(\omega - \omega_C)\,\tau/2} - \frac{\sin\,[(\omega + \omega_C)\,\tau/2]}{(\omega + \omega_C)\,\tau/2}\right\}$$

$$(2\text{-}99)$$

4. *The Unit Step Function.*   The unit step function shown in Fig. 2-20 is defined as follows:

$$f(t) = 1, \qquad t > 0$$
$$= 0, \qquad t < 0$$

Since this function does not satisfy Dirichlet's condition, namely, $\int_{-\infty}^{\infty} |f(t)|\,dt$ is not finite, we shall calculate first the frequency spectrum for

$$f(t) = 1, \qquad 0 < t < T$$
$$= 0, \qquad \text{elsewhere}$$

and then approximate the unit step function by making $T \to \infty$.

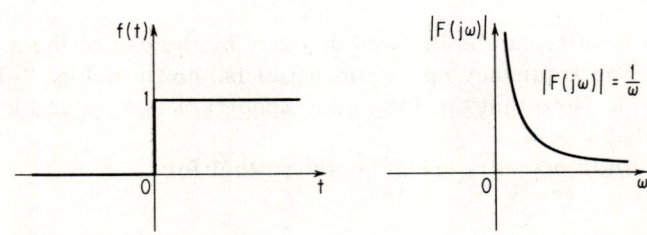

Fig. 2-20. The unit step time function and its frequency spectrum.

$$F(j\omega) = \int_{-\infty}^{\infty} f(t)e^{-j\omega t}\,dt = \int_{0}^{T} e^{-j\omega t}\,dt = \frac{1}{j\omega}(1 - e^{-j\omega T})$$

Hence $\quad f(t) = \frac{1}{2\pi} \int_{-\infty}^{\infty} \frac{1}{j\omega}\left(1 - e^{-j\omega T}\right) e^{j\omega t}\,d\omega$

$$= \frac{1}{2\pi} \int_{-\infty}^{\infty} \frac{1}{j\omega} e^{j\omega t}\,d\omega - \frac{1}{2\pi} \int_{-\infty}^{\infty} \frac{1}{j\omega} e^{j\omega(t-T)}\,d\omega$$

By making T large, the function f(t) will approximate the unit step function. Now for $T \gg 0$, we have

$$-\frac{1}{2\pi} \int_{-\infty}^{\infty} \frac{1}{j\omega} e^{j\omega(t-T)}\,d\omega \rightarrow -\frac{1}{2\pi} \int_{-\infty}^{\infty} \frac{e^{-j\omega T}}{j\omega}\,d\omega = \frac{1}{2}$$

which is a standard integral. Therefore

$$f(t) = \frac{1}{2} + \frac{1}{2\pi} \int_{-\infty}^{\infty} \frac{1}{j\omega} e^{j\omega t}\,d\omega \qquad\qquad (2\text{-}100)$$

This result can be checked, since

$$\frac{1}{2\pi} \int_{-\infty}^{\infty} \frac{e^{j\omega t}}{j\omega}\,d\omega = -\pi, \qquad t < 0$$

$$= +\pi, \qquad t > 0$$

Therefore

$$f(t) = 1, \qquad t > 0$$

$$= 0, \qquad t < 0$$

From Eq. (2-100), it follows that

$$F(j\omega) = \frac{1}{j\omega} \qquad\qquad (2\text{-}101)$$

It should be noted here that the function f(t) is given by Eq. (2-100) and not by the usual expression, as in Eq. (2-65). The reason is that in the derivation of $F(j\omega)$ we have neglected to account for the d-c term in the time function which, as we shall see later on, has an impulse function as its Fourier transform. It will be shown that the complete expression for $F(j\omega)$ is

$$F(j\omega) = \pi\,\delta(\omega) + \frac{1}{j\omega} \tag{2-102}$$

where $\pi\,\delta(\omega)$ is the impulse function of strength π at the origin. Equation (2-100) can be derived more simply from the exponential decay function which is given below.

5. *The Exponential Decay Function.* This function, shown in Fig. 2-21a, is defined as follows:

$$f(t) = e^{-\alpha t}, \qquad t > 0$$
$$= 0, \qquad\quad t < 0$$

$$F(j\omega) = \int_{-\infty}^{\infty} f(t)e^{-j\omega t}\,dt$$

$$= \int_{0}^{\infty} e^{-(\alpha + j\omega)t}\,dt = \frac{1}{\alpha + j\omega} \tag{2-103}$$

and $$|F(j\omega)| = \frac{1}{\sqrt{\alpha^2 + \omega^2}} = A(\omega) \tag{2-104}$$

The amplitude and phase of $F(j\omega)$ are plotted in Fig. 2-21b. Note that when $\alpha \rightarrow 0$, the exponential decay function approaches a step function, and $F(j\omega) \rightarrow 1/j\omega$ as before.

Similarly, it can be shown that the Fourier transform of the exponential rise function is given by

$$F(j\omega) = \frac{\alpha}{\omega\sqrt{\alpha^2 + \omega^2}} \tag{2-105}$$

and $$f(t) = \frac{1}{2} + \frac{1}{2\pi} \int_{-\infty}^{\infty} \frac{\alpha}{j\omega(\alpha + j\omega)} e^{j\omega t}\,d\omega \tag{2-106}$$

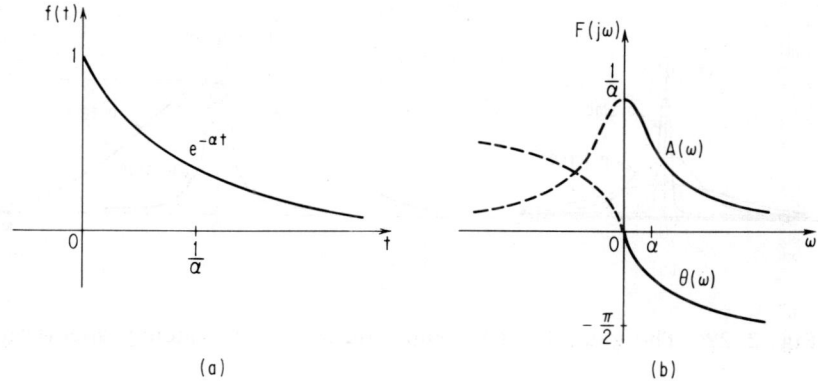

Fig. 2-21. The exponential decay time function.

6. *The Gaussian Distribution Function.* This function is very basic in noise theory, and the general form for one variable is

$$f(t) = \frac{1}{\sqrt{2\pi}\,\sigma}\, e^{-t^2/2\sigma^2} \tag{2-107}$$

$$F(j\omega) = \int_{-\infty}^{\infty} f(t) e^{-j\omega t}\, dt = \frac{1}{\sqrt{2\pi}\,\sigma} \int_{-\infty}^{\infty} e^{-(t^2/2\sigma^2 + j\omega t)}\, dt$$

$$= \frac{e^{-\omega^2\sigma^2/2}}{\sqrt{2\pi}\,\sigma} \int_{-\infty}^{\infty} e^{-(1/2\sigma^2)(t + j\omega\sigma^2)^2}\, dt \tag{2-108}$$

Put $\quad \dfrac{1}{\sqrt{2}\,\sigma}\,(t + j\omega\sigma^2) = x$

and $\quad \dfrac{1}{\sqrt{2}\,\sigma}\, dt = dx$

Hence $\quad F(j\omega) = \dfrac{e^{-\omega^2\sigma^2/2}}{\sqrt{\pi}} \displaystyle\int_{-\infty}^{\infty} e^{-x^2}\, dx = e^{-\omega^2\sigma^2/2} \tag{2-109}$

since $\quad \displaystyle\int_{-\infty}^{\infty} e^{-x^2}\, dx = \sqrt{\pi}$

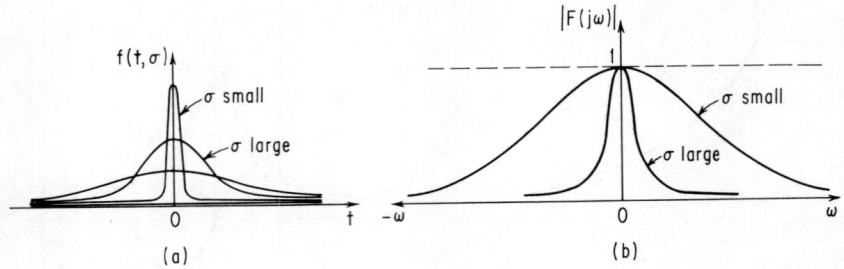

Fig. 2-22. The gaussian time function and its frequency spectrum.

Thus the spectrum of a gaussian time function is a gaussian frequency function. Note however that the position of σ^2 in the exponent is now in the numerator whereas in the time function it appears in the denominator. The physical significance is that, corresponding to a narrow f(t) as shown in Fig. 2-22a, the spectrum as shown in Fig. 2-22b is broad, and vice versa. In the limit as $\sigma \to 0$, the time function f(t) approaches the unit impulse or delta function, so that

$$\lim_{\sigma \to 0} \quad f(t, \sigma) = \delta(t)$$

while $\int_{-\infty}^{\infty} f(t)\, dt = 1$, unit area independent of σ. The corresponding Fourier spectrum is in the limit a constant equal to one, as shown in Fig. 2-22b. Since the unit impulse or delta function plays a very important role in the analysis of signal transmission through linear systems, we shall elaborate on this concept in the following section.

2.6   The Unit Impulse or Delta Function[4, 8-10]

From Fig. 2-17, it is seen that the spacing between the zeros is inversely proportional to τ, the duration time of the pulse. Thus for a pulse of extremely short duration, the zeros will be far apart, resulting in a broad spectrum. In the limiting case, consider the rectangular pulse to be one of extremely short duration and very large amplitude. In particular, let $\tau \to 0$ and $V \to \infty$, such that $V\tau = 1$ (unit area), and we get, by definition, the unit impulse or delta function.

We denote by $\delta(t - t_0)$ the function which is equal to zero for all values of t except $t = t_0$, in which case it goes to infinity.

Thus $\quad \delta(t - t_0) = 0, \qquad t \neq t_0$

$$= \infty, \qquad t = t_0 \qquad\qquad (2\text{-}110)$$

in such a way that $\int_{-\infty}^{\infty} \delta(t - t_0)\, dt = 1$.

Consider the integral $\int_{-\infty}^{\infty} f(t)\, \delta(t - t_0)\, dt$ where f(t) is continuous at t_0. From the properties of the unit impulse function, the only contribution of f(t) to the integral is at the point $t = t_0$, for

$$\int_{-\infty}^{\infty} f(t)\, \delta(t - t_0)\, dt = \int_{t_0 - \epsilon}^{t_0 + \epsilon} f(t)\, \delta(t - t_0)\, dt$$

$$= f(t_0) \int_{t_0 - \epsilon}^{t_0 + \epsilon} \delta(t - t_0)\, dt = f(t_0) \qquad (2\text{-}111)$$

since in the interval $(t_0 \pm \epsilon)$, f(t) could be considered practically as a constant quantity.

From Eq. (2-111), it follows that to integrate a function f(t) by the unit impulse function centered at the point t_0 simply implies evaluating the given function at t_0.

A more rigorous approach to the definition of the delta function may be used by defining it by its integral properties only and not as a function of time defined for every t. This approach is followed in the mathematical literature where $\delta(t - t_0)$ is defined by the integral

$$\int_{-\infty}^{\infty} f(t)\, \delta(t - t_0)\, dt = f(t_0) \qquad\qquad (2\text{-}112)$$

Thus the delta function is not defined as a function of time but is characterized only by its integral properties, as demonstrated in the following example, where the Fourier transform for the delta function $\delta(t)$ is found.

$$F(j\omega) = \int_{-\infty}^{\infty} \delta(t) e^{-j\omega t}\, dt = e^{-j\omega t}\Big|_{t=0} = 1 \qquad (2\text{-}113)$$

The spectrum of $\delta(t)$ is constant as shown in Fig. 2-23.

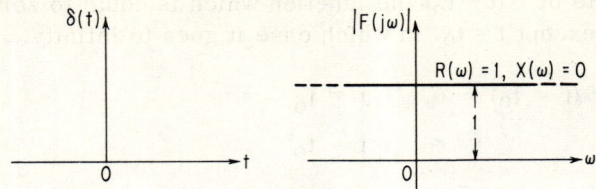

Fig. 2-23. Spectrum of delta function $\delta(t)$.

For an impulse displaced t_0 sec in time, the Fourier transform is

$$F(j\omega) = \int_{-\infty}^{\infty} \delta(t - t_0)e^{-j\omega t} \, dt = e^{-j\omega t_0} \qquad (2\text{-}114)$$

i.e., $\delta(t - t_0)$ and $e^{-j\omega t_0}$ form a transform pair.

This again gives a flat amplitude-frequency spectrum but introduces a phase factor to account for the time delay assumed.

This result can be extended to a periodic impulse function given by

$$f(t) = \sum_{n=-\infty}^{\infty} \delta(t - nT) \qquad (2\text{-}115)$$

where T is the period between the impulse functions. The corresponding Fourier spectrum is given by

$$F(j\omega) = \sum_{n=-\infty}^{\infty} e^{-jn\omega T} \qquad (2\text{-}116)$$

which consists of an infinite series of cosine terms since the sine terms will cancel out, as shown in Fig. 2-24.

In a similar manner, we may derive the time function f(t) which corresponds to the impulse function $2\pi\,\delta(\omega)$.

$$f(t) = \frac{1}{2\pi} \int_{-\infty}^{\infty} F(j\omega)e^{j\omega t} \, d\omega = \int_{-\infty}^{\infty} \delta(\omega)e^{j\omega t} = 1 \qquad (2\text{-}117)$$

Thus the Fourier transform of f(t) = 1 is an impulse frequency

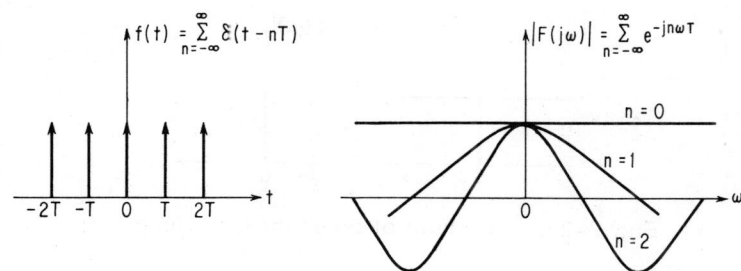

Fig. 2-24. Spectrum of periodic impulse functions.

function at the origin of strength 2π, as shown in Fig. 2-25. Also, the Fourier transform of f(t) = $e^{j\omega_0 t}$ is F(jω) = $2\pi \delta (\omega - \omega_0)$.

We have proved above that the spectrum of δ (t) is a constant quantity. We shall now prove that the spectrum of two delta functions $\delta(t - t_0)$ and $\delta(t + t_0)$ symmetrically disposed about the origin is a cosine frequency function. Given

$$f(t) = \delta(t - t_0) + \delta(t + t_0) \tag{2-118}$$

Hence
$$F(j\omega) = \int_{-\infty}^{\infty} f(t)e^{-j\omega t} dt$$

$$= \int_{-\infty}^{\infty} [\delta(t - t_0) + \delta(t + t_0)] e^{-j\omega t} dt$$

$$= e^{-j\omega t_0} + e^{j\omega t_0} = 2 \cos t_0\omega \tag{2-119}$$

as shown in Fig. 2-26; this follows also directly from Eq. (2-114).

Similarly, we may find the time function corresponding to two impulse functions $\pi\delta (\omega - \omega_0)$ and $\pi\delta (\omega + \omega_0)$ symmetrically disposed about the origin, as shown in Fig. 2-27a.

Let F(jω) = $\pi \delta (\omega - \omega_0) + \pi \delta (\omega + \omega_0)$; therefore

$$f(t) = \frac{1}{2\pi} \int_{-\infty}^{\infty} \pi [\delta(\omega - \omega_0) + \delta(\omega + \omega_0)] e^{j\omega t} d\omega$$

or
$$f(t) = \frac{1}{2} \left(e^{j\omega_0 t} + e^{-j\omega_0 t} \right) = \cos \omega_0 t \tag{2-120}$$

as shown in Fig. 2-27b. This result which follows directly from the

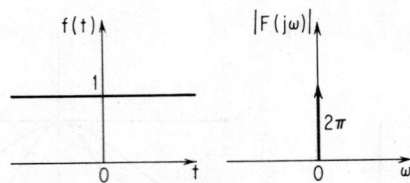

Fig. 2-25. Spectrum of constant time functions.

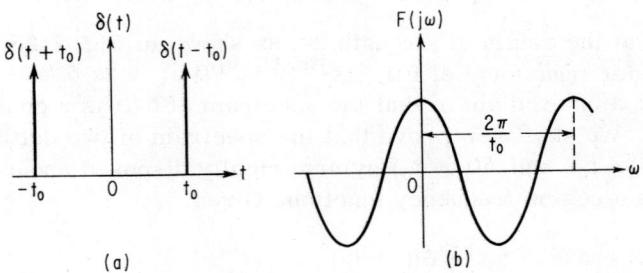

Fig. 2-26. Spectrum of two delta functions, $\delta(t - t_0)$ and $\delta(t + t_0)$.

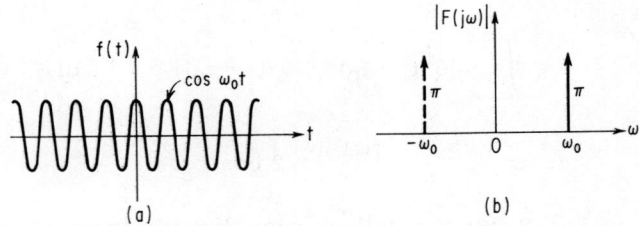

Fig. 2-27. Spectrum of a cosine time function.

definition of delta function is of fundamental importance because it provides a means of dealing with discrete frequencies at the same time as with continuous spectra. A cosine time function $\cos \omega_0 t$ may be represented in the frequency plane by two impulse functions of strength π symmetrically disposed about the origin; i.e., $\cos \omega_0 t$ is the transform of $\pi [\delta(\omega - \omega_0) + \delta(\omega + \omega_0)]$.

Similarly, if $F(j\omega) = j\pi \, \delta(\omega + \omega_0) - j\pi \, \delta(\omega - \omega_0)$, then

$$f(t) = \sin \omega_0 t \qquad\qquad (2\text{-}121)$$

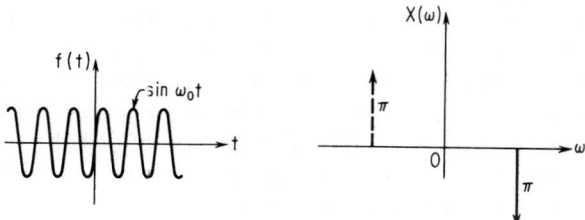

Fig. 2-28. Spectrum of a sine time function.

The frequency spectrum representing the sine function is purely imaginary and consists of two impulse functions of strength π symmetrically disposed about the origin, as shown in Fig. 2-28, so that $\sin \omega_0 t$ is the transform of $j\pi [\delta (\omega + \omega_0) - \delta (\omega - \omega_0)]$.

2.7 Line Spectra and Some Singularity Functions [8,10]

The results of the last section may be extended to a periodic function $f(t)$, of period T, namely

$$f(t + T) = f(t) \qquad (2\text{-}122)$$

Using Eq. (2-21), $f(t)$ can be expressed as a sum of exponentials; thus

$$f(t) = \sum_{n=-\infty}^{\infty} C_n e^{jn\omega_0 t} \qquad (2\text{-}123)$$

where $\quad C_n = \dfrac{1}{T} \displaystyle\int_{-T/2}^{T/2} f(t) e^{-jn\omega_0 t}\, dt \qquad (2\text{-}124)$

and $\omega_0 = 2\pi/T$; since the Fourier transform of $e^{j\omega_0 t}$ is $2\pi \delta (\omega - \omega_0)$, we can generalize this result to a periodic function. Thus, the Fourier transform of a periodic function $f(t)$ is given by

$$F(j\omega) = 2\pi \sum_{n=-\infty}^{\infty} C_n \delta (\omega - n\omega_0) \qquad (2\text{-}125)$$

which consists of a sequence of equidistant impulse functions distance ω_0 apart, as shown in Fig. 2-29.

Similarly, we shall show that the Fourier transform of a pulse train consisting of a sequence of equidistant impulse time functions $\delta(t - nT)$ can be represented by a sequence of equidistant frequency impulses of amplitude ω_0 spaced ω_0 apart, as shown in Fig. 2-30. That is to say, we shall prove that if

$$s_T(t) = \sum_{n=-\infty}^{\infty} \delta(t - nT) \tag{2-126}$$

then

$$S_T(j\omega) = \omega_0 \sum_{n=-\infty}^{\infty} \delta(\omega - n\omega_0) \tag{2-127}$$

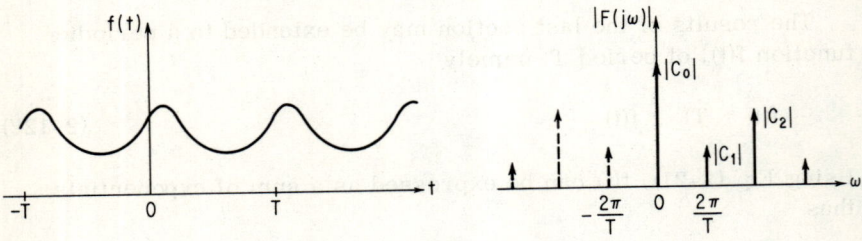

Fig. 2-29. Frequency spectrum of a periodic function.

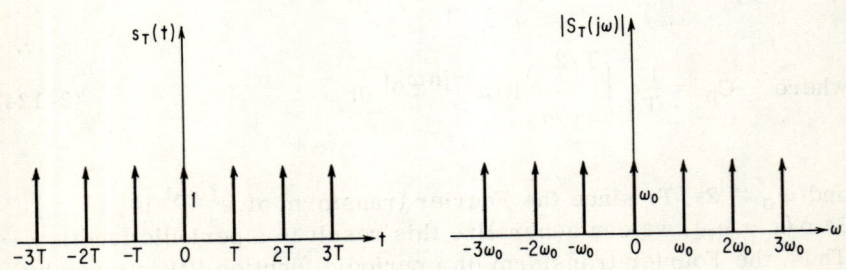

Fig. 2-30. Frequency spectrum of a pulse train of equidistant pulses

where $\omega_0 = 2\pi/T$. First we shall prove that $s_T(t)$ can be expressed in terms of the Fourier series:

$$s_T(t) = \sum_{n=-\infty}^{\infty} \delta(t - nT) = \frac{1}{T} \sum_{n=-\infty}^{\infty} e^{jn\omega_0 t} \tag{2-128}$$

This can be done by considering the periodic pulse train given by Eq. (2-38), namely,

$$f(t, \tau) = V \frac{\tau}{T} \sum_{n=-\infty}^{\infty} \frac{\sin(n\omega_0 \tau/2)}{n\omega_0 \tau/2} e^{jn\omega_0 t} \tag{2-129}$$

Let $\tau \to 0$ in such a way that the area of the pulse remains constant so that $V\tau \to 1$; hence

$$\lim_{\substack{\tau \to 0 \\ V\tau \to 1}} f(t, \tau) = s_T(t) = \frac{1}{T} \sum_{n=-\infty}^{\infty} e^{jn\omega_0 t} = \sum_{n=-\infty}^{\infty} \delta(t - nT)$$

However, since the Fourier transform of $f(t) = e^{jn\omega_0 t}$ is $F(j\omega) = 2\pi \delta(\omega - n\omega_0)$, it follows that the Fourier transform of $s_T(t)$ is given by

$$S_T(j\omega) = \frac{2\pi}{T} \sum_{n=-\infty}^{\infty} \delta(\omega - n\omega_0) = \omega_0 \sum_{n=-\infty}^{\infty} \delta(\omega - n\omega_0)$$

We can also express $S_T(j\omega)$ in terms of the Fourier series:

$$S_T(j\omega) = \omega_0 \sum_{n=-\infty}^{\infty} \delta(\omega - n\omega_0) = \sum_{n=-\infty}^{\infty} e^{-jnT\omega} \tag{2-130}$$

which follows directly from Eqs. (2-115) and (2-116).

The transform of a few additional singular functions, which will be useful in future analysis, will be discussed presently.

Consider the singular function

$$f(t) = \text{sgn } t = 1, \quad t > 0$$
$$= -1, \quad t < 1 \qquad (2\text{-}131)$$

as shown in Fig. 2-31. We shall prove that the Fourier transform of the function sgn t is equal to $2/j\omega$. Consider the function

$$f(t) = \frac{1}{2\pi} \int_{-\infty}^{\infty} \frac{2}{j\omega} e^{j\omega t} \, d\omega = \frac{1}{\pi} \int_{-\infty}^{\infty} \frac{\sin \omega t}{\omega} \, d\omega \qquad (2\text{-}132)$$

which can be shown by contour integration to equal 1 if $t > 0$ and -1 if $t < 0$; thus $f(t) = \text{sgn } t$, and its Fourier transform equals $2/j\omega$.

Another singular function which is useful in many applications is the step function u(t) which was discussed above and is shown again in Fig. 2-32a. The step function u(t) can be represented by

$$u(t) = \tfrac{1}{2} + \tfrac{1}{2} \text{ sgn } t \qquad (2\text{-}133)$$

Using the previous results, namely, Eqs. (2-117) and (2-132), the Fourier transform of u(t) is given by

$$U(j\omega) = R(\omega) + jX(\omega) = \pi \, \delta(\omega) + \frac{1}{j\omega} \qquad (2\text{-}134)$$

as shown in Fig. 2-32b. Also, the Fourier transform of $e^{j\omega_o t} u(t)$ is $[\pi \, \delta(\omega - \omega_o) + 1/j(\omega - \omega_o)]$.

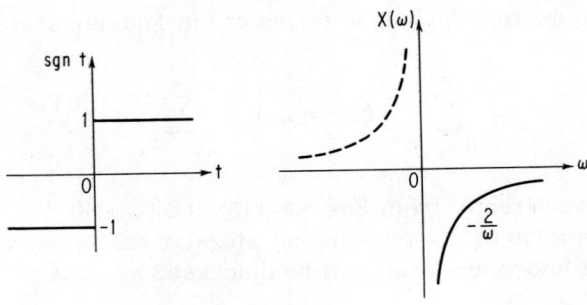

Fig. 2-31. The function sgn t.

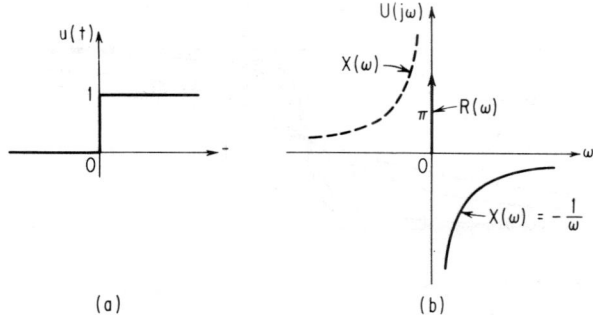

Fig. 2-32. Step function and its spectrum.

The time function $u(t)$ can be found directly by taking the inverse transform of $U(j\omega)$.

Using Eq. (2-74), we can write

$$u(t) = \frac{1}{2\pi} \int_{-\infty}^{\infty} [R(\omega) \cos \omega t - X(\omega) \sin \omega t] \, d\omega$$

$$= \frac{1}{2\pi} \int_{-\infty}^{\infty} \pi \, \delta(\omega) \cos \omega t \, d\omega + \frac{1}{2\pi} \int_{-\infty}^{\infty} \frac{\sin \omega t}{\omega} \, d\omega$$

$$= \frac{1}{2} + \frac{1}{\pi} \int_{0}^{\infty} \frac{\sin \omega t}{\omega} \, d\omega \qquad\qquad (2\text{-}135)$$

We shall presently show that the δ function can be derived by taking the first-order derivative of the unit step function. With the exception of the point $t = 0$, the unit step function $u(t)$ has a derivative everywhere which is equal to zero. To overcome the difficulty at the origin, we consider the function

$$u(t, \lambda) = \frac{1}{2} + \frac{1}{\pi} \arctan(\lambda t), \qquad |\arctan \lambda t| \le \frac{\pi}{2} \qquad (2\text{-}136)$$

This function approximates the behavior of the unit step function as $\lambda \to \infty$, as shown in Fig. 2-33.

The derivative is

$$\delta(t, \lambda) = \frac{d}{dt} u(t, \lambda) = \frac{\lambda}{\pi(\lambda^2 t^2 + 1)} \qquad\qquad (2\text{-}137)$$

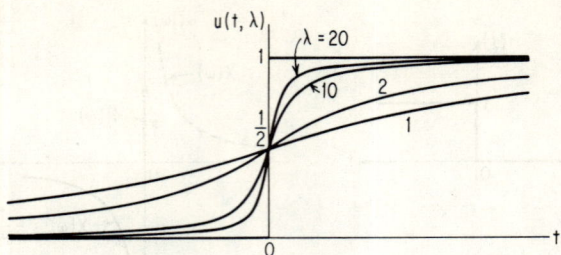

Fig. 2-33. Approximation of the unit step function.

Just as we consider

$$u(t) = \lim_{\lambda \to \infty} \left[\frac{1}{2} + \frac{1}{\pi} \arctan (\lambda t) \right] \qquad (2\text{-}138)$$

we may consider

$$\delta(t) = \frac{d}{dt} u(t) = \lim_{\lambda \to \infty} \frac{\lambda}{\pi (\lambda^2 t^2 + 1)} \qquad (2\text{-}139)$$

From the last equation, it is obvious that $\delta(t)$ equals zero for $t \neq 0$ and equals $+\infty$ for $t = 0$, since $\delta(0, \lambda) = \lambda/\pi$. The behavior of $\delta(t, \lambda)$ is shown in Fig. 2-34; as $\lambda \to \infty$, the value of λ/π increases while the peak becomes narrower, so that

$$\int_{-\infty}^{\infty} \delta(t, \lambda) \, dt = 1 \qquad (2\text{-}140)$$

2.8 Correlation Functions for Aperiodic Signals [7,10]

We shall derive the correlation functions for aperiodic signals in a manner analogous to the case of periodic signals.

By the use of the multiplication theorem [Eq. (2-93)], we consider the integral

$$\int_{-\infty}^{\infty} f_1(t) f_2(t) \, dt = \frac{1}{2\pi} \int_{-\infty}^{\infty} F_1^*(j\omega) \cdot F_2(j\omega) \, d\omega \qquad (2\text{-}141)$$

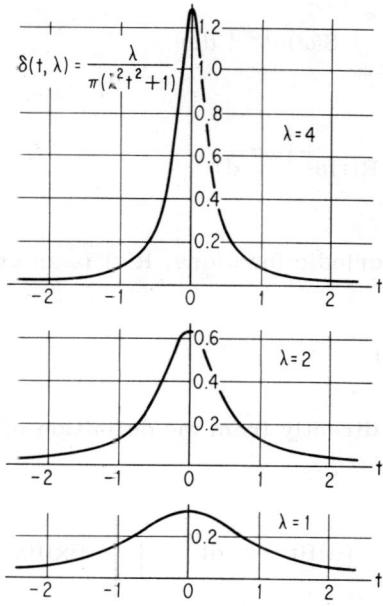

Fig. 2-34. Approximation of the delta function.

Let $f_1(t) = f(t)$ and $f_2(t) = f(t + \tau)$; by the shift theorem the Fourier transform of $f(t + \tau)$ is given by $e^{j\omega\tau} \cdot F(j\omega)$; thus Eq. (2-141) reduces to

$$\int_{-\infty}^{\infty} f(t)\, f(t + \tau)\, dt = \frac{1}{2\pi} \int_{-\infty}^{\infty} F^*(j\omega) \cdot F(j\omega) e^{j\omega\tau}\, d\omega$$

$$= \frac{1}{2\pi} \int_{-\infty}^{\infty} |F(j\omega)|^2\, e^{j\omega\tau}\, d\omega$$

$$= \frac{1}{2\pi} \int_{-\infty}^{\infty} S(\omega) e^{j\omega\tau}\, d\omega \qquad (2\text{-}142)$$

The expression $R(\tau) = \int_{-\infty}^{\infty} f(t) f(t + \tau)\, d\tau$ is defined as the auto-correlation function for an aperiodic signal $f(t)$. Thus we have shown that the autocorrelation function $R(\tau)$ and the energy density function $S(\omega)$ form a Fourier transform pair:

$$R(\tau) = \frac{1}{2\pi} \int_{-\infty}^{\infty} S(\omega)e^{j\omega\tau} d\omega \qquad\qquad (2\text{-}143)$$

and $\quad S(\omega) = \displaystyle\int_{-\infty}^{\infty} R(\tau)e^{-j\omega\tau} d\tau \qquad\qquad (2\text{-}144)$

As in the case of periodic functions, $R(\tau)$ is an even function of τ, namely,

$$R(-\tau) = R(\tau) \qquad\qquad (2\text{-}145)$$

This can be shown directly from the definition of the autocorrelation function as follows:

$$R(-\tau) = \int_{-\infty}^{\infty} f(t)f(t-\tau)\,dt = \int_{-\infty}^{\infty} f(x)f(x+\tau)\,dx = R(\tau)$$

and the Fourier transform pair can be written in the trigonometric form

$$R(\tau) = \frac{1}{\pi} \int_{0}^{\infty} S(\omega)\cos\omega\tau\,d\omega \qquad\qquad (2\text{-}146)$$

and $\quad S(\omega) = 2\displaystyle\int_{0}^{\infty} R(\tau)\cos\omega\tau\,d\tau \qquad\qquad (2\text{-}147)$

The cross-correlation function for aperiodic signals $f_1(t)$ and $f_2(t)$ is defined in a manner similar to that of periodic functions,

$$R_{12}(\tau) = \int_{-\infty}^{\infty} f_1(t)f_2(t+\tau)\,d\tau \qquad\qquad (2\text{-}148)$$

and the cross-spectral density spectrum

$$S_{12}(\omega) = F_1^*(j\omega) \cdot F_2(j\omega) \qquad\qquad (2\text{-}149)$$

The corresponding Fourier transform pair for the cross-correlation function is

$$R_{12}(\tau) = \frac{1}{2\pi} \int_{-\infty}^{\infty} S_{12}(\omega)e^{j\omega\tau} \, d\omega \qquad (2\text{-}150)$$

and

$$S_{12}(\omega) = \int_{-\infty}^{\infty} R_{12}(\tau)e^{-j\omega\tau} \, d\tau \qquad (2\text{-}151)$$

Now we shall show that

$$R_{21}(\tau) = R_{12}(-\tau) \qquad (2\text{-}152)$$

By definition

$$R_{21}(\tau) = \int_{-\infty}^{\infty} f_2(\tau)f_1(t + \tau) \, dt = \int_{-\infty}^{\infty} f_2(x - \tau)f_1(x) \, dx$$

where $x = t + \tau$, or

$$R_{21}(\tau) = \int_{-\infty}^{\infty} f_1(x)f_2\,[x + (-\tau)] \, dx = R_{12}\,(-\tau)$$

Similarly $$S_{21}(\omega) = F_1(j\omega) \cdot F_2^{*}(j\omega) = S_{12}^{*}(\omega) \qquad (2\text{-}153)$$

As stated in the introduction to this chapter, there are three types of signals which are commonly encountered in communication problems. Two of them, namely, the periodic and aperiodic signals, were discussed in this chapter, and the mathematical tools required to deal with these two types of signals were reviewed. Before proceeding, however, with the discussion of the third type, the random signal, we shall demonstrate in the next chapter the application of these mathematical techniques to the problem of transmission of signals through linear systems. The mathematical techniques for dealing with the random signal and the application to noise problems will be outlined in Chap. 4.

References

1. Cherry, C.: "Pulses and Transients in Communication Circuits," Dover Publications, Inc., New York.
2. Cuccia, C. L.: "Harmonics, Sidebands, and Transients in Communication Engineering," McGraw-Hill Book Company, New York.

3. Granlund, J.: Private Communication.
4. Stuart, R. D.: "An Introduction to Fourier Analysis," Methuen & Co., Ltd., London.
5. Guillemin, E. A.: "The Mathematics of Circuit Analysis," John Wiley & Sons, Inc., New York.
6. Goldman, S. A.: "Frequency Analysis, Modulation, and Noise," McGraw-Hill Book Company, New York.
7. Lee, Y. W.: "Statistical Theory of Communication," John Wiley & Sons, Inc., New York.
8. Papoulis, A.: "The Fourier Integral and Its Application to Linear Systems," McGraw-Hill Book Company, New York.
9. Schwartz, M.: "Information Transmission, Modulation, and Noise," McGraw-Hill Book Company, New York.
10. Solodovnikov, V. V.: "Introduction to the Statistical Dynamics of Automatic Control Systems," Dover Publications, Inc., New York.

3

SIGNAL TRANSMISSION THROUGH
LINEAR SYSTEMS

In the last chapter, we discussed the application of the Fourier
series and the Fourier integral to periodic and transient signals,
with special emphasis on the two distinct types of spectra involved,
namely, the discrete and the continuous spectra. The primary ob-
jective of this chapter is to demonstrate the application of these
powerful mathematical tools to the solution of problems relating to
the transmission of these type of signals through linear systems or
filters. The statement of linearity implies that the system response
$g(t)$ to any excitation $f(t)$ can be described by the solution of a set
of differential equations with constant coefficients, and thus the principle of superposition is also implied. Other modes of description of
the linear system which are more practical in the solution of complex problems are:

1. The frequency characteristic or transfer function which relates the output response of the system to the input as a function of
frequency.

2. The impulse or weighting function response of the linear system, i.e., the system output response produced by an impulse input.
These modes of description are simply related, and their specific
advantages for specific problems will be discussed below.

3.1 Frequency Characteristics: Transfer Function[1-3]

A linear dynamic system may be described by a differential equation of the n'th order with constant coefficients. The output g(t) resulting from an input signal f(t) is given by

$$a_n \frac{d^n g(t)}{dt^n} + a_{n-1} \frac{d^{n-1} g(t)}{dt^{n-1}} + \cdots + a_1 \frac{dg(t)}{dt} + a_0 g(t)$$

$$= b_m \frac{d^m f(t)}{dt^m} + b_{m-1} \frac{d^{m-1} f(t)}{dt^{m-1}} + \cdots + b_1 \frac{df(t)}{dt} + b_0 f(t) \qquad (3-1)$$

Consider the case where the signal f(t) represents a harmonic function of time in the form

$$f(t) = V_o \cos (\omega t + \theta_o) \qquad (3-2)$$

where V_o is the amplitude, ω is the angular frequency, and θ_o is the phase. Without loss of generality, we shall assume that $\theta_o = 0$, and we can write

$$f(t) = V_o \cos \omega t = \frac{V_o}{2} e^{j\omega t} + \frac{V_o}{2} e^{-j\omega t} \qquad (3-3)$$

which is in the form of two exponential signals.

The output response produced by the exponential input signals is obtained from the particular solution of Eq. (3-1). It can be shown that the forced oscillations produced in a linear dynamic system by a harmonic signal are also harmonic functions of time, having the same angular frequency as the signal, but differing from the latter in amplitude and in phase. The expression for the output signal is given by

$$g(t) = V_o A(\omega) \cos [\omega t + \phi(\omega)] \qquad (3-4)$$

where $A(\omega)$ and $\phi(\omega)$ are the amplitude and phase characteristics of the complex transfer function $H(j\omega)$, or

$$H(j\omega) = A(\omega) e^{j\phi(\omega)} = P(\omega) + jQ(\omega) \qquad (3-5)$$

It follows therefore that if the input and output signals consist of steady sinusoidal time functions, then $H(j\omega)$ represents the ratio of the complex amplitude of the output to that of the input.

Functions $P(\omega)$ and $Q(\omega)$ are called respectively real and imaginary frequency characteristics. The expression for $H(j\omega)$ can be derived from the differential equation of the system and is given by

$$H(j\omega) = \frac{b_m(j\omega)^m + \cdots + b_1(j\omega) + b_0}{a_n(j\omega)^n + \cdots + a_1(j\omega) + a_0} \tag{3-6}$$

Thus, the frequency characteristics of a system can be determined not only from its differential equation but also experimentally, simply by applying to the system a harmonic signal of angular frequency ω and measuring the output (allowing for the transients to be damped out). The frequency is varied from ω_1 to ω_2 over the frequency range of interest, and the ratio $A(\omega)$ of the amplitude $g(t)$ of the output signal to the amplitude of the applied signal, and the phase difference between the output and input signals $\phi(\omega)$, are plotted as functions of frequency.

3.2 Application of Fourier Transform to the Transmission of Transient Signals through Idealized Networks[1, 3, 4]

In this section, we shall establish the functional relationship between the amplitude and phase characteristics of the transfer function and the frequency spectra of the input and output signals. We shall assume a nonperiodic input time function $f(t)$ with its associated Fourier transform $F(j\omega)$ and we shall find the transient response $g(t)$. We have established in Sec. 3.1 that an input signal of the form $e^{j\omega t}$ produces in a linear system a response of the form $H(j\omega)e^{j\omega t}$. Since the signal $f(t)$ can be expressed as the inverse transform of the frequency spectrum $F(j\omega)$, it may be considered as an infinite sum of signals of the form

$$\frac{1}{2\pi} F(j\omega)e^{j\omega t} \, d\omega$$

each signal being modified in amplitude and phase in going through the linear system by the complex factor $H(j\omega)$. As a consequence of the linearity of the system and the applicability of the principle of superposition to it, the input signal $f(t)$ produces an output signal

$$g(t) = \frac{1}{2\pi} \int_{-\infty}^{\infty} H(j\omega) \cdot F(j\omega)e^{j\omega t} \, d\omega = \frac{1}{2\pi} \int_{-\infty}^{\infty} G(j\omega)e^{j\omega t} \, d\omega \tag{3-7}$$

where $G(j\omega) = H(j\omega) \cdot F(j\omega) = \int_{-\infty}^{\infty} g(t)e^{-j\omega t} \, dt$ $\tag{3-8}$

Equation (3-8) will be used to study the dependence of the output signal on the network amplitude and phase characteristics. It states in fact that the Fourier transform $G(j\omega)$ of the output signal $G(t)$ is equal to the product of the transfer function $H(j\omega)$ of the system and the Fourier transform $F(j\omega)$ of the input signal $f(t)$. This implies that if the steady-state response of the linear system to the range of sinusoidal excitation is known, then its response to any other signal can be determined uniquely.

The transfer function $H(j\omega)$ of a linear system, or as it is sometimes called, the system function, is often specified by its attenuation $\alpha(\omega)$ and phase shift or phase lag $\theta(\omega)$, defined by

$$\alpha(\omega) = -\ell n\, A(\omega), \qquad \theta(\omega) = -\phi(\omega)$$

Hence $H(j\omega) = e^{-\alpha(\omega)} \cdot e^{-j\theta(\omega)} = e^{-j[\alpha(\omega) + \theta(\omega)]}$ (3-9)

The application of Eq. (3-8) may be illustrated with the following problem. Consider a low-pass filter with the following characteristics:

$$H(j\omega) = A(\omega)e^{-j\theta(\omega)}$$

where $A(\omega) = K, \qquad |\omega| \leq \omega_c$

$\qquad\qquad = 0, \qquad$ elsewhere

and $\qquad \theta(\omega) = t_0\omega$

as shown in Fig. 3.1. While these characteristics are physically not realizable, nevertheless such an idealized filter will serve to illustrate the application of Eq. (3-8) to a filter problem. Let the input signal $f(t)$ be a single rectangular pulse; then the problem is to find the transient response $g(t)$.

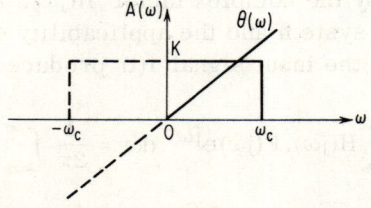

Fig. 3-1. Amplitude and phase characteristics of low-pass filter.

Using Eq. (2-96), the transform of the input signal is

$$F(j\omega) = V\tau \frac{\sin{(\omega\tau/2)}}{\omega\tau/2} \qquad\qquad (3\text{-}10)$$

and applying Eq. (3-8), we obtain the transform of the output signal

$$G(j\omega) = V\tau \frac{\sin{(\omega\tau/2)}}{\omega\tau/2} \cdot K e^{-jt_0\omega}, \qquad -\omega_c < \omega < \omega_c \qquad (3\text{-}11)$$

$$= 0, \qquad\qquad\qquad\qquad \text{elsewhere}$$

Therefore $g(t) = \dfrac{1}{2\pi} \displaystyle\int_{-\infty}^{\infty} G(j\omega) e^{j\omega t}\, d\omega$

$$= \frac{KV\tau}{2\tau} \int_{-\infty}^{\infty} \frac{\sin{(\omega\tau/2)}}{\omega\tau/2} e^{j\omega(t-t_0)}\, d\omega$$

This reduces to

$$g(t) = \frac{KV\tau}{\pi} \int_{0}^{\omega_c} \frac{\sin{(\omega\tau/2)}}{\omega\tau/2} \cos{\omega(t-t_0)}\, d\omega$$

$$= \frac{KV\tau}{2\pi} \int_{0}^{\omega_c} \left[\frac{\sin{\omega(t-t_0+\tau/2)}}{\omega\tau/2} - \frac{\sin{\omega(t-t_0-\tau/2)}}{\omega\tau/2} \right] d\omega$$

$$(3\text{-}12)$$

These integrals can be expressed in terms of the well-known sine integral by change of variables, and we obtain

$$g(t) = \frac{KV}{\pi} \left(\int_{0}^{\omega_c(t-t_0+\tau/2)} \frac{\sin x}{x}\, dx - \int_{0}^{\omega_c(t-t_0-\tau/2)} \right.$$

$$\left. \times \frac{\sin x}{x}\, dx \right)$$

$$= \frac{KV}{\pi} \left\{ S_i\left[\omega_c\left(t-t_0+\frac{\tau}{2}\right)\right] - S_i\left[\omega_c\left(t-t_0-\frac{\tau}{2}\right)\right] \right\}$$

$$(3\text{-}13)$$

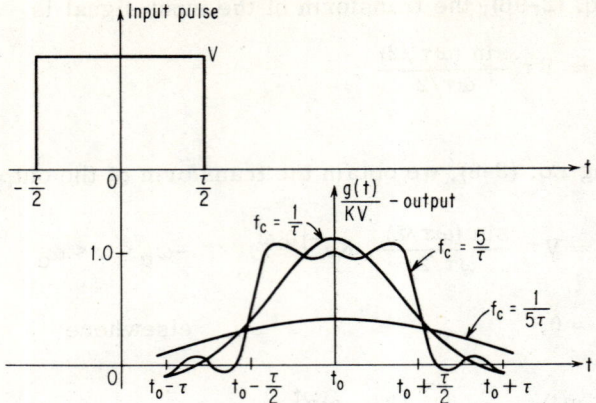

Fig. 3-2. Response of low-pass filter.

where $\quad S_i(x) = \displaystyle\int_0^x \frac{\sin \eta}{\eta}\, d\eta$ $\qquad\qquad$ (3-14)

is the well-known sine integral of x.

The response of the idealized low-pass filter is shown in Fig. 3-2 for different filter bandwidths. Note that the output is displaced t_0 sec from the input and resembles the input for $f_c \gg 1/\tau$, where $f_c = \omega_c/2\pi$ is the cutoff frequency of the low-pass filter, and τ is the width of rectangular input pulse.

3.3 Distortionless Transmission[4-7]

The problem of distortionless transmission is a very basic one in many fields of communication. Distortionless transmission implies an operation where the output signal g(t) looks like the input except for a constant time delay and possible change of scale; i.e., the output is a delayed replica of the input signal. Mathematically, such an operation is given by

$$g(t) = Kf(t - t_0), \qquad K = \text{constant} \qquad\qquad (3\text{-}15)$$

and t_0 is the constant time delay of the filter.

The transfer function $H(j\omega)$ of the filter which will satisfy the requirements for distortionless transmission may be determined as follows. Using Eq. (3-8), we write

$$G(j\omega) = F(j\omega).E(j\omega) = \int_{-\infty}^{\infty} g(t)e^{-j\omega t} \, dt$$

$$= K \int_{-\infty}^{\infty} f(t - t_0)e^{-j\omega t} \, dt \qquad (3-16)$$

Put $t - t_0 = \tau$, $dt = d\tau$;   hence

$$G(j\omega) = K \int_{-\infty}^{\infty} f(\tau)e^{-j\omega(\tau + t_0)} \, d\tau = Ke^{-j\omega t_0} \int_{-\infty}^{\infty} f(\tau)e^{-j\omega \tau} \, d\tau$$

$$= Ke^{-j\omega t_0} F(j\omega) \qquad (3-17)$$

so that $H(j\omega) = Ke^{-j\omega t_0}$ \qquad (3-18)

which represents a linear network with a constant-amplitude response and linear phase shift. A more general expression for distortionless transmission is

$$H(j\omega) = Ke^{-j(\omega t_0 \pm n\pi)} \qquad (3-19)$$

where n is an integer including zero. The condition for distortionless transmission is still satisfied because if n is even, $e^{\pm jn\pi} = 1$, while if n is odd, $e^{\pm jn\pi} = -1$ and in both cases, the amplitude response is constant.

Thus, the conditions for distortionless transmission are:

1. $|H(j\omega)| = K$, a constant \qquad (3-20)

2. $\theta(\omega) = \omega t_0 \pm n\pi$ \qquad (3-21)

which means that for ideal transmission, the gain must be constant independent of frequency, while the phase shift should be a straight line passing through zero or a multiple of π at $\omega = 0$. The slope of this line is $d\theta(\omega)/d\omega = t_0$, the time delay of the network. The phase shift at zero frequency $\pm n\pi$ is known as the linear phase-shift intercept, where n is an integer. If n is not an integer, the output signal will not be a replica of the input, and this will result in distortion. For instance, if $n = 1/2$, the cosine components become sine, and vice versa, which may cause extreme modification of the waveshape.

1. *Distortionless Transmission of a Bandpass Filter.* If the spectrum of the transmitted signal is limited to a band of frequencies, the above conditions should hold over that band only. This can

be shown as follows: let the input signal consist of two parts, of $f_1(t)$ and $f_2(t)$, where $f_1(t)$ is to be transmitted and $f_2(t)$ to be rejected by the network. We further suppose that the frequency of $f_1(t)$ is band-limited in the range $\omega_1 < \omega < \omega_2$, where ω_1 and ω_2 are the lower and upper limits of the frequency range of the bandpass network. The problem is to find the characteristic of $H(j\omega)$ of the network for distortionless transmission. Given

$$f(t) = f_1(t) + f_2(t) \tag{3-22}$$

and $\quad F(j\omega) = \int_{-\infty}^{\infty} f(t)e^{-j\omega t}\, dt = F_1(j\omega) + F_2(j\omega) \tag{3-23}$

For distortionless transmission, the output signal $g(t)$ should be of the form

$$g(t) = Kf_1(t - t_0) \tag{3-24}$$

As before we have

$$F(j\omega) \cdot H(j\omega) = \int_{-\infty}^{\infty} g(t)e^{-j\omega t}\, dt \tag{3-25}$$

so that $[F_1(j\omega) + F_2(j\omega)] H(j\omega) = K \int_{-\infty}^{\infty} f_1(t - t_0)e^{-j\omega t}\, dt \tag{3-26}$

If we substitute $t - t_0 = \tau$, the equation reduces to

$$[F_1(j\omega) + F_2(j\omega)] H(j\omega) = Ke^{-j\omega t_0} \cdot F_1(j\omega) \tag{3-27}$$

Since this equation is valid for all frequencies, it must be valid for all values of ω at which $F_1(j\omega) \neq 0$ and $F_2(j\omega) = 0$. Consider the frequency range $-\omega_2 < \omega < -\omega_1$ and $\omega_1 < \omega < \omega_2$ in which $F_2(j\omega) = 0$ and $F_1(j\omega) \neq 0$, so that

$$F_1(j\omega) \cdot H(j\omega) = Ke^{-j\omega t_0} \cdot F_1(j\omega) \tag{3-28}$$

It follows therefore that in the passband

$$H(j\omega) = Ke^{-j\omega(t_0 \pm n\pi)} \tag{3-29}$$

and outside the passband

$$H(j\omega) = 0$$

Thus for an ideal bandpass filter, we require

$$H(j\omega) = K, \qquad \omega_1 \le \omega \le \omega_2 \qquad\qquad (3\text{-}30)$$

a constant-amplitude characteristic in the transmission band, and infinite attenuation outside the band. The phase shift must be a straight line in the passband

$$\theta(\omega) = \omega t_0 \pm n\pi \qquad\qquad (3\text{-}31)$$

The phase shift at $\omega = 0$ is $n\pi$, but at the mid-band frequency ω_c, the phase shift can have any value. This is illustrated in Fig. 3-3, for both the low-pass and bandpass filter networks.

We have just established that a distortionless linear system must have a flat amplitude response and a linear phase-shift characteristic over the range of the significant spectral bandwidth of the input signal. In practice, it may not be possible to fulfill these requirements, and a certain amount of signal distortion will be introduced in the output signal. If the filter characteristics are

$$A(\omega) \neq K \qquad \theta(\omega) = \omega t_c \pm n\pi$$

the filter is called amplitude-distorted; if

$$A(\omega) = K \qquad \theta(\omega) \neq \omega t_0 \pm n\pi$$

it is phase-distorted. The effect of amplitude and phase distortion will be discussed in the following section.

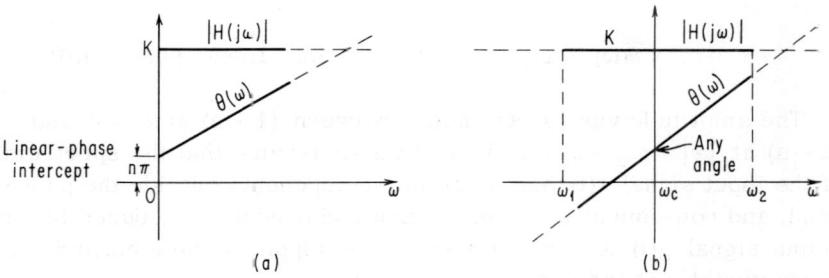

Fig. 3-3. Ideal characteristics for distortionless transmission: (a) Low-pass; (b) bandpass.

3.4   Distortion Analysis Using the Method of Paired Echoes[4, 7, 8]

We have seen that for distortionless transmission, the amplitude response of the network in the passband should be constant and the phase response linear. We shall consider now the effect on the output wave of a network which is not flat in the passband, or whose phase is not linear.

1. *First-order Amplitude Distortion.*   We shall consider first the case where the passband is not flat, as shown in Fig. 3-4. Since the

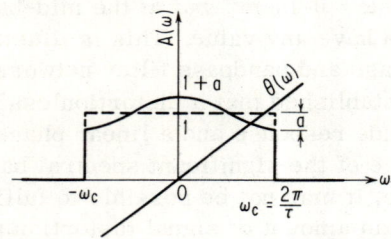

Fig. 3-4. Amplitude characteristic of bandpass network.

amplitude characteristic of the network must be an even function of ω, we shall assume the characteristic of the bandpass filter to be given by

$$A(\omega) = \left|H(j\omega)\right| = 1 + a \cos \frac{\tau}{2} \omega, \qquad -\frac{2\pi}{\tau} < \omega < \frac{2\pi}{\tau} \qquad (3\text{-}32)$$

$$= 0, \qquad\qquad\qquad \text{elsewhere}$$

$$\theta(\omega) = \omega t_o, \qquad\qquad \text{i.e., linear phase shift}$$

The amplitude varies, as shown, between $(1 + a)$ at $\omega = 0$ and $(1 - a)$ at $\left|\omega\right| = \omega_c = 2\pi/\tau$. We shall also assume that the spectrum of the input signal $f(t)$ has negligible components outside the passband, and consequently, the distortion components introduced in the output signal $g(t)$ are due entirely to the ripple in the amplitude characteristic of the network.

The response $g(t)$ will be obtained as follows. The transfer characteristic of the network is

$$H(j\omega) = A(\omega)e^{-j\theta(\omega)} = \left(1 + a\cos\frac{\tau}{2}\omega\right)e^{-j\omega t_0}$$

$$= e^{-j\omega t_0} + \frac{1}{2}ae^{-j\omega(t_0 - \tau/2)}$$

$$+ \frac{1}{2}ae^{-j\omega(t_0 + \tau/2)} \qquad (3\text{-}33)$$

Hence the output spectrum $G(j\omega)$ is given by

$$G(j\omega) = F(j\omega) \cdot H(j\omega)$$

$$= F(j\omega)\left(e^{-j\omega t_0} + \frac{1}{2}ae^{-j\omega(t_0 - \tau/2)} + \frac{1}{2}ae^{-j\omega(t_0 + \tau/2)}\right)$$

$$(3\text{-}34)$$

From the last expression, we note that the output signal $g(t)$ consists of three terms which by the use of the shift theorem are

$$g(t) = f(t - t_0) + \frac{a}{2}f\left(t - t_0 + \frac{\tau}{2}\right) + \frac{a}{2}f\left(t - t_0 - \frac{\tau}{2}\right) \qquad (3\text{-}35)$$

This is plotted in Fig. 3.5, where it can be seen that in addition to the main signal which is delayed by t_0, the response consists of two symmetrically spaced smaller signals of amplitude $\frac{1}{2}a$. The accompanying signals which are delayed $\pm\tau/2$ from the main signals are termed paired echoes and represent the distortion in the transmitted signal.

In the general case, it is possible to represent the amplitude characteristics of the transfer function of the bandpass by means of a Fourier series where each term will give rise to a pair of echoes. The resulting distortion terms in the output signal will consist of a set of symmetrical paired echoes.

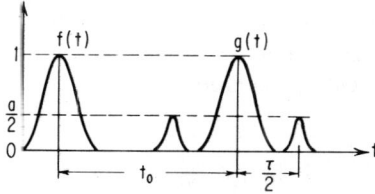

Fig. 3-5. Response of bandpass filter with cosine variation of amplitude characteristic.

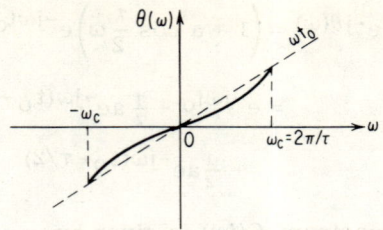

Fig. 3-6. Nonlinear phase characteristic of bandpass network.

2. *First-order Phase Distortion.*   Now we shall consider the case of a nonlinear phase characteristic, as shown in Fig. 3.6. Since the phase characteristic is an odd function, it can be represented by a sine series. As a first approximation, we shall consider $\theta(\omega)$ to be of the form

$$\theta(\omega) = \omega t_0 - b \sin \frac{\tau}{2}\omega \qquad (3\text{-}36)$$

Since $A(\omega) = |H(j\omega)| = K, \qquad -\omega_c < \omega < \omega_c$

we have $H(j\omega) = Ke^{-j[\omega t_0 - b \sin (\tau/2)\omega]}$ $\qquad (3\text{-}37)$

Therefore $G(j\omega) = F(j\omega)H(j\omega) = KF(j\omega)e^{-j[\omega t_0 - b \sin (\tau/2)\omega]}$

$$(3\text{-}38)$$

Now from the theory of Bessel functions we have

$$e^{jb \sin \theta} = \sum_{n=-\infty}^{\infty} J_n(b)e^{jn\theta} \qquad (3\text{-}39)$$

where the J's are the Bessel functions of the first kind. If b is assumed to be small such that the higher-order Bessel functions can be neglected, then $G(j\omega)$ reduces to

$$G(j\omega) = KF(j\omega)e^{-j\omega t_0} [J_0(b) + (e^{j(\tau/2)\omega} - e^{-j(\tau/2)\omega})J_1(b)]$$

$$= KJ_0(b)F(j\omega)e^{-j\omega t_0} + KJ_1(b)F(j\omega)e^{-j\omega(t_0 - \tau/2)}$$

$$- KJ_1(b)F(j\omega)e^{-j\omega(t_0 + \tau/2)} \qquad (3\text{-}40)$$

Now applying the shift theorem, we obtain for the output transient g(t) the following expression:

$$g(t) = KJ_0(b)f(t - t_0) + KJ_1(b)f\left(t - t_0 + \frac{T}{2}\right) - KJ_1(b)f\left(t - t_0 - \frac{T}{2}\right)$$

(3-41)

If we further approximate for small b,

$$J_0(b) \doteq 1 \quad \text{and} \quad J_1(b) \doteq \tfrac{1}{2} b$$

the transient response g(t) (for K = 1) becomes

$$g(t) = f(t - t_0) + \tfrac{1}{2}bf\left(t - t_0 + \frac{T}{2}\right) - \tfrac{1}{2}bf\left(t - t_0 - \frac{T}{2}\right)$$

(3-42)

As before, the principal component $f(t - t_0)$ is accompanied by a pair of echoes of decreased amplitude. Note however that the echoes are one leading and one lagging, making the pair skew-symmetric, as shown in Fig. 3-7.

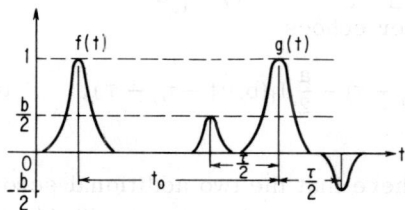

Fig. 3-7. Response of bandpass network with a nonlinear phase characteristic.

3. *Combined First-order Amplitude and Phase Distortion.* In this case, the transfer characteristics of the bandpass filter are given by

$$H(j\omega) = A(\omega) = 1 + a \cos \frac{T}{2}\omega$$

(3-43)

$$\theta(\omega) = \omega t_0 - b \sin \frac{T}{2}\omega$$

Proceeding as before, we obtain the following expression for the output signal $g(t)$:

$$g(t) = J_0(b)f(t - t_0) + J_1(b)f\left(t - t_0 + \frac{\tau}{2}\right) - J_1(b)f\left(t - t_0 - \frac{\tau}{2}\right)$$

$$+ \frac{a}{2}J_0(b)f\left(t - t_0 + \frac{\tau}{2}\right) + \frac{a}{2}J_0(b)f\left(t - t_0 - \frac{\tau}{2}\right)$$

$$+ \frac{a}{2}J_1(b)f(t - t_0 + \tau) - \frac{a}{2}J_1(b)f(t - t_0) + \frac{a}{2}J_1(b)f(t - t_0)$$

$$- \frac{a}{2}J_1(b)f(t - t_0 - \tau) \tag{3-44}$$

As shown in Fig. 3-8, the output $g(t)$ consists of the following components:

(1) $J_0(b)f(t - t_0)$,   main undistorted transmitted signal delayed by time delay of network t_0

(2) $\left[J_1(b) + \frac{a}{2}J_0(b)\right] f\left(t - t_0 + \frac{\tau}{2}\right) + \left[\frac{a}{2}J_0(b) - J_1(b)\right]f\left(t - t_0 - \frac{\tau}{2}\right)$
four first-order echoes

(3) $\frac{a}{2}J_1(b)f(t - t_0 + \tau) - \frac{a}{2}J_1(b)f(t - t_0 - \tau)$,   two second-order echoes

It should be noted here that the two additional second-order echoes which appear in the output as given by Eq. (3-44) canceled each other out only because the same constant $\omega_c = 2\pi/\tau$ was assumed for both amplitude and phase distortion.

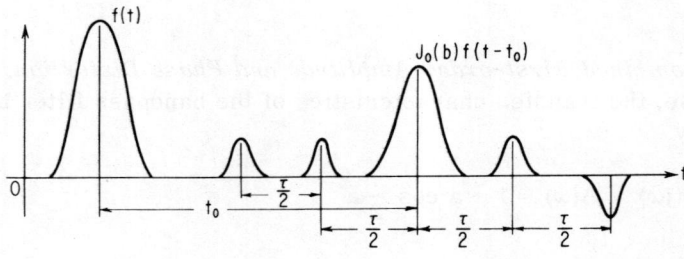

Fig. 3-8. Combined amplitude and phase distortion echoes.

3.5 Impulse Response of Linear Systems and the Convolution Integral[1,3,4]

The general problem of finding the system response g(t) corresponding to a transient input function f(t) can be solved from a knowledge of the system response to a family of simple singularity functions such as the unit impulse function or step function. In particular, we shall find the impulse response to a unit impulse function which is denoted by h(t). Thus, if

$$f(t) = \delta(t)$$

then $g(t) = h(t)$

Also, if $f(t) = \delta(t - t_0)$

then $g(t) = h(t - t_0)$

The relationship between the impulse response h(t) and the system transfer function $H(j\omega)$ can be derived as follows:
Since the Fourier transform of the impulse function is

$$F(j\omega) = \int_{-\infty}^{\infty} \delta(t)e^{-j\omega t}\, dt = 1 \tag{3-45}$$

and $G(j\omega) = H(j\omega)F(j\omega)$, it follows, therefore, that the Fourier transform of the impulse response h(t) due to an impulse time function $\delta(t)$ is

$$G(j\omega) = H(j\omega)$$

or $H(j\omega)$ and h(t) are Fourier transforms of one another. Therefore,

$$H(j\omega) = \int_{-\infty}^{\infty} h(t)e^{-j\omega t}\, dt \tag{3-46}$$

and $$h(t) = \frac{1}{2\pi} \int_{-\infty}^{\infty} H(j\omega)e^{j\omega t}\, d\omega \tag{3-47}$$

with $$H(j\omega) = P(\omega) + jQ(\omega) = A(\omega)e^{j\phi(\omega)} \tag{3-48}$$

we have from Eq. (2-74)

$$h(t) = \frac{1}{\pi} \int_0^\infty A(\omega) \cos [\omega t + \phi(\omega)] \, d\omega \qquad (3\text{-}49)$$

and from Eq. (2-77) we have

$$h(t) = \frac{2}{\pi} \int_0^\infty P(\omega) \cos \omega t \, d\omega = -\frac{2}{\pi} \int_0^\infty Q(\omega) \sin \omega t \, d\omega,$$

$$t > 0 \qquad (3\text{-}50)$$

For example, we shall find the impulse response of the ideal low-pass filter of Sec. 3.2:

$$H(j\omega) = K e^{-j\omega t_0}, \qquad |\omega| \le \omega_c$$

$$= 0, \qquad\qquad \text{elsewhere}$$

Let $f(t) = \delta(t)$. Therefore $F(j\omega) = 1$ and $G(j\omega) = H(j\omega)$. The impulse response

$$h(t) = \frac{1}{\pi} \int_0^\infty A(\omega) \cos [\omega t + \phi(\omega)] \, d\omega = \frac{K}{\pi} \int_0^\infty \cos \omega(t - t_0) \, d\omega$$

$$= \frac{K\omega_c}{\pi} \cdot \frac{\sin \omega_c(t - t_0)}{\omega_c(t - t_0)} \qquad (3\text{-}51)$$

with $h_{max} = K\omega_c/\pi$, as shown in Fig. 3-9.

This result can be derived directly from the duality theorem discussed in Chap. 2, since $h(t)$ is the inverse transform of $H(j\omega)$.

For the response of a system to a unit step, the reader is referred to Papoulis,[4] where many interesting applications are given.

The physical significance of Eqs. (3-46) and (3-47) is that the output frequency spectrum must be just the frequency spectrum of the network itself. Or, if we measure the output frequency spectrum of any linear system with an impulse applied at the input, we can determine the network transfer function. This implies that an impulse function may be used as a test signal in order to determine the transfer function of the linear network. The network transfer function is then measured by sweeping the desired range with a narrow-band receiver at the output of the network.

In practice, a periodic series of very narrow pulses might be em-

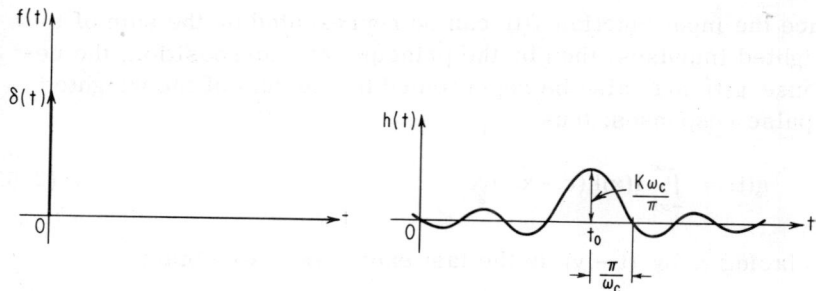

Fig. 3-9. Response of a low-pass filter to a unit impulse.

ployed so that a repetitive pattern of the frequency spectrum would
be obtained. The spectrum of a series of narrow pulses contains
discrete frequencies (multiples of the repetition frequency); but with
a slow enough repetition rate, the spectral lines will be spaced
closely enough to approximate the continuous spectrum of a single
narrow pulse or impulse as discussed in Chap. 2.

1. *Convolution Integral.*   We shall now show how to derive the
transient response $g(t)$ caused by a signal $f(t)$ when the impulse
response $h(t)$ of the linear system is known.

Consider an arbitrary time function $f(t)$, as shown in Fig. 3-10.
This function may be represented by a series of impulses of varying weight. In particular, consider the impulse function of area
$f(x)\,dx$ applied at $t = x$. Using the integral definition of the delta
function, the impulse function at $t = x$ is given by $f(x)\,dx\,\delta(t-x)$,
and the response is by definition $f(x)\,dx\,h(t-x)$. The input function
$f(t)$ is considered to be a linear superposition of these impulse functions; thus

$$f(t) = \int_{-\infty}^{\infty} f(x)\,\delta(t-x)\,dx \tag{3-52}$$

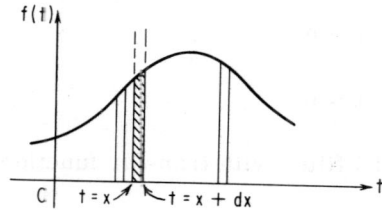

Fig. 3-10. Impulse representation of a function.

Since the input function $f(t)$ can be represented by the sum of the weighted impulses, then by the principle of superposition, the response $g(t)$ may also be represented by the sum of the weighted impulse responses; thus

$$g(t) = \int_{-\infty}^{\infty} f(x)h(t-x) \; dx \tag{3-53}$$

Replacing x by $(t-y)$ in the last expression, we obtain

$$g(t) = -\int_{\infty}^{-\infty} f(t-y)h(y) \; dy = \int_{-\infty}^{\infty} f(t-y)h(y) \; dy \tag{3-54}$$

Hence $\quad g(t) = \int_{-\infty}^{\infty} f(t-x)h(x) \; dx = \int_{-\infty}^{\infty} f(x)h(t-x) \; dx \tag{3-55}$

This result is referred to as the superposition integral or as the convolution of $f(t)$ and $h(t)$. This is sometimes denoted by

$$g(t) = h(t) * f(t) \tag{3-56}$$

Convolution in the time domain occurs in communication systems whenever a signal is transmitted through a fixed linear network, while convolution in the frequency domain occurs whenever a carrier wave is amplitude-modulated. This follows from the fact that the Fourier transform of a product of two time functions equals the convolution of their corresponding spectra, as shown in Chap. 2. However, as we have noted in Chap. 2, amplitude modulation of a carrier is equivalent to shifting or translation of the spectrum of the modulating signal by the carrier frequency. This concept will be used extensively in the next section.

We shall demonstrate the use of the convolution integral in the solution of a simple problem. Assume that $f(t)$ is a simple exponential defined for positive time only:

$$f(t) = e^{-at}, \qquad t > 0$$

$$f(t) = 0, \qquad t < 0 \tag{3-57}$$

$f(t)$ is applied to a RC filter with transfer function

$$H(j\omega) = \frac{1/RC}{j\omega + 1/RC}$$

The problem is to find the output signal g(t). Applying the convolution integral, we have

$$g(t) = \int_{-\infty}^{\infty} f(\tau)h(t - \tau) \, d\tau \tag{3-58}$$

Since $\qquad\qquad f(\tau) = e^{-a\tau}, \qquad\qquad\qquad \tau > 0 \tag{3-59}$

therefore $\quad h(t - \tau) = \dfrac{1}{RC} e^{-(t - \tau)/RC}, \qquad t > \tau \tag{3-60}$

[This follows from the result $h(t) = (1/RC)e^{-t/RC}$ for the RC network.] Since $f(t) = 0$, $t < 0$ and $h(t - \tau) = 0$, $\tau > t$, we obtain

$$g(t) = \int_{-\infty}^{\infty} f(\tau)h(t - \tau) \, d\tau = \frac{1}{RC} \int_{0}^{t} e^{-a\tau} \cdot e^{-(t-\tau)/RC} \, d\tau$$

$$= \frac{e^{-t/RC}}{RC} \cdot \frac{e^{\tau(1/RC - a)}}{1/RC - a} \bigg|_{0}^{t} = \frac{e^{-at} - e^{-t/RC}}{1 - aRC} \tag{3-61}$$

We shall show now that the convolution integral satisfies the transform relations of Eq. (3-8), namely,

$$G(j\omega) = F(j\omega) \cdot H(j\omega)$$

Taking the Fourier transform of both sides of Eq. (3-55), we obtain

$$G(j\omega) = \int_{-\infty}^{\infty} e^{-j\omega t} \left[\int_{-\infty}^{\infty} f(x)h(t - x) \, dx \right] dt \tag{3-62}$$

and assuming absolute convergence of the integrals, we interchange the order of integration, so that

$$G(j\omega) = \int_{-\infty}^{\infty} f(x) \left[\int_{-\infty}^{\infty} e^{-j\omega t}h(t - x) \, dt \right] dx \tag{3-63}$$

But $\quad \displaystyle\int_{-\infty}^{\infty} e^{-j\omega t}h(t - x) \, dt = e^{-j\omega x} \cdot H(j\omega) \tag{3-64}$

Therefore $\quad G(j\omega) = H(j\omega) \displaystyle\int_{-\infty}^{\infty} f(x)e^{-j\omega x} \, dx = H(j\omega) \, F(j\omega) \tag{3-65}$

This result could have been derived directly from Eq. (2-84), whose physical interpretation is that the spectrum of a convolution of two time functions equals the product of the spectra, and keeping in mind that $H(j\omega)$ is the transform of $h(t)$.

3.6 Bandpass Filters and Equivalent[4,9]

Bandpass filters are very commonly used in modulation systems, and consequently their response to modulated signals has an important bearing on the system fidelity for reproducing the message function without appreciable distortion. In this section, special techniques will be developed for analyzing the response of such filters to band-limited signals, leading to simple results. Specifically, it will be shown how to analyze high-frequency problems by means of low-frequency analogues. First we shall prove the low-pass band-pass transformation theorem, and then we shall show that the impulse response h(t) of the bandpass filter is a modulated signal that can readily be derived from the response of an appropriately chosen low-pass filter. Finally, it will be shown that the response of the bandpass filter of center frequency ω_0 to an input signal of the form

$$f(t) = f_\ell(t) \cos \omega_0 t \qquad (3-66)$$

where $f_\ell(t)$ is a slowly varying envelope function, is derived from the response of a low-pass filter to the envelope $f_\ell(t)$.

1. *Equivalent Low-pass Filter*. We consider a symmetrical bandpass filter whose transfer function $H(j\omega)$ is as shown in Fig. 3-11a. From Figure 3-11b it is seen that

$$H(j\omega) = H_1(j\omega) + H_2(j\omega) \qquad (3-67)$$

The transfer function $H_\ell(j\omega)$ of Fig. 3-11c is obtained by shifting $H_1(j\omega)$ to the left by ω_0.

$$H_\ell(j\omega) = H_1[j(\omega + \omega_0)] \qquad H_\ell[j(\omega - \omega_0)] = H_1(j\omega) \qquad (3-68)$$

This is called the equivalent low-pass filter. Similarly, $H_\ell(j\omega)$ can also be obtained by shifting $H_2(j\omega)$ to the right:

$$H_\ell(j\omega) = H_2[j(\omega - \omega_0)] \qquad H_\ell[j(\omega + \omega_0)] = H_2(j\omega) \qquad (3-69)$$

Now we shall prove that the impulse response h(t) of a symmetrical bandpass filter of center frequency ω_0 is an amplitude-modulated signal,

$$h(t) = 2h_\ell(t) \cos \omega_0 t \qquad (3-70)$$

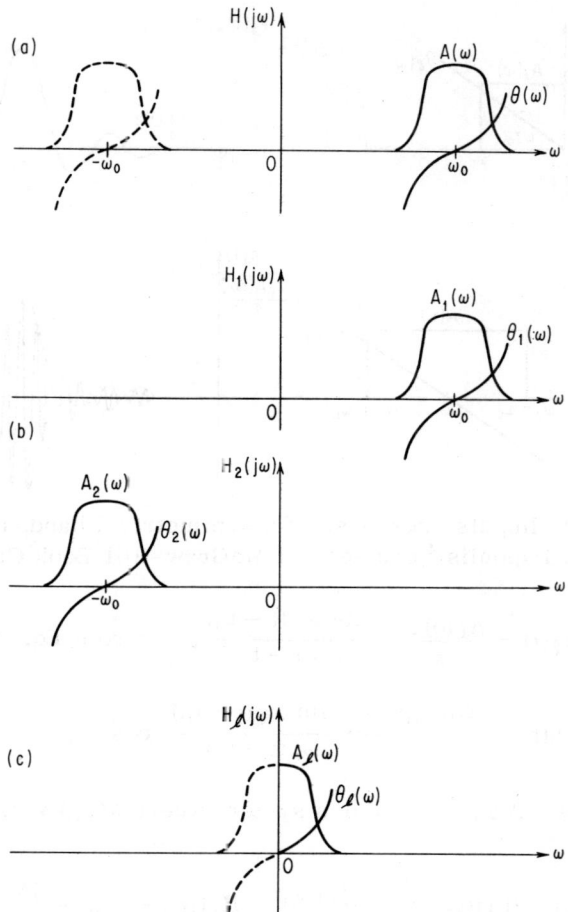

Fig. 3-11. Equivalent low-pass filter. (From A. Papoulis,[4] courtesy of McGraw-Hill Book Company.)

where $h_\ell(t)$ is the impulse response of the equivalent low-pass filter. From Eqs. (3-67) to (3-69) we obtain

$$H(j\omega) = H_\ell[j(\omega - \omega_0)] + H_\ell[j(\omega + \omega_0)] \qquad (3-71)$$

Therefore $\quad h(t) = h_\ell(t)e^{j\omega_0 t} + h_\ell(t)e^{-j\omega_0 t} = 2h_\ell(t) \cos \omega_0 t \qquad (3-72)$

This result can be used to find the impulse response of an ideal bandpass filter of center frequency ω_0 and bandwidth $2\omega_c$, which is shown in Fig. 3-12 with its equivalent low-pass filter. Since

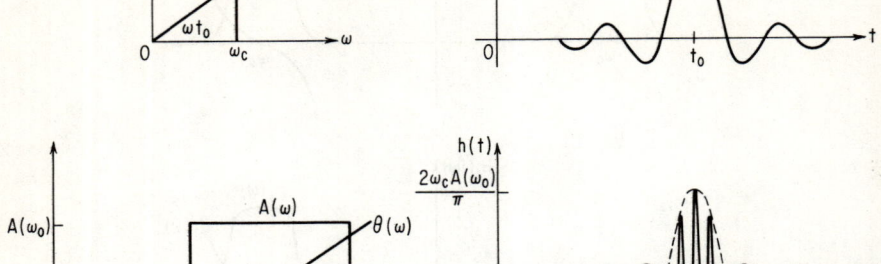

Fig. 3-12. Impulse response of a symmetrical bandpass filter.
(From A. Papoulis,[4] courtesy of McGraw-Hill Book Company.)

$$h_\ell(t) = \frac{A_\ell(0)\,\omega_c}{\pi}\;\frac{\sin \omega_c(t - t_0)}{\omega_c(t - t_0)}, \qquad \text{from Eq. (3-51)}$$

therefore

$$h(t) = \frac{2A(\omega_0)\omega_c}{\pi} \cdot \frac{\sin \omega_c(t - t_0)}{\omega_c(t - t_0)}\; \cos \omega_0 t \tag{3-73}$$

where $A_\ell(0) = A(\omega_0) = K$. For a system where $\theta(\omega_0) \neq 0$, it can be shown[4] that

$$H_\ell j(\omega) = H_1[j(\omega + \omega_0)]e^{j\,\theta(\omega_0)} = H_2[j(\omega - \omega_0)]\,e^{-j\,\theta(\omega_0)} \tag{3-74}$$

and

$$h(t) = 2h_\ell(t)\,\cos\,[\omega_0 t - \theta(\omega_0)] \tag{3-75}$$

2. *Response to a Modulated Input.* Now we shall find the response $g(t)$ of the symmetrical bandpass filter to an amplitude-modulated input:

$$f(t) = f_\ell(t)\,\cos\,\omega_0 t \tag{3-76}$$

We shall assume that the spectral components of the envelope $f_\ell(t)$ are narrow-band in the sense that the Fourier transform of $f_\ell(t)$ is $F_\ell(j\omega) = 0$, $|\omega| > \omega_0$. We shall show that

$$g(t) = g_\ell(t)\,\cos\,[\omega_0 t - \theta(\omega_0)] \tag{3-77}$$

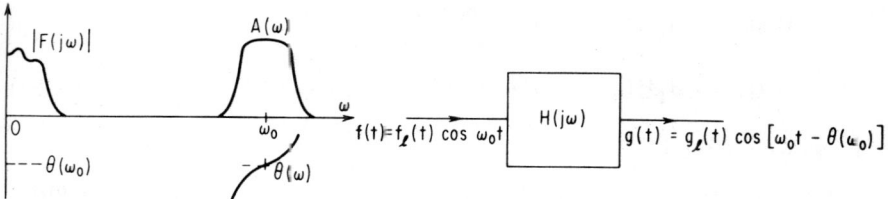

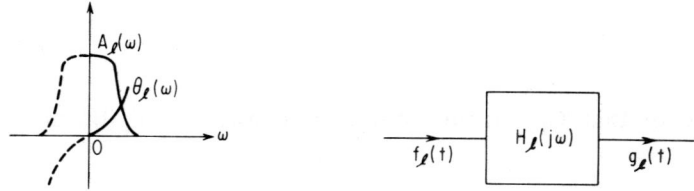

Fig. 3-13. Response to a modulated input. (From A. Papoulis,[4] courtesy of McGraw-Hill Book Company.)

where $g_\ell(t)$ is the response of the equivalent low-pass filter to $f_\ell(t)$. (See Fig. 3-13.) The Fourier transform of $f(t)$ is given by

$$F(j\omega) = \frac{F_\ell[j(\omega + \omega_0)] + F_\ell[j(\omega - \omega_0)]}{2} \qquad (3\text{-}78)$$

which follows from Eq. (3-76) and the shift theorem. Now, since

$$G(j\omega) = F(j\omega) \cdot H(j\omega)$$

we obtain $$G(j\omega) = \frac{F_\ell[j(\omega + \omega_0)] + F_\ell[j(\omega - \omega_0)]}{2}$$

$$\times [H_1(j\omega) + H_2(j\omega)] \qquad (3\text{-}79)$$

Also because the spectrum of $F_\ell(j\omega) = 0$ for $|\omega| > \omega_0$, it follows that

$$F_\ell[j(\omega - \omega_0)] \, H_2(j\omega) = 0 \qquad \text{and} \qquad F_\ell[j(\omega + \omega_0)] \, H_1(j\omega) = 0$$

so that $$G(j\omega) = \frac{F_\ell[j(\omega - \omega_0)] \, H_1(j\omega) + F_\ell[j(\omega + \omega_0)] \, H_2(j\omega)}{2} \qquad (3\text{-}80)$$

but from Eq. (3-74) we have

$$H_1(j\omega) = H_\ell[j(\omega - \omega_o)]\, e^{-j\theta(\omega_o)}$$

and $$H_2(j\omega) = H_\ell[j(\omega + \omega_o)]\, e^{j\theta(\omega_o)} \qquad (3\text{-}81)$$

Therefore $G(j\omega) =$

$$\frac{F_\ell[j(\omega-\omega_o)]H_\ell[j(\omega-\omega_o)]e^{-j\theta(\omega_o)} + F_\ell[j(\omega+\omega_o)]H_\ell[j(\omega+\omega_o)]e^{j\theta(\omega_o)}}{2}$$

$$(3\text{-}82)$$

We note however that $G_\ell(j\omega)$, the Fourier transform of $g_\ell(t)$, is given by

$$G_\ell(j\omega) = F_\ell(j\omega)\, H_\ell(j\omega) \qquad (3\text{-}83)$$

so that $$G(j\omega) = \frac{G_\ell[j(\omega - \omega_o)]e^{-j\theta(\omega_o)} + G_\ell[j(\omega + \omega_o)]e^{j\theta(\omega_o)}}{2}$$

$$(3\text{-}84)$$

Taking the inverse transforms of both sides, we obtain

$$g(t) = \frac{g_\ell(t)e^{j[\omega_o t - \theta(\omega_o)]} + g_\ell(t)e^{-j[\omega_o t - \theta(\omega_o)]}}{2}$$

$$= g_\ell(t)\cos[(\omega_o t - \theta(\omega_o))] \qquad (3\text{-}85)$$

We have just shown that the response of a symmetrical bandpass filter to the high-frequency modulated signal $f(t) = f_\ell(t)\cos\omega_o t$ can be expressed in the form $g(t) = g_\ell(t)\cos[\omega_o t - \theta(\omega_o)]$, where $g_\ell(t)$ is the response of the equivalent low-pass filter to the envelope function $f_\ell(t)$ which describes the instantaneous amplitude of the excitation.

A simplified alternative method of finding the response $g(t)$ of a symmetrical bandpass filter to the amplitude-modulated input $f(t) = f_\ell(t)\cos\omega_o t$ will now be given.

Let the input to a bandpass filter be represented by

$$f(t) = f_\ell(t)e^{j\omega_o t} \qquad (3\text{-}86)$$

and hence the output $g(t)$ is given by

$$g(t) = \frac{1}{2\pi} \int_{-\infty}^{\infty} F(j\omega)H(j\omega)e^{j\omega t} \, d\omega$$

$$= \frac{1}{2\pi} \int_{-\infty}^{\infty} F_\ell[j(\omega - \omega_c)] H(j\omega)e^{j\omega t} \, d\omega \qquad (3\text{-}87)$$

Put $\omega - \omega_0 = \omega_1$, or $\omega = \omega_1 + \omega_0$; hence

$$g(t) = \left\{ \frac{1}{2\pi} \int_{-\infty}^{\infty} F_\ell(j\omega_1)H[j(\omega_1 + \omega_0)] e^{j\omega_1} \, d\omega_1 \right\} e^{j\omega_0 t}$$

$$= \left[\frac{1}{2\pi} \int_{-\infty}^{\infty} F_\ell(j\omega_1)H_\ell(j\omega_1)e^{j\omega_1} \, d\omega_1 \right] e^{j\omega_0 t}$$

$$= g_\ell(t)e^{j\omega_0 t} \qquad (3\text{-}88)$$

The response of an unsymmetrical filter to an amplitude-modulated signal $f(t) = f_\ell(t) \cos \omega_0 t$ is of practical interest to single-sideband transmission. This will be discussed in Chap. 5, where it will be shown that the process of filtering introduces a quadrature component in the output signal $g(t)$, causing $g(t)$ to become amplitude- and phase-modulated. Similarly, it will be shown that the impulse response $h(t)$ of such a filter is also amplitude- and phase-modulated.

The quadrature components of the single-sideband signal will be shown to form a Hilbert transform pair. As we shall discover later on, Hilbert transforms play an important role in communication theory, specifically in its application to single sideband. In the following section, it will be shown that the real and imaginary components of the transfer function $H(j\omega)$ also form a Hilbert transform pair. In Chap. 5, some of the properties of Hilbert transforms will be discussed in connection with the concept of the analytic signal and the problem of compatible single sideband detection.

3.7 Relationship between the Frequency Characteristics: Hilbert Transforms[1, 4, 9-12]

In this section we shall examine the relationship between the frequency characteristics of the transfer function $H(j\omega)$. We shall show that under certain conditions, the transfer function $H(j\omega) = P(\omega) + jQ(\omega)$ can be determined if only one frequency characteristic, either $P(\omega)$ or $Q(\omega)$, is known; similarly, $H(j\omega)$ can be de-

rived if either the amplitude characteristic $A(\omega)$ or the phase characteristic $\phi(\omega)$ is known.

From Eq. (3-50), we note that the real and imaginary components
of $H(j\omega)$ are not independent, but are related by the equation

$$\int_0^\infty P(\omega) \cos \omega t \, d\omega = -\int_0^\infty Q(\omega) \sin \omega t \, d\omega, \qquad t > 0 \qquad (3\text{-}89)$$

From this relationship, we shall presently show that

$$P(\omega) = -\frac{1}{\pi} \int_{-\infty}^\infty \frac{Q(u)}{u - \omega} \, du \qquad (3\text{-}90)$$

and

$$Q(\omega) = \frac{1}{\pi} \int_{-\infty}^\infty \frac{P(u)}{u - \omega} \, du \qquad (3\text{-}91)$$

where Cauchy's principal value is used in the improper integral.
This value is obtained through approaching the point $u = \omega$ symmetrically from both sides, i.e.,

$$\int_{-\infty}^\infty = \lim_{\epsilon \to 0} \left(\int_{-\infty}^{\omega-\epsilon} + \int_{\omega+\epsilon}^\infty \right) \qquad (3\text{-}92)$$

These relations, which are known as Hilbert transforms, will be of
great use in the discussion of single-sideband modulation in Chap. 5.

To prove Eqs. (3-90) and (3-91), we resolve the impulse response
$h(t)$ into its even and odd components, $h_e(t)$ and $h_o(t)$, respectively,
as shown in Fig. 3-14. Since $h(t) = 0$ for $t < 0$, we obtain

$$h_e(t) = h_o(t), \qquad t > 0 \qquad (3\text{-}93)$$

and

$$h_e(t) = -h_o(t), \qquad t < 0 \qquad (3\text{-}94)$$

These relations can be expressed in the following form:

$$h_o(t) = h_e(t) \, \text{sgn} \, t \qquad (3\text{-}95)$$

$$h_e(t) = h_o(t) \, \text{sgn} \, t \qquad (3\text{-}96)$$

where sgn t is the sign function of Fig. 2-31. We shall show now
that the Fourier transforms of $h_e(t)$ and $h_o(t)$ are $P(\omega)$ and $jQ(\omega)$,
respectively.

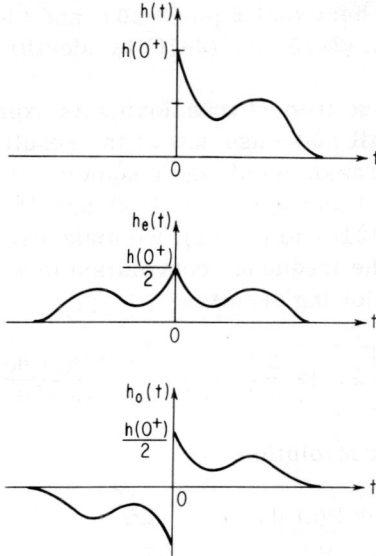

Fig. 3-14. Resolution of the impulse response into even and odd components.

Since
$$H(j\omega) = \int_{-\infty}^{\infty} h(t)e^{-j\omega t} \, dt \qquad (3\text{-}97)$$

therefore
$$P(\omega) + jQ(\omega) = \int_{-\infty}^{\infty} h_e(t) \cos \omega t \, dt - j \int_{-\infty}^{\infty} h_0(t) \sin \omega t \, dt \qquad (3\text{-}98)$$

or
$$P(\omega) = \int_{-\infty}^{\infty} h_e(t) \cos \omega t \, dt = 2 \int_{0}^{\infty} h_e(t) \cos \omega t \, dt \qquad (3\text{-}99)$$

and
$$Q(\omega) = -\int_{-\infty}^{\infty} h_0(t) \sin \omega t \, dt = -2 \int_{0}^{\infty} h_0(t) \sin \omega t \, dt \qquad (3\text{-}100)$$

Similarly
$$h_e(t) = \frac{1}{2\pi} \int_{-\infty}^{\infty} P(\omega) \cos \omega t \, d\omega = \frac{1}{\pi} \int_{0}^{\infty} P(\omega) \cos \omega t \, d\omega \qquad (3\text{-}101)$$

and
$$h_0(t) = -\frac{1}{2\pi} \int_{-\infty}^{\infty} Q(\omega) \sin \omega t \, d\omega = -\frac{1}{\pi} \int_{0}^{\infty} Q(\omega) \sin \omega t \, d\omega \qquad (3\text{-}102)$$

It should be noted here that Eqs. (3-101) and (3-102) can be derived directly from Eqs. (2-75) and (2-76) by identifying $f_e(t) \equiv h_e(t)$ and $f_o(t) \equiv h_o(t)$.

In order to prove Hilbert transforms as expressed by Eqs. (3-90) and (3-91), we shall make use now of the result derived in Chap. 2 that the Fourier transform of sgn t equals $2/j\omega$. Since the Fourier transforms of $h_e(t)$ and $h_o(t)$ are $P(\omega)$ and $Q(\omega)$, respectively, as given by Eqs. (3-101) and (3-102), we make use of Eqs. (3-95) and (3-96) and apply the frequency convolution theorem [Eq. (2-83)]. We obtain then the following results:

$$jQ(\omega) = \frac{1}{2\pi}\left[P(\omega) * \frac{2}{j\omega}\right] = \frac{1}{2\pi}\int_{-\infty}^{\infty} \frac{2P(u)\,du}{j(\omega - u)} \tag{3-103}$$

where $*$ denotes convolution

or $\quad Q(\omega) = \frac{1}{\pi}\int_{-\infty}^{\infty} \frac{P(u)\,du}{u - \omega} \tag{3-104}$

Similarly from Eq. (3-96), we derive

$$P(\omega) = \frac{1}{2\pi}\left[jQ(\omega) * \frac{2}{j\omega}\right] = \frac{-1}{\pi}\int_{-\infty}^{\infty} \frac{Q(u)\,du}{u - \omega} \tag{3-105}$$

Let us now derive a similar relationship between the amplitude and phase characteristics. From Eq. (3-5), we obtain

$$\ln H(j\omega) = \ln A(\omega) + j\phi(\omega)$$

and hence by direct substitution, we obtain the following results:

$$\ln A(\omega) = \frac{-1}{\pi}\int_{-\infty}^{\infty} \frac{\phi(u)\,du}{u - \omega} \tag{3-106}$$

and $\quad \phi(\omega) = \frac{1}{\pi}\int_{-\infty}^{\infty} \frac{\ln A(u)\,du}{u - \omega} \tag{3-107}$

It should be noted here that Hilbert transforms apply only to physically realizable systems, the transfer functions of which are analytic and bounded in the lower half-plane.

The following example will illustrate the application of Hilbert transforms.

Let $P(\omega) = \pi\, \delta(\omega)$, an impulse frequency function of strength π; the problem is to find $Q(\omega)$. From Eq. (3-91), we obtain

$$Q(\omega) = \frac{1}{\pi} \int_{-\infty}^{\infty} \frac{P(u)}{u - \omega}\, du = \frac{1}{\pi} \int_{-\infty}^{\infty} \frac{\pi\, \delta(u)\, du}{u - \omega} = \frac{-1}{\omega}$$

as shown in Fig. 2-32.

Hilbert transforms can also be derived in a different form. From Eqs. (3-101) and (3-102), we obtain

$$\int_{-\infty}^{\infty} P(\omega)\, \cos \omega t\, d\omega = - \int_{-\infty}^{\infty} Q(\omega)\, \sin \omega t\, d\omega, \qquad \text{for } t \geq 0$$

$$(3\text{-}108)$$

since $\quad h_e(t) = h_o(t), \qquad t > 0$

and $\quad h_e(t) = -h_o(t), \qquad t < 0$

we can rewrite Eq. (3-108) to be valid for all values of t as follows:

$$\int_{-\infty}^{\infty} P(\omega)\, \cos \omega t\, d\omega = - \int_{-\infty}^{\infty} Q(\omega)\, \sin \omega |t|\, d\omega, \qquad -\infty < t < \infty$$

$$(3\text{-}109)$$

From Eq. (3-99), we obtain

$$P(\omega) = \int_{-\infty}^{\infty} h_e(t)\, \cos \omega t\, dt = \frac{1}{2\pi} \int_{-\infty}^{\infty} \cos \omega t\, dt \int_{-\infty}^{\infty} P(u)\, \cos ut\, du$$

$$= \frac{-1}{2\pi} \int_{-\infty}^{\infty} \cos \omega t\, dt \int_{-\infty}^{\infty} Q(u)\, \sin u|t|\, du$$

$$= \frac{-2}{\pi} \int_{0}^{\infty} \cos \omega t\, dt \int_{0}^{\infty} Q(u)\, \sin ut\, du \qquad (3\text{-}110)$$

The inverse of this equation can be shown to be[10]

$$Q(\omega) = \frac{-2}{\pi} \int_{0}^{\infty} \sin \omega t\, dt \int_{0}^{\infty} P(u)\, \cos ut\, du \qquad (3\text{-}111)$$

The relations of Eqs. (3-110) and (3-111) are also known as Hilbert transforms. The use of these relations will be illustrated by the following example.[10] Let

$$H(j\omega) = \frac{1}{1 + j\omega} = \frac{1}{1 + \omega^2} - j\frac{\omega}{1 + \omega^2} \qquad (3\text{-}112)$$

Therefore $\quad P(\omega) = \dfrac{1}{1 + \omega^2}$ \hfill (3-113)

and $\qquad Q(\omega) = \dfrac{-\omega}{1 + \omega^2}$ \hfill (3-114)

We shall demonstrate that the imaginary component of $H(j\omega)$ can be derived from its real component

$$Q(\omega) = \frac{-2}{\pi} \int_0^\infty \sin \omega t \, dt \int_0^\infty \frac{1}{1 + u^2} \cos ut \, du$$

$$= \frac{-2}{\pi} \int_0^\infty \frac{\pi}{2} e^{-t} \sin \omega t \, dt$$

$$= \frac{-\omega}{1 + \omega^2} \qquad (3\text{-}115)$$

in agreement with Eq. (3-114). Similarly,

$$P(\omega) = \frac{-2}{\pi} \int_0^\infty \cos \omega t \, dt \int_0^\infty \frac{-u}{1 + u^2} \sin ut \, du$$

$$= \frac{2}{\pi} \cdot \frac{\pi}{2} \int_0^\infty e^{-t} \cos \omega t \, dt$$

$$= \frac{1}{1 + \omega^2} \qquad (3\text{-}116)$$

in agreement with Eq. (3-113).

As stated previously, the application of Hilbert transforms to the theory of single sideband will be given in Chap. 5.

3.8   The Principle of Stationary Phase[4, 7, 13]

In our future discussions, we shall encounter expressions for modulated signals in the form

$$I = \int_{t_1}^{t_2} V(t) \cos \phi(t) \, dt \qquad (3\text{-}117)$$

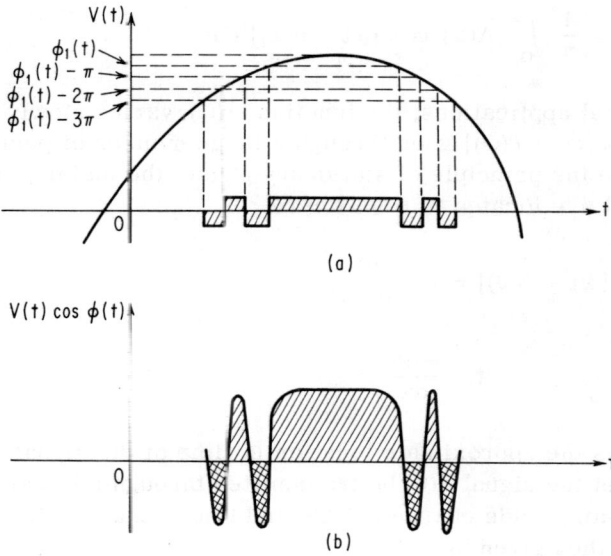

Fig. 3-15. Graphical illustration of the principle of stationary phase. (After S. Goldman,[7] courtesy of McGraw-Hill Book Company.)

where $V(t)$ is a slowly varying function of t compared to $\cos \phi(t)$. In particular, we shall assume that while $\phi(t)$ changes by 2π, the fractional change of the function $V(t)$ is small. As seen from Fig. 3-15, the main contribution to the value of the integral is derived from the regions where $\phi(t)$ has stationary values. Otherwise, those regions in which $\cos \phi(t)$ is negative will tend to cancel the positive contributions from the regions where $\cos \phi(t)$ is positive. The stationary values of $\phi(t)$ are given by the equation

$$\frac{d}{dt}[\phi(t)] = 0 \qquad\qquad (3\text{-}118)$$

The principle of stationary phase refers to the proposition that in an integral of the type of Eq. (3-117), there is a general cancellation of positive and negative portions of the integral except for the ranges of stationary phase. The application of this principle to several practical problems is illustrated in the following examples.

1. *Location of a Signal.* Consider a real function of time $f(t)$ which can be represented according to Eq. (2-74) by the Fourier integral

$$f(t) = \frac{1}{\pi} \int_0^\infty A(\omega) \cos [\omega t + \theta(\omega)] \, d\omega \qquad (3\text{-}119)$$

In practical applications, the function $A(\omega)$ varies slowly with ω while $\cos [\omega t + \theta(\omega)]$ goes through a large number of periods. According to the principle of stationary phase, the major portions of the signal are located in time where

$$\frac{d}{d\omega} [\omega t + \theta(\omega)] = t + \frac{d\theta}{d\omega} = 0 \qquad (3\text{-}120)$$

or $\qquad\qquad\qquad t = \dfrac{-d\theta}{d\omega} \qquad (3\text{-}121)$

This gives the approximate location in time of the signal.

Now let the signal $f(t)$ be transmitted through a linear network of constant-amplitude characteristic and linear phase shift. The output signal is then given by

$$g(t) = \frac{K}{\pi} \int_0^\infty A(\omega) \cos [\omega t + \theta(\omega) - \omega t_0] \, d\omega \qquad (3\text{-}122)$$

and again according to the principle of stationary phase, the major portions of the signal are located in time where

$$\frac{d}{d\omega} [\omega t + \theta(\omega) - \omega t_0] = t - t_0 + \frac{d\theta}{d\omega} = 0 \qquad (3\text{-}123)$$

or $\qquad\qquad\qquad t = t_0 - \dfrac{d\theta}{d\omega} \qquad (3\text{-}124)$

The locations of both the original signal and the transmitted signal are shown in Fig. 3-16.

2. *Spectral Distribution of Frequency Components.* The principle of stationary phase can be used to locate the main frequency ranges in the distribution of the spectral components of a signal. As an example, we consider a frequency-modulated wave which is given by

$$i(t) = e^{j[\omega_c t + \Delta\omega \int_0^t s(t) \, dt]} \qquad (3\text{-}125)$$

where ω_c is the unmodulated carrier angular frequency, $\Delta\omega$ is the

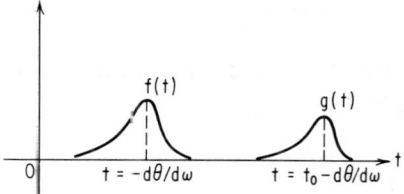

Fig. 3-16. Location of signals using the principle of stationary phase.

peak angular frequency deviation, and s(t) is the modulating information signal. A fairly complete treatment of frequency modulation is given in subsequent chapters, starting with Chap. 7, where it
is shown that the frequency spectrum of the frequency-modulated
signal occupies theoretically an infinite range. In practice, however,
most of the energy of the significant sidebands will be confined to
certain limits. The significant frequency range of i(t) can be determined qualitatively under certain assumptions as follows. Let

$$i(t) = e^{j\omega_c t} \cdot g(t) \tag{3-126}$$

where $g(t) = e^{j\Delta\omega \int_0^t s(t)dt}$ $\qquad$ (3-127)

Let $G(j\omega)$ represent the Fourier transform of g(t). Therefore

$$G(j\omega) = \int_0^T g(t)e^{-j\omega t}\,dt = \int_0^T e^{j[\Delta\omega \int_0^t s(t)dt - \omega t]}\,dt \tag{3-128}$$

where T can be made as great as desired.
We now suppose that, in the epoch $0 \leq t \leq T$, the expression
$\left| \Delta\omega \int_0^T s(t)\,dt \right|$ becomes very large compared with 2π. On this
assumption, it follows from the principle of stationary phase that
for a fixed value of ω, the important contributions to the last integral occur for those values of the integration variable t for which

$$\frac{d}{dt}\left[\Delta\omega \int_0^t s(t)\,dt - \omega t\right] = 0 \tag{3-129}$$

or $\qquad\qquad\qquad\qquad \omega = \Delta\omega s(t) \qquad\qquad\qquad\qquad$ (3-130)

Consequently, the significant sidebands of the spectrum $G(j\omega)$ are
confined to the range

$$\Delta\omega s(t)_{min} \leq \omega \leq \Delta\omega s(t)_{max}$$

or the frequency spectrum of $i(t)$ lies mainly in the range $(\omega_c \pm \Delta\omega)$, provided $|s(t)|_{max} = 1$.

3.9 Group Velocity[4, 7, 14]

In the last two examples, we have used the principle of stationary phase for the location of the principal portions of a signal both in time and in the frequency spectrum. In this section, we shall show that this principle can also be used to locate a signal in space, thus leading to the important concept of group velocity. Consider the propagation of plane waves through space or down a transmission line. A plane wave of a single frequency can be represented analytically by the expression

$$g_1(t) = A \cos \omega \left(t - \frac{x}{V} \right) \tag{3-131}$$

where A is the amplitude, x the distance traveled in time t, and V the velocity of propagation, which is more correctly known as the phase velocity. If the phase velocity is a function of the medium, then in case of several waves of slightly different wavelength, dispersion will exist in the medium which will result in reinforcement and interference along the distance traveled by the waves. The velocity with which the regions of reinforcement or interference advance is known as the group velocity. To find this velocity, consider a complex signal $g(t)$ which consists of a band of frequencies from ω_1 to ω_2, traveling through the dispersive medium, the velocity of propagation differing for each frequency.

The complex signal $g(t)$ may be represented by the Fourier integral

$$g(t) = \frac{1}{\pi} \int_{\omega_1}^{\omega_2} A(\omega) \cos \left[\omega \left(t - \frac{x}{V} \right) \right] d\omega \tag{3-132}$$

By the principle of stationary phase, the major portion of the composite signal $g(t)$ is located where

$$\frac{d}{d\omega} \left[\omega \left(t - \frac{x}{V} \right) \right] = t - \frac{x}{V} + \frac{\omega x}{V^2} \frac{dV}{d\omega} = 0 \tag{3-133}$$

From this, we obtain for the location of the signal, the expression

$$x = \frac{t}{(1/V) - (\omega/V^2)\, dV/d\omega} \tag{3-134}$$

If we denote the group velocity by V_g, then from the relation

$$x = V_g t$$

we obtain the expression

$$\frac{1}{V_g} = \frac{1}{V} - \frac{\omega}{V^2}\frac{dV}{d\omega} = \frac{d}{d\omega}\left(\frac{\omega}{V}\right) \tag{3-135}$$

By the substitution of

$$\frac{\omega}{V} = \frac{2\pi}{\lambda}$$

where λ is the wavelength, it can readily be shown that

$$V_g = V - \lambda\,\frac{dV}{d\lambda} \tag{3-136}$$

The group velocity V_g can be considered as the velocity at which the energy of the disturbance is propagated. In case the phase velocity is independent of the frequency, then $V_g = V$, which corresponds to a nondispersive medium.

3.10   Phase Delay and Envelope Delay[4, 5, 7]

It has been shown in Sec. 3.3 that for distortionless transmission, the phase characteristic must have a linear slope, so that the ratio $\theta(\omega)/\omega$ is constant for all frequencies of the entire signal and thus represents a constant time delay. However in case of nonuniform slope, the transmitted signal will be distorted, depending on the degree of the nonlinearity of the phase characteristic. In order to estimate the degree of distortion, we consider the relative phase shift between adjacent frequency components of frequencies ω and $(\omega + \delta\omega)$ having phase shifts $\theta(\omega)$ and $\theta(\omega) + \delta\theta(\omega)$, as shown in Fig. 3-17. The signal corresponding to these components will be the "beat" wave shown in Fig. 3-18, having the envelope angular fre-

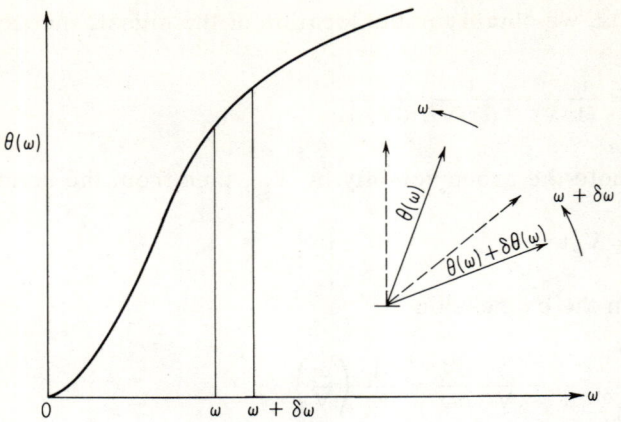

Fig. 3-17. Nonuniform phase-envelope delay.

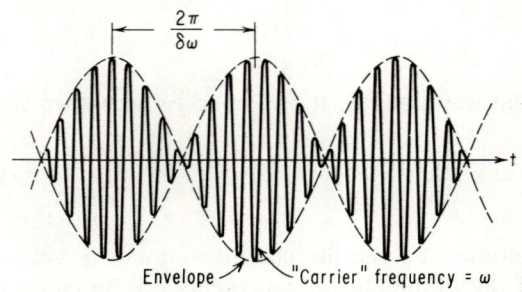

Fig. 3-18. Beat wave.

quency $\delta\omega$. Since the relative shift of the two adjacent signals is $\delta\theta(\omega)$, the time taken for the envelope to reach its peak value will be $\delta\theta(\omega)/\delta\omega$ and is called the "envelope delay" time. Note that this equals the slope of the tangent to the phase curve at the frequency ω. Curves of envelope delay plotted against frequency are more indicative of transient distortion than are curves of phase delay $\theta(\omega)/\omega$.

The distinction between phase delay and envelope delay corresponds to that between phase velocity and group velocity. Let the phase characteristic of Fig. 3-17 represent the phase-shift-frequency characteristic of a transmission line, and let $B(\omega)$ denote the phase shift per unit length of line, which will generally be a function of ω. At any one frequency ω, the phase velocity is then $\omega/B(\omega)$, but the velocity with which a composite signal corresponding to a narrow spectrum ($\omega \pm \delta\omega/2$) is transmitted is equal to

$\delta\omega/\delta B(\omega)$, which is the reciprocal of the slope of the tangent to the phase-shift curve at the point ω. The group velocity $\delta\omega/\delta B(\omega)$ is thus the reciprocal of the group (or envelope) delay time $\delta B(\omega)/\delta\omega$. These concepts of group and phase velocity are very useful in the analysis of the transmission of narrow-band signals through band-pass filters. Let

$$f(t) = V(t) \cos \omega_c t \qquad (3\text{-}137)$$

represent a frequency- or amplitude-modulated signal which is applied to the input of a bandpass filter of constant-amplitude signal; then it can be shown that the output signal $g(t)$ is given by[4]

$$g(t) = V(t - t_{gr}) \cos \omega_c(t - t_o) \qquad (3\text{-}138)$$

where $t_o = \dfrac{\theta(\omega_c)}{\omega_c}$,   carrier delay $(3\text{-}139)$

and $t_{gr} = \dfrac{\delta\theta(\omega)}{\delta\omega}\bigg|_{\omega=\omega_c}$,   group delay $(3\text{-}140)$

so that the group delay evaluated at the carrier frequency equals the delay of the envelope $V(t)$.

References

1. Solodovnikov, V. V.: "Introduction to the Statistical Dynamics of Automatic Control Systems," Dover Publications, Inc., New York.
2. James, H. M., N. B. Nichols, and R. S. Phillips: "Theory of Servomechanisms," MIT Radiation Lab. Series, vol. 25, McGraw-Hill Book Company, New York.
3. Schwartz, M.: "Information Transmission, Modulation, and Noise," McGraw-Hill Book Company, New York.
4. Papoulis, A.: "The Fourier Integral and Its Applications," McGraw-Hill Book Company, New York.
5. Cherry, C.: "Pulses and Transients in Communication Circuits," Dover Publications, Inc., New York.
6. Peskin, E.: "Transient and Steady-state Analysis of Electric Networks," D. Van Nostrand Company, Inc., Princeton, N. J.
7. Goldman, S.: "Frequency Analysis, Modulation, and Noise," McGraw-Hill Book Company, New York.
8. Stuart, R. D.: "An Introduction to Fourier Analysis," Methuen & Co., Ltd., London.

9. Baghdady, E. J. (ed.): "Lectures of Communication System Theory," McGraw-Hill Book Company, New York.
10. Lee, Y. W.: "Statistical Theory of Communication," John Wiley & Sons, Inc., New York.
11. Guillemin, E. A.: "The Mathematics of Circuit Analysis," John Wiley & Sons, Inc., New York.
12. Guillemin, E. A.: "The Theory of Linear Physical Systems," John Wiley & Sons, Inc., New York.
13. Carson, J. R., and T. C. Fry: Variable-frequency Electric Circuit Theory, BSTJ, vol. 16, pp. 513-540, October, 1937.
14. Page, L.: "Introduction to Theoretical Physics," D. Van Nostrand Company, Inc., Princeton, N. J.

4

MATHEMATICAL BACKGROUND II:
RANDOM SIGNAL THEORY—APPLICATION TO
NOISE PROBLEMS

In Chap. 2, we developed mathematical tools for handling periodic and aperiodic (transient) functions which are commonly encountered in the study of communication systems. We have noted that periodic and aperiodic functions of time are completely determined in the time domain for every value of the independent variable, and can be specified over the entire range of frequencies by a transformation from the time domain to the frequency domain.

In this chapter, we shall outline mathematical techniques for dealing with random signals and random phenomena which play an important role in the design of communication systems. No attempt will be made to present a rigorous treatment of random signal theory. The purpose of this chapter is to enable the uninitiated reader to gain sufficient mathematical background for proper understanding of the material to be covered in the chapters to follow. To this end, we shall review only the basic principles of the theory of probability and statistics and show their application to random phenomena. The concept of generalized harmonic analysis will be used to show how to relate the statistical description of a random process to its spectrum.

A random phenomenon is one in which the fluctuations of the quantity under observation as a function of time cannot be precisely predicted. All information-carrying functions (messages) in the form of fluctuating quantities, such as voltages, pressures, and temperatures, are characterized by not being subject to precise prediction. Similarly, thermal noise, shot noise, and other forms of disturbance in the transmission of a message have the same characteristic. However, these random phenomena show statistical regularities as the number of observations is increased under similar conditions. Since the theory of probability is the mathematical theory of phenomena which show statistical regularity, the methods of treating random phenomena are based on such concepts. The description and analysis of messages and noise using concepts which have been borrowed from the probability theory are therefore very useful in modern communication theory.

In the following, we shall describe random signals or processes by the use of probability functions, autocorrelation functions, and spectral densities. It will be shown that the last two are uniquely related by the Fourier transform.

4.1 Random Variable and Probability Distribution Functions[1-4]

Consider the outcome or result of a certain experiment which is not deterministic but random and exhibits statistical regularity. Let this random experiment be repeated under similar conditions a large number of times, and the outcome of this experiment be given by a single real quantity which is called the random variable. This quantity is known by other names such as chance variable, stochastic variable, and variate. Associated with a particular outcome of the experiment in a particular trial is a value x_k, and the random variable x has a probability $P(x = x_k)$ of assuming the value x_k. The values $x_1, x_2, \ldots, x_k$, each defining a particular outcome or result of an experiment, correspond to sample points. The aggregate of all the sample points associated with the outcome of the experiment defines a sample space. A sample point may be thought of as a possible outcome of the experiment performed under a given set of conditions, and a sample space is simply the set of all possible outcomes which could be realized when an operation is performed under the given set of conditions.

The random variable x is a function defined on the sample space where it may assume all possible values of x_k; the frequency function $P(x = x_k)$ or more simply $P(x_k)$ denotes the probability of the random variable x assuming a value x_k.

A random variable may assume discrete values, or it may assume a continuous range of values. If the random variable x can assume, in any finite interval, a finite number of distinct values x_1, x_2, ... , x_k with probability of occurrence $P(x_1)$, $P(x_2)$, ..., $P(x_k)$, x is called a discrete random variable. If the range of variation of a random variable is continuous, the variable is called a continuous random variable. Examples of a discrete random variable are the outcome of tossing a coin with two possible events, head or tail, or of tossing a die with six possible events. Continuous random variables are exemplified by the value of a thermal noise voltage at the output of an amplifier at a specified instant of time. Because of our interest in the statistics of noise and signals, we shall confine ourselves mainly to the study of continuous random variables.

1. *Discrete Random Variables: Probability Distribution Func- tions.* As stated above, a discrete random variable can take on only a finite number of values in any finite interval. With each value x_k there is associated a probability $P(x_k)$; the complete set of proba- bilities $P(x_k)$ is called the probability distribution function of the discrete random variable x. From the above definition, we have

$$P(x \leq X) = \sum_{x_k \leq X} P(x_k) \qquad (4\text{-}1)$$

and $\quad P(x \leq \infty) = \sum_{\text{all } k} P(x_k) = 1 \qquad (4\text{-}2)$

In the outcome of many experiments we are very often concerned with the probability of the joint occurrence of two events or the joint probability. In this case, the discrete random variable is two- dimensional, and the joint probability distribution function is given by

$$P(x \leq X, \; y \leq Y) = \sum_{x_k \leq X} \sum_{y_m \leq Y} P(x_k, y_m) \qquad (4\text{-}3)$$

where $P(x_k, y_m)$ denotes the probability of occurrence of the joint event x_k and y_m. It follows that

$$P(x \leq +\infty, \; y \leq -\infty) = \sum_{\text{all } k} \sum_{\text{all } m} P(x_k, y_m) = 1 \qquad (4\text{-}4)$$

The probability of occurrence of event x_k irrespective of the out- come of y is simply $P(x_k)$; thus

$$P(x_k) = \sum_{\text{all } m} P(x_k, y_m) \tag{4-5}$$

and similarly $\quad P(y_m) = \sum_{\text{all } k} P(x_k, y_m) \tag{4-6}$

2. Discrete Random Variables: Conditional Probabilities and Statistical Independence. The probability of occurrence of event y_m, given that x_k has occurred, is called the conditional probability and is denoted by $P(y_m | x_k)$; similarly $P(x_k | y_m)$ is the conditional probability of x_k, given that y_m has occurred.

In order to derive the relation between the conditional probabilities and joint probabilities of the combined events, and the elementary or unconditional probabilities of the individual events, we consider the following experiment. Let the random experiment of the joint occurrence of events x_k and y_m of the random variables x and y be repeated N times, and assume that event x_k occurs $n(x_k)$ times, event y_m occurs $n(y_m)$ times, and that the joint event (x_k, y_m) occurs $n(x_k, y_m)$ times.

By definition of conditional probability,

$$P(y_m | x_k) = \frac{n(x_k, y_m)}{n(x_k)} = \frac{n(x_k, y_m)/N}{n(x_k)/N} \tag{4-7}$$

and making use of the definition of joint probability, this can be written in the form

$$P(y_m | x_k) = \frac{P(x_k, y_m)}{P(x_k)} \tag{4-8}$$

Similarly $\quad P(x_k | y_m) = \dfrac{P(x_k, y_m)}{P(y_m)} \tag{4-9}$

from which it follows that

$$P(x_k, y_m) = P(x_k) \cdot P(y_m | x_k) = P(y_m) \cdot P(x_k | y_m) \tag{4-10}$$

Thus we have shown that the probability of the joint occurrence of events x_k and y_m is equal to the product of the unconditional probability of event x_k and the conditional probability of event y_m, given that event x_k has occurred. Similarly, $P(x_k, y_m)$ is equal to the product of the unconditional probability of event y_m and the conditional probability of event x_k, given that event y_m has occurred.

In case the probability of occurrence of event y_m is independent of the occurrence of event x_k, then y_m is said to be statistically independent of x_k. In this case, the conditional probability of occurrence of event y_m is simply equal to its unconditional probability. Hence

$$P(y_m \mid x_k) = P(y_m) \tag{4-11}$$

and similarly $P(x_k \mid y_m) = P(x_k)$ (4-12)

From Eq. (4-10) it follows that if events x_k and y_m are statistically independent, then

$$P(x_k, y_m) = P(x_k)\, P(y_m) \tag{4-13}$$

which means that the probability of occurrence of the joint event (x_k, y_m) is equal to the product of their unconditional probabilities.

From Eqs. (4-4) and (4-13) we find that

$$P(x \le X,\ y \le Y) = \sum_{x_k \le X} \ \sum_{y_m \le Y} P(x_k) P(y_m) \tag{4-14}$$

when x and y are statistically independent. Also from Eq. (4-1) we obtain the relation

$$P(x \le X,\ y \le Y) = P(x \le X) P(y \le Y) \tag{4-15}$$

which is valid for all values of X and Y.

3. *Continuous Random Variables: Probability Density Functions.* In dealing with continuous random variables, we shall find it useful to introduce the concept of a probability density function p(x), which is defined as the derivative of the probability distribution function,

$$p(X) = \lim_{\Delta X \to 0} \frac{P(X - \Delta X/2 \le x \le X + \Delta X/2)}{\Delta X} \tag{4-16}$$

where $P(X - \Delta X/2 \le x \le X + \Delta X/2)$ defines the probability that x lies between $(X - \Delta X/2)$ and $(X + \Delta X/2)$. The probability that the random variable x will lie in the range x_1 to x_2 is then

$$P(x_1 \le x \le x_2) = \int_{x_1}^{x_2} p(x)\ dx = P(x_2) - P(x_1) \tag{4-17}$$

In particular, the probability that x will lie somewhere in its allowable range of variation must be 1. Since in general x can range from $-\infty$ to $+\infty$, it is convenient to write

$$\int_{-\infty}^{\infty} p(x) \; dx = 1 \qquad\qquad (4\text{-}18)$$

It should be noted that $p(x) \; \Delta x$ is the probability that x will be found in the range $(x \pm \Delta x/2)$ and that from the definition of probability density function, $p(x)$ is a nonnegative function. From Eq. (4-17), it follows that the probability that the random variable is less than some value X is given by

$$p(x \leq X) = \int_{-\infty}^{X} p(x) \; dx \qquad\qquad (4\text{-}19)$$

The function $P(x \leq X)$ or simply $P(X)$ is called the cumulative distribution function; it is a positive increasing monotonic function since $p(x)$ is nonnegative.

In particular, $P(x \leq -\infty) = 0$ and $P(x \leq +\infty) = 1$.

If $P(x)$ possesses a first derivative, we have

$$p(x) = \frac{dP(x)}{dx} \qquad\qquad (4\text{-}20)$$

Typical probability density and probability distribution functions are shown in Fig. 4-1 and Fig. 4-2.

4. *Continuous Random Variables: Joint Probability Density Functions.* Similarly, a joint probability distribution function is defined by $P(x \leq X, \; y \leq Y)$, the probability that the random variable $x \leq X$ and that the random variable $y \leq Y$. In this case, the joint

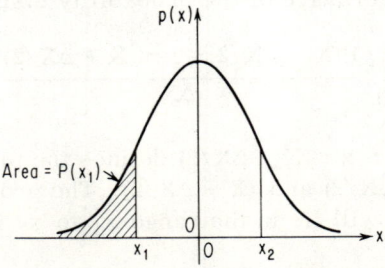

Fig. 4-1. Probability density function.

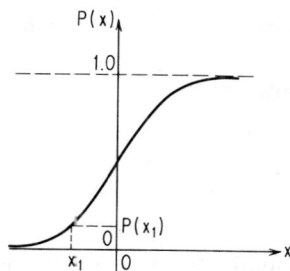

Fig. 4-2. Probability distribution function.

sample space is two-dimensional, the xy plane, and the joint probability distribution function is the probability that a result of an experiment will correspond to a sample falling in the quadrant $(-\infty \leq x \leq X, \ -\infty \leq y \leq Y)$ of that sample space. From the definition of joint probability, it follows that

$$P(x \leq -\infty, \ y \leq Y) = 0 = P(x \leq X, \ y \leq -\infty) \qquad (4\text{-}21)$$

and

$$P(x \leq +\infty, \ y \leq +\infty) = 1 \qquad (4\text{-}22)$$

Also

$$P(x \leq X, \ y \leq +\infty) = P(x \leq X) \qquad (4\text{-}23)$$

$$P(x \leq +\infty, \ y \leq Y) = P(y \leq Y) \qquad (4\text{-}24)$$

The joint probability density function of two coordinates $p(x,y)$ is defined as

$$P(x_1 \leq x \leq x_2, \ y_1 \leq y \leq y_2) = \int_{x_1}^{x_2} \int_{y_1}^{y_2} p(x,y) \ dx \ dy \qquad (4\text{-}25)$$

The joint probability distribution function $P(x \leq X, \ y \leq Y)$ gives the probability that the random variables x and y are less than the specified values X and Y in terms of the probability density function. Hence

$$P(x \leq X, \ y \leq Y) = \int_{-\infty}^{Y} \int_{-\infty}^{X} p(x,y) \ dx \ dy \qquad (4\text{-}26)$$

The expression $p(x,y) \ dx \ dy$ may be interpreted as the probability that a sample point falls in an incremental area $dx \ dy$ about the point (x,y) in the sample space. In particular, the probability that a sample point falls anywhere in the sample space, which is a certainty, is given by

$$P(x \leq +\infty, \ y \leq +\infty) = \int_{-\infty}^{\infty} \int_{-\infty}^{\infty} p(x,y) \ dx \ dy = 1 \qquad (4\text{-}27)$$

From Eq. (4-19) it follows that

$$P(x \leq X) = P(x \leq X, \ y \leq +\infty) = \int_{-\infty}^{X} dx \int_{-\infty}^{\infty} p(x,y) \ dy \qquad (4\text{-}28)$$

which is simply the probability that $x \leq X$, irrespective of the value of y. This corresponds to a sample point falling anywhere in the half-plane $(-\infty = \leq x - X)$. Similarly

$$P(y \leq Y) = P(x \leq +\infty, \ y \leq Y) = \int_{-\infty}^{Y} dy \int_{-\infty}^{\infty} p(x,y) \ dx \qquad (4\text{-}29)$$

By taking the derivatives of Eqs. (4-28) and (4-29), we obtain the relations

$$p(X) = \int_{-\infty}^{\infty} p(X,y) \ dy \qquad (4\text{-}30)$$

and $\quad p(Y) = \int_{-\infty}^{\infty} p(x,Y) \ dx \qquad (4\text{-}31)$

which correspond to Eqs. (4-5) and (4-6) of the discrete random variable case.

5. *Statistically Independent Continuous Random Variables.* The random variables x and y are said to be statistically independent if the experiments that characterize them are statistically independent. We have shown in the case of statistically independent discrete random variables [Eq. (4-15)] that the relation

$$P(x \leq X, \ y \leq Y) = P(x \leq X)P(y \leq Y) \qquad (4\text{-}32)$$

is valid for all values of X and Y. We may extend this definition to continuous independent random variables, and assuming that the various probability densities exist, Eq. (4-32) may be written in the form

$$\int_{-\infty}^{Y} \int_{-\infty}^{X} p(x,y) \ dx \ dy = \int_{-\infty}^{X} p(x) \ dx \int_{-\infty}^{Y} p(y) \ dy \qquad (4\text{-}33)$$

Taking the partial derivatives of both sides first with respect to X and then to Y, we obtain

$$p(X,Y) = p(X) \cdot p(Y) \qquad (4\text{-}34)$$

or simply $p(x,y) = p(x) \cdot p(y)$ (4-35)

Random variables x and y which satisfy Eq. (4-35) for all values of x and y are called statistically independent random variables.

4.2 Statistical Averages and Moments[1-3, 5, 6]

The probability functions discussed previously provide us with information as to the probability that a random variable will occupy a specified portion of its range. They are used to describe the average behavior of random variables. In this section, it will be shown that the various averages of random variables or of functions of random variables can be determined through the use of the probability functions.

To demonstrate the method used, let us determine first the average value of a discrete random variable. Let the total number of trials or experiments be N, and let the outcome of the trials be x_1, x_2, ... , x_k. Associated with these discrete values are the numbers of successful trials N_1, N_2, ... , N_k where

$$N_1 + N_2 + \cdots + N_k = N$$ (4-36)

The weighted average of x is obtained by weighting each value of x by the number of times it appears and dividing by the total number of trials N.

$$\tilde{x} = \text{av. } x = \frac{N_1x_1 + N_2x_2 + \cdots + N_kx_k}{N}$$ (4-37)

For a large number of trials, N_k/N is the relative frequency of occurrence of x_k, or its probability $P(x_k)$. The statistical average x can be written as

$$\tilde{x} = \sum_{n=1}^{k} x_n P(x_n)$$ (4-38)

where $\sum_{n=1}^{k} P(x_n) = 1$

In a similar manner, the mean square value of x, or its second moment, is given by

$$\tilde{x^2} = \text{av. } x^2 = \frac{N_1 x_1^2 + N_2 x_2^2 + \cdots + N_k x_k^2}{N} \qquad (4\text{-}39)$$

or for a large number of trials,

$$\tilde{x^2} = \sum_{n=1}^{k} x_n^2 P(x_n) \qquad (4\text{-}40)$$

In general, let $F(x)$ represent some specified function of the discrete random variable x, and suppose we wish to find the average value of $F(x)$. The usual way of calculating the average value of $F(x)$ is to divide the weighted sum of the values obtained for $F(x)$ by the number of observations. We write

$$\text{av. } F(x) = \sum_{n=1}^{k} F(x_n) P(x_n) \qquad (4\text{-}41)$$

In exactly the same manner, if $F(x)$ is a function of the continuous random variable, we may calculate the weighted average of $F(x)$ by noting that $p(x) \Delta x$ is the probability that x will be found in the range $(x \pm \Delta x/2)$.

It follows therefore that

$$\text{av. } F(x) = \sum_{n=1}^{k} F(x_n) p(x_n) \Delta x \qquad (4\text{-}42)$$

In the limit, as $\Delta x \to 0$,

$$\text{av. } F(x) = \int_{-\infty}^{\infty} F(x) p(x) \, dx \qquad (4\text{-}43)$$

The expression, mathematical expectation of $F(x)$, abbreviated as $E[F(x)]$ or $<[F(x)]>$, is occasionally used instead of av. $F(x)$.

Of special interest is the average of $F(x) = x^n$, where n is a positive integer. We define the average of x^n as the n'th moment of the distribution function and denote it by m_n.

$$m_n = \text{av. } x^n = \int_{-\infty}^{\infty} x^n p(x) \, dx \qquad (4\text{-}44)$$

The average value of x, m_1, is the ordinary arithmetic mean and is frequently called the first moment. The average value of x^2, m_2, is the mean square.

In cases in which the random variable x is a voltage or current, m_1 represents the constant cr d-c component, and m_2 is proportional to the mean power. Frequently, we are interested only in the a-c component of the measured values. The corresponding power averages are called central moments and are defined by

$$\mu_n = \text{av. } (x - m_1)^n = \int_{-\infty}^{\infty} (x - m_1)^n p(x) \, dx \qquad (4\text{-}45)$$

Evidently from the last equation, it follows that $\mu_1 = 0$. The most important central moment is μ_2, which is defined as the variance cr dispersion of the distribution. From the definitions

$$\mu_2 = \int_{-\infty}^{\infty} (x - m_1)^2 p(x) \, dx$$

$$= \int_{-\infty}^{\infty} x^2 p(x) \, dx - 2m_1 \int_{-\infty}^{\infty} xp(x) \, dx + m_1^2 \int_{-\infty}^{\infty} p(x) \, dx$$

$$= m_2 - 2m_1^2 + m_1^2 = m_2 - m_1^2 \qquad (4\text{-}46)$$

The square root of the variance σ is called the standard deviation of the random variable x. Therefore

$$\sigma = \sqrt{\mu_2}$$

or $\qquad \sigma^2 + m_1^2 = m_2 \qquad (4\text{-}47)$

The standard deviation σ is the rms value of the a-c component, and the variance is the mean square proportional to the mean power of the a-c component.

It is to be noted that although random functions cannot be predicted exactly and must be considered statistically, a knowledge of the distribution functions of the random variable enables us to determine the arithmetic mean and the rms value of the random variable.

The concept of averages of a probability density distribution may be extended to joint probability distributions. Thus,

$$\text{av. } F(x,y) = \int_{-\infty}^{\infty} \int_{-\infty}^{\infty} F(x,y) p(x,y) \, dx \, dy \qquad (4\text{-}48)$$

In the special case in which the random variables x and y are independent, this reduces to

$$\text{av. } F(x,y) = \int_{-\infty}^{\infty} \int_{-\infty}^{\infty} F(x,y)\, p(x)\, p(y)\ dx\ dy \tag{4-49}$$

1. *Average and Average Square of the Sum of Random Variables.* We shall be often concerned with a random variable which results from the sum of two or more random variables. Let the random variable z be the sum of two random variables x and y,

$$z = x + y \tag{4-50}$$

We shall calculate the mean and the mean square of the resultant random variable z in terms of its components x and y. By definition [Eq. (4-48)], the average value of z is given by

$$\tilde{z} = \text{av. } z = \int_{-\infty}^{\infty} \int_{-\infty}^{\infty} (x+y)\, p(x,y)\ dx\ dy \tag{4-51}$$

where $p(x,y)$ is the joint probability density of the random variables x and y. This may be expressed in the form

$$\tilde{z} = \int_{-\infty}^{\infty} x\ dx \int_{-\infty}^{\infty} p(x,y)\ dy + \int_{-\infty}^{\infty} y\ dy \int_{-\infty}^{\infty} p(x,y)\ dx \tag{4-52}$$

From Eqs. (4-31) and (4-30), we obtain

$$\int_{-\infty}^{\infty} p(x,y)\ dx = p(y)$$

and

$$\int_{-\infty}^{\infty} p(x,y)\ dy = p(x)$$

so that

$$\tilde{z} = \int_{-\infty}^{\infty} x\, p(x)\ dx + \int_{-\infty}^{\infty} y\, p(y)\ dy = \tilde{x} + \tilde{y} \tag{4-53}$$

We have just shown that the mean of the sum of two random variables is the sum of their means; the same procedure can be extended to the sum of more than two random variables.

To calculate the mean square of z, we start with the definition of Eq. (4-48).

$$\text{av. } z^2 = \tilde{z}^2 = \widetilde{(x+y)}^2 = \int_{-\infty}^{\infty} \int_{-\infty}^{\infty} (x+y)^2\, p(x,y)\ dx\ dy$$

$$= \int_{-\infty}^{\infty} x^2\ dx \int_{-\infty}^{\infty} p(x,y)\ dy + \int_{-\infty}^{\infty} y^2\ dy$$

$$\times \int_{-\infty}^{\infty} p(x,y)\ dx$$

$$+ 2 \int_{-\infty}^{\infty} \int_{-\infty}^{\infty} xyp(x,y) \, dx \, dy$$

$$= \widetilde{x^2} + \widetilde{y^2} + 2\widetilde{xy} \qquad (4\text{-}54)$$

This result states that the mean of the square of the sum of x plus y equals the sum of the mean squares of x and y plus twice the mean of their product. It should be noted here that the above results apply to random variables which may or may not be statistically independent. We shall now show that the mean of the product of two statistically independent random variables is the product of their mean values.

2. *Average of the Product of Independent Random Variables.* Again using Eq. (4-48), we define the mean of the product of xy by the relation

$$\text{av. } (xy) = \widetilde{xy} = \int_{-\infty}^{\infty} \int_{-\infty}^{\infty} xyp(x,y) \, dx \, dy \qquad (4\text{-}55)$$

But since x and y are independent random variables, Eq. (4-49) is applicable and we obtain

$$\widetilde{xy} = \int_{-\infty}^{\infty} xp(x) \, dx \int_{-\infty}^{\infty} yp(y) \, dy = \widetilde{x} \cdot \widetilde{y} \qquad (4\text{-}56)$$

and Eq. (4-54) reduces to

$$\text{av. } (x + y)^2 = \widetilde{(x + y)}^2 = \widetilde{x^2} + \widetilde{y^2} + 2\widetilde{x} \cdot \widetilde{y} \qquad (4\text{-}57)$$

3. *Variances of a Sum of Random Variables.* Using the result of Eq. (4-46), we may extend it to the random variable $z = x + y$, where x and y are also random variables. Thus

$$\mu_{2,z} = \widetilde{(z - \widetilde{z})^2} \qquad (4\text{-}58)$$

From Eqs. (4-53) and (4-54), we have

$$\widetilde{x + y} = \widetilde{x} + \widetilde{y} \quad \text{and} \quad \widetilde{(x + y)^2} = \widetilde{x^2} + \widetilde{y^2} + 2\widetilde{xy}$$

By direct substitution, we obtain

$$\mu_{2,z} = \widetilde{[(x - \widetilde{x})] + (y - \widetilde{y})]^2} = \widetilde{(x - \widetilde{x})^2} + \widetilde{(y - \widetilde{y})^2} - \widetilde{2(x - \widetilde{x})(y - \widetilde{y})}$$

$$(4\text{-}59)$$

In particular, if x and y are statistically independent, then

$$\overbrace{(x - \tilde{x})(y - \tilde{y})} = \overbrace{(x - \tilde{x})} \cdot \overbrace{(y - \tilde{y})}$$

But $\overbrace{x - \tilde{x}} = 0$, and $\overbrace{y - \tilde{y}} = 0$, by definition. Therefore

$$\mu_{2,z} = \mu_{2,x} + \mu_{2,y} \qquad\qquad (4\text{-}60)$$

or $\qquad \sigma_z^2 = \sigma_x^2 + \sigma_y^2$

That is to say, if x and y are independent variables, the variance of their sum is equal to the sum of their variances.

4.3 Transformation of Probability Density Functions[3-5]

In practical applications, it is often required to determine the probability density function of a random variable which has passed through a system, linear or nonlinear, such as amplifiers, rectifiers, etc. Generally, let the output y of the device be related to the input x, where x is a random variable, by the equation

$$y = F(x) \qquad\qquad (4\text{-}61)$$

Our problem is to determine the probability density function q(y) of the random variable y in terms of the known probability density p(x).

Let x vary between x_1 and x_2 as shown in Fig. 4-3, and correspondingly let y vary between y_1 and y_2. Assuming a one-to-one correspondence between x and y, we can state that

$$P(x_1 \leq x \leq x_2) = P(y_1 \leq y \leq y_2)$$

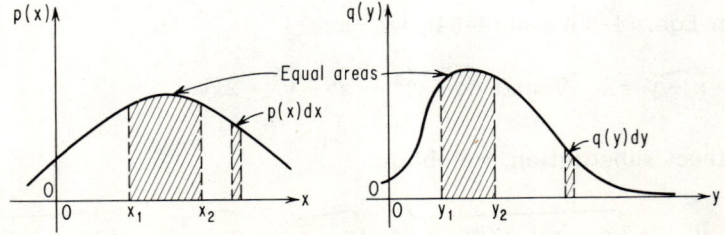

Fig. 4-3. Transformation of probability density functions.

or $$\int_{x_1}^{x_2} p(x) \, dx = \int_{y_1}^{y_2} q(y) \, dy \qquad (4\text{-}62)$$

i.e., the shaded areas under the curves of Fig. 4-3 are equal.

Assuming that Eq. (4-61) can be solved for x, we denote the inverse of $y = F(x)$ by $x = f(y)$ where it is also assumed that $f(y)$ is a single-valued function of y. Then by direct substitution, we obtain

$$\int_{x_1}^{x_2} p(x) \, dx = \int_{y_1}^{y_2} p[f(y)] \, f'(y) \, dy = \int_{y_1}^{y_2} q(y) \, dy \qquad (4\text{-}63)$$

and it follows therefore that

$$q(y) = p \, [f(y)] \, f'(y) \qquad (4\text{-}64)$$

or $\quad q(y) \, dy = p(x) \, dx \qquad (4\text{-}65)$

Equation (4-65) can be derived from Eq. (4-62) since in the limit as $x_1 \rightarrow x_2$, then $y_1 \rightarrow y_2$, and $q(y) \, dy = p(x) \, dx$.

Thus by the use of Eq. (4-65), the probability density of the new random variable y can be determined in terms of the known probability density function $p(x)$ or the original random variable x and the given transformation.

As an example of the application of Eq. (4-65), we consider the uniformly distributed function $p(x) = 1$ in the range $0 \le x \le 1$, as shown in Fig. 4-4, and proceed to find the probability density function of the output of a square-law detector with the characteristic $y = x^2$. Using Eq. (4-65), we have

$$q(y) \frac{dy}{dx} = p(x) = 1 \qquad (4\text{-}66)$$

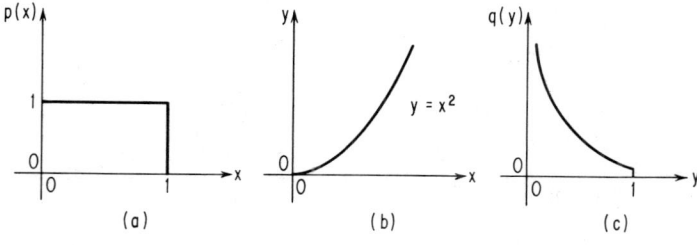

Fig. 4-4. Example of a transformation of probability density function: (a) Input variable; (b) transformation characteristic; (c) output variable.

Therefore, $q(y) = \dfrac{1}{2\sqrt{y}}$, $0 \leq y \leq 1$ (4-67)

as shown in Fig. 4-4.

As an additional example illustrating the application of Eq. (4-65) to the problem of transformation of probability density functions, we shall consider now the case of gaussian-distributed noise passed through a square-law detector with the characteristic $y = x^2$. Using the symmetrical form, the gaussian density function is given by

$$p(x) = \dfrac{1}{\sqrt{2\pi}\ \sigma}\ e^{-x^2/2\sigma^2},\qquad -\infty < x < \infty \tag{4-68}$$

As we shall see later on, the gaussian distribution function plays a very important role in the theory of noise, and it will be discussed more fully in the next section. Our present problem is to find the density function $q(y)$ of the instantaneous power at the output of the square-law detector corresponding to the gaussian-distributed input noise voltage $p(x)$.

With reference to Fig. 4-5, we note that both positive and negative values of the random variable x contribute to the detector output y. It follows that

$$P(y_1 \leq y \leq y_2) = P(x_1 \leq x \leq x_2) + P(-x_1 \leq x \leq -x_2) \tag{4-69}$$

Since the gaussian function is symmetrical about $x = 0$, we obtain the following result:

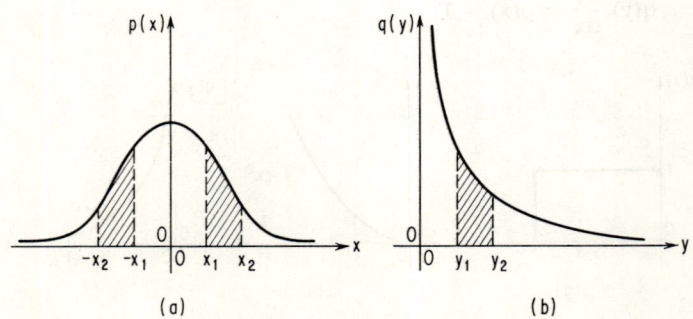

Fig. 4-5. Gaussian noise at output of square-law device: (a) Input distribution; (b) output distribution.

$$\int_{y_1}^{y_2} q(y)\ dy = 2\int_{x_1}^{x_2} p(x)\ dx \qquad (4\text{-}70)$$

or
$$q(y) = 2p\,[f(y)]\ f'(y) \qquad (4\text{-}71)$$

In this case, $y = x^2 \qquad f(y) = \sqrt{y} \qquad f'(y) = \dfrac{1}{2\sqrt{y}}$

and from Eq. (4-71) we get

$$q(y) = \frac{e^{-y/2\sigma^2}}{\sqrt{2\pi y}\ \sigma}, \qquad 0 \le y < \infty \qquad (4\text{-}72)$$

which is the distribution of the noise power at the output of the square-law detector.

4.4 Typical Examples of Distribution Functions[1-5]

1. *The Uniform or Rectangular Distribution.* The random variable x is uniformly distributed with density function $p(x) = K$ between $x = a$ and $x = b$, as shown in Fig. 4-6a. Since $\int_a^b p(x)\ dx = 1$ the constant $K = \dfrac{1}{b-a}$. The first moment

$$m_1 = \int_a^b xp(x)\ dx = \frac{1}{b-a}\int_a^b x\ dx = \frac{a+b}{2} \qquad (4\text{-}73)$$

as expected. The second moment

$$m_2 = \int_a^b x^2 p(x)\ dx = \frac{1}{b-a}\int_a^b x^2\ dx = \frac{b^2 + ab + a^2}{3} \qquad (4\text{-}74)$$

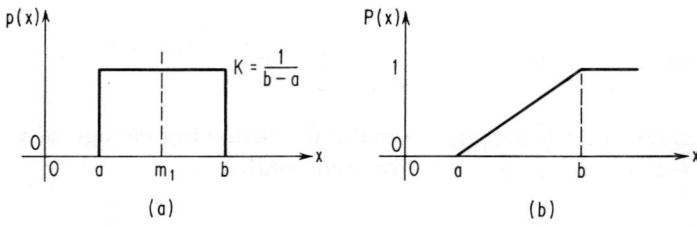

(a) (b)

Fig. 4-6. Uniform distribution function: (a) Probability density function; (b) cumulative distribution function.

The variance $\quad \sigma^2 = m_2 - m_1^2 = \dfrac{(b-a)^2}{12}$ $\qquad$ (4-75)

The standard deviation σ is therefore

$$\sigma = \frac{b-a}{2\sqrt{3}} \qquad (4\text{-}76)$$

The cumulative distribution function $P(x)$ is shown in Fig. 4-6b.

$$P(x) = \begin{cases} 0, & x < a \\[2mm] \dfrac{x-a}{b-a}, & a < x < b \\[2mm] 1, & x \geq b \end{cases} \qquad (4\text{-}77)$$

It will be shown in Chap. 20 that these results are applicable to the problem of "quantizing noise," arising from the errors in conversion of a signal in analogue form to a digital approximation in a pulse-code-modulation (PCM) system.

2. *The Gaussian or Normal Distribution.* One of the most important distributions in noise theory is the gaussian or normal distribution whose general form for one variable is

$$p(x) = \frac{1}{\sqrt{2\pi}\,\sigma}\, e^{-(x-x_0)^2/2\sigma^2}, \qquad -\infty < x < \infty \qquad (4\text{-}78)$$

shown in Fig. 4-7a. The parameters have been adjusted to render

$$\int_{-\infty}^{\infty} p(x)\ dx = 1$$

The first moment or average value of x is

$$m_1 = \int_{-\infty}^{\infty} xp(x)\ dx = \frac{1}{\sqrt{2\pi}\,\sigma} \int_{-\infty}^{\infty} xe^{-(x-x_0)^2/2\sigma^2}\ dx \qquad (4\text{-}79)$$

To evaluate this integral, we make the following change of variables: $y = (x - x_0)/\sqrt{2\sigma^2}$, $dx = \sqrt{2}\,\sigma\ dy$, and obtain

$$m_1 = \frac{1}{\sqrt{\pi}} \int_{-\infty}^{\infty} (\sqrt{2}\,\sigma y + x_0)e^{-y^2}\ dy$$

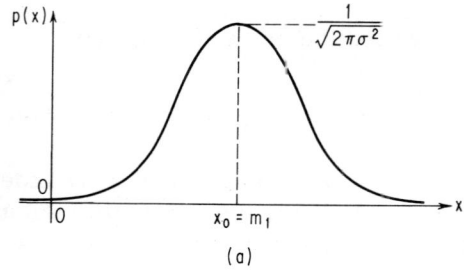
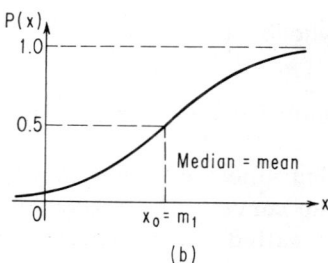

(a) (b)

Fig. 4-7. Gaussian distribution function: (a) Probability density function; (b) cumulative distribution function.

Since the function $y e^{-y^2}$ is odd, $\int_{-\infty}^{\infty} y e^{-y^2} dy = 0$, and m_1 reduces to

$$m_1 = \frac{x_o}{\sqrt{\pi}} \int_{-\infty}^{\infty} e^{-y^2} dy = x_o \tag{4-80}$$

Similarly $\quad m_2 = \int_{-\infty}^{\infty} x^2 p(x) \, dx = x_o^2 + \sigma^2 \tag{4-81}$

Therefore the variance of the distribution

$$\mu_2 = m_2 - m_1^2 = \sigma^2 \tag{4-82}$$

where $m_1 = x_o$ is the mean, σ^2 is the variance, and σ is the standard deviation. Thus, the normal distribution is completely defined by the mathematical expectation and variance of the random variable. From Eq. (4-78), it is seen that the gaussian density function is symmetrical about $x = x_o$ where $p(x)$ is peaked and equal to $1/\sqrt{2\pi}\,\sigma$.

The cumulative distribution function or the probability that the random variable x is less than some value X is plotted in Fig. 4-7b and is given by

$$P(x \le X) = \int_{-\infty}^{X} p(x) \, dx = \int_{-\infty}^{X} \frac{e^{-(x-x_o)^2/2\sigma^2}}{\sqrt{2\pi}\,\sigma} \, dx = \frac{1}{2}\left(1 + \text{erf}\,\frac{X - x_o}{\sigma}\right)$$

$$= \Phi\left(\frac{X - x_o}{\sigma}\right)$$

$$\tag{4-83}$$

where $\text{erf } u \equiv \dfrac{2}{\sqrt{2\pi}} \int_0^u e^{-t^2/2} \, dt$, called the error function and

$\Phi(u) = \dfrac{1}{\sqrt{2\pi}} \int_{-\infty}^u e^{-t^2/2} \, dt$ is the probability integral. It is to be noted

that since $p(x)$ is symmetrical about $x = x_0$, and the total area under the curve is 1, it follows that $P(x_0) = 0.5$. The 0.5 probability point is called the median, which coincides for the gaussian function with the average value.

We shall have many occasions to apply the gaussian distribution function to the study of the statistics of noise and to the problem of detection of signals in the presence of noise.

3. *The Rayleigh Distribution Function.* The Rayleigh distribution is encountered in many applications of statistics. Specifically it is used in the study of narrow-band gaussian random processes and in the analysis of short-term fading which characterizes tropospheric scatter propagation. The probability density function is given by

$$p(x) = \frac{x}{\sigma^2} e^{-x^2/2\sigma^2}, \qquad 0 \le x \le +\infty \tag{4-84}$$

$$= 0, \qquad\qquad -\infty \le x \le 0$$

The average value $\tilde{x}$ is given by

$$\tilde{x} = \int_0^\infty x p(x) \, dx = \sqrt{\frac{\pi}{2}} \, \sigma \tag{4-85}$$

and the mean square value $\tilde{x^2}$ is

$$\tilde{x^2} = \int_0^\infty x^2 p(x) \, dx = 2\sigma^2 \tag{4-86}$$

The cumulative distribution function is

$$P(x \le X) = \int_0^X p(x) \, dx = 1 - e^{-X^2/2\sigma^2} \tag{4-87}$$

From Eq. (4-87), the median x_m is given by

$$0.5 = e^{-x_m^2/2\sigma^2}$$

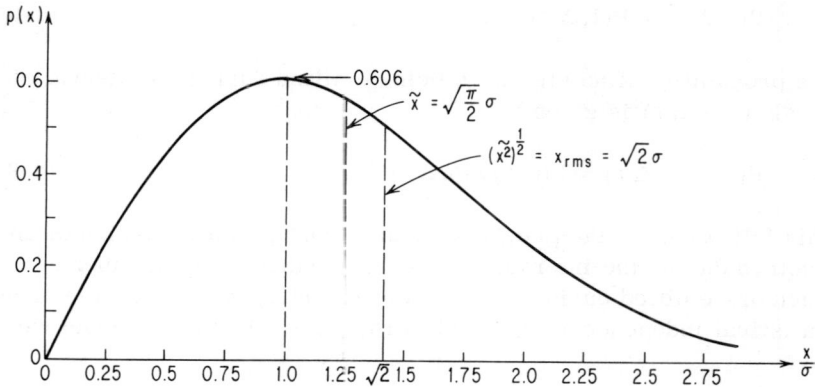

Fig. 4-8. Rayleigh probability density function.

or $x_m^2 = 1.386\sigma^2$ (4-88)

The function $p(x)$ is plotted in Fig. 4-8.

4. *The Poisson Distribution.* The Poisson distribution characterizes several random processes which are frequently encountered
in communication theory, such as shot noise in a thermionic vacuum
tube, transmission of telegraph signals, etc. It is an example of a
discrete probability distribution, since it is defined only for discrete
values of the random variable. In the following derivation of the
Poisson distribution, we shall consider as an illustration the statistical properties of shot noise, which is caused by fluctuations in a
stream of electrons from the cathode to the anode. The problem is
to determine the probability $P(K,\tau)$ that exactly K electrons are
emitted from the cathode during a time interval of length τ. The
following analysis is based on the reasonable assumption that in the
temperature-limited case, the probability of emission of an electron
during a given interval is statistically independent of the number of
electrons emitted previously, and that the probability is proportional
to the time interval $\Delta\tau$, provided $\Delta\tau$ is small. Thus, the probability of one electron being emitted in $\Delta\tau$ is given by

$$P(1,\Delta\tau) = \mu\,\Delta\tau$$ (4-89)

where μ is a constant to be determined. Since $\Delta\tau$ is small, we
further assume that the probability of the emission of more than one
electron is negligible; thus for small $\Delta\tau$, we obtain the following
approximate relation:

$$P(0,\Delta\tau) + P(1,\Delta\tau) = 1 \qquad\qquad (4\text{-}90)$$

The probability of no electrons being emitted during an interval of length $(\tau + \Delta\tau)$ is given by

$$P(0,\ \tau + \Delta\tau) = P(0,\tau)P(0,\Delta\tau) \qquad\qquad (4\text{-}91)$$

This follows from the previous assumption that the emission of an electron during the interval $\Delta\tau$ is independent of the number of electrons emitted during τ, and consequently, the expression for statistical independence is valid. From Eqs. (4-89) and (4-90) we have

$$P(0,\Delta\tau) = 1 - \mu\Delta\tau$$

and Eq. (4-91) reduces to

$$\frac{P(0,\ \tau + \Delta\tau) - P(0,\tau)}{\Delta\tau} = -\mu P(0,\tau) \qquad\qquad (4\text{-}92)$$

As $\Delta\tau \rightarrow 0$, this difference equation becomes the differential equation

$$\frac{dP(0,\tau)}{d\tau} = -\mu P(0,\tau)$$

which has the solution $\qquad P(0,\tau) = e^{-\mu\tau} \qquad\qquad (4\text{-}93)$

with the initial conditions $\quad P(0,0) = \lim_{\Delta\tau \rightarrow 0} P(0,\Delta\tau) = 1$

The last result follows from Eqs. (4-89) and (4-90). Thus, the probability of no electrons being emitted in τ sec is given by $e^{-\mu\tau}$, where μ is a constant yet to be determined.

We return now to our original problem, to determine the probability of exactly K electrons being emitted during an interval of length $(\tau + \Delta\tau)$. For $\Delta\tau$ small, there must be either one electron or none emitted in $\Delta\tau$; therefore

$$P(K,\ \tau + \Delta\tau) = P(K - 1,\ \tau)P(1,\Delta\tau) + P(K,\tau)P(0,\Delta\tau) \qquad (4\text{-}94)$$

Using the results obtained previously for $P(1,\Delta\tau)$ and $P(0,\Delta\tau)$, we have

$$\frac{P(K,\ \tau + \Delta\tau) - P(K,\tau)}{\Delta\tau} + \mu P(K,\tau) = \mu P(K-1,\ \tau) \tag{4-95}$$

As $\Delta\tau \rightarrow 0$, we obtain the differential equation

$$\frac{dP(K,\tau)}{d\tau} + \mu P(K,\tau) = \mu P(K-1,\ \tau) \tag{4-96}$$

as a recursion equation relating $P(K,\tau)$ to $P(K-1,\ \tau)$. Since $P(K,0) = 0$, the solution to this first-order linear differential equation is

$$P(K,\tau) = \mu e^{-\mu\tau} \int_0^\tau e^{\mu t} P(K-1,\ t)\ dt \tag{4-97}$$

The last equation enables us to determine $P(K,\tau)$ from $P(K-1,\ \tau)$ by the following continuation process. We take $K = 1$, to obtain $P(1,\tau)$ from Eq. (4-97).

$$P(1,\tau) = (\mu\tau)e^{-\mu\tau} \tag{4-98}$$

From this result, we determine $P(2,\tau)$ and then $P(3,\tau)$ and so on. The final general result, namely, the probability that exactly K electrons will be emitted during an interval of τ sec is

$$P(K,\tau) = \frac{(\mu\tau)^K}{K!} e^{-\mu\tau}, \qquad K = 0, 1, 2, \dots \tag{4-99}$$

which is the Poisson probability distribution.

The constant μ can be evaluated from the average number of electrons emitted during the interval of length τ sec. Since the possible number of electrons emitted during that interval ranges from zero to infinity, we obtain

$$\text{av. } K = E_\tau[K] = \sum_{K=0}^\infty KP(K,\tau) = e^{-\mu\tau} \sum_{K=0}^\infty \frac{K(\mu\tau)^K}{K!}, \qquad K \geq 0$$

Since the $K = 0$ term is zero, we can write

$$E_\tau[K] = \mu\tau\ e^{-\mu\tau} \sum_{K=1}^\infty \frac{(\mu\tau)^{K-1}}{(K-1)!}$$

But $\displaystyle\sum_{K=1}^\infty \frac{(\mu\tau)^{K-1}}{(K-1)!} = 1 + \mu\tau + \frac{(\mu\tau)^2}{2!} + \cdots + = e^{\mu\tau}$

Hence $E_T[K] = \mu\tau$ (4-100)

If we define $\overline{n} = E_T[K]/\tau$ as the average number of electrons emitted per second, then $\overline{n} = \mu$, and $P(K,\tau)$ becomes

$$P(K,\tau) = \frac{(\overline{n}\tau)^K e^{-\overline{n}\tau}}{K!}$$ (4-101)

Note that for $K = 1$ and $\overline{n}\,\Delta\tau \rightarrow 0$, this equation reduces to $P(1,\Delta\tau) = \overline{n}\,\Delta\tau$, which checks with Eq. (4-89). We also note that

$$\sum_{K=0}^{\infty} P(K,\tau) = \sum_{K=0}^{\infty} \frac{(\mu\tau)^K}{K!} e^{-\mu\tau} = e^{-\mu\tau} \sum_{K=0}^{\infty} \frac{(\mu\tau)^K}{K!} = e^{-\mu\tau} \cdot e^{\mu\tau} = 1$$

as expected.

4.5 Distribution of Sum of Independent Random Variables: the Characteristic Function [1, 3, 5, 6]

In many physical problems, it is often required to determine the distribution function of the sum of a number of independent random variables in terms of the statistics of the individual components. As a specific example, it is often required to find the distribution function of the total noise power of tandem tropospheric communication hops when the distributions of each hop are known.

Consider first the simplest case of finding the probability density function of two independent random variables. Let $z = x + y$; given the probability density functions $p_1(x)$ and $p_2(y)$, the problem is to determine the probability density function $p(z)$ of the random variable z. This problem can readily be solved by the use of Fourier transforms of the probability density functions $p_1(x)$ and $p_2(y)$. In statistics, the Fourier transform of the probability density function is called the characteristic function and is defined by

$$C_x(jv) = \text{av.} \; (e^{jvx}) = \int_{-\infty}^{\infty} e^{jvx} p_1(x) \, dx$$ (4-102)

Similarly $C_y(jv) = \text{av.} \; (e^{jvy}) = \int_{-\infty}^{\infty} e^{jvy} p_2(y) \, dy$ (4-103)

The characteristic function of the probability density $p(z)$ is by definition equal to

$$C_z(jv) = \int_{-\infty}^{\infty} \int_{-\infty}^{\infty} e^{jv(x+y)} p(x,y) \, dx \, dy$$ (4-104)

where $p(x,y)$ is the joint probability density function of the random variables x and y. Since x and y are independent random variables, Eq. (4-104) is equal to

$$C_z(jv) = \int_{-\infty}^{\infty} e^{jvx} p_1(x) \, dx \int_{-\infty}^{\infty} e^{jvy} p_2(y) \, dy = C_x(jv) \cdot C_y(jv)$$

$$(4-105)$$

We can derive now the relation among the probability densities of the random variables x, y, and z by taking the Fourier transform of both sides of Eq. (4-105).

$$p(z) = \frac{1}{2\pi} \int_{-\infty}^{\infty} C_z(jv) e^{-jvz} \, dv = \frac{1}{2\pi} \int_{-\infty}^{\infty} C_x(jv) C_y(jv) e^{-jvz} \, dv$$

$$= \frac{1}{2\pi} \int_{-\infty}^{\infty} C_y(jv) e^{-jvz} \, dv \int_{-\infty}^{\infty} p_1(x) e^{jvx} \, dx$$

$$= \frac{1}{2\pi} \int_{-\infty}^{\infty} p_1(x) \, dx \int_{-\infty}^{\infty} C_y(jv) e^{-jv(z-x)} \, dv \qquad (4-106)$$

But $\quad \dfrac{1}{2\pi} \displaystyle\int_{-\infty}^{\infty} C_y(jv) e^{-jv(z-x)} \, dv = p_2(z-x)$ $\qquad\qquad$ (4-107)

Hence $\quad p(z) = \displaystyle\int_{-\infty}^{\infty} p_1(x) p_2(z-x) \, dx$ $\qquad\qquad$ (4-108)

which is the familiar convolution integral. Thus the probability density function of the sum of two independent random variables is given by the convolution of their respective probability density functions.

This process may be extended to n independent random variables. Let $z = x_1 + x_2 + \cdots + x_n$:

$$C_z(jv) = C_{x_1}(jv) \cdot C_{x_2}(jv) \cdot \cdots \cdot C_{x_n}(jv) \qquad (4-109)$$

and $\quad p(z) = \dfrac{1}{2\pi} \displaystyle\int_{-\infty}^{\infty} C_z(jv) e^{-jvz} \, dv$ $\qquad\qquad$ (4-110)

By repeating the convolution of Eq. (4-108), the probability density $p(z)$ of the sum of n independent random variables can be expressed in terms of the probability densities of the components.

The following examples will serve to illustrate the application of characteristic functions to the calculation of distribution functions of sums of random variables.

1. *Symmetrical Uniform Distributions.* Consider the symmetrical uniform distribution shown in Fig. 4-9a.

$$p(x) = \frac{1}{x_o}, \qquad \frac{-x_o}{2} < x < \frac{x_o}{2}$$

$$= 0, \qquad \text{elsewhere}$$

The characteristic function

$$C_x(jv) = \int_{-x_o/2}^{x_o/2} \frac{e^{jvx}}{x_o} \, dx = \frac{\sin(vx_o/2)}{vx_o/2} \qquad (4\text{-}111)$$

which is shown plotted in Fig. 4-9b; the familiar $(\sin x)/x$ function is the Fourier transform of a rectangular pulse. The probability density function of the sum of n uniformly distributed variables over the same range is

$$p(z) = \frac{1}{2\pi} \int_{-\infty}^{\infty} C_z(jv)e^{-jvz} \, dv = \frac{1}{2\pi} \int_{-\infty}^{\infty} \left[\frac{\sin(vx_o/2)}{vx_o/2} \right]^n$$

$$\times e^{-jvz} \, dv \qquad (4\text{-}112)$$

where $z = nx$.

It is to be noted that when $v = 0$, the integral has the value of

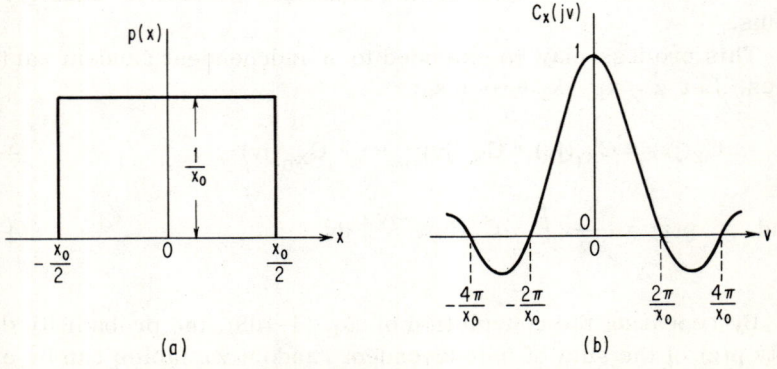

(a)   (b)

Fig. 4-9. Symmetrical uniform distribution: (a) Probability density function; (b) characteristic function.

unity, and for large n it damps out rapidly away from the origin. The characteristic function $C_Z(jv)$ may be expanded in a power series about $v = 0$ as follows:

$$C_Z(jv) = \frac{\sin^n(vx_0/2)}{(vx_0/2)^n} = \left[1 - \frac{(vx_0)^2}{24} + \cdots\right]^n$$

$$\doteq 1 - \frac{n}{24}(vx_0)^2 + \cdots \tag{4-113}$$

This power series may be approximated by the exponential function

$$e^{-(n/24)(vx_0)^2} \doteq 1 - \frac{n}{24}(vx_0)^2 + \cdots$$

so that $\qquad C_Z(jv) \doteq e^{-(n/24)(vx_0)^2} \tag{4-114}$

where n is large and v is small. Comparing Eq. (4-114) with Eq. (2-109), we note that $C_Z(jv)$ is a gaussian-type function whose Fourier transform is given by Eq. (2-107), so that, for n large,

$$p(z) \doteq \frac{1}{\sqrt{2\pi}\,\sigma}\,e^{-z^2/2\sigma^2} \tag{4-115}$$

where $\quad \sigma^2 = \dfrac{n(x_0)^2}{12} \tag{4-116}$

It should be noted here that the standard deviation of the uniform distribution is $x_0/\sqrt{12}$, so that for n large, the resultant of n distributions tends toward a gaussian distribution with $\sqrt{n}$ times the standard deviation of the original distribution. This result is as expected from the important theorem in statistics called the central-limit theorem, which states that the distribution of the sum of n independent random variables approaches the gaussian distribution for n large.

2. *Gaussian Distributions.* The problem is to calculate the probability density function of the sum of n gaussian-distributed variables having mean values of $a_1, a_2, \ldots, a_n$ and standard deviations $\sigma_1, \sigma_2, \ldots, \sigma_n$. The general form of the gaussian distribution for one variable is, from Eq. (4-78),

$$p(x) = \frac{1}{\sqrt{2\pi}\,\sigma}\, e^{-(x - x_0)^2/2\sigma^2} \tag{4-117}$$

Therefore the characteristic function of a gaussian distribution is

$$C_x(jv) = \frac{1}{\sqrt{2\pi}\,\sigma}\int_{-\infty}^{\infty} e^{jvx - (x - x_0)^2/2\sigma^2}\,dx$$

$$= e^{jx_0v - \sigma^2 v^2/2} \cdot \frac{1}{\sqrt{2\pi}\,\sigma}\int_{-\infty}^{\infty} e^{-(x - x_0 - j\sigma^2 v)^2/2\sigma^2}\,dx \tag{4-118}$$

The integral in Eq. (4-118) is the integral of the function $e^{-z^2/2\sigma^2}$, where z is a complex variable, in the complex plane along a line parallel to the real axis, namely, the line $y = -j\sigma^2 v$, which can be shown to be equal to the integral of the same function along the real axis, that is, the integral $\int_{-\infty}^{\infty} e^{-x^2/2\sigma^2}\,dx$, but this integral has the value of $\sqrt{2\pi}\,\sigma$. It follows therefore that the characteristic function of the normal distribution is

$$C_x(jv) = e^{jx_0v - \sigma^2 v^2/2} \tag{4-119}$$

This result may have also been derived by using the same method as used in Chap. 2 where it has been shown that the Fourier transform of a gaussian distribution is also gaussian.

Therefore, the probability density function $p(z)$ of the random variable $z = x_1 + x_2 + \cdots + x_n$ is given by

$$p(z) = \frac{1}{2\pi}\int_{-\infty}^{\infty} e^{-(v^2/2)(\sigma_1^2 + \sigma_2^2 + \cdots + \sigma_n^2) + jv(a_1 + a_2 + \cdots + a_n) - jvz}\,dv$$

$$= \frac{1}{\sqrt{2\pi}\,\sigma}\, e^{-(z - a)/2\sigma^2} \tag{4-120}$$

where $a = a_1 + a_2 + \cdots + a_n \tag{4-121}$

$$\sigma^2 = \sigma_1^2 + \sigma_2^2 + \cdots + \sigma_n^2 \tag{4-122}$$

Thus we have obtained the important result that the resultant distribution of the sum of n gaussian-distributed variables is also gaussian, whose mean value equals the sum of the means, and where variance equals the sum of the variances.

3. *Random-phase Distributions.*   Consider the distribution of the resultant of n independent sine waves of random phase, a problem encountered frequently in noise theory. Let

$$x_k = A_k \sin \theta_k \tag{4-123}$$

and let $p(x_k)$ denote the probability density function of x_k, with θ_k uniformly distributed over 2π rad from $-\pi$ to $+\pi$ and A_k an arbitrary constant. The problem is to find the probability density function $p(z)$ where $z = \sum_{k=1}^{n} x_k$. First we shall find the probability density function for x_k; we note that

$$q(\theta_k) = 0, \qquad \theta < -\pi \text{ and } \theta > \pi \tag{4-124}$$

$$= \frac{1}{2\pi}, \qquad -\pi < \theta < \pi$$

and that

$$\theta_k = f(x_k) = \sin^{-1} \frac{x_k}{A_k} \tag{4-125}$$

$$f'(x_k) = \frac{1}{\sqrt{A_k^2 - x_k^2}} \tag{4-126}$$

Using the equation $p(x) = q[f(x)]\, f'(x)$, we obtain for $p(x)$ the following result:

$$p(x_k) = \frac{1}{\pi \sqrt{A_k^2 - x_k^2}}, \qquad -A_k \leq x \leq A_k \tag{4-127}$$

$$= 0, \qquad\qquad x \text{ elsewhere}$$

The density function $p(x_k)$ has been multiplied by a factor of 2 because each value of x_k in the range $-A_k$ to A_k corresponds to two values of θ_k. The probability density functions $q(\theta_k)$ and $p(x_k)$ are plotted in Fig. 4-10a and b.

The characteristic function of $p(x_k)$ is

$$C_k(jv) = \frac{1}{\pi} \int_{-A_k}^{A_k} \frac{e^{jvx_k}\, dx}{\sqrt{A_k^2 - x_k^2}} = J_0(A_k v) \tag{4-128}$$

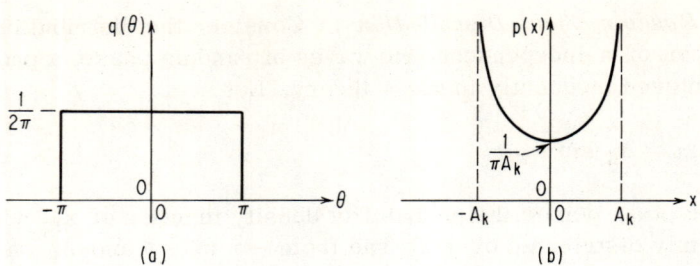

Fig. 4-10. Phase uniformly distributed: (a) $q(\theta)$; (b) $p(x)$: $x_k = A_k \sin \theta_k$.

where $J_0(A_k v)$ is the zero-order Bessel function of the first kind. The characteristic function $C_z(jv)$ for the sum of n random variables is

$$C_z(jv) = \prod_{k=1}^{n} J_0(A_k v) \tag{4-129}$$

and consequently $\quad p(z) = \dfrac{1}{2\pi} \displaystyle\int_{-\infty}^{\infty} C_z(jv) e^{-jvz} \, dv \tag{4-130}$

Using the same approach as in the case of n uniform distributions, we expand $J_0(A_k v)$ in a power series, assuming that $A_k v \ll 1$:

$$J_0(A_k v) \doteq 1 - \frac{(A_k v)^2}{4} = e^{-A_k^2 v^2/4} \tag{4-131}$$

From Eq. (4-129), we obtain

$$C_z(jv) = e^{-\frac{1}{2}\sigma^2 v^2} \tag{4-132}$$

where $\quad \sigma^2 = \frac{1}{2}(A_1^2 + A_2^2 + \cdots + A_n^2)$

Equation (4-132) represents the characteristic function of a normal distribution with zero average value and variance σ^2. It follows therefore that

$$p(z) = \frac{1}{\sqrt{2\pi}\,\sigma} e^{-z^2/2\sigma^2} \tag{4-133}$$

Since $A_k^2/2$ represents the average power in a sine wave of amplitude A_k, we conclude that σ^2 is the total mean power of the sum of n random sine waves. The result of Eq. (4-133) is another example of the central-limit theorem. We shall have occasion to make use of this result in the study of noise problems, where it will be shown that a noise voltage expressed as a sum of a large number of sine waves of random phases has a random instantaneous amplitude which is normally distributed.

4.6 Stationary Random Processes[1-3, 6-8]

In communication problems, we are interested not in a single random time function but rather in an aggregate or ensemble of random functions which either are generated by similar sources or may be obtained from a single source upon repeated experiments under essentially the same conditions. We shall presently elaborate on the two methods of obtaining the statistical properties of the aggregate of the random functions.

In our discussion of probability functions, we were concerned with the frequency of occurrence of a given outcome in a large number of experiments repeated under similar conditions. The relative frequency of occurrence can be determined from the performance of a large number of identical experiments, either simultaneously or repeated in time succession. In the first method, which is often called the ensemble method, we assume that there are available a large number of systems equipped with identical instruments and that a set of instantaneous values are obtained at the same time. In the second method, which is sometimes called extension-in-time method, the probability density and distribution functions are derived from observations on one system over a considerable period of time. The statistics derived from these two methods will generally be functions of time and not necessarily identical.

The statistical properties of random functions using the ensemble method can be derived experimentally in the following manner. As an example, we shall consider the measurements of noise voltage across resistors. The ensemble will consist of a collection of waveforms, one for each resistor, representing the variation of noise voltage as a function of time. The noise voltage will be characterized by certain random functions $x_1(t)$, $x_2(t)$, ... , differing from one another as shown in Fig. 4-11. At a given instant $t = t_1$, a set of voltage measurements can be obtained, and from these the statistical averages or ensemble averages can be derived as follows. We can determine, for example, from these measurements, the fraction of the total number of functions $x_i(t)$ which at the instant of time t_1 has a value ranging between x and $x + dx$. This fraction depends on t

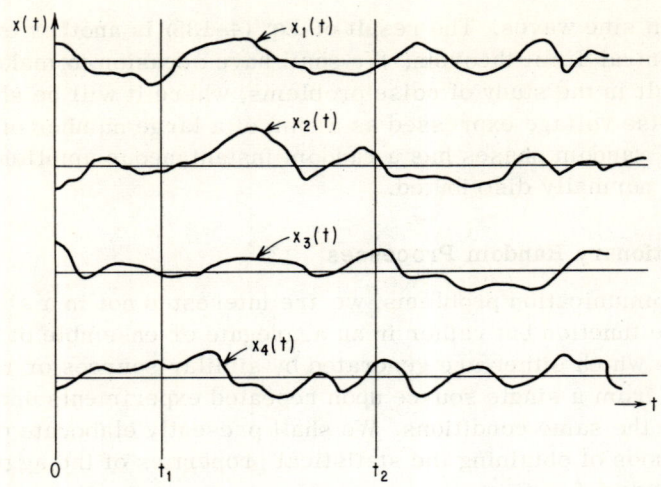

Fig. 4-11. An ensemble of voltage waveforms from a large number of identical noise sources.

and is proportional to Δx, for small values of Δx. We designate this fraction by $p_1(x,t) \Delta x$, where $p_1(x,t)$ is called the first probability distribution function. We may also consider all possible pairs of values x_1 and x_2, observed in two different instants t_1 and t_2. The fractional portion of the pairs of values for which x lies between x_1 and $(x_1 + \Delta x_1)$ for $t = t_1$, and between x_2 and $(x_2 + \Delta x_2)$ for $t = t_2$, is designated by $p_2(x_1,t_1;x_2,t_2) \Delta x_1 \Delta x_2$ where $p_2(x_1,t_1;x_2,t_2)$ is called the second probability density function. Similarly, we may continue the process and define higher probability density functions. The specification of such a set of experiments made under identical conditions, with the corresponding random variables and probability functions, is said to define a random process. Thus we note that the random process may be characterized by a complete set of probability distribution functions which define it in a statistical sense.

From the above, we conclude that the probability density functions p_1, p_2, ... , characterizing a random process, can be determined experimentally provided we have at our disposal a large number of similar systems. However, the assumption of the availability of a large number of systems required for the ensemble method has serious practical limitations, and consequently, the extension-in-time method is more attractive in practice. As stated previously, in this method we determine the frequency of occurrence from a large number of observations derived from the performance of a large number of identical experiments repeated in time succession. This may be accomplished in the following manner. Consider one observation of very long duration being recorded as the output of a single system under test. The record can then be cut into pieces of length T, and the different pieces may then be considered as the aggregate of observations on one system in successive time periods, from which the probability distributions can be determined. The first probability density $p_1(x)$ will be independent of time; the second one will depend only on the time difference $(t_2 - t_1)$; etc. Thus, the random process consisting of the different records of the ensemble of observations is characterized by the following functions:

$p_1(x)\ dx$,   probability of finding x between x and $(x + dx)$

$p_2(x_1,x_2,\tau)\ dx_1\ dx_2$,   joint probability of finding a pair of values of x in the ranges $(x_1, x_1 + dx_1)$ and $(x_2, x_2 + dx_2)$, which are a time interval $\tau = t_2 - t_1$ apart from each other

and so on.

1. *Stationary Random Process.* From the experiments described above, using the two methods of observation, we may derive two kinds of averages, namely, ensemble averages and time averages. The ensemble average $\tilde{x}$ can be evaluated from the first probability distribution function which is derived from observations made on many similar systems at the same time.

$$\tilde{x} = \int_{-\infty}^{\infty} x p_1(x,t)\ dx \qquad\qquad (4\text{-}134)$$

In general, this average is a function of the time of observations. The time average, which we shall denote by $\bar{x}$, as distinguished from the ensemble average, represents the mean value of the observations made on one of these systems for a large number of successive instants of time.

$$\bar{x} = \lim_{T \to \infty} \frac{1}{2T} \int_{-T}^{T} x(t)\ dt \qquad\qquad (4\text{-}135)$$

The time average will of course be independent of time, but will, in general, differ for the various systems $x_1(t)$, $x_2(t)$, ..., of the ensemble.

A random process is said to be stationary if the ensemble averages are independent of the time at which the averages for the group are computed, and also if the time averages are independent of the functions making up this group. Thus a stationary random process from the ensemble point of view is defined as one in which the statistics derived from observations on the ensemble members at any two distinct instants of time are the same. A stationary random process from the extension-in-time point of view is defined as one in which the statistics measured from observations made on one system over a long interval of time are independent of the time in which the observation time interval is located.

Although a random process may be stationary, the ensemble average is in general not equal to the time average of a given sample function. However, many types of random processes have this property; namely, the time average of a sample function is equal to the ensemble average. A stationary random process is defined as "ergodic" if all types of ensemble averages are interchangeable with the corresponding time averages,

$$\tilde{x} = \bar{x} \qquad (4\text{-}136)$$

Thus in "ergodic" processes, the averages across an ensemble are equal to the averages over time of a single function of infinite extent. This can be extended to higher-order moments of the function $x(t)$. In the case of an ergodic process, we have the n'th-order moment

$$\tilde{x^n} = \int_{-\infty}^{\infty} x^n p_1(x,t) \, dx \qquad (4\text{-}137)$$

equal to the time average of x^n

$$\bar{x^n} = \lim_{T \to \infty} \frac{1}{2T} \int_{-T}^{T} [x(t)]^n \, dt \qquad (4\text{-}138)$$

so that $\tilde{x^n} = \bar{x^n}$ \hfill (4-139)

In general, in an ergodic process, the two methods of observations which we used above will yield the same statistics.

As discussed above, the ensemble of random variables $x_k(t)$ may be obtained from simultaneous observations of a large number of systems or from observations of a single system over a sufficiently long interval of time. This may be accomplished by cutting up the

total record in many pieces, each of interval T, where T is very
large compared with all periods occurring in the process under consideration.   As stated previously, only in an ergodic process are the
statistical properties of the stationary processes obtained by the two
methods of observation the same.

   2. *Autocorrelation Functions.*   The autocorrelation function plays
a very important role in the harmonic analysis of random functions.
We shall show that the same role played in the theory of aperiodic
function by the Fourier transform pair, consisting of the impulse response h(t) and the associated transfer function $H(j\omega)$, is played in
the theory of random processes by the correlation function and spectral density.   If f(t) is a member function of an ensemble, its autocorrelation function is given by the expression

$$R(\tau) = \lim_{T \to \infty} \frac{1}{2T} \int_{-T}^{T} f(t)f(t + \tau) \, dt \tag{4-140}$$

This is a time average and is sometimes abbreviated to

$$R(\tau) = \overline{f(t) \, f(t + \tau)} \tag{4-141}$$

In ensemble terms, the autocorrelation function is given by

$$R(\tau) = \text{av.} \{[f(t) - \widetilde{f(t)}] \, [f(t + \tau) - \widetilde{f(t + \tau)}]\} \tag{4-142}$$

Confining ourselves to a random process where $\widetilde{f(t)} = 0$, this reduces
to

$$R(\tau) = \text{av.} \, [f(t) \, f(t + \tau)] = \int_{-\infty}^{\infty} \int_{-\infty}^{\infty} f_1 f_2 p_2 (f_1, f_2, \tau) \, df_1 df_2 \tag{4-143}$$

and if the process is "ergodic," then the time average equals the
ensemble average.

   In order to appreciate the importance of the correlation function
in the harmonic analysis of an ensemble of random functions, let us
consider the physical meaning of the concept of correlation function.
If at the time t, the value of f(t) is large, then the probability is
small that at the instant (t + τ), where τ is sufficiently small, the
value of f(t + τ) will be equal to zero. However, for τ large, the
function f(t + τ) may have any arbitrary value. Thus, it is clear that
the behavior of a random quantity f(t) is characterized not only by
its value at time t, but also by the mutual relation between the values of the function f(t) at instants t and (t + τ). The correlation

function characterizes to some extent the statistical relation between the values of the function $f(t)$ at time t and $(t + \tau)$, where τ can be both positive and negative. In fact, in cases where the values of $f(t)$ and $f(t + \tau)$ can be regarded as being statistically independent, then the mean value of the product is equal to the product of the mean value; i.e.,

$$\overline{f(t)\ f(t + \tau)} = \overline{f(t)}\ \ \overline{f(t + \tau)} = 0 \qquad (4\text{-}144)$$

which leads to $R(\tau) = 0$. Even in case of strong statistical dependence of the values of $f(t)$ and $f(t + \tau)$, the statistical relation between $f(t)$ and $f(t + \tau)$ becomes weaker as τ is increased, and as $|\tau| \rightarrow \infty$, it approaches zero, so that the values become statistically independent as $|\tau| \rightarrow \infty$. Since for $\tau = 0$, the correlation is maximum, we can express the measure of correlation between $f(t)$ and $f(t + \tau)$ by the ratio

$$\rho = \frac{R(\tau)}{R(0)} \qquad (4\text{-}145)$$

at instants t and $(t + \tau)$. The correlation function gives a measure of this relation between $f(t)$ and $f(t + \tau)$.

Some of the properties of the autocorrelation function are as follows:

The autocorrelation function for a periodic or an aperiodic function has been shown to be an even function of τ. For a random function, it is also the case, since

$$R(-\tau) = \lim_{T \rightarrow \infty} \frac{1}{2T} \int_{-T}^{T} f(t)f(t - \tau)\ dt$$

$$= \lim_{T \rightarrow \infty} \frac{1}{2T} \int_{-T - \tau}^{T - \tau} f(x)f(x + \tau)\ dx \qquad (4\text{-}146)$$

where $x = t - \tau$. By displacing both limits of integration in the same direction by τ, the interval $2T$ has not been changed. Therefore,

$$R(-\tau) = \lim_{T \rightarrow \infty} \frac{1}{2T} \int_{-T}^{T} f(x)f(x + \tau)\ dx = R(\tau) \qquad (4\text{-}147)$$

The mean square of f(t) or the mean power is given by

$$R(0) = \lim_{T \to \infty} \frac{1}{2T} \int_{-T}^{T} f^2(t) \, dt = \overline{f^2(t)} \tag{4-148}$$

which is also known as the variance of the random process.

3. *Power Spectral Density.* The concept of spectral density has already been introduced in connection with periodic and aperiodic functions. However, since for random functions, the signal goes on indefinitely and consequently the Fourier integral and energy spectral density cannot be defined, it is useful to introduce the concept of power spectral density on the assumption that the average power of the random signal is finite. This concept is of extreme importance in the study of random functions because like the correlation function, it has the same form for any of the random functions f(t), which characterize the stationary random process under consideration.

The concept of power spectral density may be introduced as an extension of the energy density function $|F(j\omega)|^2$, which has been introduced in the discussion of aperiodic functions. In the case of random functions, we consider a finite segment $f_T(t)$ of the random function f(t) extending from $t = -T$ to $t = +T$. Thus

$$f_T(t) = f(t), \qquad -T \le t \le T$$

and $f_T(t) = 0,$ for all other values of t

Let $F_T(j\omega)$ denote the Fourier transform of the function $f_T(t)$; hence

$$F_T(j\omega) = \int_{-\infty}^{\infty} f_T(t) e^{-j\omega t} \, dt = \int_{-T}^{T} f(t) e^{-j\omega t} \, dt \tag{4-149}$$

The spectral density $S(\omega)$ of the function f(t) is defined by the expression*

$$S(\omega) = \lim_{T \to \infty} \left[\text{av.} \; \frac{|F_T(j\omega)|^2}{2\pi T} \right]$$

$$\tag{4-150}$$

We shall show now that the mean square value of the random function f(t) is given by

---

*For a rigorous discussion of the definition of $S(\omega)$ see References (1) and (3).

$$R(0) = \overline{f^2(t)} = \frac{1}{2\pi} \int_{-\infty}^{\infty} S(\omega) \, d\omega \tag{4-151}$$

Consider
$$\int_{-\infty}^{\infty} |F_T(j\omega)|^2 \, d\omega = \int_{-\infty}^{\infty} F_T(-j\omega) \cdot F_T(j\omega) \, d\omega$$

$$= \int_{-\infty}^{\infty} F_T(-j\omega) \, d\omega \int_{-\infty}^{\infty} f_T(t) e^{-j\omega t} \, dt$$

or, changing the order of integration,

$$\int_{-\infty}^{\infty} |F_T(j\omega)|^2 \, d\omega = \int_{-\infty}^{\infty} f_T(t) \, dt \int_{-\infty}^{\infty} F_T(-j\omega) e^{-j\omega t} \, d\omega$$

$$= 2\pi \int_{-\infty}^{\infty} f_T^2(t) \, dt$$

Hence
$$\int_{-\infty}^{\infty} f_T^2(t) \, dt = \int_{-T}^{T} f_T^2(t) \, dt = \frac{1}{2\pi} \int_{-\infty}^{\infty} |F_T(j\omega)|^2 \, d\omega \tag{4-152}$$

By definition
$$\overline{f^2(t)} = \lim_{T \to \infty} \frac{1}{2T} \int_{-T}^{T} f_T^2(t) \, dt$$

$$= \lim_{T \to \infty} \frac{1}{2\pi} \int_{-\infty}^{\infty} \frac{1}{2T} |F_T(j\omega)|^2 \, d\omega$$

Hence
$$R(0) = \overline{f^2(t)} = \frac{1}{2\pi} \int_{-\infty}^{\infty} S(\omega) \, d\omega \tag{4-153}$$

The physical significance of the function $S(\omega)$ is as follows. The average power of $f_T(t)$ over the interval 2T is given by

$$P_{av.} = \frac{1}{2T} \int_{-T}^{T} f_T^2(t) \, dt = \frac{1}{2\pi} \int_{-\infty}^{\infty} \frac{1}{2T} |F_T(j\omega)|^2 \, d\omega \tag{4-154}$$

In the limit when $T \to \infty$, the expression for the average power becomes

$$P_{av} = \frac{1}{2\pi} \int_{-\infty}^{\infty} S(\omega) \, d\omega \tag{4-155}$$

From the definition of spectral density, we note that $S(\omega)$ is real and an even function of ω. From Eq. (4-155), it is seen that the two-sided spectral density $S(\omega)$ may be physically interpreted as the average power output of a narrow-band filter divided by its band-width (positive and negative frequencies). For

$$P_{av.} = \int_{-\infty}^{\infty} S(\omega) \, df \doteq 2S(\omega) \, \Delta f \tag{4-156}$$

where Δf is the bandwidth in cycles per second of the narrow-band filter.

4. *Band-limited White Noise.* A random signal whose power spectral density is constant, independent of frequency, is referred to as impulse or white noise. White noise is not physically realizable because the concept of constant power spectral density over the entire frequency range implies infinite power as seen from the relation

$$P_{av.} = R(0) = \frac{1}{2\pi} \int_{-\infty}^{\infty} S(\omega) \, d\omega = \frac{1}{2\pi} \int_{-\infty}^{\infty} N_0/2 \, d\omega \rightarrow \infty$$

where $S(\omega) = N_0/2$ watts/cps, a constant.

Nevertheless, it is a useful concept where the power spectral density is constant over a limited frequency range and zero outside, such as the noise power in a receiver IF. In this case, we have band-limited white noise where the two-sided spectral density $S(\omega)$ is given by

$$S(\omega) = N_0/2, \qquad \omega_1 < |\omega| < \omega_2$$

$$= 0, \qquad \text{otherwise}$$

This type of noise is physically realizable because its mean square value is finite,

$$P_{av} = \frac{2}{2\pi} \int_{\omega_1}^{\omega_2} N_0/2 \, d\omega = BN_0 \tag{4-157}$$

where $B = (\omega_2 - \omega_1)/2\pi$, the IF bandwidth.

In practice the transfer function $H(j\omega)$ of the linear network is not constant, and consequently we must consider in the calculation of P_{av} the equivalent-noise bandwidth. The equivalent-noise bandwidth of a network is defined as the bandwidth of an ideal rectangular network having the same maximum gain and passing the same average power from a white-noise source as the actual network under consideration. It will be shown later that the output spectral density $S_0(\omega)$ of a signal at the output of a linear dynamic system is related to the input spectral density $S_i(\omega)$ by the relation

$$S_0(\omega) = |H(j\omega)|^2 \, S_i(\omega) = N_0/2 \, |H(j\omega)|^2, \text{ for white noise.} \quad (4\text{-}158)$$

Relating the output noise powers of the actual and ideal networks as shown in Fig. 4-12, we have

$$P_{av} = \frac{1}{2\pi} \int_{-\infty}^{\infty} N_0/2 \, |H(j\omega)|^2 \, d\omega = N_0 \, |H(j\omega_0)|^2 \cdot B_n$$

Hence

$$B_n = \frac{\dfrac{1}{2\pi} \int_{-\infty}^{\infty} |H(j\omega)|^2 \, d\omega}{|H(j\omega_0)|^2} \quad (4\text{-}159)$$

The equivalent-noise bandwidth is often related to the half-power bandwidth of the network, thus yielding a direct relationship between the 3-db response and the output-noise powers. For example, it can be shown that the equivalent-noise bandwidth of a single-tuned circuit is given by

$$B_n = (\pi/2) B_{3db} \quad (4\text{-}160)$$

and of two stagger-tuned circuits

$$B_n = \frac{\pi}{2\sqrt{2}} B_{3db} \quad (4\text{-}161)$$

where B_{3db} is the overall 3-db bandwidth.

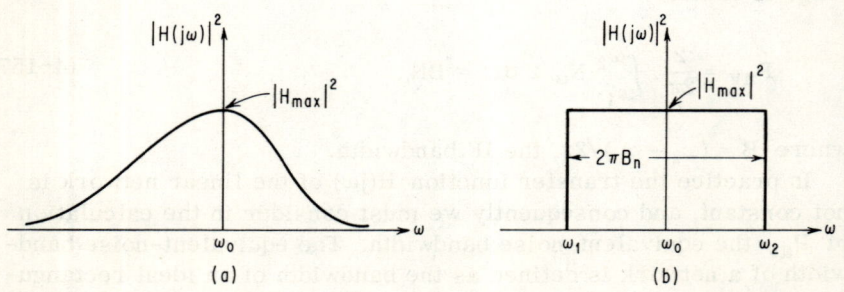

Fig. 4-12. Equivalent-noise bandwidth. (a) Response of actual network; (b) response of ideal network.

5. *Relation between Autocorrelation Function and Spectral
Density: The Wiener-Khinchin Theorem*.   It has been mentioned
earlier that in the theory of stationary random processes, the correlation function and the spectral density form a Fourier transform
pair analogous to the transform pair of the impulse response h(t)
and the transfer function H(jω) associated with aperiodic functions.
We shall show now that

$$R(\tau) = \frac{1}{2\pi} \int_{-\infty}^{\infty} S(\omega) e^{j\omega\tau} d\omega = \frac{1}{\pi} \int_{0}^{\infty} S(\omega) \cos \tau\omega \, d\omega \qquad (4\text{-}162)$$

and $$S(\omega) = \int_{-\infty}^{\infty} R(\tau) e^{-j\omega\tau} \, d\tau = 2 \int_{0}^{\infty} R(\tau) \cos \omega\tau \, d\tau \qquad (4\text{-}163)$$

The cosine transforms are obtained because $R(\tau)$ is a real and even
function of τ, so that its transform $S(\omega)$ is also a real and even
function. Let us introduce the function

$$R_T(\tau) = \frac{1}{2T} \int_{-T}^{T} f_T(t) f_T(t + \tau) \, dt \qquad (4\text{-}164)$$

The Fourier transform of the function $R_T(\tau)$ is

$$\int_{-\infty}^{\infty} R_T(\tau) e^{-j\omega\tau} \, d\tau = \frac{1}{2T} \int_{-\infty}^{\infty} e^{-j\omega\tau} \, d\tau \int_{-\infty}^{\infty} f_T(t) f_T(t + \tau) \, dt$$

$$= \frac{1}{2T} \int_{-\infty}^{\infty} d\tau \int_{-\infty}^{\infty} f_T(t) f_T(t+\tau) e^{-j\omega(t+\tau)} e^{j\omega t} \, dt$$

$$= \frac{1}{2T} \int_{-\infty}^{\infty} f_T(t) e^{j\omega t} \, dt \int_{-\infty}^{\infty} f_T(t+\tau) e^{-j\omega(t+\tau)} \, d\tau$$

Since $$F_T^*(j\omega) = F_T(-j\omega)$$

this reduces to

$$\int_{-\infty}^{\infty} R_T(\tau) e^{-j\omega\tau} \, d\tau = \frac{1}{2T} F_T(j\omega) \cdot F_T^*(j\omega) \qquad (4\text{-}165)$$

Noting that $R_T(\tau)$ is an even function of τ, we finally obtain

$$2 \int_{0}^{\infty} R_T(\tau) \cos \omega\tau \, d\tau = \frac{1}{2T} \left| F_T(j\omega) \right|^2 \qquad (4\text{-}166)$$

or in the limit

$$\lim_{T \to \infty} 2 \int_0^\infty R_T(\tau) \cos \omega \tau \; d\tau = \lim_{T \to \infty} \frac{1}{2T} \; | \; F_T(j\omega) \; |^2$$

From the last equation, we derive that

$$S(\omega) = 2 \int_0^\infty R(\tau) \cos \omega \tau \; d\tau \qquad (4\text{-}167)$$

which implies that $S(\omega) = S(-\omega)$.

As stated earlier, this relationship between the correlation function and the spectral density has great significance in the theory of stationary processes. It permits the determination of the spectral density $S(\omega)$ from a given correlation function $R(\tau)$ and conversely.

6. *Application of Wiener-Khinchin Theorem to Band-limited White Noise.* As an example let us consider the application of the Wiener-Klinchin theorem to white noise, which as we have noted previously is defined to have a constant power spectral density $S(\omega)$ = $N_0/2$ watts/cps over the entire frequency range. The correlation function is

$$R(\tau) = \frac{1}{2\pi} \int_{-\infty}^\infty S(\omega) e^{j\omega \tau} \; d\omega = \frac{N_0/2}{2\pi} \int_{-\infty}^\infty e^{j\omega \tau} \; d\omega = N_0/2 \; \delta(\tau)$$

$$(4\text{-}168)$$

where $\displaystyle \int_{-\infty}^\infty \delta(\tau) \; d\tau = 1$

As discussed previously, while white noise is not physically realizable, this concept is useful where the power spectral density is constant over a limited frequency range and zero outside. As an example we shall find first the correlation function of the noise process at the output of an ideal lowpass filter with cut-off angular frequency ω_b.

$$R(\tau) = \frac{1}{2\pi} \int_{-\infty}^\infty S(\omega) \; e^{j\omega \tau} \; d\omega = \frac{N_0/2}{2\pi} \int_{-\omega_b}^{\omega_b} e^{j\omega \tau} \; d\omega = \frac{N_0 \omega_b}{2\pi} \frac{\sin \omega_b \tau}{\omega_b \tau}$$

$$(4\text{-}169)$$

As a second example we shall find the correlation function of narrow-band noise. Consider the application of white noise to a narrow-band filter centered at angular frequency ω_c and of band-width $\Delta\omega$ ($\Delta\omega \ll \omega_c$). For the sake of simplicity assume the filter to be of the ideal rectangular type, such that

$$H(j\omega) = K \qquad \omega_c - \frac{\Delta\omega}{2} < \omega < \omega_c + \frac{\Delta\omega}{2}$$

$$H(j\omega) = 0 \qquad \text{elsewhere}$$

As noted previously the output spectral density $S_o(\omega)$ is related to the input spectral density by

$$S_o(\omega) = |H(j\omega)|^2 \cdot S_i(\omega)$$

Hence $\quad S_o(\omega) = (N_o/2)K^2 \qquad \omega_c - \frac{\Delta\omega}{2} < |\omega| < \omega_c + \frac{\Delta\omega}{2}$

$$S_o(\omega) = 0 \qquad\qquad \text{elsewhere}$$

The autocorrelation function of the narrow-band noise process at the filter output is

$$R(\tau) = \frac{1}{2\pi} \int_{-\infty}^{\infty} S_o(\omega)e^{j\omega\tau}\, d\omega = \frac{(N_o/2)K^2}{2\pi} \int_{-\omega_c - \Delta\omega/2}^{-\omega_c + \Delta\omega/2} e^{j\omega\tau}\, d\omega$$

$$+ \frac{(N_o/2)K^2}{2\pi} \int_{\omega_c - \Delta\omega/2}^{\omega_c + \Delta\omega/2} e^{j\omega\tau}\, d\omega$$

$$= \frac{N_o K^2}{2\pi} \int_{\omega_c - \Delta\omega/2}^{\omega_c + \Delta\omega/2} e^{j\omega\tau}\, d\omega = N_o K^2 \frac{\Delta\omega}{2\pi} \frac{\sin \Delta\omega\tau/2}{\Delta\omega\tau/2}$$

$$\times \cos \omega_c\tau \qquad\qquad\qquad\qquad\qquad (4\text{-}17\text{C})$$

As shown in Figure 4-13, $R(\tau)$ resembles an amplitude modulated carrier whose envelope follows the familiar $(\sin x)/x$ relation, becoming zero at $\tau = 1/B$ [$B = (\Delta\omega/2\pi)$] and oscillating with decreasing intensity for larger values of τ.

7. *Application of Wiener-Khinchin Theorem to Gaussian Noise Spectrum.* In this case the amplitude response of the transfer func-

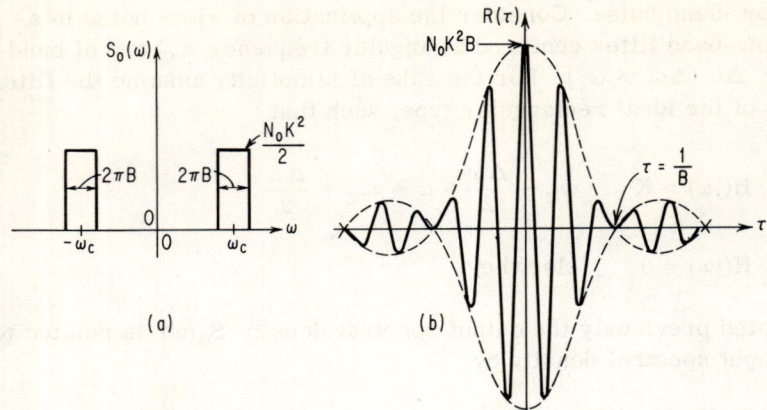

Fig. 4-13. Power spectrum and correlation function of narrow-band white noise. (a) Power spectrum; (b) correlation function. (After M. Schwartz,[5] courtesy of McGraw-Hill Book Company.)

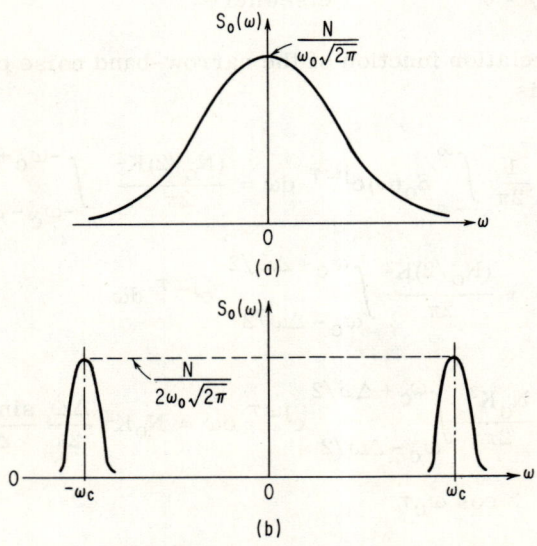

Fig. 4-14. Gaussian power spectrum. (a) Low-pass; (b) bandpass.

tion of the linear network is not constant but has a gaussian frequency selectivity. First we shall consider the gaussian low-pass filter shown in Fig. 4-14a. The power spectral density at the filter output is

$$S_0(\omega) = \frac{N e^{-\omega^2/2\omega_0^2}}{\sqrt{2\pi}\,\omega_0} \tag{4-171}$$

where N is the mean total power, and at $\omega = \omega_0$ the spectral density is down one-quarter neper or 2.17 db relative to the density at zero frequency. This spectral distribution simulates the more realistic gradual cutoff response of a low-pass filter encountered in physical networks than the sharp cutoff of the ideal filter. The autocorrelation function for the gaussian low-pass spectrum is

$$R(\tau) = \frac{1}{2\pi} \int_{-\infty}^{\infty} S_0(\omega)\, e^{j\omega\tau}\, d\omega = \frac{N}{(2\pi)^{3/2}\,\omega_0} \int_{-\infty}^{\infty} e^{-\omega^2/2\omega_0^2}$$

$$\times\, e^{j\omega\tau}\, d\omega = \frac{N}{2\pi}\, e^{-(\omega_0\tau)^2/2} \tag{4-172}$$

The gaussian bandpass spectrum is shown in Fig. 4-14b; its power spectral density is given by

$$S_0(\omega) = \frac{N}{2\omega_0\sqrt{2\pi}}[e^{-(\omega-\omega_c)/2\omega_0^2} + e^{-(\omega+\omega_c)^2/2\omega_0^2}] \tag{4-173}$$

and its autocorrelation function is

$$R(\tau) = \frac{N}{2\pi}\, e^{-(\omega_0\tau)^2/2} \cos\omega_c\tau \tag{4-174}$$

The narrow-band case in which $\omega_0 \ll \omega_c$ is the one most frequently of interest. The parameter $2f_0$ ($= 2\,\omega_0/2\pi$) can readily be shown to be equal to 1.7B, where 2B is the 3 db-bandwidth of the bandpass filter. Such a gaussian filter is a good approximation for the response of a sharply tuned multistage amplifier to white noise.

At this point, we should compare the relationship between the correlation function and the spectral density which we have just derived for a random function, with the analogous result derived in the analysis of periodic functions as given by the transform pair of Eqs. (2-51) and (2-52). In the case of periodic functions, the autocorrelation function is periodic and the power spectrum discrete, while in the case of random functions, the autocorrelation function is aperiodic and the spectrum continuous.

8. *Autocorrelation and Spectral Density Function of a Random*

Wave with a Periodic Component. We shall derive now an expression for the spectral density for the case when the random wave $f(t)$ has a periodic component of angular frequency ω_0. In this case, the power density spectrum consists of a continuous curve for the random component and a series of impulses for the periodic component. At the fundamental and harmonic frequencies of the periodic component, the impulses are infinite in amplitude but of finite area and can be represented by the well-known delta functions.

Consider the periodic function

$$f(t) = \sum_{n=-\infty}^{\infty} C_n e^{jn\omega_0 t} = C_0 + 2\sum_{n=1}^{\infty} |C_n| \cos(n\omega_0 t - \theta_n)$$

(4-175)

and its autocorrelation

$$R(\tau) = \sum_{n=-\infty}^{\infty} |C_n|^2 e^{jn\omega_0 \tau}$$

(4-176)

The power spectrum $S(\omega)$ is therefore given by

$$S(\omega) = \int_{-\infty}^{\infty} R(\tau) e^{-j\omega\tau}\, d\tau = \sum_{n=-\infty}^{\infty} |C_n|^2 \int_{-\infty}^{\infty} e^{-j(\omega - n\omega_0)\tau}\, d\tau$$

(4-177)

Since the Fourier transform of $e^{jn\omega_0 \tau}$ is equal to $2\pi\, \delta(\omega - n\omega_0)$, it follows that

$$S(\omega) = \sum_{n=-\infty}^{\infty} |C_n|^2 \cdot 2\pi\, \delta(\omega - n\omega_0)$$

(4-178)

In the case of a simple sinusoid plus a d-c term, we have

$$f(t) = a_0 + a_1 \cos(\omega_0 t - \theta)$$

where $C_0 = a_0$, $C_1 = a_1/2$, and $C_{-1} = a_1/2$. Hence

$$S(\omega) = 2\pi\left[a_0^2\, \delta(\omega) + \frac{a_1^2}{4}\, \delta(\omega - \omega_0) + \frac{a_1^2}{4}\, \delta(\omega + \omega_0) \right]$$

(4-179)

Thus, the spectral density of the random function plus the periodic components will consist of a continuous part and a certain number

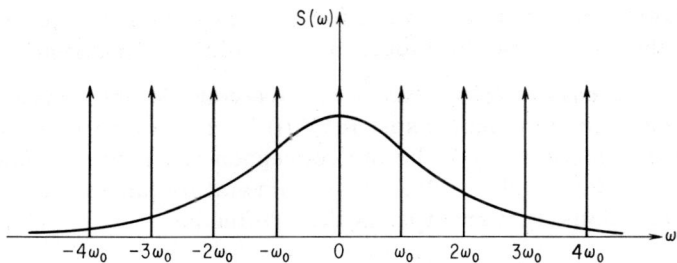

Fig. 4-15. Power density spectrum of a random wave with a periodic component.

of delta functions for discrete frequency values, as shown in Fig. 4-15.

9. *Autocorrelation Function of Signal plus Noise.* Consider a sample function $x(t)$ of an ergodic random process which is the sum of a desired signal $s(t)$ and a noise component $n(t)$ which is statistically independent of $s(t)$.

$$x(t) = s(t) + n(t) \tag{4-180}$$

The autocorrelation function of $x(t)$ is

$$R_x(\tau) = \overline{[x(t)x(t-\tau)]} = \overline{\{[s(t)+n(t)]\,[s(t-\tau)+n(t-\tau)]\}}$$

$$= \overline{\{[s(t)s(t-\tau)] + [s(t-\tau)n(t)] + [n(t-\tau)s(t)]}$$

$$+ \overline{[n(t)n(t-\tau)]\}} \tag{4-181}$$

since the noise and signal are assumed to be statistically independent; hence

$$R_x(\tau) = \overline{[s(t)s(t-\tau)]} + \overline{[s(t-\tau)n(t)]} + \overline{[n(t-\tau)s(t)]}$$

$$+ \overline{[n(t)n(t-\tau)]} \tag{4-182}$$

If either the noise or the signal has a zero mean, the last equation reduces to

$$R_x(\tau) = \overline{[s(t)s(t-\tau)]} + \overline{[n(t)n(t-\tau)]} = R_s(\tau) + R_n(\tau)$$

$$\tag{4-183}$$

or the autocorrelation function of the sum of signal plus noise is equal to the sum of the individual autocorrelation functions.

10. *Cross-correlation Functions.*   We consider two ensembles of stationary random processes; let $f_1(t)$ be a member function for the first ensemble and $f_2(t)$ be the corresponding member function for the second ensemble. The cross-correlation function of the random functions $f_1(t)$ and $f_2(t)$ is then defined as

$$R_{12}(\tau) = \lim_{T \to \infty} \frac{1}{2T} \int_{-T}^{T} f_1(t) f_2(t + \tau) \, dt \tag{4-184}$$

As in the case of periodic functions, the order of the subscripts of the cross-correlation function is important, so that

$$R_{21}(\tau) = \lim_{T \to \infty} \frac{1}{2T} \int_{-T}^{T} f_2(t) f_1(t + \tau) \, dt \tag{4-185}$$

and $R_{12}(\tau) = R_{21}(-\tau)$ \hfill (4-186)

The last relation can be shown as follows:

$$R_{12}(-\tau) = \lim_{T \to \infty} \frac{1}{2T} \int_{-T}^{T} f_1(t) f_2(t - \tau) \, dt$$

$$= \lim_{T \to \infty} \frac{1}{2T} \int_{-T}^{T} f_2(x) f_1(x + \tau) \, dx = R_{21}(\tau) \tag{4-187}$$

Similarly, the cross-power density spectrum of the random functions $f_1(t)$ and $f_2(t)$ is defined as

$$S_{12}(\omega) = \lim_{T \to \infty} \frac{1}{2T} F_{1,T}^{*}(j\omega) \cdot F_{2,T}(j\omega) \tag{4-188}$$

where $F_{1,T}(j\omega) = \int_{-T}^{T} f_{1,T}(t) e^{-j\omega t} \, dt$

and $F_{2,T}(j\omega) = \int_{-T}^{T} f_{2,T}(t) e^{-j\omega t} \, dt$

From the relations $F_{1,T}(j\omega) = F_{1,T}^*(-j\omega)$

$$F_{2,T}(j\omega) = F_{2,T}^*(-j\omega)$$

it follows that

$$S_{12}(\omega) = S_{12}^*(-\omega) = S_{21}(-\omega) \qquad (4\text{-}189)$$

The transform pair of the cross-correlation functions is

$$R_{12}(\tau) = \frac{1}{2\pi} \int_{-\infty}^{\infty} S_{12}(\omega) e^{j\omega\tau}\, d\omega \qquad (4\text{-}190)$$

and $S_{12}(\omega) = \int_{-\infty}^{\infty} R_{12}(\tau) e^{-j\omega\tau}\, d\tau \qquad (4\text{-}191)$

The function $S_{12}(\omega)$ is in general complex, and consequently it has no simple physical interpretation. To clarify its meaning consider the random process

$$f(t) = s(t) + n(t) \qquad (4\text{-}192)$$

where $s(t)$ and $n(t)$ are statistically correlated. Hence

$$R_f(\tau) = \overline{f(t) \cdot f(t-\tau)} = \overline{[s(t) + n(t)][s(t-\tau) + n(t-\tau)]}$$

$$= R_s(\tau) + R_n(\tau) + R_{sn}(\tau) + R_{ns}(\tau) \qquad (4\text{-}193)$$

and $S_f(\omega) = S_s(\omega) + S_n(\omega) + S_{sn}(\omega) + S_{ns}(\omega) \qquad (4\text{-}194)$

But $S_{ns}(\omega) = S_{sn}^*(\omega)$

Hence $S_f(\omega) = S_s(\omega) + S_n(\omega) + 2R_e[S_{sn}(\omega)] \qquad (4\text{-}195)$

The first two terms of Eq. (4-195) represent the spectral power density in the absence of correlation between $s(t)$ and $n(t)$; the term $2R_e[S_{sn}(\omega)]$ can be interpreted as the "interference intensity" caused by the statistical relation between the random processes $s(t)$ and $n(t)$.

11. *The Correlation Function and Spectral Density of a Random Telegraph Signal.*   The following example will serve to illustrate the calculation of the correlation function and spectral density of a

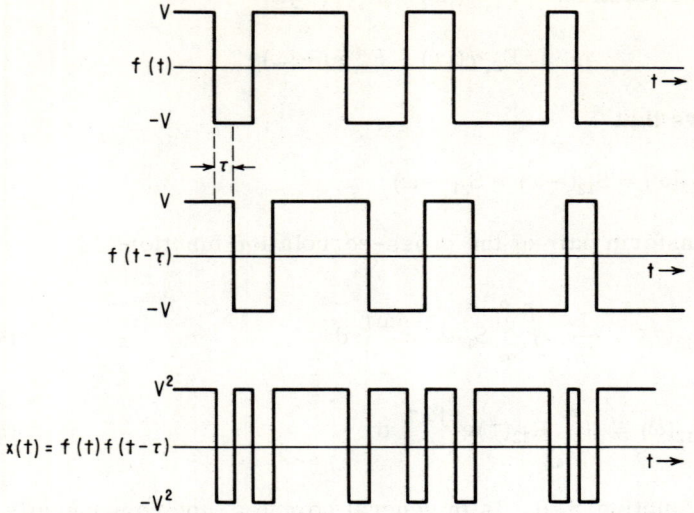

Fig. 4-16. Graph of $f(t)$, $f(t - \tau)$, and $x(t) = f(t) \times f(t - \tau)$.

random function. Let $f(t)$ represent a random telegraph signal which is equal to either $+V$ or $-V$ so that it is of the form of a flat-top wave, as shown in Fig. 4-16. Assume that the number of transitions or zeros obey a Poisson distribution. This assumption is justified if we assume that if on the average, there are μ changes of sign per second, the probability of a change of sign in the interval $(t, t + dt)$ is $\mu\,dt$ and is independent of the number of transitions outside the interval $(t, t + dt)$. Thus, the probability of obtaining exactly N zeros or transitions in T sec is given by

$$P(N, T) = \frac{(\mu T)^N}{N!}\, e^{-\mu T} \qquad\qquad (4\text{-}196)$$

where μ is the average number of transitions or zero crossings per second.

    To compute the correlation function, we consider the average value of the product $x(t) = f(t)f(t - \tau)$. This product equals $+V^2$ or $-V^2$, depending on whether the functions have like or unlike signs. In the first case, there are an even number, including zero, of changes of sign in the interval $(t, t + \tau)$ and in the second case, there are an odd number of changes of sign. Thus average value of $f(t)f(t - \tau)$ is given by

$$\text{av. } x(t) = \sum_n x_n P(x_n) = V^2 \sum_n P(x_n)$$

or   av. $x(t) = \overline{f(t)f(t - \tau)} = V^2 \times$ probability of an even number of
changes of sign in the interval $(t, \ t+\tau)$
$- V^2 \times$ probability of an odd number of
changes of sign in the interval $(t, \ t+\tau)$

$$(4\text{-}197)$$

Since by assumption, the probability of a change of sign during the
interval τ is independent of what happens outside that interval, it
follows that the assumption will remain valid for any interval, irrespective of its starting point. Hence, the probabilities in Eq.
(4-197) are independent of t and can be obtained from Eq. (4-196)
by setting $T = \tau$. Thus, Eq. (4-197) reduces to

$$\overline{f(t)f(t - \tau)} = V^2[P(0) + P(2) + P(4) + \cdots] - V^2[P(1) + P(3) + P(5) + \cdots]$$

$$= V^2 e^{-\mu\tau}\left[1 - \frac{\mu\tau}{1!} + \frac{(\mu\tau)^2}{2!} - \cdots\right] = V^2 e^{-2\mu|\tau|}$$

$$(4\text{-}198)$$

From Eq. (4-141), this equals the correlation function for $f(t)$, so
that

$$R(\tau) = V^2 e^{-2\mu|\tau|} \qquad\qquad (4\text{-}199)$$

which is shown plotted in Fig. 4-17.

The corresponding spectral density $S(\omega)$ is from Eq. (4-167)
given by

$$S(\omega) = 2V^2 \int_0^\infty \epsilon^{-2\mu\tau} \cos \omega\tau \ d\tau = \frac{4V^2\mu}{\omega^2 + 4\mu^2} \qquad (4\text{-}200)$$

Correlation functions and spectral density functions of this type are
often encountered in practice. For example, the spectral density of

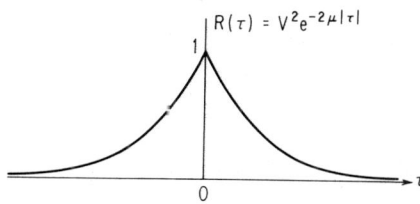

Fig. 4-17. Autocorrelation function $f(t)$ shown in Fig. 4-16.

a current flowing through an inductance and a resistance in series in response to a very wideband thermal noise-white noise is in the form of Eq. (4-200).

4.7   Response of Linear System to Random Function[2, 6, 8]

In Chap. 3, we discussed the response of a linear dynamic system to an aperiodic function, having an impulse response $h(t)$ and a transfer function $H(j\omega)$. In this section, we shall study the response of such a linear system to an input signal of which only certain statistical properties are known. We shall suppose that only the spectral density $S_i(\omega)$ of the input random function, or its time equivalent, the autocorrelation function $R_i(\tau)$, is known. The problem is to derive expressions for the output spectral density $S_0(\omega)$ in terms of $S_i(\omega)$ and the transfer function $H(j\omega)$ of the linear network. We shall develop similar expressions for the output cross-power density spectrum of two linearly transformed random functions.

1. *The Correlation Function and Spectral Density of a Signal at the Output of a Linear Dynamic System.*   Let the input signal $x_i(t)$ be a stationary random signal having a correlation function $R_i(\tau)$ and a spectral density $S_i(\omega)$. The problem is to find the correlation function $R_0(\tau)$ and the spectral density $S_0(\omega)$ of the output signal $x_0(t)$. By definition, the autocorrelation function of the output is given by

$$R_0(\tau) = \lim_{T \to \infty} \frac{1}{2T} \int_{-T}^{T} x_0(t + \tau)x_0(t) \ dt \qquad (4\text{-}201)$$

but

$$x_0(t) = \int_{-\infty}^{\infty} x_i(t - \lambda)h(\lambda) \ d\lambda \qquad (4\text{-}202)$$

Therefore, by substitution

$$R_0(\tau) = \lim_{T \to \infty} \frac{1}{2T} \int_{-T}^{T} dt \left[\int_{-\infty}^{\infty} x_i(t - \lambda) \cdot h(\lambda) \ d\lambda \right.$$

$$\left. \times \int_{-\infty}^{\infty} x_i(t + \tau - \eta) \cdot h(\eta) \ d\eta \right]$$

Interchanging the order of integration gives

$$R_0(\tau) = \lim_{T \to \infty} \int_{-\infty}^{\infty} d\lambda \int_{-\infty}^{\infty} d\eta h(\eta) h(\lambda)$$

$$\times \left[\frac{1}{2T} \int_{-T}^{T} x_-(t + \tau - \eta) x_i(t - \lambda) \, dt \right]$$

But $\quad \lim_{T \to \infty} \dfrac{1}{2T} \displaystyle\int_{-T}^{T} x_i(t + \tau - \eta) x_i(t - \lambda) \, dt$

$$= \lim_{T \to \infty} \frac{1}{2T} \int_{-T}^{T} x_i(t + \tau + \lambda - \eta) x_i(t) \, dt = R_i(\tau + \lambda - \eta)$$

so that $R_0(\tau) = \displaystyle\int_{-\infty}^{\infty} h(\lambda) \, d\lambda \int_{-\infty}^{\infty} h(\eta) R_i(\tau + \lambda - \eta) \, d\eta$ (4-203)

This equation represents the relationship between the correlation functions of the input and output signals of a linear dynamic system whose impulse response is h(t).

Let us now prove the corresponding relationship between the spectral densities of the input and output signals. Since the correlation function and spectral density form a Fourier pair, we have

$$S_0(\omega) = \int_{-\infty}^{\infty} R_0(\tau) e^{-j\omega\tau} \, d\tau \tag{4-204}$$

By substitution of Eq. (4-203) into (4-204), we have

$$S_0(\omega) = \int_{-\infty}^{\infty} d\tau \int_{-\infty}^{\infty} d\lambda \int_{-\infty}^{\infty} e^{-j\omega\tau} h(\eta) h(\lambda) R_i(\tau + \lambda - \eta) \, d\eta$$

$$= \int_{-\infty}^{\infty} d\tau \int_{-\infty}^{\infty} d\lambda \int_{-\infty}^{\infty} e^{-j\omega(\tau + \lambda - \eta)} \cdot e^{j\omega\lambda} \cdot e^{-j\omega\eta}$$

$$\times R_i(\tau + \lambda - \eta) \cdot h(\lambda) \cdot h(\eta) \cdot d\eta$$

$$= \int_{-\infty}^{\infty} h(\eta) e^{-j\omega\eta} \, d\eta \cdot \int_{-\infty}^{\infty} h(\lambda) e^{j\omega\lambda} \, d\lambda$$

$$\times \int_{-\infty}^{\infty} R_i(\tau + \lambda - \eta) e^{-j\omega(\tau + \lambda - \eta)} \, d\tau \tag{4-205}$$

But $\quad \int_{-\infty}^{\infty} h(\eta) e^{-j\omega\eta} \, d\eta = H(j\omega), \qquad$ transfer function of the network

and $\quad \int_{-\infty}^{\infty} h(\lambda) e^{j\omega\lambda} \, d\lambda = H(-j\omega), \qquad$ complex conjugate of $H(j\omega)$

Hence $\quad S_0(\omega) = |H(j\omega)|^2 \int_{-\infty}^{\infty} R_i(\tau) e^{-j\omega\tau} \, d\tau = |H(j\omega)|^2 \cdot S_i(\omega)$

$$(4-206)$$

This is the relationship we set out to prove. The physical significance is that since each spectral component is modified by $H(j\omega)$, the power spectrum is modified by $|H(j\omega)|^2$.

Since $R_0(\tau)$ and $S_0(\omega)$ form a Fourier pair, namely,

$$R_0(\tau) = \frac{1}{2\pi} \int_{-\infty}^{\infty} S_0(\omega) e^{j\omega\tau} \, d\omega$$

it follows from Eq. (4-206) that

$$R_0(\tau) = \frac{1}{2\pi} \int_{-\infty}^{\infty} |H(j\omega)|^2 S_i(\omega) e^{j\omega\tau} \, d\omega \qquad (4-207)$$

In particular, the variance of the output is

$$R_0(0) = \frac{1}{2\pi} \int_{-\infty}^{\infty} |H(j\omega)|^2 S_i(\omega) \, d\omega = \overline{x_0^2(t)} \qquad (4-208)$$

This equation relates the average power of the random function at the output of the filter to the transfer function $H(j\omega)$ of the filter and the input spectral density $S_i(\omega)$. As an illustration, consider a bandpass filter such that $H(j\omega) = 1$ in the range $\omega_1 < \omega < \omega_1 + \Delta\omega$, $-\omega_1 - \Delta\omega < \omega < -\omega_1$, and 0 otherwise. From Eq. (4-208) we obtain

$$\overline{x_0^2(t)} = \frac{1}{2\pi} \int_{-\infty}^{\infty} |H(j\omega)|^2 S_i(\omega) \, d\omega$$

$$= S_i(\omega) \frac{\Delta\omega}{2\pi} + S_i(-\omega) \frac{\Delta\omega}{2\pi} = 2S_i(\omega) \frac{\Delta\omega}{2\pi} = 2S_i(\omega) \, \Delta f$$

From this result, it is apparent that $2S_i(\omega)$ equals the average power of the random function in a unit frequency band; this is the reason why $S_i(\omega)$ is identified with the two-sided power spectral density of the random function.

Equation (4-208) can be derived directly as follows:

$$x_o(t) = \int_{-\infty}^{\infty} h(\tau)x_i(t-\tau)\, d\tau \tag{4-209}$$

and the average output power $\overline{x_o^2(t)}$ is given by

$$\begin{aligned}
\overline{x_o^2(t)} &= \int_{-\infty}^{\infty}\int_{-\infty}^{\infty} h(\tau)h(\sigma)\overline{x_i(t-\tau)x_i(t-\sigma)}\, d\tau\, d\sigma \\
&= \int_{-\infty}^{\infty}\int_{-\infty}^{\infty} h(\tau)h(\sigma)R_i(\tau-\sigma)\, d\tau\, d\sigma \\
&= \frac{1}{2\pi}\int_{-\infty}^{\infty}\int_{-\infty}^{\infty} h(\tau)h(\sigma)\int_{-\infty}^{\infty} e^{j\omega(\tau-\sigma)} S_i(\omega)\, d\omega \\
&= \frac{1}{2\pi}\int_{-\infty}^{\infty} S_i(\omega)\, d\omega \int_{-\infty}^{\infty} e^{j\omega\tau}\, h(\tau)\, d\tau \int_{-\infty}^{\infty} e^{-j\omega\sigma} h(\sigma)\, d\sigma \\
&= \frac{1}{2\pi}\int_{-\infty}^{\infty} H(j\omega)H(-j\omega)S_i(\omega) = \frac{1}{2\pi}\int_{-\infty}^{\infty} |H(j\omega)|^2 S_i(\omega)\, d\omega
\end{aligned} \tag{4-210}$$

2. *Cross-correlation of Linearly Transformed Random Functions.* Consider two random functions $x_1(t)$ and $x_2(t)$ to be applied to two linear networks whose transfer function are $H_1(j\omega)$ and $H_2(j\omega)$, respectively, as shown in Fig. 4-18.

The cross-correlation of the output random functions $x_a(t)$ and $x_b(t)$ of the linear system is

$$R_{ab}(\tau) = \lim_{T\to\infty} \frac{1}{2T}\int_{-T}^{T} x_a(t)x_b(t+\tau)\, dt \tag{4-211}$$

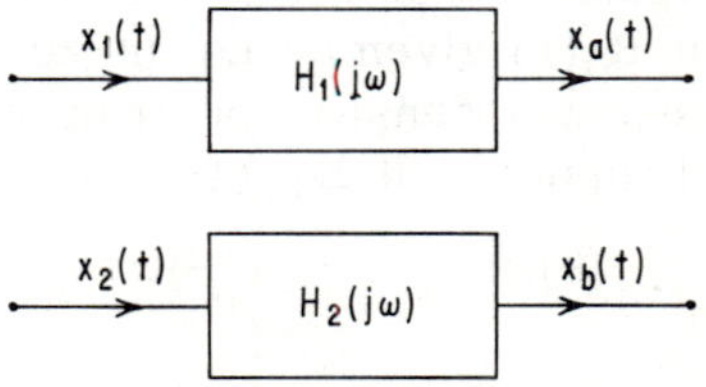

Fig. 4-18. Random functions under linear transformations.

In order to express the output cross-correlation function $R_{ab}(\tau)$ in terms of the system characteristics and the input cross-correlation function, we proceed as follows:

$$x_a(t) = \int_{-\infty}^{\infty} h_1(\nu)x_1(t - \nu) \, d\nu$$

$$x_b(t) = \int_{-\infty}^{\infty} h_2(\sigma)x_2(t - \sigma) \, d\sigma$$

(4-212)

Hence Eq. (4-211) reduces to

$$R_{ab}(\tau) = \lim_{T \to \infty} \frac{1}{2T} \int_{-T}^{T} dt \int_{-\infty}^{\infty} h_1(\nu)x_1(t - \nu) \, d\nu$$

$$\times \int_{-\infty}^{\infty} h_2(\sigma)x_2(t + \tau - \sigma) \, d\sigma$$

(4-213)

Inverting the order of integration, we obtain

$$R_{ab}(\tau) = \int_{-\infty}^{\infty} h_1(\nu) \, d\nu \int_{-\infty}^{\infty} h_2(\sigma) \, d\sigma$$

$$\times \lim_{T \to \infty} \frac{1}{2T} \int_{-T}^{T} x_1(t - \nu)x_2(t + \tau - \sigma) \, dt$$

(4-214)

Since $R_{12}(\tau + \nu - \sigma) = \lim_{T \to \infty} \dfrac{1}{2T} \displaystyle\int_{-T}^{T} x_1(t - \nu)x_2(t + \tau - \sigma) \, dt$

Eq. (4-214) assumes the form

$$R_{ab}(\tau) = \int_{-\infty}^{\infty} h_1(\nu) \, d\nu \int_{-\infty}^{\infty} h_2(\sigma)R_{12}(\tau - \nu - \sigma) \, d\sigma$$

(4-215)

This result is analogous to that obtained above for the output auto-correlation function $R_0(\tau)$ given by Eq. (4-203).

The output cross-power density spectrum can now be obtained by taking the Fourier transform of Eq. (4-215).

$$S_{ab}(\omega) = \int_{-\infty}^{\infty} R_{ab}(\tau)e^{-j\omega t} \, d\tau = \int_{-\infty}^{\infty} e^{-j\omega \tau} \, d\tau \int_{-\infty}^{\infty} h_1(\nu) \, d\nu$$

$$\times \int_{-\infty}^{\infty} h_2(\sigma)R_{12}(\tau + \nu - \sigma) \, d\sigma$$

(4-216)

and by introducing a change of variable $\mu = \tau + \nu - \sigma$, we obtain

$$S_{ab}(\omega) = \int_{-\infty}^{\infty} e^{-j\omega(\mu + \sigma - \nu)} \, d\mu \int_{-\infty}^{\infty} h_1(\nu) \, d\nu \int_{-\infty}^{\infty} h_2(\sigma) R_{12}(\mu) \, d\sigma$$

$$= \left[\int_{-\infty}^{\infty} h_1(\nu) e^{j\omega\nu} \, d\nu \right] \left[\int_{-\infty}^{\infty} h_2(\sigma) e^{-j\omega\sigma} \, d\sigma \right]$$

$$\times \left[\int_{-\infty}^{\infty} R_{12}(\mu) e^{-j\omega\sigma} \, d\mu \right] \qquad (4\text{-}217)$$

But
$$H_1(j\omega) = \int_{-\infty}^{\infty} h_1(t) e^{-j\omega t} \, dt \qquad (4\text{-}218)$$

$$H_2(j\omega) = \int_{-\infty}^{\infty} h_2(t) e^{-j\omega t} \, dt$$

are the transfer functions of the linear networks, and since

$$S_{12}(\omega) = \int_{-\infty}^{\infty} R_{12}(\mu) e^{-j\omega\mu} \, d\mu \qquad (4\text{-}219)$$

is the cross-power density spectrum of the input random functions, hence

$$S_{ab}(\omega) = H_1(j\omega) H_2^*(j\omega) S_{12}(\omega) \qquad (4\text{-}220)$$

which is analogous to Eq. (4-206), namely

$$S_0(\omega) = |H(j\omega)|^2 S_i(\omega)$$

4.8 Mathematical Representation of Random Noise: the Gaussian Random Process [1-13]

In the study of noise problems in radio receivers, we encounter a special type of random process which is characterized by the fact that all the basic probability distribution functions are gaussian distributions. In this section, the theory of the gaussian random process will be given, with special application to the solution of noise problems.

Consider an ensemble of stationary random functions with a normal distribution. Within the time interval $-T/2$ to $T/2$ we assume that any member of this ensemble, say $x_m(t)$, can be expressed as a Fourier series:

$$x_m(t) = \sum_{k=1}^{\infty} (a_{mk} \cos k\omega_o + b_{mk} \sin k\omega_o t) \qquad (4\text{-}221)$$

where $\omega_o = 2\pi f_o$, $f_o = 1/T$, and

$$a_{mk} = \frac{2}{T} \int_{-T/2}^{T/2} x_m(t) \cos k\omega_o t \, dt$$

$$\hspace{8cm} (4\text{-}222)$$

$$b_{mk} = \frac{2}{T} \int_{-T/2}^{T/2} x_m(t) \sin k\omega_o t \, dt$$

In this series, there is no constant term since it is assumed, without loss of generality, that the mean value of the member function $x_m(t)$ is zero. All members of the ensemble will have a Fourier series expansion of the same form but with different amplitudes of coefficients. These coefficients a_{mk} and b_{mk} will therefore, generally speaking, be different for the various members of the ensemble. Furthermore, we shall assume that these coefficients are mutually independent, and have gaussian distributions with mean values equal to zero and with variances which may depend on the order k, but which are the same for all a_{mk} and b_{mk} for the same value of k. It follows therefore that the entire ensemble can be represented by

$$x(t) = \sum_{k=1}^{\infty} (a_k \cos k\omega_o t + b_k \sin k\omega_o t) \hspace{3cm} (4\text{-}223)$$

where a_k and b_k are random variables with the assumption that

$$\overline{a_k} = \overline{b_k} = 0$$

$$\overline{a_k a_\ell} = \overline{b_k b_\ell} = \sigma_k^2 \delta_{k\ell} \hspace{5cm} (4\text{-}224)$$

$$\overline{a_k b_\ell} = 0$$

where $\delta_{k\ell} = 1$ for $k = \ell$; $\delta_{k\ell} = 0$ for $k \neq \ell$; and σ_k^2 is the variance of a_k or b_k.

The joint probability distribution function of the coefficients is

$$p(a_1, a_2, \ldots; b_1, b_2, \ldots) = \prod_{k=1}^{\infty} \frac{1}{2\pi\sigma_k^2} e^{-(a_k^2 + b_k^2)/2\sigma_k^2} \hspace{2cm} (4\text{-}225)$$

Let us evaluate the mean square value of $x(t)$; we have

$$\overline{x^2(t)} = \sum_{k=1}^{\infty} (\overline{a_k^2} \cos^2 k\omega_o t + \overline{b_k^2} \sin^2 k\omega_o t)$$

$$= \sum_{k=1}^{\infty} \sigma_k^2 = \frac{1}{2T} \sum_{k=1}^{\infty} S_T(k\omega_0) \doteq \frac{1}{2\pi} \int_{-\infty}^{\infty} S(\omega) \, d\omega = \sigma^2$$

(4-226)

Let us consider now the second probability distribution function $p_2(x_1, x_2, \tau)$, giving the probability of finding a pair of values of x in the intervals $(x_1, x_1 + dx)$, and $(x_2, x_2 + dx_2)$ in the instants of time t_1 and t_2, where $t_2 - t_1 = \tau$. Since $x(t_1)$ and $x(t_2)$ are both linear functions of a_k and b_k, we shall obtain a two-dimensional normal distribution; the mean square values $\overline{x^2(t_1)}$ and $\overline{x^2(t_2)}$ are given by Eq. (4-226),

$$\overline{x(t_1)x(t_2)} = \sum_{k=1}^{\infty} (\overline{a_k^2} \cos k\omega_0 t_1 \cos k\omega_0 t_2 + \overline{b_k^2} \sin k\omega_0 t_1 \sin k\omega_0 t_2)$$

$$= \sum_{k=1}^{\infty} \sigma_k^2 \cos k\omega_0 \tau \doteq \frac{1}{2\pi} \int_{-\infty}^{\infty} S(\omega) \cos \tau\omega \, d\omega$$

$$= R(\tau) = \rho(\tau)\sigma^2$$

(4-227)

where $\rho(\tau)$ is the normalized correlation function. As expected, since the process is stationary, the correlation depends only on the time difference τ.

The normalized correlation function $\rho(\tau)$ corresponds to the normalized spectral density

$$S_\sigma(\omega) = \frac{1}{\sigma^2} S(\omega) = \frac{S(\omega)}{\frac{1}{2\pi} \int_{-\infty}^{\infty} S(\omega) \, d\omega}$$

(4-228)

Similarly, it can be shown[1] that all probability distribution functions are gaussian and depend only on σ^2 and $\rho(\tau)$.

1. *The Narrow-band Gaussian Random Process.* A random process is said to be a narrow-band random process if the width $\Delta\omega$ of the significant region of its spectral density is small compared with the center angular frequency ω_c of that region. As an example, consider the passage of white noise through a bandpass filter. The output can be considered as a sample function of such a random process, which is expressible in the form

$$n(t) = V(t) \cos [\omega_c t + \phi(t)]$$

(4-229)

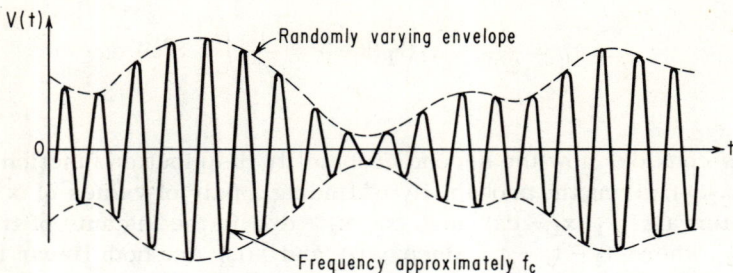

Fig. 4-19. Noise at the output of a narrow-band filter.

When viewed in an oscilloscope, the output will appear to be a sinusoidal wave with slowly varying envelope $V(t)$ and slowly varying phase $\phi(t)$ as shown in Figure 4-19. In the following, we shall determine some of the statistical properties of the envelope and phase of $n(t)$ when the narrow-band random process is a stationary gaussian random process.

It has been shown earlier that the distribution of the resultant of n independent sine waves of random phase approaches a gaussian distribution for n large. As pointed out by Bennett[3] and Rice[12] this result suggests a method of using a Fourier series representation as a mathematical model for the narrow-band gaussian process. In this presentation we approximate the continuous noise power spectrum with an equivalent discrete power spectrum; thus the narrow-band noise process $n(t)$ is approximated by

$$n(t) = \sum_{n=1}^{\infty} A_n \cos (\omega_n t + \theta_n) = \sum_{n=1}^{\infty} \sqrt{2S'(f_n)\,\Delta f} \, \cos (\omega_n t + \theta_n)$$

(4-230)

where $\omega_n = 2\pi f_n = \dfrac{2\pi n}{T}$, $f = \dfrac{1}{T}$, $S'(f_n)$ denotes the noise power spectral density (one-sided, assuming positive frequencies only), and θ_n a uniformly distributed random variable. In this mathematical model the continuous noise power $S'(f_n)\,\Delta f$ in a slot of Δf cps is represented by the discrete power spectrum $A_n^2/2$, and the ensemble of the sum of sinusoids of random phase approaches the required gaussian ensemble as $\Delta f \rightarrow 0$.

In order to emphasize the narrow-band characteristics of $n(t)$, let the spectrum of $n(t)$ be centered about the carrier angular frequency ω_c and symmetrical with respect to ω_c. Put $\omega_n = (\omega_n - \omega_c) + \omega_c$ and expand Eq. (4-230) in terms of the frequencies $(\omega_n - \omega_c)$ and ω_c. Then we can express $n(t)$ in the form

$$n(t) = x_c(t) \cos \omega_c t - x_s(t) \sin \omega_c t \tag{4-231}$$

where $x_c(t) = \sum_{n=1}^{\infty} \sqrt{2S'(f_n)\,\Delta f} \cos [(\omega_n - \omega_c)t + \theta_n]$

and $x_s(t) = \sum_{n=1}^{\infty} \sqrt{2S'(f_n)\,\Delta f} \sin [(\omega_n - \omega_c)t + \theta_n]$ (4-232)

Since the range of frequencies included in $n(t)$ is assumed small as compared with ω_c, the random variables $x_c(t)$ and $x_s(t)$ are slowly varying functions compared with $\cos \omega_c t$. Furthermore, since they are defined as the sum of independent sinusoids of random phase, it follows that the random functions $x_c(t)$ and $x_s(t)$ are independent gaussian variables. The mean-squared value of $x_c(t)$ and $x_s(t)$ as well as $n(t)$ is obtained directly from Eqs. (4-232) and (4-231), namely:

$$\overline{n^2(t)} = \frac{\overline{x_c^2(t)}}{2} + \frac{\overline{x_s^2(t)}}{2} \tag{4-233}$$

and $\overline{x_c^2(t)} = \overline{x_s^2(t)} = \sum_n S'(f_n)\Delta f = \sigma^2 = N$

Hence, in the limit as $T \to \infty$

$$\overline{n^2(t)} = \overline{x_c^2(t)} = \overline{x_s^2(t)} = \int_0^{\infty} S'(f)\,df = N \tag{4-234}$$

where N denotes the total noise powers in the narrow-band gaussian process $n(t)$.

The spectral density of a narrow-band white gaussian process is shown in Fig. 4-20. In this case

$$\overline{n^2(t)} = \overline{x_c^2(t)} = \overline{x_s^2(t)} = N_0 \frac{\Delta \omega}{2\pi} \tag{4-235}$$

where $S'(f) = N_0$ watts/cps is the one-sided noise power spectral density which is constant, throughout the significant region of the narrow-band process.

We shall find now the probability distribution of the envelope and phase of the narrow-band noise process $n(t)$. From Eqs. (4-229) and (4-231) we obtain

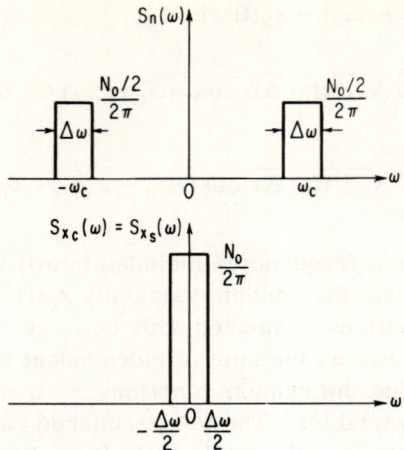

Fig. 4-20. Spectral density of narrow-band white gaussian process.

$$x_C(t) = V(t) \cos \phi(t)$$

$$x_S(t) = V(t) \sin \phi(t)$$

(4-236)

and

$$V(t) = \sqrt{x_C^2(t) + x_S^2(t)}$$

(4-237)

$$\phi(t) = \tan^{-1} \frac{x_S(t)}{x_C(t)}$$

(4-238)

The two random variables $x_C(t)$ and $x_S(t)$ have a gaussian distribution, namely:

$$p(x_C) = \frac{1}{\sqrt{2\pi}\,\sigma}\, e^{-x_C^2/2\sigma^2}$$

(4-239)

$$p(x_S) = \frac{1}{\sqrt{2\pi}\,\sigma}\, e^{-x_S^2/2\sigma^2}$$

and because they are independent the joint probability density function of $x_C(t)$ $x_S(t)$ evaluated at a given instant of time is given by

$$p(x_C, x_S) = \frac{1}{2\pi\sigma^2}\, e^{-(x_C^2 + x_S^2)/2\sigma^2} = \frac{1}{2\pi\sigma^2}\, e^{-V^2/2\sigma^2}$$

(4-240)

where $x_C \equiv x_C(t)$, $x_S \equiv x_S(t)$, $V \equiv V(t)$, $\phi \equiv \phi(t)$

The last equation will be used to determine the statistical properties of the envelope and phase random variables V and ϕ, respectively. Given the distributions of the two random variables x_c and x_s, the problem is to find the distributions of two other variables V and ϕ.

Equating probability density functions, we have

$$p(x_c, x_s) \, dx_c \, dx_s = q(V, \phi) \, dV \, d\phi \tag{4-241}$$

Transforming differential areas, we obtain

$$dx_c \, dx_s = V \, dV \, d\phi \tag{4-242}$$

so that $q(V, \phi) = \dfrac{V}{2\pi \sigma^2} \, e^{-V^2/2\sigma^2}$, $V \geq 0$ and $0 \leq \phi \leq 2\pi$

$$= 0, \qquad\qquad\qquad \text{otherwise} \tag{4-243}$$

The probability density function $V(t)$ for the envelope only can be obtained by integrating the last equation over ϕ from 0 to 2π.

$$q(V) = \int_0^{2\pi} q(V, \phi) \, d\phi = \frac{V}{\sigma^2} \, e^{-V^2/2\sigma^2}, \qquad V \geq 0$$

$$= 0, \qquad\qquad\qquad \text{otherwise} \tag{4-244}$$

This is the Rayleigh probability density function which has been discussed in Sec. 4.4 [Eq. (4-84)]; the peak of this distribution is at $V = \sigma$ and is equal to $(1/\sigma)e^{-1/2}$. It is easily seen that $q(V)$ is properly normalized, so that $\int_0^\infty q(V) \, dV = 1$.

The probability density function of $\phi(t)$ is obtained by integrating $q(V, \phi)$ over all values V; this gives

$$q(\phi) = \frac{1}{2\pi}, \qquad 0 \leq \phi \leq 2\pi$$

$$= 0, \qquad \text{elsewhere} \tag{4-245}$$

The random phase angle $\phi(t)$ is uniformly distributed. It follows therefore from Eqs. (4-241) to (4-245) that

$$q(V, \phi) = q(V) \, q(\phi) \tag{4-246}$$

so that $V(t)$ and $\phi(t)$ are independent random variables.

2. *Envelope of Sine Wave plus Narrow-band Gaussian Process.* In this section, we shall derive expressions for the probability density functions of the envelope and phase angle of the sum of a sine wave and a narrow-band gaussian noise. The envelope results will be used in a subsequent chapter in the discussion of envelope detection in amplitude-modulated signals. Similarly, the probability density function of the phase angle of the sum will be of interest in the discussion of phase modulation.

Consider a carrier wave and white gaussian noise applied to a narrow-band IF amplifier. Let the carrier wave be given by

$$e_c(t) = A_c \cos[\omega_c t + \psi(t)] = a(t) \cos \omega_c t - b(t) \sin \omega_c t \quad (4\text{-}247)$$

where $a(t) = A_c \cos \psi(t)$, $b(t) = A_c \sin \psi(t)$,

$$A_c = \sqrt{a^2(t) + b^2(t)} = \text{constant}$$

and $\psi(t)$ is a random variable uniformly distributed over the interval $(0 - 2\pi)$ and independent of the gaussian process. Let $e_n(t)$ represent a sample function of a stationary narrow-band gaussian process which, using the notation of Eq. (4-229), is given by

$$e_n(t) = V(t) \cos[\omega_c t + \phi(t)] = x_c(t) \cos \omega_c t - x_s(t) \sin \omega_c t$$

$$(4\text{-}248)$$

The resultant signal $e_r(t) = e_c(t) + e_n(t)$ can be expressed in terms of an envelope and phase, namely,

$$e_r(t) = R(t) \cos[\omega_c t + \theta(t)] = X_c(t) \cos \omega_c t - X_s(t) \sin \omega_c t$$

$$(4\text{-}249)$$

where $X_c(t) = a(t) + x_c(t)$, $X_s(t) = b(t) + x_s(t)$,

and $\quad \theta(t) = \tan^{-1} \dfrac{X_s(t)}{X_c(t)} = \tan^{-1} \dfrac{b(t) + x_s(t)}{a(t) + x_c(t)}$

The envelope $R(t)$ is given by

$$R(t) = \sqrt{X_c^2(t) + X_s^2(t)}$$

We shall proceed now by finding the probability density function of the envelope $R(t)$ which, as mentioned previously, will be of interest in the process of envelope detection.

Since $x_c(t)$ and $x_s(t)$ are independent gaussian random variables with zero mean values and variances σ^2, $X_c(t)$ and $X_s(t)$ are also independent gaussian variables with means $a(t)$ and $b(t)$, respectively. Thus

$$p_1(X_c) = \frac{1}{\sqrt{2\pi}\,\sigma}\, e^{-(X_c - a)^2/2\sigma^2} \qquad\qquad (4\text{-}250)$$

$$p_2(X_s) = \frac{1}{\sqrt{2\pi}\,\sigma}\, e^{-(X_s - b)^2/2\sigma^2} \qquad\qquad (4\text{-}251)$$

$$p_3(\psi) = \frac{1}{2\pi}, \qquad 0 \le \psi \le 2\pi \qquad\qquad (4\text{-}252)$$

where $a \equiv a(t)$, $b \equiv b(t)$, $X_c \equiv X_c(t)$, and $X_s \equiv X_s(t)$

Furthermore, the joint probability density function

$$p(X_c,X_s,\psi) = p_1(X_c)p_2(X_s)p_3(\psi)$$

$$= \frac{1}{4\pi^2\sigma^2}\, e^{-(1/2\sigma^2)[(X_c - a)^2 + (X_s - b)^2]} \qquad\qquad (4\text{-}253)$$

We make now the following transformation:

$$R^2 = X_c^2 + X_s^2$$

$$X_c = R\cos\theta$$

$$X_s = R\sin\theta$$

so that $\quad p(X_c,X_s,\psi)\, dX_c\, dX_s = q(R,\theta,\psi)\, dR\, d\theta \qquad\qquad (4\text{-}254)$

But the differential area

$$dX_c\, dX_s = R\, dR\, d\theta \qquad\qquad (4\text{-}255)$$

Therefore $\quad q(R,\theta,\psi) = \dfrac{R}{4\pi^2\sigma^2}\, e^{-(1/2\sigma^2)[R^2 + A_c^2 - 2RA_c\cos(\theta - \psi)]}$

$$\qquad\qquad (4\text{-}256)$$

The probability density function $q(R)$ of the envelope $R(t)$ is obtained by integrating the joint density function $q(R,\theta,\psi)$ over all values of θ and ψ; thus

$$q(R) = \int_0^{2\pi} d\psi \int_0^{2\pi} q(R,\theta,\psi) \; d\theta$$

$$= \frac{R}{4\pi^2 \sigma^2} e^{-(R^2 + A_c^2)/2\sigma^2} \int_0^{2\pi} d\psi \int_{-\psi}^{2\pi - \psi} e^{(RA_c/\sigma^2) \cos \phi} \; d\phi$$

$$(4\text{-}257)$$

where $\phi = \theta - \psi$

But the integral

$$I_0(x) = \frac{1}{2\pi} \int_0^{2\pi} e^{x \cos \phi} \; d\phi = \sum_{n=0}^{\infty} \frac{x^{2n}}{2^{2n}} (n!)^2 \qquad (4\text{-}258)$$

is available in tabulated form where $I_0(x)$ is the Bessel function of zero order and imaginary argument. It follows therefore that the probability density function of the envelope of the sum of a sine wave and a narrow-band gaussian random process is given by

$$q(R) = \frac{R}{\sigma^2} e^{-(R^2 + A_c^2)/2\sigma^2} \cdot I_0(R \; A_c/\sigma^2), \qquad R(t) \geq 0$$

$$= 0, \qquad\qquad\qquad\qquad\qquad \text{otherwise} \;\; (4\text{-}259)$$

This equation can be expressed in terms of the input carrier-to-noise ratio, for the input carrier power

$$C_i = \frac{A_c^2}{2} \qquad\qquad\qquad\qquad (4\text{-}260)$$

and the input noise power $N_i = \sigma^2$ so that the input $(C/N)_i = z$ $= A_c^2/2\sigma^2$. Therefore $\qquad\qquad\qquad\qquad\qquad\qquad (4\text{-}261)$

$$q(R) = \frac{R}{\sigma^2} e^{-(R^2/2\sigma^2 + z)} \cdot I_0\left(\frac{R}{\sigma} \sqrt{2z}\right) \qquad\qquad (4\text{-}262)$$

For small-input carrier-to-noise ratio, $z \ll 1$, $I_0(x) = 1 + \frac{x^2}{4} + \cdots$ $\doteq e^{x^2/4}$, and the probability density function reduces to

$$q(R) \doteq \frac{R}{\sigma^2} e^{-R^2/2\sigma^2} \qquad\qquad\qquad (4\text{-}263)$$

which is the Rayleigh density function of Eq. (4-244). For large-input carrier-to-noise ratio, $z \gg 1$; i.e., the argument x of the modified Bessel function is large, and $I_0(x)$ can be expanded in an asymptotic series as follows:

$$I_0(x) = \frac{e^x}{\sqrt{2\pi x}} \left(1 + \frac{1}{8x} + \cdots \right) \doteq \frac{e^x}{\sqrt{2\pi x}}$$

(4-264)

for $x \gg 1$ or $RA_c \gg \sigma^2$. Hence

$$q(R) \doteq \frac{R}{\sqrt{2\pi} \, RA_c \sigma^2} e^{-(R - A_c)^2/2\sigma^2} = \frac{1}{\sigma} \left(\frac{R}{2\pi A_c}\right)^{\frac{1}{2}} e^{-(R - A_c)^2/2\sigma^2}$$

$$\doteq \frac{1}{\sqrt{2\pi} \, \sigma} e^{-(R - A_c)^2/2\sigma^2}$$

(4-265)

This function peaks sharply about the point $R = A_c$. Hence for large-input carrier-to-noise ratio, the probability density function of the envelope R(t) is gaussian with mean value of A_c. Plots of q(R) for various values of z are illustrated in Fig. 4-21; we note here the transition from a Rayleigh distribution for $z = 0$ to a normal distribution about the mean A_c for $z \gg 1$.

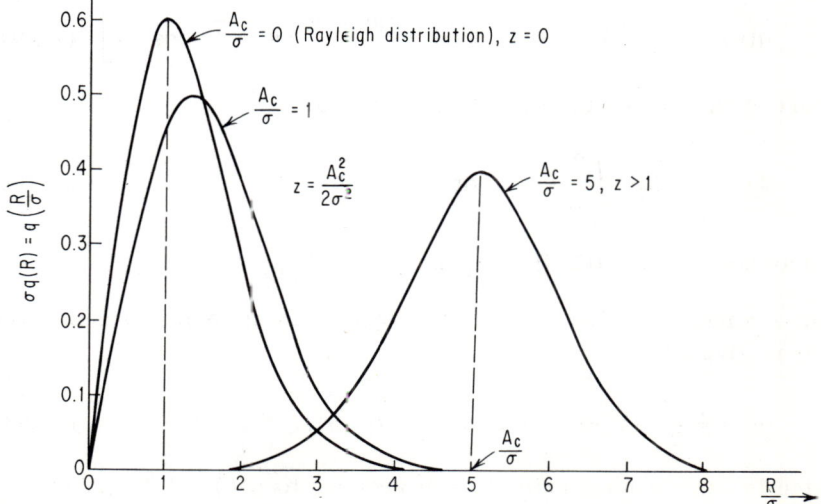

Fig. 4-21. Probability density function of the envelope of narrow-band noise plus a sinusoid.

As mentioned previously, the probability density function of the phase $q(\theta)$ will be of interest in the discussion of phase or frequency modulation. To derive $q(\theta)$ we integrate Eq. (4-256) over the range of the envelope $R(t)$ and obtain the joint probability density function $q(\theta,\psi)$,

$$q(\theta,\psi) = \frac{1}{4\pi^2\sigma^2} \int_0^\infty \text{Re}^{-(1/2\sigma^2)[R^2+A_c^2-2RA_c\cos(\theta-\psi)]}\, dR$$

To evaluate this integral, put $\phi = \theta - \psi$ and on completing the square, this equation can be expressed in the following form:

$$q(\theta,\psi) = \frac{1}{4\pi^2\sigma^2}\, e^{-A_c^2\sin^2\phi/2\sigma^2} \cdot \int_{-\infty}^\infty \text{Re}^{-(R-A_c\cos\phi)^2/2\sigma^2}\, dR$$

An approximate expression for the joint probability density function of the phase angles θ and ψ is given by Davenport and Root[1] for the case when $A_c\cos(\theta-\psi) \gg \sigma$:

$$q(\theta,\psi) = \frac{A_c\cos(\theta-\psi)}{(2\pi)^{3/2}\cdot\sigma} \cdot e^{-(A_c^2/2\sigma^2)\sin^2(\theta-\psi)}, \quad 0 \le \theta; \psi \le 2\pi$$

$$(4\text{-}266)$$

A somewhat simpler expression is obtained by Hancock[4] for $\psi = 0$,

$$q(\theta) = \frac{e^{-z}}{2\pi}\left[1 + \sqrt{4\pi z}\,\cos\theta\, e^{z\cos^2\theta} \cdot \Phi(\sqrt{2z}\,\cos\theta)\right] \quad (4\text{-}267)$$

where $\Phi(x)$ denotes the probability integral

$$\Phi(x) = \frac{1}{\sqrt{2\pi}}\int_{-\infty}^x e^{-t^2/2}\, dt \qquad (4\text{-}268)$$

and as before $z = (C/N)_i = A_c^2/2\sigma^2$

For z small, i.e., low carrier-to-noise ratio, a series expansion of $q(\theta)$ is given by

$$q(\theta) \doteq \frac{1}{2\pi}(1 + \sqrt{\pi z}\,\cos\theta + z\cos 2\theta), \qquad z < 0.1 \qquad (4\text{-}269)$$

which in case of no carrier ($z = 0$) reduces to $q(\theta) = 1/2\pi$ as expected. For large carrier-to-noise ratio, the probability integral may be approximated by

$$\Phi(x) \doteq 1 - \frac{e^{-x^2/2}}{\sqrt{2\pi}\,x}, \qquad x > 3 \tag{4-270}$$

Using this approximation in Eq. (4-267), it may be shown that

$$q(\theta) \doteq \sqrt{z/\pi}\ \cos\theta \cdot e^{-z\sin^2\theta}, \qquad 1 > \cos\theta > \frac{2.5}{\sqrt{z}} \tag{4-271}$$

$$q(\theta) \doteq 0, \qquad\qquad \frac{-2.5}{\sqrt{z}} > \cos\theta > -1 \tag{4-272}$$

In the vicinity of $\theta = 0$ and for large z, the probability density $q(\theta)$ may therefore be approximated by

$$q(\theta) \doteq \sqrt{z/\pi} = e^{-z\theta^2} \tag{4-273}$$

A gaussian distribution with zero mean and variance $\sigma_\theta^2 = \dfrac{1}{2z}$.

The probability density function $q(\theta)$ is plotted in Fig. (4-22), where it is seen that for zero carrier, the probability density function is rectangular and for large carrier-to-noise ratio, in the vicinity of $\theta = 0$, the curve is approximated by a gaussian distribution.

Having reviewed the basic mathematical prerequisites for the study of modulation theory, we shall start in the next chapter with the topic of "Linear Modulation Systems." As might be expected, additional techniques will be introduced, as required, in the discussion of some special modulation topics. These, together with the mathematical background provided so far, should be adequate for our future discussions.

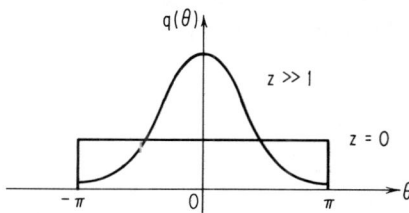

Fig. 4-22. Probability density function of phase of sine wave plus noise.

References

 1. Davenport, W. B., Jr., and W. L. Root: "An Introduction to the Theory of Random Signals and Noise," McGraw-Hill Book Company, New York.
 2. Solodovnikov, V. V.: "Introduction to the Statistical Dynamics of Automatic Control Systems," Dover Publications, Inc., New York.
 3. Bennett, W. R.: Methods of Solving Noise Problems, *Proc. IRE,* May, 1956.
 4. Hancock, J. C.: "An Introduction to the Principles of Communication Theory," McGraw-Hill Book Company, New York.
 5. Schwartz, M.: "Information Transmission, Modulation, and Noise," McGraw-Hill Book Company, New York.
 6. Lee, Y. W.: "Statistical Theory of Communication," John Wiley & Sons, Inc., New York.
 7. Lawson, J. L. , and G. E. Uhlenbeck: "Threshold Signals," *MIT Radiation Lab. Series,* vol. 24, McGraw-Hill Book Company, New York.
 8. Wainstein, L. A., and V. D. Zubakov: "Extraction of Signals from Noise," Prentice-Hall, Inc., Englewood Cliffs, N. J.
 9. Harman, W. W.: "Principles of the Statistical Theory of Communication," McGraw-Hill Book Company, New York.
10. Nichols, J. H., and N. R. Phillips: "Theory of Servomechanism," *Radiation Lab. Series,* vol. 25, McGraw-Hill Book Company, New York.
11. Blackman, R. B., and J. W. Tukey: The Measurement of Power Spectra from the Point of View of Communication Engineering, *BSTJ,* part 1, January, 1958; part 2, March, 1958.
12. Rice, S. O.: Mathematical Analysis of Random Noise, *BSTJ,* July, 1944, pp. 282-332; January, 1945, pp. 46-157.
13. Rice, S. O.: Statistical Properties of a Sine Wave Plus Random Noise, *BSTJ,* January, 1948, pp. 109-157.

5

LINEAR MODULATION SYSTEMS

In continuous-wave (C-W) modulation systems, a radio-frequency sinusoid is used as a carrier in which the amplitude, phase, or frequency is modulated by the information-carrying signals or messages. The modulating signal may consist of one message only, or more often it is a composite of frequency- or time-division multiplexed messages. In a linear modulation process, the frequency components of the modulating signal are translated to occupy a different position in the spectrum. This is accomplished by effectively multiplying together the time functions that describe the modulating signal and the carrier. In a nonlinear modulation process, new components are generated in the resulting spectrum of the modulated signal which do not have a one-to-one correspondence in the original spectra involved. This chapter is devoted to a discussion of linear modulation operations and systems. The process of linear detection or demodulation will be discussed in Chap. 6.

5.1 Linear Modulation [1-3]

The process of linear modulation is essentially a multiplication process, where the time functions that describe the modulating sig-

171

nal and carrier are multiplied together. It will be shown that this operation is equivalent to a symmetrical translation of the baseband spectrum through a distance ω_c, the carrier angular frequency, about the frequency axis. Such a modulation system is shown in its general form in Fig. 5-1, where the resulting product is further processed by a linear operation such as filtering. As will be shown, modulation methods such as amplitude modulation, vestigial-sideband modulation, single-sideband and double-sideband modulation are all special forms of linear modulation.

At the receiver, a demodulation or detection process is used to derive the original modulation signal from the incoming modulated carrier. This may be accomplished by multiplying the incoming modulated signal by a locally generated carrier whose frequency and phase are identical to the carrier component of the incoming signal. Such an operation is often called synchronous detection or product demodulation and is usually followed by low-pass filtering.

In practice, simpler detectors are often employed in linear modulation systems, such as envelope detectors or square-law detectors. However, as we shall see later on, the system performance under severe conditions of reception is superior with synchronous detection as compared to other types. We have already noted earlier that the purpose of the carrier is to transfer the intelligence spectrum to a frequency region which is more suitable for propagation. The modulation function must be limited to those operations which will yield modulated signals amenable to subsequent demodulation. As will be discussed later, the detection techniques which are currently available are coherent detection, envelope detection, and phase or frequency detection; consequently, the modulation functions must operate either on the amplitude phase or envelope of the modulated signal, as exemplified by linear and exponential modulation. In this chapter, only linear modulation will be discussed; exponential modulation will be discussed in Chap. 7.

Linear modulation techniques have been in use from the earliest days of radio. We shall discuss the general characteristics of the

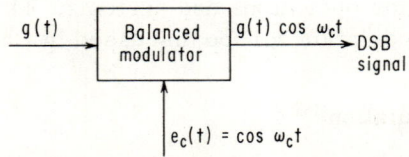

Fig. 5-1. Double-sideband transmission system.

various forms of linear modulation systems, not necessarily in the historical order of invention, but in a logical order of development to emphasize the derivation of these systems from the basic double-sideband system (DSB), which is shown in Fig. 5-1.

The overall system performance of the various linear communication systems will be arrived at by analyzing first the transmitter and receiver performance separately, and then considering their combined performance from an overall system point of view. This will finally lead to a comparison of the various linear modulation systems, and to a judicious assessment of the various system advantages and disadvantages.

5.2 Double-sideband (DSB) Transmission [1-4]

This is shown schematically in Fig. 5-1, where the carrier signal is assumed to be a sinusoid and the message signal g(t) is considered to have a zero d-c component, and consequently no carrier component appears in the output. By modifying g(t) to contain a d-c term, the DSB signal can be transformed into an AM signal, as will be discussed in the next section.

Let $G(j\omega)$ represent the Fourier transform of the message signal g(t) which is confined to the region $|\omega| \leq \omega_b$; g(t) is thus a baseband signal, band-limited to $-\omega_b \leq \omega \leq \omega_b$. The product

$$e(t) = g(t) \cos \omega_c t \qquad (5-1)$$

represents a modulated carrier, where the carrier frequency $\omega_c \gg \omega_b$. This modulation process is commonly called double-sideband suppressed carrier (DSB-SC) or simply double sideband (DSB). The spectral components of the DSB signal e(t) are given by its Fourier transform.

$$E(j\omega) = \tfrac{1}{2}G[j(\omega - \omega_c)] + \tfrac{1}{2} G[j(\omega + \omega_c)] \qquad (5-2)$$

This result is obtained by the use of the shift theorem [Eq. (2-81)]. Figure 5-2a illustrates the magnitude of a typical baseband spectrum $G(j\omega)$, and Fig. 5-2b shows the corresponding DSB spectrum $E(j\omega)$, illustrating the presence of upper and lower sidebands. Note that the Fourier spectrum of the modulation signal is translated symmetrically $\pm\omega_c$ about the origin.

The process of product modulation as represented by Eq. (5-1) can be generalized mathematically as the product of two time functions

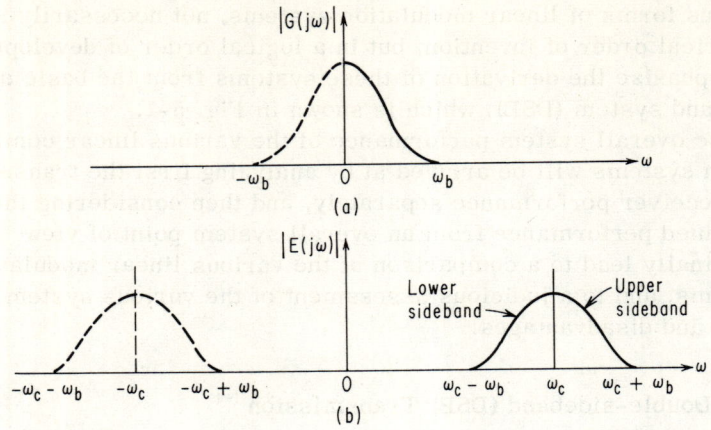

Fig. 5-2. Baseband signal and double-sideband spectra.

$$e(t) = s[g(t)] \cdot c(t) \tag{5-3}$$

where s represents a functional operation on the message signal g(t), and c(t) is a function of time representing the carrier. By the use of the convolution theorem [Eq. (2-83)] we obtain for the resulting spectrum $E(j\omega)$ the following expression:

$$E(j\omega) = \frac{1}{2\pi} \int_{-\infty}^{\infty} S[j(\omega - \omega_1)]C(j\omega_1)d\omega_1 \tag{5-4}$$

If the carrier is a narrow-band waveform, as might be used in noise modulation, then the convolution expressed in Eq. (5-4) results in a spreading of the signal band, as well as its translation to the vicinity of the carrier frequency.

Considering again the harmonic carrier as given by Eq. (5-1), as a result of simple multiplication of the intelligence signal and carrier wave, the amplitude and phase of the carrier are modulated in direct proportion to the magnitude of the modulating signal. This results in an RF envelope, which follows faithfully the waveform of the modulating signal, as seen from Fig. 5-3. This direct relationship between the amplitude of the modulating signal and the envelope of the modulated signal is characteristic of DSB and AM only.

It is well known that in any radio system, the system performance is determined by the average signal power, while the peak transmitter power is a limiting design factor. The relationship between peak power and average power of the transmitted signal

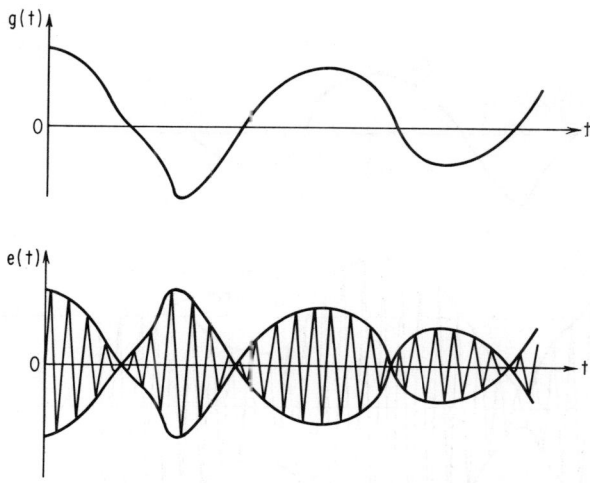

Fig. 5-3. Double-sideband waveforms.

for linearly modulated systems may be established from the study
of the transmitted waveform, which is shown for example in Fig.
5-3 for DSB. A comparison based on the relationship between peak
power and average power will be made later for DSB, AM, and SSB.

5.3   Amplitude Modulation (AM) Transmission [1-3]

Amplitude modulation is one of the oldest forms of modulation.
It may be derived from DSB by adding a d-c term to the modulating
signal g(t), as in Eq. (5-1). The resulting AM waveform is

$$e(t) = [1 + g(t)]\cos \omega_c t \tag{5-5}$$

shown in Fig. 5-4. It is to be noted that the RF modulated signal is
in phase with the modulation signal, similar to the case of DSB, the
only difference being the addition of the d-c term. It is obvious
from the waveform in Fig. 5-4 that as the result of the addition of
the d-c term, the peak envelope voltage has been doubled, and consequently, the peak envelope power of the AM wave is four times as
great at the peak envelope power of the DSB wave. Since the average
intelligence in both AM and DSB systems is the same, we may conclude that under equal peak power limitations, DSB has a 6-db advantage in intelligence power over AM, independent of the waveform
of the modulation signal.

In order to derive the frequency spectrum of an AM signal, we
shall first rewrite Eq. (5-5) as follows:

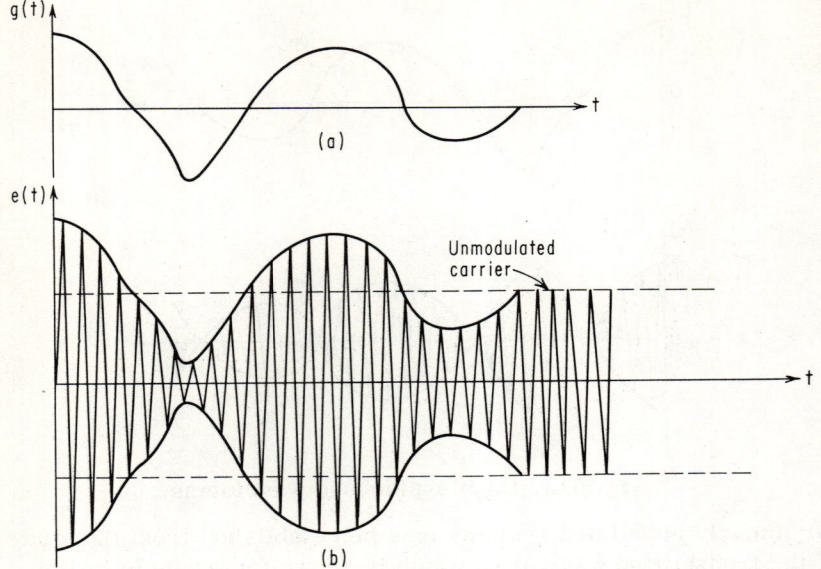

Fig. 5-4. Amplitude modulation of a carrier: (a) Modulating signal; (b) amplitude modulated carrier.

$$e(t) = [1 + m_a s(t)]\cos \omega_c t \qquad\qquad (5\text{-}6)$$

where the factor m_a is known as the "modulation index," "modulation factor," or degree of modulation; the percentage modulation is given by $100m_a$ per cent. A basic assumption is made that the modulating signal $s(t)$ varies slowly compared with the carrier and that $|m_a s(t)| \leq 1$, to ensure an undistorted envelope. We shall presently consider two types of modulating signals.

1. *Periodic Modulating Signal.* The modulating signal is band-limited with no d-c component,

$$s(t) = \sum_{n=-M}^{M} C_n e^{jn\omega t} = 2 \sum_{n=1}^{M} |C_n| \cos (n\omega t + \theta_n) \qquad (5\text{-}7)$$

where $\omega = 2\pi/T$, T is the period of the modulation, and $M\omega \ll \omega_c$. The output signal $e(t)$ becomes

$$e(t) = \cos \omega_c t + m_a \sum_{n=1}^{M} |C_n| \{\cos [(\omega_c + n\omega)t + \theta_n]$$

$$+ \cos \left[(\omega_c - n\omega)t - \theta_n \right] \} \tag{5-8}$$

The corresponding frequency spectrum is shown in Fig. 5-5. As in the case of DSB, the spectrum of the modulating signal is translated symmetrically about the carrier ω_c. However, in addition to the sidebands which contain the amplitude and phase of the original signal, the carrier is also transmitted. The carrier component conveys no intelligence, generally speaking, except the frequency, phase, and amplitude of the unmodulated signal. The presence of the carrier, however, in AM as in other types of linear modulation permits simplified receiving techniques to be employed.

2. *Nonperiodic Modulating Signal.* Let s(t) represent a nonperiodic signal which is band-limited to ω_M, where $\omega_M \ll \omega_c$. The modulated signal e(t) is given by

$$e(t) = [1 + m_a s(t)] \cos \omega_c t$$

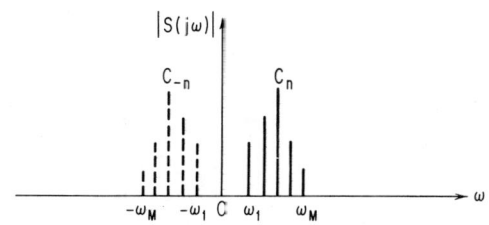

(a) Modulating signal spectrum

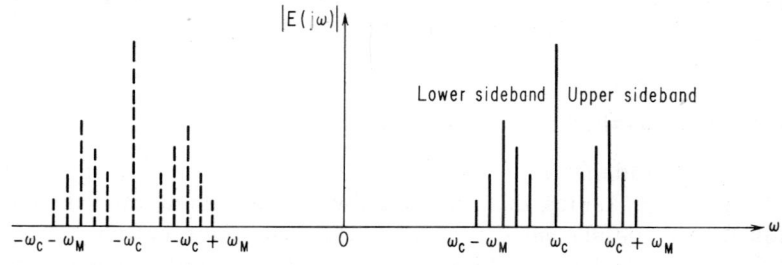

(b) Amplitude modulated spectrum

Fig. 5-5. AM spectrum-periodic modulating signal.

and its Fourier transform $E(j\omega)$ is

$$E(j\omega) = \pi\,\delta(\omega - \omega_c) + \pi\,\delta(\omega + \omega_c) + \frac{m_a}{2}\,S[j(\omega - \omega_c)]$$

$$+ \frac{m_a}{2}\,S[j(\omega + \omega_c)] \qquad (5-9)$$

The Fourier spectrum of the AM signal is shown in Fig. 5-6, where it is seen that in the process of amplitude modulation, the Fourier spectrum $S(j\omega)$ of the modulating signal has been translated from symmetry about $\omega = 0$ to symmetry about ω_c.

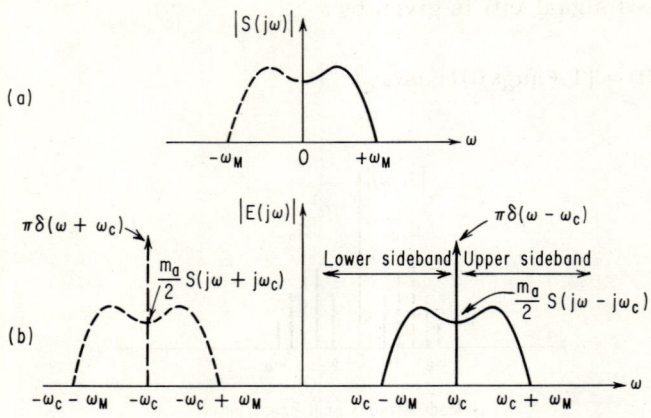

Fig. 5-6. Spectrum of AM wave: (a) Spectrum of modulating signal; (b) spectrum of amplitude-modulated wave.

5.4 Vestigial-sideband Transmission [1,2]

Vestigial-sideband modulation is derived from a conventional DSB signal by passing the output of the product modulator through a filter $H_V(j\omega)$, as shown in Fig. 5-7. The transfer function $H_V(j\omega)$ of the filter treats the two sidebands of the DSB signal in such a manner as to attenuate one sideband differently from the other. The output signal can best be analyzed by the use of the impulse response $h_V(t)$ of the system function $H_V(j\omega)$. By making use of the convolution integral as given in Eq. (3-55), the output signal $e_0(t)$ of the vestigial-sideband transmitter is given by

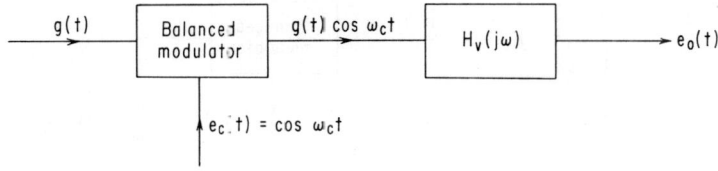

Fig. 5-7. Vestigial-sideband transmission system.

$$e_0(t) = \int_{-\infty}^{\infty} h_v(\tau)g(t - \tau) \cos [\omega_c(t - \tau)]d\tau$$

$$= \left[\int_{-\infty}^{\infty} h_v(\tau) \cos \omega_c\tau \cdot g(t - \tau)d\tau \right] \cos \omega_c t$$

$$+ \left[\int_{-\infty}^{\infty} h_v(\tau) \sin \omega_c\tau \cdot g(t - \tau)d\tau \right] \sin \omega_c t \qquad (5\text{-}10)$$

From the last equation, it is seen that the process of filtering has introduced a quadrature component in the output signal. In fact, the vestigial-sideband output signal consists now of a DSB-modulated cosine carrier and a DSB-modulated sine carrier. The low-frequency modulating components in the bracketed terms in Eq. (5-10) may be considered as the result of passing the modulating signal $g(t)$ through networks whose impulse response is $h_i(t)$ and $h_q(t)$, respectively, and system functions $H_i(j\omega)$ and $H_q(j\omega)$, where

$$h_i(t) = h_v(t) \cos \omega_c t \qquad (5\text{-}11)$$

$$h_q(t) = h_v(t) \sin \omega_c t \qquad (5\text{-}12)$$

$H_i(j\omega)$ and $H_q(j\omega)$ are the Fourier transform of $h_i(t)$ and $h_q(t)$, respectively. Again making use of the shift theorem, [Eq. (2-81)], we obtain the following equations for the system functions of two hypothetical networks:

$$H_i(j\omega) = \frac{1}{2} \left\{ H_v[j(\omega - \omega_c)] + H_v[j(\omega + \omega_c)] \right\} \qquad (5\text{-}13)$$

$$H_q(j\omega) = \frac{1}{2j} \left\{ H_v[j(\omega - \omega_c)] - H_v[j(\omega + \omega_c)] \right\} \qquad (5\text{-}14)$$

The physical significance of these expressions is that the process

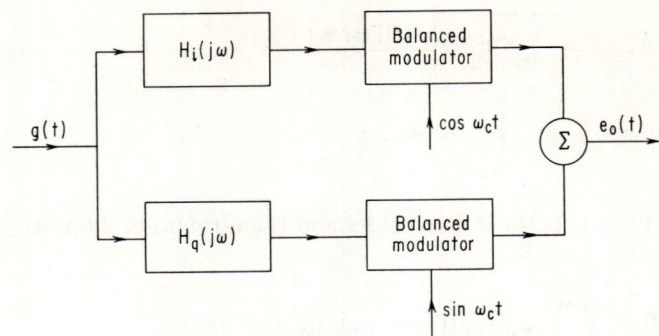

Fig. 5-8. Equivalent vestigial-sideband transmission system.

of vestigial-sideband modulation by the use of the filter network $H_v(j\omega)$, as shown in Fig. 5-7, may be replaced by an equivalent vestigial-sideband system shown in Fig. 5-8, whose output signal $e_0(t)$ is identical to that of Fig. 5-7. We shall see later that such an equivalent modulation system is used to produce single-sideband (SSB) modulation, which may be thought of as a limiting form of vestigial-sideband modulation.

5.5 Single-sideband (SSB) Transmission [1,2,4,5]

Single-sideband transmission may be produced in the same manner as vestigial sideband, by using a high-pass filter $H_S(j\omega)$ which completely eliminates all signals on one side of the carrier frequency, as shown in Fig. 5-9. The transfer function $H_S(j\omega)$ of the ideal high-pass filter is defined by

$$H_S(j\omega) = [\tfrac{1}{2} + \tfrac{1}{2} \text{ sgn } (\omega - \omega_c)] + [\tfrac{1}{2} - \tfrac{1}{2} \text{ sgn } (\omega + \omega_c)] \qquad (5\text{-}15)$$

where sgn ω is the signum function discussed in Chap. 2. The output spectrum $E_S(j\omega)$ is obtained by multiplying Eqs. (5-2) and (5-15),

$$E_S(j\omega) = H_S(j\omega)E(j\omega) = \tfrac{1}{2} G [j(\omega - \omega_c)][\tfrac{1}{2} + \tfrac{1}{2} \text{ sgn } (\omega - \omega_c)]$$

$$+ \tfrac{1}{2} G [j(\omega + \omega_c)][\tfrac{1}{2} - \tfrac{1}{2} \text{ sgn } (\omega + \omega_c)] \qquad (5\text{-}16)$$

The two missing products are zero because the terms in each product occur on nonoverlapping frequency intervals. The frequency characteristic of the high-pass filter $H_S(j\omega)$ and the spectrum of the single-sideband output signal are shown in Fig. 5-10.

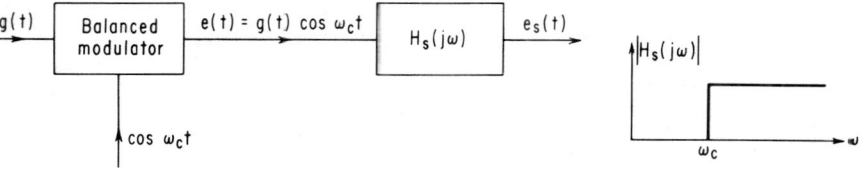

Fig. 5-9. Single-sideband transmission system.

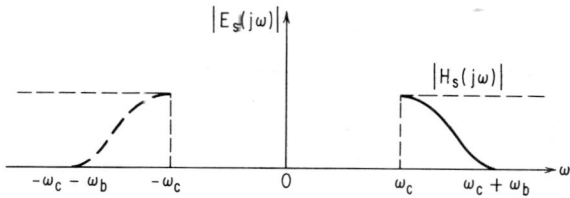

Fig. 5-10. Single-sideband spectrum and high-pass filter.

We shall return to Eq. (5-16) in connection with the concept of analytic signal and Hilbert transforms as applied to SSB, which will be discussed in Chap. 6.

The equations which were derived for vestigial-sideband modulation are also applicable to this case, since SSB may be considered as an extreme example of vestigial modulation. However, in order to emphasize the appearance of the quadrature component mentioned previously (which is characteristic for SSB), it would be more instructive to examine the SSB system by assuming a periodic wave for the message function g(t), thereby making use of the Fourier series.

Consider the modulation signal g(t) to consist of a band-limited square wave, as given in Fourier series form by Eq. (2-30).

$$g(t) = \frac{4V}{\pi} \sum_{n=1,3,5,\ldots}^{N} \frac{\sin (n\pi/2)}{n} \cos n\omega_s t \qquad (5\text{-}17)$$

where $\omega_s = 2\pi/T$. Such a modulation function g(t) may be generated by passing a perfect square wave through a low-pass filter whose bandwidth is limited to the N'th harmonic of the square wave. We are interested in studying the effects on the transmitted SSB signal which are caused by bandwidth variations of the modulation signal. This can be accomplished by varying the magnitude of N in Eq. (5-17) and studying the SSB output signal $e_s(t)$. Figure 5-11 depicts the bandwidth-limited modulating square wave with the overshoot at

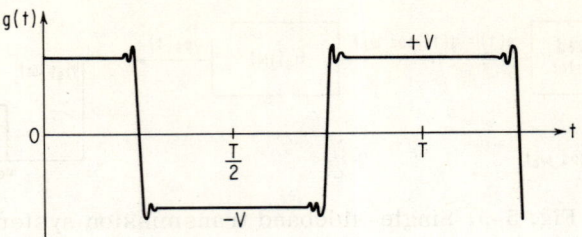

Fig. 5-11. Square-wave g(t) signal.

the transition time, which is recognizable as the Gibbs phenomenon associated with terminated Fourier series. The DSB output of the product modulator e(t) is given by

$$e(t) = g(t) \cos \omega_c t = \left[\frac{4V}{\pi} \sum_{n=1,3,5,\ldots}^{N} \frac{\sin (n\pi/2)}{n} \cos n\omega_s t \right] \cos \omega_c t$$

$$= \frac{2V}{\pi} \sum_{n=1,3,5,\ldots}^{N} \frac{\sin (n\pi/2)}{n} \cos (\omega_c + n\omega_s)t$$

$$+ \frac{2V}{\pi} \sum_{n=1,3,5,\ldots}^{N} \frac{\sin (n\pi/2)}{n} \cos (\omega_c - n\omega_s)t \qquad (5\text{-}18)$$

The DSB signal as represented by Eq. (5-18) may be transformed into a SSB by removing one of the summation terms by the use of the single-sideband filter $H_s(j\omega)$. The resulting SSB signal $e_s(t)$ is given by

$$e_s(t) = \frac{2V}{\pi} \sum_{n=1,3,5,\ldots}^{N} \frac{\sin (n\pi/2)}{n} \cos (\omega_c + n\omega_s)t \qquad (5\text{-}19)$$

By comparing the SSB signal as represented by this equation with the original modulating signal g(t) as given by Eq. (5-17), we note that the original frequency components have been translated by an amount ω_c, the SSB carrier frequency. However, the true consequences of this simple translation may be appreciated by expanding the SSB expression of Eq. (5-19) into the following expression:

$$e_S(t) = \left[\frac{2V}{\pi} \sum_{n=1,3,5,\ldots}^{N} \frac{\sin (n\pi/2)}{n} \cos n\omega_s t \right] \cos \omega_c t$$

$$- \left[\frac{2V}{\pi} \sum_{n=1,3,5,\ldots}^{N} \frac{\sin (n\pi/2)}{n} \sin n\omega_s t \right] \sin \omega_c t \qquad (5\text{-}20)$$

The first term of Eq. (5-20) is identical to the DSB signal e(t) of Eq. (5-18); the second term, however, represents the quadrature component resulting from the filtering action of $H_S(j\omega)$. It is to be noted that these new quadrature components of the modulation voltage shown in the second pair of brackets of Eq. (5-20) are similar to the original frequency component except for a 90° phase shift of each Fourier component. The effect of these quadrature components on the envelope of the SSB signal can be evaluated at the times of crossover of the original square wave by setting t = T/4 in Eq. (5-20). At this instant of time, the inphase component of the SSB has zero amplitude, and the quadrature component has its maximum amplitude. This is seen clearly in Eq. (5-21), which was derived from Eq. (5-20) by setting t = T/4. Thus in the vicinity of t = T/4,

$$e_S\left(t = \frac{T}{4}\right) = - \frac{2V}{\pi} \left[1 + \frac{1}{3} + \frac{1}{5} + \cdots + \frac{1}{N} \right] \sin \omega_c t \qquad (5\text{-}21)$$

Assuming that $\omega_c \gg \omega_s$, Eq. (5-21) represents the SSB signal for a short period near the peak of the SSB envelope. Since the harmonic series in the square brackets is divergent, it is obvious that the larger the N, the greater the peak envelope of the SSB signal, and at least theoretically, an infinite amount of peak power would be required to transmit a perfect square using SSB transmission. In practice, we are dealing with finite bandwidths, N is limited to a finite value, and so is the peak envelope power. But the resulting SSB envelope is not simply related to the waveform of the modulating signal as in the case of DSB and AM.

In addition to the mathematical analysis of SSB and its envelope relationship to the original modulating signal, the following simple physical reasoning will explain why a simple translation of the frequency components of the square wave by a fixed amount ω_c has such a drastic effect upon the resulting envelope. The frequency components of the modulating signal are harmonically related to the fundamental frequency of the modulating signal. But in the process of translation, the harmonic relationship is lost, and the resulting SSB signal is no longer representative of a periodic function. Whereas in the original signal, the more harmonic components that are included in the series, the more closely g(t) is approximated, when the same components are translated by ω_C, the larger their number, the larger the peak envelope becomes, as shown in Eq. (5-21).

As mentioned earlier, in the majority of communication systems, system performance is directly related to the average signal power, while the peak transmitter power is the limiting design factor. In order to compare the merits of SSB vs. DSB, we specify equal sideband power for both systems and proceed to rate the systems on the basis of the resulting peak transmitter power. From our previous discussion of the square-wave modulating signal, it is obvious that no meaningful comparison can be made between the two systems unless the waveform of the modulating signal is specified. Such is not the case when a comparison is made between DSB and AM since their envelopes are simply related to the modulation signal, independently of the modulation waveform.

1. *The Computation of Average-to-Peak Power Ratio in SSB.*[6] The following computation of transmitter average-to-peak power ratio as a function of the modulating waveform will serve to illustrate the importance of the waveform of the modulating signal when a comparison is made between the merits of linear modulation systems.

Let the modulating signal be represented by the periodic function $\sin^\nu x$, $0 < x \le \pi$, and $-\sin^\nu x$, $\pi \le x \le 2\pi$, where x is a linear function of time and $0 \le \nu \le 1$. As seen from Fig. 5-12, the waveform is sinusoidal when $\nu = 1$, and as $\nu \to 0$, the waveform becomes square.

Since the SSB modulation process shifts the modulating signal in the frequency domain without altering its total energy content, the average power of the SSB is simply equal to the modulating signal average power. Hence

$$P_{av} = \frac{1}{\pi} \int_0^\pi \sin^{2\nu} x \, dx = \frac{1}{\sqrt{\pi}} \frac{\Gamma(\nu + \frac{1}{2})}{\Gamma(\nu + 1)} , \qquad 0 < \nu \le 1 \qquad (5-22)$$

The calculation of SSB peak power is more complex for the following reasons. Before modulation, the peak power of the modulating signal is unity for all values of ν where $0 < \nu \le 1$ because the frequency components of the modulating signal are harmonically related and add in phase to produce the unity peak amplitude. As

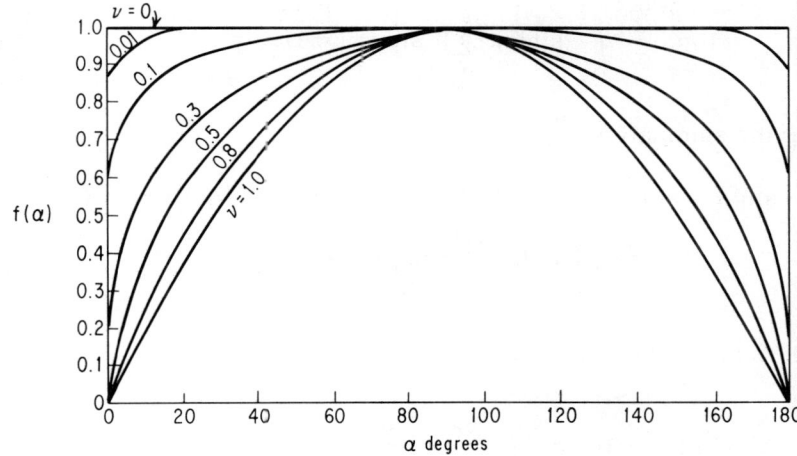

Fig. 5-12. Plot of $f(\alpha) = \sin^\nu \alpha$ for various values of ν. (After W. K. Squires and E. Bedrosian,[6] Proc. IRE.)

pointed out earlier, the harmonic relationship is lost in the process of translation, and consequently, the frequency components will all add "in phase" at some time. The peak amplitude after modulation will therefore be equal to the Fourier summation, or

$$A = \sum_{n=1}^{\infty} |a_n(\nu)| \tag{5-23}$$

where

$$a_n(\nu) = \frac{2}{\pi} \int_0^\pi \sin^\nu x \, \sin nx \, dx \tag{5-24}$$

Since, from symmetry, the even terms vanish and the odd terms are positive, then

$$A = \sum_{n=1}^{\infty} |a_n| = \sum_{n=1}^{\infty} a_{2n-1} = \sum_{n=1}^{\infty} \frac{2}{\pi} \int_0^\pi \sin^\nu x \, \sin (2n-1)x \, dx$$

Changing the order of integration, we obtain

$$A = \lim_{N\to\infty} \sum_{n=1}^{N} \frac{2}{\pi} \int_0^\pi \sin^\nu x \, \sin (2n-1)x \, dx$$

$$= \lim_{N \to \infty} \frac{2}{\pi} \int_0^\pi \sin^\nu x \sum_{n=1}^\infty \sin (2n - 1)x \, dx \qquad (5-25)$$

Using the summation

$$\sin \alpha + \sin (\alpha + \delta) + \sin (\alpha + 2\delta) + \cdots + \sin [\alpha + (n - 1)\delta]$$

$$= \frac{\sin \{\alpha + [(n - 1)/2]\delta\} \sin (n\delta/2)}{\sin (\delta/2)} \qquad (5-26)$$

and letting $\delta = 2\alpha$, we obtain

$$\sum_{n=1}^N \sin (2n - 1)\alpha = \frac{\sin^2 N\alpha}{\sin \alpha} = \frac{1 - \cos 2N\alpha}{2 \sin \alpha} \qquad (5-27)$$

which when substituted in Eq. (5-25) gives

$$A = \frac{1}{\pi} \int_0^\pi \sin^{\nu-1} x \, dx - \frac{1}{\pi} \lim_{N \to \infty} \int_0^\pi \sin^{\nu-1} x \cos 2Nx \, dx$$

$$(5-28)$$

Now it can be shown that if $f(x)$ is integrable over (a, b), then as $\lambda \to \infty$, $\int_a^b f(x) \cos \lambda x \, dx \to 0$. Since $\sin^{\nu-1} x$ is integrable in $(0, \pi)$ for $0 < \nu \leq 1$, the right-hand integral in Eq. (5-28) vanishes in the limit, and the left-hand integral reduces to

$$A = \frac{1}{\sqrt{\pi}} \frac{\Gamma(\nu/2)}{\Gamma\left(\dfrac{\nu + 1}{2}\right)}, \qquad 0 < \nu \leq 1 \qquad (5-29)$$

The ratio P_{av}/A^2 gives the SSB average-to-peak power ratio as a function of ν. This is plotted in Fig. 5-13 with the corresponding ratios for AM and DSB.

It can be concluded from Fig. 5-13 that for small values of ν, i.e., when the modulating signal is approaching a square, the SSB ratio compares poorly with either AM or DSB; in fact, the SSB ratio can be approximated in this region by ν^2, and for a perfect square, the SSB average power output is zero for any finite peak-power limitation. For modulating signals represented by moderately rectangular

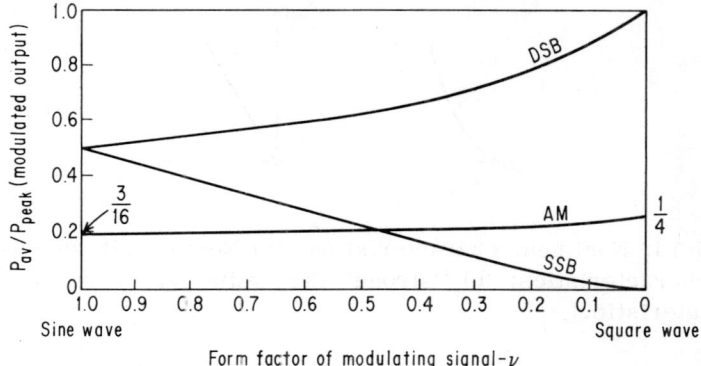

Fig. 5-13. Average-to-peak power relations as a function of modulating signal. (After W. K. Squires and E. Bedrosian,[6] Proc. IRE.)

waveforms, it can be concluded that DSB is superior to AM, which is considerably better than SSB.

5.6 Amplitude Modulators [2,3,7]

Since the mathematical expression for AM appears as a product function, an obvious way of producing such a signal is to develop a device in which the output is proportional to the product of the two input functions. Modulators possessing this property are commonly referred to as product modulators.

In the analysis of modulators and demodulators, we shall distinguish among linear, nonlinear, and switching systems. Generally, linear systems are governed by linear differential equations that have constant coefficients, and cannot be used as modulators for they do not generate new frequencies. However, adding a switch to a linear system which switches the system in a specified manner from one linear condition to another, independently of the signal, does not change the basic characteristics of a linear system. Such a system is still governed by linear differential equations, but the coefficients of the equations now vary with time according to the prescribed switching action. The system is now a linear time-varying system. Such a system can be used for modulation and will generate new frequencies. Modulation or generation of new frequencies is also possible by using devices whose static terminal characteristics are nonlinear. The differential equations governing the system become nonlinear differential equations. Two typical kinds

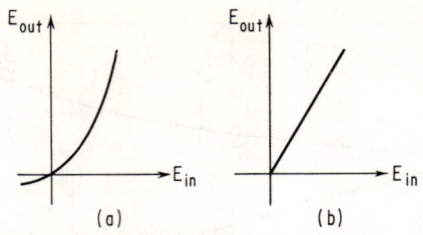

Fig. 5-14. Nonlinear characteristics: (a) Nonlinearity due to curvature characteristics; (b) "strong" discontinuity, piecewise-linear characteristic.

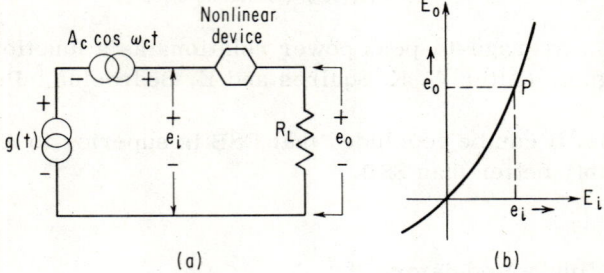

Fig. 5-15. Square-law modulator: (a) simple circuit; (b) terminal characteristics. (From M. Schwartz,[3] courtesy of McGraw-Hill Book Company.)

of nonlinear characteristics are shown in Fig. 5-14. Figure 5-15a represents the characteristic of a typical square-law diode or of a triode in certain portions of its characteristics. Figure 5-15b represents an approximation to a linear rectifier (this device is not linear), and also the characteristics of a switching device.

Simple modulators having the nonlinear characteristics of Fig. 5-14 will presently be analyzed.

1. *Square-law Modulator.* Square-law modulators are widely used, and all employ devices in which the output is proportional to the square of the input. A simple form of this type of modulator is shown in Fig. 5-15. Let P represent a fixed operating point, and e_0 and e_i represent the incremental output and input variations, respectively. Expanding e_0 in a power series of e_i,

$$e_0 = a_1 e_i + a_2 e_i^2 + a_3 e_i^3 + \cdots \qquad (5\text{-}30)$$

From Fig. 5-15a, $e_i = [A_c \cos \omega_c t + g(t)]$, and retaining just two terms of the power series, we obtain

$$e_0 = a_1[A_c \cos \omega_c t + g(t)] + a_2[A_c \cos \omega_c t + g(t)]^2$$

$$= a_1 g(t) + a_2 A_c^2 \cos^2 \omega_c t + a_2 g^2(t) + a_1 A_c \cos \omega_c t \left[1 + \frac{2a_2}{a_1} g(t)\right]$$

Unwanted terms   Amplitude-modulated terms

(5-31)

The unwanted terms consist of harmonics of the modulating signal and the carrier and can be filtered out; the remaining terms contain the desired AM signal of the form $K[1 + m_a g(t)] \cos \omega_c t$. It should be noted here that for proper filtering, the lower sidebands of the amplitude-modulated terms should not overlap with the second harmonic of the modulating signal.

2. *Piecewise-linear Modulator-Rectifier-type Modulator.* A typical circuit with piecewise-linear characteristics is shown in Fig. 5-16. In an ideal linear rectifier, the flow of current is propor-

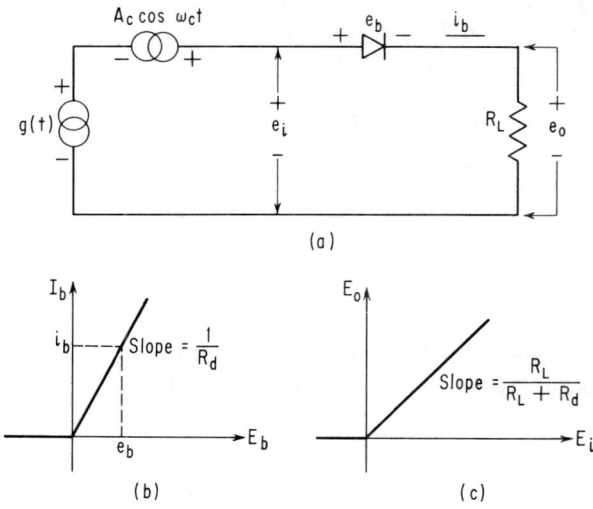

Fig. 5-16. Piecewise-linear modulator: (a) Simple modulator; (b) diode characteristic; (c) circuit characteristic. (From M. Schwartz,[3] courtesy of McGraw-Hill Book Company.)

tional to the applied voltage, provided the voltage is positive, and no current flows when the voltage is negative.

Let the modulating signal g(t) in series with a sinusoidal carrier $A_c \cos \omega_c t$ be applied to the rectifier, and let the carrier amplitude A_c be much greater than the maximum modulating signal g(t). The applied signal $e_i(t)$ is then

$$e_i(t) = A_c \cos \omega_c t + g(t), \qquad A_c \gg g(t) \tag{5-32}$$

The output signal

$$e_o(t) \doteq \frac{R_L e_i(t)}{R_L + R_d} = b e_i(t), \qquad A_c \cos \omega_c t > 0$$
$$\tag{5-33}$$
$$= 0, \qquad A_c \cos \omega_c t < 0$$

where R_L = load resistance, and R_d = diode forward resistance.

The output signal e_o varies periodically between two values $be_i(t)$ and zero at the carrier frequency; this is equivalent to switching the input signal $e_i(t)$ between two regions of the diode operation. The nonlinear device has thus been converted into a linear switching device, or expressed mathematically, the nonlinear equation has been replaced by a time-varying one.

Rewriting Eq. (5-33), we have

$$e_o(t) \doteq [A_c \cos \omega_c t + g(t)]S(t) \tag{5-34}$$

with $S(t) = V, \qquad -\frac{1}{4} T < t < \frac{1}{4} T, \ \omega_c = \frac{2\pi}{T}$

$\qquad\qquad = 0, \qquad t \text{ elsewhere}$

where S(t) may be considered as a periodic square-wave switching function, as shown in Fig. 5-17. Using Eq. (2-21), we obtain

$$S(t) = \sum_{n=-\infty}^{\infty} C_n e^{jn\omega_c t} = C_o + 2\sum_{n=1}^{\infty} |C_n| \cos(n\omega_c t - \theta_n) \tag{5-35}$$

where $C_n = \frac{V}{2} \frac{\sin(n\omega_c T/4)}{n\omega_c T/4} = \frac{V}{2} \frac{\sin(n\pi/2)}{n\pi/2}$

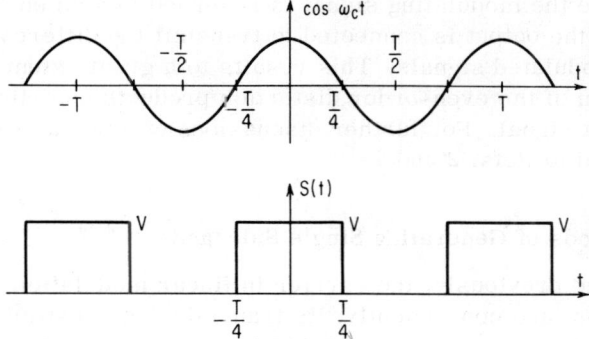

Fig. 5-17. Switching function.

or $\quad S(t) = V\left[\dfrac{1}{2} + \displaystyle\sum_{n=1}^{\infty} \dfrac{\sin\,(n\pi/2)}{n\pi/2}\,\cos\,n\omega_c t\right]$ $\qquad$ (5-36)

Finally $e_0(t) = [A_c\,\cos\,\omega_c t + g(t)]S(t)$

$$= V\left(\left\{\dfrac{1}{2}\,g(t) + \dfrac{2}{\pi}\,A_c\,\cos^2\,\omega_c t\right.\right.$$

$$+ \sum_{n=3}^{\infty} \dfrac{\sin\,(n\pi/2)}{n\pi/2}\,[g(t) + A_c\,\cos\,\omega_c t]\,\cos\,n\omega_c t\Big\}$$

$$+ \dfrac{A_c}{2}\,\cos\,\omega_c t\left[1 + \dfrac{4}{\pi A_c}\,g(t)\right]\right) \qquad (5\text{-}37)$$

The high-frequency components can be filtered out by the use of a bandpass filter, and the output $e_0(t)$ reduces to an AM signal:

$$e_0(t) \doteq K[1 + m_a g(t)]\,\cos\,\omega_c t \qquad (5\text{-}38)$$

where $\quad K = \dfrac{VA_c}{2} \qquad m_a = \dfrac{4}{\pi A_c}$

This modulation process may be considered as sampling the modulating wave briefly at regular intervals at the carrier rate, and applying the ensemble of samples to the input of a bandpass filter having a center frequency coincident with the carrier frequency.

In practical applications, some form of balanced modulators are used where the modulating signal is reversed in sign on one modulator, and the output is connected to transmit the difference between the two modulated signals. This results to a great extent in the elimination of the even-order distortion products contributed by the modulating signal. For further discussions of modulators, the reader is referred to Refs. 2 and 7.

5.7   Methods of Generating Single Sidebands [2,3,8-10]

As noted previously, the carrier in linear modulation carries no information, and consequently, its transmission constitutes a waste of power. Another advantage of SSB is the conservation of frequency spectrum, since only one sideband is used to transmit the intelligence. The high-frequency portion of the spectrum between 2 and 25 Mc which is readily suited for long-range communication is increasingly overcrowded, and its economical frequency utilization is therefore highly desirable. However, the elimination of the carrier gives rise to a number of rather difficult problems at the receiver in providing a suitable demodulating carrier. These problems include the necessity of having highly stable frequency control of transmitter and receiver and the use of complex receiver automatic-frequency-control techniques. In this section, we shall describe a number of techniques for SSB generation and compare some of their advantages and disadvantages.

1. *Frequency Discrimination.* In this method, the single-sideband modulator consists basically of a suitable amplitude modulator, followed by a filter capable of passing the desired sideband and attenuating all other components sufficiently to meet the requirements of the system. It is often necessary in practical systems to resort to a multiple-modulation process in which the first carrier frequency is so chosen as to place a practical type of highly discriminating filter in its optimum frequency range. The output of this filter goes to a second modulator supplied with a carrier frequency that is so chosen as to place the selected sideband in the desired interval for transmission over the medium. By this process, the separation between the wanted and unwanted sidebands at the final filter can be greatly increased. Typical carrier telephone systems employing frequency discrimination are K, L, and M systems of the Bell Telephone System.

2. *Phase-shift Method of SSB Generation.* The problem of generating SSB signals with the required sideband suppression and channel bandwidth represents a challenge to the designer of SSB equipment.

The phase-shift method of generating SSB signals provides a means for extending the useful bandwidth of single-sideband systems, with increased attenuation of the unwanted sidebands without the use of sharp cutoff filters.

In this system, two double-sideband signals are generated in balanced modulators and added together, as shown in Fig. 5-18. Because the upper and lower sidebands of an AM signal differ in the sign of their phase angles, phase discrimination may be used to cancel one sideband of the DSB system. This can be demonstrated mathematically for a sinusoidal modulating signal as follows. The DSB output of a balanced modulator is of the form

$$e_1(t) = \cos \omega_m t \cdot \cos \omega_c t$$

$$= \tfrac{1}{2} [\cos (\omega_c + \omega_m)t + \cos (\omega_c - \omega_m)t] \qquad (5\text{-}39)$$

where ω_m is the angular frequency of the modulating signal. A SSB signal is of the form $\cos (\omega_c - \omega_m)t$. This implies that we must add to $e_1(t)$ an expression of the form

$$e_2(t) = \tfrac{1}{2} [\cos (\omega_c - \omega_m)t - \cos (\omega_m + \omega_c)t]$$

$$= \sin \omega_m t \cdot \sin \omega_c t \qquad (5\text{-}40)$$

This indicates that a product modulator or a balanced modulator is again needed with the modulating signal phase-shifted to $\sin \omega_m t$, multiplied by the phase-shifter carrier $\sin \omega_c t$. This simplified approach is helpful in understanding the need for two balanced modulators for the generation of SSB modulation. A more sophisticated way of looking at the phase-shift method of SSB generation is to compare Fig. 5-18 with the equivalent vestigial-sideband transmission system of Fig. 5-8. It becomes obvious that in SSB, one system function $H_i(j\omega)$ is a constant, independent of frequency, while the second one, namely, $H_q(j\omega)$, is a simple phase-shift network of 90°, constant over the frequency range of the modulating signal $g(t)$. In practice, it is difficult to realize constant phase-shift networks over a wide enough modulating-signal frequency range. If, however, a phase-shifting network is included in each modulating-signal path, the required differential phase shift can be maintained within any desired tolerance over the modulating frequency range. This is shown in Fig. 5-19 where the phase-shifting networks are identified by α and β. The analysis of the phase-shift

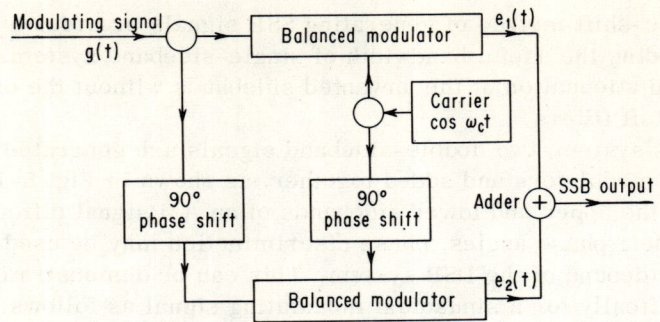

Fig. 5-18. Phase-shift method of generating SSB.

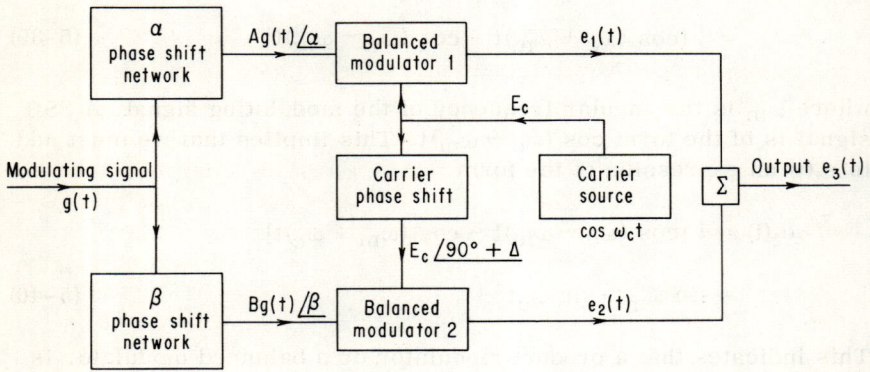

Fig. 5-19. Phase-shift method of generating SSB using phase-shifting networks. (After D. E. Norgaard,[9] Proc. IRE.)

method of generating SSB using phase-shifting networks is given by Norgaard.[9]

In conclusion, it should be noted that no bandpass filters are required in the phase-shift method of SSB generation. However, since the operation of any modulator depends upon an inherent nonlinearity, the circuits which follow the modulators should thus be designed for adequate harmonic suppression.

3. *A Third Method of Generation of SSB Signals.* In the phasing method described previously, the degree to which the undesired sideband may be suppressed depends upon accurate balancing and requires very careful control of amplitudes and phases. The design and construction of a wideband 90° phase-shifting network is quite an art to achieve and often presents practical difficulties.

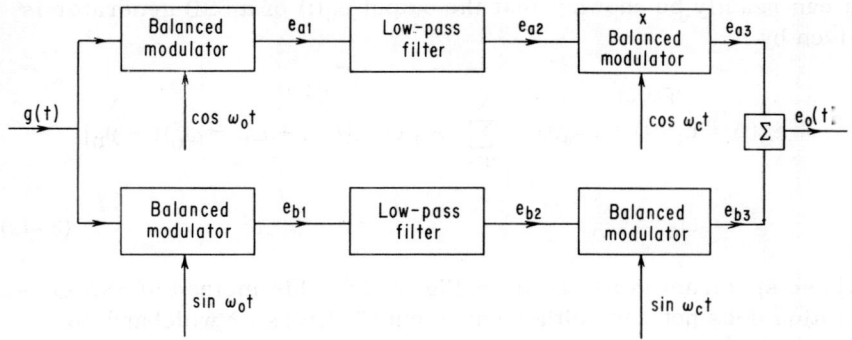

Fig. 5-20. Single-sideband generator. (After D. K. Weaver,[10] Proc. IRE.)

A third method of SSB generation as described by Weaver [10] is shown in Fig. 5-20. This method is suitable, as in the phasing method, for generation of a SSB signal at any frequency in the IF or RF spectrum. It differs basically from either the conventional filter or the phasing method in that no sharp cutoff filters or wide-band 90° phase-shifting networks are required.

Let the modulating signal $g(t)$ be band-limited as shown in Fig. 5-21 and be given by

$$g(t) = \sum_{n=1}^{N} A_n \cos (\omega_n t + \phi_n) \tag{5-41}$$

where $\omega_L \leq \omega_n \leq \omega_L + 2\pi B$ and $\omega_0 = \omega_L + \pi B$

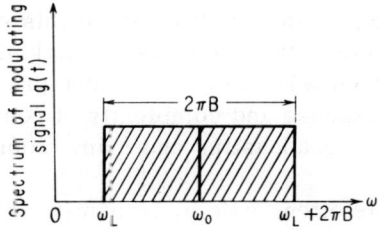

Fig. 5-21. Input-signal spectrum.

It can readily be shown [10] that the output $e_0(t)$ of a SSB generator is given by

$$e_0(t) = e_{a3}(t) + e_{b3}(t) = \sum_{n=1}^{N} A_n \cos \left[(\omega_c + \omega_n - \omega_0)t + \phi_n \right],$$

$$\omega_c \gg \omega_0 \tag{5-42}$$

whose spectrum is as shown in Fig. 5-22. This method of SSB generation does not need either sharp cutoff filters or wideband 90° phase-difference networks. Note that the frequency normally referred to as the carrier corresponds to $(\omega_c - \omega_0)$ and that the angular frequency ω_c is in the center of the SSB spectrum.

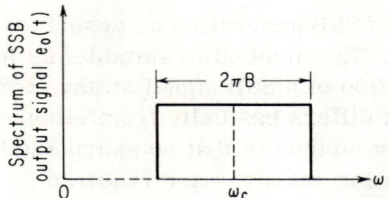

Fig. 5-22. Spectrum of output signal.

5.8 Compatible Single-sideband (CSSB) Transmission [11-17]

In our previous discussions of linear modulation systems, we have noted that AM and DSB systems are less efficient than SSB systems from the point of view of power utilization and conservation of frequency spectrum. The severe shortage of spectrum space available, especially in the broadcast band, makes it very attractive to explore the feasibility of replacing the existing AM systems with suppressed-carrier SSB systems.

The number of SSB transmitters in operation has greatly increased during the past decade because of this serious spectrum shortage and because of the other operational advantages which were discussed previously. However, there are many services that cannot justify the expense and complexity of conventional SSB equipment: for example, broadcasting and many mobile communication systems.

A compromise is often made by transmitting the full carrier in addition to the single sideband so that a simple linear envelope detector may be used to demodulate the signal. However, the original modulating signal appears distorted at the output of the envelope detector. This distortion is easily tolerated in speech communication systems but would be undesirable in other applications.

In order to effect an orderly and economically feasible transition from AM to SSB communication in many existing services, it appears necessary to design the new SSB equipment in such a way that it is compatible with existing AM equipment.

A CSSB system has recently been defined by CCIR [11] in the following manner: "A single-sideband transmission is considered to be compatible if it can be received on the existing conventional double-sideband receivers without any modifications whatsoever and with satisfactory quality of reception." It follows from this definition of compatibility that a perfect CSSB signal should have the following characteristics:

1. An envelope which follows faithfully the modulating signal with minimum nonlinear distortion.

2. A spectrum characteristic limited to one sideband only. Since these specifications are incompatible, it is obvious that in a CSSB system these requirements can be met only approximately.

The original CSSB which was proposed by Kahn [12,13] has been shown, according to an analysis by Costas, [14] to transmit spurious components outside the desired frequency band. These components are spurious in the sense that they do not belong to a true single-sideband signal, but they are very necessary in the CSSB system in order that the output wave have an envelope which may be detected with an AM receiver without distortion. However, as discussed previously, a truly compatible signal should have an envelope which reproduces the modulating message signal with minimum distortion and a spectrum which is limited to one sideband only about the carrier.

While these requirements are theoretically incompatible, Kahn proposed a modified CSSB system [15] designed to meet approximately these requirements. Since a distortion-free envelope is of primary importance for broadcast application, it appears that only the second requirement can be relaxed in practical application. Consequently, in the modified CSSB system, the emphasis is to produce a practically distortion-free envelope plus a phase-modulated (PM) signal whose energy is practically concentrated in one sideband only. This concept of considering a SSB signal as a hybrid modulated wave in amplitude and phase will be discussed in the following section in connection with the concept of analytic signal and the application of Hilbert transforms to SSB.

5.9   The Concept of Analytic Signal and Its Application to SSB [4,5,16,18-23]

The problem of CSSB modulation systems has stimulated interest in the theoretical aspects of simultaneous amplitude and phase modulation by the use of Hilbert transforms and the concept of the "analytic signal." In the following, we shall introduce the concept of analytic signal and discuss its application to the compatibility problems in SSB transmission. In this connection, we shall make use of Hilbert transforms in representing certain modulated signals, with special emphasis on the compatibility problem. Finally, we shall discuss some of the proposed CSSB systems in the light of our theoretical analysis.

1. *The Concept of Analytic Signal.* We have seen earlier that a SSB signal can be regarded as the resultant of quadrature modulation of a carrier by a pair of signals in phase quadrature. Let $s(t)$ represent an arbitrary message function, and $\sigma(t)$ its harmonic conjugate (obtained from $s(t)$, ideally by a network whose amplitude response is unity, and whose phase response is a constant 90° lag at all frequencies). Then it follows that the modulated wave

$$f(t) = s(t) \cos \omega_c t - \sigma(t) \sin \omega_c t \qquad (5\text{-}43)$$

is an upper-sideband signal with no spectral components below the carrier angular frequency ω_c. In fact, any band-limited wave (e.g., bandpass noise) can be expressed in the form of Eq. (5-43), if ω_c is chosen to be below the band edge.

Equation (5-43) can be rewritten in the form

$$f(t) = \sqrt{s(t)^2 + \sigma(t)^2} \, \cos \left[\omega_c t + \tan^{-1} \frac{\sigma(t)}{s(t)} \right]$$

$$= \alpha(t) \cos [\omega_c t + \phi(t)] \qquad (5\text{-}44)$$

regarding the single-sideband signal as a hybrid amplitude- and phase-modulated wave. In a conventional SSB system, the message is normally conveyed by the inphase component $s(t)$, detection being accomplished by synchronous demodulation with a carrier generated locally at the receiver. It is conceivable, however, that the message could be conveyed by the envelope $\alpha(t)$ if a corresponding phase $\phi(t)$ could be found that would yield a single-side spectrum for the hybrid wave. Reception would then be accomplished by a conventional envelope detector, and a single-sideband transmission could be made compatible with standard AM receivers.

Since the signals s(t) and $\sigma(t)$ are uniquely related to each other, it is apparent that the envelope $\alpha(t)$ and the phase $\phi(t)$ must be related also. The compatibility problem consists in the study of conditions on the envelope $\alpha(t)$ that yield to a realizable phase function $\phi(t)$, and in determining the relation that must hold between them. The solution is most easily obtained through the concept of the analytic signal.

Basically, the analytic signal is a complex function of the variable whose real and imaginary parts form a Hilbert pair. When applied to our problem of SSB, the analytic signal is a function of time given by

$$\psi(t) = s(t) + j\sigma(t) = \alpha(t)e^{j\phi(t)} \tag{5-45}$$

where the function $\sigma(t)$ is related to s(t) by the Hilbert transform relation

$$\sigma(t) = \frac{1}{\pi} \int_{-\infty}^{\infty} \frac{s(\tau)}{t - \tau} \, d\tau = \widehat{s(t)} \tag{5-46}$$

where $\widehat{s(t)}$ denotes Hilbert transform of s(t), and

$$s(t) = -\frac{1}{\pi} \int_{-\infty}^{\infty} \frac{\sigma(\tau)}{t - \tau} \, d\tau \tag{5-47}$$

where the Cauchy principal value of the integral at $\tau = t$ is taken as discussed in Sec. 3.7.

Equivalent formulas to Eq. (5-46) are

$$\widehat{s(t)} = \frac{1}{\pi} \int_{-\infty}^{\infty} \frac{s(t + \tau)}{-\tau} \, d\tau \tag{5-48}$$

and $$\widehat{s(t)} = \frac{1}{\pi} \int_{0+}^{\infty} \frac{s(t + \tau) - s(t - \tau)}{\tau} \, d\tau \tag{5-49}$$

Since the Hilbert transform of a Hilbert transform is the negative of the original function, therefore

$$\widehat{\widehat{s(t)}} = -s(t)$$

In practice, the message function is identified with either the real or
the imaginary part of the analytic signal. The analytic signal representation of modulated signals will be shown to possess several interesting properties; since the real and imaginary parts of the
analytic signal form a Hilbert pair, we shall discuss the analytic
signal through the properties of Hilbert transforms.

2. *Properties of Hilbert Transforms.* We have noted already that
$\widehat{\widehat{s(t)}} = -s(t)$. Consider the Hilbert transform of $s(t) = \cos(\omega t + \varphi)$.
Using Eq. (5-48), we obtain

$$\widehat{s(t)} = \frac{1}{\pi} \int_{-\infty}^{\infty} \frac{\cos(\omega t + \omega \tau + \varphi)}{-\tau} \, d\tau$$

$$= -\frac{1}{\pi} \int_{-\infty}^{\infty} \frac{1}{\tau} [\cos(\omega t + \varphi) \cos \omega \tau] d\tau$$

$$+ \frac{1}{\pi} \int_{-\infty}^{\infty} \frac{1}{\tau} [\sin(\omega t + \varphi) \sin \omega \tau] d\tau$$

The first integral is zero since $(\cos \omega \tau)/\tau$ is odd, and we obtain

$$\widehat{s(t)} = \frac{\sin(\omega t + \varphi)}{\pi} \int_{-\infty}^{\infty} \frac{\sin \omega \tau}{\tau} \, d\tau = \sin(\omega t + \varphi) \qquad (5\text{-}50)$$

since $\int_{-\infty}^{\infty} \frac{\sin x}{x} \, dx = \pi$. We have thus proved that the Hilbert

transform of $\cos(\omega t + \psi)$ is $\sin(\omega t + \psi)$. Similarly, it can be shown
that if

$$s(t) = \sin(\omega t + \varphi) \qquad (5\text{-}51)$$

then $\widehat{s(t)} = -\cos(\omega t + \varphi) \qquad (5\text{-}52)$

Next we consider the spectral properties of the Hilbert transform.
Let $S(j\omega)$ denote the Fourier transform of $s(t)$, so that

$$S(j\omega) = \int_{-\infty}^{\infty} s(t) e^{-j\omega t} \, dt \qquad (5\text{-}53)$$

We shall show that the Fourier transform of $s(t)$, namely, $\widehat{S(j\omega)}$, is given by

$$\widehat{S(j\omega)} = -jS(j\omega) \text{ sgn } \omega \tag{5-54}$$

From Eq. (5-46), the function $\widehat{s(t)}$ may be considered as the convolution of $s(t)$ and $1/\pi t$. Therefore, the Fourier transform of $s(t)$ is the product of the Fourier transform of $s(t)$ and that of $1/\pi t$. That is,

$$\widehat{S(j\omega)} = \begin{cases} -jS(j\omega), & \omega > 0 \\ 0, & \omega = 0 \\ jS(j\omega), & \omega < 0 \end{cases} \tag{5-55}$$

or $\widehat{S(j\omega)} = -jS(j\omega) \text{ sgn } \omega$

Thus for positive frequencies, the spectrum of $\sigma(t)$ is identical to that of $s(t)$ except for the multiplying factor $(-j)$ corresponding to a 90° phase lag. Now let $\Psi(j\omega)$ denote the Fourier transform of the analytic signal $\psi(t)$,

$$\Psi(j\omega) = S(j\omega) + j\widehat{S(j\omega)} = S(j\omega) + S(j\omega) = 2S(j\omega), \qquad \omega > 0$$
$$= S(j\omega) - S(j\omega) = 0, \qquad\qquad \omega < 0 \tag{5-56}$$

a very interesting result. It states that the Fourier transform of the analytic signal vanishes for negative frequencies while that of its complex conjugate vanishes for positive frequencies. The non-physical signal $\psi(t)$ has been called the analytic signal because as a function of a complex variable z, $\psi(z)$ is analytic in the upper-half z plane. By comparing Eqs. (5-44) and (5-45), it can be seen that the magnitude and phase of the complex signal $\psi(t)$ are identical to the envelope and phase of the single-sideband wave. Thus a study of single sideband can be made through the analytic signal without reference to the arbitrary carrier frequency ω_c. The envelope $\alpha(t)$, phase $\phi(t)$, and signal $s(t)$ may be defined by

$$\psi(t) = \alpha(t)e^{j\phi(t)} = s(t) + j\widehat{s(t)} \tag{5-57}$$

and $s(t) = \alpha(t) \cos \phi(t)$ \hfill (5-58)

where $\alpha(t) = \sqrt{s(t)^2 + s(t)^2}$ \hfill (5-59) and $\quad \phi(t) = \tan^{-1} \dfrac{\widehat{s(t)}}{s(t)}$ (5-60)

3. *Single-sideband Modulation.* In this section, we shall illustrate the application of Hilbert transforms and the concept of analytic signal to SSB modulation. In Sec. 5.5, we derived an expression for the output spectrum of a SSB modulator, which is reproduced by Eq. (5-61) with change of notation $g(t) \equiv s(t)$.

$$E_S(j\omega) = H_S(j\omega)E(j\omega) = \tfrac{1}{2} S[j(\omega - \omega_c)][\tfrac{1}{2} + \tfrac{1}{2} \operatorname{sgn}(\omega - \omega_c)]$$
$$+ \tfrac{1}{2} S[j(\omega + \omega_c)][\tfrac{1}{2} - \tfrac{1}{2} \operatorname{sgn}(\omega + \omega_c)]$$
(5-61)

This expression may be rewritten as

$$E_s(j\omega) = \frac{1}{4} \left\{ S[j(\omega - \omega_c)] + S[j(\omega + \omega_c)] \right\}$$
$$+ \frac{j}{4} S[j(\omega - \omega_c)][-j \operatorname{sgn}(\omega - \omega_c)]$$
$$- \frac{j}{4} S[j(\omega + \omega_c)][-j \operatorname{sgn}(\omega + \omega_c)]$$
(5-62)

Using the results of Eq. (5-55), namely,

$$\widehat{S(j\omega)} = -jS(j\omega) \operatorname{sgn} \omega$$

we derive an expression for the output signal $e_S(t)$ by taking the inverse transform of $E_S(j\omega)$.

$$e_S(t) = \frac{s(t)}{4} e^{j\omega_c t} + \frac{s(t)}{4} e^{-j\omega_c t} - \frac{\widehat{s(t)}}{4j} e^{j\omega_c t} + \frac{\widehat{s(t)}}{4j} e^{-j\omega_c t}$$
$$= \frac{s(t)}{2} \cos \omega_c t - \frac{\widehat{s(t)}}{2} \sin \omega_c t$$
(5-63)

Thus, the SSB product modulator can be represented by the sum of two product modulators as shown in Fig. 5-23, where the carriers are in phase quadrature and the modulation signals form a Hilbert pair. This is the well-known phase-shift method of SSB generation discussed in Sec. 5.5 and shown in Fig. 5-18.

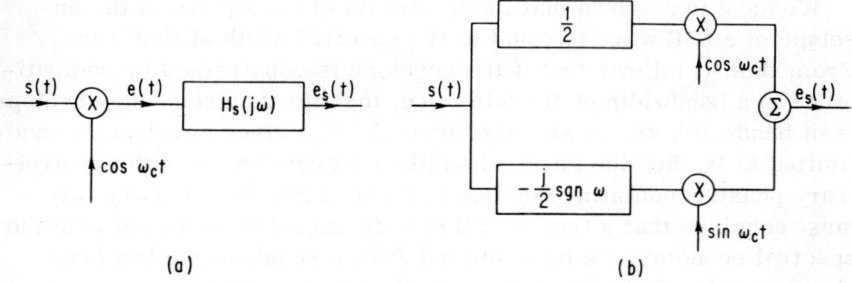

Fig. 5-23. SSB modulator and its equivalent.

4. *The Compatibility Problem.* The compatibility problem in SSB communication can be studied from the relationship of the magnitude and phase of the analytic signal $\psi(t)$, which should be identical to the envelope and phase of the SSB signal. From Eq. (5-57) we obtain

$$\alpha^2(t) = |\psi(t)|^2 \tag{5-64}$$

By taking Fourier transform of both sides, we write

$$\int_{-\infty}^{\infty} \alpha^2(t) e^{-j\omega t} \, dt = \int_{-\infty}^{\infty} |\psi(t)|^2 e^{-j\omega t} \, dt$$

$$= \frac{1}{2\pi} \int_{-\infty}^{\infty} \psi(t) e^{-j\omega t} \int_{-\infty}^{\infty} \Psi^*(j\omega_1) e^{-j\omega_1 t} \, d\omega_1 \, dt$$

$$= \frac{1}{2\pi} \int_{-\infty}^{\infty} \Psi^*(j\omega_1) \int_{-\infty}^{\infty} \psi(t) e^{-j(\omega+\omega_1)t} dt \, d\omega_1$$

$$= \frac{1}{2\pi} \int_{-\infty}^{\infty} \Psi^*(j\omega_1) \Psi[j(\omega_1 + \omega)] d\omega_1$$

$$= \frac{2}{\pi} \int_{0}^{\infty} S^*(j\omega_1) S[j(\omega_1 + \omega)] \, d\omega_1, \qquad \omega > 0 \tag{5-65}$$

It should be noted here that this result can be obtained directly from Eq. (2-83). Assuming the signal $\psi(t)$ [also $s(t)$] to be band-limited to B cycles/sec, then

$$\int_{-\infty}^{\infty} \alpha^2(t) e^{-j\omega t} \, dt = \frac{2}{\pi} \int_{0}^{(2\pi B - \omega)} S^*(j\omega_1) S[j(\omega_1 + \omega)] d\omega_1, \qquad \omega > 0 \tag{5-66}$$

and the right-hand side vanishes for $\omega > 2\pi B$.

We have just shown that the bandwidth of the square of the envelope of a SSB wave is equal to the spectral width of that wave. From this, it follows that if the envelope is constrained by compatibility to a bandwidth of B cycles/sec, then the square of the envelope is of bandwidth 2B, or stated differently, if a given envelope is band-limited to B, then the spectral width of a hybrid wave with any arbitrary phase-modulating function must be $\geq$ 2B. In summary, we must conclude that a truly CSSB system cannot be achieved with the spectral economy of a conventional SSB; a result which has been shown previously in the case of single-tone modulation.

Although it is theoretically impossible to generate a SSB signal which can be detected by a linear envelope detector, it is possible to generate a SSB signal whose envelope is the square root of the message function which can be detected without distortion by a square-law envelope detector. Such a SSB system, originated by Powers, [16] has recently been described in the literature. [23] In this new SSB system, the carrier is amplitude-modulated with the square root of the desired message function and at the same time, and it is phase-modulated with the Hilbert transform of the logarithm of the waveform. Although neither the amplitude nor the phase components are, in general, band-limited, the transmitted signal is nevertheless band-limited, as discussed above.

References

1. Costas, J. P.: Linear Modulation Systems, Lecture Delivered during the MIT Special Summer Program on Reliable Long-range Radio Communication, 1960.
2. Black, H. S.: "Modulation Theory," D. Van Nostrand Company, Inc., Princeton, N. J.
3. Schwartz, M.: "Information Transmission, Modulation, and Noise," McGraw-Hill Book Company, New York.
4. Bedrosian, E.: The Analytic Signal Representation of Modulated Waveforms, *Proc. IRE*, October, 1962.
5. Kuo, F. F., and S. L. Freeny: Hilbert Transforms and Modulation Theory, *Proc. NEC*, vol. 18, pp. 51-58, 1962.
6. Squires, W. K., and E. Bedrosian: The Computation of Single-sideband Peak Power (correspondence), *Proc. IRE*, January, 1960, pp. 123-124.
7. "Handbook of Line Communication," vol. 1, published by The Royal Signals, London, England, 1947.
8. Honey, J. F. and D. K. Weaver: An Introduction to Single-sideband Communications, *Proc. IRE*, SSB issue, December, 1956.

9. Norgaard, D. E.: The Phase-shift Method of Single-sideband Signal Generation, *Proc. IRE*, December, 1956.
10. Weaver, D. K.: A Third Method of Generation and Detection of Single-sideband Signals, *Proc. IRE*, December, 1956.
11. CCIR, 9th Plenary Assembly, Los Angeles, Calif.: "Resolutions, Questions, and Study Programmers," vol. 2, pp. 165-166, 1959.
12. Kahn, L. R.: A Compatible Single-sideband, presented at 2d Annual Symposium on Aeronautical Communications, Utica, N. Y., October, 1956.
13. Kahn, L. R.: A Compatible Single-sideband Modulation System, *Proc. Radio Club Am.*, vol. 34, pp. 1-9, March, 1958.
14. Costas, J. P.: A Mathematical Analysis of the Kahn Compatible Single-sideband System, *Proc. IRE*, vol. 47, pp. 1396-1401, July, 1958.
15. Kahn, L. R.: Compatible Single-sideband, *Proc. IRE*, October, 1961, pp. 1503-1527.
16. Powers, K. H.: The Compatibility Problem in SSB Transmission, *Proc. IRE*, August, 1960, pp. 1431-1435.
17. Granlund, J.: Interference in Frequency-modulated Reception, *MIT Res. Lab. Electron. Tech. Rept.* 42, Jan. 20, 1949.
18. Dugundji, J.: Envelopes and Pre-envelopes of Real Waveforms, *IRE Trans. Inform. Theory*, vol. IT-4, pp. 53-57, March, 1958.
19. Oswald, J. R. V.: The Theory of Analytic Band Limited Signals Applied to Carrier Systems, *IRE Trans. Circuit Theory*, vol. CT-3, pp. 244-251, 1956.
20. Logan, B. F. and M. R. Schroeder: A Solution to Problem of Compatible Single-sideband Transmission, *IRE Trans. Inform. Theory*, vol. IT-8, No. 5, September, 1962.
21. Gabor, D.: Theory of Communication, *J. IEE London*, vol. 93, part 3, pp. 429-457, November, 1946.
22. Ville, J.: Théorie et Application de la Notion de Signal Analytique, *Cables et Trans.*, vol. 2, pp. 61-74, January, 1948.
23. Von Urff, C. A. and F. I. Zoris: The Square-law Single-sideband System, *IRE Trans. Commun. Systems*, September, 1962.

6

LINEAR DEMODULATION OR DETECTION

The process of separating a modulating signal from a modulated carrier is called demodulation or detection. Demodulation is basically the inverse of modulation and requires nonlinear or linear time-varying parameter devices. There is a distinction, however, between the method of demodulating DSB or SSB signals and the method of demodulating normal AM signals.

In DSB or SSB detection, the carrier must be supplied at the receiver before the modulation process can take place. This method of detection was referred to earlier as coherent or synchronous detection, since the detector must have a carrier wave that is synchronous with that used at the transmitter. The coherence may be accomplished by transmitting a pilot carrier for the synchronization purpose. The precision frequency control required of both SSB transmitters and receivers is the basic limitation to their widespread use at the present time. The degree of precision frequency control required in the 2- to 30-Mc frequency range is 1 part in 10^7. Another stumbling block in the path of widespread use of SSB transmission is the problems involved in achieving the compatibility of AM and SSB systems. Some of these problems will be discussed later.

In conventional AM systems, coherent detection is not necessary because all the information is contained only in the amplitude of the envelope and not in both the amplitude and phase. Consequently, in normal AM detection, the modulating signal is recovered by applying the modulated carrier to a half-wave rectifier, whose output is then filtered to provide the desired modulating signal. Thus the advantage of normal AM is the ease with which reception is accomplished. The disadvantage is the large amount of power wasted in the carrier component.

In the following sections, we shall discuss in greater detail the various linear modulation receiving systems. In our discussion, we shall consider only the effect of additive stationary, white, gaussian noise upon system performance and ignore interference due to other transmissions or intentional jamming. These simplifying assumptions will enable us to gain a basic understanding of the operation of the various receiving systems. The output signal-to-noise ratio for the various linear modulation systems will be derived first by using a simplified approach which is applicable to both cases of either high-input carrier-to-noise ratio or very low ratio. A more general treatment of a more complex nature will then be presented, which is applicable to a wider range of carrier-to-noise ratio and produces more general results.

In order to make a direct comparison of the efficiencies of the different receiving systems, we shall assume the same average intelligence or sideband power for all systems under consideration. Similarly, we shall assume the same white-noise power density in the communication channel of the various receiving systems. Finally, the transmission and reception efficiencies will be combined to produce a measure of relative performance of the overall linear modulation systems.

6.1 Double-sideband (DSB) Reception [1-5]

In DSB reception the incoming signal $e_r(t)$ is multiplied by a locally generated signal which is phase-synchronized with the carrier component of the received signal $e_r(t)$. A basic DSB receiver is shown in Fig. 6-1, where the resulting product is filtered by a low-pass filter to produce the output signal $e_d(t)$ of the receiver.

The multiplication process shown in Fig. 6-1 is accomplished in practice by using what is commonly known as a product detector or demodulator. Product detectors take many different forms, and the details of their design will not be discussed here. However, it is important to note certain results that are obtained when product detection is used. In contrast to normal AM or FM detection, for example, where the detector in such cases has only one input, the

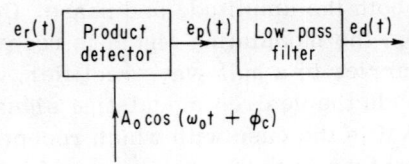

Fig. 6-1. Block diagram of double-sideband (DSB) receiver.

product detector requires two inputs: the received signal and the locally generated signal.

1. *Spectral Analysis of DSB Reception.*   Let the linearly modulated received carrier be given by

$$e_r(t) = A_c g(t) \cos (\omega_0 t + \phi_0) \tag{6-1}$$

and the locally generated oscillation be denoted by

$$e_0(t) = A_0 \cos (\omega_0 t + \phi_0) \tag{6-2}$$

The output of the product demodulator is therefore given by

$$
\begin{aligned}
e_p(t) &= A_0 A_c g(t) \cos (\omega_c t + \phi_c) \cos (\omega_0 t + \phi_0) \\
&= \frac{A_0 A_c g(t)}{2} \{[\cos (\omega_c - \omega_0)t + \phi_c - \phi_0] \\
&\quad + [\cos (\omega_c + \omega_0)t + \phi_c + \phi_0]\} \tag{6-3}
\end{aligned}
$$

Assuming that the local oscillator is properly synchronized with the transmitted carrier, Eq. (6-3) is simplified to

$$e_p(t) = \frac{A_0 A_c}{2} [g(t) + g(t) \cos (2\omega_c t + 2\phi_c)] \tag{6-4}$$

From Eq. (6-4), it is seen that the output consists of the original message function which is extracted by the low-pass filter and a modulated carrier of double the frequency of the local carrier oscillator.

In order to appreciate the spectral significance of the operation of the product demodulator, we make use of Fourier transform analysis. Using Eq. (5-2), we derive the Fourier transform of $e_r(t)$

$$E_r(j\omega) = \frac{A_c}{2} \left\{ e^{j\phi_c} G[j(\omega - \omega_c)] + e^{-j\phi_c} G[j(\omega + \omega_c)] \right\} \tag{6-5}$$

The Fourier transform of the local oscillator signal $\cos(\omega_c t + \phi_c)$ is obtained by the use of Eq. (2-120),

$$\pi\left[e^{j\phi_c}\,\delta(\omega-\omega_c) + e^{-j\phi_c}\,\delta(\omega+\omega_c)\right] \tag{6-6}$$

The Fourier transform of the product of the two time functions $e_r t \cdot A_0 \cos(\omega_c t + \phi_c)$ is given by the convolution of their Fourier transforms. Thus

$$E_p(j\omega) = \frac{A_c A_0}{2} G(j\omega) + \frac{A_c A_0}{4}\left\{e^{j2\phi_c} G[j(\omega-2\omega_c)]\right.$$

$$\left. + e^{-j2\phi_c} G[j(\omega+2\omega_c)]\right\} \tag{6-7}$$

Thus, in the absence of distortion, a repetition of the modulating process applied to the modulated wave $e_r(t)$ produces two results. One is the recovery of a spectrum proportional to the original spectrum of the modulating wave $g(t)$. The other is a pair of spectra displaced about $\pm 2\omega_c$. If the carrier frequency is high enough, the spectra are sufficiently separated so that the modulating wave can be recovered by filtering.

2. *Signal-to-Noise Improvement in DSB Demodulation.* In order to analyze the effect of the noise contribution upon the demodulated output signal $e_d(t)$, we shall assume that the modulation signal is band-limited and that the voltage $e_r(t)$ delivered to the DSB receiver has been passed through an appropriate bandpass filter. In practice, such predetector filtering is not necessary, as the low-pass post-detector filter can provide all the necessary selectivity. We assume the use of a predetector filter so that the noise component of $r(t)$ may be expressed in terms of the two-phase representation of the form $n(t) = x_c(t)\cos\omega_c t - x_s(t)\sin\omega_c t$, where $n(t)$ represents the bandpass gaussian noise voltage, and $x_c(t)$ and $x_s(t)$ represent low-pass gaussian voltages which are gaussian random variables independent of each other, provided that the noise spectrum is symmetrical about the center angular frequency ω_c. The concept of band-limited noise in communication has already been discussed in Chap. 4. It has been shown that the average power of $x_c(t)$ is equal to the average power of $x_s(t)$, which in turn is equal to the average power of the original noise voltage $n(t)$, as is indicated by Eq. (6-8).

$$\overline{x_c^2(t)} = \overline{x_s^2(t)} = \overline{n^2(t)} \tag{6-8}$$

Thus the total receiver input voltage $e_r(t)$ may be written in terms of a signal component and a noise component as follows:

$$e_r(t) = g(t) \cos \omega_c t + n(t) = g(t) \cos \omega_c t + x_c(t) \cos \omega_c t$$
$$- x_s(t) \sin \omega_c t \qquad (6-9)$$

where for sake of simplicity $A_o = A_c = 1$, and $\phi_c = 0$.

From Eq. (6-9), it is seen that one component of the noise contribution has the same carrier phase as that of the desired signal, while the second noise component is in phase quadrature with the desired signal. Since the locally generated carrier is in phase synchronism with the carrier component of $e_r(t)$, the output of the low-pass filter will be

$$e_d(t) = g(t) + x_c(t) \qquad (6-10)$$

Thus the process of synchronous detection has removed one of the noise terms by virtue of its phase-sensitive property. Comparing Eqs. (6-10) with (6-9), we note that the output $e_d(t)$ of the DSB receiver has 3 db better signal-to-noise power ratio than the input $e_r(t)$. In other words, the process of synchronous or coherent detection yields a 3-db gain from pre- to postdectector. This situation may be expressed mathematically relating the signal-to-noise output $(S/N)_o$ to the signal-to-noise input $(S/N)_i$ by the expression

$$\left(\frac{S}{N}\right)_{o \, (DSB)} = 2 \left(\frac{S}{N}\right)_{i \, (DSB)} \qquad (6-11)$$

The detection gain in DSB receiver has been explained from the point of view of the quadrature representation of the noise voltage in the predetector portion of the receiver. This effect can also be explained by considering that the sideband voltages of the DSB signal add coherently at the postdetector while the noise voltages add incoherently at that point. Consequently, the noise components from the upper- and lower-sideband frequency range add as power, while the upper- and lower-sideband components of signal add in phase, to give four times the signal power recovered from one sideband alone.

Another very important observation can be made from Eq. (6-10), namely, that there exists no threshold effect if product detection is used in DSB and AM receivers. The output signal-to-noise is, of course, proportional to the input signal-to-noise, but the desired

signal is always there, even under adverse receiving conditions.
This signal preservation property of synchronous detection may
turn out to be of great value in future communication systems where
signal processing requires that the desired signal be present at all
times.

3. *Phase and Frequency Synchronization of Local Oscillator.*
The effect of lack of phase synchronization on the detected signal
can be seen from Eq. (6-3) by putting $\omega_c = \omega_0$ and $\phi_c \neq \phi_0$. The detected output after filtering is then given by

$$e_d(t) = \frac{A_c A_0 g(t)}{2} \cos (\phi_c - \phi_0) = kg(t) \cos \psi \qquad (6\text{-}12)$$

where $\psi = \phi_c - \phi_0$ represents the phase difference between the
transmitted carrier and the locally generated oscillator. Thus the
message signal amplitude is maximum when $\psi = 0$, that is, when the
local carrier is in phase with the original modulated incoming carrier. If $\psi = \pi/2$, the recovered components from the upper and
lower sidebands cancel, and no signal is received.

A method of deriving the control signals in a DSB receiver for
eliminating both phase and frequency errors in the receiver local
oscillator is shown in Fig. 6-2. In this system, the local oscillator
signal is split into two quadrature components; these two quadrature
signals feed separate product detectors as shown. The filtered outputs of these two product detectors are then in turn multiplied together to produce an output signal which is proportional to local
oscillator phase error. If the local oscillator is properly synchronized to the phase of the incoming signal, the upper low-pass

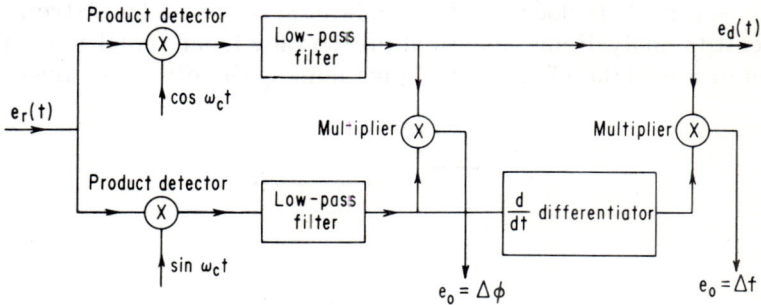

Fig. 6-2. DSB receiver with phase and frequency error-signal
deviation.

filter in Fig. 6-2 will contain the desired modulation voltage g(t),
while the lower low-pass filter output will be zero, due to the quadrature relationship of the corresponding local oscillator signal and
the incoming DSB signal. Under these conditions, multiplication of
the two low-pass filter outputs will yield no control signal. If we
now assume a small error in the phase of the local oscillator signal, the output voltage from the upper low-pass filter will be reduced somewhat in amplitude, but otherwise no change in this voltage will take place. The output of the lower low-pass filter will now
show some signal voltage g(t), and this voltage will be either in
phase with the signal voltage from the upper filter or in exact
phase opposition to the upper-filter output voltage, depending upon
the sign of the phase error. Thus a d-c voltage will be produced at
the output of the first multiplier following the low-pass filters,
whose polarity will depend upon the polarity of the phase error, and
whose magnitude will depend, at least for small phase errors, upon
the magnitude of this phase error. This control voltage may be used
to adjust the local oscillator signal and thereby to remove the phase
error. A phase-control loop of the type just described will not only
control local oscillator phase but, over a narrow range, will also
control oscillator frequency. If one of the low-pass filter output
voltages is differentiated and multiplied by the remaining low-pass
filter output voltage, a d-c voltage will result which is proportional
to frequency error. Thus, both phase- and frequency-control voltages may be obtained from a DSB receiver and may be used appropriately to remove the corresponding errors.

6.2   AM Reception[1,4,5]

A very common AM detector is shown in Fig. 6-3. This very
common circuit is deceptively simple in appearance but extremely
difficult to analyze mathematically, when the input signal is composed of a modulated AM signal plus noise. The difficulty arises

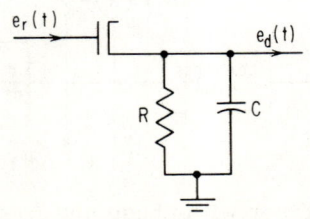

Fig. 6-3. Envelope detector.

from the nonlinear device which forms part of the circuit, and from
the storage property of the diode load circuit. The mathematical
difficulties can be avoided by assuming that the combination of nonlinear device and RC load provides an output of $e_d(t)$ which is equal
to the envelope of $e_r(t)$, the input voltage. The behavior of the envelope detector circuit will presently be examined for very good
and very poor input signal-to-noise ratios.

1. *Signal-to-Noise Improvement in AM Detection.* If predetector
filtering is provided, the voltage input to the envelope detector may
be represented by Eq. (6-13), where the noise is expressed in terms
of inphase and quadrature components as was done previously. If a
good signal-to-noise ratio is assumed, the phasor diagram of Fig.
6-4 may be used.

$$e_r(t) = A_c [1 + g(t)] \cos \omega_c t + x_c(t) \cos \omega_c t - x_s(t) \sin \omega_c t$$

$$(6-13)$$

This diagram is drawn for the angular frequency ω_c of the AM carrier signal and shows the phase relationships between the AM wave
and the two components of the noise wave. It should be kept in
mind that the two noise-modulation components $x_c(t)$ and $x_s(t)$
are independent low-frequency gaussian random variables. Since
a good signal-to-noise ratio has been assumed, the rms values
of $x_c(t)$ and $x_s(t)$ will at all times be small compared to A_c,
and their amplitudes most of the time. If we further assume
that the modulating voltage $g(t)$ is small compared to unity most of
the time, we may express the output voltage of the envelope detector by considering only the projection of the resultant vector along
the axis of the carrier signal. The output voltage from the envelope

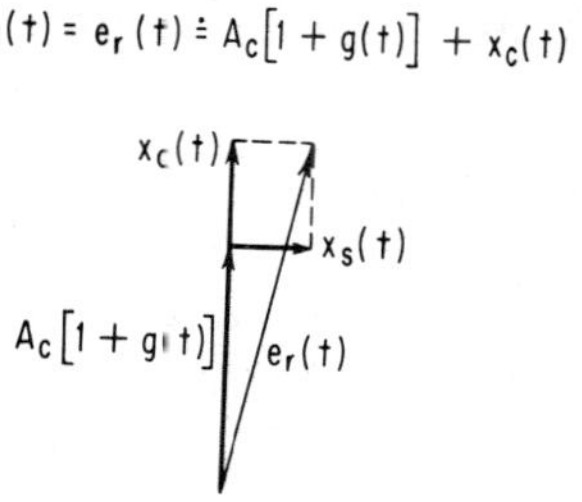

$$e_d(t) = e_r(t) \doteq A_c[1 + g(t)] + x_c(t)$$

Fig. 6-4. AM detector phasor diagram for high $\left(\dfrac{S}{N}\right)_i$.

detector will then be given approximately by Eq. (6-14). Here again we see the elimination of quadrature noise effects.

$$e_d(t) = e_r(t) \doteq A_c [1 + g(t)] + x_c(t) \tag{6-14}$$

The same result can be derived algebraically for a single modulating frequency ω_m as follows. From Eq. (6-13), we have

$$e_r(t) = A_c (1 + m_a \cos \omega_m t) \cos \omega_c t + x_c(t) \cos \omega_c t - x_s(t) \sin \omega_c t$$

$$= R(t) \cos [\omega_c t + \psi(t)] \tag{6-15}$$

where
$$R(t) = \sqrt{[A_c (1 + m_a \cos \omega_m t) + x_c(t)]^2 + x_s^2(t)} \tag{6-16}$$

and
$$\tan \psi(t) = \frac{x_s(t)}{A_c (1 + m_a \cos \omega_m t) + x_c(t)} \tag{6-17}$$

The response of the envelope detector is $R(t)$. In the case in which noise is small relative to the carrier amplitude, a power-series expansion for $R(t)$ yields

$$R(t) = A_c (1 + m_a \cos \omega_m t) \left[1 + \frac{2 x_c(t)}{A_c (1 + m_a \cos \omega_m t)} \right.$$

$$\left. + \frac{x_c^2(t) + x_s^2(t)}{A_c^2 (1 + m_a \cos \omega_m t)^2} \right]^{\frac{1}{2}} \doteq A_c (1 + m_a \cos \omega_m t) + x_c(t)$$

$$\tag{6-18}$$

the same result as given by Eq. (6-14). It is now quite simple to calculate pre- and postdetector average power signal-to-noise ratios and relate them as shown in Eq. (6-19).

$$\left(\frac{S}{N}\right)_{o\,(AM)} = 2 \left(\frac{S}{N}\right)_{i\,(AM)} \tag{6-19}$$

This equation shows that the AM detector, under good signal-to-noise ratio conditions, is as efficient as the synchronous or coherent detector discussed previously. Thus if the transmission medium is stable and if the signal-to-noise ratio at the detector input is good, envelope detection of the AM signal is as good as synchronous detection. As a matter of fact, under the conditions just assumed, the

envelope detector is essentially a product detector. If the carrier-
to-noise ratio is very good, the envelope detector gates or samples
the input voltage at times determined by the carrier signal itself.
This action is caused by the d-c voltage stored in the RC diode load,
which tends to give to the diode a switching action controlled by the
positive peaks of the input carrier signal. Such gating or sampling
action can be shown to be equivalent to product detection, provided
appropriate auxiliary filtering is used.

2. *Comparison of System Performance of DSB and AM Systems.*
From Eq. (6-19), we note that the system performance of an AM
system is identical to that of DSB except for the power wasted by
the carrier; that is, if a DSB and an AM system have equal power
in the sidebands, then the output signal-to-noise ratios are equal.

If a DSB and an AM system have the same average total transmitted powers, then the output signal-to-noise ratios differ by a
factor of 3; that is:

$$\frac{(S/N)_o \text{ (DSB)}}{(S/N)_o \text{ (AM)}} = 3 \qquad (6\text{-}20)$$

The factor 3 arises from the fact that for equal total powers, the
AM system has only one-third the signal power of the DSB system.

As stated previously, in many communication systems, it is the
peak power rather than the average power which is the design limitation. Consequently, we shall presently derive expressions for the
output signal-to-noise $(S/N)_o$ for the case of DSB and AM in terms
of peak power of the system. Let P_{s1} denote the signal power in one
sideband and P_{n1} the noise power. The average total signal sideband
power $P_s = 2P_{s1}$ on the assumption that the signal power is equal
in the two sidebands. Similarly, it is assumed that the noise power
is uniformly distributed over the two sidebands so that $P_{n1} = P_{n2}$
$= P_n$. As discussed previously, the average noise components from
the upper and lower sidebands add on a power basis in the detection
process, while the upper- and lower-sideband components of the
signal add in phase to give four times the signal power recovered
from one sideband alone. Hence for DSB and AM, we obtain at the
output of the detector

$$\left(\frac{S}{N}\right)_o = \frac{4 P_{s1}}{2P_n} = \frac{P_s}{P_n} \qquad (6\text{-}21)$$

Let the peak power of the system be denoted by P_p. For a double-

sideband suppressed-carrier system $P_p = 2P_s$, because the upper
and lower sidebands sometimes add nearly in phase to give four
times the power of one sideband. Hence, from Eq. (6-21) we obtain

$$\left(\frac{S}{N}\right)_{o\,(DSB)} = \frac{P_s}{P_n} = \frac{P_p}{2P_n} \qquad (6\text{-}22)$$

Consider now the case of AM. Let A_c denote the peak amplitude of
the unmodulated transmitter carrier; therefore, for 100 per cent
amplitude modulation, the peak envelope of the modulated carrier
equals $2A_c$, and the peak average power $P_p = 2A_c^2$, the average being
taken over a carrier cycle at the peak of the modulating signal. Now
the average transmitted power taken over the cycle of the modulating signal is

$$P_{av} = \frac{A_c^2}{2} + \frac{A_c^2}{8} + \frac{A_c^2}{8} = \frac{3A_c^2}{4} \qquad (6\text{-}23)$$

Since the peak average power $P_p = 2A_c^2$, therefore

$$P_p = \frac{8}{3} P_{av} = 8P_s \qquad (6\text{-}24)$$

where $P_s = A_c^2/4$ is the total signal power or $P_s = P_{av}/3$. Using
Eq. (6-21), we obtain

$$\left(\frac{S}{N}\right)_{o\,(AM)} = \frac{P_s}{P_n} = \frac{P_p}{8P_n} \qquad (6\text{-}25)$$

We conclude therefore that from the point of view of peak power
limitation DSB has a 4:1 advantage over AM.

3. *Signal-to-Noise Ratio in AM for Poor-input Carrier-to-Noise.*
We have shown above that for good-input carrier-to-noise ratios,
the envelope AM detector is as efficient as the product detector.
In case of poor carrier-to-noise ratio, we shall use a different expression for the noise voltage $n(t)$. In the previous example, $n(t)$
was represented in terms of its inphase and quadrature components;
in the present case, it is more convenient to represent $n(t)$ in
terms of the slowly varying envelope $V(t)$ and slowly varying phase
$\phi(t)$, as in Eq. (4-229). We recall here that $V(t)$ and $\phi(t)$ are independent random variables, $V(t)$ being Rayleigh-distributed and $\phi(t)$
uniformly distributed from 0 to 2π. Thus

$$n(t) = V(t) \cos [\omega_c t + \phi(t)] \qquad\qquad (6-26)$$

We now draw a phasor diagram for the AM detector input voltage
r(t), but this time we let the noise voltage of Eq. (6-26) be the reference for the diagram. This results in Fig. 6-5. To the vector V(t)
we have added the AM signal $A_c [1 + g(t)]$ with the angle ϕ between
them. The resultant vector has an amplitude $e_R(t)$ as shown.

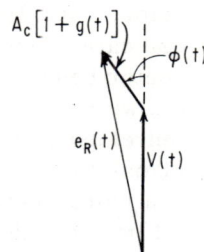

Fig. 6-5. AM detector phasor diagram for poor $\left(\dfrac{S}{N}\right)_i$.

Now, since V(t) is a random variable, there will be times when
V(t) is not large compared to $A_c [1 + g(t)]$. Here again, we assume
that the signal-to-noise ratio is low, so that most of the time V(t)
will be much greater than the quantity $A_c [1 + g(t)]$. Under these
conditions, the envelope voltage $e_R(t)$ may be obtained by a summation of the vector projections along the axis of V(t), and the quadrature components with respect to this axis can be ignored. When
this is done, the output voltage $e_d(t)$ from the envelope detector will
be given by Eq. (6-27).

$$e_d(t) = e_R(t) \doteq V(t) + A_c [1 + g(t)] \cos \phi(t)$$

$$= V(t) + A_c \cos \phi(t) + A_c g(t) \cos \phi(t) \qquad (6-27)$$

This equation shows that the modulating voltage g(t) no longer appears as a part of the detector output voltage but that g(t) is modulated or multiplied by the cosine of the noise angle ϕ. Since $\phi(t)$ is
a uniformly distributed random variable over the interval 0 to 2π,
we can see from Eq. (6-27) that the detector output does not in fact
contain the intelligence information g(t) at all. This signal mutila-
tion encountered under poor carrier-to-noise ratio conditions in
envelope detection accounts for the so-called threshold effect in
AM. It should be clearly understood, however, that this threshold

effect is a distinct property of the envelope detector, and no AM threshold will be noted if synchronous or coherent detection is used.

Both the envelope detector and the linear rectifier distort the recovered signal when overmodulation occurs. Such distortion is avoided in a product demodulator, which recovers the signal by multiplication of the AM wave by the unmodulated carrier. We shall summarize AM detection as follows: Under good signal-to-noise operation, the conventional envelope detector behaves essentially as a product or coherent detector, and a linear relationship between output and input signal-to-noise ratio results. For poor signal-to-noise ratios, the envelope detector behaves very badly, and under very poor reception conditions, we may lose the intelligence signal completely in the process of envelope detection. The intermediate case between good and poor signal-to-noise ratio will be discussed in a subsequent section where a more thorough mathematical treatment of envelope detection will be presented.

6.3 Single-sideband Reception[1,3]

The simplest method of SSB reception is the so-called filtering method shown schematically in Fig. 6-6. The received signal $e_r(t)$ is first passed through a predetector bandpass filter whose bandwidth is equal to the modulation bandwidth. The output of this bandpass filter is multiplied in a product detector by the local oscillator signal, and the output voltage $e_d(t)$ is obtained directly from this product detector. Since the action of the product detector may be considered to be one of frequency translation, that is, all the signal and noise components are translated from the radio frequency to baseband frequency, therefore the relationship between the input and output signal-to-noise ratios for the SSB receiver is given by

$$\left(\frac{S}{N}\right)_{o\,(SSB)} = \left(\frac{S}{N}\right)_{i\,(SSB)} \tag{6-28}$$

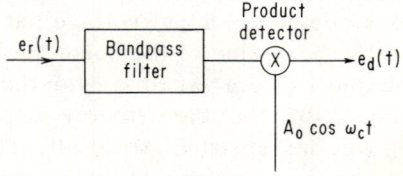

Fig. 6-6. Block diagram of SSB receiver.

When comparing this result with that obtained with DSB or good
signal-to-noise ratio AM, we note that although in DSB and AM we
have $(S/N)_o = 2 (S/N)_i$, the noise power admitted into the system is
double that of a SSB, and, consequently, the ratio $(S/N)_i$ in a SSB is
twice as good as in the other systems, and therefore $(S/N)_o$ (SSB)
$= (S/N)_o$ (AM or DSB).

The same result can be arrived at from the point of view of coherence of the sidebands. Although the noise bandwidth in SSB is
one-half that for DSB, the signal-to-noise ratios are the same, because overlapping of the sideband signals no longer occurs, and the
advantage of coherent addition of the sidebands as in the case of
DSB is lost. Thus the translated signal spectrum has the same
power as the received spectrum, in the same manner that the translated output noise is equal to the input noise.

We have just noted that the SSB receiver translates the received
SSB modulated carrier back down to baseband. It is thus obvious
that a frequency difference between the transmitter and receiver
local oscillators will result as a frequency change in the modulating
signal. The question naturally arises as to the frequency tolerance
of the local oscillator used in the SSB receiver. The degree of stability will depend on the type of signal transmitted. In case of voice
transmission, a maximum frequency error of 20 cps may be tolerated in practice, and independent stable oscillators are used in the
modulator and demodulator. In transmission systems where the
waveform of the modulating signal is to be preserved, not only must
the local oscillator in the SSB receiver have the proper frequency,
but it must also have the proper phase. A pilot carrier or pilot tone
is normally used in such cases to provide the proper frequency and
phase control.

6.4   Comparison of Linear Modulation Systems [1,3,4]

On the basis of our previous discussions and system performance, we may summarize our findings as follows:

1. A DSB system shows a 6-db power advantage over an AM system, regardless of the modulating waveform; that is, for equal
sideband power, 6 db of additional peak power capability is required
for AM as compared to DSB.

2. Power comparisons of AM and DSB with SSB are impossible
to make in general, since the waveform of the modulating signal is
an all-important factor. With sinusoidal modulation, SSB is superior
to both DSB and AM. However, with square-wave modulation, DSB
and AM can both be superior to SSB.

3. For a given peak transmitter power limitation, a DSB system is a 6-db $(S/N)_o$ advantage over an AM system.

4. The use of product detection in SSB and DSB receivers as compared to envelope detection in AM receivers is of no advantage if the medium is stable and if the input signal-to-noise ratio is good. Where product detection begins to contribute to the operational superiority of SSB and DSB systems over AM systems is at marginal or poor signal-to-noise conditions. Even under such conditions, the use of product detection results in a linear relationship between the output signal-to-noise ratio and the input signal-to-noise ratio, and an intelligence signal which may be enhanced by further processing. This is to be contrasted with the threshold effect in envelope AM detection which can completely lose the intelligence signal when the input signal-to-noise ratio becomes unfavorable.

The performance of the linear modulation system in case of selective fading in the transmission medium is also of great interest. Selective fading is quite often encountered in high-frequency links, for example. Such selective fading is roughly equivalent to the insertion of a filter in the transmission path, having attenuation peaks and nulls which keep changing in frequency location and in height or depth. In AM reception, such selective fading can be extremely damaging. For example, the carrier component of the AM signal may be removed by such selective fading, and only the sidebands may at times be available for envelope detection. This causes a very serious nonlinear distortion in output waveform and renders, quite often, the output signal completely unintelligible, even though the signal-to-noise ratio may be excellent. In contrast, such selective fading does not produce this type of distortion or degradation when product detection is employed as in SSB or DSB. Some intelligence frequencies are attenuated and others are amplified by the selective fading, but no new distortion products are introduced as in the case of AM. For voice transmission, this filtering effect in SSB and DSB is not normally serious with regard to loss of intelligibility, whereas the loss of carrier in the AM case is extremely serious because of the envelope detection process which is employed. The trouble with AM in a perturbed medium is normally not lack of power or signal-to-noise ratio; rather it is a distortion which renders the output signal unintelligible, even though it may be heard loud and clear. Thus the vast improvement in high-frequency performance which can be achieved by using SSB or DSB systems as compared to AM systems has really very little to do with the transmission mode employed. The great difference in performance which is noted is almost entirely explained by the use of product detection in SSB and DSB, as compared to envelope detection in AM.

6.5 Signal-to-Noise Ratio in Linear Modulation Systems[2-5]

This problem was discussed previously, using a simplified approach, and formulas were derived for the output signal-to-noise ratios $(S/N)_o$, for SSB, DSB, and AM. In this section, a more extensive analysis will be made which will also be applicable later on to exponential modulation systems. We shall consider first product demodulation for a general linear modulation signal, and then envelope detection for an AM signal.

1. *Product Demodulation.* First we consider double-sideband suppressed carrier. Let the received modulated carrier be given by the expression

$$e_{DSB}(t) = A_c g(t) \cos (\omega_c t + \phi_c) \tag{6-29}$$

which represents double-sideband suppressed-carrier (DSB) linear modulation, where $g(t)$ is the message function, A_c the carrier amplitude, and ω_c the carrier angular frequency. The incoming modulated carrier is multiplied by the locally generated oscillator

$$e_o(t) = A_o \cos (\omega_c t + \phi_o) \tag{6-30}$$

The IF amplifier preceding the detector is band-limited to $(f_c \pm f_m)$, where f_m is the highest-frequency component of $g(t)$, and the product demodulator is followed by a low-pass filter f_m cycles/sec wide. The detected signal output of the demodulator is

$$e_d(t) = k_d A_o A_c g(t) \cos (\phi_c - \phi_o) \tag{6-31}$$

where k_d is a constant of the demodulator. The mean square value of the signal component is

$$\overline{e_d^2(t)} = k_d^2 A_o^2 A_c^2 \overline{g^2(t)} \cos^2 (\phi_c - \phi_o) \tag{6-32}$$

where $\overline{g^2(t)} = \frac{1}{T} \int_o^T g^2(t)\, dt$, T being the basic period of $g(t)$

Now consider multiplication of the local oscillator by the noise components. At the detector input, the noise has a one-sided power

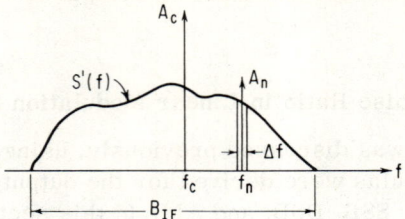

Fig. 6-7. Noise spectrum of IF.

spectral density of $S'(f)$ watts/cps. In order to calculate the mean square value of the output noise, we approximate the continuous noise power spectrum with an equivalent discrete power spectrum as shown in Fig. 6-7. The noise appearing at the mean frequency f_n in a slot which is $(f_n - f_c)$ cycles away from the center of IF and occupying a width of Δf cycles/sec can be approximated by a sine wave of the same average power and the same frequency. Thus the equivalent sine wave is given by

$$n(t) = A_n \cos (\omega_n t + \phi_n) \qquad (6-33)$$

where $A_n = \sqrt{2S'(f_n)\,\Delta f}$, $\omega_n = 2\pi f_n = \omega_c + 2\pi n\,\Delta f$, and $f_n - f_c = n\,\Delta f$.

The phase angle ϕ_n is a random variable that is uniformly distributed over $0 \le \phi \le 2\pi$.

The total noise within the IF bandwidth B_{IF} can be expressed by

$$e_n(t) = \sum_{n\,\Delta f = -B_{IF}/2}^{B_{IF}/2} A_n \cos (\omega_n t + \phi_n) \qquad (6-34)$$

and the corresponding noise component at the output of the product detector is

$$e_{n(d)}(t) = k_d A_o \sum_{n\,\Delta f = -B_{IF}/2}^{B_{IF}/2} A_n \cos (2\pi n\,\Delta ft + \phi_n - \phi_o) \qquad (6-35)$$

Because of the assumed random nature of the ϕ_n's and their statistical independence, the noise components add incoherently. That is to say, the mean square value of the sum equals the sum of the

mean square values of the components. The mean square value of
the output noise is, therefore,

$$\overline{e_n^2(t)} = k_d^2 A_o^2 \int_{-f_m}^{f_m} S'(f_c + f) \, df \tag{6-36}$$

where $B_{IF} = 2f_m$, f_m denoting the highest modulation frequency and
the output signal-to-noise power ratio is given by $\overline{e_d^2(t)}/\overline{e_n^2(t)}$, or

$$\left(\frac{S}{N}\right)_o = \frac{\overline{g^2(t)} \, A_c^2 \cos^2 (\phi_c - \phi_o)}{\int_{-f_m}^{f_m} S'(f_c + f) \, df} \tag{6-37}$$

Assuming the noise to be white and of mean square density of N_0
watts/cps, Eq. (6-37) reduces to

$$\left(\frac{S}{N}\right)_o = 2\overline{g^2(t)} \left(\frac{A_c^2/2}{2N_0 f_m}\right) \cos^2 (\phi_c - \phi_o) \tag{6-38}$$

If we assume also that the IF passband is limited to $f_c \pm f_m$ and
having the characteristics of an ideal rectangular filter, then by
the use of Eq. (6-29) the input signal-to-noise to the demodulator
$(S/N)_i$ is given by

$$\left(\frac{S}{N}\right)_i = \frac{A_c^2/2}{2N_0 f_m} \cdot \overline{g^2(t)} \tag{6-39}$$

and $$\frac{(S/N)_o}{(S/N)_i} = 2 \cos^2 (\phi_c - \phi_o) \tag{6-40}$$

which represents a maximum improvement of 3 db when the local
oscillator is in phase with the incoming carrier; this result is in
agreement with Eq. (6-11), which was derived from elementary considerations.

Next we consider product demodulation of an AM signal given by

$$e_{AM}(t) = A_c [1 + m_a g(t)] \cos (\omega_c t + \phi_c) \tag{6-41}$$

where $m_a = 1$ for 100 per cent modulation, and $|g(t)| \le 1$.

First we find the input carrier power. The mean square value
of $e_{AM}(t)$ is given by

$$\overline{e_{AM}^2(t)} = \frac{A_c^2}{2} + A_c^2 m_a \overline{g(t)} + \frac{A_c^2}{2} m_a^2 \overline{g^2(t)} \tag{6-42}$$

$$= \frac{A_c^2}{2} \left[1 + m_a^2 \overline{g^2(t)} \right] \quad , \text{ if } \overline{g(t)} = 0 \tag{6-43}$$

Thus if $g(t)$ has no d-c component, the carrier-to-noise ratio at the input to the demodulator is

$$\left(\frac{S}{N}\right)_i = \frac{A_c^2/2}{2N_o f_m} \left[1 + m_a^2 \overline{g^2(t)} \right] \tag{6-44}$$

Now we shall find the output signal power. As before, the received modulated carrier is multiplied by the locally generated oscillator signal, and the demodulated output is

$$e_d(t) = k_d A_o A_c \cos(\phi_c - \phi_o) + k_d A_o A_c m_a g(t) \cos(\phi_c - \phi_o) \tag{6-45}$$

The d-c component is removed at the output of the detector, and the output signal-to-noise power ratio is

$$\left(\frac{S}{N}\right)_o = 2 \frac{A_c^2/2}{2N_o f_m} \overline{g^2(t)} \, m_a^2 \cos^2(\phi_c - \phi_o) \tag{6-46}$$

Therefore for product demodulation of AM, we have

$$\frac{(S/N)_o}{(S/N)_i} = \frac{2 \overline{g^2(t)} m_a^2 \cos^2(\phi_c - \phi_o)}{1 + m_a^2 \overline{g^2(t)}} \tag{6-47}$$

which is maximum for $m_a = 1$ and $\phi_c = \phi_o$.

In a SSB receiver with product demodulation, we have

$$\frac{(S/N)_o}{(S/N)_i} = \cos^2(\phi_c - \phi_o) \tag{6-48}$$

which is equal to unity for $\phi_c = \phi_o$. It is important to note here that in product demodulation, we do not experience threshold effect as already discussed previously.

2. *Envelope Detection of AM.* In this analysis, we shall limit our discussion to the extreme cases in which the carrier is either much stronger than the noise or much weaker. The intermediary cases require elaborate mathematical analysis and will be discussed separately in the next section.

Consider first the case of a carrier much stronger than the noise; the output signal-to-noise power ratio is first calculated by ignoring the contribution of noise. Thus from Eq. (6-41) we obtain

$$(S)_o = k_d^2 m_a^2 A_c^2 \overline{g^2(t)} \tag{6-49}$$

The contribution of the noise is calculated next, ignoring the message modulation; this is valid for this case of a strong carrier. The effect of the noise component contained in slot Δf centered at $f_n = f_c + n \Delta f$ is to add directly to the carrier to give the sum

$$A_c \cos (\omega_c t + \phi_c) + A_n \cos (\omega_n t + \phi_n) \tag{6-50}$$

where $A_n = \sqrt{2S'(f_n) \Delta f} \ll A_c$.

The resultant envelope $R(t)$ can be obtained from Fig. 6-8, namely,

$$R(t) = A_c \sqrt{1 + 2\frac{A_n}{A_c} \cos [(\omega_n - \omega_c)t + \phi_n - \phi_c] + \frac{A_n^2}{A_c}}$$

$$\doteq A_c \left\{ 1 + \frac{A_n}{A_c} \cos [(\omega_n - \omega_c)t + \phi_n - \phi_c] \right\}, \qquad A_n \ll A_c \tag{6-51}$$

and the phase angle $\theta(t)$ is given by

$$\tan \theta(t) = \frac{A_n \sin [(\omega_n - \omega_c)t + \phi_n - \phi_c]}{A_c + A_n \cos [(\omega_n - \omega_c)t + \phi_n - \phi_c]} \tag{6-52}$$

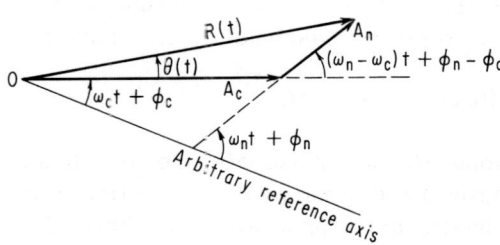

Fig. 6-8. Phasor diagram of unmodulated carrier and small noise component.

The sum of one noise component at frequency f_n and the carrier at frequency f_c gives rise to a hybrid modulation. The phase term $\theta(t)$ will be discussed in Chap. 14 in the analysis of FM noise. At the present, we are interested in the amplitude term which will be detected by the envelope detector. In the absence of the other noise components, the noise element considered will yield an output signal whose mean square value is given by $k_d^2 A_n^2/2$, provided

$$|\omega_n - \omega_c| \leq 2\pi f_m$$

Since the various component noise contributions combine incoherently because of their independent random phasing, the mean square value of the total noise output becomes

$$(N)_o = k_d^2 \int_{-f_m}^{f_m} S'(f_c + f) \, df \tag{6-53}$$

and the output signal-to-noise mean power ratio is given by

$$\left(\frac{S}{N}\right)_o = \frac{m_a^2 A_c^2 \overline{g^2(t)}}{\int_{-f_m}^{f_m} S'(f_c + f) \, df} \tag{6-54}$$

For white noise, we have $S'(f) = N_0$ watts/cps, and

$$\left(\frac{S}{N}\right)_o = 2 \frac{A_c^2/2}{2 N_0 f_m} \, m_a^2 \overline{g^2(t)} \tag{6-55}$$

which is identical to the result of Eq. (6-46) for AM product demodulation with $\phi_c = \phi_0$. We conclude therefore that for high-input signal-to-noise ratio, the envelope AM detector is as efficient as the product AM detector.

The same result can be derived by representing the total noise present within the receiver passband by the sinusoid

$$e_n(t) = V(t) \cos [\omega_c t + \phi(t)] \tag{6-56}$$

where the envelope $V(t)$ and phase $\phi(t)$ are time functions that vary slowly for a narrow-band noise process relative to $\cos \omega_c t$. For a narrow-band gaussian noise process, as we have shown in Chap. 4, $V(t)$ is Rayleigh-distributed, and $\phi(t)$ is uniformly distributed over $0 \leq \phi \leq 2\pi$. The mean square value of the noise within the band-limited IF is given by

$$(N)_i = \frac{\overline{V^2(t)}}{2} = \int_{-f_m}^{f_m} S'(f_c + f) \, df = \sigma^2 \qquad (6-57)$$

where σ^2 is the variance of the noise process.

This presentation, which has been more fully discussed in Chap. 4, will also be used in the analysis of poor carrier-to-noise ratio, as we shall see later on.

Using a vectorial presentation similar to Fig. 6-8, we can readily show that the envelope $R(t)$ of the resultant sinusoid of the total noise and the AM signal combined is given by

$$R(t) = \sqrt{V^2(t) + 2V(t)A_c [1 + m_a g(t)] \cos [\phi(t) - \phi_c] + A_c^2 [1 + m_a g(t)]^2}$$

$$(6-58)$$

Since in the present case of high carrier-to-noise ratio, it is assumed that $V(t) \ll A_c [1 + mg(t)]$ most of the time, the envelope described by Eq. (6-58) is closely approximated by

$$R(t) \doteq A_c [1 + m_a g(t)] + V(t) \cos [\phi(t) - \phi_c] \qquad (6-59)$$

The d-c term in this expression is removed by the low-pass filter following the envelope detector, and consequently the mean square value of the detected signal is

$$(S)_o = k_d^2 m_a^2 A_c^2 \overline{g^2(t)} \qquad (6-60)$$

and the mean square value of the total noise output is

$$(N)_o = k_d^2 \frac{\overline{V^2(t)}}{2} \qquad (6-61)$$

The output signal-to-noise mean power ratio is therefore given by

$$\left(\frac{S}{N}\right)_o = \frac{m_a^2 A_c^2 \overline{g^2(t)}}{\int_{-f_m}^{f_m} S'(f_c + f) \, df} \qquad (6-62)$$

as in Eq. (6-54).

Next we consider the case of poor input signal-to-noise ratio; namely, the noise is much stronger than the modulated carrier and $V(t) \gg A_c [1 + mg(t)]$ most of the time. In this case, the envelope of signal plus noise in Eq. (6-58) is closely approximated by

$$V(t) + A_c [1 + m_a g(t)] \cos [\phi(t) - \phi_c] \qquad (6\text{-}63)$$

Closer examination of Eq. (6-63) reveals a very interesting result; namely, the message function $g(t)$ is multiplied by a random variable $\cos [\phi(t) - \phi_c]$, and consequently it may be lost in the noise. This results in a threshold effect which exists only in envelope detection and does not exist when a product demodulator is used in the detection of the AM signal.

6.6 Signal-to-Noise Ratios in Envelope Detector (Linear-amplitude Detector)[4-10]

In the last section, the signal-to-noise ratio at the output of the linear envelope detector was determined for the extreme cases of very high and very low input signal-to-noise ratio. In this section, we shall present an approach for finding the output signal-to-noise ratio which is applicable to the general case of a given input signal-to-noise ratio, including the two extreme cases dealt with previously. It will also be shown that the results derived previously can be obtained from the general result as a limiting process. To simplify the problem, we consider the input to the detector to consist of an unmodulated carrier plus narrow-band gaussian noise. The problem is then to determine the output of the detector when the input is either pure noise or noise plus carrier. The envelope detector is characterized by the following relation between the output amplitude I due to an input signal V,

$$\begin{aligned} I &= 0, & V < 0 \\ &= kV, & V > 0 \end{aligned} \qquad (6\text{-}64)$$

where k is constant. At the output of the detector, the carrier component is removed by a low-pass filter, and the output will then consist of a d-c signal which represents the envelope of the unmodulated carrier and superimposed on it a random low-frequency signal due to the noise,

$$I = I_{DC} + I_{AC} = \frac{kR(t)}{\pi} \qquad (6\text{-}65)$$

where $R(t)$ is the instantaneous envelope amplitude of the modulated carrier by the noise. The factor $1/\pi$ results from the fact that the ratio of mean power in the positive half-cycle of the carrier to the mean power under the modulation envelope over a full carrier cycle is $1/\pi^2$.

1. *D-C Output.* From Eq. (6-65), it follows that for a carrier of amplitude A_C in the absence of noise, the output signal I is given by

$$I = I_{DC} = \frac{kA_C}{\pi} \tag{6-66}$$

To find the d-c component I_{DC} due to carrier plus noise, we need to find the average value of I in Eq. (6-65). We start with Eq. (4-262), which represents the probability density of the amplitude R(t) of the noise-modulated carrier at the input of the envelope detector,

$$q(R) = \frac{R}{\sigma^2} e^{-(R^2/2\sigma^2 + z)} \cdot I_0 \left(\frac{R}{\sigma} \sqrt{2z}\right) \tag{6-67}$$

where σ^2 is the variance of the noise process, and z equals the detector input carrier-to-noise, namely,

$$z = \left(\frac{C}{N}\right)_i = \frac{A_C^2}{2\sigma^2} \tag{6-68}$$

The average or mean value of the total output I due to carrier and noise is simply

$$I_{DC} = \frac{k}{\pi} \overline{R} = \frac{k}{\pi} \int_0^\infty R q(R)\, dR \tag{6-69}$$

which is the first moment of the distribution function. Since in the calculation of the detector $(S/N)_0$, we shall make use of a number of moments of q(R), let us first find the n'th moment and then derive the first moment from the general case. The n'th moment is given by

$$\overline{R^n} = \int_0^\infty R^n q(R)\, dR = \int_0^\infty \frac{R^{n+1}}{\sigma^2} e^{-(R^2/2\sigma^2 + z)} \cdot I_0 \left(\frac{R}{\sigma} \sqrt{2z}\right) dR \tag{6-70}$$

This formidable integral is available in tabulated form:

$$\overline{R^n} = (2\sigma^2)^{\frac{n}{2}} \left(\frac{n}{2}\right)!\; {}_1F_1\left(-\frac{n}{2},\, 1;\, -z\right) \tag{6-71}$$

where $_1F_1(-n/2, 1; -z)$ is the confluent hypergeometric function defined by

$$_1F_1(a, b; \mu) = 1 + \frac{a}{b}\mu + \frac{a(a+1)\mu^2}{b(b+1)2!} + \frac{a(a+1)(a+2)\mu^3}{b(b+1)(b+2)3!} \qquad (6\text{-}72)$$

In our particular case, $a = -n/2$, $b = 1$, $\mu = -z$, and we obtain

$$_1F_1\left(-\frac{n}{2}, 1; -z\right) = 1 + \frac{n}{2}z - \frac{n/2(1-n/2)}{2!}z^2 + \cdots \qquad (6\text{-}73)$$

We proceed now with finding the average or mean value I_{DC} of the total output due to carrier plus noise. This is obtained from Eq. (6-71) by putting n = 1:

$$I_{DC} = \overline{s_0 + n_0} = \frac{k}{\pi}\overline{R} = \frac{k}{\pi}(2\sigma^2)^{\frac{1}{2}}(\tfrac{1}{2})! \, _1F_1(-\tfrac{1}{2}, 1; -z) \qquad (6\text{-}74)$$

The d-c output due to noise alone in the absence of a carrier is obtained by setting $A_c = 0$ or $z = 0$. Hence

$$\overline{n_0} = \frac{k}{\pi}\sqrt{2\sigma^2}\,(\tfrac{1}{2})! = \frac{k}{\pi}\sqrt{\pi/2}\,\sigma = \frac{k\sigma}{\sqrt{2\pi}} \qquad (6\text{-}75)$$

It is of interest to note here that in the absence of a carrier $R(t) = V(t)$, where the noise envelope $V(t)$ is Rayleigh-distributed; the average value of $V(t)$ can be obtained directly from Eq. (4-85), namely,

$$\overline{R(t)} = \overline{V(t)} = \sqrt{\pi/2}\,\sigma$$

which checks with Eq. (6-75).

The d-c output due to carrier alone is

$$\overline{s_0} = \frac{k\sigma}{\sqrt{2\pi}}\,[_1F_1(-\tfrac{1}{2}, 1; -z) - 1] \qquad (6\text{-}76)$$

For the general case of carrier plus noise, Eq. (6-74) can be shown[8] to expand to

$$I_{DC} = \overline{s_0 + n_0} = \frac{k\sigma}{\sqrt{2\pi}}\,e^{-A_c^2/4\sigma^2}\left[\left(1 + \frac{A_c^2}{2\sigma^2}\right)I_0\left(\frac{A_c^2}{4\sigma^2}\right) + \frac{A_c^2}{2\sigma^2}I_1\left(\frac{A_c^2}{4\sigma^2}\right)\right]$$

$$(6\text{-}77)$$

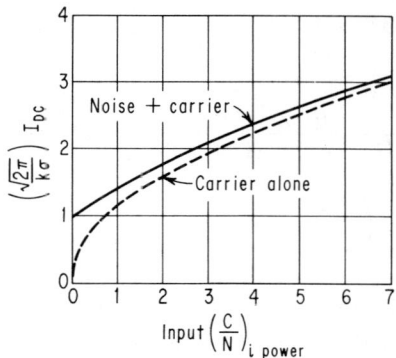

Fig. 6-9. DC output of a linear detector (σ = constant).

where I_1 is the modified Bessel function of the first kind and first order. A plot of I_{DC} (normalized) versus $(C/N)_i$ is shown in Fig. 6-9.

2. *A-C Output: Output Signal-to-Noise Power Ratio* $(S/N)_0$.   The output signal power S_0 is given by

$$(S)_0 = \overline{s_0^2} = \overline{s_0}^{\cdot 2} = \frac{k^2\sigma^2}{2\pi} \, [_1F_1 \, (-\tfrac{1}{2}, \, 1; \, -z) - 1]^2 \qquad (6\text{-}78)$$

The mean square noise voltage is given by the variance; thus

$$(N)_0 = \frac{k^2}{\pi^2} \, \overline{I_{AC}^2} = \frac{k^2}{\pi^2} \, (\overline{I^2} - I_{DC}^2) = \frac{k^2}{\pi^2} \, [\overline{R^2} - \overline{R}^2] = \frac{2k^2\sigma^2}{\pi^2} \, (1 + z)$$

$$- \frac{k^2}{2\pi} \, \sigma^2 \, [_1F_1 \, (-\tfrac{1}{2}, \, 1; \, -z)]^2 \qquad (6\text{-}79)$$

Finally, the output signal-to-noise power ratio is given by

$$\left(\frac{S}{N}\right)_0 = \frac{[_1F_1 \, (-\tfrac{1}{2}, \, 1; \, -z) - 1]^2}{(4/\pi)(1 + z) - [_1F_1 \, (-\tfrac{1}{2}, \, 1; \, -z)]^2} \qquad (6\text{-}80)$$

where $_1F_1 \, (-\tfrac{1}{2}, \, 1; \, -z) \doteq 1 + \tfrac{1}{2} z - \tfrac{1}{16} z^2 + \cdots \qquad (6\text{-}81)$

Having obtained the expression for the output signal-to-noise power ratio for the general case, it will be of interest to derive from it expressions for both limiting cases discussed above. First we consider the case of low-input carrier-to-noise ratio, i.e., $z < 1$. The confluent hypergeometric function may be approximated by

$$_1F_1 \left(-\tfrac{1}{2}, 1; -z\right) \doteq 1 + \frac{z}{2} \tag{6-82}$$

and $\quad \left(\dfrac{S}{N}\right)_O = \dfrac{(\tfrac{1}{2} z)^2}{(4/\pi)(1 + z) - (1 + z/2)^2} \doteq 0.916z^2$

or $\quad \left(\dfrac{S}{N}\right)_O = 0.916 \left(\dfrac{C}{N}\right)_i^2 \tag{6-83}$

Thus, in the case of poor $(C/N)_i$, there is a deterioration in the output signal-to-noise ratio, and consequently a linear envelope detector should not be used in this instance.

For large-input carrier-to-noise ratio, $z \gg 1$, and

$$_1F_1 \left(-\tfrac{1}{2}, 1; -z\right) = \frac{2\sqrt{z}}{\sqrt{\pi}} \left(1 + \frac{1}{4z}\right), \qquad z > 10 \tag{6-84}$$

The output signal-to-noise ratio becomes

$$\left(\frac{S}{N}\right)_O \doteq \frac{(2\sqrt{z/\pi} - 1)^2}{(4/\pi)(1 + z) - (4/\pi)\, z(1 + 1/4z)^2} \tag{6-85}$$

or $\quad \left(\dfrac{S}{N}\right)_O \doteq 2z = 2 \left(\dfrac{C}{N}\right)_i$

which is identical with the result obtained earlier in Eq. (6-55). The effect of the carrier level on the output noise can be derived from Eq. (6-79). For $z = 0$, the output noise power in the absence of a carrier is simply

$$(N)_O = \frac{k^2}{\pi^2} \left(2\sigma^2 - \frac{\pi}{2}\, \sigma^2\right) = \frac{k^2}{\pi^2}\, (0.43\sigma^2) \tag{6-86}$$

For large z, the approximation of Eq. (6-84) may be used, and the result is

$$(N)_O = \frac{k^2}{\pi^2} \left[2\sigma^2 (1 + z) - \frac{\pi}{2}\, \sigma^2 \cdot \frac{4}{\pi}\, z\left(1 + \frac{1}{4z}\right)^2\right]$$

$$= \frac{k^2}{\pi} \left[2\sigma^2 (1 + z) - 2\sigma^2 \left(z + \tfrac{1}{2}\right)\right] \doteq \frac{k^2}{\pi^2}\, \sigma^2 \tag{6-87}$$

Thus in the presence of a strong carrier, the output noise power is increased by a factor of 2.33 over the no-carrier conditions; this phenomenon partly accounts for the increase in noise heard on an ordinary AM receiver when a carrier is tuned in.

As before, we can derive Eq. (6-86) directly from the results derived in Chap. 4 for the Rayleigh distribution. In the absence of a carrier, Eq. (6-79) reduces to

$$(N)_o = \frac{k^2}{\pi^2} (\overline{V^2} - \overline{V}^2) = \frac{k^2}{\pi^2} \left(2\sigma^2 - \frac{\pi}{2} \sigma^2 \right) = \frac{k^2}{\pi^2} (0.43\sigma^2)$$

A plot of Eq. (6-79) is shown in Fig. 6-10, where the a-c power output, namely, $(N)_o$, is plotted as a function of $(C/N)_i$ normalized to k^2/π^2 $(0.43\sigma^2)$. If the carrier is amplitude-modulated, then the a-c signal power output of the linear detector is given by k^2/π^2 $(m_a^2 A_c^2/2)$, which is shown plotted in curve 2 for $m_a = 0.5$.

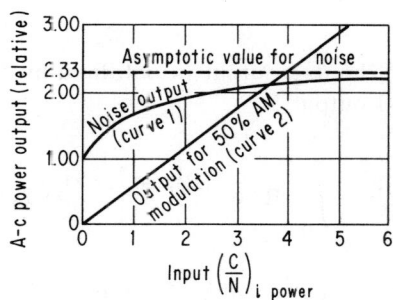

Fig. 6-10. AC power output of a linear detector (σ = constant).

6.7 Square-law Amplitude Detector[8,9]

In this section, we shall calculate the $(S/N)_o$ when a square-law amplitude detector is used. This type of detector is characterized by the following relation of the output amplitude I as a function of the input amplitude V:

$$I = aV^2 + bV + c \tag{6-88}$$

where a, b, and c are constants. As before, we consider the input signal to consist either of pure random noise or noise plus a carrier. Using the results of Sec. 4-8, the input may be considered a modulated carrier, given by

$$V = R(t) \cos [\omega_c t + \theta(t)] \tag{6-89}$$

where $R(t)$ is the envelope which, in case of pure noise, reduces to the random variable $V(t)$ Rayleigh-distributed, and $\theta(t)$ is a random phase. The output of the detector is then

$$I = aR^2 \cos^2 [\omega_c t + \theta(t)] + bR \cos [\omega_c t + \theta(t)] + c = \frac{aR^2}{2}$$

$$+ \frac{aR^2}{2} \cos [2\omega_c t + 2\theta(t)] + bR \cos [\omega_c t + \theta(t)] + c \tag{6-90}$$

The components of the carrier frequency and its second harmonic are removed by a low-pass filter, and the detector output reduces to

$$I = \frac{aR^2}{2} + c \tag{6-91}$$

    1. *D-C Output.*   The d-c output is obtained by finding the average value of the total output I.

$$I_{DC} = \frac{a}{2} \overline{R^2} + c = \frac{a}{2\sigma^2} \int_0^\infty R^3 e^{-(R^2/2\sigma^2 + z)} \cdot I_0 \left(\frac{R}{\sigma} \sqrt{2z} \right) + c$$

$$= a\sigma^2 (1 + z) + c = a\sigma^2 + a(A_c^2/2) + c \tag{6-92}$$

Thus we note that the carrier and noise contribute independently to the output power, each giving a d-c output which is a linear function of its input power. It is this property of the square-law detector which makes it useful as a power-measuring device.

    2. *A-C Output: Output Signal-to-Noise Power Ratio.*   The power $(N)_0$ of the low-frequency output of the detector can be found as before by subtracting the d-c power from the total mean output power, i.e.,

$$(N)_0 = \overline{I^2_{AC}} = \overline{I^2} - I^2_{DC} \tag{6-93}$$

Now $\overline{I^2} = \overline{\left(\frac{aR^2}{2} + c \right)^2} = \frac{a^2 \overline{R^4}}{4} + ac\overline{R^2} + c^2$

which is shown by Goldman[8] to reduce to

$$\overline{I^2} = a^2\left(2\sigma^4 + 2\sigma^2 A_c^2 + \frac{A_c^4}{4}\right) + ac(2\sigma^2 + A_c^2) + c^2 \qquad (6\text{-}94)$$

The a-c output is then

$$(N)_0 = \overline{I_{AC}^2} = \overline{I^2} - I_{DC}^2 = a^2\sigma^4 + \sigma^2 A_c^2 = a^2\sigma^4\left[1 + 2\left(\frac{C}{N}\right)_i\right] \qquad (6\text{-}95)$$

From Eq. (6-95), we conclude that the noise power output of a square-law detector is a linear function of the input $(C/N)_i$ power ratio. This is contrasted with the linear-amplitude detector, where it has been shown in Eq. (6-87) that for large $(C/N)_i$ the output noise stabilizes at a maximum value of σ^2. Similarly, we note that for $C/N_i \rightarrow 0$ the output noise power is proportional to σ^4, compared to σ^2 in the linear case.

References

1. Costas, J. P.: Linear Modulation Systems, Lecture Delivered during the MIT Special Summer Program on Reliable Long-range Radio Communication, 1960.
2. Baghdady, E. J. (ed.): "Lectures on Communication Theory," chap. 19, McGraw-Hill Book Company, New York, 1961.
3. Black, H. S.: "Modulation Theory," D. Van Nostrand Company, Inc., Princeton, N. J.
4. Bennett, W. R.: "Electrical Noise," McGraw-Hill Book Company, New York.
5. Hancock, J. C.: "An Introduction to the Principles of Communication Theory," McGraw-Hill Book Company, New York.
6. Schwartz, M.: "Information Transmission, Modulation, and Noise," McGraw-Hill Book Company, New York.
7. Davenport, W. B., Jr., and W. L. Root: "An Introduction to the Theory of Random Signals and Noise," McGraw-Hill Book Company, New York.
8. Goldman, S.: "Frequency Analysis, Modulation, and Noise," McGraw-Hill Book Company, New York.
9. King, N. H.: An Introduction to the Effect of Some Detectors on Noisy Signal, *AWA Tech. Rev.*, no. 3, 1962.
10. Lawson, J. L., and G. E. Uhlenbeck: "Threshold Signals," MIT Radiation Lab. Series, vol. 24, McGraw-Hill Book Company, New York.

7

EXPONENTIAL MODULATION: BASIC PRINCIPLES AND SPECTRAL DISTRIBUTION

We have investigated in the last two chapters the effect of slowly varying the amplitude of a sinusoidal carrier in accordance with the information to be transmitted. The desired information was found to be concentrated in sidebands about the carrier frequency; and by choosing the carrier frequency high enough, information transmission by means of linear modulation of the carrier was shown to be practicable. We have also noted that many information channels may be transmitted simultaneously, by means of frequency-multiplexing techniques.

Linear modulation is, however, not the only means of modulating a sine-wave carrier; the phase and frequency of a sine wave can also be modulated in accordance with the information-bearing signal. We shall show later, by comparing the information-handling capacities of different systems, that this type of nonlinear modulation provides better discrimination against noise and interfering signals than linear modulation; however, it will be shown that wider bandwidth is needed to obtain this improved response. We shall see that the ability to exchange bandwidth occupancy in the transmission medium for improved noise performance is a general property of information transmission systems, whether of the sine-wave-carrier or pulse-carrier type.

In a nonlinear modulation process, the message time function g(t) is undergoing a nonlinear transformation before multiplication by the carrier. The exponential transformation is the most practical nonlinear transformation because the original message function g(t) can readily be recovered in the demodulation process. Exponential modulation may be defined as an operation whereby the carrier function

$$e_c(t) = A_c e^{j(\omega_c t - \phi_c)} \tag{7-1}$$

is multiplied by the transformed message function which is of the form $e^{jm_p g(t)}$, to produce an angle-modulated carrier given by

$$e_{exp}(t) = A_c e^{j[\omega_c t + m_p g(t) + \phi_c]} \tag{7-2}$$

Here A_c is the amplitude of the unmodulated carrier, ω_c is the unmodulated carrier angular frequency, m_p is a constant usually referred to as the phase-modulation index, and ϕ_c is a constant phase angle.

This chapter will be devoted to the spectral analysis of exponentially modulated carrier. Starting with single-tone modulation, the results will then be extended to multitone signals, including complex periodic modulating signals. This analysis will be generalized to cover the theory of multitone combined amplitude and frequency modulation. The spectral distribution of double modulation such as FM/FM signal which is frequently used in telemetry systems will also be covered. Finally, consideration will be given to the problem of transmission bandwidth of multitone frequency modulation, which is of great practical interest in the design of multichannel FM systems. However, before we embark on the theory of exponential modulation, we must first define some fundamental terms which are frequently encountered in FM, namely, instantaneous phase, frequency, and amplitude.

7.1 Definitions: Instantaneous Phase, Frequency, and Amplitude[1-5]

Consider a real time function f(t) which describes the message signal and assume that this function may be represented by the real or imaginary part of the complex time function F(t) given by

$$F(t) = A(t)e^{j\theta(t)} \tag{7-3}$$

Furthermore, if A(t) contains none of the zeros of f(t), then $\frac{d\theta(t)}{dt}$ is by definition the "instantaneous frequency." The amplitude function A(t) is the "instantaneous amplitude," and its magnitude $|A(t)|$ is the "envelope." The following discussions will clarify the physical significance of these definitions.

Consider the real time function

$$f(t) = A(t) \cos \theta(t) = A(t) \cos [\omega t + \psi(t)] \tag{7-4}$$

In amplitude modulation, A(t) is of the form

$$A(t) = A_0[1 + m_a g(t)] \tag{7-5}$$

Similarly, we may modulate the phase in the form

$$\psi(t) = \psi_0[1 + m_p g(t)] \tag{7-6}$$

and thus get phase modulation. In order to obtain an expression for frequency modulation, it would, however, be erroneous to substitute in Eq. (7-4) the expression

$$\omega = \omega_0[1 + m_f g(t)], \qquad m_f = \text{constant} \tag{7-7}$$

for this would lead to a physical absurdity, as we shall presently show. For consider the flow of a frequency-modulated current through an inductance L; let the current be given by

$$I = f(t) = A_0 \cos \{\omega_0[1 + m_f g(t)]t + \psi_0\} \tag{7-8}$$

The voltage across the inductance L is V = L dI/dt; therefore

$$V = -LA_0\omega_0[1 + m_f g(t) + m_f t g'(t)] \sin\{\omega_0[1 + m_f g(t)]t + \psi_0\} \tag{7-9}$$

which leads to a physical absurdity since $V \rightarrow \infty$ as $t \rightarrow \infty$. Following Helmholtz and Van der Pol,[2] we think of angular frequency as the time rate of change of phase, just as angular velocity is the time derivative of angular position. Therefore, we rewrite Eq. (7-4) as follows:

$$f(t) = A_0 \cos [\int_0^t \omega(t) \, dt + \psi_0] \tag{7-10}$$

This form makes it possible to extend the expression for a simple harmonic motion to the case of frequency modulation

$$f(t) = A_0 \cos \left\{ \int_0^t \omega_0[1 + m_f g(t)]dt + \psi_0 \right\}$$

$$= A_0 \cos[\omega_0 t + m_f \omega_c \int_0^t g(t) \, dt + \psi_0] \qquad (7\text{-}11)$$

Thus we define instantaneous frequency $\omega_i(t) = d\theta(t)/dt$, or

$$\omega_i(t) = \frac{d}{dt} \left\{ \int_0^t \omega_0[1 + m_f g(t)]dt + \psi_0 \right\} = \omega_0[1 + m_f g(t)] \qquad (7\text{-}12)$$

which is identical to Eq. (7-7).

An alternative definition of instantaneous frequency[3] essentially counts the density of zero crossings per unit interval of time. When applied to frequency-modulated signals, it gives the same results as the previous definition. Consider a sinusoidal signal $\cos \omega_0 t$; in a period $T = 2\pi/\omega_0$ the number of zero crossings is two, and the number of crossings in π sec must equal $2\pi/T = \omega_0$, or stated differently, the angular frequency of a sinusoid is equal to the number of zero crossings in a time interval of π sec. This definition may be extended to the case of a time function $s(t)$ as follows:

The instantaneous frequency of $s(t)$ is defined at the time t as the ratio of the number of zeros of $s(t)$ in the interval of time between $(t - \tau/2)$ and $(t + \tau/2)$ divided by τ/π, or as the mean density of zero crossings averaged over the time interval τ/π.

As an example, consider the signal $\cos [\omega_0 t + s(t)]$. If this function has consecutive zeros at $t = t_1$ and $t = t_2$, then

$$\omega_0(t_1 - t_2) + s(t_1) - s(t_2) = \pi \qquad (7\text{-}13)$$

Assume that $s'(t)$ changes slowly compared to $\cos \omega_0 t$ so that $s(t_1) - s(t_2) = s'(t_1)(t_1 - t_2)$, and therefore

$$t_1 - t_2 = \frac{\pi}{\omega_0 + s'(t)} \qquad (7\text{-}14)$$

If we define τ in such a way that $s'(t)$ is practically constant during the time interval τ, then the number of zeros within the time τ is

$$\frac{\tau}{t_1 - t_2} = \frac{\tau[\omega_0 + s'(t)]}{\pi} \qquad (7\text{-}15)$$

It follows therefore from the above definition that the instantaneous frequency is

$$\omega_i(t) = \omega_o + s'(t) \tag{7-16}$$

As stated above, the same result can be obtained using the first definition, namely,

$$\omega_i(t) = \frac{d\theta(t)}{dt} \qquad \text{where } \theta(t) = \omega_o t + s(t)$$

The first definition of instantaneous frequency is applicable to an ordinary sinusoid-carrier frequency-modulated signal whose amplitude does not contribute to zero crossings. The second one is more suitable for use in the analysis of frequency-modulated signals whose amplitude does contribute to zero crossings. In our future discussion, we shall make use only of the first definition of instantaneous frequency.

1. *Application to a Complex Signal.* We shall presently show how to apply the definition of instantaneous frequency to a function f(t) which consists of several sinusoidally varying time functions that have different frequencies and amplitudes. If

$$f(t) = \sum_{n=1}^{N} A_n(t) \cos(\omega_n t + \theta_n), \tag{7-17}$$

then using the same presentation as in Eq. (7-3), we define F(t) as a corresponding sum of exponentials

$$F(t) = \sum_{n=1}^{N} A_n(t) e^{j(\omega_n t + \theta_n)} = \sum_{n=1}^{N} A_n(t) e^{j\phi_n(t)} = A(t) e^{j\theta(t)} \tag{7-18}$$

where $\quad A(t) = \sqrt{[\text{Re } F(t)]^2 + [\text{Im } F(t)]^2} \tag{7-19}$

and $\quad \theta(t) = \text{Im}\,[\ell n\, F(t)] \tag{7-20}$

This is equivalent to finding the resultant phasor by adding the component phasors vectorially in the complex plane, using an arbitrary reference axis for the measurement of the phase angles $\phi_n(t)$.

Let us apply these definitions to a problem which will come up in the study of interference in frequency modulation. The problem is to compute the instantaneous frequency of the resultant vibration $f(t)$ caused by two frequency-modulated signals, the amplitude of the stronger being 1 and that of the weaker ρ. The angular frequencies of the modulating signals are p and q, respectively, and their modulation indices m_1 and m_2. The two frequency-modulated components are

$$\cos \phi_1(t) = \cos (\omega_1 t + m_1 \sin pt) \tag{7-21}$$

$$\rho \cos \phi_2(t) = \rho \cos (\omega_2 t + m_2 \sin qt + \psi_0) \tag{7-22}$$

and $\quad f(t) = \cos \phi_1(t) + \rho \cos \phi_2(t)$ $\hspace{3cm}$ (7-23)

From Fig. 7-1, we can readily see that the instantaneous amplitude $A(t)$ of the resultant is given by

$$A(t) = [1 + \rho^2 + 2\rho \cos \phi(t)]^{\frac{1}{2}} \tag{7-24}$$

where $\quad \phi(t) = \phi_2(t) - \phi_1(t)$ $\hspace{3cm}$ (7-25)

and the phase angle $\theta(t)$ is given by

$$\theta(t) = \tan^{-1} \frac{\sin \phi_1 + \rho \sin \phi_2}{\cos \phi_1 + \rho \cos \phi_2} = \phi_1 + \tan^{-1} \frac{\rho \sin \phi}{1 + \rho \cos \phi} \tag{7-26}$$

The resultant is given by

$$f(t) = A(t) \cos \theta(t) = A(t) \cos [\omega_1 t + m_1 \sin pt + \psi(t)] \tag{7-27}$$

where $\quad \tan \psi(t) = \dfrac{\rho \sin \phi}{1 + \tau \cos \phi}$ $\hspace{3cm}$ (7-28)

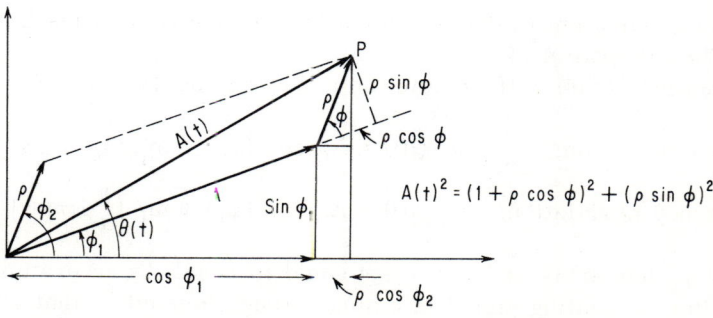

Fig. 7-1. Addition of two characteristic vectors.

The instantaneous frequency

$$\omega_i(t) = \frac{d\theta}{dt} = \omega_1 + m_1 p \cos pt + \frac{\rho \cos \phi + \rho^2}{1 + 2\rho \cos \phi + \rho^2} \cdot \frac{d\phi}{dt} \qquad (7-29)$$

The first two terms in Eq. (7-29) give the instantaneous frequency in the absence of a second signal, and the third term represents the effect of the second signal on the instantaneous frequency. We shall return to analyze this result in greater detail in our future discussions of interference problems in frequency modulation.

7.2   Fundamental Considerations of Exponential Modulation [1,6-12]

We shall explore the properties of exponential modulation by analyzing the following expression:

$$e_{exp}(t) = A_c \cos [\omega_c t + \phi_c + \psi(t)] \qquad (7-30)$$

where $\psi(t) = m_p g(t)$,   phase modulation $(7-31)$

and $\psi(t) = m_f \int_0^t g(\tau)d\tau$,   frequency modulation $(7-32)$

where g(t) is the modulating signal or message function, and $\psi(t)$ is the instantaneous phase angle which is modulated by the message function. We note here that in phase modulation, the instantaneous phase of the modulated signal depends linearly upon the modulating signal g(t); while in frequency modulation, the instantaneous frequency varies linearly with g(t), namely,

$$\omega_i(t) = \omega_c + m_f g(t) \qquad (7-33)$$

which is equivalent to the statement that the phase depends linearly upon the integral of g(t).

Equation (7-30) will assume the following special forms:

Phase modulation: $e_{PM}(t) = A_c \cos [\omega_c t + m_p g(t)]$ $(7-34)$

Frequency modulation: $e_{FM}(t) = A_c \cos [\omega_c t + m_f \int_0^t g(\tau)d\tau]$ $(7-35)$

where ϕ_c has arbitrarily been set equal to zero. As an example, when the modulating signal is a single-tone sinusoid so that g(t) = $\cos \omega_m t$, we obtain the following expressions:

Phase modulation: $e_{PM}(t) = A_c \cos(\omega_c t + m_p \cos \omega_m t)$

$$(7\text{-}36)$$

Frequency modulation: $e_{FM}(t) = A_c \cos\left(\omega_c t + \dfrac{m_f}{\omega_m} \sin \omega_m t\right)$

$$(7\text{-}37)$$

This is illustrated in Fig. 7-2 where the modulating signal is a sinusoid; amplitude modulation of the carrier is also included for comparison. It should be noted here that a distinction can be made between the phase- and frequency-modulated carrier only when compared with the modulating signal, as illustrated in Fig. 7-2. We

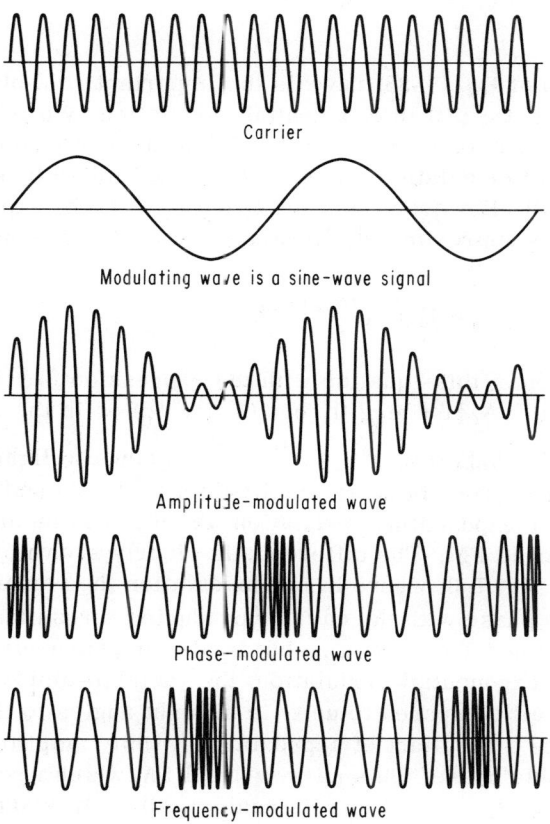

Carrier

Modulating wave is a sine-wave signal

Amplitude-modulated wave

Phase-modulated wave

Frequency-modulated wave

Fig. 7-2. Amplitude, phase, and frequency modulation of a sine-wave carrier by a sine-wave signal.

also note that, for phase modulation, the peak phase deviation or modulation index m_p is a constant, independent of the frequency of the modulating signal, and for frequency modulation, the peak phase deviation is given by m_f/ω_m, which is inversely proportional to the frequency of the modulating signal.

The general case of exponential modulation or, as it is sometimes called, angle modulation can best be analyzed by using the exponential form of the frequency-modulated carrier as given by Eq. (7-2) and expanding it in a power series of $\psi(t) = m_p g(t)$.

$$e_{exp}(t) = A_c e^{j(\omega_c t + \phi_c)} \cdot e^{j\psi(t)}$$

$$= A_c e^{j(\omega_c t + \phi_c)} \left[1 + j\psi(t) - \frac{1}{2!} \psi^2(t) - j\frac{1}{3!} \psi^3(t) + \cdots \right]$$

$$(7-38)$$

Examination of Eq. (7-38) reveals the important role played by the modulation factor $\psi(t)$ in exponential modulation. When $|\psi(t)|_{max}$ is not small compared to unity, we have nonlinear modulation since, in this type of modulation, the carrier is multiplied by higher powers of $\psi(t)$. However, in case $|\psi(t)|_{max} \ll 1$, the exponential modulation is approximately linear, and Eq. (7-38) reduces to

$$e_{exp} \doteq A_c[1 + j\psi(t)] e^{j(\omega_c t + \phi_c)}$$

$$(7-39)$$

A comparison of this equation with the expression for linear modulation $e_{AM}(t) = A_c[1 + m_a g(t)] \cos (\omega_c t + \phi_c)$ or in the exponential form $e_{AM}(t) = A_c[1 + m_a g(t)] \cdot e^{j(\omega_c t + \phi_c)}$ brings to light the very important difference between amplitude modulation and the approximately linear exponential modulation which is commonly referred to as narrow-band FM. The difference is clearly shown by the phasor diagram in Fig. 7-3, where in AM modulation the variable-amplitude phasor is in phase with the carrier producing a resultant which varies between $1 \pm m_a|g(t)|_{max}$ but is also in phase with the carrier; however, in exponential modulation, the variable-amplitude phasor which is small compared to unity is at right angles to the carrier and produces a resultant of approximately unity amplitude and small variable phase. This narrow-band FM which results from restricting $|\psi(t)|_{max} \ll 1$ and its analogy with AM modulation led Armstrong[6] to the invention of the first method of generating crystal-controlled or indirect FM, as we shall see later in Chap. 12 entitled Generation of FM Waves.

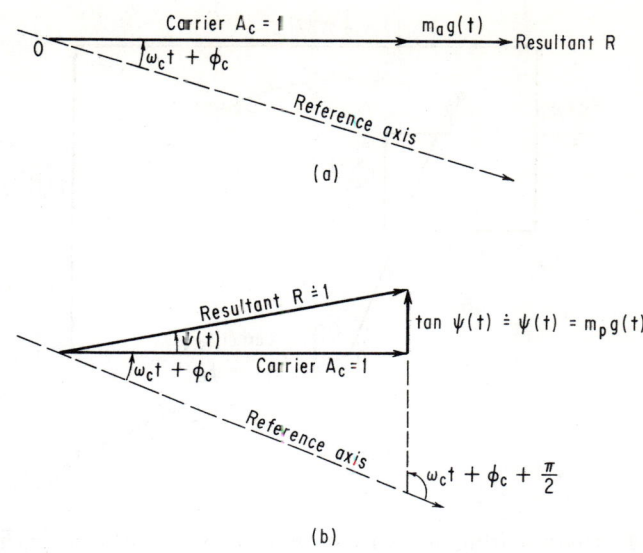

Fig. 7-3. Phasor diagrams for comparing narrow-band FM with AM.

In the general case, the so-called wideband FM in which $|\psi(t)|_{max}$ is not small compared to unity, higher powers of $\psi(t)$ contribute to the modulation process, as shown in Eq. (7-38) and represented by the phasor diagram of Fig. 7-4. We note here that the odd powers of $\psi(t)$ are in quadrature with the carrier, while the even powers are parallel to the carrier, and that the resultant R is of constant amplitude and of variable phase $\psi(t)$ with the carrier phasor A_c.

The spectral analysis of exponential modulation can best be performed by considering specific examples of $\psi(t)$. However, since higher powers of $\psi(t)$ contribute to the modulation process, it is obvious that the resultant spectrum will be of unlimited bandwidth as the index n in the power series of $\psi(t)$ becomes large. In order to obtain a clearer insight in the spectral distribution of an angular-modulated signal, we shall presently discuss specific examples of single-tone and multitone angular modulation.

7.3   Single-tone Angular Modulation [1,7,8,12]

1. *Phase Modulation.* Consider Eq. (7-30) with $\psi(t) = m_p g(t)$ $= \beta \sin \omega_m t$.

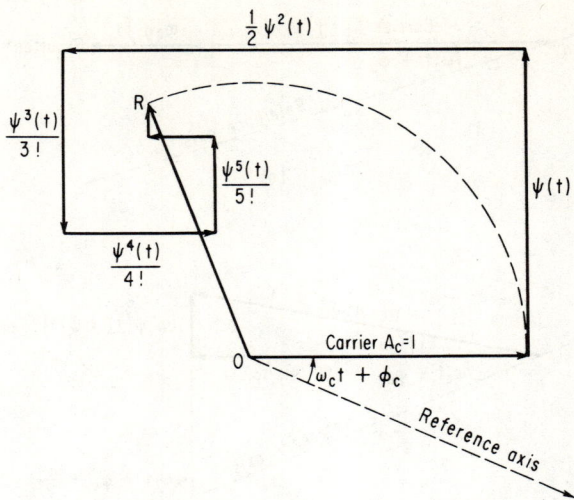

Fig. 7-4. Phasor diagram of exponentially modulated signal for large phase deviation.

Therefore $e_{PM}(t) = A_c(\cos \omega_c t + \phi_c + \beta \sin \omega_m t)$ (7-40)

where β is the peak phase deviation of the modulated signal of angular frequency ω_m. Without loss of generality, we set $\phi_c = 0$; therefore

$$e_{PM}(t) = A_c \cos (\omega_c t + \beta \sin \omega_m t)$$ (7-41)

and the instantaneous frequency $\omega_i(t)$ is equal to

$$\omega_i(t) = \omega_c + \omega_m \beta \cos \omega_m t = \omega_c + \Delta\Omega \cos \omega_m t$$ (7-42)

where $\Delta\Omega$ is the maximum frequency deviation of the modulated carrier frequency; it follows that $\Delta\Omega = \beta\omega_m$.

Following our previous discussion of narrow-band and wideband exponential modulation, we shall consider first the case of small phase deviation (narrow-band PM) where $\beta \ll 1$, and then the general case of large phase deviation (wideband PM).

Small Phase Deviation (Narrow-band PM).

$$e_{PM}(t) = A_c \cos (\omega_c t + \beta \sin \omega_m t), \qquad \beta \ll 1$$

$$e_{PM}(t) = A_c [\cos \omega_c t \cos (\beta \sin \omega_m t) - \sin \omega_c t \sin (\beta \sin \omega_m t)]$$

Since $\beta \ll 1$, therefore $\cos(\beta \sin \omega_m t) \doteq 1$ and $\sin(\beta \sin \omega_m t) \doteq \beta \sin \omega_m t$. Therefore

$$e_{PM}(t) \doteq A_c(\cos \omega_c t - \beta \sin \omega_m t \sin \omega_c t)$$

$$= A_c \cos \omega_c t - \frac{A_c \beta}{2} \cos(\omega_c - \omega_m)t + \frac{A_c \beta}{2} \cos(\omega_c + \omega_m)t$$

$$(7-43)$$

From Eq. (7-43), we note that the spectrum of narrow-band PM consists of a carrier and two sidebands which are 180° out of phase.

Equation (7-43) can readily be derived from Eq. (7-39), which corresponds to the restricted case of exponential modulation, by setting $\psi(t) = \beta \sin \omega_m t$.

$$e_{PM}(t) \doteq \text{Re}\left\{A_c e^{j\omega_c t}[1 + \psi(t)]\right\}$$

$$= \text{Re}\left[A_c e^{j\omega_c t}\left(1 - \frac{\beta}{2} e^{-j\omega_m t} + \frac{\beta}{2} e^{j\omega_m t}\right)\right]$$

$$= A_c \cos \omega_c t - \frac{A_c \beta}{2} \cos(\omega_c - \omega_m)t$$

$$+ \frac{A_c \beta}{2} \cos(\omega_c + \omega_m)t \qquad (7-44)$$

The corresponding equation for AM is

$$e_{AM}(t) = A_c \cos \omega_c t + \frac{A_c m_a}{2} \cos(\omega_c - \omega_m)t$$

$$+ \frac{A_c m_a}{2} \cos(\omega_c + \omega_m)t \qquad (7-45)$$

It is seen that the fundamental difference between the last two equations lies in the sign of the lower sideband. The rotating vectors describing these two cases are drawn in Fig. 7-5, where, in case of amplitude modulation, the sideband pair yields a resultant R_1 which adds to the carrier and produces amplitude modulation. However, the change in the sign of the lower sideband is instrumental in shifting the sideband resultant R_1 by 90°, so that the new total R is out of phase with respect to the carrier and produces a phase shift whose magnitude is $\psi(t)$. Thus, it is possible to take an AM wave and, by means of a suitable network, shift one sideband by 180° (or both sidebands by 90°) and produce phase modulation. This is the

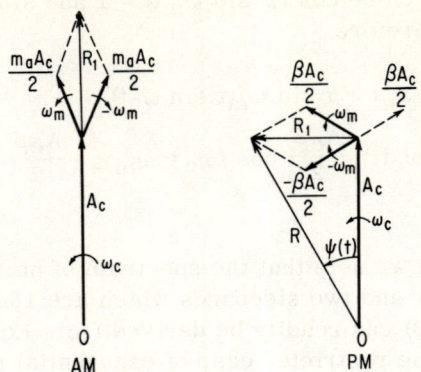

Fig. 7-5. Vector representation of AM and narrow-band PM.

basis of the indirect FM system. Note that the representation of Fig. 7-5 is identical to that of Fig. 7-3 where the case of restricted exponential modulation was discussed.

Large Phase Deviation (Wideband PM)

$$e_{PM}(t) = A_c \cos(\omega_c t + \beta \sin \omega_m t), \qquad \beta \text{ not small}$$

$$= A_c [\cos \omega_c t \cos(\beta \sin \omega_m t) - \sin \omega_c t \sin(\beta \sin \omega_m t)]$$

$$= A_c \cos \omega_c t \left[J_0(\beta) + 2 \sum_{\lambda=1}^{\infty} J_{2\lambda}(\beta) \cos 2\lambda \omega_m t \right]$$

$$- A_c \sin \omega_c t \left[2 \sum_{\nu=1}^{\infty} J_{2\nu-1}(\beta) \sin(2\nu - 1)\omega_m t \right]$$

$$= A_c \left[\cos \omega_c t \sum_{n=-\infty}^{\infty} J_n(\beta) \cos n\omega_m t \right.$$

$$\left. - \sin \omega_c t \sum_{n=-\infty}^{\infty} J_n(\beta) \sin n\omega_m t \right] \qquad (7\text{-}46)$$

These equations were derived from the well-known Fourier series expansions

$$\cos(\beta \sin \omega_m t) = J_0(\beta) + 2J_2(\beta) \cos 2\omega_m t + 2J_4(\beta) \cos 4\omega_m t + \cdots$$

$$= J_0(\beta) + 2 \sum_{\lambda=1}^{\infty} J_{2\lambda}(\beta) \cos 2\lambda \omega_m t \qquad (7\text{-}47)$$

$$\sin (\beta \sin \omega_m t) = 2J_1(\beta) \sin \omega_m t + 2J_3(\beta) \sin 3\omega_m t + \cdots$$

$$= 2 \sum_{\nu=1}^{\infty} J_{2\nu-1}(\beta) \sin (2\nu - 1)\omega_m t \qquad (7\text{-}48)$$

Finally, the spectral distribution for wideband modulation is

$$e_{PM}(t) = A_c \{J_0(\beta) \cos \omega_c t - J_1(\beta)[\cos (\omega_c - \omega_m)t - \cos (\omega_c + \omega_m)t]$$

$$+ J_2(\beta) [\cos (\omega_c - 2\omega_m)t + \cos (\omega_c + 2\omega_m)t]$$

$$- J_3(\beta) [\cos (\omega_c - 3\omega_m t) - \cos (\omega_c + 3\omega_m)t] + \cdots \} \qquad (7\text{-}49)$$

We thus have a time function consisting of a carrier and an infinite number of sidebands, whose amplitudes are proportional to $J_n(\beta)$ spaced at frequencies $\pm\omega_m$, $\pm2\omega_m$, etc., away from the carrier, as shown in Fig. 7-6.

Noting that $J_n(\beta) = (-1)^n J_{-n}(\beta)$, we may express it in a more compact form:

$$e_{PM}(t) = A_c \sum_{n=-\infty}^{\infty} J_n(\beta) \cos (\omega_c + n\omega_m)t \qquad (7\text{-}50)$$

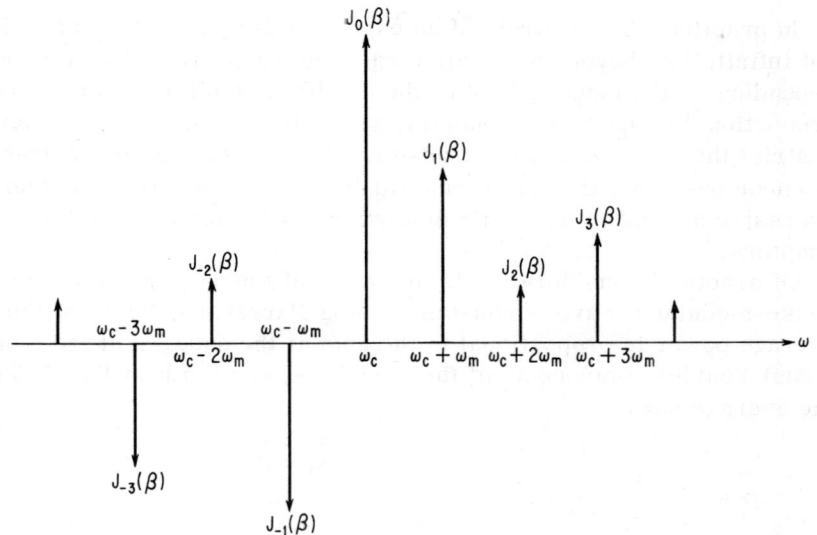

Fig. 7-6. Composition of FM wave into sidebands.

or $e_{PM}(t) = Re\left(A_c e^{j(\omega_c t + \beta \sin \omega_m t)}\right)$

$$= Re\left[A_c \sum_{n=-\infty}^{\infty} J_n(\beta) e^{j(\omega_c + n\omega_m)t}\right] \qquad (7\text{-}51)$$

The last result can be derived in a more elegant way directly from Eq. (7-38) which represents the general case of exponential modulation:

$$e_{PM}(t) = Re\ A_c\left(e^{j(\omega_c t + \phi_c)} \cdot e^{j\psi(t)}\right)$$

$$= Re\ A_c\left(e^{j\omega_c t} \cdot e^{j\beta \sin \omega_m t}\right)$$

$$= Re\ A_c\left[e^{j\omega_c t} \sum_{n=-\infty}^{\infty} J_n(\beta) e^{jn(\omega_m t)}\right]$$

$$= A_c \sum_{n=-\infty}^{\infty} J_n(\beta) \cos(\omega_c + n\omega_m)t \qquad (7\text{-}52)$$

where we used the identity

$$e^{j\beta \sin \omega_m t} = \sum_{n=-\infty}^{\infty} J_n(\beta) e^{jn\omega_m t} \qquad (7\text{-}53)$$

In practice, the spectrum of an exponentially modulated signal is not infinite for, beyond a certain frequency range from the carrier, depending on the magnitude of β, the sideband amplitudes which are proportional to $J_n(\beta)$ are becoming negligibly small. Thus we may restrict the bandwidth of the transmitting and receiving equipment to encompass only the significant sidebands without introducing an excessive amount of harmonic distortion, as we shall see in later chapters.

Of practical consideration is the fact that the average power in a phase-modulated wave is constant. Using Parseval's theorem, the average power is proportional to the sum of the squares of the individual Fourier components of the modulated wave. From Eq. (7-50) the average power

$$P = \frac{A_c^2}{2} \sum_{-\infty}^{\infty} J_n^2(\beta) = \frac{A_c^2}{2} \qquad (7\text{-}54)$$

This result is obtained from the properties of Bessel functions, namely,

$$\sum_{n=-\infty}^{\infty} J_n^2(\beta) = 1 \qquad\qquad (7\text{-}55)$$

2. *Frequency Modulation.* We have noted earlier in Eqs. (7-31) and (7-33) that the instantaneous phase in PM depends linearly upon the modulating signal g(t), while in FM the instantaneous frequency depends linearly on g(t). From Eq. (7-35), the instantaneous frequency is given by

$$\omega_i(t) = \frac{d\phi(t)}{dt} = \omega_c + m_f g(t) \qquad\qquad (7\text{-}56)$$

and assuming a sinusoidal modulating signal $\cos \omega_m t$ with a maximum frequency deviation $\Delta\Omega$ we obtain

$$\omega_i(t) = \omega_c + \Delta\Omega \cos \omega_m t, \qquad \Delta\Omega \ll \omega_c \qquad\qquad (7\text{-}57)$$

where the frequency deviation $\Delta\Omega$ is independent of the modulating frequency and is proportional to the amplitude of the modulating signal.

The instantaneous phase angle $\theta(t)$ for this special case is given by

$$\theta(t) = \int_0^t \omega_i(t)\ dt = \omega_c t + \frac{\Delta\Omega}{\omega_m} \sin \omega_m t + \theta_0 \qquad\qquad (7\text{-}58)$$

θ_0 may be taken as zero by referring to an appropriate phase reference so that the frequency-modulated carrier is given by

$$e_{FM}(t) = A_c \cos (\omega_c t + \beta \sin \omega_m t) \qquad\qquad (7\text{-}59)$$

where β is inversely proportional to the modulating angular frequency ω_m, namely, $\beta = \Delta\Omega/\omega_m$, which is of the same form as Eq. (7-41) for PM. It should be emphasized, however, that while Eqs. (7-59) and (7-41) are of the same form, nevertheless, the respective modulation indices β assume different significance. In PM the maximum phase deviation β is constant, independent of the modulating angular frequency ω_m, but in FM the maximum frequency deviation $\Delta\Omega$ is constant, independent of ω_m.

Since the expressions representing phase- and frequency-modulated signals are identical as given by Eqs. (7-59) and (7-41), it follows therefore that for a single modulating frequency the spectral representations are identical. However, it is of interest to

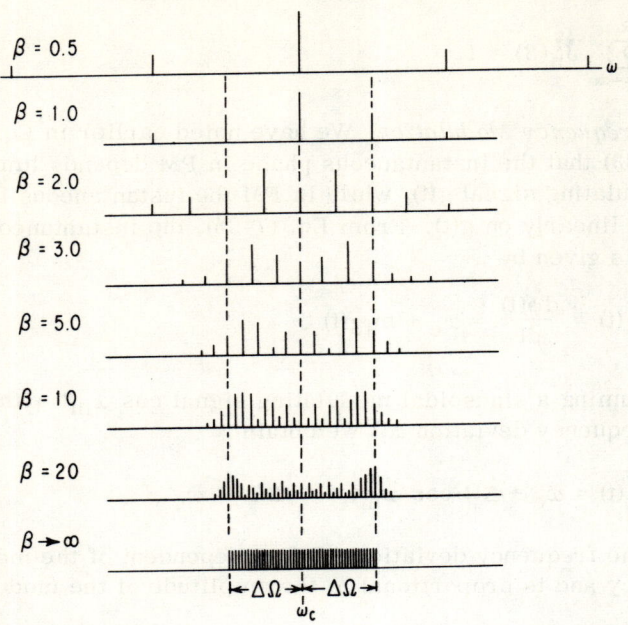

Fig. 7-7. FM spectrum of single-tone modulation bandwidth vs. modulation index β.

examine the spectral behavior as ω_m is varied, $\Delta\Omega$ being held constant. As shown in Fig. 7-7, as $\beta \rightarrow \infty$, the number of sidebands increases and the spectral components become more and more confined to the band between $\omega_c \pm \Delta\Omega$.

The bandwidth is equal to $2\Delta f_c(\Delta\Omega = 2\pi \Delta f_c)$ only for very large modulation index. For smaller values of β, we determine the bandwidth by counting the significant number of sidebands. The word "significant" is usually taken to mean those sidebands which have a magnitude at least 1 per cent of the magnitude of the unmodulated carrier.

3. *Variation of Amplitude of Carrier and Sidebands as a Function of Modulation Index β.* We have noted before that for $\beta \ll 1$ we have essentially a carrier and one pair of significant sidebands. As β increases, the number of significant sidebands increases while the total average power of the sidebands plus the carrier remains constant, equal to $A_c^2/2$. Since the amplitudes of the carrier and sidebands are proportional to the Bessel function $J_n(\beta)$ where n = 0, ± 1, ± 2, ..., the variation of the amplitudes as a function of the modulation index β can be obtained from the plot of the Bessel functions of

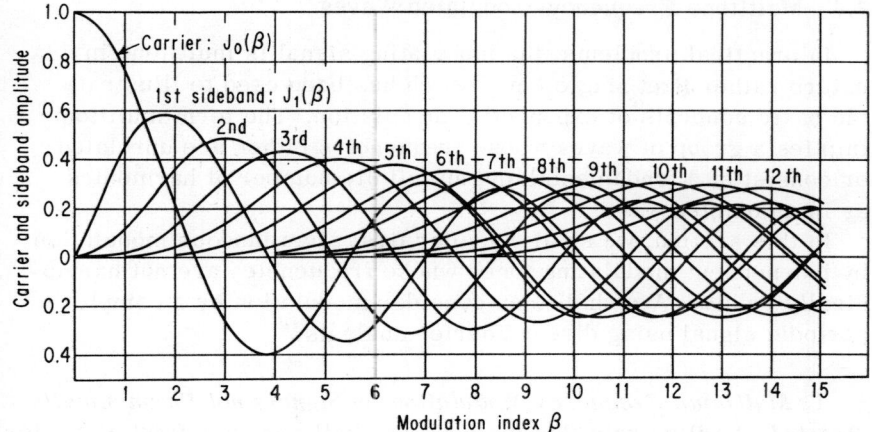

Fig. 7-8. Plot of Bessel functions of first kind as a function of argument β.

the first kind, as shown in Fig. 7-8. The behavior of the carrier amplitude is determined from the zero-order Bessel function $J_0(\beta)$. As β is increased, the value of the Bessel function drops off rapidly until, at $\beta = 2.404$, the amplitude is zero. As seen from the plot, the zero-order Bessel function is oscillatory with decreasing peak amplitude, the spacing between zeros approaching asymptotically the constant value π. This follows from the approximation

$$J_0(\beta) \doteq \frac{\cos(\beta - \pi/4)}{\sqrt{\pi\beta/2}} \qquad (7\text{-}60)$$

The fact that the carrier can go to zero for single-tone modulation can be used to measure the frequency deviation of the modulated carrier by increasing the amplitude of the modulating signal and observing the corresponding zeros of the carrier amplitude. These can be observed with a sharply tuned selective receiver which is capable of rejecting the sidebands and accepting the carrier only. In this experiment, the modulating frequency is kept constant so that the deviations corresponding to the zero points can be plotted as a function of the modulating voltage, and thus a calibration curve is obtained. It should be noted that the zeros corresponding to $J_1(\beta)$, $J_2(\beta)$, etc., can also be used to provide additional points on the calibration curve, provided the receiver is tuned to the corresponding sidebands.

7.4   Multitone Frequency-modulated Waves [7,9-11,16]

In practical problems, the modulating signal is multitone in nature rather than single tone, which has been used to illustrate the basic concepts of exponential modulation. The term multitone implies a group of waves whose frequencies either are unrelated or consist of a fundamental and an infinite number of harmonics as in a complex wave.

In this section, we shall consider first simultaneous modulation by independent modulating tones whose frequencies are not harmonically related. We shall then consider modulation by a complex periodic signal using direct Fourier analysis.

1. *Multitone Frequency Modulation by Signals not Harmonically Related.* To illustrate this method, we shall consider first a carrier being frequency-modulated by two tones. Let the instantaneous angular frequency of the modulated carrier be given by

$$\omega_i(t) = \omega_c + \Delta\Omega_1 \cos \omega_1 t + \Delta\Omega_2 \cos \omega_2 t \qquad (7\text{-}61)$$

The frequency-modulated signal is of the form

$$e(t) = A_c e^{j(\omega_c t + \beta_1 \sin \omega_1 t + \beta_2 \sin \omega_2 t)}$$

$$= A_c \left[\sum_{n=-\infty}^{\infty} J_n(\beta_1) e^{jn\omega_1 t} \right] \cdot \left[\sum_{m=-\infty}^{\infty} J_m(\beta_2) e^{jm\omega_2 t} \right] e^{j\omega_c t} \qquad (7\text{-}62)$$

where the modulation index of the first modulating signal $\beta_1 = \Delta\Omega_1/\omega_1$, and of the second modulating signal $\beta_2 = \Delta\Omega_2/\omega_2$.

Equation (7-62) can be expanded to give an expression for the sideband amplitudes, as follows:

$$e(t) = A_c \sum_{n,m=-\infty}^{\infty} C_{n,m} e^{j(\omega_c + n\omega_1 + m\omega_2)t} \qquad (7\text{-}63)$$

Equation (7-63) yields four types of sideband terms: the carrier, a set of sidebands corresponding to the modulating wave ω_1, a second set corresponding to ω_2, and a third set resulting from the sum and beat frequencies of the two modulating waves.

The amplitude of these spectral lines are listed below:

1 Carrier: $J_0(\beta_1) J_0(\beta_2) A_c e^{j\omega_c t}$

2 Sidebands due to ω_1: $J_n(\beta_1) J_0(\beta_2) A_c e^{j(\omega_c \pm n\omega_1)t}$

3 Sidebands due to ω_2: $J_m(\beta_2)J_0(\beta_1)A_c e^{j(\omega_c \pm m\omega_2)t}$

4 Beat frequencies at
$\omega_c \pm n\omega_1 \pm m\omega_2$: $J_n(\beta_1)J_m(\beta_2)A_c e^{j(\omega_c \pm n\omega_1 \pm m\omega_2)t}$

In two-tone modulation the amplitudes of the spectral components are generally reduced, unlike single-tone frequency modulation with the same index, this being due to the fact that each amplitude is determined by the product of two Bessel functions.

The method of analysis can be generalized to the case of N-tone modulation as follows. As before, the instantaneous frequency is

$$\omega_i(t) = \omega_c + \Delta\Omega_1 \cos \omega_t + \Delta\Omega_2 \cos \omega_2 t + \cdots + \Delta\Omega_N \cos \omega_N t \tag{7-64}$$

where the frequencies ω_1, ω_2, ..., ω_N are not harmonically related. The frequency-modulated carrier is given by

$$e(t) = A_c e^{j \int_0^t \omega_i(t)dt} = A_c e^{j\theta(t)} \tag{7-65}$$

so that $\theta(t) = \omega_c t + \beta_1 \sin \omega_1 t + \beta_2 \sin \omega_2 t + \cdots + \beta_N \sin \omega_N t$
$$\tag{7-66}$$

Hence

$$e(t) = A_c \left[\sum_{n_1=-\infty}^{\infty} J_{n_1}(\beta_1)e^{jn_1\omega_1 t} \right] \left[\sum_{n_2=-\infty}^{\infty} J_{n_2}(\beta_2)e^{jn_2\omega_2 t} \right] \cdots$$

$$\times \left[\sum_{n_N=-\infty}^{\infty} J_{n_N}(\beta_N)e^{jn_N\omega_N t} \right] e^{j\omega_c t} \tag{7-67}$$

or $$e(t) = A_c \left[\prod_{m=1}^{N} \sum_{n_m=-\infty}^{\infty} J_{n_m}(\beta_m)e^{jn_m\omega_m t} \right] e^{j\omega_c t} \tag{7-68}$$

The series expansion of Eq. (7-68) is given in Ref. 16.

2. *Frequency Modulation by a Periodic Complex Signal.* Complex-periodic-wave frequency and phase modulation may be analyzed by using direct Fourier analysis. Consider the FM wave given by

$$e(t) = A_c \cos [\omega_c t + \psi(t)] \tag{7-69}$$

where $\psi(t)$ represents the phase angle which is modulated by the complex periodic signal. Hence

$$\omega_i(t) = \omega_c + \frac{d\psi(t)}{dt}$$

is the instantaneous frequency of the modulated carrier. Thus

$$e(t) = A_c \, [\cos \, \omega_c t \, \cos \, \psi(t) - \sin \, \omega_c t \, \sin \, \psi(t)] \qquad (7\text{-}70)$$

Expand in the Fourier series:

$$\cos \, \psi(t) = \sum_{n=0}^{\infty} (a_n \sin n\omega_m t + b_n \cos n\omega_m t)$$

$$\sin \, \psi(t) = \sum_{n=0}^{\infty} (c_n \sin n\omega_m t + d_n \cos n\omega_m t) \qquad (7\text{-}71)$$

where ω_m is the angular repetition rate of the modulating function. Substituting in Eq. (7-70), we obtain

$$e(t) = A_c \cos \, \omega_c t \sum_{n=0}^{\infty} (a_n \sin n\omega_m t + b_n \cos n\omega_m t)$$

$$- A_c \sin \, \omega_c t \sum_{n=0}^{\infty} (c_n \sin n\omega_m t + d_n \cos n\omega_m t) \qquad (7\text{-}72)$$

from which we can derive the amplitudes of the sidebands. Using the complex notation, we may express it in the following compact form:

$$e(t) = A_c \sum_{n=-\infty}^{\infty} C_n e^{j(\omega_c + n\omega_m)t} \qquad (7\text{-}73)$$

where $\quad C_n = \dfrac{1}{T_m} \int_{-T_m/2}^{T_m/2} e^{j\psi(t)} \cdot e^{-jn\omega_m t} \, dt \qquad (7\text{-}74)$

As an illustration, let us apply this method to the square-wave frequency modulation which is shown in Fig. 7-9. The instantaneous frequency can be described as follows:

$$\omega_i(t) = \omega_c - \Delta\omega, \qquad -\frac{T_m}{2} < t < -\frac{T_m}{4}$$

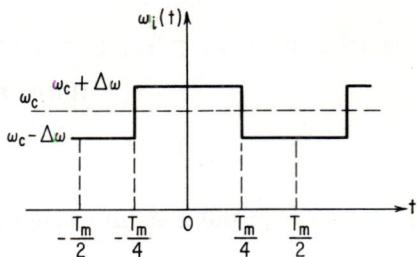

Fig. 7-9. Frequency modulation by square wave.

$$\omega_i(t) = \omega_c + \Delta\omega, \qquad -\frac{T_m}{4} < t < \frac{T_m}{4}$$

$$\omega_c - \Delta\omega, \qquad \frac{T_m}{4} < t < \frac{T_m}{2} \qquad (7\text{-}75)$$

The corresponding instantaneous phase variation may be described as

$$\psi(t) = \begin{cases} -\dfrac{\Delta\omega}{\omega_m}(\pi + \omega_m t), & -\dfrac{T_m}{2} < t < -\dfrac{T_m}{4} \\[2ex] \Delta\omega t, & -\dfrac{T_m}{4} < t < \dfrac{T_m}{4} \\[2ex] \dfrac{\Delta\omega}{\omega_m}(\pi - \omega_m t), & \dfrac{T_m}{4} < t < \dfrac{T_m}{2} \end{cases} \qquad (7\text{-}76)$$

The coefficients C_n can now be evaluated by the use of Eq. (7-74); the Fourier series of the modulated carrier is given by [7,11]

$$e(t) = A_c \sum_{n=-\infty}^{\infty} \frac{2\beta}{\pi(\beta^2 - n^2)} \sin(\beta - n)\frac{\pi}{2} \cdot \cos(\omega_c + n\omega_m)t \qquad (7\text{-}77)$$

$$= \frac{2A_c}{\pi\beta} \sin\frac{\pi\beta}{2} \cos\omega_c t \qquad \text{(carrier)}$$

$$+ \frac{2\beta A_c}{\pi(\beta^2 - 1^2)} \cos\frac{\beta\pi}{2} [\cos(\omega_c - \omega_m)t - \cos(\omega_c + \omega_m)t]$$

$$\text{(first sideband pair)}$$

$$- \frac{2\beta A_c}{\pi(\beta^2 - 2^2)} \sin\frac{\beta\pi}{2} [\cos(\omega_c - 2\omega_m)t + \cos(\omega_c + 2\omega_m)t]$$

$$\text{(second sideband pair)}$$

$$- \frac{2\beta A_c}{\pi(\beta^2 - 3^2)} \cos \frac{\beta\pi}{2} [\cos(\omega_c - 3\omega_m)t - \cos(\omega_c + 3\omega_m)t]$$

(third sideband pair)

$$+ \cdots$$

(7-78)

where $\beta = \Delta\omega/\omega_m$.

From Eq. (7-77), we see that the absolute magnitude of the n'th term is

$$|A_n| = \frac{2\beta}{\pi} \frac{\sin[(\beta - n)\pi/2]}{\beta^2 - n^2} A_c$$

(7-79)

By inspection of Eq. (7-79), we arrive at the following results:

1. When β is an odd integer, the amplitudes of the odd sidebands are equal to zero.

2. When β is an even integer, the amplitudes of the even sidebands are equal to zero.

3. When $\beta = n$,

$$|A_{\beta=n}| = \lim_{\beta \to n} \frac{2\beta}{n} \frac{\sin[(\beta - n)\pi/2]}{\beta^2 - n^2} A_c$$

$$= \frac{\beta}{\beta + n} A_c \lim_{\beta \to n} \frac{\sin[(\beta - n)\pi/2]}{(\beta - n)\pi/2} \to \frac{A_c}{2}$$

$$|A_{\beta=n}|^2 = \frac{A_c^2}{4}$$

(7-80)

indicating that, for the special case $\beta = n$, one-half the spectral energy is found in the sidebands which are displaced from the carrier angular frequency by $n\omega_m$.

The more general aspects of Eq. (7-77) are illustrated in Fig. 7-10 where $\Delta\omega$ is kept constant and ω_m is varied to illustrate the square-wave spectrum amplitudes. For low $\beta(\omega_m > \Delta\omega)$, the energy is now distributed among the carrier and the first pair of sidebands. As β is increased, more and more energy is contributed by the sidebands in the vicinity of $(\omega_c \pm \Delta\omega)$. As $\beta \to \infty$, which corresponds to the wave frequency having an infinitely long half period at $(\omega_c + \Delta\omega)$ and an infinitely long half period at $(\omega_c - \Delta\omega)$, the energy and amplitude spectra will consist of only two spectral lines, indicating that all the energy is at $(\omega_c \pm \Delta\omega)$.

In working with the complicated mathematical expressions often involved in complex-wave and multitone spectrum analysis, it is extremely important to note that the spectral contributions of an

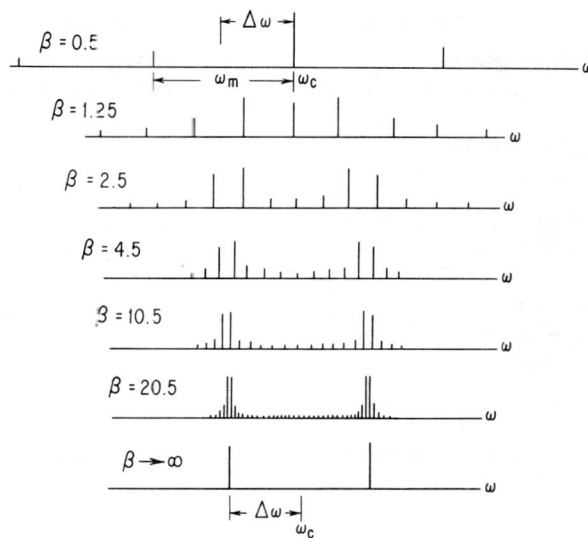

Fig. 7-10. Spectrum behavior of a square-wave FM wave as β is increased with $\Delta\omega$ constant

FM wave are functions of the rate of change of frequency during the entire cycle; the larger the frequency of an FM wave remains in a certain range of frequencies, the greater will be the spectral contribution in that frequency range. This is clearly shown in Fig. 7-11 where several complex-wave FM waveforms are shown with typical corresponding spectra resulting from large modulation indices.

7.5 Simultaneous Amplitude and Angular Modulation [7, 10, 13-15]

In a practical FM communication system, the generation of FM carrier is often accompanied by simultaneous amplitude modulation of sufficient magnitude to affect the spectral distribution of the modulated carrier.

For single-tone amplitude plus frequency modulation, we can write the following expression:

$$e_{AM-FM}(t) = A_c(1 + m_a \cos \omega_m t)e^{j(\omega_c t + \beta \sin \omega_m t)} \qquad (7-81)$$

As noted earlier, we distinguish between phase and frequency modulation only in comparison with the modulating signal which in this case is given by $A_c m_a \cos \omega_m t$. Since in this instance the instantaneous frequency of the modulated carrier is proportional to $\cos \omega_m t$, in

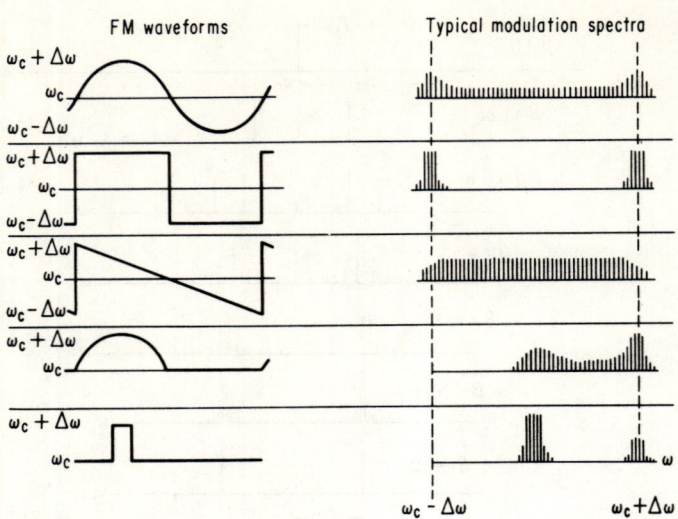

Fig. 7-11. FM waveforms and corresponding typical modulation spectra. (From C. L. Cuccia,[7] courtesy of McGraw-Hill Book Company.)

phase with the modulating signal, the result is frequency modulation. In contrast with Eq. (7-81), the expression describing simultaneous single-tone amplitude and phase modulation is given by

$$e_{AM-PM}(t) = A_c(1 + m_a \cos \omega_m t)e^{j(\omega_c t + \beta \cos \omega_m t)} \qquad (7\text{-}82)$$

where it is seen that the instantaneous phase of the modulated carrier is proportional to the modulating signal. We shall presently show that the spectral distribution of the two expressions are not identical, which is in contrast to pure FM and PM where the sidebands are identical.

1. *Simultaneous Amplitude and Frequency Modulation.* To derive the spectral components of Eq. (7-81), we express first the AM function in exponential form; thus Eq. (7-81) can be expressed in the form

$$e_{AM-FM}(t) = A_c e^{j(\omega_c t + \beta \sin \omega_m t)} + \frac{m_a A_c}{2} e^{j[(\omega_c + \omega_m)t + \beta \sin \omega_m}$$

$$+ \frac{m_a A_c}{2} e^{j[(\omega_c - \omega_m)t + \beta \sin \omega_m t]} \qquad (7\text{-}83)$$

Using the identity of Eq. (7-53), namely,

$$e^{j\beta \sin\omega_m t} = \sum_{n=-\infty}^{\infty} J_n(\beta) e^{jn\omega_m t}$$

we obtain

$$e_{AM-FM}(t) = A_c \sum_{n=-\infty}^{\infty} J_n(\beta) e^{j(\omega_c + n\omega_m)t}$$

$$+ \frac{m_a A_c}{2} \sum_{n=-\infty}^{\infty} J_n(\beta) e^{j[\omega_c + (n+1)\omega_m]t}$$

$$+ \frac{m_a A_c}{2} \sum_{n=-\infty}^{\infty} J_n(\beta) e^{j[\omega_c + (n-1)\omega_m]t} \qquad (7-84)$$

From Eq. (7-84), we obtain directly the expression for the sidebands as follows:

Carrier: $A_c e^{j\omega_c t}$

Upper n'th sideband:

$$\left\{ J_n(\beta) + \frac{m_a}{2} [J_{n+1}(\beta) + J_{n-1}(\beta)] \right\} A_c e^{j(\omega_c + n\omega_m)t}$$

Odd, lower, sideband ($n = 1, 3, \ldots$):

$$\left\{ -J_n(\beta) + \frac{m_a}{2} [J_{n+1}(\beta) + J_{n-1}(\beta)] \right\} e^{j(\omega_c - n\omega_m)t}$$

Even, lower, sideband ($n = 2, 4, \ldots$):

$$\left[J_n(\beta) - \frac{m_a}{2} [J_{n+1}(\beta) + J_{n-1}(\beta)] \right] e^{j(\omega_c - n\omega_m)t} \qquad (7-85)$$

Examination of the last results reveals the following interesting facts:

1. The carrier remains unchanged.
2. The spectrum is asymmetrical with respect to the carrier.
3. It is possible to cancel some lower sidebands so as to generate an approximate SSB spectrum by proper choice of m_a and β.

The last result is of importance, since it provides a simple means of obtaining a single-sideband spectrum by modulating the frequency as well as the amplitude of a carrier. It should be pointed out, however, that this method is applicable only to a low modulation index, since only the first lower sideband can be made to vanish identically for all values of β. This can readily be shown by the use of the recursion formula

$$nJ_n(\beta) = \frac{\beta}{2}[J_{n+1}(\beta) + J_{n-1}(\beta)] \tag{7-86}$$

and the relation for the lower sideband [Eq. (7-85)]. By choosing $m_a = \beta$, we note that the amplitude of the first lower sideband is identically zero for all values of β. For further discussion of the spectral distribution of such a system, the reader is referred to Golden and Schroeder [14] and Chakarabarti. [15]

2. *Simultaneous Amplitude and Phase Modulation.* Using the same method as before, we obtain

$$e_{AM-PM}(t) = A_c e^{j(\omega_c t + \beta \cos \omega_m t)} + \frac{m_a A_c}{2} e^{j[(\omega_c + \omega_m)t + \beta \cos \omega_m t]}$$

$$+ \frac{m_a A_c}{2} e^{j[(\omega_c - \omega_m)t + \beta \cos \omega_m t]} \tag{7-87}$$

Using the relation

$$e^{j\beta \cos \omega_m t} = \sum_{n=-\infty}^{\infty} j^n J_n(\beta) e^{jn\omega_m t} \tag{7-88}$$

we get

$$e_{AM-PM}(t) = A_c \sum_{n=-\infty}^{\infty} j^n J_n(\beta) e^{j(\omega_c + n\omega_m)t}$$

$$+ \frac{m_a A_c}{2} \sum_{n=-\infty}^{\infty} j^n J_n(\beta) e^{j[\omega_c + (n+1)\omega_m]t}$$

$$+ \frac{m_a A_c}{2} \sum_{n=-\infty}^{\infty} j^n J_n(\beta) e^{j[\omega_c + (n-1)\omega_m]t} \tag{7-89}$$

This can be regrouped to illustrate the amplitudes of the upper and lower sidebands as follows:

$$e_{AM} = PM(t) = A_c \left(\left[J_0(\beta) + jm_a J_1(\beta) \right] e^{j\omega_c t} \right.$$

$$+ \left\{ \frac{m_a}{2} \left[J_0(\beta) - J_2(\beta) \right] + j J_1(\beta) \right\} \left[e^{j(\omega_c + \omega_m)t} + e^{j(\omega_c - \omega_m)t} \right]$$

$$+ \left\{ -J_2(\beta) + j \frac{m_a}{2} \left[J_1(\beta) - J_3(\beta) \right] \right\}$$

$$\times \left[e^{j(\omega_c + 2\omega_m)t} + e^{j(\omega_c - 2\omega_m)t} \right]$$

$$+ \left\{ -\frac{m_a}{2} \left[J_2(\beta) - J_4(\beta) \right] - j J_3(\beta) \right\}$$

$$\times \left[e^{j(\omega_c + 3\omega_m)t} + e^{j(\omega_c - 3\omega_m)t} \right]$$

$$+ \left\{ J_4(\beta) - j \frac{m_a}{2} \left[J_3(\beta) - J_5(\beta) \right] \right\}$$

$$\left. \times \left[e^{j(\omega_c + 4\omega_m)t} + e^{j(\omega_c - 4\omega_m)t} \right] + \cdots \right) \qquad (7\text{-}90)$$

From Eq. (7-90), the following conclusions are derived:
1. The carrier vector is modified by a quadrature component.
2. The spectrum remains symmetrical.
Thus, we see that it is impossible to cancel lower sidebands as in amplitude plus frequency modulation.

7.6 Generalized Theory of Multitone Combined Amplitude and Frequency Modulation [10]

In this section, we shall discuss the general case of combined amplitude and frequency modulation, where the modulating signal causing amplitude modulation may be different from that causing frequency modulation.

Consider first a carrier whose angular frequency is ω_c, being amplitude-modulated with a signal of frequency Ω_1 and simultaneously frequency-modulated with a signal of angular frequency ω_1. The result is

$$e(t) = A_c [1 + m_1 \cos (\Omega_1 t + \theta_1)] \cos [\omega_c t + \beta_1 \sin (\omega_1 t + \phi_1)]$$

$$= A_c [1 + m_1 \cos (\Omega_1 t + \theta_1)] \sum_{n=-\infty}^{\infty} J_n(\beta_1) \cos [(\omega_c + n\omega_1)t + n\phi_1]$$

$$= A_c \sum_{n=-\infty}^{\infty} J_n(\beta_1) \cos \left[(\omega_c + n\omega_1)t + n\phi_1 \right]$$

$$+ \frac{m_1 A_c}{2} \sum_{n=-\infty}^{\infty} J_n(\beta_1) \{ \cos \left[(\omega_c + n\omega_1 + \Omega_1)t + n\phi_1 + \theta_1 \right]$$

$$+ \cos \left[(\omega_c + n\omega_1 - \Omega_1)t + n\phi_1 - \theta_1 \right] \} \tag{7-91}$$

which reduces to Eq. (7-84) for $\Omega_1 = \omega_1$ as expected.

Equation (7-91) indicates that an amplitude- and frequency-modulated signal spectrum consists of the sum of three individual spectra: (1) the center spectrum results from the frequency-modulated carrier by the modulating signal whose frequency is ω_1, (2) the lower spectrum is the frequency-modulated signal spectrum decreased in amplitude by $m_1/2$ and shifted downward in frequency by Ω_1, (3) the upper spectrum is the frequency-modulated signal spectrum decreased in amplitude by $m_1/2$ but shifted upward in frequency by Ω_1. The same result may be arrived at by considering first the carrier to be frequency-modulated by ω_1, and then each sideband of the resulting spectrum to be amplitude-modulated by Ω_1. If Ω_1 is an integer multiple of ω_1, some of the resulting sidebands overlap. In this case, the resultant sideband is a vector summation of the individual sidebands with due regard to ϕ_1 and θ_1.

Equation (7-91) can readily be extended to the multitone combined amplitude- and frequency-modulation case. The results for a two-tone and a multitone case are:

1. *Two-tone Combined Amplitude and Frequency Modulation*

$$e(t) = A_c[1 + m_1 \cos (\Omega_1 t + \theta_1) + m_2 \cos (\Omega_2 + \theta_2)]$$

$$\times \cos \left[\omega_c t + \beta_1 \sin (\omega_1 t + \phi_1) + \beta_2 \sin (\omega_2 t + \phi_2) \right]$$

$$= A_c[1 + m_1 \cos (\Omega_1 t + \theta_1) + m_2 \cos (\Omega_2 t + \theta_2)] \sum_{n_1=-\infty}^{\infty} \sum_{n_2=-\infty}^{\infty}$$

$$\times J_{n_1}(\beta_1) \, J_{n_2}(\beta_2) \cos \left[(\omega_c + n_1\omega_1 + n_2\omega_2)t + n_1\phi_1 + n_2\phi_2 \right]$$

$$= A_c \sum_{n_1=-\infty}^{\infty} \sum_{n_2=-\infty}^{\infty} J_{n_1}(\beta_1) J_{n_2}(\beta_2)$$

$$\times \cos \left[(\omega_c + n_1\omega_1 + n_2\omega_2)t + n_1\phi_1 + n_2\phi_2 \right]$$

$$+ \frac{m_1 A_C}{2} \sum_{n_1=-\infty}^{\infty} \sum_{n_2=-\infty}^{\infty} J_{n_1}(\beta_1) J_{n_2}(\beta_2)$$

$$\times \{ \cos \left[(\omega_C + n_1\omega_1 + n_2\omega_2 + \Omega_1)t + n_1\phi_1 + n_2\phi_2 + \theta_1 \right]$$

$$+ \cos \left[(\omega_C + n_1\omega_1 + n_2\omega_2 - \Omega_1)t + n_1\phi_1 + n_2\phi_2 - \theta_1 \right] \}$$

$$+ \frac{m_2 A_C}{2} \sum_{n_1=-\infty}^{\infty} \sum_{n_2=-\infty}^{\infty} J_{n_1}(\beta_1) J_{n_2}(\beta_2)$$

$$\times \{ \cos \left[(\omega_C + n_1\omega_1 + n_2\omega_2 + \Omega_2)t + n_1\phi_1 + n_2\phi_2 + \theta_2 \right]$$

$$+ \cos \left[(\omega_C + n_1\omega_1 + n_2\omega_2 - \Omega_2)t + n_1\phi_1 + n_2\phi_2 - \theta_2 \right] \} \qquad (7\text{-}92)$$

As before, the combined amplitude- and frequency-modulation
process may be considered in two steps in tandem. The carrier is
first frequency-modulated by the two signals $\cos (\omega_1 t + \phi_1)$ and
$\cos (\omega_2 t + \phi_2)$, and then the resulting sidebands are amplitude-
modulated by the signals $\cos (\Omega_1 t + \theta_1)$ and $\cos (\Omega_2 t + \theta_2)$, with the
corresponding indices of modulation m_1 and m_2.

2. *Multitone Combined Amplitude and Frequency Modulation*

$$e(t) = A_C \left[1 + \sum_{j=1}^{\ell} m_j \cos (\Omega_j t + \theta_j) \right] \cos \left[\omega_C t + \sum_{i=1}^{k} \beta_i \sin (\omega_i t + \phi_1) \right]$$

$$= A_C \left[1 + \sum_{j=1}^{\ell} m_j \cos (\Omega_j t + \theta_j) \right]$$

$$\times \prod_{i=1}^{k} \sum_{n_i=-\infty}^{\infty} \left[J_{n_i}(\beta_i) \cos \left(\omega_C + \sum_{i=1}^{k} n_i \omega_i \right) t + \sum_{i=1}^{k} n_i \phi_1 \right]$$

$$= A_C \prod_{i=1}^{k} \sum_{n_i=-\infty}^{\infty} \left[J_{n_i}(\beta_i) \cos \left(\omega_C + \sum_{i=1}^{k} n_i \omega_1 \right) t + \sum_{i=1}^{k} n_i \phi_1 \right]$$

$$+ \frac{A_C}{2} \sum_{j=1}^{\ell} \prod_{i=1}^{k} \left[\sum_{n_i=-\infty}^{\infty} m_j J_{n_i}(\beta_i) \right] \left\{ \cos \left[\left(\omega_C + \sum_{i=1}^{k} n_i \omega_i + \Omega_j \right) t \right. \right.$$

$$+ \sum_{i=1}^{k} n_1\phi_i + \theta_j \Bigg] + \cos \left[\left(\omega_c + \sum_{i=1}^{k} n_i\omega_i - \Omega_j \right) t \right.$$

$$\left. + \sum_{i=1}^{k} n_i\phi_i - \theta_j \Bigg] \right\} \tag{7-93}$$

It is to be noted that the average power during modulation for the multitone case is not constant and is given by

$$P = \frac{A_c^2}{2} \left(1 + \sum_{j=1}^{\ell} \frac{m_j^2}{2} \right) \tag{7-94}$$

7.7 Spectral Distribution of an FM/FM Signal [18]

This type of double modulation is frequently employed in telemetry systems where the transmitted carrier is frequency-modulated with the output of one or several subcarrier oscillators, which in turn have been frequency-modulated by data signals. In the following, we shall derive an expression for the spectral distribution of an FM/FM signal for one subcarrier only.

Let ω_c = carrier angular frequency

ω_s = subcarrier angular frequency

ω_m = modulating angular frequency

The subcarrier signal is given by

$$e_s(t) = A_s \cos \left[\omega_s t + \phi_s + \frac{\Delta\omega_s}{\omega_m} \sin (\omega_m t + \phi_m) \right]$$

$$= A_s \cos [\omega_s t + \phi_s + \beta_s \sin (\omega_m t + \phi_m)] \tag{7-95}$$

where $\Delta\omega_s$ = peak frequency deviation of subcarrier

and β_s = peak phase deviation of subcarrier

The instantaneous frequency of the carrier wave is given by

$$\omega_i(t) = \omega_c + \Delta\omega \cos [\omega_s t + \phi_s + \beta_s \sin (\omega_m t + \phi_m)] \tag{7-96}$$

where $\Delta\omega$ = maximum frequency deviation of carrier.

Let the modulated carrier wave be represented by

$$e(t) = A_c \cos \theta(t) \tag{7-97}$$

where $\theta(t) = \int_0^t \omega_i(t) \, dt$

$$= \int_0^t \{\omega_c + \Delta\omega \cos [\omega_s t + \phi_s + \beta_s \sin (\omega_m t + \phi_m)]\} \, dt$$

$$= \omega_c t + \phi_c + \Delta\omega \int_0^t \cos [\omega_s t + \phi_s + \beta_s \sin (\omega_m t + \phi_m)] \, dt$$

$$(7\text{-}98)$$

Evaluate the integral

$$I = \int_0^t \cos [\omega_s t + \phi_s + \beta_s \sin (\omega_m t + \phi_m)] \, dt$$

$$I = \frac{1}{\omega_s} \sin [\omega_s t + \phi_s + \beta_s \sin (\omega_m t + \phi_m)]$$

$$- \frac{\Delta\omega_s}{\omega_s} \int_0^t \cos (\omega_m t + \phi_m)$$

$$\times \cos [\omega_s t + \phi_s + \beta_s \sin (\omega_m t + \phi_m)] \, dt \qquad (7\text{-}99)$$

In practice $\omega_s \gg \omega_m$, and the term $\cos (\omega_m t + \phi_m)$, which is a slowly varying function compared to $\cos \omega_s t$, may be treated as a constant in front of $\cos (\omega_s t + \phi_s + \beta_s \sin \omega_m t + \phi_m)$ during the integration period.

Therefore

$$I = \frac{1}{\omega_s} \sin [\omega_s t + \phi_s + \beta_s \sin (\omega_m t + \phi_m)] - \frac{\Delta\omega_s}{\omega_s} \cos (\omega_m t + \phi_m) \cdot I$$

or

$$I \left[1 + \frac{\Delta\omega_s}{\omega_s} \cos (\omega_m t + \phi_m) \right]$$

$$= \frac{1}{\omega_s} \sin [\omega_s t + \phi_s + \beta_s \sin (\omega_m t + \phi_m)]$$

Since $\dfrac{\Delta\omega_s}{\omega_s} \ll 1$, we can neglect the term $\left(\dfrac{\Delta\omega_s}{\omega_s}\right) \cos (\omega_m t + \phi_m)$

and $\quad I = \dfrac{1}{\omega_s} \sin [\omega_s t + \phi_s + \beta_s \sin (\omega_m t + \phi_m)] \qquad (7\text{-}100)$

Finally, by substituting this result in Eq. (7-98), we obtain

$$\theta(t) = \omega_c t + \phi_c + \frac{\Delta\omega}{\omega_s} \sin[\omega_s t + \phi_s + \beta_s \sin(\omega_m t + \phi_m)]$$

$$(7\text{-}101)$$

and the modulated carrier wave of Eq. (7-97) is given by

$$e(t) = A_c \cos\theta(t)$$

$$= A_c \cos\{\omega_c t + \phi_c + \beta \sin[\omega_s t + \phi_s + \beta_s \sin(\omega_m t + \phi_m)]\}$$

$$(7\text{-}102)$$

where $\beta = \Delta\omega/\omega_s$. This can be written in the complex form

$$e(t) = A_c \text{Re}\left(e^{j(\omega_c t + \phi_c)} \cdot e^{j\beta \sin[\omega_s t + \phi_s + \beta_s \sin(\omega_m t + \phi_m)]}\right)$$

$$= A_c \text{Re}\, e^{j(\omega_c t + \phi_c)} \times \sum_{p=-\infty}^{\infty} J_p(\beta) e^{jp(\omega_s t + \phi_s)}$$

$$\times \sum_{q=-\infty}^{\infty} J_q(p\beta_s)\, e^{jq(\omega_m t + \phi_m)}$$

$$(7\text{-}103)$$

Finally, we obtain the expression

$$e(t) = A_c \sum_{p=-\infty}^{\infty} \sum_{q=-\infty}^{\infty} J_p(\beta) J_q(p\beta_s)$$

$$\times \cos[(\omega_c + p\omega_s + q\omega_m)t + \phi_c + p\phi_s + q\phi_m]$$

$$(7\text{-}104)$$

In case of n subcarriers, it can be shown that Eq. (7-104) assumes the form

$$e(t) = A_c \prod_{k=1}^{n} \sum_{p=-\infty}^{\infty} \sum_{q=-\infty}^{\infty} J_p(\beta_k) J_q(p\beta_{sk})$$

$$\times \cos\left[\left(\omega_c + p\sum_{k=1}^{n} \omega_{sk} + q\sum_{k=1}^{n} \omega_{mk}\right)t + \phi_c + p\sum_{k=1}^{n} \phi_{sk} + q\sum_{k=1}^{n} \phi_{mk}\right]$$

$$(7\text{-}105)$$

where n subcarriers of frequencies ω_{s1}, ω_{s2}, ..., ω_{sn} are respectively modulated by the data signals of frequencies ω_{m1}, ω_{m2}, ..., ω_{mn}.

7.8   Bandwidth Considerations in Multitone FM [19,20]

The problem of bandwidth design is of paramount practical importance in the design of communication systems. The bandwidth of the bandpass filters for the transmission and reception of the modulated carrier should be adequate enough to encompass the significant sidebands only.

As noted previously, the theoretically infinitely wide frequency band of a FM signal is in practice limited to a finite number of significant sidebands compatible with the distortion and intermodulation specifications, the number of significant sidebands being a function of the modulation index β and the peak deviation $\Delta\omega$. For $\beta < 1$, the spectrum is limited to the carrier and one pair of significant sidebands only of amplitudes A_c and $\beta A_c / 2$, respectively, because $J_0(\beta) \doteq 1$ and $J_1(\beta) \doteq \beta/2$. The second and subsequent sidebands can be neglected because, for small β, we have $J_n(\beta) \doteq \beta^n / 2^n n!$. For $\beta \gg 1$ or $\Delta\omega \gg \omega_m$, the required angular bandwidth is given by $2\Delta\omega$, as shown in Fig. 7-7. It should be noted here that the same conclusion has been reached in the discussion of the principle of stationary phase in Chap. 3, where it has been shown that the spectrum of an FM wave lies mainly in the range $(\omega_c \pm \Delta\omega)$, provided $\beta > 2\pi$.

In case of multitone modulation, it is evident from Eq. (7-68) that, as the number of modulating signals is increased, the amplitudes of the sidebands decrease rapidly as expected, since the total energy of the modulated wave is constant and independent of the form of the modulating signal. Since the sidebands extend from zero to infinite frequencies, it is impossible to assign a definite bandwidth for the frequency-modulated wave, but it is obvious that in general, beyond a certain frequency range from the carrier, the sidebands' amplitudes are negligibly small, and consequently may be neglected. In fact, it follows from Eq. (7-68) that if each value of β_m is fairly large compared to unity, the sidebands which lie much outside the band $\omega_c \pm \sum_{m=1}^{N} \beta_m \omega_m$ can be neglected.

If the component waves of the complex signal have a low modulation index, the spectrum will consist of a pair of sidebands for each component wave of frequencies $(\omega_c \pm \omega_m)$. The bandwidth is thus twice the highest modulation frequency. In some practical applications, the complex signal will have a low-frequency component of high modulation index and a high-frequency wave of low modulation index. In this case, the IF bandwidth is approximately given by

$$B_{IF} = 2\Delta F + 2f_m \qquad\qquad (7\text{-}106)$$

where ΔF is the peak frequency deviation of the carrier due to the low-frequency component and f_m is the frequency of the highest-frequency component of the complex modulating signal. Thus, the bandwidth equals twice the maximum deviation of the low-frequency wave plus twice the maximum modulating frequency.

For intermediate values of β there are available several approximate expressions for use in the design of multiplex communication systems. An approximate formula for the IF bandwidth required for such systems which is commonly used is

$$B_{IF} = 2(\Delta F + 2f_m) = 2\Delta F\left(1 + \frac{2}{\beta}\right) \qquad (7\text{-}107)$$

where ΔF = peak frequency deviation for the system
and $\quad f_m$ = highest baseband frequency
A plot of Eq. (7-107) is shown in Fig. 7-12, where the bandwidth occupied by the significant sideband is plotted (normalized for ΔF) vs. the modulation index β. We note that for $\beta \gg 1$, $B_{IF} \rightarrow 2\Delta F$ as expected.

Additional information on the transmission bandwidth and energy distribution of a frequency-modulated carrier by a frequency-division multiplex signal is given in Ref. 20.

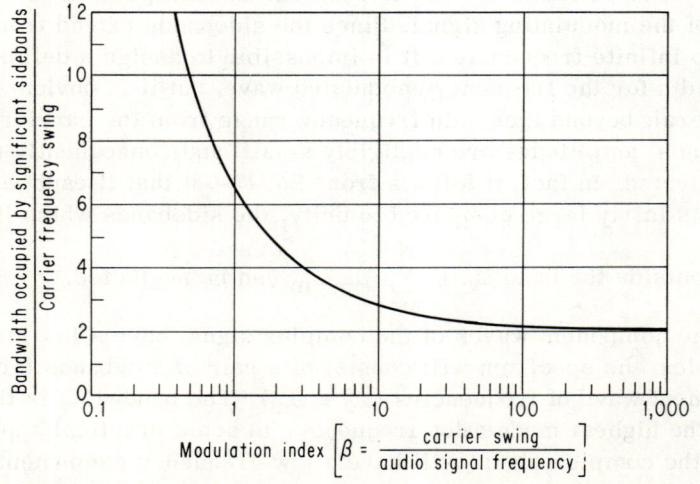

Fig. 7-12. Significant bandwidth (normalized) vs modulation index β. (From C. E. Tibbs,[19] courtesy of Chapman & Hall, Ltd.)

References

1. Baghdady, E. J. (ed.): "Lectures on Communication System Theory," chap. 19, McGraw-Hill Book Company, New York, 1961.
2. Van der Pol, B.: The Fundamental Principles of Frequency Modulation, *J. IEE London*, vol. 93, part 3, pp. 153-158, May, 1946.
3. Stumpers, F. L. H. M.: Theory of Frequency Modulation Noise, *Proc. IRE*, vol. 36, pp. 1081-902, September, 1948.
4. Granlund, J.: Interference in Frequency-modulated Reception, *MIT Res. Lab. Electron. Tech. Rept.* 42, Jan. 20, 1949.
5. Stumpers, F. L. H. M.: Interference Problems in Frequency Modulation, *Philips Res. Rept.* 2, pp. 136-160, 1947.
6. Armstrong, E. H.: A Method of Reducing Disturbances in Radio Signaling by a System of Frequency Modulation, *Proc. IRE*, vol. 24, pp. 689-740, May, 1936.
7. Cuccia, C. L.: "Harmonics, Sidebands, and Transients in Communication Engineering," McGraw-Hill Book Company, New York.
8. Schwartz, M.: "Information Transmission, Modulation, and Noise," McGraw-Hill Book Company, New York.
9. Black, H. S.: "Modulation Theory," D. Van Nostrand Company, Inc., Princeton, N. J.
10. Giacoletto, L. J.: Generalized Theory of Multitone Amplitude and Frequency Modulation, *Proc. IRE*, vol. 35, July, 1947.
11. Corrington, M. S.: Variation of Bandwidth with Modulation Index in Frequency Modulation, *Proc. IRE*, vol. 35, October, 1947.
12. Panter, P. F.: A Review of the Sideband Theory of FM with Particular Reference to Distortion Problems, ITT Federal Laboratories Internal Report, 1947.
13. Chakarabarti, N. B.: Combined AM and PM for a One-sided Spectrum (correspondence), *Proc. IRE*, vol. 47, p. 1663, September, 1959.
14. Golden, R. M., and M. R. Schroeder: Discussion of Combined AM and PM for a One-sided Spectrum (correspondence), *Proc. IRE*, vol. 48, p. 1049, June, 1961.
15. Chakarabarti, N. B.: Reply to Correspondence (Ref. 14), *Proc. IRE*, p. 1049, June, 1961.
16. Vellat, T.: Beitrag zur Theorie der Seitenbänder bei Frequenz-modulation, *Elek. Nachr.-Tech.*, vol. 18, Heft 7, July, 1941.
17. Cherry, E. C., and R. S. Rivlin: Non-linear Distortion with Particular Reference to the Theory of Frequency Modulated Waves, part 1, *Phil. Mag.*, vol. 32, p. 265, October, 1941.

18. Marcus, J.: "La Modulation de Fréquence," Editions Eyrolles, Paris, 1960.
19. Tibbs, C. E., and G. G. Johnstone: "Frequency Modulation Engineering," John Wiley & Sons, Inc., New York.
20. Smith, J. R. W., and J. L. Slow: Energy Distribution in a Wave Frequency Modulated by a Multichannel Telephone Signal, *A.T.E. Journal*, vol. 12, No. 3, July, 1956.

8

DISTORTION OF FREQUENCY-MODULATED SIGNALS THROUGH LINEAR SYSTEMS: FOURIER METHOD

When a frequency-modulated signal is impressed upon a linear network which has transmission characteristics which are nonlinear functions of frequency, the output will in general exhibit some amplitude modulation with distortion, in addition to distortion components in the phase of the FM signal. While the unwanted amplitude modulation may be removed to a large extent by means of limiters, the distortion components of the phase angle will appear in the output of the frequency discriminator, which is proportional to the derivative of the phase angle of the FM signal.

The distortion arises first from the fact that the sidebands of the FM signal are attenuated unequally, owing to the nonlinear amplitude characteristic of the network and, secondly, owing to the nonlinear phase-frequency characteristic of the network. Hence in the design of IF amplifiers, for example, it is necessary to provide as wide a passband as is consistent with permissible FM distortion limitations and adequate channel selectivity.

The problem of distortion introduced in an FM signal has attracted attention at a very early stage and has since been treated by numerous workers in the field (see references at end of chapter).

Essentially, two methods of approach are available in the solution of problems dealing with the response of a linear system to
variable-frequency excitations. One approach is commonly referred
to as the steady-state or Fourier method. This spectral approach
treats the linear system as a selector of certain frequency bands.
First, the impressed signal is analyzed into its Fourier components.
Then the spectrum of the steady-state response of the network is
obtained by multiplying each input spectral component by the value
that the transfer function of the network assumes at the frequency
of the input component. In this approach, each FM sideband is
weighted by the value of the network transfer function at the respective sideband frequency, and the output FM wave is considered to be
the vector sum of the weighted sidebands.[1] This method is concep-
tually simple and straightforward. It is often the final resort in
checking the validity and accuracy of results obtained by other
methods. Unfortunately, when the number of significant spectral
components is large, the computation becomes extremely laborious,
and, in perspective, the significance of the results may be lost in a
maze of complex numbers. [2]

Medhurst [3] and Assadourian [4] have applied the Fourier method
directly to the solution of network problems. In each case, a worka-
ble form of solution has been obtained by restricting attention to
small-order distortion, or, more correctly, to small divergence of
the amplitude and phase response from linearity. In Medhurst's
treatment, the divergence of the network response from linearity
is small but arbitrary. Assadourian considers the special case
where the divergence has the form of a sinusoidal ripple.

Under conditions where the divergence of the network response
from linearity is not small, the standard method of approach has
been to approximate the response by a finite power series or
trigonometrical series, and to use sufficient terms to represent the
response within a specified accuracy over a given frequency interval. This method would be described more appropriately as a
pseudo-Fourier method, because the final result is closely related
to the asymptotic series of Carson and Fry [5] by consideration of
sidebands. The pseudo-Fourier method has been used by several
workers in the field; [6-8] their methods lead to results of any desired accuracy but are limited in value by the complexity.

The Fourier method does not show clearly the relation between
the distortion and the characteristic of the network. This relation
is much more clearly brought out by the dynamic or asymptotic
method developed by Carson and Fry, which yields the result in the
form of a power series. In this method of approach, the network is

viewed as a system which, by virtue of its energy-storage elements, will exercise an inertia or sluggishness that sets a limit on the kind of frequency (or amplitude) changes that will be reproduced in its response.

In this chapter, we shall discuss several analytical techniques for finding the response of a linear network to variable-frequency excitation by the use of the steady-state or Fourier method. Four-terminal networks having linear or nonlinear amplitude-frequency and phase-shift-frequency characteristics will be considered and it will be shown that the response is, in general, a hybrid wave in which both the frequency-modulation and amplitude-modulation terms are distorted. However, in the case of linear phase-shift-frequency characteristic it will be shown that the output is an undistorted FM wave, except for a constant delay in the carrier and the modulating signal. For a linear amplitude-frequency characteristic, the output is a hybrid wave in which both the frequency- and amplitude-modulation terms are substantially undistorted, provided the carrier frequency is large compared with the frequency deviation.

Several methods of analysis of small-order distortion due to variations in attenuation and phase will be outlined, and in some instances a comparison will be made between the results obtained, using the first-order theory and the quasi-stationary approximation. Since this quasi-stationary method will be used quite often in our future discussions, it behooves us to elaborate on it at this time.

The quasi-stationary approximation is intimately connected with the definition of the phase-frequency characteristic. We recall that the phase-frequency characteristic is obtained by passing through the network single-frequency tones over a certain frequency range and measuring the corresponding phase shift introduced by the network, after sufficient time has elapsed to allow transients to die out. In the quasi-stationary mode, the frequency varies continuously but slowly enough so that the transients have no measurable effect on the phase variations. In particular, if the frequency varies slowly in a sinusoidal manner, the additional carrier phase variation is then obtained by inserting this sinusoid into the phase transmission characteristics.

The conditions of validity of the quasi-stationary approximation are of great importance, and will depend on individual circumstances; however, in the limit when the index of modulation β is very large, the quasi-stationary conditions for validity are always met.

8.1   Fourier Method [9-12]

As discussed in Chap. 3, a linear dynamic system may be described by a differential equation with constant coefficients. Consider a network whose amplitude-frequency and phase-frequency characteristics vary nonlinearly with frequency. Let the differential equation of the network be given by

$$\left[a_r \left(\frac{d}{dt}\right)^r + \cdots + a_1 \left(\frac{d}{dt}\right) + a_0 \right] e(t) = \left[b_s \left(\frac{d}{dt}\right)^s + \cdots \right.$$

$$\left. + b_1 \left(\frac{d}{dt}\right) + b_0 \right] i(t)$$

or operationally $f\left(\dfrac{d}{dt}\right) e(t) = g\left(\dfrac{d}{dt}\right) i(t)$ (8-1)

where $e(t)$ is the output voltage resulting from the input current $i(t)$, which is frequency-modulated. The FM current $i(t)$ is given by

$$i(t) = A_c e^{j(\omega_c t + \beta \sin \omega_m t)} = A_c \sum_{n=-\infty}^{\infty} J_n(\beta) e^{j(\omega_c + n\omega_m t)} \qquad (8-2)$$

where $\beta = \Delta\omega/\omega_m$ is the modulation index.

The steady-state response is

$$e(t) = A_c \sum_{n=-\infty}^{\infty} J_n(\beta) \frac{g[j(\omega_c + n\omega_m)]}{f[j(\omega_c + n\omega_m)]} \cdot e^{j(\omega_c + n\omega_m)t}$$

$$= A_c \sum_{n=-\infty}^{\infty} J_n(\beta) Z_n(j\omega) \, e^{j(\omega_c + n\omega_m)t} \qquad (8-3)$$

where $Z_n(j\omega) = Z_n[j(\omega_c + n\omega_m)] = X_n + jY_n = P_n e^{j\theta_n}$

$$P_n = \sqrt{X_n^2 + Y_n^2} \qquad \text{and} \qquad \tan\theta_n = \frac{Y_n}{X_n} \qquad (8-4)$$

$Z_n(j\omega)$ is the impedance of the network at the angular frequency $\omega = \omega_c + n\omega_m$. The sidebands of the output signal will, in general, be modified in amplitude and phase by the impedance characteristic, namely, the n'th sideband will be given by

$$A_c P_n J_n(\beta) \cos [(\omega_c + n\omega_m)t + \theta_n]$$

The output is given by

$$e(t) = \text{Re} \left\{ A_c \sum_{n=-\infty}^{\infty} P_n J_r(\beta) e^{j[(\omega_c + n\omega_m)t + \theta_n]} \right\}$$

$$= \text{Re} \left[R(t) e^{j[\omega_c t + \theta(t)]} \right]$$

or $e(t) = R(t) \cos [\omega_c t + \theta(t)]$ \hfill (8-5)

where $R(t)$ contains amplitude-modulation terms of frequencies
which are harmonically related to the modulating frequency ω_m,
and $\theta(t)$ equals the original modulation signal and its harmonic distortion terms. Equation (8-5) represents a "hybrid" wave in which
both the frequency-modulation and amplitude-modulation terms are
distorted. As stated above, the amplitude modulation can be minimized by means of limiters, and consequently we shall consider the
distortion components of the phase angle $\theta(t)$ only.

The phase angle of the output wave is

$$\phi(t) = \omega_c t + \theta(t) = \omega_c t + \tan^{-1} \frac{\sum_{n=-\infty}^{\infty} P_n J_n(\beta) \sin (n\omega_m t + \theta_n)}{\sum_{n=-\infty}^{\infty} P_n J_n(\beta) \cos (n\omega_m t + \theta_n)}$$

$$= \omega_c t + \tan^{-1} \frac{\sum_{n=-\infty}^{\infty} J_n(\beta) (X_n \sin n\omega_m t + Y_n \cos n\omega_m t)}{\sum_{n=-\infty}^{\infty} J_n(\beta) (X_n \cos n\omega_m t - Y_n \sin n\omega_m t)} \tag{8-6}$$

Equation (8-6) is the most general form of the output phase angle of
an FM signal which has passed through a network whose amplitude
and phase characteristics are known.

The instantaneous angular frequency $\omega_i(t)$ of the output signal is

$$\omega_i(t) = \frac{d\phi(t)}{dt} = \omega_c - \frac{d\theta(t)}{dt} \tag{8-7}$$

$$\text{where } \tan \theta(t) = \frac{B}{A} = \frac{\displaystyle\sum_{n=-\infty}^{\infty} P_n J_n(\beta) \sin (n\omega_m t + \theta_n)}{\displaystyle\sum_{n=-\infty}^{\infty} P_n J_n(\beta) \cos (n\omega_m t + \theta_n)}$$

$$= \frac{\displaystyle\sum_{n=-\infty}^{\infty} I_n \sin (n\omega_m t + \theta_n)}{\displaystyle\sum_{n=-\infty}^{\infty} I_n \cos (n\omega_m t + \theta_n)} \tag{8-8}$$

where I_n represents the amplitude of the n'th sideband after passing through the network.

$$\frac{d\theta(t)}{dt} = \frac{AB' - BA'}{A^2 + B^2} \tag{8-9}$$

$$\text{where} \quad A^2 = \sum_{n,\,m=-\infty}^{\infty} I_n I_m \cos (n\omega_m t + \theta_n) \cos (m\omega_m t + \theta_m)$$

$$B^2 = \sum_{n,\,m=-\infty}^{\infty} I_n I_m \sin (n\omega_m t + \theta_n) \sin (m\omega_m t + \theta_m)$$

$$\text{and} \quad A^2 + B^2 = \sum_{n,\,m=-\infty}^{\infty} I_n I_m \left[\cos (n-m)\,\omega_m t + \overline{\theta_n - \theta_m}\right]$$

$$\text{Also} \quad B' = \sum_{n=-\infty}^{\infty} n\omega_m I_n \cos (n\omega_m t + \theta_n)$$

$$AB' = \sum_{n,\,m=-\infty}^{\infty} n\omega_m I_n I_m \cos (n\omega_m t + \theta_n) \cos (m\omega_m t + \theta_m)$$

$$\text{and} \quad A' = -\sum_{n=-\infty}^{\infty} n\omega_m I_n \sin (n\omega_m t + \theta_n)$$

$$BA' = -\sum_{n,\,m=-\infty}^{\infty} n\omega_m I_n I_m \sin (n\omega_m t + \theta_n) \sin (m\omega_m t + \theta_m)$$

$$AB' - BA' = \sum_{n,\, m=-\infty}^{\infty} n\omega_m I_n I_m \cos\left[(n-m)\,\omega_m t + \overline{\theta_n - \theta_m}\right]$$

Finally, $\dfrac{d\theta(t)}{dt} = \dfrac{\displaystyle\sum_{n,\, m=-\infty}^{\infty} n\omega_m I_n I_m \cos\left[(n-m)\,\omega_m t + \overline{\theta_n - \theta_m}\right]}{\displaystyle\sum_{n,\, m=-\infty}^{\infty} I_n I_m \cos\left[(n-m)\,\omega_m t + \overline{\theta_n - \theta_m}\right]}$

$$(8\text{-}10)$$

The output of the FM discriminator is proportional to $d\theta(t)/dt$. Equation (8-10) may be expressed in terms of the rectangular components of P_n, namely:

$$\frac{d\theta}{dt} = \frac{\left\{\begin{array}{l}\displaystyle\sum_{n,\, m=-\infty}^{\infty} n\omega_m J_n(\beta)\, J_m(\beta)\,[(X_n X_m + Y_n Y_m)\cos(n-m)\,\omega_m t \\ \qquad\qquad - (X_n Y_m - X_m Y_n)\sin(n-m)\,\omega_m t]\end{array}\right\}}{\left\{\begin{array}{l}\displaystyle\sum_{n,\, m=-\infty}^{\infty} J_n(\beta)\, J_m(\beta)\,[(X_n X_m + Y_n Y_m)\cos(n-m)\,\omega_m t \\ \qquad\qquad + (X_n Y_m - X_m Y_n)\sin(n-m)\,\omega_m t]\end{array}\right\}}$$

$$(8\text{-}11)$$

In many practical problems, the amplitude-frequency characteristic of the network is an even function of the frequency while the phase-frequency characteristic is odd; i.e., $X_n = X_{-n}$ and $Y_n = -Y_{-n}$. This means that $P_n = P_{-n}$ and $\theta_n = -\theta_{-n}$; and since $J(\beta)_{-n} = (-1)^n J_n(\beta)$, we have $I_{-n} = (-1)^n I_n$, and Eq. (8-6) reduces to

$$\phi(t) = \omega_c t + \tan^{-1}\frac{2\displaystyle\sum_{\nu=0}^{\infty} I_{2\nu+1}\sin\left[(2\nu+1)\,\omega_m t + \theta_{2\nu+1}\right]}{I_0 + 2\displaystyle\sum_{\lambda=1}^{\infty} I_{2\lambda}\cos(2\lambda\omega_m t + \theta_{2\lambda})} \qquad (8\text{-}12)$$

$$= \omega_c t$$

$$+ \tan^{-1} \frac{2 \sum_{\nu=0}^{\infty} J_{2\nu+1}(\beta)[X_{2\nu+1} \sin(2\nu+1)\omega_m t + Y_{2\nu+1}\cos(2\nu+1)\omega_m t]}{J_0(\beta)X_0 + 2\sum_{\lambda=1}^{\infty} J_{2\lambda}(\beta)[X_{2\lambda}\cos 2\lambda\omega_m - Y_{2\lambda}\sin 2\lambda\omega_m t]}$$

$$(8\text{-}13)$$

It may easily be seen that the distortion components corresponding to the symmetrical case, as given by Eqs. (8-12) and (8-13), consist of odd harmonics of the modulating frequency. For by changing the variable $\omega_m t$ into $(\omega_m t + \pi)$, we obtain

$$\theta(\omega_m t + \pi) = \tan^{-1} \frac{2\sum_{\nu=0}^{\infty} I_{2\nu+1}\sin[(2\nu+1)(\omega_m t + \pi) + \theta_{2\nu+1}]}{I_0 + 2\sum_{\lambda=1}^{\infty} I_{2\lambda}\cos[2\lambda(\omega_m t + \pi) + \theta_{2\lambda}]}$$

$$= -\theta(\omega_m t) \qquad (8\text{-}14)$$

Hence the expression must be of the form

$$\theta(t) = \sum_{\nu=0}^{\infty} A_{2\nu+1}\sin(2\nu+1)\omega_m t + \sum_{\nu=0}^{\infty} B_{2\nu+1}\cos(2\nu+1)\omega_m t$$

$$(8\text{-}15)$$

In practical problems when the modulation index is small, the intensity of the sidebands diminishes rapidly, owing to the nature of the Bessel coefficients and the amplitude characteristic of the network; hence, the number of significant sidebands in Eq. (8-11) is rather small. The distortion components may then be calculated with the aid of harmonic analysis using, for example, the 12-point schedule or others. The general formula (8-11) does not give the harmonic components of the output signal in explicit form, except in special cases which will be discussed below.

8.2 Effect of Linear Phase-shift Characteristic of Network upon an FM Signal [9, 12]

Before we embark on the problem of distortion due to the non-linearities of the amplitude and frequency characteristics of the network, we shall determine the effect of the linear phase-shift characteristic upon the time delay of the modulating signal. We shall show later that the imparted time delay of the modulating signal plays a vital role in the design of CW ranging systems.

Let the FM signal as given by Eq. (8-2) be applied to the network having a linear phase-shift-frequency characteristic given by $\theta_n = \theta_0 + n\omega_m T_0$, where T_0 = constant, and zero attenuation at all frequencies. Hence Eq. (8-6) becomes

$$\phi(t) = (\omega_c t + \theta_0) + \tan^{-1} \frac{\displaystyle\sum_{n=-\infty}^{\infty} J_n(\beta) \sin [n\omega_m(t + T_0) + \theta_0]}{\displaystyle\sum_{n=-\infty}^{\infty} J_n(\beta) \cos [n\omega_m(t + T_0) + \theta_0]} \qquad (8\text{-}16)$$

First let us consider the case when $\theta_0 = 0$:

$$\phi(t) = \omega_c t + \tan^{-1} \frac{\displaystyle\sum_{n=-\infty}^{\infty} J_n(\beta) \sin n\omega_m t'}{\displaystyle\sum_{n=-\infty}^{\infty} J_n(\beta) \cos n\omega_m t'}$$

$$= \omega_c t + \tan^{-1} [\tan (\beta \sin \omega_m t')] = \omega_c t + \beta \sin \omega_m(t + T_0) \qquad (8\text{-}17)$$

where $t' = t + T_0$, which means a delay of T_0 sec in the modulating signal.

For the analysis of the general case where $\theta_0 \neq 0$, we consider the frequency-modulated signal given by Eq. (8-2). Since the attenuation is constant or zero at all frequencies, the sidebands are modified in passing through the network in phase only. The output is thus given by

$$\text{Output} = A_c \sum_{n=-\infty}^{\infty} J_n(\beta) \sin (\omega_c t + n\omega_m t + \theta_n)$$

$$= A_c \sin [\omega_c t - \theta_0 + \beta \sin \omega_m(t + T_0)] \qquad (8\text{-}18)$$

where $\quad \theta_n = \theta_0 + n\omega_m T_0 = \eta_0 + \omega_c T_0 + n\omega_m T_0 \qquad$ (8-19)

The output can be written in the form

$$\text{Output} = A_c \sin [\omega_c(t + T_0) + \eta_0 + \beta \sin \omega_m(t + T_0)] \qquad (8\text{-}20)$$

where $\eta_0 = \pm n\pi$, $n = 0, 1, 2, 3, \ldots$ $\qquad\qquad$ (8-21)

We conclude therefore that a linear phase shift introduces no distortion components but only a time delay of T_0 sec in the modulation
signal and the carrier, and an additional phase shift η_0 in the carrier. We shall see later in Chap. 11 how this important result is
applied to ranging systems.

8.3   Analysis of Small-order Distortion Due to Variations in Attenuation and Phase [3,13]

The problem of evaluation of the distortion introduced in a FM
wave in passing through a linear network using the Fourier method
is very complex when a general solution is sought. As we have
noted in Sec. 8.1, the expression for the derivative of the phase
angle of the output [Eq. (8-10)] contains a double-infinitive series
in the numerator and denominator and is unmanageable when the
number of significant sidebands is large. In the following, several
methods will be given which are applicable to small-order distortion arising from small variations in attenuation and phase and applicable to simple types of modulating signals.

1. *First-order Distortion by the Method of Medhurst.*[3]   Let $s(t)$
be the modulating signal (expressed as a phase modulation), and ω_c
the angular carrier frequency. Then the input modulated wave

$$e_i(t) = A_c \sin [\omega_c t + s(t)]$$

$$= A_c \cos s(t) \sin \omega_c t + A_c \sin s(t) \cos \omega_c t \qquad (8\text{-}22)$$

When the modulated wave passes through a passive network whose
amplitude and phase response vary nonlinearly with frequency, the
output wave will, in general, contain both frequency-modulation distortion and amplitude modulation. Let Δ denote the distortion terms
introduced in the modulating signal $s(t)$, and assuming that Δ is
small, the output modulated wave with distortion becomes

$$e_0(t) = A_c(1 + \Delta_3) \sin [\omega_c t + s(t) + \Delta]$$

$$= A_C(1 + \Delta_3) [\cos s(t) - \Delta \sin s(t)] \sin \omega_c t$$
$$+ A_C(1 + \Delta_3) [\sin s(t) + \Delta \cos s(t)] \cos \omega_c t \qquad (8\text{-}23)$$

to the first order of Δ. The term $(1 + \Delta_3)$ represents an amplitude modulation introduced by the distorting network which may also be assumed to be small, so that $\Delta_3 \ll 1$.

The output wave can also be expressed in the following form by operating upon the right-hand side of Eq. (8-22):

$$e_o(t) = A_C [\cos s(t) + \Delta_1] \sin \omega_c t + A_C [\sin s(t) + \Delta_2] \cos \omega_c t$$
$$(8\text{-}24)$$

where Δ_1 and Δ_2 are small. Then from (8-23) and (8-24), we obtain

$$(1 + \Delta_3) [\cos s(t) - \Delta \sin s(t)] = \cos s(t) + \Delta_1$$
$$(1 + \Delta_3) [\sin s(t) + \Delta \cos s(t)] = \sin s(t) + \Delta_2$$

To the first order in the Δ's, these equations become

$$\Delta_3 \cos s(t) - \Delta \sin s(t) = \Delta_1$$
$$\Delta_3 \sin s(t) + \Delta \sin s(t) = \Delta_2$$

Eliminating Δ_3, it is found that

$$\Delta = \Delta_2 \cos s(t) - \Delta_1 \sin s(t) \qquad (8\text{-}25)$$

This basic result will be applied to several simple distortion problems. Consider the simple case of $s(t) = \beta \sin \omega_m t$, where ω_m is the angular frequency of the modulating signal, and β the index of modulation. From Eq. (8-22), the spectral components are

$$e_i(t) = A_C J_0(\beta) \sin \omega_c t + A_C \sum_{n=1}^{\infty} J_n(\beta) \{\sin [(\omega_c + n\omega_m)t]$$
$$+ (-1)^n \sin [(\omega_c - n\omega_m)]t\} \qquad (8\text{-}26)$$

Let the amplitude and phase characteristics of the network be represented by $1 + \rho(\omega)$ (voltage ratio) and $\phi(\omega)$ rad, respectively. Then the distorted spectrum (for $A_C = 1$) is given by

$$e_0(t) = J_0(\beta)\,[1 + \rho(\omega_c)]\,\sin[\omega_c t + \phi(\omega_c)]$$

$$+ \sum_{n=1}^{\infty} J_n(\beta)\,\{[1 + \rho(\omega_c + n\omega_m)]$$

$$\times \sin[(\omega_c + n\omega_m)t + \phi(\omega_c + n\omega_m)]$$

$$+ (-1)^n\,[1 + \rho(\omega_c - n\omega_m)]\,\sin[(\omega_c - n\omega_m)t$$

$$+ \phi(\omega_c - n\omega_m)]\} \tag{8-27}$$

In the following, the assumption is made that the departure of $\phi(\omega)$ from linearity is small and that $\rho(\omega)$ is small. Expanding Eq. (8-27) to first order in ρ and ϕ, the distorted spectral components are given by

$$e_0(t) = \left\{\cos\,(\beta\,\sin\,\omega_m t) + J_0(\beta)\,\rho(\omega_c) + \sum_{n=1}^{\infty} J_n(\beta)\right.$$

$$\times\,[\rho(\omega_c + n\omega_m) + (-1)^n\rho(\omega_c - n\omega_m)]\cos n\omega_m t - \sum_{n=1}^{\infty} J_n(\beta)$$

$$\left.\times\,[\phi(\omega_c + n\omega_m) - (-1)^n\phi(\omega_c - n\omega_m)]\sin n\omega_m t\right\}\sin\omega_c t$$

$$+ \left\{\sin\,(\beta\,\sin\,\omega_m t) + \sum_{n=1}^{\infty} J_n(\beta)\right.$$

$$\times\,[\rho(\omega_c + n\omega_m) - (-1)^n\rho(\omega_c - n\omega_m)]\sin n\omega_m t + J_0(\beta)\phi(\omega_c)$$

$$+ \sum_{n=1}^{\infty} J_n(\beta)\,[\phi(\omega_c + n\omega_m) + (-1)^n\phi(\omega_c - n\omega_m)]$$

$$\left.\times\,\cos n\omega_m t\right\}\cos\omega_c t \tag{8-28}$$

This has the same form as Eq. (8-24); consequently, using Eq. (8-25), the distortion components of the phase-modulated signal are

$$\Delta = \left[J_0(\beta) + 2\sum_{n=1}^{\infty} J_{2n}(\beta)\cos 2n\omega_m t\right]\left\{\sum_{n=1}^{\infty} J_n(\beta)\,[\rho(\omega_c + n\omega_m)\right.$$

$$-(-1)^n \rho(\omega_c - n\omega_m)] \sin n\omega_m t + J_0(\beta)\phi(\omega_c) + \sum_{n=1}^{\infty} J_n(\beta)$$

$$\times [\phi(\omega_c + n\omega_m) + (-1)^n \phi(\omega_c + n\omega_m)] \cos n\omega_m t \Bigg\}$$

$$- 2 \sum_{n=1}^{\infty} J_{2n-1}(\beta) \sin(2n-1)\omega_m t \left\{ J_0(\beta)\rho(\omega_c) + \sum_{n=1}^{\infty} J_n(\beta) \right.$$

$$\times [\rho(\omega_c + n\omega_m) + (-1)^n \rho(\omega_c - n\omega_m)] \cos n\omega_m - \sum_{n=1}^{\infty} J_n(\beta)$$

$$\times [\phi(\omega_c + n\omega_m) - (-1)^r \phi(\omega_c - n\omega_m)] \sin n\omega_m t \Bigg\} \qquad (8\text{-}29)$$

Any desired harmonic can be picked out from this equation. It should be noted here that these terms represent the harmonics of the phase-modulation distortion; the harmonic distortion components of the output signal are obtained by differentiating Eq. (8-29) with respect to time. The ratio between the second-harmonic distortion and the fundamental is

$$\frac{2}{\beta} \left\{ \left[\sum_{n=1}^{\infty} J_n(J_{n-2} - J_{n+2}) \rho(\omega_c + n\omega_m) \right. \right.$$

$$\left. - \sum_{n=1}^{\infty} J_n(J_{n-2} - J_{n+2}) \rho(\omega_c - n\omega_m) \right]^2$$

$$+ \left[2J_0 J_2 \phi(\omega_c) + \sum_{n=1}^{\infty} J_n(J_{n-2} + J_{n+2}) \phi(\omega_c + n\omega_m) \right.$$

$$\left. \left. + \sum_{n=1}^{\infty} J_n(J_{n-2} + J_{n-2}) \phi(\omega_c - n\omega_m) \right]^2 \right\}^{-\frac{1}{2}} \qquad (8\text{-}30)$$

and that between the third-harmonic distortion and the fundamental is

$$\frac{3}{\beta} \left\{ \left[-2J_0 J_3 \rho(\omega_c) + \sum_{n=1}^{\infty} J_n(J_{n-3} - J_{n+3}) \rho(\omega_c + n\omega_m) \right. \right.$$

$$+ \sum_{n=1}^{\infty} J_n(J_{n-3} + J_{n+3}) \rho(\omega_c - n\omega_m) \Bigg]^2$$

$$+ \Bigg[\sum_{n=1}^{\infty} J_n(J_{n-3} + J_{n+3}) \phi(\omega_c - n\omega_m)$$

$$- \sum_{n=1}^{\infty} J_n(J_{n-3} + J_{n+3}) \phi(\omega_c - n\omega_m) \Bigg]^2 \Bigg\}^{\frac{1}{2}} \qquad (8\text{-}31)$$

It should be noted here that Eqs. (8-30) and (8-31) are applicable not only to continuous analytical functions of phase and amplitude characteristic but also to discontinuous functions, provided the values are known at the sideband positions.

The N'th harmonic component of the phase-modulation distortion is given by

$$\cos{(N\omega_m t)} \sum_{n=-\infty}^{\infty} J_n(J_{n-N} + J_{n+N}) \phi(\omega_c + n\omega_m)$$

$$+ \sin{(N\omega_m t)} \sum_{n=-\infty}^{\infty} J_n(J_{n-N} - J_{n+N}) \rho(\omega_c + n\omega_m) \quad (8\text{-}32)$$

This expression is derived on the assumption that both ϕ and ρ are less than unity over the range of significant sidebands.

2. *Extended First-order Theory of Medhurst.* [13]   The condition that both ϕ and ρ should be small, which is required for Eq. (8-32) to be valid, may sometimes impose a rather severe restriction on the application of the formula. Medhurst has extended the first-order theory to the case where ϕ and ρ are not small. He retains the terms $\cos \phi$ and $\sin \phi$ without approximations but restricts his analysis to the case where second- and higher-order terms in the Δ's may be neglected. Starting with the basic expression of the undistorted FM wave, namely, $\sin{(\omega_c t + \beta \sin \omega_m t)}$ the sidebands are modified in amplitude and phase in passing through the distorting network to produce an output wave given by

$$e_o(t) = \sum_{n=-\infty}^{\infty} J_n(\beta) \, [1 + \rho(\omega_c + n\omega_m)] \sin{[(\omega_c + n\omega_m)t}$$

$$+ \phi(\omega_c + n\omega_m)] \qquad (8\text{-}33)$$

From this equation, the following expressions for Δ_1 and Δ_2 are derived

$$\Delta_1 = \sum_{n=-\infty}^{\infty} J_n(\beta)\Big(\{[1 + \rho(\omega_c + n\omega_m)]\cos[\phi(\omega_c + n\omega_m)] - 1\}$$

$$\times \cos n\omega_m t - [1 + \rho(\omega_c + n\omega_m)]\sin\phi(\omega_c + n\omega_m)\sin n\omega_m t\Big)$$
(8-34)

and $\quad \Delta_2 = \sum_{n=-\infty}^{\infty} J_n(\beta)\Big(\{[1 + \rho(\omega_c + n\omega_m)]\cos\phi(\omega_c + n\omega_m) - 1\}$

$$\times \sin n\omega_m + [1 + \rho(\omega_c + n\omega_m)]\sin\phi(\omega_c + n\omega_m)\cos n\omega_m t\Big)$$

(8-35)

These expressions are inserted in Eq. (8-25), and after some algebraic manipulation, the following expression is derived for the harmonic-distortion terms of the phase modulation. The N'th harmonic component of the phase-modulation distortion is given by

$$\cos(N\omega_m t) \sum_{n=-\infty}^{\infty} J_n(J_{n-N} + J_{n+N}) \{[1 + \rho(\omega_c + n\omega_m)]$$

$$\times \sin\phi(\omega_c + n\omega_m)\} + \sin(N\omega_m t)\sum_{n=-\infty}^{\infty} J_n(J_{n-N} - J_{n+N})$$

$$\times \{[1 + \rho(\omega_c + n\omega_m)]\cos\phi(\omega_c + n\omega_m) - 1\}$$
(8-36)

This expression for the extended first-order distortion reduces to the expression for small distortion as given by Eq. (8-32) if we take ϕ sufficiently small and neglect the product of ρ and ϕ.

The extended theory has a wider range of application than the approximate theory, because under certain conditions, the Δ's remain small when ϕ is not small.

3. *Calculation of Distortion for Low Modulation Index.* When the index of modulation β is small, the number of significant sidebands is not too large, and the distortion components may then be evaluated directly using Eqs. (8-12) and (8-33). The spectrum of the distorted wave is then

$$e_0(t) = \sum_{n=-\infty}^{\infty} [1 + \rho(\omega_c + n\omega_m)] J_n(\beta) \sin [(\omega_c + n\omega_m)t$$

$$+ \phi(\omega_c + n\omega_m)] \tag{8-37}$$

and the distorted phase modulation according to Eq. (8-12) is

$$\theta(t) = \tan^{-1}\left(\frac{\left\{\begin{array}{l}2\displaystyle\sum_{n=1}^{\infty}\{1 + \rho[\omega_c + (2n-1)\omega_m]\}J_{2n-1}(\beta)\\ \times \sin\{(2n-1)\omega_m t + \phi[\omega_c + (2n-1)\omega_m]\}\end{array}\right\}}{\left\{\begin{array}{l}1 + \rho(\omega_c) + 2\displaystyle\sum_{n=1}^{\infty}[1 + \rho(\omega_c + 2n\omega_m)]J_{2n}(\beta)\\ \times \{\cos[2n\omega_m t + \phi(\omega_c + 2n\omega_m)]\}\end{array}\right\}}\right) \tag{8-38}$$

This equation is valid only for the symmetrical case when $\rho(\omega)$ and $\phi(\omega)$ are respectively even and odd functions about $\omega = \omega_c$. This formula is exact, and in case of a large number of significant sidebands, a digital computer may be used to extract the fundamental and harmonics by numerical Fourier analysis.

In our particular case of low modulation index, since the number of significant sidebands is limited to a relatively small number, the results may be obtained by ordinary numerical Fourier analysis. The most straightforward method of deciding whether sufficient sidebands have been taken in Eq. (8-38) is to add a further one, re-evaluate the results, and compare the revised fundamental and harmonics with the original results.

In case of large modulation index, the modulation distortion can be evaluated using the quasi-stationary approach when the network characteristics are known analytically. The extended first-order approach is particularly applicable to the intermediate region.

8.4   Distortion Due to Small Sinusoidal Variations of Transmission Characteristics [4,16]

In the previous section, expressions were derived for the harmonic distortion arising from the nonlinearity of the amplitude and phase characteristics of the network, by expressing these non-linearities by the first few terms of a power series. In practice, however, adequate presentation of these characteristics by power series tends to require a substantial number of high-order terms, leading to excessively elaborate distortion formulas.

In this section, we shall derive expressions for the distortion due to a small sinusoidal ripple on either phase and/or amplitude characteristic, using the method of echo distortion. It will be shown that the distortion can be expressed as the product of the distortion due to a single echo and a factor involving the ripple wavelength as the modulating frequency. This method of analysis has already been introduced in Chap. 3, where the method of paired echoes was used in the distortion analysis. In the following, however, we shall first derive an expression for the distortion arising from a single echo and then show its application to the case of sinusoidal ripple.

1. *Analysis of Single Echo.* Consider one component of the spectrum of a modulated signal having an angular frequency ω and amplitude unity. This component is combined with its echo, $a \cos(\omega t - \tau)$, where a is the relative amplitude, and τ is the delay time. The resultant is given by

$$e_r(t) = \cos \omega t + a \cos \omega(t - \tau) = A(\omega) \cos[\omega t + \phi(\omega)] \qquad (8\text{-}39)$$

Using Eqs. (7-24) and (7-26), we obtain

$$A(\omega) = \sqrt{1 + 2a \cos \omega\tau + a^2} = 1 + a \cos \omega\tau, \qquad a < 1 \qquad (8\text{-}40)$$

and $\quad \phi(\omega) = \tan^{-1} \dfrac{-a \sin \omega\tau}{1 + a \cos \omega\tau} \doteq -a \sin \omega\tau, \qquad a < 1 \qquad (8\text{-}41)$

so that the addition of the echo produces, to first order, the same distortion effect as passage through a network whose phase and amplitude characteristics both consist of sinusoidal ripples.

This formula can now be extended to a phase-modulated signal, $\cos[\omega_c t + s(t)]$, where s(t) represents the phase modulation due to the modulation signal, which may consist of a number of tones. The sum of signal and echo will be of the form

$$e_r(t) = \cos[\omega_c t - s(t)] + a \cos[\omega_c(t - \tau) + s(t - \tau)]$$
$$= R(t) \cos[\omega_c t + s(t) + \psi(t)] \qquad (8\text{-}42)$$

where ω_c is the angular carrier frequency. For sufficiently small τ, we have

$$\psi(t) \doteq -a \sin[\omega_c \tau + s(t) - s(t-\tau)] = -a \sin(\omega_c \tau) \cos[s(t) - s(t-\tau)]$$

$$- a \cos(\omega_c \tau)[\sin s(t) - s(t-\tau)] \tag{8-43}$$

Let the modulation signal $s(t)$ consist of a number of tones, so that

$$s(t) = \sum_s f(\omega_s) \sin(\omega_s t + \phi_s)$$

Therefore
$$s(t) - s(t-\tau) = \sum_s f(\omega_s) \sin(\omega_s t - \phi_s) - \sum_s f(\omega_s)$$

$$\times \sin(\omega_s t - \omega_s \tau + \phi_s) = 2 \sum_s f(\omega_s)$$

$$\times \sin\left(\frac{\omega_s \tau}{2}\right) \cos[\omega_s(t - \frac{\tau}{2}) + \phi_s] \tag{8-44}$$

Suppose that
$$- \cos[s(t) - s(t-\tau)] = \sum_n \frac{1}{a} D_s(\omega_n)$$

$$\times \cos[\omega_n(t - \frac{\tau}{2}) + \psi_n] \tag{8-45}$$

and
$$- \sin[s(t) - s(t-\tau)] = \sum_m \frac{1}{a} D_c(\omega_m)$$

$$\times \cos[\omega_m(t - \frac{\tau}{2}) + \zeta_m] \tag{8-46}$$

Then from Eq. (8-44), the phase-modulation distortion becomes

$$\sin(\omega_c \tau) \sum_n D_s(\omega_n) \cos[\omega_n(t - \frac{\tau}{2}) + \psi_n] + \cos(\omega_c \tau) \sum_m D_c(\omega_m)$$

$$\times \cos[\omega_m(t - \frac{\tau}{2}) + \zeta_m] \tag{8-47}$$

where D_s and D_c depend on the delay time.

2. *Sinusoidal Ripple on Group-delay Characteristic.* Let the phase characteristic be of the form

$$\phi(\omega) = -a \sin \omega \tau \tag{8-48}$$

so that the corresponding group-delay characteristic is

$$\frac{d\phi(\omega)}{d\omega} = -a\tau \cos \omega\tau \qquad (8\text{-}49)$$

Consider a single component of the frequency-modulated signal, using Eq. (8-39). The output signal is

$$e_d(t) = A(\omega) \cos (\omega t - a \sin \omega\tau) \doteq A(\omega) [\cos \omega t - \frac{a}{2} \cos \omega(t + \tau)$$

$$+ \frac{a}{2} \cos \omega(t - \tau)] \qquad (8\text{-}50)$$

Thus when a is small, the assumed phase characteristic is equivalent to two echoes, one advanced and one retarded. Note here that a similar result was obtained in Chap. 3 [Eq. (3-42)].

Following the same method as in the case of the single echo, the phase-modulation distortion is given by

$$\psi(t) = -\frac{a}{2} \sin (\omega_c\tau) \cos[s(t) - s(t + \tau)] + \frac{a}{2} \cos (\omega_c\tau) \sin [s(t) - s(t + \tau)]$$

$$- \frac{a}{2} \sin (\omega_c\tau) \cos [s(\tau) - s(t - \tau)] - \frac{a}{2} \cos (\omega_c\tau) \sin [s(t) - s(t - \tau)]$$

$$(8\text{-}51)$$

In case of multiple-tone modulation signal, we have

$$s(t) - s(t - \tau) = 2 \sum_s f(\omega_s) \sin (\omega_s \frac{\tau}{2}) \cos [\omega_s(t - \frac{\tau}{2}) + \phi_s] \qquad (8\text{-}52)$$

and $\quad s(t) - s(t + \tau) = -2 \sum_s f(\omega_s) \sin (\omega_s \frac{\tau}{2}) \cos [\omega_s(t + \frac{\tau}{2} + \phi_s]$

$$(8\text{-}53)$$

Using the results of Eqs. (8-45) and (8-46), the corresponding functions for the advanced echo can be obtained, and finally it can be shown that the phase-modulation distortion becomes

$$\sin (\omega_c\tau) \sum_n D_s(\omega_n) \cos \left(\omega_n \frac{\tau}{2}\right) \cos (\omega_n t + \psi_n) + \cos (\omega_c\tau)$$

$$\times \sum_m D_c(\omega_m) \cos\left(\omega_m \frac{\tau}{2}\right) \cos (\omega_m t + \zeta_m) \qquad (8\text{-}54)$$

3. *Sinusoidal Ripple an Amplitude Characteristic.*   The amplitude characteristic is given by

$$A(\omega) = 1 + a \cos \omega\tau, \qquad a \ll 1 \qquad (8\text{-}55)$$

The analysis is similar to the previous case. The phase-modulation
distortion can be shown to be equal to

$$\sin (\omega_c\tau) \sum_n D_s(\omega_n) \sin\left(\omega_n \frac{\tau}{2}\right) \sin (\omega_n t + \psi_n) + \sin (\omega_c\tau) \sum_m D_c(\omega_m)$$

$$\times \sin\left(\omega_m \frac{\tau}{2}\right) \sin (\omega_m t + \zeta_m) \qquad (8\text{-}56)$$

8.5   Nonlinear Distortion of Complex Frequency-modulated Signal: Exact Solution Using Steady-state Approach [19]

The previous analysis of nonlinear distortion was confined to
special cases of a frequency-modulated carrier in which the modulating signal consisted of a single sinusoidal component, and for
particular forms of transmission network. In this section, the analysis is more general; it covers the case where the modulating signal is complex, and is not limited to the condition that the carrier
frequency be large compared with the modulating frequencies.

In this analysis, the nonlinear amplitude-frequency characteristic will be expressed as a Fourier series over a given frequency
range; by taking a sufficient number of terms, the amplitude-frequency characteristic can be represented over any desired range of
frequencies, and with any degree of accuracy. Following closely the
work of Cherry and Rivlin, [19] we shall use a Fourier series representation of the nonlinear phase-shift-frequency characteristic and
restrict ourselves to small departures from linearity in this characteristic. This does not impose a serious limitation since in practice the tolerable amount of crosstalk in multiplex systems is small,
and consequently, the practically important situations are characterized by small phase variations.

1. *The Application of a FM Complex Signal to a Linear Network: Linear Amplitude-Frequency Characteristics.*   Consider a frequency-modulated signal given by

$$e_i(t) = A_c \sin\left[\omega_c t + \sum_{s=1}^{N} \beta_s \cos(\omega_s t + \phi_s)\right] \qquad (8-57)$$

applied to the input terminals of a four-terminal network having the linear amplitude-frequency characteristic

$$A(\omega) = c_0 + c_1\omega \qquad (8-58)$$

Furthermore, assume that the network introduces zero phase shift at all frequencies. It has been shown by Cherry and Rivlin that the applied frequency-modulated signal of Eq. (8-57) can be expressed in the form

$$e_i(t) = \sum_{\alpha_s=-\infty}^{\infty} A_c \left[\prod_{s=1}^{N} J_{\alpha_s}(\beta_s)\right] \cos\left[\omega_c t + (\alpha-1)\frac{\pi}{2} + \sum_{s=1}^{N} \alpha_s \psi_s\right] \qquad (8-59)$$

with the notation $\psi_s = \omega_s t + \phi_s$, $\alpha = \sum_{s=1}^{N} \alpha_s$. It should be noted here that Eq. (8-59) is the more general expression of N-tone frequency modulation which was discussed earlier and given by Eq. (7-68).

In passing through the network, each spectral component of Eq. (8-59) is modified by the amplitude characteristic of Eq. (8-58) to produce an output

$$e_0(t) = \sum_{\alpha_s=-\infty}^{\infty} A_c \left[\prod_{s=1}^{N} J_{\alpha_s}(\beta_s)\right]\left[c_0 + c_1\left(\omega_c + \sum_{s=1}^{N} \alpha_s\omega_s\right)\right]$$

$$\times \cos\left[\omega_c t + (\alpha-1)\frac{\pi}{2} + \sum_{s=1}^{N} \beta_s \cos\psi_s\right] \qquad (8-60)$$

which, after laborious mathematical operations, reduces to

$$e_0(t) = A_c\left[(c_0 + c_1\omega_c) - c_1\sum_{s=1}^{N} \beta_s\omega_s \sin\psi_s\right]$$

$$\times \sin\left(\omega_c t + \sum_{s=1}^{N} \beta_s \cos\psi_s\right) \qquad (8-61)$$

We conclude therefore that for a linear amplitude-frequency characteristic, the output signal is a "hybrid" wave in which the resulting amplitude-modulation terms are undistorted and the frequency-modulation terms are identical to those for the input.

2. *The Application of a FM Complex Signal to a Linear Network: Linear Phase-shift—Frequency Characteristic.* We consider now a network having a linear phase-shift—frequency characteristic given by

$$\psi_0 = \eta_0 + \eta_1 \omega \qquad (8\text{-}62)$$

and zero attenuation at all frequencies. In any physical network, η_0 will be zero or an integral multiple of π. Applying the frequency-modulated signal of Eq. (8-57) to this network, each spectral component of Eq. (8-59) is modified linearly in phase to produce an output signal given by

$$
\begin{aligned}
e_0(t) &= \sum_{\alpha_s = -\infty}^{\infty} A_c \left[\prod_{s=1}^{N} J_{\alpha_s}(\beta_s) \right] \cos \left[\omega_c t + (\alpha - 1)\frac{\pi}{2} + \sum_{s=1}^{N} \alpha_s \psi_s \right. \\
&\quad \left. + \eta_0 + \eta_1 \left(\omega_c + \sum_{s=1}^{N} \alpha_s \omega_s \right) \right] = \sum_{\alpha_s = -\infty}^{\infty} A_c \left[\prod_{s=1}^{N} J_{\alpha_s}(\beta_s) \right] \\
&\quad \times \cos \left[\omega_c t + (\alpha - 1)\frac{\pi}{2} + \eta_0 + \eta_1 \omega_c + \sum_{s=1}^{N} \alpha_s (\psi_s + \eta_1 \omega_s) \right] \\
&= A_c \sin \left[\omega_c (t + \eta_1) + \eta_0 + \sum_{s=1}^{N} \beta_s \cos (\psi_s + \eta_1 \omega_s) \right] \qquad (8\text{-}63)
\end{aligned}
$$

From Eq. (8-63), it follows that the carrier has undergone a phase shift of $(\eta_0 + \eta_1 \omega_c)$ and the modulating signal has been delayed in time by η_1; this result is identical to that derived in Sec. 8.2 for a single-tone modulating signal.

3. *The Application of a FM Complex Signal to a Linear Network: Linear Amplitude Frequency and Phase-shift—Frequency Characteristics.* A network having linear amplitude-frequency and phase-shift—frequency characteristics can be considered to consist of two networks in tandem, the first having the linear phase-shift-frequency characteristic of Eq. (8-62) and producing zero attenuation for all frequencies, the second having the linear amplitude-frequency

characteristic of Eq. (8-58) and introducing zero phase shift at all frequencies. Applying the frequency-modulated signal of Eq. (8-57) to the first network, the output signal, by the use of Eq. (8-63), is then given by

$$e_0(t) = A_c \sin \left[\omega_c t + \eta_0 + \eta_1 \omega_c + \sum_{s=1}^{N} \beta_s \cos (\psi_s + \eta_1 \omega_s) \right] \qquad (8\text{-}64)$$

Now let us apply this signal to the second network having the amplitude-frequency characteristic of Eq. (8-58). The output is obtained from Eq. (8-61) by substituting $(\omega_c t + \eta_0 + \eta_1 \omega_c)$ for $\omega_c t$ and $(\psi_s + \eta_1 \omega_s)$ for ψ_s. The result will be found to consist of two components: $e_{01}(t) + e_{02}(t)$ where

$$e_{01}(t) = (c_0 + c_1 \omega_c) A_c \sin \left[\omega_c t + \eta_0 + \eta_1 \omega_c + \sum_{s=1}^{N} \beta_s \cos (\psi_s + \eta_1 \omega_s) \right]$$

$$(8\text{-}65)$$

which is the applied FM signal, modified in amplitude by the factor $(c_0 + c_1 \omega_c)$ and delayed as in Eq. (8-63), and

$$e_{02}(t) = c_1 \omega_c A_c \left[\sum_{s=1}^{N} k_s \sin (\psi_s + \eta_1 \omega_s) \right] \sin \left[\omega_c t + \eta_0 + \eta_1 \omega_c \right.$$

$$\left. + \sum_{s=1}^{N} \beta_s \cos (\psi_s + \eta_1 \omega_s) \right], \qquad k_s = \frac{\beta_s \omega_s}{\omega_c} \qquad (8\text{-}66)$$

which is a "hybrid" wave consisting of the applied frequency-modulated wave, delayed in time, and modulated in amplitude by the modulating signal

$$\sum_{s=1}^{N} k_s (\psi_s + \eta_1 \omega_s)$$

which is the original modulating function delayed by time η_1. As mentioned above, such a network followed by an envelope detector can be used to detect the original modulating signal. It is of interest to note here that several types of frequency discriminators employ this same basic principle of detection; this will be more fully investigated in Chap. 13.

4. *The Application of a FM Complex Signal to a Linear Network: Nonlinear Amplitude-Frequency and Phase-shift-Frequency Characteristics.* We have just seen that a network having both amplitude and phase-shift characteristics as linear functions of frequency can be used as an ideal discriminator for frequency-modulated signals. If these characteristics are no longer linear, then the network will no longer behave as a perfect discriminator; the departure from linearity will produce distortion in the envelope of the output wave.

The nonlinear amplitude characteristic may be expressed over a given frequency range B to any desired degree of approximation by the following relation

$$A(\omega) = c_0 + c_1\omega + \sum_{n=1}^{K} (g_n \cos \frac{2\pi n \omega}{B} + h_n \sin \frac{2\pi n \omega}{B}) \qquad (8\text{-}67)$$

The frequency range B must be taken sufficiently large to include all significant sideband components in the output. This is determined from a consideration of the "bandwidth" of the applied signal, together with the form of the characteristic.

In a similar manner, the phase-shift-frequency characteristic may be expressed to any desired degree of approximation by the relation

$$\psi_0 = \eta_0 + \eta_1\omega + \sum_{n-1}^{K} (\lambda_n \cos \frac{2\pi n \omega}{B} + \mu_n \sin \frac{2\pi n \omega}{B}) \qquad (8\text{-}68)$$

The Effect of Nonlinear Amplitude-Frequency Characteristic. First we shall consider the case of nonlinear amplitude-frequency characteristic given by Eq. (8-67) and zero delay at all frequencies. The output signal is then given by

$$e_0(t) = \sum_{\alpha_s = -\infty}^{\infty} A_c \left[\prod_{s=1}^{N} J_{\alpha_s} (\beta_s) \right] \left\{ c_0 + c_1 (\omega_0 + \sum_{s=1}^{N} \alpha_s \omega_s) \right.$$

$$+ \sum_{n=1}^{K} \left[g_n \cos \frac{2\pi n}{B} \left(\omega_0 + \sum_{s=1}^{N} \alpha_s \omega_s \right) \right.$$

$$\left. \left. + h_n \sin \frac{2\pi n}{B} \left(\omega_0 + \sum_{s=1}^{N} \alpha_s \omega_s \right) \right] \right\} \cos \left[\omega_c t + (\alpha - 1)\frac{\pi}{2} + \sum_{s=1}^{N} \alpha_s \psi_s \right]$$

$$(8\text{-}69)$$

This is shown [19] to produce the following result:

$$e_0(t) = A_c \left[(c_0 + c_1\omega_c) - c_1 \sum_{s=1}^{N} \beta_s \omega_s \sin \psi_s \right.$$

$$\times \sin \left(\omega_c t + \sum_{s=1}^{N} \beta_s \cos \psi_s \right) + \sum_{n=1}^{K} \left(\frac{1}{2} g_n A_c \right.$$

$$\times \left\{ \sin \left[\omega_c t + \frac{2\pi n}{B}\omega_c + \sum_{s=1}^{N} \beta_s \cos \left(\psi_s + \frac{2\pi n}{B}\omega_s \right) \right] \right.$$

$$+ \sin \left[\omega_c t - \frac{2\pi n}{B}\omega_c + \sum_{s=1}^{N} \beta_s \cos \left(\psi_s - \frac{2\pi n}{B}\omega_s \right) \right] \right\} - \frac{1}{2} h_n A_c$$

$$\times \left\{ \cos \left[\omega_c t + \frac{2\pi n}{B}\omega_c + \sum_{s=1}^{N} \beta_s \cos \left(\psi_s + \frac{2\pi n}{B}\omega_s \right) \right] \right.$$

$$\left. \left. - \cos \left[\omega_c t - \frac{2\pi n}{B}\omega_c + \sum_{s=1}^{N} \beta_s \cos \left(\psi_s - \frac{2\pi n}{B}\omega_s \right) \right] \right\} \right) \quad (8\text{-}70)$$

Thus, we note that the output consists of an undistorted frequency-modulated wave and a hybrid wave identical with Eq. (8-61), corresponding to the linear term of the amplitude-frequency characteristic, plus a distortion term which consists of "echoes" of the applied signal delayed and advanced in time from the applied signal. Each component in the Fourier series given by Eq. (8-67), representing the nonlinearity of the amplitude-frequency characteristic, is giving rise to one pair of "echoes" delayed and advanced by the time $2\pi n/B$. This is very similar to Eq. (3-35) where the output is given by the original signal and two echoes.

The Effect of Nonlinear Phase-shift-Frequency Characteristic. We shall now consider the case of the nonlinear phase-shift-frequency characteristic given by Eq. (8-68) and zero attenuation at all frequencies. Equation (8-68) may be rewritten in the form

$$\psi_0 = \eta_0 + \eta_1 \omega + \sum_{n=1}^{K} \left[\nu_n \cos\left(\frac{2\pi n}{B}\omega + \chi_n \right) \right] \quad (8\text{-}71)$$

where $\quad \nu_n^2 = \lambda_n^2 + \mu_n^2$ and $\chi_n = \tan^{-1}\left(-\dfrac{\mu_n}{\lambda_n}\right)$

Since the linear phase-shift component of Eq. (8-71) introduces a constant delay only and no distortion in the applied signal, we can therefore neglect in our analysis the term $(\eta_0 + \eta_1\omega)$ and consider the remaining terms only. The output signal is then given by

$$
\begin{aligned}
e_0(t) = A_c \sum_{\alpha_s=-\infty}^{\infty} &\left[\prod_{s=1}^{N} J_{\alpha_s}(\beta_s)\right] \cos\left\{\omega_c t + (\alpha-1)\frac{\pi}{2} + \sum_{s=1}^{N} \alpha_s\psi_s \right. \\
&\left. + \sum_{n=1}^{K} \nu_n \cos\left[\frac{2\pi n}{B}\left(\omega_c + \sum_{s=1}^{N}\alpha_s\omega_s\right) + \chi_n\right]\right\}
\end{aligned}
\tag{8-72}
$$

As stated previously, we restrict our analysis to small phase variations from linearity so that only first-order terms of ν_n are retained. The output $e_0(t)$ can then be shown to reduce to

$$
\begin{aligned}
e_0(t) = A_c \sin&\left(\omega_c t + \sum_{s=1}^{N}\beta_s\cos\psi_s\right) + \sum_{n=1}^{K}\frac{1}{2}\nu_n \\
\times &\left\{\cos\left[\omega_c t + \frac{2\pi n}{B}\omega_c + \chi_n + \sum_{s=1}^{N}\beta_s\cos\left(\psi_s + \frac{2\pi n}{B}\omega_s\right)\right]\right. \\
&\left. + \cos\left[\omega_c t - \frac{2\pi n}{B}\omega_c - \chi_n + \sum_{s=1}^{N}\beta_s\cos\left(\psi_s - \frac{2\pi n}{B}\omega_s\right)\right]\right\}
\end{aligned}
\tag{8-73}
$$

As before, the output is seen to consist of an undistorted frequency-modulated component plus a pair of "echoes" arising from each term of the phase-shift characteristic; this result is identical to Eq. (3-42) except for the constant delay which was deliberately omitted.

References

1. Brown, R. F.: Frequency-modulation Distortion in Linear Networks, *Proc. IEE London*, paper 2196R, January, 1957.
2. Baghdady, E. G.: Theory of Low-distortion Reproduction of FM Signals in Linear Systems, *IRE Trans. Circuit Theory*, vol. CT-5, p. 202, September, 1958.

3. Medhurst, R. G.: Harmonic Distortion of FM Waves by Linear
    Networks, *Proc. IEE London*, paper 1650, vol. 101, part 3,
    p. 171, May, 1954.
4. Assadourian, F.: Distortion of a Frequency-modulated Signal
    by Small Loss and Phase Variations, *Proc. IRE*, vol. 40, p. 172,
    1952.
5. Carson, J. R., and T. C. Fry: Variable-frequency Electric Cir-
    cuit Theory, *BSTJ*, vol. 16, pp. 513-540, October, 1937.
6. Franz, W. J.: The Transmission of a Frequency-modulated
    Wave through a Network, *Proc. IRE*, vol. 34, p. 114P, 1946.
7. Gold, B.: The Solution of Steady State Problems in FM, *Proc.
    IRE*, vol. 37, p. 1264, 1949.
8. Collins, R. H. P., and J. K. Skwirzynski: The Distortion of FM
    Signals in Passive Networks, *Marconi Rev.*, vol. 17, 4th quar-
    ter, pp. 113-136, 1954.
9. Panter, P. F.: A Review of the Sideband Theory of FM with
    Particular Reference to Distortion Problems, ITT Federal
    Laboratories Internal Report, 1947.
10. Kulp, M.: "Spektra und Klirrfactoren Frequenz und Amplituden
    modulierter Schwingungen," Terl I, ENT (19), 72 (1942); Heft 5,
    II, ENT (19), 96 (1942), Heft 6, III (19), 126, (1942), Heft 7.
11. Stumpers, F. W. L. M.: Distortion of Frequency Modulated
    Signals in Electrical Networks, *Commun. News*, vol. 9, p. 82,
    1948.
12. Guttinger, P.: Der Einfluss von Dampfungs und Phasen verzer-
    rungen auf Frequenzmodulierte Wellen, *Assoc. Suisse Elec-
    triciens Bull.*, vol. 36, May, 1945.
13. Medhurst, R. G.: Fundamental and Harmonic Distortion of
    Waves Frequency-modulated with a Single Tone, *Proc. IEE
    London*, paper 3182, March, 1960.
14. Gerlach, A. A.: Distortion-bandpass Considerations in Angular
    Modulation, *Proc. IRE*, vol. 38, p. 1203, 1950.
15. Gladwin, A. S.: The Distortion of Frequency-modulated Waves
    by Transmission Networks, *Proc. IRE*, vol. 35, p. 1436, 1947.
16. Medhurst, R. G., and G. F. Small: Distortion in Frequency-
    modulation Systems Due to Small Sinusoidal Variations of
    Transmission Characteristics, *Proc. IRE*, vol. 40, pp. 1608-
    1612, November, 1956.
17. Van der Pol, B.: The Fundamental Principles of Frequency
    Modulation, *J. IEE London*, vol. 93, part 3, pp. 153-158, May,
    1946.
18. Bloch, A.: Modulation Theory, *J. IEE London*, vol. 91, part 3,
    p. 31, 1944.

19. Cherry, E. C., and R. S. Rivlin: Non-linear Distortion, with
     Particular Reference to the Theory of Frequency Modulated
     Waves, part 1, *Phil. Mag.*, vol. 32, p. 265, October, 1941.
20. Cherry, E. C., and R. S. Rivlin: Non-linear Distortion, with
     Particular Reference to the Theory of Frequency Modulated
     Waves, part 2, *Phil. Mag.*, vol. 33, p. 272, 1942.
21. Skwirzynski, J. K.: The Linear Distortion of FM Signals in
     Band-pass Filters for Large Modulation Frequencies, *Marconi
     Rev.*, vol. 17, 4th quarter, p. 101, 1954.

9

DISTORTION OF FREQUENCY-MODULATED SIGNALS THROUGH LINEAR SYSTEMS: DYNAMIC OR ASYMPTOTIC METHOD

The problem of calculating the distortion of FM signals by linear networks using the dynamic approach was first considered by Carson and Fry[1] in their classic paper entitled Variable-frequency Electric Circuit Theory. This appears to be the earliest publication of a detailed mathematical analysis of the dynamic response of a linear system to a variable-frequency excitation. Various other important contributions have since been published, among which we note the work of Van der Pol,[2] Stumpers,[3] and Baghdady.[4]

The basic viewpoint taken by Carson and Fry is that the dynamic response can be broken up into two components: the quasi-stationary component and the distortion component. As stated in the last chapter, the quasi-stationary component represents the part of the response that can be obtained formally from conventional sinusoidal steady-state circuit theory by substituting the variable instantaneous frequency for the assumed constant frequency. In general, owing to the fact that the system response cannot follow the excitation, the complete solution requires a correction term. This correction term embodies the effect of the system sluggishness upon its response to a given FM signal.[4]

Carson and Fry derived the correction term as an infinite series expansion of the distorted output signal in terms of derivatives of the network transfer characteristics, but the convergence properties of the series were obscured by the complexity of the analysis. Van der Pol and Stumpers followed up with a more direct expansion of the filter response into the desired form of expression. Stumpers has shown that the Carson and Fry series is asymptotic.

In this chapter, we shall first analyze the problem of distortion of pure frequency-modulated signals through linear systems, using the dynamic approach. We shall also derive expressions for the dynamic response of a linear system to a "hybrid" signal consisting of both frequency and amplitude modulation. The conditions for quasi-stationary response and the application to the problem of FM-to-AM conversion will also be investigated in some detail.

9.1 Development of the Expansions Representing the Dynamic Response [1-5]

The dynamic response of a linear network to a frequency-modulated signal is generally derived by two methods. One method starts with the convolution integral, and the second method makes use of the Fourier transform. In the following, we shall derive the expansions representing the dynamic response by the use of both methods. In the convolution integral method, we shall follow closely the derivation of Baghdady.[4]

1. *Convolution Integral Approach.* Consider a linear system that is characterized by its unit-impulse response h(t). Let this system be excited by a FM signal

$$i(t) = e^{j[\omega_c t + \theta(t)]} = e^{j\int_0^t \omega_i(\tau)\,d\tau} \tag{9-1}$$

where

$$\omega_i(t) = \omega_c + \frac{d\theta(t)}{dt} = \omega_c + s(t) \tag{9-2}$$

is the instantaneous frequency of the excitation, ω_c is the frequency of the unmodulated carrier and s(t) carries the specification of the frequency modulation or the message function.

Denote the response, after the initial transients have died out, by

$$e(t) = E(t)\, e^{j[\omega_c t + \theta(t)]} \tag{9-3}$$

Using Eq. (3-55), the expression for the convolution integral, the response e(t) may also be expressed in the form

$$e(t) = \int_0^\infty h(\tau)i(t - \tau) \, d\tau \qquad (9-4)$$

Substitution from Eqs. (9-3) and (9-1) yields

$$E(t) = e(t)e^{-j[\omega_c t + \theta(t)]} = \int_0^\infty h(\tau)e^{-j[\omega_c \tau + \theta(t) - \theta(t - \tau)]} \, d\tau \qquad (9-5)$$

which can be rearranged by the use of Eq. (9-2) as

$$E(t) = \int_0^\infty h(\tau)\left[e^{-j\theta(t)} \cdot e^{j\theta(t - \tau)}\right]e^{-j\omega_c \tau} \, d\tau \qquad (9-6)$$

$$= \int_0^\infty h(\tau)\left[e^{j[\theta(t - \tau) - \theta(t) + \tau \theta'(t)]}\right]e^{-j\omega_i(t)\tau} \, d\tau \qquad (9-7)$$

$$= \int_0^\infty h(\tau)e^{-j\omega \tau} \cdot g(t,\tau) \, d\tau \qquad (9-8)$$

where $g(t,\tau)$ denotes each bracketed quantity in the integrands of Eqs. (9-6) and (9-7). The function $g(t,\tau)$ can be expanded in a Taylor series in powers of $(-\tau)$,

$$g(t,\tau) = \sum_{n=0}^\infty a_n(-\tau)^n \qquad (9-9)$$

and Eqs. (9-6) and (9-7) are then both given by

$$E(t) = \int_0^\infty h(\tau)\left[\sum_{n=0}^\infty a_n(-\tau)^n\right]e^{-j\omega \tau} \, d\tau \qquad (9-10)$$

Two useful expansions of the integral of Eq. (9-10) are possible; in the first expansion ω denotes the carrier frequency, $\omega = \omega_c$, while in the second expansion ω denotes the instantaneous frequency, $\omega = \omega_i(t)$. As will be shown later, the first expansion brings out the discriminator action, while the second is more amenable to the analysis of the conditions for the validity of the quasi-stationary approximation.

From Eq. (9-10), we now obtain under certain general conditions of convergence

$$E(t) = \sum_{n=0}^{\infty} a_n \left[\int_0^{\infty} (-\tau)^n h(\tau) e^{-j\omega\tau} \, d\tau \right] \qquad (9\text{-}11)$$

Let $Z(j\omega)$ denote the transfer function of the network, and since the Fourier transform of the impulse response $h(\tau)$ equals the transfer function, we obtain

$$Z(j\omega) = \int_0^{\infty} h(\tau) e^{-j\omega\tau} \, d\tau \qquad (9\text{-}12)$$

the quantity in brackets in Eq. (9-11) is recognized as equal to $d^n Z(j\omega)/d(j\omega)^n$, and therefore Eq. (9-11) reduces to

$$E(t) = \sum_{n=0}^{\infty} a_n \frac{d^n Z(j\omega)}{d(j\omega)^n} \qquad (9\text{-}13)$$

The arrangements of terms as indicated in Eq. (9-6) thus leads to the Carson and Fry expansion [$\omega = \omega_c$ in Eq. (9-13)], whereas the one in Eq. (9-7) leads to the Van der Pol-Stumpers' expansion [$\omega = \omega_i(t)$ in Eq. (9-13)]. Table 9-1 summarizes both expansions and gives the detailed form of $g(t,\tau)$ for each expansion. Note that the complex amplitude $E(t)$ in Eq. (9-3) may be expressed in the form

$$E(t) = P(t) + jQ(t) = R(t) e^{j\psi(t)} \qquad (9\text{-}14)$$

where $R(t) = \sqrt{P(t)^2 + Q(t)^2}$ and $\psi(t) = \tan^{-1}[Q(t)/P(t)]$; $R(t)$ represents the amplitude-modulation distortion while the frequency distortion component is contained in $\psi(t)$. The total response to the excitation is given by

$$e(t) = R(t) \, e^{j[\omega_c t + \int_0^t s(t) \, dt + \psi(t)]} \qquad (9\text{-}15)$$

which is generally a "hybrid" wave, as discussed in Chap. 8. From Table 9-1 we note also that the coefficients a_n in Eq. (9-9) equal $(1/n!) B_n(t)$ in one case and $(1/n!) C_n(t)$ in the other, where $B_n(t)$ and $C_n(t)$ are defined in Table 9-2.

2. *Validity of the Carson and Fry Expansion.*   From Tables 9-1 and 9-2, the distorted output signal is given by

$$e(t) = e^{j[\omega_c t + \theta(t)]} \left[Z(j\omega_c) + \sum_{n=1}^{\infty} \frac{1}{n!} B_n(t) Z^{(n)}(j\omega_c) \right] \qquad (9\text{-}16)$$

Table 9-1.   Development of the Expansion for the Complex
Amplitude E(t) of the Steady-state Network Response
to a Variable-frequency Excitation

$$i(t) = e^{j[\omega_c t + \theta(t)]}$$

Carson and Fry expansion	Van der Pol-Stumpers' expansion

1. $E(t) = \displaystyle\int_0^\infty h(\tau) e^{-j[\omega_c \tau + \theta(t) - \theta(t - \tau)]} \, d\tau = \int_0^\infty h(\tau) e^{-j\omega\tau} \cdot g(t,\tau) \, d\tau$

$h(t)$ = impulse response of linear passive network

$\omega = \omega_c$ and $g(t,\tau) = e^{-j\theta(t)} \cdot e^{j\theta(t - \tau)}$ $\bigg|$ $\omega = \omega_i(t) = \omega_c + \theta'(t)$ and

$g(t,\tau) = e^{j[\theta(t - \tau) - \theta(t) + \tau\theta'(t)]}$

2. Assume that $\theta(t)$ has finite derivatives of all orders for all values of t.
Thus,

$\theta(t - \tau) = \displaystyle\sum_{n=0}^\infty \frac{\theta^{(n)}(t)}{n!} \cdot (-\tau)^n$ will converge uniformly and absolutely for

all values of t and τ. Note also that $e^x = \displaystyle\sum_{n=0}^\infty \frac{1}{n!} x^n$ has an infinite

range of uniform and absolute convergence.

3. Therefore

$g(t,\tau) = \displaystyle\sum_{n=0}^\infty \frac{1}{n!} B_n(t)(-\tau)^n$ $\qquad$ $g(t,\tau) = \displaystyle\sum_{n=0}^\infty \frac{1}{n!} C_n(t)(-\tau)^n$

converge uniformly for all values of t and τ.
$B_n(t)$ and $C_n(t)$ are given in Table 9-2.

4. Substitute from step 3 into step 1, interchange summation and integration,
and obtain

$E(t) = \displaystyle\sum_{n=0}^\infty \frac{1}{n!} B_n(t)$ $\qquad\qquad$ $E(t) = \displaystyle\sum_{n=0}^\infty \frac{1}{n!} C_n(t)$

$\times \displaystyle\int_0^\infty (-\tau)^n h(\tau) e^{-j\omega_c \tau} \, d\tau$ $\qquad$ $\times \displaystyle\int_0^\infty (-\tau)^n h(\tau) e^{j\omega_i(t)\tau} \, d\tau$

Table 9-1. (Continued)

5. From Eq. (9-12)

$$Z(j\omega) = \int_0^\infty h(\tau)e^{-j\omega\tau}\,d\tau; \text{ therefore } Z^{(n)}(j\omega) = \frac{d^n Z(j\omega)}{d(j\omega)^n}$$

$$= \int_0^\infty (-\tau)^n h(\tau)e^{-j\omega\tau}\,d\tau$$

6. Substitute from step 5 into step 4 to obtain

$$E(t) = \sum_{n=0}^\infty \frac{1}{n!}\,B_n(t)\,Z^{(n)}(j\omega_c) \qquad\qquad E(t) = \sum_{n=0}^\infty \frac{1}{n!}\,C_n(t)\,Z^{(n)}[j\omega_i(t)]$$

$$\text{where } Z^{(n)}(j\omega_c) = \frac{d^n Z(j\omega)}{d(j\omega)^n}\bigg|_{\omega=\omega_c} \qquad\qquad Z^{(n)}[j\omega_i(t)] = \frac{d^n Z(j\omega)}{d(j\omega)^n}\bigg|_{\omega=\omega_i(t)}$$

Equation (9-16) is sometimes given in a slightly different form, namely,

$$e(t) = e^{j\left[\omega_c t + \int_0^t s(\mu)\,d\mu\right]}\left[Z(j\omega_c) + \sum_{n=1}^\infty \frac{(j)^n}{n!}\,B_n'(t)\,Z^{(n)}(j\omega_c)\right]$$

$$B_n(t) \equiv j^n B_n'(t) \quad (9\text{-}17)$$

where $s(t)$ is the variable part of the instantaneous angular frequency. Equation (9-17), which is the basic formula derived by Carson and Fry, may not be convergent, and, as stated earlier, Stumpers has shown[3] that this expansion is of asymptotic nature and diverges in most practical cases. This is due to the peculiar structure of the operators B_n. Thus:

$$B_1' = s(t) \qquad\qquad B_3' = s^3 - 3jss' \;\Big|\; -s''$$

$$B_2' = s^2 - js' \qquad\qquad B_4' = s^4 - 6js^2 s' \;\Big|\; -4ss'' - 3s'^2 + js''' \quad (9\text{-}18)$$

$$\text{and} \quad B_n' = \left(s - j\frac{d}{dt}\right)\left(s - j\frac{d}{dt}\right)\cdots\left(s - j\frac{d}{dt}\right)s = \left[s - j\frac{d}{dt}\right]^{n-1} s,$$

$$\text{operationally, } s \equiv s(t) \qquad\qquad (9\text{-}19)$$

Table 9-2. The Functions $B_n(t)$ and $C_n(t)$ of Table 9-1

n	$B_n(t) \equiv e^{-j\,\theta(t)} \cdot \dfrac{d^n}{dt^n}\, e^{j\,\theta(t)}$	$\dfrac{1}{n!}\,C_n(t)$
0	1	1
1	$j\,\dfrac{d\theta}{dt}$	0
2	$-\left(\dfrac{d\theta}{dt}\right)^2 + j\,\dfrac{d^2\theta}{dt^2}$	$j\,\dfrac{\theta''(t)}{2!}$
3	$\ldots\ldots\ldots\ldots\ldots\ldots$	$j\,\dfrac{\theta'''(t)}{3!}$
4	$B_{n+1} = \left(j\,\dfrac{d\theta}{dt} + \dfrac{d}{dt}\right) B_n$	$j\,\dfrac{\theta^{IV}}{4!} + \dfrac{1}{2!}\left(j\,\dfrac{\theta''}{2!}\right)^2$
5		$j\,\dfrac{\theta^{V}}{5!} + \left(j\,\dfrac{\theta''}{2!}\right)\left(j\,\dfrac{\theta'''}{3!}\right)$
6		$j\,\dfrac{\theta^{VI}}{6!} + \left(j\,\dfrac{\theta''}{2!}\right)\left(j\,\dfrac{\theta^{IV}}{4!}\right)$ $+ \dfrac{1}{2!}\left(j\,\dfrac{\theta'''}{3!}\right)^2 + \dfrac{1}{3!}\left(j\,\dfrac{\theta''}{2!}\right)^3$

Carson and Fry then use the following approximation:

$$B'_n \doteq s^n - \frac{j(n-1)n}{2}\,s's^{n-2} \tag{9-20}$$

which, as we see, includes only the first two terms of each B_n, i.e., in the expressions of the B's (9-18), the terms which are to the left of the vertical dashed line. An approximate expression for the response $e(t)$, as given by Eq. (9-17), can then be derived by expanding $Z[\,j\omega_i(t)]$ as a power series of $s(t)$ as follows:

$$Z[j\omega_i(t)] = Z(j\omega_c) + \sum_{n=1}^{\infty} \frac{s^n}{n!}\, Z^{(n)}(j\omega_c) \tag{9-21}$$

Assuming that the series is convergent and using Eq. (9-20), we obtain by direct substitution the following approximate expression for the response $e(t)$:

$$e(t) = e^{j[\omega_c t + \int_0^t s(t)dt]} \cdot Z(\overline{j\omega}) \tag{9-22}$$

where $\quad Z(\overline{j\omega}) = Z(j\omega_i) + j\,\dfrac{s'(t)}{2}\,Z^{(2)}(j\omega_i) \tag{9-23}$

and $\omega_i(t) = \omega_c + s(t)$.

The expression (9-22) is of great importance in the further history of the problem. It is identical with the formulas used subsequently by Stumpers, Van der Pol, and hence by many authors following Van der Pol, as we shall see later on. The first term of Eq. (9-23) is recognized as the quasi-stationary approximation. Some very useful results may be obtained by means of this approximation, but there is always a question as to the range of its validity. In fact, the quasi-stationary approximation is only true in the limit when the modulation frequency tends to zero. We can also state generally that Eq. (9-23) is true only as far as the first order of the modulating frequency.

3. *Derivation of the Carson and Fry Expansion Using the Fourier Integral Approach.*[1] Carson and Fry have also derived the expansion of Eq. (9-17) using the Fourier transform method. Using their approach with some modification, we write as before

$$i(t) = e^{j[\omega_c t + \int_0^t s(\mu)d\mu]} = e^{j\omega_c t}\,g(t) \tag{9-24}$$

where $\quad g(t) = e^{j\int_0^t s(\mu)d\mu}$

The Fourier transform of $g(t)$ is

$$G(j\omega) = \frac{1}{2\pi} \int_0^T g(t)e^{-j\omega t}\,dt \tag{9-25}$$

and $\quad g(t) = \int_{-\infty}^{\infty} G(j\omega)e^{j\omega t}\,d\omega$

where $g(t)$ is defined for $0 \le t \le T$ and zero for $t > T$. The Fourier transform of $i(t)$ is $I(j\omega)$; therefore from Eq. (9-24)

$$I(j\omega) = G(j\omega - j\omega_c) \tag{9-26}$$

and the response $\quad e(t) = \int_{-\infty}^{\infty} Z(j\omega) G(j\omega - j\omega_c) e^{j\omega t} \, d\omega \tag{9-27}$

If a change of variable $\omega' = \omega - \omega_c$ is made in Eq. (9-27) and ω' is then replaced by ω in the result, Eq. (9-27) becomes

$$e(t) = e^{j\omega_c t} \int_{-\infty}^{\infty} Z(j\omega + j\omega_c) G(j\omega) e^{j\omega t} \, d\omega \tag{9-28}$$

Assume that $Z(j\omega + j\omega_c)$ can be expanded in a Taylor series in the vicinity of $\omega = \omega_c$:

$$Z(j\omega + j\omega_c) = Z(j\omega_c) + \sum_{n=1}^{\infty} \frac{(j\omega)^n}{n!} Z^{(n)}(j\omega_c) \tag{9-29}$$

Therefore $\quad e(t) = e^{j\omega_c t} \left[Z(j\omega_c) \int_{-\infty}^{\infty} G(j\omega) e^{j\omega t} \, d\omega \right.$

$$\left. + \sum_{n=1}^{\infty} \frac{1}{n!} Z^{(n)}(j\omega_c) \int_{-\infty}^{\infty} (j\omega)^n \, G(j\omega) e^{j\omega t} \, d\omega \right] \tag{9-30}$$

In order to express Eq. (9-30) in a more suitable form, we note that

$$\int_{-\infty}^{\infty} (j\omega)^n \, G(j\omega) e^{j\omega t} \, d\omega = \frac{d^n g(t)}{dt^n} \tag{9-31}$$

From Eq. (9-31), we have

$$\int_{-\infty}^{\infty} (j\omega) G(j\omega) e^{j\omega t} \, d\omega = \frac{dg(t)}{dt} = e^{j \int_0^t s(\mu) d\mu} \cdot js(t) \tag{9-32}$$

$$\int_{-\infty}^{\infty} (j\omega)^2 \cdot G(j\omega) e^{j\omega t} \, d\omega = \frac{d^2 g(t)}{dt^2} = e^{j \int_0^t s(\mu) d\mu} \cdot \left(j \frac{ds}{dt} - s^2 \right)$$

$$= e^{j \int_0^t s(\mu) d\mu} \cdot j^2 \left[s(t) - j \frac{d}{dt} \right] s(t) \tag{9-33}$$

$$\int_{-\infty}^{\infty} (j\omega)^3 G(j\omega) e^{j\omega t}\, d\omega = \frac{d^3 g(t)}{dt^3} = e^{j\int_0^t s(\mu)d\mu} \cdot j^3 \left[s(t) - j\,\frac{d}{dt} \right]^2 s(t)$$

$$(9\text{-}34)$$

In these results, the expression $(s - j\, d/dt)^{n-1}\, s$ is exactly equal to B_n' of Eq. (9-19), and consequently Eq. (9-30) reduces to Carson and Fry's expansion as given by Eq. (9-17).

4. *Analysis of Van der Pol-Stumpers' Expansion.*[3-5] An important property of the Van der Pol-Stumpers' expansion was demonstrated by Stumpers. Stumpers assumed a periodic frequency modulation s(pt) of fundamental repetition frequency p rad/sec and showed that the Carson and Fry and the Van der Pol-Stumpers' expansions are expressible in the form

$$E(pt) = \sum_{n=o}^{\infty} G_n(pt) p^n \tag{9-35}$$

where E(pt) is the complex amplitude of the response. In this form, the two expansions are divergent, but asymptotic as the fundamental repetition frequency of the modulation p goes to zero. The functions $G_n(pt)$ are presented in Table 9-3.

Table 9-3. The Coefficients $G_n(pt)$ in Stumpers'
Asymptotic Series

n	$G_n(\phi), \qquad \phi \equiv pt$
0	$Z(j\omega_i)$
1	$j\,\dfrac{1}{2!}\, s'(\phi) Z^{(2)}(j\omega_i)$
2	$j\,\dfrac{1}{3!}\, s''(\phi) Z^{(3)}(j\omega_i) + \dfrac{1}{2!}\left[j\,\dfrac{s'(\phi)}{2!} \right]^2 Z^{(4)}(j\omega_i)$
3	$j\,\dfrac{s'''(\phi)}{4!}\, Z^{(4)}(j\omega_i) + \left[j\,\dfrac{s'(\phi)}{2!} \right]\left[j\,\dfrac{s''(\phi)}{3!} \right] Z^{(5)}(j\omega_i) + \dfrac{1}{3!}\left[j\,\dfrac{s'(\phi)}{2!} \right]^3$
	$\times\, Z^{(6)}(j\omega_i)$

$$G_n(\phi) = \sum \text{ coefficients of } p^n \text{ in the Van der Pol-Stumpers'}$$
$$\text{expansion after substitution of } \theta^{(\nu)}(t) = p^{\nu-1} s^{(\nu-1)}(\phi)$$

$$(9\text{-}36)$$

$$i(t) = e^{j[\omega_c t + \int_0^t s(pt)\,dt]}$$

$$e(t) = e^{j[\omega_c t + \int_0^t s(pt)\,dt]} \cdot \left\{ Z(j\omega_i) + \frac{jps'(pt)}{2} Z^{(2)}(j\omega_i) \right.$$

$$\left. + p^2 \left[\frac{js''(pt) Z^{(3)}(j\omega_i)}{6} - \frac{s'(pt)^2}{8} Z^{(4)}(j\omega_i) \right] + \cdots \right\} \qquad (9\text{-}37)$$

The expansion (9-35) is readily obtainable from the expansion in Table 9-2 by substituting

$$\theta^{(n)}(t) = p^{(n-1)} s(pt)$$

and then grouping the results in terms of ascending powers of p. Therefore, we may conclude that, when the frequency modulation of the excitation is periodic, the complex amplitude of the forced response can be expanded into an asymptotic series about p = 0, where p is the fundamental angular frequency of the modulation.

Returning briefly to Stumpers' asymptotic expansion of Eq. (9-37), we can derive from this series the instantaneous frequency and amplitude by expressing the transfer impedance in terms of its modulus and phase, namely,

$$Z(j\omega) = A(\omega)e^{j\phi(\omega)} \qquad\qquad (9\text{-}38)$$

Applying this to Eq. (9-37), Stumpers derives the following expression for the instantaneous frequency $\omega_i(t)$:

$$\omega_i(t) = \omega_c + s(pt) + p\phi's' + p^2 \left[\frac{s''}{2} \left(-\frac{A''}{A} + \phi'^2 \right) \right.$$

$$\left. + \frac{s'^2}{2} \left(\frac{A''A' - A'''A}{A^2} + 2\phi'\phi'' \right) \right] + \cdots \qquad (9\text{-}39)$$

$$\text{where} \quad s' = \frac{ds(pt)}{d(pt)}, \qquad \phi' = \frac{d\phi(\omega_i)}{d\omega_i}, \qquad A' = \frac{dA(\omega_i)}{d\omega_i}$$

The instantaneous amplitude is given by

$$R(t) = A(\omega_i) + p\left(A'\phi' + \frac{\phi''A}{2}\right) \tag{9-40}$$

In the series of Eqs. (9-39) and (9-40), we call the terms in p of zero degree the "static terms" and those of the first degree, the "quasi-stationary" terms; together they constitute the quasi-stationary approximation.

5. *Conditions for Quasi-stationary Response.* [4] It has been stated earlier that the quasi-stationary response can be obtained formally from conventional steady-state theory by substituting the variable instantaneous frequency for the assumed constant frequency; i.e.,

$$e(t) = Z[j\omega_i(t)] e^{j\int_o^t \omega_i(t)dt} \tag{9-41}$$

The actual response, after the initial transients have died out, is / given by [Eq. (9-3)],

$$e(t) = E(t) e^{j\int_o^t \omega_i(t)dt} \tag{9-42}$$

Baghdady[4] has shown that the actual response can be expressed by

$$e(t) = [Z(j\omega_i) + R_{1e}(t)] e^{j\int_o^t \omega_i(t)dt}$$

where the correction term $R_{1e}(t)$ satisfies the condition

$$|R_{1e}(t)| \le \frac{1}{2}|\theta''(t)|_{max} \cdot \left|\frac{d^2Z(j\omega_i)}{d(j\omega_i)^2}\right| \tag{9-43}$$

This represents an upper bound on the magnitude of the error incurred in using only $Z(j\omega_i)$ to represent $E(t)$.* From the relation $E(t) = Z(j\omega_i) + R_{1e}(t)$, the relative error in the assumption that $E(t) = Z(j\omega_i)$ is bounded by

$$\epsilon_{max} = \frac{1}{2}|\theta''(t)|_{max} \cdot \left|\frac{Z''(j\omega_i)}{Z(j\omega_i)}\right| \ll 1 \tag{9-44}$$

When applied to a single-tuned high-Q circuit, it is shown that

*The validity of this bound has been questioned by H. E. Rowe in Trans. IRE Circuit Theory, vol. CT-9, September, 1962.

$$\epsilon_{max} = \frac{\omega_m}{(BW)/2} \cdot \frac{\Delta\Omega}{(BW)/2} \ll 1 \tag{9-45}$$

where BW is the bandwidth, in radians per second, between the half-power points; ω_m is the modulating angular frequency; and $\Delta\Omega$ peak frequency deviation.

9.2   Response of a Linear System to an AM-FM Signal[6]

It has been mentioned previously that the effect of the transmission medium upon a pure FM signal is to produce, under certain conditions, a "hybrid" signal consisting of both frequency- and amplitude-modulation terms. As we have noted previously, some of the factors which contribute to the variation in the amplitude and phase of the FM signal are: limited bandwidth, nonuniform frequency response, and interference caused by noise and crosstalk.

Expressions were derived in the last section for the dynamic response of a linear system to a pure frequency-modulated wave. However, in practical applications, the input signal may be a "hybrid" consisting of both frequency and amplitude modulation. In the following, we shall derive expressions for the response of a linear system to a signal modulated simultaneously in amplitude and frequency. The derivation of the formula for the response is a generalization of the Fourier transform method used by Carson and Fry, for a pure FM input, which was developed previously.

Following the notation of Eq. (9-24), the input signal is given by

$$i(t) = A(t)\, e^{j[\omega_c t + \int_0^t s(\tau)d\tau]} = e^{j\omega_c t}\, g(t) \tag{9-46}$$

where $\quad g(t) = A(t)\, e^{j\int_c^t s\, dt}, \quad s \equiv s(t) \tag{9-47}$

It is to be noted here that in the case of pure FM signal $g(t) = A_0 e^{j\int_0^t s\, dt}$, where $A_0 = $ constant. Following the same derivations as given in Sec. 9.1, let us examine Eq. (9-31), namely:

$$\int_{-\infty}^{\infty} (j\omega)^n\, G(j\omega)e^{j\omega t}\, d\omega = \frac{d^n g(t)}{dt^n} = \frac{d^n}{dt^n}\left[A(t)e^{j\int_0^t s\, dt}\right] \tag{9-48}$$

We calculate the derivatives of the left-hand side of Eq. (9-48):

$$\int_{-\infty}^{\infty} (j\omega) G(j\omega) e^{j\omega t} \, d\omega = e^{j \int_0^t s \, dt} (A' + jsA) \tag{9-49}$$

$$\int_{-\infty}^{\infty} (j\omega)^2 G(j\omega) e^{j\omega t} \, d\omega = e^{j \int_0^t s \, dt} \left[A'' + 2jsA' + j^2 \left(s - j \frac{d}{dt} \right) s \right] \tag{9-50}$$

$$\int_{-\infty}^{\infty} (j\omega)^3 G(j\omega) e^{j\omega t} \, d\omega = e^{j \int_0^t s \, dt} \Bigg\{ A''' + 3jsA''$$
$$+ 3j^2 \left[\left(s - j \frac{d}{dt} \right) s \right] A' + j^3 \left[\left(s - j \frac{d}{dt} \right)^2 s \right] A \Bigg\} \tag{9-51}$$

Note that Eqs. (9-49) to (9-51) reduce to Eqs. (9-32) to (9-34), for the case of pure FM, namely, when $A(t)$ is a constant.

In these results, $(s - j \, d/dt)^n \, s^1$ is exactly the operator denoted as B_n' in Eq. (9-19), namely

$$B_n'(t) = \left(s - j \frac{d}{dt} \right)^{n-1} s = \left(s - j \frac{d}{dt} \right) \left(s - j \frac{d}{dt} \right) \cdots \left(s - j \frac{d}{dt} \right) s$$

where the operator $(s - j \, d/dt)$ must be applied $(n - 1)$ times. The preceding special cases of Eq. (9-48) can now be expressed in terms of $B_n(t)$. For example,

$$\int_{-\infty}^{\infty} (j\omega)^3 G(j\omega) e^{j\omega t} = e^{j \int_0^t s \, dt} [A''' + 3j B_1 A'' + 3j^2 B_2 A' + j^3 B_3 A] \tag{9-52}$$

This result is in a very interesting form. The expression in the brackets looks like a binomial expansion. In fact, if we make the symbolical substitutions

$$B = j B_1', \qquad B^2 = j^2 B_2' \qquad \text{and} \qquad B^3 = j^3 B_3'$$

we can write the last result in the form

$$\int_{-\infty}^{\infty} (j\omega)^3 G(j\omega) e^{j\omega t} \, d\omega = e^{j \int_0^t s \, dt} (A''' + 3BA'' + 3B^2 A' + B^3 A^0)$$
$$= e^{j \int_0^t s \, dt} \cdot (A + B)^3$$

where $A^o = A$, because the exponent refers to differentiation. The previous result can be generalized by mathematical induction to give

$$\int_{-\infty}^{\infty} (j\omega)^n G(j\omega) e^{j\omega t}\, d\omega = e^{j \int_0^t s(\tau)d\tau} (A + B)^n$$

$$= e^{j \int_0^t s\, dt} \left[A^{(n)} + nA^{(n-1)}B \right.$$

$$+ \frac{n(n-1)}{2!} A^{(n-2)}B^2 + \cdots$$

$$\left. + nA^1 B^{n-1} + A^o B^n \right]$$

$$= e^{j \int_0^t s\, dt} \left[A^{(n)} + nj B_1' A^{(n-1)} \right.$$

$$+ \frac{n(n-1)}{2!} j^2 B_2' A^{(n-2)} + \cdots$$

$$\left. + nj^{n-1} B_{n-1}' A^1 + j^n B_n' A^o \right] \qquad (9\text{-}53)$$

Finally, the response of the network to a signal modulated simultaneously in amplitude and frequency is obtained by substituting Eq. (9-53) into (9-30), which is rewritten as Eq. (9-54).

$$e(t) = e^{j\omega_c t} \left[Z(j\omega_c) \int_{-\infty}^{\infty} G(j\omega) e^{j\omega t}\, d\omega \right.$$

$$\left. + \sum_{n=1}^{\infty} \frac{1}{n!} Z^{(n)}(j\omega_c) \int_{-\infty}^{\infty} (j\omega)^n \cdot G(j\omega) e^{j\omega t}\, d\omega \right]$$

$$= e^{j(\omega_c t + \int_0^t s\, dt)} \cdot \left\{ Z(j\omega_c)A(t) + \sum_{n=1}^{\infty} \frac{[A(t) + B]^n}{n!} Z^{(n)}(j\omega_c) \right\}$$

$$(9\text{-}54)$$

Equation (9-54) can be interpreted as the generalization of the basic result of Carson and Fry for the FM case as given in Eq. (9-17); it reduces to Eq. (9-17) when A(t) is a constant. In case of AM only, Eq. (9-54) reduces to

$$e_{AM}(t) = e^{j\omega_c t} \left[Z(j\omega_c)A(t) + \sum_{n=1}^{\infty} \frac{A(t)^{(n)}}{n!} Z^{(n)}(j\omega_c) \right] \qquad (9\text{-}55)$$

If A(t) has the special form

$$A(t) = I(1 + m_a \sin pt) \tag{9-56}$$

then Eq. (9-55) should be identical with the steady-state result obtained from sideband analysis, as will be shown presently. Consider

$$i(t) = I(1 + m_a \sin pt)\, e^{j\omega_c t} = I\left(e^{j\omega_c t} + \frac{m_a}{2j}\, e^{j(\omega_c + p)t}\right.$$

$$\left. - \frac{m_a}{2j}\, e^{j(\omega_c - p)t}\right) \tag{9-57}$$

The output e(t) may be obtained by multiplying $I(t) \cdot Z(j\omega)$; therefore

$$e(t) = I\left[Z(j\omega_c)e^{j\omega_c t} + \frac{m_a}{2j}\, Z(j\omega_c + jp)e^{j(\omega_c + p)t} - \frac{m_a}{2j}\right.$$

$$\left. \times\, Z(j\omega_c - jp)e^{j(\omega_c - p)t}\right] \tag{9-58}$$

Now expand $Z(j\omega_c \pm jp)$ in a power series about $\omega = \omega_c$. Therefore

$$Z(j\omega_c \pm jp) = Z(j\omega_c) + \sum_{n=1}^{\infty} \frac{(\pm jp)^n}{n!}\, \frac{d^n Z(j\omega)}{d(j\omega)^n}\bigg|_{\omega = \omega_c} \tag{9-59}$$

If this series is substituted into Eq. (9-58), we obtain an expression identical to Eq. (9-55), which represents the response of the network to a pure AM wave.

Let us consider again Eq. (9-54), which we derived previously for the response of the linear system to a signal modulated simultaneously in amplitude and frequency. If we rewrite this equation in the form

$$e(t) = R(t)e^{j[\omega_c t + \int_0^t s\, dt + \psi(t)]} \tag{9-60}$$

we see that the effect of the linear system upon the input i(t) is in general to change both the instantaneous frequency and amplitude. No matter whether the input is originally a FM, an AM, or an AM-FM signal, the output is generally an AM-FM or a hybrid modulated signal whose amplitude is $R(t)$, and the instantaneous frequency of the response $[\omega_i(t)]_r$ is given by

$$[\omega_i(t)]_r = \omega_c + s(t) + \frac{d\psi(t)}{dt} \tag{9-61}$$

9.3 Theory of FM-to-AM Conversion [4,7]

The general theory of detection of frequency-modulated waves will be treated in Chap. 13. In our present analysis, the FM discriminator will be considered as a network that is designed to convert the instantaneous-frequency variations of an amplitude-limited wave into instantaneous amplitude variations, which are then detected by an envelope detector. It will be shown in Chap. 13 that the operation of many FM discriminators such as off-resonance, Travis, Foster-Seeley, ratio detector, etc., is based on this principle.

Consider an amplitude-limited frequency-modulated carrier given by Eq. (9-1), applied to a linear network designed to operate as a discriminator. From Eq. (9-3) and (9-16), we find that the complex amplitude of the response $E(t)$ can be represented by

$$E(t) = \left[Z(j\omega_c) + \sum_{n=1}^{\infty} \frac{1}{n!} B_n(t) Z^n(j\omega_c) \right] = R(t)e^{j\psi(t)} \tag{9-62}$$

The envelope $R(t)$ contains the message function and generally some distortion terms introduced by the nonlinearities of the amplitude-frequency and phase-shift-frequency characteristics of the network. For a network designed for FM-to-AM conversion with little distortion, we take $n = 1$ in Carson and Fry's expansion of Eq. (9-62) and obtain

$$E(t) = Z(j\omega_c) + jZ'(j\omega_c)\,\theta'(t) + R_{2e}(t) \tag{9-63}$$

where $R_{2e}(t)$ represents the error incurred in approximating the complex amplitude by two terms in the Carson and Fry expansion. It can be shown[4] that

$$|R_{2e}(t)| \le \tfrac{1}{2} \left| \sqrt{[\theta'(t)]^4 + [\theta''(t)]^2} \right|_{max} \cdot \left| \frac{d^2 Z(j\omega_c)}{d(j\omega_c)^2} \right| \tag{9-64}$$

Assuming an ideal envelope detector, the detected output is then given by

$$R(t) = |E(t)| = |Z(j\omega_c)| \cdot \left| 1 + j\,\frac{Z'(j\omega_c)}{Z(j\omega_c)}\,\frac{d\theta(t)}{dt} + \frac{R_{2e}(t)}{Z(j\omega_c)} \right|$$

$$(9-65)$$

For good sensitivity and low distortion, we require that $jZ'(j\omega_c)/Z(j\omega_c)$ be real, that $Z'(j\omega_c)$ be high, and that the relative error

$$\epsilon = \frac{1}{2}\left| \sqrt{\left(\frac{d\theta}{dt}\right)^4 + \left(\frac{d^2\theta}{dt^2}\right)^2}\,\right|_{max} \cdot \left| \frac{Z''(j\omega_c)}{Z(j\omega_c)} \right| \ll 1 \qquad (9-66)$$

As an illustration, let

$$\theta(t) = \beta \sin \omega_m t, \qquad \beta = \frac{\Delta\omega}{\omega_m}$$

and let the network be a high-Q parallel-resonant circuit (off-resonance discriminator) whose half-bandwidth between half-power points is given by α rad/sec. If x_c represents the deviation of ω_c from the resonant frequency in units of α, then

$$\varepsilon = \left(\frac{\Delta\omega}{\alpha}\right)^2 (1 + x_c^2)^{-1} < \left(\frac{\Delta\omega}{\alpha}\right)^2, \qquad \beta > 1$$

$$(9-67)$$

$$= \frac{\omega_m \Delta\omega}{\alpha^2}(1 + x_c^2)^{-1} < \left(\frac{\omega_m}{\alpha}\right)\left(\frac{\Delta\omega}{\alpha}\right), \qquad \beta < 1$$

We shall conclude this section with a few examples which will illustrate the application of the expressions derived above to the problem of FM-to-AM conversion.

1. *Ideal Discriminator.* Such an ideal discriminator is, for example, an inductance L in the plate circuit of a vacuum tube, followed by an envelope detector, as shown in Fig. 9-1. The impedance of the network may be described as

$$Z(j\omega) = jX_c[1 + a(\omega - \omega_c)]$$

where $\qquad X_c = \omega_c L \qquad$ and $\qquad a = \dfrac{1}{\omega_c}$

Therefore $\quad Z(j\omega_c) = jX_c$

$$Z'(j\omega_c) = aX_c$$

$$Z''(j\omega_c) = 0$$

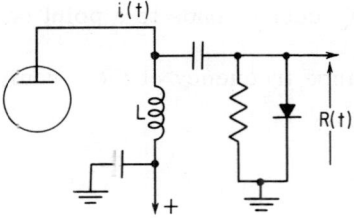

Fig. 9-1. Ideal discriminator.

If a FM wave of the form $i(t) = e^{j \int_0^t \omega_i(t) dt}$, where $\omega_i(t) = \omega_c + \theta'(t)$ $= \omega_c + \Delta\omega \cos pt$, is passed through the inductance, the output voltage after rectification is obtained by the use of Eq. (9-65), giving the result

$$R(t) = |Z(j\omega_c)| \left| 1 + j \frac{Z'(j\omega_c)}{Z(j\omega_c)} \theta'(t) \right|$$

$$= |jX_c| \left| 1 + j \frac{aX_c}{jX_c} \Delta\omega \cos pt \right| = X_c \left(1 + \frac{\Delta\omega}{\omega_c} \cos pt \right)$$

which contains the initial modulating voltage without distortion. This discriminator is ideal because the correction term $R_{2e}(t)$ is zero, and, at least in theory, no distortion terms appear in the output.

2. *Off-resonance Discriminator.* The circuit often used in practice as a discriminator consists of a parallel-tuned circuit in the plate circuit of a tube, followed by an envelope detector, as shown in Fig. 9-2. This is called an off-resonance discriminator because the

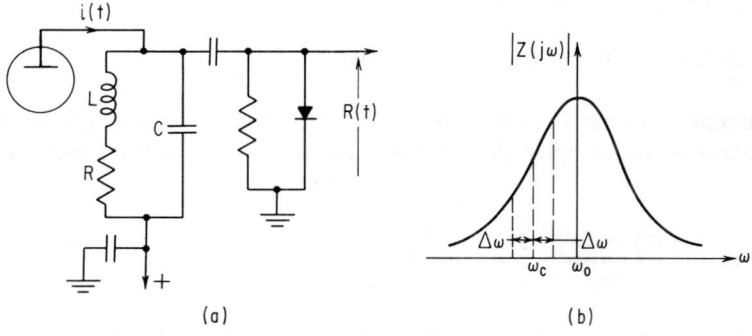

(a)   (b)

Fig. 9-2. Off-resonance discriminator.

carrier frequency ω_c corresponds to a point on the slope of the resonance curve.

Denote the resonance frequency of the circuit by ω_0; the impedance $Z(j\omega)$ is then

$$Z(j\omega) = \frac{Z_0}{1 + jx}$$

where Z_0 is the resonance impedance or the dynamic resistance of the parallel tuned circuit, and

$$x = Q\left(\frac{\omega}{\omega_0} - \frac{\omega_0}{\omega}\right)$$

$$x_c = Q\left(\frac{\omega_c}{\omega_0} - \frac{\omega_0}{\omega_c}\right), \qquad \omega_c = \text{carrier angular frequency}$$

Let us expand $Z(j\omega)$ about ω_c or $Z(jx)$ about x_c, where $x - x_c < 1$, which is the condition for convergence. The result is

$$R(t) = \frac{R}{\sqrt{1 + x_c^2}}\left[1 - \frac{2x_c}{1 + x_c^2}\left(\frac{\Delta\omega}{\omega_c}Q\right)\cos pt + \frac{2x_c^2 - 1}{(1 + x_c^2)^2}\left(\frac{\Delta\omega}{\omega_c}Q\right)^2\right.$$

$$\left. \times \cos 2pt + \frac{3x_c^2 - 2x_c^3}{(1 + x_c^2)^3}\left(\frac{\Delta\omega}{\omega_c}Q\right)^3 \cos 3pt + \cdots\right]$$

where the calculation is carried out to third-order terms. The amplitude of the harmonics are clearly given in terms of the deviation $\Delta\omega$ and Q. The distortion is a function of the operating point x_c. Notice that at the point of inflection

$$2x_c^2 - 1 = 0 \qquad \text{or} \qquad x_c = 0.707$$

the second harmonic vanishes, and the third harmonic is given by the ratios of the amplitudes of the third harmonic to the fundamental, or

$$d_3 = \frac{(3x_c^2 - 2x_c^3)/(1 + x_c^2)^3}{2x_c/(1 + x_c^2)}\left(\frac{\Delta\omega}{\omega_c}Q\right)^2$$

Evaluated at $x_c = \frac{1}{2}$, the result is $d_3 = \frac{4}{9}[(\Delta\omega/\omega_c)Q]^2$. Additional examples of distortion problems in FM discriminators will be given in Chap. 13.

9.4 Application of the Dynamic Method to the Problem of Distortion of FM Signals through Linear Systems [2,3,7]

In this section, we shall illustrate the use of the expansions which were derived above for the response of several linear networks to frequency-modulated signals. The results will, in some instances, be compared with those obtained using the quasi-stationary approximation.

1. *Distortion in a Single-tuned Circuit.* We apply a frequency-modulated signal $i(t) = e^{j[\omega_c t + \int_0^t s(pt)dt]}$ to a single-tuned circuit, as shown in Fig. 9-3; the problem is to find the response $e(t)$ or the voltage $e(t)$ across the tuned circuit. Using Van der Pol-Stumpers' expansion [Eq. (9-37)], the response $e(t)$ is given by

$$e(t) = e^{j[\omega_c t + \int_0^t s(pt)dt]} \cdot \left[Z(j\omega_i) + \frac{jps'(pt)}{2} Z''(j\omega_i) \right.$$

$$\left. + p^2 \frac{js''(pt) Z'''(j\omega_i)}{6} + \cdots \right] \tag{9-68}$$

where $Z(j\omega_i) = \dfrac{Z_0}{1 + jx_i}$, $x_i = Q\left(\dfrac{\omega_i}{\omega_c} - \dfrac{\omega_c}{\omega_i}\right) \doteq \dfrac{2Q}{\omega_c}(\omega_i - \omega_c)$

and $Z_0 = \dfrac{L}{CR}$, the resonance impedance

It should be noted that the carrier frequency of the applied signal ω_c coincides in this example with the resonant frequency of the tuned circuit; consequently $\omega_c^2 LC = 1$. Stumpers has derived the

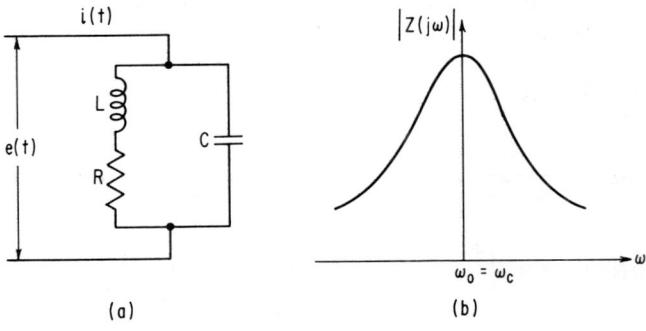

(a)   (b)

Fig. 9-3. Single-tuned circuit.

following expression for the instantaneous frequency of the response $e(t)$ for the special case where $s(t) = \Delta\omega \cos pt$.

$$[\omega_i(t)]_r = \omega_c + \Delta\omega \cos pt + \frac{p\alpha \, \Delta\omega \sin pt}{\alpha^2 + \Delta\omega^2 \cos^2 pt}$$

$$+ \frac{p^2 \, \Delta\omega \cos pt}{(\alpha^2 + \Delta\omega^2 \cos^2 pt)^3} \cdot [-\alpha^4 - 6\alpha^2 \, \Delta\omega^2 \sin^2 pt$$

$$+ \Delta\omega^4 \cos^2 pt(1 + \sin^2 pt)] \tag{9-69}$$

where $\alpha = R/2L$. Equation (9-69) can be expressed in terms of the harmonic distortion components.

$$[\omega_i(t)]_r = \omega_c + \Delta\omega \cos pt + p \sum_{n=0}^{\infty} C_{2n+1} \sin (2n + 1)pt$$

$$+ p^2 \sum_{n=0}^{\infty} D_{2n+1} \cos (2n + 1)pt \tag{9-70}$$

where $C_{2n+1} = (-1)^n 2c^{-(2n+1)} \left[(1 + c^2)^{\frac{1}{2}} - 1\right]^{2n+1}, \qquad c = \frac{\Delta\omega}{\alpha}$

$$\tag{9-71}$$

and $D_{2n+1} = (-1)^n 2(2n + 1)^2 \alpha^{-1} c^{-(2n+1)}(1 + c^2)^{-\frac{1}{2}} [(1 + c^2)^{\frac{1}{2}} - 1]^{2n+1}$

$$\tag{9-72}$$

Let us compare the result of Eq. (9-69) with the quasi-stationary solution, namely, $e(t) = Z[j\omega_i(t)]e^{j\int_0^t \omega_i(t) \, dt}$.

Expressing the impedance in terms of the amplitude and phase response, we write

$$Z(j\omega_i) = A(\omega_i)e^{j\phi(\omega_i)} \tag{9-73}$$

where $A(\omega_i) = \frac{Z_0}{\sqrt{1 + x_i^2}} \doteq Z_0 \left[1 - \frac{2Q^2}{\omega_c^2} (\omega_i - \omega_c)^2\right] \tag{9-74}$

and $\phi(\omega_i) = \tan^{-1} \left[-\frac{2Q}{\omega_c} (\omega_i - \omega_c)\right] = \tan^{-1} \left[-\frac{1}{\alpha} (\omega_i - \omega_c)\right] \tag{9-75}$

The response is therefore given by

$$e(t) = A(\omega_i) e^{j[\int_0^t \omega_i dt + \phi(\omega_i)]}$$ (9-76)

and the instantaneous frequency of the response equals the time derivative of the phase angle, or

$$[\omega_i(t)]_r = \frac{d}{dt}\left[\int_0^t \omega_i \, dt + \phi(\omega_i)\right] = \omega_i(t) + \frac{d}{dt}[\phi(\omega_i)]$$

$$= \omega_c + s(t) + \frac{d}{d\omega_i}[\phi(\omega_i)]s'(t)$$ (9-77)

In this problem, we have considered $s(t) = \Delta\omega \cos pt$, so that

$$[\omega_i(t)]_r = \omega_c + \Delta\omega \cos pt - \Delta\omega p \sin pt \frac{d\phi(\omega_i)}{d\omega_i}$$ (9-78)

From Eq. (9-75), we derive the following expression:

$$\tan \phi(\omega_i) = -\frac{1}{\alpha}(\omega_i - \omega_c)$$

Hence $$\frac{d\phi(\omega_i)}{d\omega_i} = -\frac{1}{\alpha}\frac{1}{1 + \tan^2 \phi(\omega_i)}$$

$$= -\frac{1}{\alpha}\frac{1}{1 + (1/\alpha)^2(\Delta\omega \cos pt)^2}$$

Finally $$[\omega_i(t)]_r = \omega_c + \Delta\omega \cos pt + \frac{\Delta\omega\alpha p \sin pt}{\alpha^2 + \Delta\omega^2 \cos^2 pt}$$

We note here that the distortion term using the quasi-stationary approximation is equal to the first distortion term only of Eq. (9-69).

2. *Van der Pol's Approximate Equation.* Van der Pol considered also the case when a higher approximation than the quasi-stationary solution must be taken into account. He derived the following expression for the instantaneous frequency of the response.

$$[\omega_i(t)]_r = \omega_i(t) + \frac{d\phi(\omega_i)}{dt} + \frac{d}{dt}\left\{\frac{d\omega_i(t)}{dt}\left[\phi'^2(\omega_i) - \frac{A''(\omega_i)}{A(\omega_i)}\right]\right\}$$ (9-79)

The second term represents the distortion derived from the quasi-stationary approximation, whereas the third term gives the distortion which must also be considered when the quasi-stationary solution does not suffice. In most practical applications, Eq. (9-79) can be simplified to

$$[\omega_i(t)]_r = \omega_i(t) + \frac{d\phi(\omega_i)}{dt} + \left[\phi'^2(\omega_c) - \frac{A''(\omega_c)}{A(\omega)}\right]\frac{d^2\omega_i(t)}{dt^2} \qquad (9\text{-}80)$$

Equation (9-80) shows that, whereas the quasi-stationary solution is determined by the phase characteristic of the linear network only, the last term depends also upon the amplitude characteristic of the network.

References

1. Carson, J. R., and T. C. Fry: Variable-frequency Electric Circuit Theory, *BSTJ*, vol. 16, pp. 513-540, October, 1937.
2. Van der Pol, B.: The Fundamental Principles of Frequency Modulation, *J.IEE London*, vol. 93, part 3, pp. 153-158, May, 1946.
3. Stumpers, F. L. H. M.: Distortion of Frequency Modulated Signals Electrical Networks, *Commun. News*, vol. 9, pp. 82-92, April, 1948.
4. Baghdady, E. J.: Theory of Low-distortion Reproduction of FM Signals in Linear Systems, *Trans. IRE Circuit Theory*, vol. CT-5, pp. 202-214, September, 1958.
5. Collins, R. H. P., and J. K. Skwirzynski: The Distortion of FM Signals in Passive Networks, *Marconi Rev.*, vol. 17, 4th quarter, pp. 113-136, 1954.
6. Assadourian, F.: Theory of Steady-state Response of a Linear System to an AM-FM Input, ITT Federal Laboratories Internal Report, 1947.
7. Vellat, T.: Der Empfang Frequenzmodulierter Wellen, *Elek. Nach.-Tech.*, vol. 18, p. 72, 1940.
8. Bloch, A.: Modulation Theory, *J.IEE London*, vol. 91, part 3, p. 31, 1944.

10

TRANSIENT RESPONSE IN FREQUENCY MODULATION

In the last two chapters, we have dealt with the problem of finding the response of a linear network to angle-modulation signals. Two methods of approach were used to calculate the nonlinear distortion introduced by the system. In Chap. 8, the Fourier or steady-state approach was used where the input signal was resolved into its spectral steady-state components, and each component modified by the transfer function of the network. The output was then found by synthesis of the modified components. In Chap. 9, we demonstrated the use of the quasi-stationary approach with corrections as expressed in the series of Carson and Fry or Van der Pol-Stumpers and modified by Baghdady.

We have noted earlier that the response of the network to FM signals consists actually of two parts, the quasi-steady-state response and the transient response. As shown above, the concept of quasi-stationary behavior of a network implies that the instantaneous amplitude and frequency at the output can be derived from the instantaneous amplitude and frequency at the input by multiplication by the value of the transfer function applicable to that frequency.[1] As long as the instantaneous frequency of the modulated signal varies slowly enough with time so that the network to which the

325

modulated signal is applied behaves in a quasi-stationary manner, the instantaneous frequency of the network response will be a close replica of the instantaneous frequency of the excitation. However, when the instantaneous frequency of the modulating signal changes abruptly so that quasi-stationary conditions are violated, the instantaneous input and output frequencies of the network may differ significantly.[2] The problem of finding the transient response appears to become of considerable importance, in view of the growing popularity of digital systems of modulation which, when applied to exponential modulation, would result in modulation by functions containing points of discontinuity such as step and ramp functions.

The transient response of linear networks to FM signals, assuming an instantaneous jump in the signal frequency, has been investigated by several workers.[3-6] Their work has been summarized by Weiner,[2] who has carried out extensive experimental work to verify the analytical results. His results agree with the theoretical treatment and are accounted for by a simple theoretical model, originally developed by Baghdady,[7] which explains the mechanism of the generation of FM transients in terms of the normal modes of the network. It will be shown that when the frequency of the signal applied to a single-tuned circuit is suddenly varied, the resultant output signal is composed of two parts: a transient component having a frequency equal to the natural frequency of the circuit and a steady-state component produced by the applied signal. These two components can be added vectorially to give a resultant whose instantaneous frequency is obtained by finding the time derivative of its phase angle. The resultant frequency of the response will be seen to consist also of two parts: a mean frequency which varies exponentially, the rate of variation being governed by the damping factor of the circuit, and an alternating or "flutter" component whose frequency of variation is equal to the difference between the final frequency and the natural frequency of the circuit.

In this chapter, we shall first discuss the theoretical model of Baghdady as presented by Weiner, which will be followed by experimental results in support of the theoretical model. We shall then develop analytical expressions for the transient response of simple linear networks to frequency and phase steps, using the method of Laplace transforms.

10.1 Theoretical Model to Simplify FM Transient Analysis[2, 7]

The analysis of FM transients can be greatly simplified by the use of the theoretical model which was developed by Baghdady.[7] We

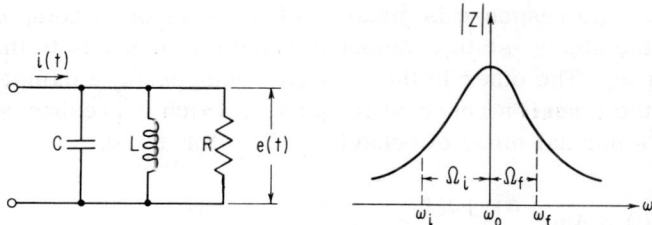

Fig. 10-1. Parallel-resonance circuit and its resonance curve.

shall illustrate its application by considering the response of a single-tuned high-Q circuit to a frequency step. The frequency step occurring at $t = 0$ can be represented by the application of a signal $e^{j\omega_i t}$ at $t \ll 0$, its removal at $t = 0$, and the application of a signal $e^{j\omega_f t}$ at $t = 0$, where ω_i and ω_f are fixed angular frequencies of the initial and final excitation, respectively. Consider the parallel-resonant circuit as shown in Fig. 10-1; the impedance of this circuit, assuming high Q is

$$Z(j\Omega) = \frac{R}{1 + j\,\Omega/\alpha} = \frac{R}{1 + jx} \tag{10-1}$$

where $\Omega = \omega - \omega_0$ is the deviation of the frequency of excitation from the resonant frequency ω_0, and $\alpha = 1/2RC$ is the damping constant or the half-bandwidth of the tuned circuit. The assumed excitation is denoted by

$$i(t) = e^{j \int_0^t [\omega_i + u_{-1}(t)(\omega_f - \omega_i)]\, dt} \tag{10-2}$$

where $u_{-1}(t)$ is a unit step function occurring at $t = 0$, defined as

$$u_{-1}(t) = 0, \qquad t < 0$$

$$= 1, \qquad t > 0$$

For $t < 0$, the response $e(t)$ is found from steady-state considerations to be

$$e(t) = \frac{R}{1 + j\,\Omega_i/\alpha}\ e^{j\omega_i t} = Z(j\Omega_i)e^{j\omega_i t} \tag{10-3}$$

For $t > 0$, i.e., after the frequency of excitation has switched from

ω_i to ω_f, the response is made up of two sinusoidal components. One is the steady-state component which corresponds to the new excitation ω_f. The other is the transient component, a damped oscillation at the circuit's resonant frequency, which corresponds to the circuit's normal mode of behavior. Thus for $t > 0$,

$$e(t) = A_T' e^{-\alpha t} e^{j\omega_o t} + \frac{R}{1 + j\,\Omega_f/\alpha}\,e^{j\omega_f t}$$

$$= A_T' e^{-\alpha t}\, e^{j\omega_o t} + Z(j\Omega_f)\,e^{j\omega_f t} \tag{10-4}$$

Since the two expressions for Eqs. (10-3) and (10-4) must be identical at $t = 0$, we have

$$A_T' = Z(j\Omega_i) - Z(j\Omega_f) \tag{10-5}$$

Therefore, the response for $t > 0$ is given by

$$e(t) = [Z(j\Omega_i) - Z(j\Omega_f)]\, e^{-\alpha t} e^{j\omega_o t} + Z(j\Omega_f)e^{j\omega_f t} \tag{10-6}$$

Equation (10-6) reveals that there are two damped oscillations at the circuit's resonant frequency. One oscillation is caused by the shift in frequency from ω_i to ω_o of the energy which the circuit possessed prior to $t = 0$. The other oscillation, equal in magnitude but opposite in phase to the steady-state component at $t = 0_+$, results from the restriction that the steady-state component of the response must equal zero at $t = 0_+$.

Normalizing $e(t)$ with respect to $Z(j\Omega_f)e^{j\omega_f t}$, we have from Eq. (10-6) for $t > 0$:

$$e'(t) = \frac{e(t)}{Z(j\Omega_f)e^{j\omega_f t}} = 1 + \left[\frac{Z(j\Omega_1)}{Z(j\Omega_f)} - 1\right] e^{-\alpha t} e^{-j\Omega_f t} = 1 + A_T e^{-\alpha t}$$

$$\times e^{-j\Omega_f t} \tag{10-7}$$

The phasor diagram corresponding to Eq. (10-7) is shown in Fig. 10-2. Note that the unit vector is stationary while the vector $A_T e^{-\alpha t}$ decays exponentially with time and rotates at an angular frequency Ω_f; $\phi = \phi_T - \Omega_f t$, where ϕ_T = phase angle of complex quantity A_T $= |A_T|e^{j\phi_T}$. The instantaneous frequency of $e'(t)$ is given by the

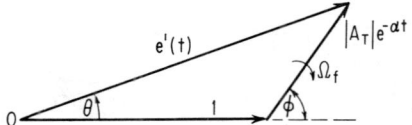

Fig. 10-2. Phasor diagram for e'(t) to explain FM transients.

time derivative of θ and equals the instantaneous frequency of
e(t) minus ω_f. The phasor diagram for A(t) is shown in Fig. 10-3.

The phasor models of Figs. 10-2 and 10-3 can be used to explain
all the amplitude and frequency transient phenomena of the response
of a single-tuned circuit to a series of frequency steps. Consider,
for example, the situation shown in Fig. 10-4 where (O – O') represents the unit vector. As the phasor, $A_Te^{-\alpha(t)}$ decays exponentially,
the phasor e'(t) wobbles back and forth with a period (approximately)
equal to $2\pi/\Omega_f$, and the tip of e'(t) traces out a spiral. At point A,
e'(t) is at rest, and the instantaneous frequency of e(t) equals ω_f.
As e'(t) rotates toward B, the time rate of change of the phase angle
θ increases, and an overshoot in the instantaneous frequency of e'(t)
results. The maximum overshoot occurs at point B. The time rate
of change of θ decreases as e'(t) rotates from B. At point C, e'(t) is
again at rest. As e'(t) rotates toward D, $d\theta/dt$ continues to decrease,
and the instantaneous frequency of e'(t) falls below ω_f. The instantaneous frequency reaches its minimum value at point D. As e'(t)
rotates from D, $d\theta/dt$ begins to increase once again. At point E,
$d\theta/dt = 0$, and one cycle of ringing in the instantaneous-frequency
response of e'(t) is completed. The cycle is repeated for points E,
F, G, and H and continues to repeat until $A_Te^{-\alpha t}$ is negligible compared to the unit vector. Since such points as B and F occur at increasingly greater distances from O, overshoots in the frequency
response of e'(t) occur with successively smaller magnitudes.

The occurrence of undershoots in the instantaneous frequency of
the response is explained with the use of Fig. 10-5. For the case
shown, the phasor $A_Te^{-\alpha t}$ is sufficiently large with respect to the

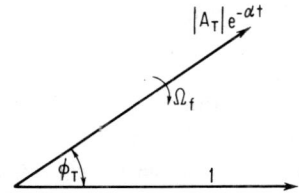

Fig. 10-3. Phasor diagram for A_T.

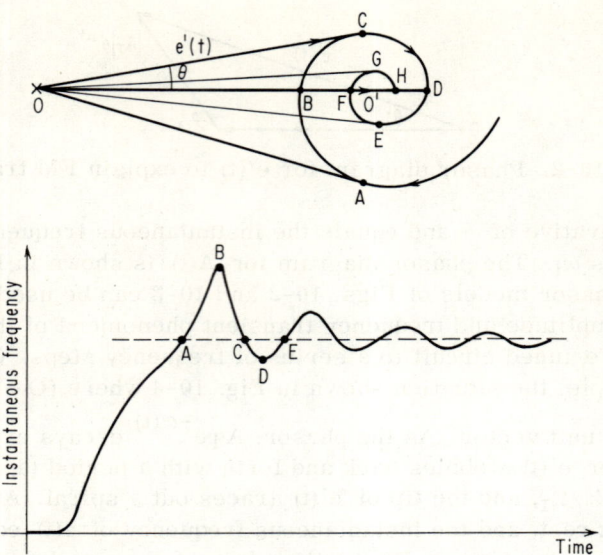

Fig. 10-4. Behavior of e′(t) which results in occurrence of instantaneous-frequency and instantaneous-amplitude transients in the tuned-circuit response.

unit phasor at t = 0, and the frequency ω_f is sufficiently high to enable the tip of the e′(t) phasor to encircle the origin O twice as $A_T e^{-\alpha t}$ revolves at the angular frequency ω_f. Since the instantaneous frequency of the resultant of two signals of different frequencies always overshoots in the direction of the frequency of the stronger signal (as will be shown in Chap. 11), the instantaneous frequency of e′(t) undershoots at such points as A and B, which occur to the left of O, and overshoots at such points as C and D, which occur to the right of O. Note that the undershoots, if they occur, must always precede the overshoots; also, successive undershoots have magnitudes that increase in value, whereas successive overshoots have magnitudes that decrease in value. When the tip of the e′(t) phasor passes through the origin, the magnitude of e′(t) equals zero, and the instantaneous frequency of the response is transitional between an overshoot and an undershoot.

When $\omega_f = \omega_o$, $\Omega_f = 0$ and the phasor $A_T e^{-\alpha t}$ is stationary with respect to the unit phasor, as shown in Fig. 10-6. Thus, the response to a frequency jump that terminates at ω_o consists of a smooth exponential rise caused by the exponential decay of $A_T e^{-\alpha t}$. When ω_f approximately equals ω_o, Ω_f is very small. Thus, $\phi = \phi_T - \Omega_f t$ is

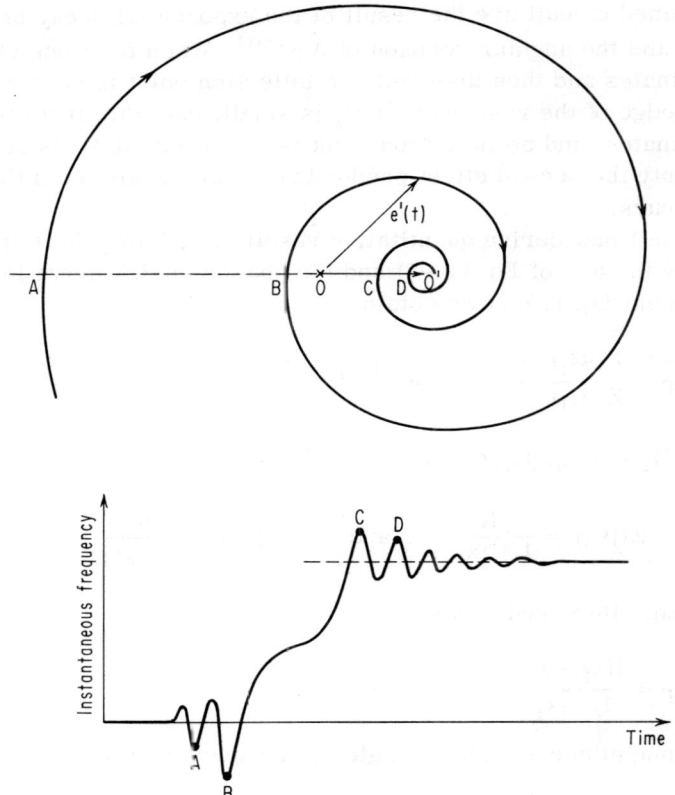

Fig. 10-5. Behavior of $e'(t)$ which results in the occurrence of both undershoots and overshoots in the instantaneous-frequency response of $e(t)$.

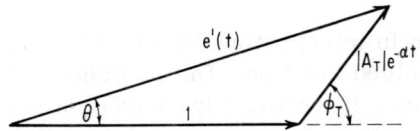

Fig. 10-6. When $\omega_f = \omega_o$, $\Omega_f = 0$ and the transients result only from the exponential decay of $A_T e^{-\alpha t}$.

approximately constant as $A_T e^{-\alpha t}$ decays, and the response is close to an exponential rise.

The inflection point that occurs, for certain values of Ω_f, in the leading edge of the instantaneous-frequency response is caused by the fact that the transients in the frequency-step response of the single-tuned circuit are the result of the exponential decay of $A_Te^{-\alpha t}$ and the angular rotation of $A_Te^{-\alpha t}$. When first one effect predominates and then the other, an inflection point is seen on the leading edge of the response. If Ω_f is small, only the first effect predominates, and no inflection point is observed. If Ω_f is very large, only the second effect predominates, and again, no inflection point occurs.

We shall now derive quantitative results regarding the FM transients by the use of Eq. (10-7) and the phasor model shown in Fig. 10-2. From Eq. (10-7) we obtain

$$A_T = \frac{Z(j\Omega_i)}{Z(j\Omega_f)} - 1 = |A_T|e^{j\phi_T} \qquad (10\text{-}8)$$

Letting $\Omega_i/\alpha = x_i$, $\Omega_f/\alpha = x_f$,

we have $\quad Z(j\Omega_i) = \dfrac{R}{1 + jx_i} \quad$ and $\quad Z(j\Omega_f) = \dfrac{R}{1 + jx_f}$

Hence Eq. (10-8) reduces to

$$A_T = \frac{j(x_f - x_i)}{1 + jx_i} \qquad (10\text{-}9)$$

and the magnitude and phase angle of A_T are given by

$$|A_T| = \frac{|x_f - x_i|}{\sqrt{1 + x_i^2}} \qquad (10\text{-}10)$$

and $\quad \phi_T = \text{ctn}^{-1}x_i \qquad (10\text{-}11)$

Care must be taken in interpreting Eq. (10-11) since the inverse cotangent is a multivalued function. The principal value of ϕ_T is determined by x_i alone. However, a knowledge of both x_i and x_f is necessary to place ϕ_T in its proper quadrant. This point is emphasized by noting that the vertical and horizontal components of A_T are given by

$$|A_T|_V = \frac{x_f - x_i}{1 + x_i^2}$$

$$\qquad\qquad\qquad\qquad (10\text{-}12)$$

$$|A_T|_H = \frac{x_i(x_f - x_i)}{1 + x_i^2}$$

Having obtained ϕ_T, it is an easy matter to determine t_1, the time at which the peak of the first overshoot occurs. Figure 10-7 illustrates the possible situations which may exist at $t = 0$. Considering ϕ_T as a positive quantity, irrespective of the quadrant in which ϕ_T appears, $t_1 = (\pi + \phi_T)/\Omega_f$ for Fig. 10-7a and b, while $t_1 = (\pi - \phi_T)/\Omega_f$ for Fig. 10-7c and d. Normalizing with respect to α, we have either

$$\tau_1 = \alpha t_1 = \frac{\pi + \phi_T}{x_f} \quad \text{or} \quad \frac{\pi - \phi_T}{x_f} \tag{10-13}$$

as the orientations at $t = 0$ may require. Obviously, the peak of the n'th overshoot occurs for either

$$\tau_n = \frac{\pi(2n - 1) + \phi_T}{x_f} \quad \text{or} \quad \frac{\pi(2n - 1) - \phi_T}{x_f} \tag{10-14}$$

The instantaneous frequency of $e'(t)$ is obtained by use of the relation

$$\frac{d\theta}{dt} = \text{Im}\left[\frac{1}{e'(t)} \quad \frac{de'(t)}{dt} \right] \tag{10-15}$$

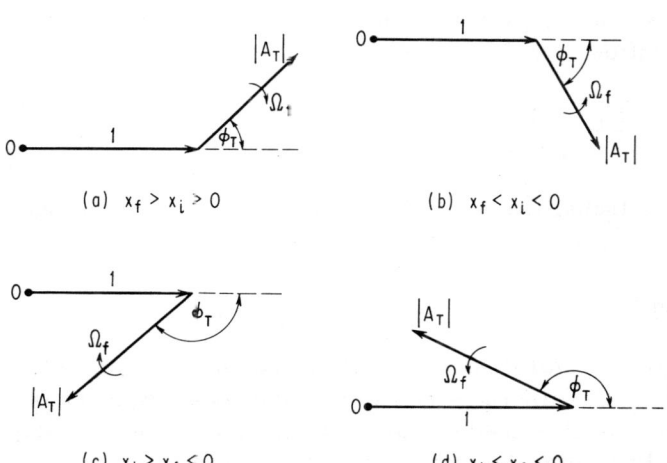

(a) $x_f > x_i > 0$ (b) $x_f < x_i < 0$

(c) $x_i > x_f < 0$ (d) $x_i < x_f < 0$

Fig. 10-7. Possible situations which may exist at $t = 0$.

where $e'(t) = 1 + |A_T| e^{-\alpha t} \cdot e^{j\phi}$, and θ and ϕ are shown in Fig. 10-2. The resulting expression is

$$\frac{d\theta}{dt} = \frac{|A_T| e^{-\alpha t} \left[-\Omega_f (\cos \phi + |A_T| e^{-\alpha t}) - \alpha \sin \phi \right]}{1 + 2|A_T| e^{-\alpha t} \cdot \cos \phi + |A_T|^2 \cdot e^{-2\alpha t}} \quad (10\text{-}16)$$

Note that this expression reduces to $(\Omega_i - \Omega_f)$ at $t = 0$, as it should, since the instantaneous frequency of the response of $e(t)$ at that time equals ω_i. At the time of the peak of the n'th overshoot, ϕ equals plus or minus $\pi(2n - 1)$. Thus, the instantaneous frequency of $e'(t)$ at the n'th overshoot is given by

$$\omega_P = \frac{d\theta}{dt}\bigg|_{t=t_n} = \frac{\Omega \omega_f |A_T|}{e^{t_n} - |A_T|} \quad (10\text{-}17)$$

Remembering that the instantaneous frequency of $e(t)$ equals the instantaneous frequency of $[e'(t) + \omega_f]$, the fractional overshoot of the instantaneous frequency of $e(t)$ at the n'th peak is

$$a_n = \frac{(\omega_p + \omega_f) - \omega_f}{\omega_f - \omega_i} = \frac{1}{x_f - x_i} \frac{x_f |A_T|}{e^{\tau_n} - |A_T|} \quad (10\text{-}18)$$

If the instantaneous frequency of the response undershoots rather than overshoots at the n'th peak, the fractional undershoot is given by the relation

$$b_n = \frac{(\omega_p + \omega_f) - \omega_i}{\omega_f - \omega_i} = a_n + 1 \quad (10\text{-}19)$$

Note that a transition between overshoot and undershoot occurs when

$$|A_T| = e^{\tau_n} \quad (10\text{-}20)$$

1. *Experimental Results of FM Transients According to Weiner.*[2,7] Oscillograms of the instantaneous frequency of the response of a single-tuned circuit to a series of frequency steps are shown in Fig. 10-8. In oscillograms a through k, the input frequency steps start at the resonance frequency of the filter. For x_f less than 0.80, the response resembles a rising exponential whose rise time is approximately independent of x_f (see Fig. 10-8a). A

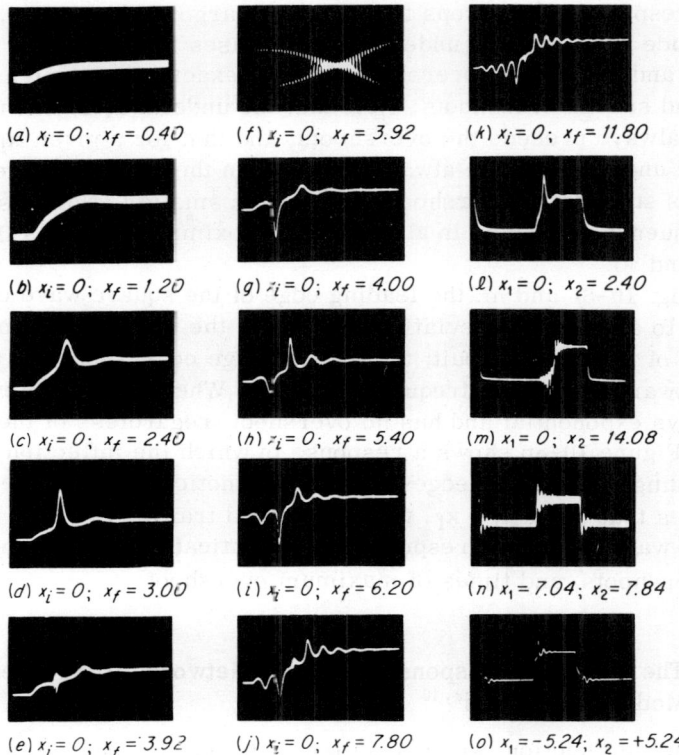

Fig. 10-8. Oscillograms of response of a single-tuned circuit to a step-modulated sinusoid. (From E. J. Baghdady,[7] courtesy of McGraw-Hill Book Company.)

small overshoot is first noticeable (not shown here) for $x_f = 0.80$. As x_f becomes greater than 0.80, the magnitude of the overshoot increases with a corresponding decrease in the rise time of the response and the time of maximum overshoot. An inflection point is also apparent on the leading edge of the transient (see Fig. 10-8b to d). For $x_f = 3.92$, the overshoot is caught in the act of turning into an undershoot, and at the instant of transition, the instantaneous amplitude of the response drops to zero. These two effects are shown in Fig. 10-8e and f, respectively. For $x_f = 4.00$, the "overshoot" has definitely become an undershoot (Fig. 10-8g). With a further increase in x_f, the magnitude of the first undershoot decreases in value whereas the magnitude of the second overshoot increases (Fig. 10-8g and h). For $x_f = 6.20$, the second overshoot becomes an undershoot (Fig. 10-8i), and the instantaneous amplitude of the response again drops to zero. For larger values of x_f, the magnitude of the second undershoot decreases in value. The third, fourth, and successive overshoots behave exactly the same as the first and second overshoots. Note that the undershoots, when they occur, always precede the overshoots, and that for a given x_f, successive undershoots are always larger than the preceding ones, whereas successive overshoots are always smaller. Note also that the frequency of ringing in all cases (approximately) equals Ω_f (Fig. 10-8j and k).

In Fig. 10-8ℓ and m, the leading edge of the square wave corresponds to a frequency deviation away from the center frequency ($x_i = 0$) of the tuned circuit; the trailing edge corresponds to a deviation toward the center frequency ($x_f = 0$). When $x_f = 0$, the response is always exponential and has no overshoot, regardless of the value of x_i. Figure 10-8n shows a response in which the inflection point in the leading and trailing edges is almost unnoticeable. Figure 10-8o indicates that when $x_i = x_f$, the leading and trailing edges of the square-wave frequency response have identical rise times, percentage overshoots, and times of maximum overshoot.

10.2 The Transient Response of Linear Networks to Frequency-Modulated Signals[9],[10]

In the last section, we have shown that a sudden change in the input signal frequency will set up a train of damped oscillations at the resonant frequency of the circuit. Furthermore, if the final frequency of the applied voltage is different from the resonant frequency of the circuit, there will be present, for an interval of time, two signals which will give a resultant of varying amplitude and frequency. A general analytical expression for the resultant instantaneous angular frequency can be obtained using the method outlined in Chap. 7. The vector addition of the two signals is shown in Fig. 10-9, where the final signal is given by $e^{j\omega_f t}$, and the damped signal is given by $\rho = A_T e^{-\alpha t} \cdot e^{j(\omega_0 t + \phi)}$, ϕ being the phase displacement of the two signals. From the diagram, we obtain directly an expression for the angle of the resultant signal $R(t)$,

$$\tan \theta = \frac{\rho \sin (\omega_0 t + \phi) + \sin \omega_f t}{\rho \cos (\omega_0 t + \phi) + \cos \omega_f t} \tag{10-21}$$

Differentiating with respect to t, we obtain

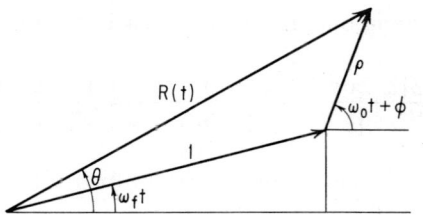

Fig. 10-9. Vector addition of two signals of different frequencies.

$$\sec^2 \theta \; \frac{d\theta}{dt} = \frac{\omega_o \rho \cos (\omega_o t + \phi) - \alpha \rho \sin (\omega_o t + \phi) + \omega_f \cos \omega_f t}{\rho \cos (\omega_o t + \phi) + \cos \omega_f t}$$

$$+ \frac{\{[\omega_o \rho \sin (\omega_o t + \phi) + \alpha \rho \cos (\omega_o t + \phi) + \omega_f \sin \omega_f t] \times [\rho \sin (\omega_o t + \phi) + \sin \omega_f t]\}}{[\rho \cos (\omega_o t + \phi) + \cos \omega_f t]^2}$$

$$(10\text{-}22)$$

Now
$$\sec^2 \theta = \frac{\cos \omega_f t + \rho (\cos \omega_o t + \phi)^2}{[\cos \omega_f t + \rho \cos (\omega_o t + \phi)]^2 + [\sin \omega_f t + \rho \sin (\omega_o t + \phi)]^2}$$

$$(10\text{-}23)$$

The instantaneous angular frequency $\omega_i(t)$ is given by

$$\omega_i(t) = \frac{d\theta}{dt} = \frac{\{\omega_o \rho^2 + \omega_f + \rho(\omega_o + \omega_f) \cos [(\omega_o - \omega_f)t + \phi] - \alpha \rho \sin [(\omega_o - \omega_f)t + \phi]\}}{1 + \rho^2 + 2\rho \cos [(\omega_o - \omega_f)t + \phi]}$$

$$(10\text{-}24)$$

Let $\omega_f = \omega_o + \overline{\omega}$, and this equation reduces to

$$\omega_i(t) = \omega_o + \frac{\overline{\omega} + \rho[\overline{\omega} \cos (-\overline{\omega}t + \phi) - \alpha \sin (-\overline{\omega}t + \phi)]}{1 + \rho^2 + 2\rho \cos (-\overline{\omega}t + \phi)}$$

$$= \omega_o + \Delta_i(t) \qquad\qquad (10\text{-}25)$$

so that $\Delta_i(t)$ gives the difference between the instantaneous resultant frequency and the natural frequency of the network.

The frequency components of $\Delta_i(t)$ will presently be analyzed.

$$\Delta_i(t) = \frac{\bar{\omega} + \rho[\bar{\omega} \cos(-\bar{\omega}t + \phi) - \alpha \sin(-\bar{\omega}t + \phi)]}{(1 + \rho^2)[1 + 2\rho/(1 + \rho^2) \cos(-\bar{\omega}t + \phi)]}$$

$$= \{\bar{\omega} + \rho[\bar{\omega} \cos(-\bar{\omega}t + \phi) - \alpha \sin(-\bar{\omega}t + \phi)]\}(1 + \rho^2)^{-1}$$

$$\times \left[1 + \frac{2\rho}{1 + \rho^2} \cos(-\bar{\omega}t + \phi)\right]^{-1} \tag{10-26}$$

Since $\rho < 1$ and $1 + \rho^2 > 2\rho$, Eq. (10-26) can be expanded into an infinite series of terms of angular frequency $\bar{\omega}$ and its higher harmonics called flutter terms, and a flutterless series given by

$$\Delta_i(t) = \frac{\bar{\omega}}{1 + \rho^2} - \frac{(\bar{\omega} - 2\alpha \sin \phi \cos \phi)\rho^2}{(1 + \rho^2)^2}$$

$$\times \left[1 + \frac{\rho^2}{2(1 + \rho^2)^2} + \frac{\rho^4}{4(1 + \rho^2)^4} + \cdots\right] \tag{10-27}$$

so that the instantaneous-frequency component without flutter is given by

$$\Delta_i(t) = \frac{\bar{\omega}}{1 + \rho^2} - \frac{2\rho^2(\bar{\omega} - 2\alpha \sin \phi \cos \phi)}{2 + 3\rho^2 + 2\rho^4} \tag{10-28}$$

It will be seen that this frequency component is affected by both the damping factor α and the initial phase angle.

The results derived above will be applied now to the investigation of the transient response of a few simple circuits to a frequency-modulated signal.

1. *Resistance-Capacitance and Resistance-Inductance Circuits.*

Consider the simple transfer function $F(s) = \dfrac{s}{s + 1/RC}$; the transient is of the form $Ae^{-(1/RC)t}$. This is nonoscillatory and decays exponentially so that ω_0 and ϕ equal zero. The resultant instantaneous frequency, using Eq. (10-24) and setting ω_0 and ϕ equal to zero, is found to be

$$\omega_i(t) = \omega_f + \rho(\omega_f \cos \omega_f t + \alpha \sin \omega_f t)(1 + 2\rho \cos \omega_f t)^{-1} \tag{10-29}$$

It is to be noted that the instantaneous frequency consists of the final steady angular frequency ω_f plus a transient frequency term which decays. If this transient frequency term is expanded, two distinct series in ρ will be obtained, one giving the steadily decaying part, and the second consisting of sine and cosine terms of the angles $\omega_f t$, $2\omega_f t$, $3\omega_f t$, etc., which gives the flutter frequency components. As seen in Fig. 10-10, the steadily decaying part is due to the fact that the successive periodic times T_1, T_2, T_3, etc., are varying.

2. *Series-tuned Circuit.*   Let the impedance of the series-tuned circuit be given by

$$Z(s) = R + sL + \frac{1}{sC} \qquad (10\text{-}30)$$

We apply a voltage $v(t)$ across the circuit given by

$$v(t) = e^{j\omega_i t} \qquad (10\text{-}31)$$

If the amplitude remains constant but its angular frequency suddenly changes to ω_f at time $t = \tau$, the Laplace transform of $v(t)$ is

$$V(s) = \int_0^\tau e^{(j\omega_i - s)t}\, dt + \int_\tau^\infty e^{(j\omega_2 s)t}\, dt \qquad (10\text{-}32)$$

or $\qquad V(s) = \dfrac{1}{s - j\omega_i} - \dfrac{1}{s - j\omega_i} e^{(j\omega_i - s)\tau} + \dfrac{1}{s - j\omega_f} e^{(j\omega_f - s)\tau}$

$$\qquad (10\text{-}33)$$

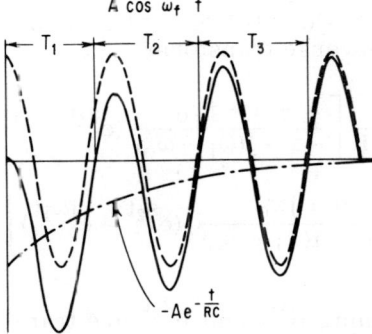

Fig. 10-10. Cosine curve added to exponential curve.

The first term on the right-hand side of this equation represents the application of a signal of angular frequency ω_i at time $t = 0$, the second represents the removal at time $t = \tau$, and the third term represents the application of a signal of equal amplitude and angular frequency ω_f at time $t = \tau$. Since the circuit is linear, the principle of superposition holds. The current due to the first term is

$$I_1(s) = \frac{1}{s - j\omega_i} \frac{1}{Z(s)}$$

(10-34)

$$\text{Now} \quad A(s) = \frac{L}{s}(s - s_1)(s - s_2)$$

(10-35)

where $\quad s_1 = -\alpha + \sqrt{\alpha^2 - \omega_0^2} \qquad s_2 = -\alpha - \sqrt{\alpha^2 - \omega_0^2}$

and $\qquad \alpha = \dfrac{R}{2L}$, $\quad \omega_0^2 = \dfrac{1}{LC}$

For the oscillatory circuit $\alpha < \omega_0$, and

$$s_1 = -\alpha + j\sqrt{\omega_0^2 - \alpha^2} \qquad s_2 = -\alpha - j\sqrt{\omega_0^2 - \alpha^2}$$

Therefore, using partial fractions, Eq. (10-34) is given by

$$I_1(s) = \frac{1}{L(s_1 - s_2)}\left[\frac{s_1}{s_1 - j\omega_i}\left(\frac{1}{s - s_1} - \frac{1}{s - j\omega_i}\right) \right.$$

$$\left. - \frac{s_2}{s_2 - j\omega_i}\left(\frac{1}{s - s_2} - \frac{1}{s - j\omega_i}\right) \right]$$

(10-36)

Assuming that $\alpha \ll \omega_0$, then $s_1 = -\alpha_1 + j\omega_0$ and $s_2 = -\alpha - j\omega_0$, and we obtain by the inverse transform

$$i_1(t) = \frac{1}{2j\omega_0 L}\left[\frac{-\alpha + j\omega_0}{-\alpha - j(\omega_i - \omega_0)}(e^{s_1 t} - e^{j\omega_i t}) \right.$$

$$\left. + \frac{\alpha + j\omega_0}{-\alpha - j(\omega_i + \omega_0)}(e^{s_2 t} - e^{j\omega_i t}) \right]$$

(10-37)

The terms containing $e^{s_1 t}$ and $e^{s_2 t}$ are transient terms due to the application of the original signal at time $t = 0$; these eventually will die away, and we are left with

$$i_1(t) = \frac{e^{j\omega_i t}}{2j\omega_0 L} \left[\frac{\alpha + j\omega_0}{\alpha + j(\omega_i + \omega_0)} - \frac{\alpha - j\omega_0}{\alpha + j(\omega_i - \omega_0)} \right] \qquad (10\text{-}38)$$

The second term of Eq. (10-33) can be derived from this equation if we measure time from the instant the frequency jump occurs, by making $\tau = 0$; thus the second term of Eq. (10-33) becomes $-1/(s - j\omega_i)$ and produces a current $i_2(t)$ similar to that of Eq. (10-37) but negative, so that

$$i_2(t) = -\frac{1}{2j\omega_0 L} \left[\frac{-\alpha + j\omega_0}{-\alpha - j(\omega_i - \omega_0)} (e^{s_1 t} - e^{j\omega_i t}) \right.$$

$$\left. + \frac{\alpha + j\omega_0}{-\alpha - j(\omega_i - \omega_0)} (e^{s_2 t} - e^{j\omega_i t}] \right] \qquad (10\text{-}39)$$

Adding Eqs. (10-38) and (10-39), we are left with only the transient terms, so that

$$i_1(t) + i_2(t) = \frac{1}{2j\omega_c L} \left[\frac{-\alpha + j\omega_0}{\alpha + j(\omega_i - \omega_0)} e^{s_1 t} + \frac{\alpha + j\omega_0}{\alpha + j(\omega_i + \omega_0)} e^{s_2 t} \right]$$

$$(10\text{-}40)$$

The third term of Eq. (10-33) will also produce a current given by an expression similar to that of Eq. (10-37) with ω_i changed to ω_f. If we write $e^{s_1 t} = e^{-\alpha t} \cdot e^{j\omega_0 t}$ and $e^{s_2 t} = e^{-\alpha t} e^{-j\omega_0 t}$, then

$$i(t) = \frac{1}{2j\omega_0 L} \left\{ \left[\left(\frac{1 - j\,\omega_c/\alpha}{1 + j(\omega_f - \omega_0)/\alpha} - \frac{1 - j\,\omega_0/\alpha}{1 + j(\omega_i - \omega_0)/\alpha} \right) e^{j\omega_0 t} \right.\right.$$

$$\left. + \left(\frac{1 + j\,\omega_0/\alpha}{1 + j(\omega_i + \omega_0)/\alpha} - \frac{1 + j\,\omega_0/\alpha}{1 + j(\omega_f + \omega_0)/\alpha} \right) e^{-j\omega_0 t} \right] e^{-\alpha t}$$

$$\left. + \left(\frac{1 + j\,\omega_0/\alpha}{1 + j(\omega_f + \omega_0)/\alpha} - \frac{1 - j\,\omega_0/\alpha}{1 + j(\omega_f - \omega_0)/\alpha} \right) e^{j\omega_f t} \right\} \qquad (10\text{-}41)$$

Let $\omega_0/2\alpha = Q$, and assume that Q is large so that $1 - j\,\omega_0/\alpha \doteq -2jQ$; also let $\omega_i + \omega_0 \gg \omega_i - \omega_0$, and $\omega_f + \omega_0 \gg \omega_f - \omega_0$. Therefore

$$i(t) \doteq \frac{Q}{\omega_o L} \left[\left(\frac{1}{1 + j\,(\omega_i - \omega_o)/\alpha} - \frac{1}{1 + j\,(\omega_f - \omega_o)/\alpha} \right) e^{-\alpha t}\, e^{j\omega_o t} \right.$$

$$\left. + \frac{1}{1 + j\,(\omega_f - \omega_o)/\alpha}\, e^{j\omega_f t} \right] \tag{10-42}$$

Put $\omega_i - \omega_o = x_i$ and $\omega_f - \omega_o = x_f$; also since $Q/\omega_o L = 1/R$, Eq. (10-42) reduces to

$$i(t) \doteq \frac{e^{j\omega_o t}}{R(1 + j\,x_f/\alpha)} \left[\frac{j(x_f - x_i)/\alpha}{1 + j\,x_i/\alpha}\, e^{-\alpha t} + e^{j x_f t} \right] \tag{10-43}$$

The instantaneous frequency is determined by the use of Eq. (10-25); we obtain for time $t = 0$,

$$\omega_i(t) = \omega_o + \frac{\alpha^2 x_f + x_i^2 x_f + (x_f - x_i)(x_f x_i - \alpha^2)}{x_f^2 + \alpha^2} = \omega_o + x_i$$

$$\tag{10-44}$$

so that $\omega_i(t) = \omega_i \qquad t = 0$

This shows that there is not any sudden frequency jump, but a smooth change from ω_i to ω_f.

It is of interest to note here that if $\omega_f = \omega_o$, i.e., $x_f = 0$, both components of the response, as given by Eq. (10-43), have the same angular frequency ω_o. However, owing to the initial phase angle between the two components and the decaying amplitude of the transient term, there is no instantaneous-frequency step from ω_i to ω_o, so that the instantaneous frequency at $t = 0$ is still ω_i.

We can summarize the results of our analysis as follows: When the frequency of the signal applied to a single-tuned circuit undergoes a sudden jump, the response is composed of two parts: (1) a transient component having a frequency equal to the natural frequency of the circuit and (2) a steady-state component produced by the applied signal. The amplitude of the transient component increases with the value of the frequency jump and also with the Q of the circuit. The vectorial addition of these two components gives a resultant whose instantaneous frequency equals the time derivative of its phase angle. The instantaneous frequency of the resultant signal is also composed of two parts: (1) a mean frequency which varies exponentially, the rate of variation being governed by the decay factor of the circuit, and (2) an alternating or "flutter" component whose frequency of variation is equal to the difference between the final signal frequency and the natural frequency of the circuit. It is important to note here that the combination of these two frequency components gives an instantaneous frequency at $t = 0_+$, namely, immediately after the change of frequency of the applied signal, equal to the original frequency. Also, if the final frequency of the applied signal is equal to the natural frequency of the circuit, there will not be any "flutter" component, but there will still be a varying mean frequency due to the difference in phase between the applied signal and the transient signal.

10.3 The Transient Response of Linear Networks to Phase-Modulated Signals[11]

The study of the transient response of linear networks to phase-modulated signals is of practical interest in the analysis of the performance of digital phase-modulation systems, which have been developed in the last few years. In these systems, the modulated carrier consists of a phase-modulated RF pulse of specified width transmitted at a known repetition rate. The performance of digital phase-modulation communication systems will be analyzed in Chap. 23; in this section, the response of some linear passive networks to a phase jump is examined.

Let the input signal be denoted by $v_i(t)$; in practice, this signal is usually a sinusoid. The input signal is defined by

$$v_i(t) = e^{j\omega t}, \qquad 0 < t < t_1$$
$$= e^{j(\omega t + \phi)}, \qquad t > t_1 \qquad (10\text{-}45)$$

where the angular frequency of the sinusoid ω = constant.

The "phase jump" occurring at $t = t_1$ can be realized by the application of a signal $e^{j\omega t}$ at $t = 0$, its removal at $t = t_1$, and the application of a signal $e^{j(\omega t + \phi)}$ at $t = t_1$; the total response is found by superposition of the individual responses. In the following analysis, we are interested only in the transient response of the circuit due to the application of the phase jump at $t = t_1$; consequently, t_1 may be chosen large enough so that the transient resulting from the applied signal $e^{j\omega t}$ at $t = 0$ has decayed to a negligible value at $t = t_1$.

1. *Transient Response of RC Network: Phase Jump a Step Function.* Applying the Laplace transform to Eq. (10-45), we get

$$V_i(s) = \frac{1}{s - j\omega} - \frac{e^{(j\omega - s)t_1}}{s - j\omega} + \frac{e^{(j\omega - s)t_1 + j\phi}}{s - j\omega} \tag{10-46}$$

The transfer function of the network shown in Fig. 10-11a is

$$H(s) = \frac{V_o(s)}{V_i(s)} = \frac{1}{1 + s\tau}, \qquad \tau = CR$$

The response will consist of three terms corresponding to the terms of Eq. (10-46); each of the three terms will be considered separately.

$$V_{01}(s) = \frac{1}{1 + s\tau} \; \frac{1}{s - j\omega} = \frac{1}{1 + j\omega\tau} \left(\frac{1}{s - j\omega} - \frac{1}{s + 1/\tau} \right)$$

Therefore $\quad v_{01}(t) = \dfrac{1}{1 + j\omega\tau} \left(e^{j\omega t} - e^{-t/\tau} \right) \tag{10-47}$

Also $\quad V_{02}(s) = \dfrac{-1}{1 + s\tau} \; \dfrac{e^{(j\omega - s)t_1}}{s - j\omega} = \dfrac{-e^{j\omega t_1}}{1 + j\tau} \left(\dfrac{e^{-st_1}}{s - j\omega} - \dfrac{e^{-st_1}}{s + 1/\tau} \right)$

Therefore $\quad v_{02}(t) = \dfrac{-e^{j\omega t_1}}{1 + j\omega\tau} \left(e^{j\omega(t - t_1)} - e^{-(t - t_1)\tau} \right) \tag{10-48}$

Similarly, the third term gives

$$v_{03}(t) = \frac{e^{j(\omega t_1 + \phi)}}{1 + j\omega\tau} \left(e^{j\omega(t - t_1)} - e^{-(t - t_1)/\tau} \right) = -v_{02}(t) \; e^{j\phi}$$

$$\tag{10-49}$$

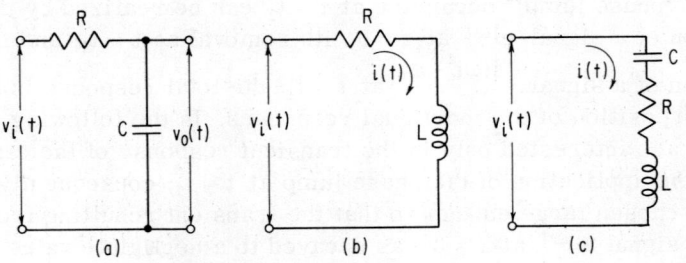

Fig. 10-11. The networks considered.

From Eq. (10-47), it is seen that the response contains a term representing the transient due to the application of $v_i(t) = e^{j\omega t}$ at $t = 0$; as stated previously, by choosing t_1 so that this transient has decayed to a negligible value at $t = t_1$, the response becomes

$$v_{01}(t) = \frac{e^{j\omega t}}{1 + j\omega\tau} \tag{10-50}$$

The total response is then

$$v_0(t) = \frac{1}{1 + j\omega\tau}\left[e^{j(\omega t + \phi)} + (1 - e^{j\phi})\, e^{j\omega_1 - (t - t_1)/\tau} \right] \tag{10-51}$$

If the applied voltage is cosinusoidal, namely,

$$v_i(t) = \cos \omega t = \mathrm{Re}\ \ e^{j\omega t}, \qquad 0 < t < t_1$$

then the response is obtained by taking the real part of Eq. (10-51).

$$\mathrm{Re}\,[v_0(t)] = \frac{1}{\sqrt{1 + (\omega\tau)^2}}\left[\cos(\omega t + \phi - \beta) + 2 \sin\frac{\phi}{2} \right.$$

$$\left. \times \sin\left(\omega t_1 + \frac{\phi}{2} - \beta\right) e^{-(t - t_1)/\tau} \right], \qquad t > t_1 \tag{10-52}$$

where $\beta = \tan^{-1} \omega\tau$.
   For $t < t_1$, we have $\phi = 0$, and

$$\mathrm{Re}\,[v_0(t)] = \frac{1}{\sqrt{1 - (\omega\tau)^2}} \cos(\omega t - \beta), \qquad \text{as expected}$$

For the purpose of examining the magnitude of the transient term, we note that the phase jump ϕ can have any value between $\pm\pi$, and t_1 can have any value between 0 and $2\pi/\omega$. The transient has a maximum value of $\pm 2/\sqrt{1 + (\omega\tau)^2}$ which occurs when $t - t_1 = 0$, $\phi = \pm\pi$ and $\omega t_1 = \beta + n\pi$. In general, for a "phase jump" ϕ, the transient term will have a maximum value $\pm 2 \sin(\phi/2)/\sqrt{1 + (\omega\tau)^2}$ which occurs when the instant of switching is such that

$$\omega t_1 = \beta - \frac{\phi}{2} + \frac{(n + 1)\pi}{2}$$

2. *Transient Response of RL Network: Phase Jump a Step Func-*

tion. In this case, we have $I(s)/V_i(s) = 1/R(1 + s\tau)$ where $\tau = L/R$, which is of the same form as the transfer function of the RC circuit, and the results will be similar.

3. *Transient Response of LCR Network: Phase Jump a Step Function.* Here

$$Z(s) = R + Ls + \frac{1}{Cs} \quad \text{and} \quad I(s) = \frac{V_i(s)}{Z(s)} \tag{10-53}$$

Thus $\quad I(s) = \dfrac{V_1(s)}{R + Ls + 1/Cs} = \dfrac{V_1(s)s}{L(s - s_1)(s - s_2)}$ \hfill (10-54)

where $\quad s_1, s_2 = -\dfrac{R}{2L} \pm \sqrt{\left(\dfrac{R}{2L}\right)^2 - \dfrac{1}{LC}} = -\alpha \pm j\sqrt{\omega_0^2 - \alpha^2}$

and $\qquad \alpha = \dfrac{R}{2L} \qquad \omega_0^2 = \dfrac{1}{LC}$

As before, let $v_i(t)$ be given as in Eq. (10-45); then

$$I(s) = \frac{s}{L(s - s_1)(s - s_2)} \left(\frac{1}{s - j\omega} - \frac{e^{(j\omega - s)t_1}}{s - j\omega} + \frac{e^{(j\omega - s)t_1 + j\phi}}{s - j\omega} \right)$$

$$\tag{10-55}$$

As before, we consider each term separately:

$$I_1(s) = \frac{s}{L(s - s_1)(s - s_2)} \; \frac{1}{s - j\omega}$$

$$= \frac{1}{L(s_1 - s_2)} \left[\frac{s_1}{s_1 - j\omega} \left(\frac{1}{s - s_1} - \frac{1}{s - j\omega} \right) \right.$$

$$\left. - \frac{s_2}{s_2 - j\omega} \left(\frac{1}{s - s_2} - \frac{1}{s - j\omega} \right) \right] \tag{10-56}$$

giving $\quad i_1(t) = \dfrac{1}{L(s_1 - s_2)} \left[\dfrac{s_1}{s_1 - j\omega} (e^{s_1 t} - e^{j\omega t}) \right.$

$$\left. - \frac{s_2}{s_2 - j\omega} (e^{s_2 t} - e^{j\omega t}) \right] \tag{10-57}$$

This expression contains transient terms due to the application of the signal $e^{j\omega t}$ at $t = 0$. As before, t_1 may be chosen so that the transient has decayed to a negligible value at $t = t_1$; hence

$$i_1(t) = \frac{e^{j\omega t}}{L(s_1 - s_2)} \left(\frac{s_1}{s_1 - j\omega} - \frac{s_2}{s_2 - j\omega} \right) \tag{10-58}$$

Similarly, we can show that

$$i_2(t) = \frac{-e^{j\omega t_1}}{L(s_1 - s_2)} \left[\frac{s_1}{s_1 - j\omega} \left(e^{s_1(t - t_1)} - e^{j\omega(t - t_1)} \right) \right.$$

$$\left. - \frac{s_2}{s_2 - j\omega} \left(e^{s_2(t - t_1)} - e^{j\omega(t - t_1)} \right) \right] \tag{10-59}$$

and $i_3(t) = -i_2(t)\, e^{j\phi}$ $\tag{10-60}$

The resultant current in the circuit

$$i(t) = i_1(t) + i_2(t) + i_3(t)$$

$$= \frac{1}{L(s_2 - s_1)} \left(\frac{s_1}{s - j\omega} - \frac{s_2}{s - j\omega} \right) e^{j(\omega t + \phi)} + (1 - e^{j\phi})$$

$$\times \left(\frac{s_1 e^{s_1(t - t_1) + j\omega t_1}}{s_1 - j\omega} - \frac{s_2 e^{s_2(t - t_1) + j\omega t_1}}{s_2 - j\omega} \right) \right] \quad t > t_1$$

$$\tag{10-61}$$

The last equation can be simplified by making the following substitutions and approximations. Let

$$\omega_0^2 - \alpha^2 = \frac{1}{LC} - \frac{R^2}{4L^2} = \omega_0^2 \left(1 - \frac{1}{4Q^2} \right) \qquad \text{where } Q = \frac{\omega_0 L}{R}$$

Assume $1/4Q^2 \ll 1$ and $\omega_0^2 - \alpha^2 \doteq \omega_0^2$; therefore $s_1 \doteq -\alpha + j\omega_0$, $s_2 \doteq -\alpha - j\omega_0$, and $s_2 - s_1 = -2j\omega_0$.
   Equation (10-61) then reduces to

$$i(t) \doteq \frac{1}{R} e^{j\omega_0 t + \phi} + \frac{e^{-\alpha(t - t_1)}}{2\omega_0 L} \left[\frac{\omega_0}{\alpha} \left(e^{j\omega_0 t} - e^{j(\omega_0 t + \phi)} \right) \right.$$

$$\left. - \frac{1}{2} \left(e^{-j\omega_0(t - 2t_1)} - e^{-j[\omega_0(t - t_1) - \phi]} \right) \right] \tag{10-62}$$

For a cosinusoidal input signal, the current i(t) is given by the real part of Eq. (10-62):

$$i(t) \doteq \frac{1}{R} \cos(\omega_0 t + \phi) + \frac{1}{2\omega_0 L} \sin\frac{\phi}{2} \cdot e^{-\alpha(t-t_1)}$$

$$\times \left\{ \sin\left(\omega_0 t + \frac{\phi}{2}\right)\left[\frac{2\omega_0}{\alpha} - \sin(2\omega_0 t_1 + \phi)\right]\right.$$

$$\left. - \cos\left(\omega_0 t + \frac{\theta}{2}\right) \cos(2\omega_0 t_1 + \phi)\right\} \qquad (10\text{-}63)$$

Since $Q = \omega_0/2\alpha$ and $Q \gg 1$, it follows that $2\omega_0\alpha \gg \sin(2\omega_0 t_1 + \phi)$, and this equation reduces to

$$i(t) \doteq \frac{1}{R} \cos(\omega_0 t + \phi) + \frac{1}{2\omega_0 L} \sin\left(\frac{\phi}{2}\right) e^{-\alpha(t-t_1)}$$

$$\times \left[\frac{2\omega_0}{\alpha} \sin\left(\omega_0 t + \frac{\phi}{2}\right) - \cos(2\omega_0 t_1 + \phi) \cos\left(\omega_0 t + \frac{\phi}{2}\right)\right]$$

$$\doteq \frac{1}{R}\left[\cos(\omega_0 t + \phi) + 2\sin\left(\frac{\phi}{2}\right) e^{-\alpha(t-t_1)} \sin\left(\omega_0 t + \frac{\phi}{2} - \delta\right)\right]$$

$$(10\text{-}64)$$

where
$$\delta = \tan^{-1}\frac{\cos(2\omega_0 t_1 + \phi)}{4Q} \qquad (10\text{-}65)$$

For phase jumps greater than a few degrees, $\delta \ll 1$; hence for practical purposes, Eq. (10-64) can be written as

$$i(t) \doteq \frac{1}{R}\left[\cos(\omega_0 t + \phi) + 2\sin\left(\frac{\phi}{2}\right) e^{-\alpha(t-t_1)} \sin\left(\omega_0 t + \frac{\phi}{2}\right)\right]$$

$$(10\text{-}66)$$

These results were verified by Ivison[11] experimentally and with an analogue computer.

References

1. Hupert, J. J.: Normalized Phase and Gain Derivatives as an Aid in Evaluation of FM Distortion, *Proc. IRE*, February, 1954.
2. Weiner, D. D.: Experimental Study of FM Transients and Quasistatic Response, S.M. Thesis, Department of Electrical Engineering, MIT, January, 1958.

3. Salinger, H.: Transients in Frequency Modulation, *Proc. IRE*, vol. 30, p. 378, 1942.
4. Bell, D. A.: Transient Response in Frequency Modulation, *Phil. Mag.*, vol. 35, p. 143, 1944.
5. Gumowski, I.: Transient Response in FM, *Proc. IRE*, vol. 42, p. 819, 1954.
6. McCoy, R. E.: FM Transient Response of Band-pass Circuits, *Proc. IRE*, vol. 42, pp. 574-579, 1954.
7. Baghdady, E. J. (ed.): "Lectures on Communication System Theory," Chap. 19, McGraw-Hill Book Company, New York, 1960.
8. Zinn, M. K.: Transient Response of an FM Receiver, *BSTJ*, 1948.
9. Cotton, S. J.: A Comparison of the Transient Response of Amplitude Modulated and Frequency Modulated Signals, *IEE London Monograph* 322R, December, 1958.
10. Cotton, S. J.: Transient Response of FM Signals, *Electron. Technol.*, November, 1961, pp. 414-419.
11. Ivison, J. M.: The Transient Response of Linear Passive Networks to Phase-modulated Signals, *Electron. Eng.*, February, 1962, pp. 87-91.

11

INTERFERENCE IN FREQUENCY-MODULATION RECEPTION: COMMON- AND ADJACENT-CHANNEL INTERFERENCE AND MULTIPATH TRANSMISSION

When two frequency-modulated transmitters are stationed so near to each other in the frequency band as to cause partial or complete overlapping of the spectra in the receiver, interference will result. The resulting interference due to other frequency-modulated transmitters may be divided into two classifications: co-channel or common-channel interference arising from transmitters operating on the same channel, and adjacent-channel interference arising from transmitters operating on different channels. When two or more interfering signals come from the same transmitter, but one is delayed with respect to the other because of longer transmission path, interference may result in the receiver, called multipath interference. It can be shown experimentally that the interference produced by two stations operating at the same frequency is less in the case of frequency modulation than in the corresponding case of amplitude modulation. It is also found that when the ratio of the desired carrier voltage to the interfering

350

voltage is high, the signal-to-noise ratio improvement due to frequency modulation is considerable. As the interfering noise voltage is increased with respect to the desired carrier-wave voltage, the improved noise suppression is obtained as long as the desired signal is several times as strong as the interfering one.

When the ratio is decreased to about 2 or 3, for wideband frequency modulation, the amount of distortion in the detected audio output of the FM receiver increases rapidly. This means that in the case of frequency modulation the signal output is either good or bad; there is only a small range for the ratio of carrier to interfering voltage that gives a noisy but tolerable signal.

The problem of common and adjacent channel interference has been analyzed by many workers in the field.[1-4] It is of particular importance in the design of FM receivers, in view of the large number of commercial and military FM transmitters in operation. In this chapter, we shall derive the basic analytical results for computing the amplitudes of the harmonics and cross-modulation frequencies produced by interference. The problem of the so-called "capture effect," where the stronger interfering signal takes over or captures the receiver, will be discussed, and particular consideration will be given to the problem of FM receiver design in order to minimize this undesirable effect.

Considerable distortion is also observed in the presence of multipath transmission in frequency-modulated systems. Crosby[5,6] was the first to note the detrimental effect of multipath transmission on frequency-modulation reception over long distances. Our discussion will be based primarily on the work of Stumpers,[3] Corrington,[7] and Sollenberger.[8] The general distortion formulas arising from multipath transmission will be derived, giving the amplitudes of the harmonics of the sinusoidal modulation signal as well as the phase error introduced in the fundamental component of the receiver output signal. This result is of particular interest to ranging systems where range information is derived from the round-trip phase shift of the modulation subcarrier.

A problem related to multipath distortion is echo distortion, which may arise in a mismatched feeder line. This form of distortion is one of the major factors that have to be considered in communication system design. Considerable work has been done in this field,[10-12] and some of the results will be reviewed in this chapter. A more extensive analysis of multiple reflections in feeder lines with special emphasis on the design of FDM-FM multichannel communication systems is given by Medhurst and Small.[12]

11.1 Interference between Two Unmodulated Carriers [1-3]

The most elementary case of interference is that produced when two unmodulated carriers, having nearly the same frequency, are added together in the IF of the receiver. This gives the usual hetero-dyne envelope as the two voltages beat together. In addition, there is a variation in the phase of the resultant, which is detected by the discriminator and appears as noise.

Let the stronger signal be represented by $E_s \cos \omega_c t$, and the interfering signal by $\rho E_s \cos (\omega_c + \omega_d)t$, where $\rho < 1$, ω_c and ω_d the angular frequencies, and $\omega_d \ll \omega_c$. The resultant signal $e_r(t)$ at the output of the IF amplifier may be derived from the vectorial addition of the unmodulated carriers as shown in Fig. 11-1.

$$e_r(t) = E_s \cos \omega_c t + \rho E_s \cos (\omega_c + \omega_d)t = A(t) \cos [\omega_c t + \theta(t)]$$

We shall now compute the envelope $A(t)$ and the instantaneous frequency

$$\omega_i(t) = \omega_c + \frac{d\theta}{dt}$$

From Fig. 11-1, we find $\quad A(t) = E_s \sqrt{1 + \rho^2 + 2\rho \cos \omega_d t}$ (11-1)

and $\quad \tan \theta = \dfrac{\rho \sin \omega_d t}{1 + \rho \cos \omega_d t}$ (11-2)

Therefore, the instantaneous frequency

$$\omega_i(t) = \omega_c + \frac{d}{dt}\left(\tan^{-1} \frac{\rho \sin \omega_d t}{1 + \rho \cos \omega_d t}\right)$$ (11-3)

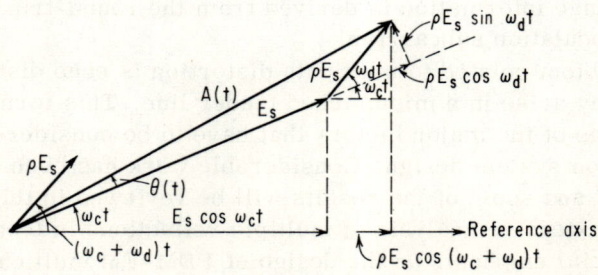

Fig. 11-1. Vectorial additions of unmodulated carriers.

The output of an envelope detector is directly proportional to the magnitude $A(t)$ of the resultant signal $e_r(t)$, while the output of a frequency discriminator is directly proportional to the instantaneous frequency $\omega_i(t)$.

1. *Fourier Series Analysis of the Envelope* $A(t)$. Since envelope detectors are almost universally used in AM receivers, it is of interest to expand the expression for the envelope $A(t)$ in a Fourier series in order to determine the harmonic components of the detected audio signal.

A Fourier analysis of $A(t)$ yields [1,12]

$$A(t) = E_s \sum_{n=o}^{\infty} A_n(\rho) \cos n\omega_d t \tag{11-4}$$

in which

$$A_0(\rho) = 1 + \left(\frac{\rho}{2}\right)^2 + \frac{1}{4}\left(\frac{\rho}{2}\right)^4 + \cdots = \text{average value of } A(t) \tag{11-5}$$

$$A_1(\rho) = \rho\left[1 - \frac{1}{2}\left(\frac{\rho}{2}\right)^2 - \frac{1}{4}\left(\frac{\rho}{2}\right)^4 - \cdots\right] \tag{11-6}$$

$$A_2(\rho) = -\left(\frac{\rho}{2}\right)^2\left[1 - \left(\frac{\rho}{2}\right)^2 - \cdots\right] \tag{11-7}$$

· ·

$$A_n(\rho) = 2(-1)^n \left[\frac{1(3)\cdots(2n-1)}{n!}\right]\left(\frac{\rho}{2}\right)^n\left[\frac{-1}{2n-1} + \frac{1}{1(n+1)}\left(\frac{\rho}{2}\right)^2\right.$$
$$\left. + \sum_{k=2}^{\infty} \frac{1(3)\cdots(2k-3)}{k!\,2^{2k}} \cdot \frac{(2n+1)(2n+3)\cdots(2n+2k-3)}{(n+1)(n+2)\cdots(n+k)}\rho^{2k}\right] \tag{11-8}$$

The d-c component or the average value of the envelope whose amplitude is $E_s A_0(\rho)$ depends primarily upon E_s, the amplitude of the stronger signal; the beat-frequency component $E_s A_1 \cos \omega_d t$ depends principally upon the amplitude ρE_s of the interfering signal, while the amplitude of the n'th harmonic of the beat frequency depends principally upon $(\rho/2)^n E_s$.

For small values of ρ,

$$A(t) \doteq E_s(1 + \rho \cos \omega_d t), \qquad \rho \ll 1 \tag{11-9}$$

As the ratio ρ is increased, the higher harmonics increase in amplitude and in the limit as $\rho \rightarrow 1$,

$$A(t) = \sqrt{2}\ E_s\sqrt{1 + \cos\ \omega_dt} = 2E_s\left|\cos\ \frac{\omega_d}{2}\,t\right| \qquad (11\text{-}10)$$

and the envelope becomes a series of rectified cosine waves. This can be expressed in terms of a harmonic series given by

$$A(t) = \frac{4E_s}{\pi}\left[1 - 2\sum_{n=1}^{\infty}\frac{(-1)^n\ \cos\ n\omega_dt}{(2n)^2 - 1}\right] \qquad (11\text{-}11)$$

2. *Fourier Series Analysis of the Instantaneous Frequency $\omega_i(t)$.*
The instantaneous frequency $\omega_i(t)$ of the resultant signal $e_r(t)$ is given by

$$\begin{aligned}\omega_i(t) &= \omega_c + \frac{d\theta}{dt} = \omega_c + \frac{d}{dt}\left(\tan^{-1}\frac{\rho\ \sin\ \omega_dt}{1 + \rho\ \cos\ \omega_dt}\right)\\[2mm] &= \omega_c + \omega_d\left(\frac{\rho\ \cos\ \omega_dt + \rho^2}{1 + \rho^2 + 2\rho\ \cos\ \omega_dt}\right)\end{aligned} \qquad (11\text{-}12)$$

The expression $d\theta/dt$ represents the instantaneous deviation of the frequency of the resultant signal from that of the stronger signal.

The Fourier analysis of the instantaneous frequency can be performed as follows:

Consider the expression

$$\tan\ \theta = \frac{\rho\ \sin\ \omega_dt}{1 + \rho\ \cos\ \omega_dt} \qquad (11\text{-}13)$$

which is represented vectorially in Fig. 11-2. From this figure, we have

$$r\ \sin\ \theta = \rho\ \sin\ \omega_dt \qquad \text{and} \qquad r\ \cos\ \theta = 1 + \rho\ \cos\ \omega_dt$$
$$(11\text{-}14)$$

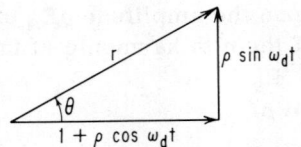

Fig. 11-2. Vectorial representation of $\tan\ \theta$.

and $\quad r = \sqrt{1 + \rho^2 + 2\rho \cos \omega_d t}$ $\qquad$ (11-15)

This may be expressed in the form

$$1 + \rho \cos \omega_d t + j\rho \sin \omega_d t = r(\cos \theta + j \sin \theta) \qquad (11-16)$$

or $\quad 1 + \rho e^{j(\omega_d t)} = re^{j\theta}$ $\qquad$ (11-17)

Taking logarithms of both sides, we obtain

$$\log\left(1 + \rho e^{j(\omega_d t)}\right) = \log r + j\theta \qquad (11-18)$$

Since

$$\log\left(1 + \rho e^{j(\omega_d t)}\right) = \rho e^{j(\omega_d t)} - \frac{\rho^2}{2} e^{2j(\omega_d t)} + \frac{\rho^3}{3} e^{3j(\omega_d t)} - \cdots,$$

$$-1 < \rho \le 1 \qquad (11-19)$$

Therefore $\quad \log r + j\theta = \rho\left(\cos \omega_d t + j \sin \omega_d t\right)$

$$- \frac{\rho^2}{2}\left(\cos 2\omega_d t + j \sin 2\omega_d t\right)$$

$$+ \frac{\rho^3}{3}\left(\cos 3\omega_d t + j \sin 3\omega_d t\right) - \cdots \qquad (11-20)$$

Equating the imaginary terms:

$$\theta = \rho \sin \omega_d t - \frac{\rho^2}{2} \sin 2\omega_d t + \frac{\rho^3}{3} \sin 3\omega_d t - \cdots \qquad (11-21)$$

and the interfering signal is proportional to

$$\frac{d\theta}{dt} = \omega_d \left(\rho \cos \omega_d t - \rho^2 \cos 2\omega_d t + \rho^3 \cos 3\omega_d t - \cdots\right)$$

$$= \omega_d \sum_{n=1}^{\infty} (-1)^{n+1} \rho^n \cos n\omega_d t \qquad (11-22)$$

There is no constant term in this expression, although nonsym-
metrical; $d\theta/dt$ has an average value equal to zero; consequently,
there is no frequency shift in the original carrier frequency ω_c.

This expansion is valid in the limits $-1 < \rho \leq 1$.
The function $d\theta/dt$ is peaked at $\cos n\omega_d t = \pm 1$.

$$\text{At } \cos n\omega_d t = 1, \quad \frac{d\theta}{dt} = \omega_d \sum_{n=1}^{\infty} (-1)^{n+1} \rho^n = \frac{\omega_d \rho}{\rho + 1} \tag{11-23}$$

$$\text{At } \cos n\omega_d t = -1, \quad \frac{d\theta}{dt} = \omega_d \sum_{n=1}^{\infty} \rho^{n+1} = \frac{\omega_d \rho}{\rho - 1} \tag{11-24}$$

Plots of $d\theta/dt$ are shown in Fig. 11-3, for $\rho = 0.8$ (solid curve) and $\rho = 1/0.8$ (dashed curve). The spike pattern will reverse polarities if $\omega_d = -\omega_d$ or if ρ becomes greater than 1.

Since the average value of $d\theta/dt$ is equal to zero, it is obvious that the average of the resultant signal over a period $2\pi/\omega_d$ sec is precisely the frequency ω_c of the stronger signal. Thus, assuming a perfect limiter, preceding the frequency discriminator, the detector output will be proportional to ($\omega_c + d\theta/dt$), and its average value will correspond to the d-c level dictated by the frequency ω_c of the stronger signal. The harmonic components of the interfering signal ω_d, as given by Eq. (11-22), will be attenuated by the low-frequency filtering and the de-emphasis circuit which usually follows the discriminator.

In the following sections, we shall examine the case of frequency-modulated interfering carriers. It should be pointed out, however,

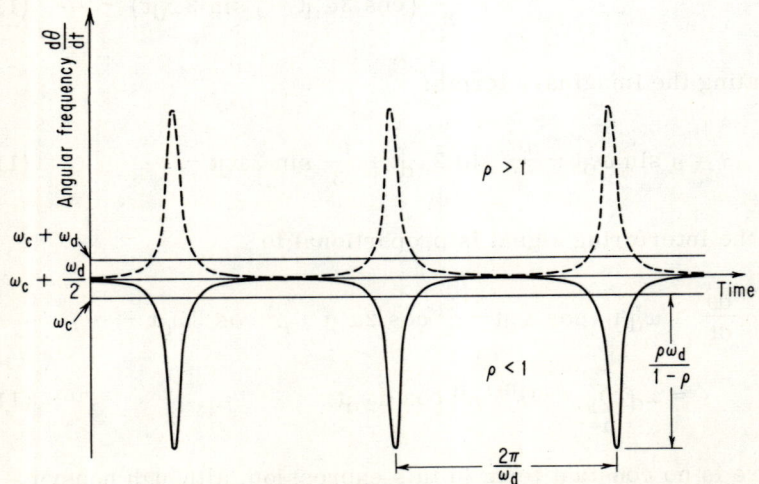

Fig. 11-3. Instantaneous frequency of resultant due to two-carrier interference: $\rho = 0.8$, solid curve; $\rho = 1/0.8$, dashed curve.

that the preceding analysis of two unmodulated carriers of slightly
different frequencies is also applicable to this case, provided the
modulation is slow in comparison with the beat frequency ω_d.

11.2   Common- and Adjacent-channel Interference [1,3]

In the previous section, we discussed the case of two unmodulated
carriers of slightly different frequency. If now the frequency of one
carrier is modulated, the problem becomes one of common- or
adjacent-channel interference, depending upon what range of frequencies the deviations of the modulated carrier cover. If the deviations of the one wave are about a mean frequency which coincides
with the frequency of the second carrier, the result is common-
channel interference. If the mean frequencies are separated by the
width of one channel, the result will be adjacent-channel interference.

Following the method by Stumpers,[3] we shall derive a general
formula which will enable us to calculate the distortion terms
arising from the interference of two carriers of equal frequency
or slightly different frequency, modulated or unmodulated. The
generalized formula will then be applied to the specific problems
of common- and adjacent-channel interference.

1. *The General Problem*. Consider the general case of two interfering signals:

$$e_1(t) = \cos \psi_1(t) = \cos (\omega_1 t + \beta_1 \sin pt) \tag{11-25}$$

$$e_2(t) = \rho \cos \psi_2(t) = \rho \cos (\omega_2 t + \beta_2 \sin qt + \psi_0) \tag{11-26}$$

where ω_1 and ω_2 are the carrier frequencies, p and q the modulating
frequencies, and $\beta_1 = \Delta\omega_1/p$, $\beta_2 = \Delta\omega_2/q$, the modulation indices.

As before (see Fig. 11-1), the resultant $e_r(t)$ is given by

$$e_r(t) = A(t) \cos \varphi(t) \tag{11-27}$$

where $A(t)$ is the instantaneous amplitude given by

$$A(t) = \sqrt{1 + \rho^2 + 2\rho \cos \psi(t)} \tag{11-28}$$

$$\psi(t) = \psi_2(t) - \psi_1(t) \tag{11-29}$$

and $\varphi(t)$ is the phase angle of the resultant signal given by

$$\varphi(t) = \psi_1(t) + \tan^{-1} \frac{\rho \sin \psi(t)}{1 + \rho \cos \psi(t)} \qquad (11\text{-}30)$$

The phase angle $\varphi(t)$ can be expressed as a Fourier series of $\psi(t)$ by making use of the results derived previously [Eq. (11-21)].

$$\varphi(t) = \psi_1(t) + \rho \sin \psi(t) - \frac{\rho^2}{2} \sin 2\psi(t) + \frac{\rho^3}{3} \sin 3\psi(t) - \cdots \qquad (11\text{-}31)$$

The instantaneous frequency is derived by differentiating Eq. (11-30).

$$\omega_i(t) = \frac{d\varphi}{dt} = \omega_1 + \Delta\omega_1 \cos pt + \frac{\rho \cos \psi + \rho^2}{1 + 2\rho \cos \psi + \rho^2} \frac{d\psi}{dt} \qquad (11\text{-}32)$$

The first terms in this result give the required signal in the absence of the interfering transmitter. The second term, namely,

$$\frac{\rho \cos \psi + \rho^2}{1 + 2\rho \cos \psi + \rho^2} \frac{d\psi}{dt}$$

represents the distortion components.

To evaluate the distortion components, we consider the function

$$f(\rho,\psi) = \frac{\rho \cos \psi + \rho^2}{1 + 2\rho \cos \psi + \rho^2} \qquad (11\text{-}33)$$

which is a periodic function of ψ of period 2π that can be represented by a Fourier series. For $\rho < 1$, we have by differentiating Eq. (11-31)

$$f(\rho,\psi) = \rho \cos \psi - \rho^2 \cos 2\psi + \rho^3 \cos 3\psi - \cdots \qquad (11\text{-}34)$$

Furthermore:

$$f(\rho,0) = \frac{\rho}{1 + \rho} \; ; \quad f\left(\rho, \frac{\pi}{2}\right) = \frac{\rho^2}{1 + \rho^2} \; ; \quad f(\rho,\pi) = \frac{-\rho}{1 - \rho}$$

The last relation is valid only when $\rho \neq 1$. As ρ approaches 1, the peak at π grows sharper and $|f(\rho,\pi)| \rightarrow \infty$.

It can be shown that the following relations are valid:

$$\lim_{\rho \to 1} f(\rho,\psi) = \tfrac{1}{2}; \qquad \psi \neq \pi$$

$$\lim_{\rho \to 1-0} f(\rho,\pi) = -\infty \qquad\qquad (11\text{-}35)$$

It is sufficient to consider the case of $\rho \leq 1$, as all other values follow from the relation

$$f(\rho,\psi) + f\left(\frac{1}{\rho},\psi\right) = 1 \qquad\qquad (11\text{-}36)$$

This relation is derived directly from Eq. (11-33) by substitution. The function $f(\rho,\psi)$ is plotted in Fig. 11-4 for various values of ρ. From Eq. (11-31), we have

$$\varphi = \psi_1 - \sum_{s=1}^{\infty} (-1)^s \rho^s \sin s\psi \qquad\qquad (11\text{-}37)$$

The expression $\sin s\psi$ may be written as

$$\sin s\psi = \operatorname{Im}\left[e^{js\psi}\right] = \operatorname{Im}\left[e^{js(\omega_2 t - \omega_1 t + \psi_0 + \beta_2 \sin qt - \beta_1 \sin pt)}\right] \qquad (11\text{-}38)$$

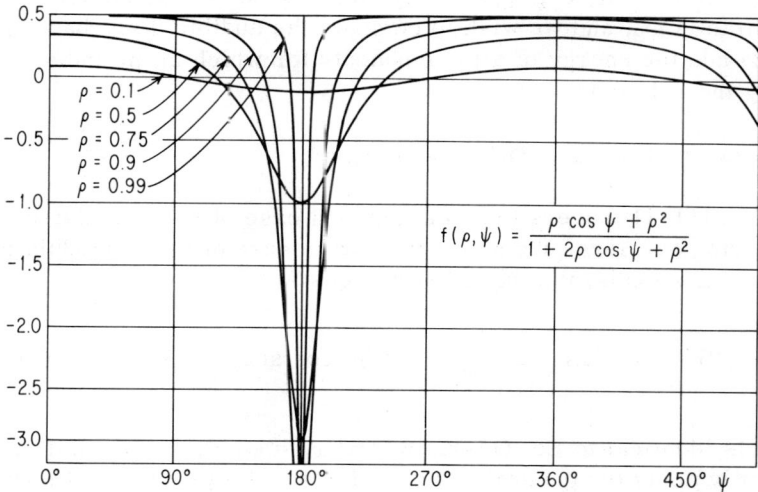

Fig. 11-4. Plot of $f(\rho,\psi)$ for various values of ρ and ψ. (From F. L. H. M. Stumpers,[3] courtesy of Philip Research Rep.)

Using the expansion

$$e^{j\beta \sin pt} = \sum_{n=-\infty}^{\infty} J_n(\beta)e^{jnpt} \tag{11-39}$$

we obtain

$$\varphi = \psi_1 - \sum_{s=1}^{\infty} \frac{(-1)^s \rho^s}{s} \sum_{n=-\infty}^{\infty} J_n(s\beta_2) \sum_{m=-\infty}^{\infty} J_m(s\beta_1)$$

$$\times \sin [(s\omega_2 - s\omega_1 - mp + nq)t + s\psi_0] \tag{11-40}$$

The resultant of the two frequency-modulated components therefore has the instantaneous frequency

$$\omega_i(t) = \frac{d\varphi}{dt} = \omega_1 + \Delta\omega_1 \cos pt - \sum_{s=1}^{\infty} \frac{(-1)^s \rho^s}{s} \sum_{m=-\infty}^{\infty} J_m(s\beta_1)$$

$$\times \sum_{n=-\infty}^{\infty} J_n(s\beta_2)(s\omega_2 - s\omega_1 - mp + nq)$$

$$\times \cos [(s\omega_2 - s\omega_1 - mp + nq)t + s\psi_0] \tag{11-41}$$

We see that if $\omega_2 \neq \omega_1$, the detected signal in general does not contain the frequency q. Hence there is no understandable crosstalk, and the interfering term gives only noise. When the audio-frequency bandwidth is ω_a, and we wish to compute the audible disturbance, we have to add the energy of all components for which m, n, and s satisfy the relation

$$-\omega_a < s(\omega_2 - \omega_1) - mp + nq < \omega_a$$

Equation (11-41) covers the most general case of two modulated interfering carriers. We note here that in case of two unmodulated interfering signals, this equation reduces to

$$\omega_i(t) = \omega_1 + (\omega_2 - \omega_1) \sum_{s=1}^{\infty} (-1)^{s+1} \rho^s \cos s(\omega_2 - \omega_1)t \tag{11-42}$$

which is identical to Eq. (11-22) where $\omega_d = \omega_2 - \omega_1$. We shall presently apply the general result to the special problems of common and adjacent channel interference.

2. *Common- and Adjacent-channel Interference; Interfering Signal Unmodulated.* This is the case when a frequency-modulated

signal and an unmodulated carrier produce the beat-rate frequency. The instantaneous frequency of the resultant signal is obtained readily from Eq. (11-41) by putting $\beta_2 = 0$ and $q = 0$:

$$\omega_i(t) = \omega_1 + \Delta\omega_1 \cos pt - \sum_{s=1}^{\infty} \frac{(-1)^s}{s} \rho^s \sum_{m=-\infty}^{\infty} J_m(s\beta_1)$$

$$\times [s(\omega_2 - \omega_1) - mp] \cos [(s\omega_2 - s\omega_1 - mp)t + s\psi_0] \quad (11\text{-}43)$$

This result was first derived by Corrington;[1] it shows that the effect of the interfering signal is to produce cross-modulation between the modulation signal of frequency p and the difference angular frequency $(\omega_2 - \omega_1)$ of the interfering unmodulated carrier and the desired modulated carrier. When $\omega_2 = \omega_1$, we have common-channel interference, and the expression for the instantaneous frequency reduces to

$$\omega_i(t) = \omega_1 + \Delta\omega_1 \cos pt + \sum_{s=1}^{\infty} \sum_{m=-\infty}^{\infty} (-\rho)^s \frac{mp}{s} J_m(s\beta)$$

$$\times \cos (mpt - s\psi_0) \quad (11\text{-}44)$$

Corrington[1] plotted the output from a FM receiver in the presence of common-channel interference, assuming a perfect limiter, for various ratios of the interfering signal voltage ρ, as shown in Fig. 11-5. When $\rho = 0$ (i.e., no interference), the output is an undistorted cosine wave. As the interference increases, the peaks and dips increase in size, until finally, in the limit, they become very narrow pulses superimposed on a cosine wave of one-half the amplitude obtained with no interference.

As ρ becomes greater than one, the interfering signal takes control, and the modulation of the desired signal is suppressed, as shown in Fig. 11-6; as ρ increases from one to infinity, the peaks and dips in the output decrease.

The effect of a de-emphasis network following the discriminator and of a low-pass filter upon the harmonic components in the output, as shown by Eq. (11-44), can be determined by computing the amplitude of each harmonic that falls within the working range. If a de-emphasis network and a low-pass audio filter are used, many of the harmonics will be attenuated or removed, and the peaks of noise are therefore reduced considerably.

3. *Common-channel Interference; Both Signals Modulated*. In this section, we shall discuss the case when the desired and inter-

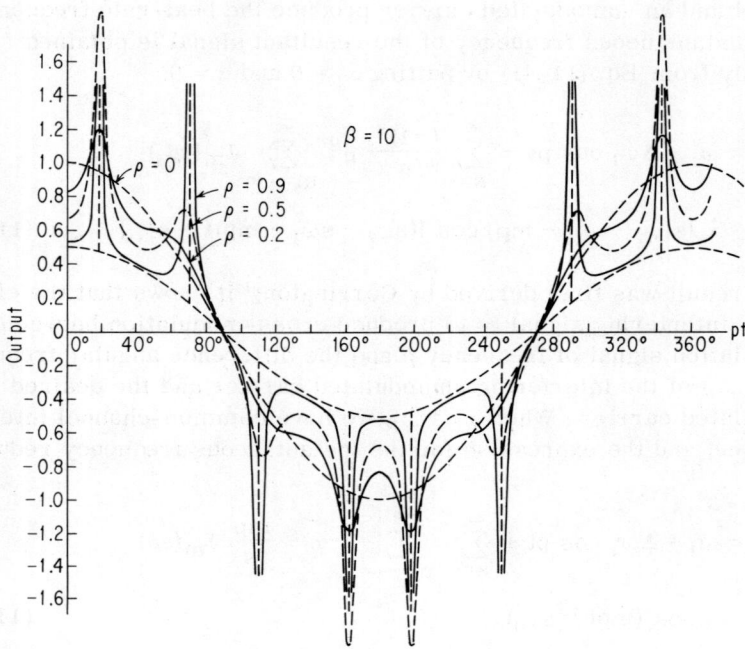

Fig. 11-5. Common-channel interference: variation of distortion as interfering signal becomes stronger. (From M. S. Corrington, [1] RCA Review.)

fering signals with a common-carrier frequency are both modulated sinusoidally. Again, using the general result of Eq. (11-41) and making $\omega_1 = \omega_2 = \omega_c$, we obtain

$$\omega_i(t) = \omega_c + \Delta\omega \cos pt - \sum_{s=1}^{\infty} \frac{(-1)^s \rho^s}{s} \sum_{m=-\infty}^{\infty} J_m(s\beta_1) \sum_{n=-\infty}^{\infty} J_n(s\beta_2)$$

$$\times (nq - mp) \cos [(nq - mp)t + s\psi_0] \qquad (11\text{-}45)$$

Here we may have crosstalk, as the frequency q occurs in the output spectrum. This result was also first derived by Corrington, [1] together with very interesting plots of the resultant audio output for various values of ρ. In these plots, Corrington shows that as $\rho \rightarrow 1$, the beat-note interference increases in amplitude, until in the limit as $\rho \rightarrow 1$ the pulses become very narrow and long, approaching an impulse function. If ρ becomes greater than one, the polarity of the pulses is reversed, and the interfering signal gains control. This

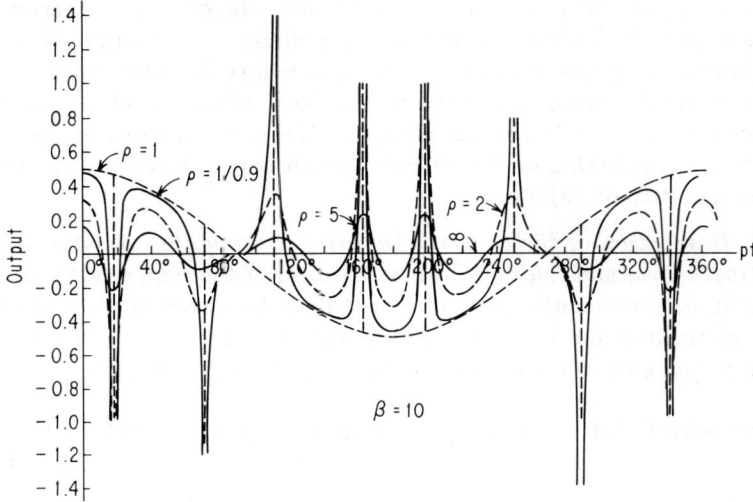

Fig. 11-6. Common-channel interference: variation of distortion as interfering signal becomes stronger. (From M. S. Corrington,[1] RCA Review.)

phenomenon whereby one of several input signals dominates in the output is known as the "capture" effect. In a later section, we shall discuss FM "capture" performance and its consideration in receiver design.

11.3   Multipath Transmission Interference[3,7,8]

It has already been mentioned earlier that multipath transmission occurs when two or more interfering signals come from the same transmitter but one is delayed with respect to the other because of longer transmission path. The problem of multipath transmission may be treated as a special case of the general problem of two interfering signals, as shown by Stumpers.[3] However, in order to obtain a clearer insight into the physical phenomenon of multipath transmission, we shall derive the distortion formulas arising from multipath transmission directly from basic considerations. We shall also derive an expression for FM multipath phase errors in CW-FM ranging systems which may arise in range measurements.

Multipath interference may introduce a phase error for the following reason. In a phase-comparison ranging system, the phase of a modulating signal which has traveled from the ground transmitter to the transponder, where it is retransmitted back to a ground receiver, is compared with the phase of the reference signal from the transmitter. The phase difference in radians between the reference signal and the received signal is $2\pi D/\lambda$, where D is the total distance traveled by the signal from the transmitter to the receiver via the transponder, and λ is the wavelength. The presence of a ground-reflected interfering signal introduces an error in the measurement of the quantity of $2\pi D/\lambda$.

1. *Derivation of the General Distortion Formulas Arising from Multipath Transmission.* In the derivation of the expression of the distortion components, it is assumed that the receiver contains a perfect limiter and a discriminator which is linear over the frequency range of interest. Referring to Fig. 11-7, we have

Direct wave: $e_1(t) = \cos \psi_1(t) = \cos [\omega_c(t - t_1) + \beta \sin p(t - t_1)]$

$$(11-46)$$

Reflected wave: $e_2(t) = \rho \cos \psi_2(t)$

$$= \rho \cos [\omega_c(t - t_2) + \beta \sin p(t - t_2) + \phi] \quad (11-47)$$

where t_1 = time required for the direct wave to reach the receiver
t_2 = time required for the reflected wave to reach the receiver
ϕ = angle of reflection of reflected signal
$\beta = \Delta\omega/p$ = modulation index
p = angular frequency of modulating signal

Therefore

$$\psi(t) = \psi_2(t) - \psi_1(t) = -\omega_c t_0 + \phi - 2\beta \cos p\left[t - t_1 - \frac{t_0}{2}\right] \sin \frac{pt_0}{2}$$

$$(11-48)$$

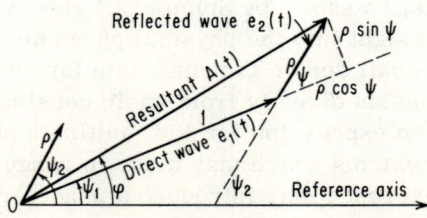

Fig. 11-7. Vector diagram showing direct wave, reflected wave, and resultant.

where $t_0 = t_2 - t_1$

t_0 is the time delay of the reflected wave relative to the direct wave. The resultant wave $e_r(t)$ is given by

$$e_r(t) = A(t) \cos \varphi(t) \tag{11-49}$$

where $A(t) = \sqrt{1 + \rho^2 + 2\rho \cos \psi(t)}$ \hfill (11-50)

$$\varphi(t) = \psi_1(t) + \tan^{-1} \frac{\rho \sin \psi(t)}{1 - \rho \cos \psi(t)} \tag{11-51}$$

The amplitude term $A(t)$ contains the variation of the resultant carrier amplitude, and the instantaneous phase of the resultant wave $\varphi(t)$ contains the original phase modulation and the resulting distortion. Thus multipath transmission resulted in amplitude modulation and distortion in the original FM signal. When the signal goes through a perfect limiter, the amplitude term is suppressed, and the output of the discriminator is proportional to the instantaneous frequency $\omega_i(t) = d\varphi(t)/dt$.

$$\frac{d\varphi(t)}{dt} = \omega_c + \Delta\omega \cos p(t - t_1) + \frac{d}{dt} \tan^{-1} \frac{\rho \sin \psi(t)}{1 + \rho \cos \psi(t)} \tag{11-52}$$

At the receiver, however, the ω_c component is removed by a balanced discriminator, leaving the demodulated output

$$\Delta\omega \cos p(t - t_1) + \frac{d}{dt} \tan^{-1} \frac{\rho \sin \psi(t)}{1 + \rho \cos \psi(t)} \tag{11-53}$$

The second term of expression (11-53) contains the harmonic distortion components of the demodulated signal. As we shall show later, the first harmonic contains a contribution to the time delay t_1 which introduces an error in the range measurements.

   Note that Eqs. (11-49) to (11-51) are identical with Eqs. (11-27), (11-28), and (11-30), respectively, except for the modified expression for $\psi(t)$ as given by Eq. (11-48). Obviously, the general result for the instantaneous frequency of the resultant waves, namely, $d\varphi(t)/dt$ as given by Eq. (11-41), is also applicable to this problem. However, since we are interested also in deriving an expression for the phase error introduced in CW-FM ranging systems, we shall derive the final result in a slightly different form, following the method of Sollenberger.[8]

Let us rewrite Eq. (11-48) as follows. Put

$$z = 2\beta \sin \frac{pt_0}{2} \tag{11-54}$$

$$B = p\left(t - t_1 - \frac{t_0}{2}\right) \tag{11-55}$$

$$\theta_0 = -\omega_c t_0 + \phi \tag{11-56}$$

Therefore $\quad \psi(t) = \theta_0 - Z \cos B \tag{11-57}$

In order to calculate the distortion components, we expand the distortion term in harmonics of the modulation frequency p. Using the results of Eq. (11-31), we obtain

$$\tan^{-1} \frac{\rho \sin \psi(t)}{1 + \rho \cos \psi(t)} = \rho \sin \psi(t) - \frac{\rho^2}{2} \sin 2\psi(t)$$
$$+ \frac{\rho^3}{3} \sin 3\psi(t) - \cdots \tag{11-58}$$

Since $\psi(t) = \theta_0 - z \cos B$, we obtain

$$\tan^{-1} \frac{\rho \sin (\theta_0 - z \cos B)}{1 + \rho \cos (\theta_0 - z \cos B)} = - \sum_{n=1}^{\infty} (-1)^n \frac{\rho^n}{n} \sin n(\theta_0 - z \cos B) \tag{11-59}$$

With the use of the identities

$$\sin n(\theta_0 - z \cos B) = \sin (n\theta_0) \cos (nz \cos B)$$
$$- \cos (n\theta_0) \sin (nz \cos B) \tag{11-60}$$

$$\cos (nz \cos B) = J_0(nz) + \sum_{m=1}^{\infty} (-1)^m J_{2m}(nz) \cos 2mB \tag{11-61}$$

$$\sin (nz \cos B) = 2 \sum_{m=0}^{\infty} (-1)^m J_{2m+1}(nz) \cos (2m + 1)B \tag{11-62}$$

it follows that

$$\tan^{-1} \frac{\rho \sin (\theta_0 - z \cos B)}{1 + \rho \cos (\theta_0 - z \cos B)}$$

$$= - \sum_{n=1}^{\infty} \frac{(-\rho)^n}{n} \sin (n\theta_0) \left[J_0(nz) + 2 \sum_{m=1}^{\infty} (-1)^m J_{2m}(nz) \cos 2mB \right]$$

$$+ \sum_{n=1}^{\infty} \frac{(-\rho)^n}{n} \cos (n\theta_0) \left[2 \sum_{m=0}^{\infty} (-2)^m J_{2m+1}(nz) \cos (2m + 1)B \right]$$

$$(11\text{-}63)$$

The only quantity containing the time is $B = p(t - t_1 - t_0/2)$. The distortion term in the output is the time derivative of Eq. (11-63), namely,

$$\frac{d}{dt} \left[\tan^{-1} \frac{\rho \sin (\theta_0 - z \cos B)}{1 + \rho \cos (\theta_0 - z \cos B)} \right]$$

$$= 2p \sum_{n=1}^{\infty} \frac{(-\rho)^n}{n} \sin (n\theta_0) \sum_{m=1}^{\infty} (-1)^m 2m J_{2m}(nz) \sin 2mB$$

$$- 2p \sum_{n=1}^{\infty} \frac{(-\rho)^n}{n} \cos (n\theta_0)$$

$$\times \sum_{m=0}^{\infty} (-1)^m (2m + 1) J_{2m+1}(nz) \sin (2m + 1)B \qquad (11\text{-}64)$$

To simplify the expression, let

$$S_m = (-1)^m 4m \sum_{n=1}^{\infty} \frac{(-\rho)^n}{n} \sin (n\theta_0) J_{2m}(nz) \qquad (11\text{-}65)$$

$$C_m = -(-1)^m 2(2m + 1) \sum_{n=1}^{\infty} \frac{(-\rho)^n}{n} \cos (n\theta_0) J_{2m+1}(nz) \qquad (11\text{-}66)$$

Therefore $\dfrac{d}{dt} \left[\tan^{-1} \dfrac{\rho \sin (\theta_0 - z \cos B)}{1 + \rho \cos (\theta_0 - z \cos B)} \right]$

$$= p \sum_{m=1}^{\infty} S_m \sin 2mB + p \sum_{m=0}^{\infty} C_m \sin (2m + 1)B \qquad (11\text{-}67)$$

Equation (11-67), which was first derived by Corrington,[7] gives the expansion of the distortion term of Eq. (11-53) in terms of harmonics of the angle B due to multipath transmission. Corrington has plotted the distorted audio output for several specific cases as shown in

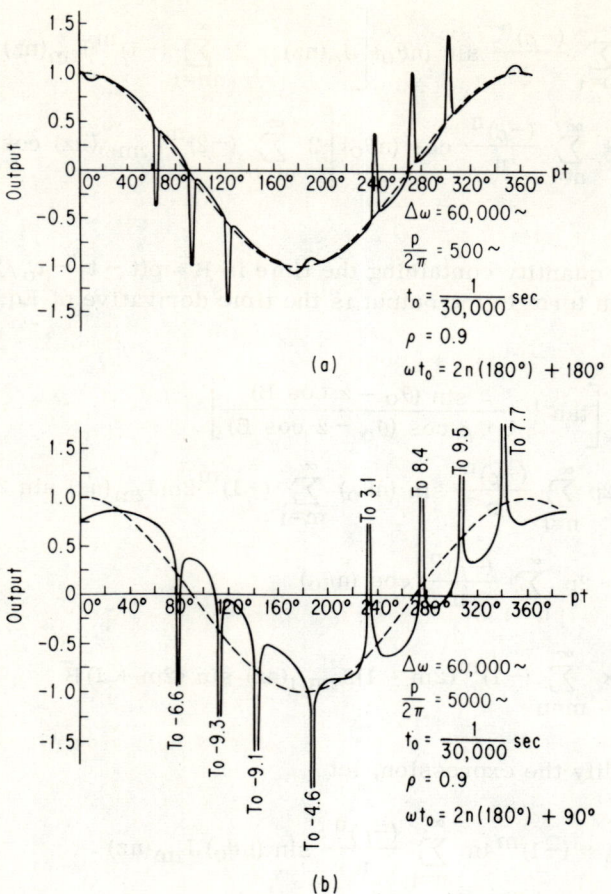

Fig. 11-8. Distorted audio output. (From M. S. Corrington,[7] Proc. IRE.)

Fig. 11-8. Superimposed on the dotted cosine wave, which represents the undistorted output, is the distorted audio output which at several instances consists of impulses corresponding to the times when $d\varphi/dt \rightarrow \infty$.

2. *Calculation of the Phase Error*. It is now assumed that the distortion components of Eq. (11-67) may be removed by filtering. The fundamental is the term in the summation of Eq. (11-67) for which m = 0, i.e., $pC_0 \sin B$. The filtered detected output of the receiver is therefore

$$e_d(t) = \Delta\omega \cos p(t - t_1) + pC_0 \sin B \qquad (11\text{-}68)$$

where
$$C_O = -2 \sum_{n=1}^{\infty} \frac{(-\rho)^n}{n} \cos(n\theta_0) J_1(nz) \tag{11-69}$$

Substituting for $B = p(t - t_1 - t_0/2)$, the output becomes

$$e_d(t) = \Delta\omega \left[\left(1 - \frac{C_O}{\beta} \sin \frac{pt_0}{2} \right) \cos p(t - t_1) \right.$$
$$\left. + \frac{C_O}{\beta} \cos \frac{pt_0}{2} \sin p(t - t_1) \right] \tag{11-70}$$

Put $A \cos \alpha = 1 - \dfrac{C_O}{\beta} \sin \dfrac{pt_0}{2}$ and $A \sin \alpha = \dfrac{C_O}{\beta} \cos \dfrac{pt_0}{2}$

$$\tag{11-71}$$

the output may be written as

$$e_d(t) = \Delta\omega A \cos[p(t - t_1) - \alpha] \tag{11-72}$$

From Eq. (11-72), it is seen that since the original phase delay to be measured is pt_1, the quantity α represents a lagging phase error. The values of α and A are readily obtained from Eq. (11-71) and are

$$A = \left[\left(1 - \frac{C_O}{\beta} \sin \frac{pt_0}{2} \right)^2 + \left(\frac{C_O}{\beta} \cos \frac{pt_0}{2} \right)^2 \right]^{\frac{1}{2}} \tag{11-73}$$

$$\alpha = \tan^{-1} \frac{(C_O/\beta) \cos(pt_0/2)}{1 - (C_O/\beta) \sin(pt_0/2)} \tag{11-74}$$

Phase Error for $\rho = 1$. The phase error when the coefficient of reflection is unity is best obtained by substituting $\rho = 1$ in expression (11-53) for the output. Using the identity

$$\tan \frac{\psi}{2} = \frac{\sin \psi}{1 + \cos \psi} \tag{11-75}$$

the output becomes

$$\Delta\omega \cos p(t - t_1) + \frac{d}{dt} \left[\frac{\psi(t)}{2} \right] \tag{11-76}$$

but from (11-57), $\psi(t) = \theta_0 - z \cos B$. Therefore

$$\frac{d}{dt}\left[\frac{\psi(t)}{2}\right] = \Delta\omega \sin p\left(t - t_1 - \frac{t_0}{2}\right) \sin \frac{pt_0}{2} \qquad (11\text{-}77)$$

The output becomes $A \cos [p(t - t_1) + \alpha]$ \hfill (11-78)

where $\quad A = \Delta\omega \cos \dfrac{pt_0}{2}$

and $\qquad \alpha = \dfrac{-pt_0}{2}$

Since t_0 is the difference between the transmit time of the direct and reflected signals, pt_0 is the phase lag of the reflected signal, and thus the phase error for $\rho = 1$ is seen to be one-half the phase lag of the reflected signal.

Phase Error for $\beta \gg 1$. An examination of Eq. (11-74) shows that the phase error is proportional to $1/\beta$ for large values of β, so that for sufficiently large β, the maximum phase error can be reduced to a predetermined value. It can be shown[8] that the phase error is always less than

$$\tan^{-1}\frac{2/\beta}{[1 - (4/\beta^2)]^{\frac{1}{2}}}$$

This formula is useful in estimating the order of magnitude of the maximum phase error for values of $\beta > 2$. For example, when $\beta = 10$, the phase error is less than $\tan^{-1}\frac{1}{5} = 11.3°$.

11.4   Echo Distortion in Long Feeder Lines [10-12]

In the last section, we dealt with the problem of distortion introduced in a FM signal due to multipath transmission. A similar phenomenon of multipath transmission occurs in long feeder lines due to a mismatch at each end of the line. The effect of the reflecting discontinuities at each end of a long feeder is to introduce distortion in the FM signal. In practice, in which a long waveguide run is composed of a number of short lengths, there will be, in addition, a series of small reflections arising from the various joins, which may however be assumed to be small compared with those existing at the ends.

In this section, the effect of the reflecting discontinuities on a frequency-modulated signal will be analyzed in terms of the harmonic distortion of the modulating signal in the output of an ideal demodulator. Only the simple case of a single modulating frequency will be discussed presently.

First we shall consider the effect of the reflecting discontinuities at each end of the feeder line only, based on the assumption that the contribution arising from the multiple reflections throughout the line is negligible compared to the effect of the mismatches at both ends. The analysis can be extended to encompass the more general case of multiple reflections arising from the discontinuities and irregularities along the line.[11,12]

1. *Calculation of Phase Distortion Terms Due to Reflections at Both Ends of the Line Only.* Consider a FM wave of the form $\cos [\omega_c t + s(t)]$, where ω_c is the carrier angular frequency, and $s(t)$ is the variable phase given by $s(t) = (\Delta\omega/p) \cos pt$, where p is the modulating angular frequency. Following the analysis of Lewin,[10] we consider the FM signal to be transmitted along the transmission line of length ℓ, which is mismatched at both ends. Depending on the mismatch at the load Z_2 (see Fig. 11-9), the signal is reflected and returns toward the generator whose internal impedance is Z_1. Assuming also a mismatch at Z_1, namely $Z_1 \neq Z_0$, the characteristic impedance of the line, a portion of the wave is reflected again toward the load Z_2. The FM signal leaving the feeder at any instant consists of several signals delayed by reflections within the feeder. The transmitted signal will, of course, consist of multiple reflections, but if the mismatches are small, the dominant interference will come from that wave that has been reflected but once from each end of the feeder before leaving it.

Let τ denote the time delay along the feeder, and V denote the group velocity of the waves; then $\tau = \ell/V$, and the transmitted signal

$$e_r(t) = \cos [\omega_c t + s(t)] + r_1 r_2 e^{-2\alpha\ell}$$
$$\times \cos [\omega_c(t - 2\tau) - s(t - 2\tau) + \phi_1 + \phi_2] \qquad (11\text{-}79)$$

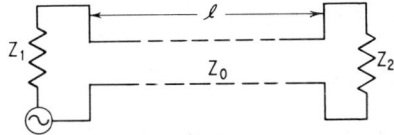

Fig. 11-9. Transmission line with terminations.

where r_1, r_2 are the reflection coefficients at each end of the feeder; ϕ_1, ϕ_2 are the phase changes occurring at each end of the feeder; and α is the attenuation constant of the line.

The reflection coefficients are given by

$$r_1 = \frac{Z_0 - Z_1}{Z_0 + Z_1} = |r_1| e^{j\phi_1}$$

$$r_2 = \frac{Z_0 - Z_2}{Z_0 + Z_2} = |r_2| e^{j\phi_2} \tag{11-80}$$

We shall presently calculate the phase distortion introduced in the composite received signal, which consists of the sum of direct and delayed signals. At the receiver, we assume an ideal limiter and a demodulator whose output is proportional to the instantaneous frequency of the signal $e_r(t)$. Equation (11-79) can be expressed in the form

$$\begin{aligned}
e_r(t) &= A \cos [\omega_c t + s(t)] + B \sin [\omega_c t + s(t)] \\
&= \sqrt{A^2 + B^2} \cos [\omega_c t + s(t) - \theta]
\end{aligned} \tag{11-81}$$

where $A = 1 + r_1 r_2 e^{-2\alpha\ell} \cos [2\omega_c\tau + s(t) - s(t - 2\tau) - \phi_1 - \phi_2]$

$$\tag{11-82}$$

$$B = r_1 r_2 e^{-2\alpha\ell} \sin [2\omega_c\tau + s(t) - s(t - 2\tau) - \phi_1 - \phi_2] \tag{11-83}$$

and $\quad \tan \theta = \dfrac{B}{A}$ $\tag{11-84}$

The amplitude modulation of the term $\sqrt{A^2 + B^2}$ is removed by limiting, so that θ represents the phase distortion due to the mismatches. Assuming small mismatches, we approximate $A \doteq 1$ and $\tan \theta \doteq \theta = B$.

The output of the demodulator is proportional to the instantaneous frequency of the amplitude-limited signal, which is $\omega_c + s'(t) - d\theta/dt$. While $s'(t) = \Delta\omega \sin pt$ represents the recovered signal, the last term contains the distortion.

$$\text{Output} = \Delta\omega \sin pt$$

$$- \frac{d}{dt} \{r_1 r_2 e^{-2\alpha\ell} [\sin 2\omega_c\tau + s(t) - s(t - 2\tau) - \phi_1 - \phi_2]\}$$

$$\tag{11-85}$$

Now

$$s(t) - s(t - 2\tau) = 2\Delta\omega\tau \cdot \sin p(t - \tau) \cdot \frac{\sin p\tau}{p\tau} = y \sin p(t - \tau) \qquad (11\text{-}86)$$

where $y = 2\Delta\omega\tau \dfrac{\sin p\tau}{p\tau}$

Expanding Eq. (11-85), we obtain for the output

$$\text{Output} = \Delta\omega \sin pt - r_1 r_2 e^{-2\alpha\ell} \sin(2\tau\omega_c - \phi_1 - \phi_2)$$

$$\times \frac{d}{dt}\{\cos[y \sin p(t - \tau)]\} + r_1 r_2 e^{-2\alpha\ell}$$

$$\times \cos(2\tau\omega_c - \phi_1 - \phi_2)\frac{d}{dt}\{\sin[y \sin p(t - \tau)]\} \qquad (11\text{-}87)$$

Equation (11-87) can be expanded by the use of the Fourier–Bessel expansion of $\sin[y \sin p(t - \tau)]$ and $\cos[y \sin p(t - \tau)]$ to yield the following harmonic distortion terms of the output.

Recovered signal $= \Delta\omega \sin pt$

Second harmonic $= 2r_1 r_2^{-2\alpha\ell}(2p)J_2(y) \sin\phi \; [\sin 2p(t - \tau)]$

Third harmonic $\; = 2r_1 r_2 e^{-2\alpha\ell}(3p)J_3(y) \sin\left(\phi + \dfrac{\pi}{2}\right) [\cos 3p(t - \tau)]$

Fourth harmonic $= 2r_1 r_2 e^{-2\alpha\ell}(4p)J_4(y) \sin(\phi + \pi) \; [\sin 4p(t - \tau)]$

$$(11\text{-}88)$$

where $\phi = 2\tau\omega_c - \phi_1 - \phi_2$

Lewin [11] and other writers [12] have extended the analysis of feeder distortion to the case of intermodulation caused by reflections at many points along a mismatched feeder for a large number of tones in a multiplex signal. Their results are of practical use in the analysis of noise allocation in communication links.

11.5   Considerations of FM Receiver Design for Rejecting Interference [2, 13-19]

In this section, we shall discuss FM receiver design criteria for good interference rejection ability. We shall show that the interference can be substantially eliminated if the receiver is provided with a properly designed limiter and discriminator. In order to support a high "capture ratio," namely, a maximum ratio of the interfering co-channel signal to the desired signal, the limiter and discriminator must be capable of accommodating the necessary amplitude and frequency excursions involved, as the amplitude of the interfering signal approaches in magnitude the desired signal. Furthermore, the stages preceding the limiter must have an essentially flat frequency response over the frequency range occupied by the desired signal.

Subsequently, we shall show that while wide banding in the limiter and discriminator is a sufficient condition for accommodating multipath or interference co-channel signals with an essentially interference-free output, there are other techniques, such as cascading a number of limiter-filter stages, which may be employed in the design of a high "capture ratio" receiver.

1. *Receiver Design Considerations Using the "Widebanding" Approach.*[2,14,15] It has been shown above that the presence of an undesired signal due to multipath or co-channel interference, within the linear passband of the receiver, introduces in the resultant signal delivered to the first limiter severe AM and FM disturbances under conditions of high-level interference.

The first important design criteria for ensuring suppression of two-path interference at arbitrarily high levels were made by Arguimbau and Granlund.[13] Their design specifications prescribe that the FM demodulator (limiter-discriminator combination) should be amplitude-insensitive and capable of delivering a rectified voltage output that is linearly related to the instantaneous-frequency variations of the resultant of the two input signals over the entire range covered by the spike pattern described by Fig. 11-3. The FM disturbance could then be suppressed or minimized after it is detected.

For proper amplitude-limiting operation, the limiter circuit must be capable of following the sharp changes in the envelope of the resultant impressed signal, which may occur at a rate equivalent to the bandwidth of the IF amplifier in cycles per second. In the discriminator, the output circuit across which the rectified voltage is taken must be capable of following the detected instantaneous frequency disturbance pattern, which may also occur at a maximum rate of one IF bandwidth.

At the basis of the "widebanding" theory lies the following argument. It has been shown above that the average frequency of the resultant signal over a period of $2\pi/\omega_d$ sec is precisely the frequency

ω_c of the stronger signal. Thus, Arguimbau and Granlund reasoned, if the average frequency is preserved at this value, the interference arising from the presence of the weaker signal will be effectively suppressed when the difference frequency ω_d lies outside the audio range. If ω_d lies in the audio range, the action of the de-emphasis network and the low-pass audio filter will reduce the interference considerably. As pointed out by Baghdady,[15] their work highlighted the following considerations in FM receiver design for rejecting interference.

1. In the linear sections, namely, the stages preceding the limiter-detector section, the bandwidth should be sufficient to accommodate a desired frequency-modulation signal over the whole range of its frequency variations. If the first stages pass the desired and undesired signals separately without modification, being linear, they will also pass the sum perfectly. Broadbanding is not necessary before limiting because only the nonlinear process of limiting broadens the spectrum of the sum.

2. Since a FM receiver should be completely insensitive to amplitude changes, the linear stages should be followed by a perfect-acting limiter to cope with amplitude ratios of the order of $(1 + \rho)/(1 - \rho)$ or $39:1$ for $\rho = 0.95$, where ρ equals the strength of the interfering signal relative to the desired stronger signal. It follows also that the IF amplifier must have sufficient gain to raise the level of the minimum amplitude for proper limiter operation.

3. To preserve the average frequency (over a period of the difference frequency ω_d) of the resultant signal at the value ω_c rad/sec, corresponding to the frequency of the stronger signal, it has been found sufficient to provide a limiter bandwidth of $[(1 + \rho)/(1 - \rho)]$ $\times$ $(BW)_{IF}$ and a frequency characteristic that is linear over the same range.

The conditions listed above are sufficient conditions for minimizing the effects of multipath or co-channel signals. Until early in 1955, it was generally believed that these conditions were both necessary and sufficient. As we shall see later on, such widebanding in the limiter and discriminator can be avoided by using other schemes such as cascading a number of limiter-filter stages. In order to evaluate properly the design implications using the "widebanding" approach, let us determine the requirements for reproducing a reasonable approximation to the disturbance pattern of Eqs. (11-11) and (11-22) in the response of a filter (that follows an ideal limiter), and for converting the FM to AM by a discriminator circuit. Baghdady has shown that for $\rho = 0.95$, a single-tuned circuit will reproduce an acceptable approximation to the disturbance pattern if $(BW)_{\lim} \geq 100\omega_d$. For a maximally flat double-tuned circuit, $(BW)_{\lim} \geq 160\omega_d$. Since ω_d may be as high as one IF bandwidth, $(BW)_{IF}$, in co-channel interference, and a few times $(BW)_{IF}$ in off-channel interference, the indicated values for $(BW)_{\lim}$ are often quite impracticable. Similarly, Baghdady arrives at the conclusion that the peak-to-peak separation of the discriminator characteristic must be approximately $2.5[(1 + \rho)/(1 - \rho)]\omega_d$. For $\rho = 0.95$, this separation, which is of the order of $100\omega_d$, will result in a discriminator characteristic of very low sensitivity.

It is obvious from the above results that wideband limiting and detecting imposes severe practical limitations on the design of the FM receiver. This difficulty can be overcome by the use of cascaded narrow-band limiters as suggested by Baghdady.[14,15,17]

2. *Cascaded Narrow-band Limiters.* In view of the difficulty of meeting the FM demodulator design requirements for faithful reproduction of the FM disturbance, the method of narrow-band limiters aims at minimizing this disturbance before it is detected. In this method, the emphasis is on predetection suppression of signals that cannot be eliminated by conventional IF filtering. This is accomplished by a repeated process of amplitude limiting followed by narrow-band filtering.

Before we present the basic mechanism of this process, it will be necessary first to consider the properties of the spectrum of the amplitude-limited resultant of the two input sinusoids. Consider the resultant signal

$$e_r(t) = E_s \cos \omega_c t + \rho E_s \cos (\omega_c + \omega_d)t \qquad (11\text{-}89)$$

to be passed through an ideal limiter that is followed by an ideal wideband filter. The filter is assumed to be sufficiently selective to make only the spectral components centered about the frequency ω_c significant with $\omega_d \ll \omega_c$, and with harmonics of ω_c and their associated sidebands completely rejected or negligible. Thus, with $A(t) \cos \phi(t)$ at the input, the signal at the output will be

$$e_\ell(t) = k_\ell \cos [\omega_c t + \theta(t)] \qquad (11\text{-}90)$$

where k_ℓ is a constant of the limiter and

$$\theta(t) = \tan^{-1} \frac{\rho \sin \omega_d t}{1 + \rho \cos \omega_d t}. \qquad (11\text{-}91)$$

The right-hand side of Eq. (11-90) can be expanded in the form

$$e_\ell(t) = k_\ell \sum_{n=-\infty}^{\infty} A_n(\rho) \cos (\omega_c - n\omega_d)t \qquad (11\text{-}92)$$

Ten-place tables of the spectral amplitude components $A_n(\rho)$ have been constructed by Granlund[2] for various values of ρ, and for n extending to fairly large values. A very close study of those tables reveals the following properties:

1. At $t = 2m\pi/\omega_d$, where m is an integer, the $A_n(\rho)$'s alternate in sign, starting with $A_{+1}(\rho)$ negative, $A_0(\rho)$ and $A_{-1}(\rho)$ positive.

2. At $t = \ell\pi/\omega_d$, where ℓ is an odd integer, all the $A_{+n}(\rho)$'s line up in the same positive direction as A_0, whereas all the $A_{-n}(\rho)$'s line up in phase opposition to $A_0(\rho)$.

3. The magnitude of $A_n(\rho)$ decreases very rapidly with n.

The most important components in the spectrum of $e_\ell(t)$ are those whose frequencies correspond to the frequencies of the two input sinusoids. The amplitude of the spectral component that has the frequency of the stronger of the two input signals is $kA_0(\rho)$. The component with amplitude $kA_{-1}(\rho)$ has the frequency of the weaker signal. The functions $A_0(\rho)$ and $A_{-1}(\rho)$ are plotted in Fig. 11-10, and they can be approximated by

$$A_0(\rho) \doteq 1 - \frac{\rho^4}{4}, \qquad \text{for } \rho < 0.7$$

and

$$A_{-1}(\rho) \doteq \frac{\rho}{2}, \qquad \text{for } \rho < 0.6$$

If we set $\quad R(\rho) = \dfrac{A_{-1}(\rho)/A_0(\rho)}{\rho} \qquad (11\text{-}93)$

then, from Fig. 11-10, it is obvious that $R(\rho) \doteq \frac{1}{2}$ for all $\rho < \frac{1}{2}$, and $R(\rho) \to 1$ only as $\rho \to 1$. This signifies that the amplitude-limiting operation reduces the weaker-to-stronger signal-amplitude ratio by a factor $R(\rho)$ that is essentially $\frac{1}{2}$ for small ρ. This effect can be utilized in various ways to achieve stronger-signal, as well as weaker-signal enhancement, such as the techniques of cascading narrow-band limiters and applying feedback or feed-forward across the limiters.[14,18,19]

An extremely effective scheme for improving the capture per-

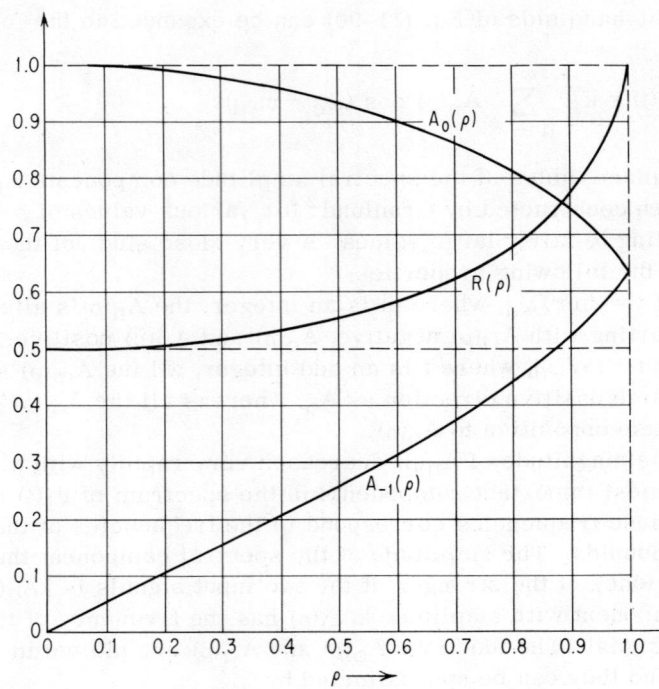

Fig. 11-10. Plots of the functions $A_0(\rho)$, $A_{-1}(\rho)$ and $R(\rho) = [A_{-1}(\rho)/A_0(\rho)]/\rho$. (From E. J. Baghdady,[14] courtesy of McGraw-Hill Book Company.)

formance of a FM receiver can be achieved by cascading several stages of amplitude limiting, each of which is followed by a bandpass filter whose bandwidth does not exceed a few times the IF bandwidth. The mechanism by which cascading narrow-band limiters reduces interference is based on the fact which we have just discussed, that each limiter spreads out the significant spectrum over a frequency range that exceeds many times the bandwidth of the IF amplifier. For example, if the signal strengths differ by 10 per cent of the amplitude of the stronger signal, then the resulting spectrum due to amplitude limiting will be of the order of 100 times the bandwidth of the IF amplifier. This spectrum can be narrowed down by placing a bandpass filter of appropriate bandwidth after the limiter without appreciably affecting the spectrum that carries the message modulation of the stronger signal. As we have seen above, the greatest achievable reduction in the relative amplitude of the weaker signal in one stage of narrow-band limiting equals one-half when $\rho < \frac{1}{2}$. This process can be repeated several times by placing a number of limiter-filter units in cascade until the proper improvement in the capture performance of the receiver is realized.

The minimum permissible limiter bandwidth is a very important parameter in the design of a FM receiver which incorporates a number of cascaded ideal narrow-band limiters. The limiter bandwidth must be wide enough to pass a sufficient number of sideband components that will add up to a resultant signal whose average frequency is equal to the frequency of the stronger of the two input carriers. It has been shown by Baghdady [15,16] that the minimum permissible bandwidth for the idealized filter associated with an amplitude limiter is one IF bandwidth for all values of $\rho \geq 0.863$, and a bandwidth of $3(\text{BW})_{\text{IF}}$ is sufficient for all $\rho \geq 0.98$. It is important to note that for a given value of limiter-filter bandwidth, $(\text{BW})_{\text{lim}}$, the effect of narrow-band limiting upon the disturbance becomes more pronounced with increasing values of the frequency difference ω_d and the interference ratio ρ. As shown previously, the greatest achievable reduction in the relative amplitude of the weaker signal in one stage of narrow-band limiting amounts to a factor of $\frac{1}{2}$ when $\rho > 0.5$. Baghdady has shown that the performance can be improved by applying feedback [14,18] or feed-forward [14,19] across the limiter.

References

1. Corrington, M. S.: Frequency Modulation Distortion Caused by Common and Adjacent Channel Interference, *RCA Rev.*, vol. 7, pp. 522-560, December, 1946.
2. Granlund, J.: Interference in Frequency-modulated Reception, *MIT Res. Lab. Electron. Tech. Rept.* 42, Jan. 20, 1949.
3. Stumpers, F. L. H. M.: Interference Problems in Frequency Modulation, *Philips Res. Rept.* 2, pp. 136-160, 1947.
4. Plusc, I.: Investigation of Frequency-modulation Signal Interference, *Proc. IRE*, October, 1947.
5. Crosby, M. A.: Frequency-modulation Propagation Characteristics, *Proc. IRE*, vol. 24, pp. 898-913, 1936.
6. Crosby, M. A.: Observations of Frequency-modulation Propagation on 26 Megacycles, *Proc. IRE*, vol. 29, pp. 398-403, 1941.
7. Corrington, M. S.: Frequency Modulation Distortion Caused by Multipath Transmission, *Proc. IRE*, vol. 34, p. 256, May, 1946.
8. Sollenberger, T. E.: Ground Reflection Phase Errors of CW-FM Tracking Systems, *Air Force Tech. Rept.* 12, January, 1952.
9. Meyers, S. T.: Non-linearity in Frequency-modulation Radio System Due to Multipath Propagation, *Proc. IRE*, May, 1946.

10. Lewin, L.: Phase Distortion in Feeders, *Wireless Eng.*, vol. 27, pp. 189–193, July, 1952.
11. Lewin, L.: Multiple Reflections in Long Feeders, *Wireless Eng.*, vol. 27, pp. 189–193, July, 1952.
12. Medhurst, R. G., and G. F. Small: An Extended Analysis of Echo Distortion in the FM Transmission of Frequency Division Multiplex, *Proc. IRE*, March, 1956.
13. Arguimbau, L. B., and J. Granlund: Trans-Atlantic Communication by Frequency Modulation, *Proc. NEC*, vol. 3, p. 644, November, 1947.
14. Baghdady, E. J. (ed.): "Lectures on Communication System Theory," chap. 19, McGraw-Hill Book Company, New York, 1961.
15. Baghdady, E. J.: Frequency-modulation Interference Rejection with Narrow-band Limiters, *Proc. IRE*, January, 1955.
16. Baghdady, E. J.: Interference Rejection in FM Receivers, *MIT Res. Lab. Electron. Tech. Rept.* 252, September, 1956.
17. Baghdady, E. J.: Theory of Stronger-signal Capture in FM Reception, *Proc. IRE*, vol. 46, pp. 728–738, April, 1958.
18. Baghdady, E. J.: Theory of Feedback around the Limiter, *IRE Natl. Conv. Record*, part 8, pp. 176–202, 1957.
19. Baghdady, E. J.: New Developments in FM Reception and Their Application to the Realization of a System of "Power-division" Multiplexing, *IRE Trans. Commun. Systems*, vol. CS-7, September, 1959.

12

GENERATION OF FREQUENCY-MODULATED SIGNALS

In this chapter, we shall discuss various methods for the generation of frequency-modulated signals. In the design of FM systems, it is important to fulfill the following conditions: (1) The carrier component of the modulated signal should not change in frequency when modulation is applied. (2) The frequency deviation of the carrier must be directly proportional to the amplitude of the modulating signal and independent of its frequency. In case of pure phase modulation, this rule applies to the resulting phase deviation. (3) The amplitude of the frequency-modulated carrier should remain constant and independent of the modulation process.

There are essentially two general methods of generating frequency-modulated signals, namely, indirect FM and direct FM.

In the indirect method of producing frequency modulation, the modulating signal is integrated first and then used to phase-modulate a crystal-controlled oscillator to produce a narrow-band FM signal. Subsequent frequency multiplication is used to increase the resulting frequency deviation to any desired level. After the desired bandwidth has been obtained, the carrier with its sidebands can be heterodyned either directly to the transmitter frequency or to some submultiple thereof. In the latter case, additional deviation and bandwidth are obtained by the subsequent multiplication to the transmitter frequency.

A more direct method for producing frequency modulation is to frequency-modulate the carrier directly in accordance with the input modulating signal. In this system, the transmitter output frequency may be modulated directly without the need for frequency multiplication. Because the carrier frequency cannot be crystal-controlled directly, frequency control can be maintained by heterodyning the carrier wave with a crystal-controlled oscillator to some low value, which is caused to operate a sensitive discriminator circuit. The output of the discriminator in turn is used to control the frequency of the carrier oscillator.

In the indirect method, there is no need for carrier frequency stabilization, but the system requires a number of frequency multipliers to produce wideband FM. In the direct method, the number of multiplier stages is smaller, but stabilization of the center frequency is more complicated.

12.1　Indirect-frequency Modulation[1-7]

This method of generation of frequency-modulated signals is used in many commercial FM transmitters. Its principle of operation is based on deriving frequency modulation from a phase modulator by integrating the modulating signal before its application to the phase modulator, as shown in the simplified block diagram of Fig. 12-1. The modulating signal is first integrated and then used to phase-modulate a crystal-controlled carrier frequency. In order to minimize the distortion which is inherent in the phase modulator, as we shall see later on, the maximum phase variation is kept small, resulting in narrow-band FM signal. The output of the phase modulator is then multiplied in frequency by the frequency multiplier to produce

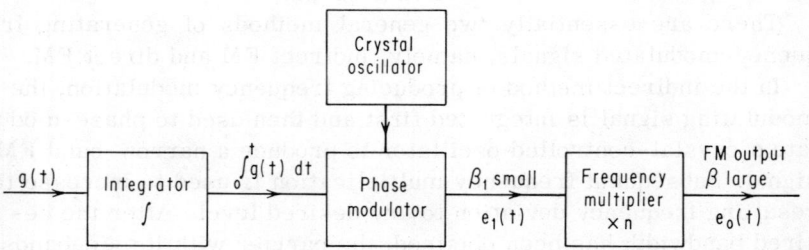

Fig. 12-1. Indirect FM modulator.

the desired wideband FM signal. The operation of the indirect FM system can be expressed mathematically as follows:

Let $e_1(t)$ denote the output of the phase-modulated signal,

$$e_1(t) = A_1 \cos [\omega_1 t + \theta_1(t)] \qquad (12\text{-}1)$$

where ω_1 is the angular frequency of the crystal oscillator, and $\theta_1(t)$ its phase angle which is being modulated by the integral of the message function $g(t)$. Thus $\theta_1(t)$ is given by

$$\theta_1(t) = k \int_0^t g(\tau) \, d\tau \qquad (12\text{-}2)$$

where k is a constant. For a sine-wave modulating signal, the output $e_1(t)$ is given by

$$e_1(t) = A_1 \cos (\omega_1 t + \beta_1 \sin \omega_m t), \qquad \beta_1 = \frac{\Delta \omega}{\omega_m} \qquad (12\text{-}3)$$

where β_1, the modulation index, is kept small to keep the distortion to a minimum.

The output of the phase modulator is then multiplied n times in frequency by the frequency multiplier to produce the final FM carrier $e_0(t)$ with constant amplitude A_0.

Therefore

$$e_0(t) = A_0 \cos [n\omega_1 t + n\theta_1(t)] \qquad (12\text{-}4)$$

which reduces, in the simple case of sine-wave modulating signal, to

$$e_0(t) = A_0 \cos (n\omega_1 t + n\beta_1 \sin \omega_m t) = A_0 \cos (\omega_c t + \beta \sin \omega_m t)$$
$$(12\text{-}5)$$

where $\beta = n\beta_1$ and $\omega_c = n\omega_1$.

In the following, we shall discuss several phase modulators which are commonly used in practice.

1. *Armstrong's Phase-shift Modulator.*   The phase-shift modulator which is illustrated in Fig. 12-2 has been used by Armstrong to produce wideband frequency modulation and is representative of the class of crystal-controlled phase-shift modulators. Its principle of operation is as follows: A crystal-controlled oscillator is ampli-

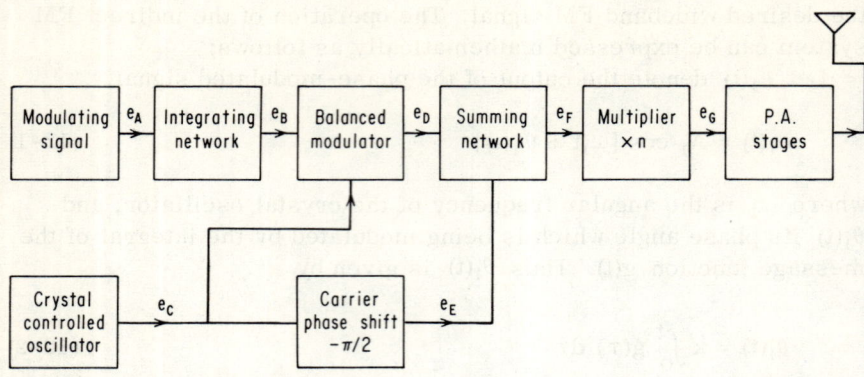

Fig. 12-2. Armstrong's basic indirect FM modulator. (From H. S. Black,[2] courtesy of D. Van Nostrand Company, Inc.)

tude-modulated in a balanced modulator whose output is combined with the carrier which has been phase-shifted by 90°. The mathematical presentation of Armstrong's modulator is as follows.[2]

Let the modulating signal be

$$e_A(t) = A_1 \cos \omega_m t \qquad (12\text{-}6)$$

and the output of the integrating network be given by

$$e_B(t) = \int_0^t A_1 \cos \omega_m t \, dt = \frac{A_1}{\omega_m} \sin \omega_m t \qquad (12\text{-}7)$$

Let the carrier voltage $e_C(t) = A_c \cos \omega_{c_1} t$ \qquad (12-8)

Since the output of the product modulator is proportional to the product of $e_B(t) \, e_C(t)$, therefore

$$e_D(t) = k e_B(t) \, e_C(t) = \frac{k A_1 A_c}{\omega_m} \sin \omega_m t \; \cos \omega_{c_1} t \qquad (12\text{-}9)$$

The carrier is also transmitted through a phase shifter whose output

$$e_E(t) = A_c \cos \left(\omega_{c_1} t - \frac{\pi}{2} \right) = A_c \sin \omega_{c_1} t \qquad (12\text{-}10)$$

The summing network produces an output voltage $e_F(t)$:

$$e_F(t) = e_D(t) + e_E(t) = A_c \left(\sin \omega_{c_1} t + \frac{kA_1}{\omega_m} \sin \omega_m t \ \cos \omega_{c_1} t \right)$$

$$= A_c \sqrt{1 + \frac{k^2 A_1^2}{\omega_m^2} \sin^2 \omega_m t} \ \sin \left[\omega_{c_1} t + \tan^{-1} \left(\frac{kA_1}{\omega_m} \sin \omega_m t \right) \right]$$

$$\doteq A_c \sin \left(\omega_{c_1} t + \frac{kA_1}{\omega_m} \sin \omega_m t \right), \qquad \frac{kA_1}{2\omega_m} \ll 1 \qquad (12\text{-}11)$$

The signal $e_F(t)$, which is narrow-band FM, is passed through a frequency multiplier which is merely a harmonic generator, and the desired harmonic is selected by filtering. As an incidental but important function, the harmonic multiplier also removes, through its nonlinear limiting action, most of any small-amplitude modulation associated with the phase-modulation process. Consequently, the output $e_G(t)$ of the frequency multiplier is a wideband frequency-modulated signal of carrier frequency $\omega_c = n\omega_{c_1}$ and of constant amplitude; its phase deviation has also increased by a factor n, $\Delta\theta_n = nkA_1/\omega_m$. Thus,

$$e_G(t) = A_c \left(\sin n\omega_{c_1} t + \frac{nkA_1}{\omega_m} \sin \omega_m t \right)$$

$$= A_c \left(\sin \omega_c t + \frac{\Delta\omega_c}{\omega_m} \sin \omega_m t \right) \qquad (12\text{-}12)$$

$$= A_c \sin (\omega_c t + \beta \sin \omega_m t)$$

where $\omega_c = n\omega_{c_1}$ and $\Delta\omega_c = nkA_1$.

The modulation index β can be controlled by the choice of n and the design factors kA_1 of the phase-shift modulator.

2. *Distortion Produced by Armstrong's Modulator*.[3] We shall show that this method of producing phase modulation has inherent distortion. This can be seen from Eq. (12-11), where the phase modulation obtained is not proportional to the integral of the modulating signal $(A_1/\omega_m) \sin \omega_m t$ but to the tangent, namely,

$$\theta(t) = \tan^{-1} \left(\frac{kA_1}{\omega_m} \sin \omega_m t \right) \doteq \frac{kA_1}{\omega_m} \sin \omega_m t = \Delta\theta \sin \omega_m t$$

$$(12\text{-}13)$$

where $\Delta\theta = kA_1/\omega_m$. Thus we see that the variable phase angle $\theta(t)$ will be proportional to the modulating signal, provided the phase shift produced is small, so that $\tan\theta \doteq \theta$. This can best be seen from the vectorial presentation of the unmodulated carrier signal and the resultant of the first sideband pair, as shown in Fig. 12-3. This vectorial presentation corresponds to the exact expression for the phase-modulated signal $e_F(t)$ of Eq. (12-11), namely,

$$e_F(t) = A_c \left[\sin\omega_{c_1}t + \frac{kA_1}{2\omega_m} \cos(\omega_{c_1}+\omega_m)\, t - \frac{kA_1}{2\omega_m} \right.$$
$$\left. \times \cos(\omega_{c_1}-\omega_m)t \right] \tag{12-14}$$

The resultant R is phase-modulated by the angle $\theta(t)$ which is proportional to $R_1 = (kA_1/\omega_m)\sin\omega_m t$, provided $\theta(t)$ is small.

The distortion which is produced by this method of modulation may be calculated as follows.

The instantaneous-frequency modulation resulting from this phase modulation is the time derivative of $\theta(t)$, where $\tan\theta = \Delta\theta\sin\omega_m t$, from Eq. (12-13). Therefore

$$\frac{d\theta(t)}{dt} = \Delta\omega_i(t) = \frac{\omega_m\,\Delta\theta\,\cos\omega_m t}{1 + (\Delta\theta)^2\,\sin^2\omega_m t} \tag{12-15}$$

The harmonic distortion as a function of $\Delta\theta$ can be determined from the Fourier expansion of Eq. (12-15). Thus

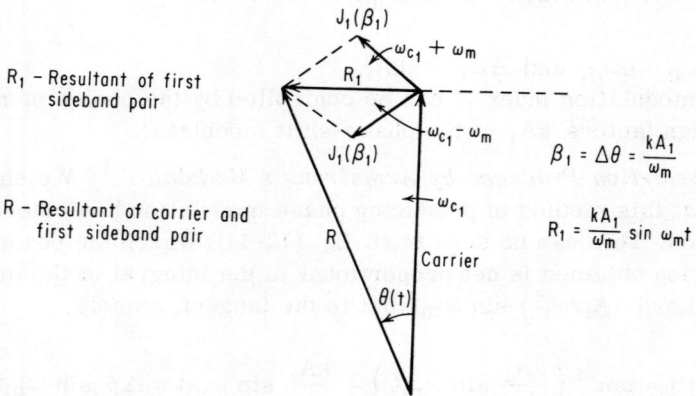

R_1 – Resultant of first sideband pair

R – Resultant of carrier and first sideband pair

$\beta_1 = \Delta\theta = \dfrac{kA_1}{\omega_m}$

$R_1 = \dfrac{kA_1}{\omega_m}\sin\omega_m t$

Fig. 12-3. Vectorial presentation of narrow-band FM.

$$\Delta \omega_i(t) = \frac{\Delta \theta \omega_m \cos \omega_m t}{1 + (\Delta \theta)^2 \sin^2 \omega_m t} = \sum_{n=1}^{\infty} A_n \cos n\omega_m t,$$

$$-\pi < \omega_m t < \pi \qquad\qquad (12\text{-}16)$$

where $\quad A_n = \dfrac{2}{\pi} \displaystyle\int_0^{\pi} \Delta \omega_i(t) \cos (n\omega_m t) \, d(\omega_m t)$

$$= \frac{2\Delta \theta}{\pi} \int_0^{\pi} \frac{\omega_m \cos \omega_m t \cos n\omega_m t}{1 + (\Delta \theta)^2 \sin^2 \omega_m t} \, d(\omega_m t)$$

$$= \frac{2\omega_m}{(\Delta \theta)^n} [\sqrt{1 + (\Delta \theta)^2} - 1]^n, \qquad \underline{n \text{ odd}} \qquad (12\text{-}17)$$

Finally, $\quad \omega_i(t) = \omega_c + \displaystyle\sum_{n=1,3,5}^{\infty} \frac{2\omega_m}{(\Delta \theta)^n} [\sqrt{1 + (\Delta \theta)^2} - 1]^n \cos n\omega_m t$

$$(12\text{-}18)$$

As seen from the last equation, the instantaneous frequency of the frequency-modulated signal will contain odd-harmonic distortion components of the modulating frequency ω_m; the most important contribution to the total harmonic distortion is derived from the third harmonic. From Eq. (12-17), it can readily be shown that for $\Delta \theta < 1$ rad, the third-harmonic distortion $D_3 = A_3/A_1$ can be approximated by $\frac{1}{4}(\Delta \theta)^2$, which amounts to 1 per cent distortion for $\Delta \theta = 0.2$ rad. Since the modulating signal is integrated before being applied to Armstrong's modulator, it follows that the largest phase deviation will occur at the lowest modulating frequency.

The following numerical example will serve to illustrate the calculations for a typical phase-modulation system.

Consider the broadcast modulation band of 50 to 15,000 cps. In keeping with the distortion requirements, we limit the maximum phase deviation $\beta_1 = \Delta \theta = 0.2$ rad, which is associated with the lowest modulating frequency of the modulating signal. Since the maximum frequency deviation of the output signal is ± 75 kc, the maximum index of modulation of the transmitted signal is therefore $\beta = \Delta f/f_m = 75,000/50 = 1,500$ rad/sec. Since $\beta = n\beta_1$, it follows therefore that $n = 1,500/0.2 = 7,500$, which means that the original crystal-oscillator frequency is to be multiplied by a factor of 7,500.

In practice, such a large frequency multiplication would result in a carrier output frequency outside the commercial band of 88 to 108 Mc. This is overcome by employing a combination of both frequency

multiplication and frequency conversion or heterodyning to produce
a wideband FM signal at the desired carrier frequency. It is important to note that the process of heterodyning does not alter the
modulation index, for this amounts to a translation of the phase angle
only, and not a multiplication.

An early type of an indirect FM transmitter is shown in the block
diagram of Fig. 12-4. Here we start with a crystal oscillator of
200 kc, which is multiplied by a factor of 64 to produce $f_o = 12,800$
kc, and then heterodyned down to $f_o = 1,875$ kc. This is finally multiplied by 48 to produce the final output frequency of 90 Mc. The frequency deviations produced at the intermediate stages are indicated
on the diagram. Note that the total frequency multiplication amounts
to $48 \times 64 = 3,072$, and that the maximum narrow-band phase deviation $\Delta\theta$ at the phase modulator, which occurs at 50 cps, is β_1
$= 1,500/3,072 = 0.488$ rad/sec.

The early broadcast transmitters employing this type of phase
modulation had certain shortcomings in inherent noise due to excessive multiplication and distortion at the lower modulating frequencies. An improved version of the Armstrong phase-shift modulator
is the double-channel modulator, which consists of two multiplier
chains derived from the same crystal-controlled oscillator and modulated in opposite phase to each other. Figure 12-5 is a block diagram of this type of modulator.

Notice that the final carrier frequency is independent of ω_{c_1} and
that the final deviation is double the total multiplication of mn. Also
the stability of the system depends only on the stability of ω_{LC}. For
the same reason, the improved modulator tends to overcome noise
and hum modulation that occurs in the first oscillator ω_{c_1}.

3. *Pulse-position or Serrasoid FM Modulator.* The first successful indirect method of producing frequency modulation was the
phase-shift method. We have seen that in order to keep the distortion to an acceptable minimum for FM broadcasting, the initial phase
shift is restricted to about $\beta_1 = 0.2$ rad. This requirement results
in a large number of multiplying stages, which in turn increase the
noise contribution. The invention of the double-channel modulator,
which is a modification of the phase-shift method, eliminated or reduced some of the difficulties by reducing the number of multipliers
and improving the inherent S/N ratio of the modulator.

The pulse-position modulator which belongs to the class of phase
modulators is capable of producing linearly a large initial peak
phase shift of $\pm 150°$ with equivalent FM distortion of a few tenths
of one per cent, and the noise originating in the modulator can be

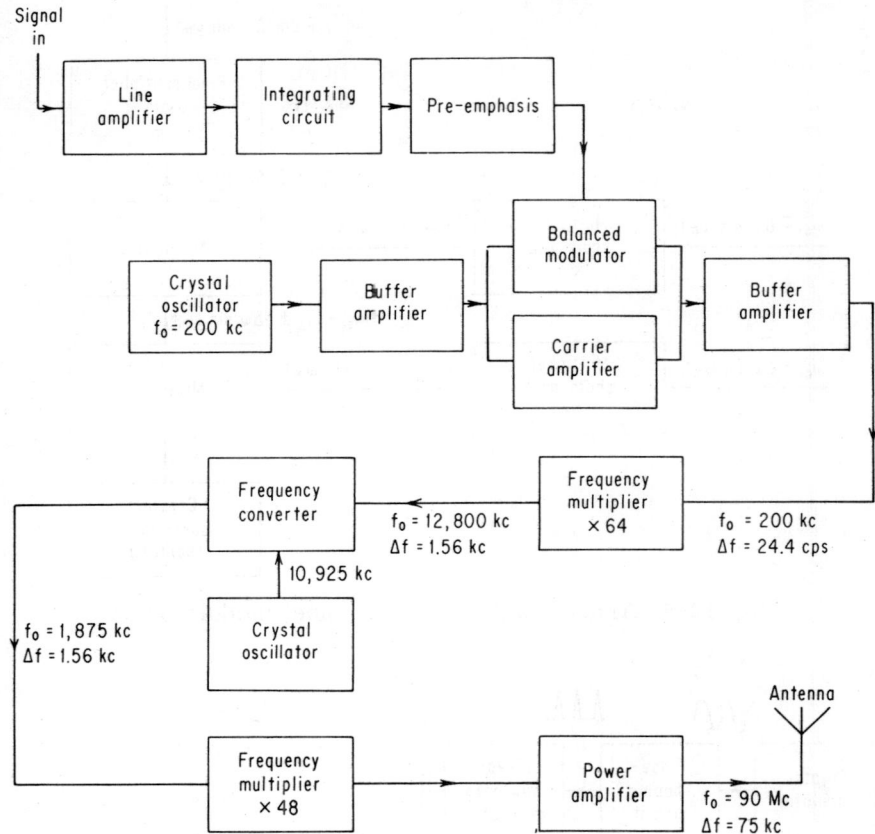

Fig. 12-4. Block diagram of an indirect FM transmitter. (From C. L. Cuccia,[3] courtesy of McGraw-Hill Book Company.)

kept to about 80 db below 100 per cent modulation. By limiting its range to $\pm 90°$ or ± 1.5 rad, the inherent distortion can be kept to a negligible minimum. For 1.5 rad and lowest modulating frequency of 50 cps, the peak deviation is given by $\beta_1 = \Delta f/50$, or $\Delta f = \pm 75$ cps. Since 100 per cent modulation in FM broadcasting requires a deviation of ± 75 kc, a frequency multiplication of 1,000 is indicated.

A particular form of this type of modulator has been developed by Day[5] and is called "serrasoid modulator"; its block diagram is shown in Fig. 12-6. The output from a crystal oscillator is passed through shaping circuits, which produce narrow positive pulses. The positive pulses trigger a linear time-base generator which generates a series of very linear sawtooth waves. The sawtooth signals are

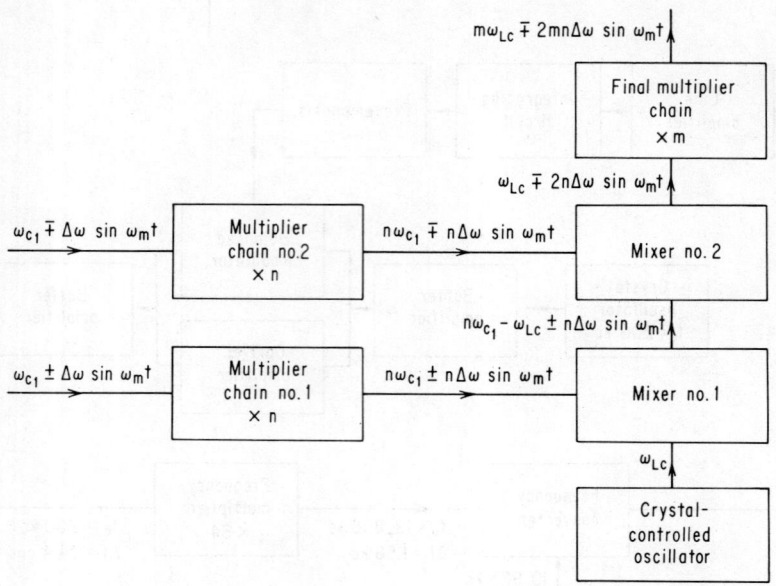

Fig. 12-5. Armstrong's double-channel modulator.

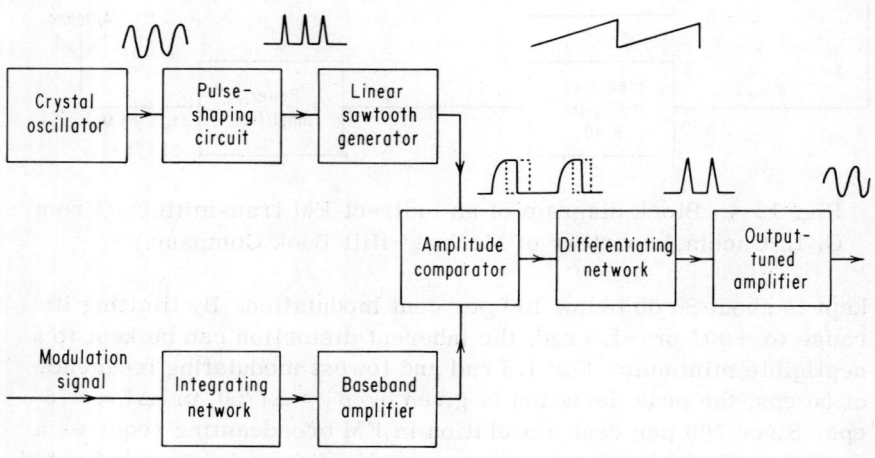

Fig. 12-6. Block diagram of pulse-position modulator.

then directly coupled to the grid of an amplitude comparator which is biased so that the conduction begins about halfway up the sawtooth. A negative-going pulse is generated each time the tube conducts. The modulating signal varies the bias so that the conduction point is moved up and down the sawtooth to produce pulse-width modulation where the leading edge of the pulses is varied. By differentiating these pulses, one obtains pulses whose time position is varied to produce phase modulation. The tuned amplifier selects a harmonic of these pulses and produces a frequency-modulated carrier. This in turn is frequency-multiplied in a chain of frequency multipliers to produce the final carrier frequency. The pulse-position modulator permits a phase variation over a range of $\pm 150°$ with excellent linearity. It is being used practically in most of FM broadcast transmitters as well as in many broadband multiplex communication systems.

12.2 Direct-frequency Modulation

In direct FM systems, the modulating signal directly varies the carrier frequency. Direct FM systems normally require a much smaller amount of frequency multiplication than the indirect systems. However, because the carrier frequency must be directly varied, crystal-controlled oscillators cannot be used, and electronic or mechanical frequency-stabilization techniques using feedback must be employed.[4]

The carrier frequency is normally generated by an oscillator whose frequency-determining circuit is a high-Q resonant circuit. Variations in the inductance or capacitance of this resonant circuit will then change the resonant frequency of the oscillator frequency. The reactance-tube modulator is commonly used to produce these variations in the tuned-circuit reactance in accordance with the modulating signal input. As an example, we may mention the reactance-tube modulator which was developed by the radio laboratories of the British Post Office for broadband radio links.[7-9] The performance of the modulator, using the well-known ladder network type of resistance-capacitance phase-shift oscillator, is adequate to meet the CCIR standards for 600-channel telephone transmission.

The reactance-tube modulator is not the only means employed to frequency-modulate a carrier in a direct manner. The Miller-effect capacitance of a tube can be used to change the effective capacitance of the oscillator tuned circuit. The capacitance cannot, however, be varied linearly over a wide range as is possible with the reactance-tube modulator, so that the frequency deviations obtainable are not as great.

Another broadband modulator commonly used is the diode reactance modulator. It provides good broadband operation, is relatively simple, and has a small number of critical alignment controls. The

diode reactance modulator operates on the principle[10] that the reactance of a series combination of a capacitor and a diode connected
across an a-c source can be varied by changing the diode bias. The
resistance change is caused by the change in angle of conduction of
the diode as its bias is changed. The modulator consists of a series
diode and capacitor combination, connected across the tank circuit
of an oscillator, with provision for supplying a bias to the diode. The
effective capacitance of the series combination can be made to vary
linearly by an audio or video signal, resulting in frequency modulation of the oscillator. This method produces less noise and distortion than the reactance-tube modulator, and the frequency deviation
can be made nearly proportional to the modulating signal. But it is
again not capable of as large a frequency deviation as the reactance-
tube modulator.

The saturable-reactor frequency modulator is another method of
producing frequency modulation.[1] It depends for its action on the
fact that the permeability of a ferrite core is changed when it is
placed in an external magnetic field, and it varies as the strength of
this magnetic field varies. Thus, if a coil is wound on the ferrite
core, its inductance will depend on the external magnetic field, and
if the external field is provided by the modulating signal, the ferrite-
cored inductance can be made to vary in accordance with the modulation. By careful design, the change in inductance can be made very
nearly linearly proportional to the modulating current causing the
external magnetic field. Inclusion of the ferrite-cored coil as a part
of an oscillator circuit then enables frequency modulation to be realized. The chief advantage of the saturable-reactor modulator,
apart from its simplicity, is that the required frequency deviation of
±100 kc/sec can be obtained over the FM broadcast band direct from
the oscillator, without the use of multiplier stages.

The klystron modulator has wide application in many wideband
microwave-relay systems. The frequency-modulated signal is generated by modulating the repeller voltage about a mean operating
point over the linear region of the frequency-repeller voltage carrier. We shall see in Chap. 15 how to minimize the nonlinearities of
the klystron characteristics by applying negative feedback to the klystron transmitter. Very often there are operational advantages in
having a modulator unit whose output is at IF rather than at the microwave carrier frequency. A method commonly used is to beat together two klystrons with frequencies separated by the IF and to apply the modulating signal to one of the klystrons. This arrangement
is being used in laboratory test equipment. The frequency stability
of the klystron modulator is poorer than with lower-frequency mod-

ulators and requires a high-gain AFC circuit to provide satisfactory stability.

1. *Generation of Frequency-modulated Waves by Varying C or L of the Oscillator Resonant Circuit.*[11]   Since in a direct FM system, the frequency variation is effected by varying the capacitance or inductance of the frequency-determining resonant circuit, it is of interest to study the effect of these variations on the frequency of the oscillator.

Consider first the case in which the capacitance $C(t)$ of the resonant circuit is varied by the modulating signal as follows:

$$C(t) = C_0 + \Delta C \sin \omega_m t \qquad\qquad (12\text{-}19)$$

where C_0 is the capacitance in the absence of modulation, and ΔC is the maximum change. The instantaneous frequency $\omega_i(t)$ is given by

$$\omega_i^2(t) = \frac{1}{LC(t)} \qquad\qquad (12\text{-}20)$$

Therefore $$2\omega_i(t)\, d\omega_i(t) = -\frac{1}{LC^2(t)}\, dC(t) \qquad\qquad (12\text{-}21)$$

or $$\frac{d\omega_i(t)}{\omega_i(t)} = \frac{-dC(t)}{2C(t)} \qquad\qquad (12\text{-}22)$$

Since the variations are considered to be small, Eq. (12-22) reduces to

$$\frac{\Delta \omega_i(t)}{\omega_0} \doteq -\frac{\Delta C(t)}{2C_0} \qquad\qquad (12\text{-}23)$$

where $\Delta C(t) = \Delta C \sin \omega_m t$. Therefore

$$\omega_i(t) = \omega_0 + \Delta \omega_i(t) \doteq \omega_0\left(1 - \frac{\Delta C}{2C_0} \sin \omega_m t\right) \qquad\qquad (12\text{-}24)$$

Similarly, if C is considered constant and the inductance L is varied, we obtain the expression

$$\omega_i(t) \doteq \omega_0\left(1 - \frac{\Delta L}{2L_0} \sin \omega_m t\right) \qquad\qquad (12\text{-}25)$$

By the substitution

$$\frac{\Delta C}{2C_O} = \frac{\Delta L}{2L_O} = -\frac{\Delta \omega}{\omega_O}$$

we obtain, for the instantaneous frequency of the oscillator which is being frequency-modulated by varying either the capacitance or the inductance of its frequency-determining resonant circuit, the relation

$$\omega_i(t) = \omega_O \left(1 + \frac{\Delta \omega}{\omega_O} \sin \omega_m t\right) = \omega_O + \Delta \omega \sin \omega_m t \qquad (12\text{-}26)$$

This can be accomplished by having the variable element of a resonant circuit in parallel with the fixed element as shown in Fig. 12-7. Let

$$\omega_O^2 L_O C_O = 1 \qquad (12\text{-}27)$$

and $\quad \omega_i^2(t) \dfrac{L_O \ell}{L_O + \ell} \cdot C_O = 1$

or $\quad \omega_i^2(t) = \dfrac{1}{L_O C_O} \cdot \dfrac{L_O + \ell}{\ell} \qquad (12\text{-}28)$

By differentiation, we obtain

$$2\omega_i(t) \; d\omega_i(t) = -\frac{1}{L_O C_O} \cdot \frac{L_O}{\ell^2} \; d\ell$$

and for small variations in ℓ we have finally

$$\frac{\Delta \omega}{\omega_O} \doteq -\tfrac{1}{2} L_O \frac{\Delta \ell}{\ell^2} \qquad (12\text{-}29)$$

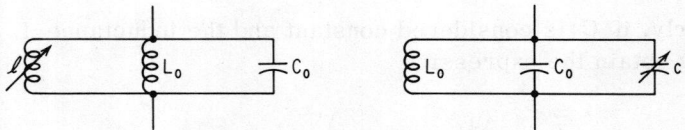

Fig. 12-7. Variable element of resonant circuit in parallel with fixed one.

Similarly, $\omega_i^2(t) = \dfrac{1}{L_0(C_0 + c)}$

or $\dfrac{1}{\omega_i^2(t)} = L_0 C_0 + L_0 c$

By differentiation $-\dfrac{2d\omega_i(t)}{\omega_i^3(t)} = L_0\, dc$

or $\dfrac{d\omega_i(t)}{\omega_i(t)} = -\dfrac{\omega_i(t)^2}{2} L_0\, dc$

Therefore $\dfrac{\Delta\omega}{\omega_0} \doteq -\dfrac{L_0\omega_0^2}{2}\Delta c = -\dfrac{\Delta c}{2C_0}$ (12-30)

2. *Basic Theory of the Reactance-tube Modulator.*[3, 4, 11, 12]   The
operation of the reactance-tube modulator is readily explained with
reference to Fig. 12-8. The plate circuit of the reactance tube is
connected in parallel to the resonant circuit of the oscillator to be
modulated. For proper operation of frequency modulation, the tube
should act like a variable reactance whose value depends upon the
grid bias, which is varied by the amplitude of the modulating signal.
We shall show that such is the case. The variations in the inductance or capacitance of the resonant circuit of the oscillator will depend on the values of Z_1 and Z_2 of the reactance-tube modulator.
Four possible variations may result. Two of these introduce an
equivalent capacitance varying with g_m of the tube; the other two
an equivalent inductance.

From Fig. 12-8, we note that the admittance Y is given by

$$Y = \frac{I}{E} = \frac{1}{Z_1 + Z_2} + \frac{g_m E_g}{E}; \qquad g_p = 0, \ R_s \gg Z_2 \qquad (12\text{-}31)$$

But $E_g = \dfrac{Z_2}{Z_1 + Z_2} E$ (12-32)

Therefore $Y = \dfrac{1 + g_m Z_2}{Z_1 + Z_2} \doteq \dfrac{g_m Z_2}{Z_1 + Z_2}, \qquad g_m Z_2 \gg 1 \qquad (12\text{-}33)$

The admittance seen across the tuned circuit is proportional to
g_m under the assumptions made.

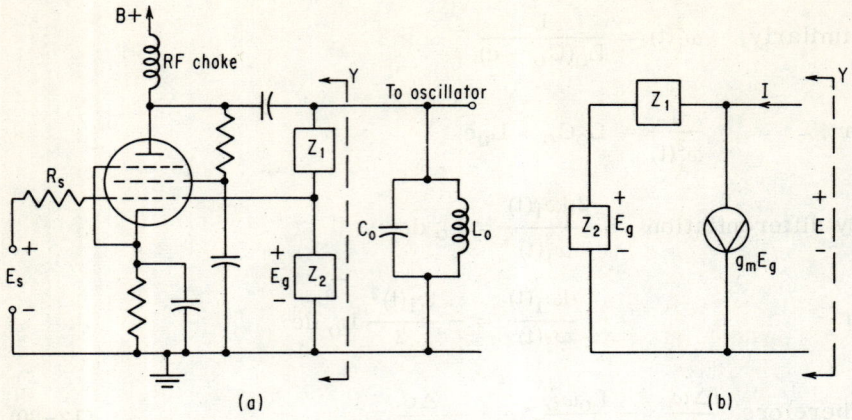

Fig. 12-8. Reactance-tube modulator analysis: (a) Generalized reactance-tube modulator; (b) simplified equivalent circuit ($g_p = 0$, $R_3 \gg Z_2$). (From M. Schwartz,[4] courtesy of McGraw-Hill Book Company.)

Case 1. $Z_1 = R$, $Z_2 = 1/j\omega C$.

$$Y = G - jB = \frac{g_m/j\omega C}{R + 1/j\omega C} = \frac{g_m}{1 + j\omega CR} = \frac{g_m(1 - j\omega CR)}{1 + (\omega CR)^2}$$

$$= \frac{1}{R_p} + \frac{1}{jX_p} \tag{12-34}$$

where $G = \dfrac{g_m}{1 + (\omega CR)^2}$ and $B = \dfrac{\omega CR\, g_m}{1 + (\omega CR)^2}$ $\qquad$ (12-35)

$$R_p = \frac{1 + (\omega CR)^2}{g_m} \qquad \text{and} \qquad X_p = \frac{1 + (\omega CR)^2}{\omega CR g_m}$$

or $L_p = \dfrac{1 + (\omega CR)^2}{g_m \omega^2 CR}$ $\qquad\qquad\qquad\qquad\qquad\qquad$ (12-36)

Thus the modulator looks like an inductance.

Case 2. $Z_1 = 1/j\omega C$, $Z_2 = R$.

Using Eq. (12-33), we obtain

$$Y = \frac{g_m R}{1/j\omega C + R} = \frac{j\omega CR\, g_m}{1 + j\omega CR} = G - jB = \frac{1}{R_p} + \frac{1}{jX_p} \qquad (12\text{-}37)$$

so that $\quad G - jB = \dfrac{(1 - j\omega CR)\, j\omega CR g_m}{1 + (\omega CR)^2} \qquad (12\text{-}38)$

where $\quad G = \dfrac{g_m(\omega CR)^2}{1 + (\omega CR)^2} \quad$ and $\quad B = -\dfrac{\omega CR g_m}{1 + (\omega CR)^2} \quad (12\text{-}39)$

The contribution of the reactance-tube modulator in parallel with the resonance circuit of the oscillator is given by

$$R_p = \frac{1 + (\omega CR)^2}{g_m(\omega CR)^2} \qquad (12\text{-}40)$$

and $\quad X_p = -\dfrac{1 + (\omega CR)^2}{g_m \omega CR} \quad$ or $\quad C_p = \dfrac{g_m\, RC}{1 + (R\omega C)^2} \qquad (12\text{-}41)$

which represents an equivalent capacitance varying with g_m.

Case 3. $Z_1 = R$, $Z_2 = j\omega L$.

$$G - jB = \frac{g_m j\omega L}{R + j\omega L} = \frac{g_m j\omega L(R - j\omega L)}{R^2 + (\omega L)^2} \qquad (12\text{-}42)$$

where $\quad G = \dfrac{g_m(\omega L)^2}{R^2 + (\omega L)^2} \quad$ and $\quad B = -\dfrac{g_m R\omega L}{R^2 + (\omega L)^2} \qquad (12\text{-}43)$

$$R_p = \frac{R^2 + (\omega L)^2}{g_m(\omega L)^2} \quad \text{and} \quad X_p = -\frac{R^2 + (\omega L)^2}{g_m R\omega L} \qquad (12\text{-}44)$$

or $\quad C_p = \dfrac{g_m R L}{R^2 + (\omega L)^2}$

which represents an equivalent capacitance.

Case 4. $Z_1 = j\omega L$, $Z_2 = R$.

$$G - jB = \frac{g_m R}{R^2 + (\omega L)^2} \qquad (12\text{-}45)$$

where $\quad G = \dfrac{g_m R^2}{R^2 + \omega^2 L^2} \quad$ and $\quad B = \dfrac{g_m \omega R L}{R^2 + \omega^2 L^2} \qquad (12\text{-}46)$

$$R_p = \frac{R^2 + (\omega L)^2}{g_m R^2} \quad \text{and} \quad X_p = \frac{R^2 + (\omega L)^2}{g_m R \cdot \omega L}$$

or $\qquad L_p = \dfrac{R^2 + (\omega L)^2}{g_m R \omega^2 L}$ $\qquad\qquad\qquad\qquad$ (12-47)

These four cases are presented in Table 12-1.

It is worth noting again that the modulating signal serves only to vary g_m and in turn the oscillator frequency. For example, consider Case 2 above.

$$Z_1 = \frac{1}{j\omega C} \quad , \quad Z_2 = R$$

With $R \ll 1/\omega_c$ and $g_m \gg 1$, the reactance-tube circuit acts as an effective capacitance $C' = g_m RC$, appearing in parallel with the oscillator tuned-circuit parameters C_0 and L_0. From Eq. (12-24), we have

$$\omega_i(t) \doteq \omega_0 \left(1 - \frac{C'}{2C_0}\right) = \omega_0 \left(1 - \frac{g_m RC}{2C_0}\right) \qquad\qquad (12\text{-}48)$$

provided $\quad g_m RC \ll C_0$

The resonance frequency varies linearly with g_m provided that the effective capacitance $g_m RC$ is small compared with capacitance C_0.

Now let $\quad g_m = g_{mo} + \Delta g_m$

With zero signal applied, $g_m = g_{mo}$, and the resonant frequency is just ω_c, the unmodulated carrier frequency; thus $\omega_c = \omega_0(1 - g_{mo}RC/2C_0)$, and $\omega_0^2 = 1/LC$. With a signal applied, we have

$$\omega_i(t) = \omega_c - \frac{\omega_0 RC}{2C_0} \Delta g_m = \omega_c - \Delta\omega$$

The frequency deviation $\quad \Delta f = \dfrac{f_0 RC}{2C_0} \Delta g_m$ $\qquad\qquad\qquad$ (12-49)

The amplitude of the modulating voltage required to effect the change Δg_m in the transconductance of the reactance modulator tube may be obtained from the transconductance-grid-voltage curve.

Table 12-1. The Equivalent Impedance Presented by a Reactance
Tube as a Function of the Phase-shifting Networks Z_1 and Z_2

		Equivalent parallel resistive and reactive components	
Z_1	Z_2	R_p	L_p or C_p
R	C	$\dfrac{1 + (\omega CR)^2}{g_m}$	$L_p = \dfrac{1 + (\omega CR)^2}{g_m \omega^2 CR}$
C	R	$\dfrac{1 + (\omega CR)^2}{g_m(\omega CR)^2}$	$C_p = \dfrac{g_m CR}{1 + (\omega CR)^2}$
R	L	$\dfrac{R^2 + \omega^2 L^2}{g_m \omega^2 L^2}$	$C_p = \dfrac{g_m LR}{R^2 + \omega^2 L^2}$
L	R	$\dfrac{R^2 + \omega^2 L^2}{g_m R^2}$	$L_p = \dfrac{R^2 + \omega^2 L^2}{g_m \omega^2 LR}$

The use of a reactance-tube modulator has serious limitations; it causes the oscillator frequency to drift, owing to changes in line voltage or similar causes, and it also introduces undesired amplitude modulation. To minimize these effects, reactance-tube modulators are used in push-pull. While unwanted frequency modulation is reduced by this method, the carrier mean frequency stability cannot be so high as that of a crystal oscillator, and some form of automatic frequency control is normally used in commercial FM transmitters in order to approach crystal stability. This is generally achieved by comparing the carrier mean frequency or a frequency derived from it with that of a crystal oscillator.

A block diagram of a typical direct FM transmitter is shown in Fig. 12-9. As stated previously, an AFC system is used to stabilize the mean frequency of the crystal oscillator.

3. *Frequency-modulated Crystal Oscillator (FMQ: Frequency-modulated Quartz).*[12] Since a conventional LC circuit does not have a frequency stability equivalent to that of a crystal, it appears that an attempt should be made to frequency-modulate a crystal oscillator. Such a circuit has been developed by the Marconi Company

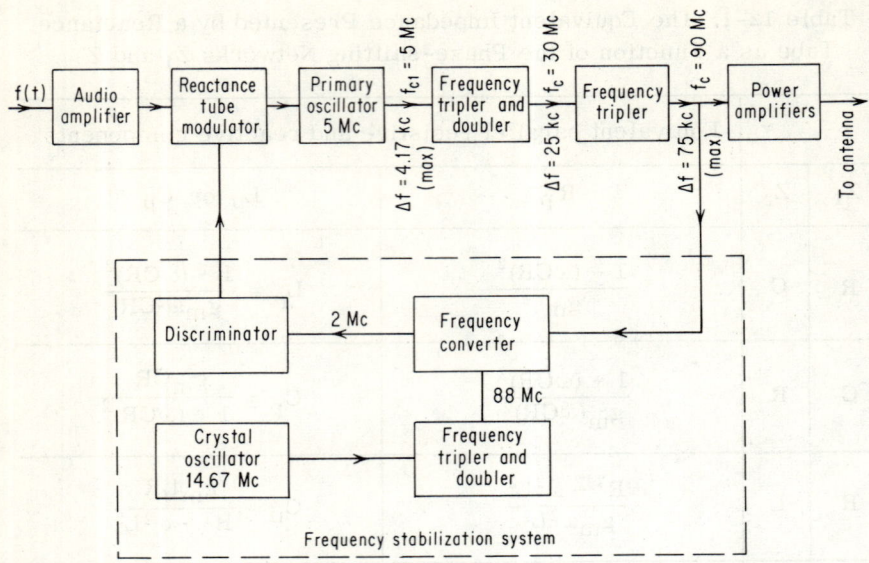

Fig. 12-9. Typical direct FM transmitter. (From M. Schwartz,[4] courtesy of McGraw-Hill Book Company.)

and is called a **FMQ modulator.** Push-pull reactance modulation is used, as shown in Fig. 12-10. It is not normally possible to frequency-modulate a crystal oscillator with satisfactory results by the use of a reactance-tube modulator because the crystal does not act as a simple resonant circuit but has a number of secondary resonances, which often fall within the maximum-deviation frequency range of its main resonance. The Marconi FMQ modulator has overcome the problem of spurious responses and produced a crystal capable of modulation. The theory of operation is as follows. With reference to Fig. 12-11a, the components L_C, R_C, and C_C are the frequency-determining parameters of the crystal; C_h, the crystal holder and stray capacitance, has very little effect on the resonant frequency of the circuit, since it is dominated by L_C and C_C. In order to effect modulation, an impedance transformation is necessary, which is obtained by the use of a quarter-wave network. The crystal operates when unmodulated at its series-resonant frequency mode, and the purpose of the $\lambda/4$ transformer is to convert the crystal series circuit into its parallel equivalent, so that the reactance modulation, which is in parallel with the crystal, can be fully effective. This is more clearly seen from the following circuit analysis. In Fig. 12-11b, the values of L and C are chosen so that

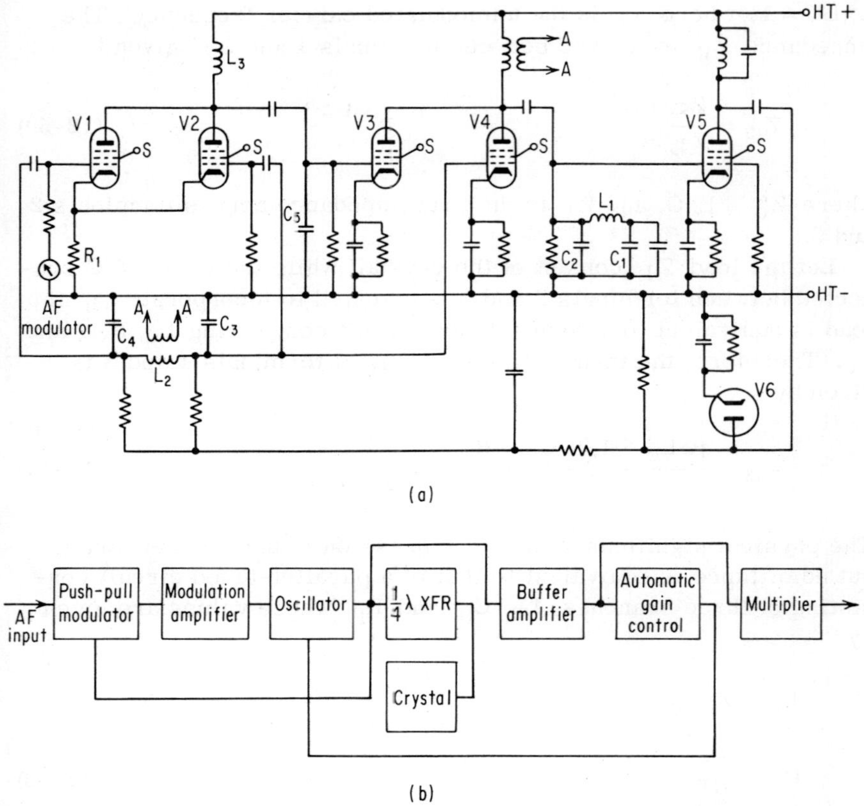

(a)

(b)

Fig. 12-10. Marconi FMQ transmitter: (a) Simplified circuit diagram of the Marconi FMQ transmitter; (b) block schematic diagram of the Marconi FMQ transmitter. (From K. R. Sturley,[1] courtesy of George Newnes Ltd., London.)

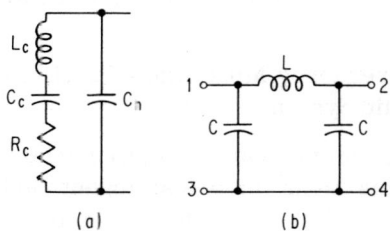

Fig. 12-11. (a) Equivalent circuit of crystal. (b) Quarter-wave π-section network.

$\omega_0^2 LC = 1$, where ω_0 is the unmodulated carrier frequency. The impedance Z_{13} measured between terminals 1 and 3 is given by

$$Z_{13} = \frac{Z_0^2}{Z_L} \tag{12-50}$$

where $Z_0^2 = L/C$, and Z_L is the load impedance across terminals 2 and 3.

Let the load Z_L consist of the crystal, while the value of C connected between terminals 2 and 3 is adjusted to incorporate C_h; the load is equivalent to a series-tuned circuit comprising L_c, R_c, and C_c. Therefore, the input admittance Y_{13} at terminals 1 and 3 is given by

$$Y_{13} = \frac{j\omega L_c + 1/j\omega C_c + R_c}{Z_0^2} \tag{12-51}$$

The physical significance of this expression is that the network input admittance is equivalent to that of a parallel-tuned circuit consisting of three branches, L_p, C_p, and R_p, whose values are given by

$$L_p = Z_0^2 C_0$$

$$C_p = \frac{L_c}{Z_0^2} \tag{12-52}$$

$$R_p = \frac{Z_0^2}{R_c}$$

Thus a reactance-tube modulator can be connected across the input terminals 1 and 3 to modulate the crystal-controlled oscillator, while its high center-frequency stability is retained.

12.3   Wideband Frequency Modulators for High-quality Multichannel Radio System [7-9,10,13]

The increasing communication requirements of both commercial and military organizations impose stringent performance characteristics on wideband modulators. The design objectives of the wideband frequency modulators are intended to meet the high-quality multichannel performance in tandem trunk communication systems employing tropospheric scatter or line-of-sight modes of propagation. The most significant performance characteristic of such a modulator is frequency response—to provide for a 600-channel system with very low intermodulation distortion and wideband frequency deviation capability.

Reactance-tube modulators do not generally meet these requirements because the maximum deviation possible with adequate linearity, and without excessive amplitude modulation, is restricted to 1 or 2 per cent of the oscillator mean frequency.

Klystrons can be used as sources of microwave frequency-modulated signals. As discussed previously, there are operational advantages in having a modulator unit whose output is at IF rather than at microwave carrier frequency. This is accomplished by beating two klystrons with frequencies separated by the IF and applying the modulating signal to one of the klystrons. At the present time, the klystron modulator seems to be the only solution to the highly linear low-noise modulator problem of meeting CCIR specifications up to 1,800 channels.

A wideband IF modulator which accepts either television or multichannel telephony was developed, as mentioned earlier, by the British Post Office Laboratories.[7-9] This IF modulator was developed from the ladder network type of resistance-capacitance phase-shift oscillator by using grounded-grid-triode tubes as the variable-impedance elements. The performance characteristics are adequate to meet CCI transmission standards when carrying either 625 line-television channels or considerably more than 600 telephone channels. The mean frequency of the oscillator is 35 Mc/sec, and it is frequency-doubled to provide the required 70-Mc/sec output signal.

The diode modulator [10] offers several advantages not found in many FM modulators. Its performance, reliability, and ease of maintenance appear to be considerably better than many other modulators. The relatively low capacitance of the diode circuit and the simplicity of the modulator circuitry make it possible to apply more than one modulator to the same oscillator without adding unduly to the total circuit complexity.

The serrasoid modulator, while extremely linear, is limited more sharply than other wideband modulators in the number of octaves of baseband frequency it is capable of handling. It does not lend itself to a universal single unit design suitable for the frequency range 250 cycles to 2.5 Mc, but is capable of covering this range by three units, namely, 60 channels—250 cycles to 300 kc, 240 channels—12 kc to 1 Mc, and 600 channels—60 kc to 2.5 Mc.

A wideband IF frequency modulator which claims to exceed the performance characteristics of radio set AN/FRC 39/56 has recently been described.[15]

References

1. Sturley, K. R.: "Frequency Modulated Radio," George Newnes Ltd., London.
2. Black, H. S.: "Modulation Theory," D. Van Nostrand Company, Inc., Princeton, N. J.
3. Cuccia, C. L.: "Harmonics, Sidebands, and Transients in Communication Engineering," McGraw-Hill Book Company, New York.
4. Schwartz, M.: "Information Transmission, Modulation, and Noise," McGraw-Hill Book Company, New York.
5. Day, J. R.: Serrasoid Modulator, *Electronics*, vol. 21, p. 72, 1948.
6. Gundlach, F. W.: Studie über den Serrasoid-Modulator, *Fernmeldetech. Z.*, Heft 6, June, 1952, pp. 256-262.
7. White, R. W., and I. A. Ravenscroft: A Frequency Modulator for Broadband Radio Relay Systems, *P. O. Elec. Engrs. J.*, vol. 48, p. 108, 1955.
8. Ravenscroft, I. A.: An Improved Frequency Modulator for Broadband Radio Relay Systems, *P. O. Elec. Engrs. J.*, vol. 50, p. 186, 1957.
9. Floyd, C. F., and R. W. White: Microwave Link Development, *Electron. Eng.*, vol. 30, no. 363, May, 1958.
10. Montgomery, G. F.: *Natl. Bur. Standards Rept.* 3593, July 1, 1956.
11. Marcus, J.: "La Modulation de fréquence," Éditions Eyrolles, Paris, 1960.
12. Tibbs, C. E., and G. G. Johnstone: "Frequency Modulation Engineering," John Wiley & Son, Inc., New York.
13. Westenborg, L. D., and H. D. Hern: "Wideband Frequency Modulator," *Conv. Rec.*, *Fifth Nat. Symp. on Global Comm.*, May, 1961, pp. 106-109.

13

DETECTION OF FREQUENCY-MODULATED SIGNALS

We have noted earlier on several occasions that when an FM signal is passed through a network whose amplitude and phase-shift characteristics are linear functions of frequency, then the output is amplitude-modulated by a precise copy of the original modulation signal. This is the basis of several well-known types of discriminators, as, for example, the off-resonance discriminator discussed in Chap. 9.

An ideal discriminator is a device whose output voltage has an instantaneous value proportional to the instantaneous frequency of the input signal. Since most practical discriminators are, in some measure, sensitive to amplitude variation of the frequency-modulated signal, a limiter is normally used ahead of the FM detector. The limiter must be capable of removing the incidental amplitude modulation of the carrier due to thermal noise and nonuniform amplitude response of the IF amplifier.

We shall show that the most common frequency detector consists of a network $Z(j\omega) = A(\omega)e^{j\theta(\omega)}$, in which the frequency modulation is first converted into an amplitude modulation, which is a copy of

405

the message function, and then detected by an envelope detector. The conversion from FM to AM is realized without distortion, provided the network has a gain characteristic $A(\omega)$ which is a linear function of ω and a phase characteristic $\theta(\omega)$ which is either constant or linear with frequency. In practice, both characteristics will depart from these ideal forms, resulting in nonlinear distortion terms in the output message function.

In practice, the frequency detector may consist of two such linear networks whose detected outputs are connected in a balanced differential manner to produce zero d-c output from an unmodulated carrier. It will be shown that with such a balanced discriminator, the limiter requirements are less severe, and that some of the distortion terms may be canceled out.

13.1 FM-to-AM Conversion [1-9]

As stated above, the more conventional discriminators make use of a network in which the frequency modulation is first converted into an amplitude modulation which is a true replica of the message function and then detected by an envelope detector. Let us examine the operation of FM-to-AM conversion more fully.

Consider the linear network for FM-to-AM conversion whose transfer characteristics are represented by Fig. 13-1 and given by

$$Z(j\omega) = A(\omega)e^{j\theta_0} \tag{13-1}$$

where

$$A(\omega) = A_c\left(1 + \rho\,\frac{\omega - \omega_c}{\Delta\omega}\right) \tag{13-2}$$

$$\theta_0 = \text{constant}$$

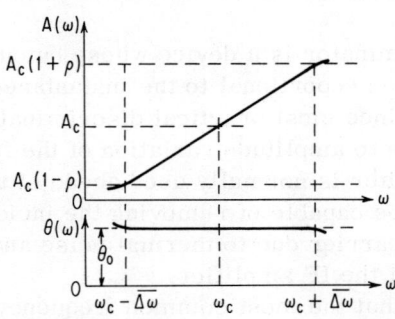

Fig. 13-1. Gain and phase characteristics of an FM to AM converting network.

and ρ is the fractional increase in gain $A(\omega)$, from $\omega = \omega_c$ to $\omega_c + \overline{\Delta\omega}$. Now

$$Z'(j\omega_c) = \frac{1}{j} A'(\omega_c) e^{j\theta_0} = \frac{A_c\rho}{j\overline{\Delta\omega}} \cdot e^{j\theta_0} \qquad (13\text{-}3)$$

$$Z''(j\omega_c) = 0$$

The response of such a network to a pure FM input $i(t)$

$$= e^{j[\omega_c t + \int_0^t s(t)dt]}, \text{ using Eq. (9-17), is}$$

$$e(t) = [Z(j\omega_c) + jB_1'(t) Z'(j\omega_c)] \, e^{j[\omega_c t + \int_0^t s(t)dt]}$$

$$= \left[A_c e^{j\theta_0} + js(t) \, \frac{A_c\rho}{j\overline{\Delta\omega}} e^{j\theta_0} \right] \cdot e^{j[\omega_c t + \int_0^t s(t)dt]}$$

or $\quad e(t) = A_c \left[1 + \frac{\rho s(t)}{\overline{\Delta\omega}} \right] e^{j[\omega_c t + \int_0^t s(t)dt + \theta_0]} \qquad (13\text{-}4)$

In the detection of $e(t)$, only its amplitude is of importance. An envelope detector following the linear network will produce an output proportional to $[1 + \rho s(t)/\overline{\Delta\omega}]$; i.e., the output will consist of a d-c term plus a term which is proportional to the modulating signal $s(t)$. As mentioned earlier, the d-c term can be removed by the use of a balanced discriminator.

In case of hybrid modulation (and imperfect limiting), the response is obtained using Eq. (9-54).

$$e(t) = \{Z(j\omega_c)A(t) + [A(t) - B]^1 Z'(j\omega_c)\} e^{j(\omega_c t + \int_0^t s(t)dt)}$$

$$= \left[A_c e^{j\theta_0} A(t) + A'(t) \, \frac{A_c\rho e^{j\theta_0}}{j\overline{\Delta\omega}} + A^0(t) \quad A_c \, \frac{js(t)}{j\overline{\Delta\omega}} \rho e^{j\theta_0} \right]$$

$$\times e^{j[\omega_c t + \int_0^t s(t)dt]}$$

$$= A_c \left[A(t) + \frac{\rho}{\overline{\Delta\omega}} A(t)s(t) + \frac{\rho}{j\overline{\Delta\omega}} A'(t) \right] e^{j[\omega_c t + \int_0^t s(t)dt + \theta_0]}$$

$$(13\text{-}5)$$

The output of the envelope detector will be proportional to the amplitude of Eq. (13-5), or

$$V = KA_c \sqrt{A^2(t)\left[1 + \frac{\rho}{\Delta\omega}s(t)\right]^2 + \frac{\rho^2}{\Delta\omega^2}[A'(t)]^2} \qquad (13\text{-}6)$$

In a balanced discriminator, the input signal $i(t) = A(t)$

$\times\, e^{j[\omega_c t + \int_0^t s(t)dt]}$ is applied simultaneously to two linear networks $Z_1(j\omega)$ and $Z_2(j\omega)$ with amplitude characteristics $A_1(\omega)$ and $A_2(\omega)$, as shown in Fig. 13-2. The response of these networks, using Eq. (13-5), is

$$e_1(t) = A_c\left[A(t) + \frac{\rho}{\Delta\omega}A(t)s(t) + \frac{\rho}{j\Delta\omega}A'(t)\right]e^{j[\omega_c t + \int_0^t s(t)dt + \theta_0]}$$

$$(13\text{-}7)$$

$$e_2(t) = A_c\left[A(t) - \frac{\rho}{\Delta\omega}A(t)s(t) - \frac{\rho}{j\Delta\omega}A'(t)\right]e^{j[\omega_c t + \int_0^t s(t)dt - \theta_0]}$$

$$(13\text{-}8)$$

The differential output of the envelope detectors, using Eq. (13-6), is given by

$$V = KA_c\left\{\sqrt{A^2(t)\left[1 + \frac{\rho}{\Delta\omega}s(t)\right]^2 + \frac{\rho^2}{\Delta\omega^2}[A'(t)]^2}\right.$$

$$\left. - \sqrt{A^2(t)\left[1 - \frac{\rho}{\Delta\omega}s(t)\right]^2 + \frac{\rho^2}{\Delta\omega^2}[A'(t)]^2}\right\} \qquad (13\text{-}9)$$

Let us compare the operation of the one channel with that of the balanced discriminators, as given by Eqs. (13-6) and (13-9). First we shall consider the detection of a pure FM signal. For the one-channel discriminator $V_{FM} = KA_c[1 + (\rho/\Delta\omega)s(t)]$, since $A'(t) = 0$. For the balanced discriminator, we note from Eq. (13-9) that the d-c term is balanced out and the sensitivity is doubled.

Next, let us consider pure AM. For a one-channel discriminator $V_{AM} = KA_c\sqrt{A^2(t) + (\rho^2/\Delta\omega^2)A'(t)^2}$; for the balanced discriminator, the output $V = 0$. Thus we conclude that the balanced FM discriminator is insensitive to amplitude modulation $A(t)$ around the center frequency ω_c, whereas the output of the one-channel detector is proportional to $\sqrt{A^2(t) + (\rho^2/\Delta\omega^2)A'(t)^2}$. But this conclusion should be qualified by stating that this result is valid only for pure amplitude modulation and not frequency modulation. The main advantage of

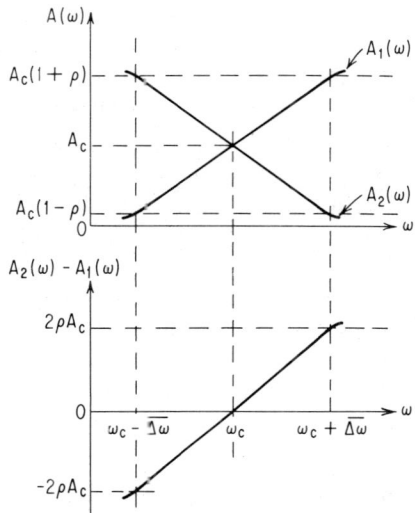

Fig. 13-2. Amplitude characteristics of a balanced discriminator.

using a balanced discriminator is the improvement in the linearity of its characteristics. In the preceding analysis, we have assumed linear amplitude characteristics of the impedance network $Z(j\omega)$. In practice, we can only approximate such ideal discriminators and, hence, the detection process is usually accompanied by distortion products of the modulating signal. By using a balanced discriminator such as a Travis discriminator, the even harmonics are canceled.

The conventional discriminators which are commonly used in commercial applications are: (1) Travis discriminator, (2) Foster-Seeley discriminator and (3) ratio detector.

The Travis discriminator has been analyzed by Zinn,[2] using the method of convolution to illustrate a generic form of a balanced frequency detector. It is a balanced discriminator consisting of two off-resonance discriminators whose rectified outputs are differentially connected. The well-known Foster-Seeley discriminator is fully covered in many publications; the first thorough analysis was given by Sturley.[3] A similar analysis is given by Johnstone[4] and others. [5,6]

The ratio detector is essentially similar to a Foster-Seeley discriminator; it differs from it in that one of the diodes is reversed in sense. The ratio detector has some limiting properties and has been specifically invented as a substitute for the use of the combination of limiters and Foster-Seeley discriminator. A very thorough analysis of the ratio detector is given by Seeley and Avins,[7] and a very illuminating comparison of its operating characteristics with those of the Foster-Seeley discriminator is made by Armstrong.[8]

A design procedure for the construction of a very linear balanced frequency discriminator of the Travis type is given by Fancourt and Skirzynski.[9] In their design, the circuit parameters are chosen in such a manner as to minimize the second and third harmonics. The design procedure is embodied in many graphs and is presented in a form which can readily be used in practice.

In the following, we shall give the theory of several types of very linear low-distortion discriminators which are frequently used for special application and whose theory of operation is not widely known.

13.2   Theory of Wideband Line Discriminator[10-13]

We have shown that the process of FM detection without distortion requires a linear network with linear amplitude characteristics. In Chap. 9, we have calculated the distortion produced by an off-resonance discriminator and showed how to minimize the distortion by operating the discriminator about the point of inflection of the resonance circuit. We have also noted above that the distortion can further be reduced by using two off-resonance circuits tuned to different frequencies and differentially connected.

In broadband FM systems, it is often required to incorporate a frequency discriminator of inherently low distortion operating at a high center frequency. While several discriminators are available, they suffer one common disadvantage; namely, the inherent distortion increases with the moaulating frequency. In the following, it will be shown that the so-called "line discriminator" has inherent low distortion, the magnitude of which is practically independent of the modulating frequency. It will be shown that the distortion is due to the nonlinearity in the amplitude-frequency response of the discriminator, while the phase-frequency response is linear and does not contribute to the distortion. Three methods of analysis will be presented: one is the quasi-stationary approach; the second method is the Fourier or steady-state approach; and the third method is the Fourier transform approach.

1. *General Theory*.   Two methods of driving a line discriminator are illustrated in Fig. 13-3a and b. The sections of coaxial lines are of equal length ℓ, the characteristic impedance being equal to

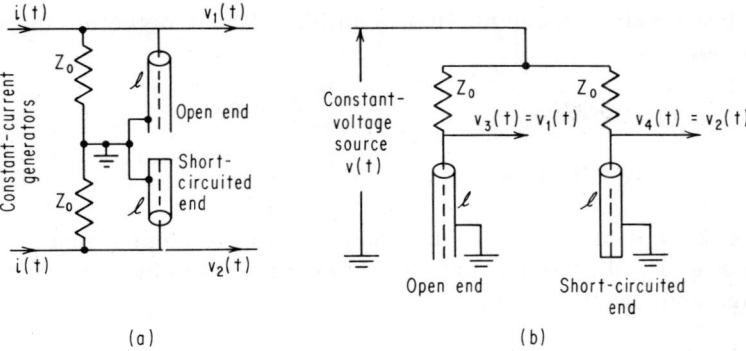

Fig. 13-3. Line discriminator connections: (a) Constant-current sources; (b) constant-voltage source.

Z_0. Frequency modulation is converted to amplitude modulation by the impedance variation in the two lines. In order to obtain a balanced system, one of the lines is open-ended and the other has its end short-circuited, so that the impedance variations are symmetrical. The impedances of the open and short-circuited lines are

$$Z_1 = -jZ_0 \cot \theta = Z_0 \frac{1 + e^{-j2\theta}}{1 - e^{-j2\theta}} \tag{13-10}$$

$$Z_2 = jZ_0 \tan \theta = Z_0 \frac{1 - e^{-j2\theta}}{1 + e^{-j2\theta}} \tag{13-11}$$

where $\theta = 2\pi \ell / \lambda$, and λ = wavelength.

In Fig. 13-3a, the two networks are fed by constant-current generators. The voltage $v_1(t)$ and $v_2(t)$ developed across the line sections are:

$$v_1(t) = \frac{-j \cot \theta}{1 - j \cot \theta} Z_0 i(t) = \frac{1}{1 + j \tan \theta} Z_0 i(t) = Z_0 i(t) \cos \theta e^{-j\theta} \tag{13-12}$$

$$v_2(t) = \frac{j \tan \theta}{1 + j \tan \theta} Z_0 i(t) = \frac{1}{1 - j \cot \theta} Z_0 i(t) = Z_0 i(t) j \sin \theta e^{-j\theta}$$
$$= Z_0 i(t) \sin \theta \, e^{j(\pi/2 - \theta)} \tag{13-13}$$

If these voltages are applied to an ideal linear detector, the rectified outputs are

$$V_1 = |\,i(t)\,|\ R \cos \theta \tag{13-14}$$

$$V_2 = |\,i(t)\,|\ R \sin \theta \tag{13-15}$$

where $Z_O = R$ and $|\,i(t)\,|$ is the amplitude of the input signal.

In Fig. 13-3b, the network is assumed to be fed by a constant-voltage source; the outputs are

$$v_3(t) = \frac{-j \cot \theta}{1 - j \cot \theta}\ v(t) = \frac{1}{1 + j \tan \theta}\ v(t) = v(t) \cos \theta\, e^{-j\theta} \tag{13-16}$$

$$v_4(t) = \frac{j \tan \theta}{1 + j \tan \theta}\ v(t) = \frac{1}{1 - j \cot \theta}\ v(t) = v(t) \sin \theta\, e^{j(\pi/2 - \theta)} \tag{13-17}$$

The detected outputs are

$$V_3 = |\,v(t)\,|\ \cos \theta \tag{13-18}$$

$$V_4 = |\,v(t)\,|\ \sin \theta \tag{13-19}$$

The impedance of the parallel combination of Fig. 13-3b is

$$Z = \frac{(R - jR \cot \theta)\,(R + jR \tan \theta)}{2R - jR \tan \theta - jR \cot \theta} = R \tag{13-20}$$

That is to say, the impedance is constant and is equal to the characteristic impedance. Thus the constant-voltage source may be replaced by a constant-current source, and since $v(t) = i(t)R$, the performances of the circuits become identical.

The difference of the rectified output in either case is given by

$$V = V_2 - V_1 = |\,v(t)\,|\ (\sin \theta - \cos \theta) \tag{13-21}$$

The lengths ℓ are chosen to render the discriminator output zero at the carrier wavelength λ_c.

This is equal to zero when $\theta_c = K\pi/4 = 2\pi\ell/\lambda_c$, where K is an odd integer. It follows therefore that the length ℓ of the line is

$$\ell = \frac{K\lambda_c}{8} \qquad \text{and} \qquad \theta = \frac{2\pi\ell}{\lambda} = \frac{K\pi}{4}\frac{\lambda_c}{\lambda}$$

Thus the length of the line should equal an odd multiple of a one-eighth wavelength.

The rectified output V is therefore given by

$$V = \sqrt{2}\ |v(t)|\ \sin\left(\theta - \frac{\pi}{4}\right) = \sqrt{2}\ |v(t)|\ \sin\ \frac{K\pi}{4}\left(\frac{\lambda_c}{\lambda} - 1\right)$$

or $\qquad V = \sqrt{2}\ |v(t)|\ \sin\ \dfrac{K\pi}{4}\left(\dfrac{\omega}{\omega_c} - 1\right)$

$$= \sqrt{2}\ |i(t)|R\ \sin\ \frac{K\pi}{4}\left(\frac{\omega}{\omega_c} - 1\right) \tag{13-22}$$

Hence the discriminator output is proportional to the characteristic impedance of the lines.

If a square-law detector is used, the detected voltages V_1 and V_2 are proportional to $\cos^2\theta$ and $\sin^2\theta$, respectively, and the discriminator output becomes

$$V = |v^2(t)|\ (\cos^2\theta - \sin^2\theta) = |v^2(t)|\ \cos 2\theta \tag{13-23}$$

Again, this is equal to zero when $2\theta_c = (2n + 1)\pi/2 = 4\pi\ell/\lambda_c$ where n is an integer, so that the discriminator again consists of an odd number of eighths of wavelength on the line.

2. *Distortion Analysis: Quasi-stationary Approach.* Consider a frequency-modulated wave whose instantaneous frequency is given by $\omega_1(t) = \omega_c + \Delta\omega \cos pt$ is to be applied to the input of a line discriminator; the output, using Eq. (13-22), is given by

$$V = \sqrt{2}\ |v(t)|\ \sin\left(\frac{K\pi}{4}\frac{\Delta\omega}{\omega_c} \cos pt\right) \tag{13-24}$$

Put $x = K\pi/4\ \Delta\omega/\omega_c$; therefore

$$V = \sqrt{2}\ |v(t)|\ \sin\ (x \cos pt)$$

$$= 2\sqrt{2}\ |v(t)|\ \sum_{n=0}^{\infty} (-1)^n J_{2n+1}(x)\ \sin\ (2n + 1)\ pt \tag{13-25}$$

Thus the fundamental component of the output is given by $2\sqrt{2}\ |v(t)|\ J_1(x)\sin pt$, and the distortion consists of odd harmonics only. The third-harmonic distortion is given by

$$D_3 = \frac{J_3(x)}{J_1(x)}$$

Since x is small, the Bessel functions may be approximated by

$$J_1(x) \doteq \frac{x}{2}\left(1 - \frac{x^2}{8}\right)$$

$$J_3(x) \doteq \frac{x^3}{48}\left(1 - \frac{x^2}{16}\right)$$

so that $\ \ D_3 \doteq \dfrac{x^2}{24}\left(1 + \dfrac{x^2}{16}\right)$ \hfill (13-26)

The distortion varies as the square of the frequency deviation and the line length.

Consider a numerical example where

$$f_c = 100 \text{ Mc} \qquad \Delta f = 10 \text{ Mc}$$

Therefore $D_3 = 0.025$ per cent for $K = 1$ and $D_3 = 0.65$ per cent for $K = 5$.

3. *Distortion Analysis: Steady-state or Fourier Spectrum Approach.* The quasi-stationary analysis assumes that the circuit is capable of following the instantaneous-frequency variations. This is evidenced by the fact that the distortion [Eq. (13-25)] is independent of the modulating frequency. In practice, the highest modulation frequency may equal the deviation, and such an assumption may not be justified. Rewriting Eqs. (13-12) and (13-13), we have

$$v_1(t) = Z_0 i(t)\ \cos\left(\frac{K\pi}{4}\ \frac{\omega}{\omega_c}\right)\ e^{-j[(K\pi/4)\omega/\omega_c]} \tag{13-27}$$

$$v_2(t) = Z_0 i(t)\ \sin\left(\frac{K\pi}{4}\ \frac{\omega}{\omega_c}\right)\ e^{-j[(K\pi/4)\omega/\omega_c - \pi/2]} \tag{13-28}$$

The linear variation of the phase corresponds to a constant delay of all components of the frequency-modulated wave and may therefore be ignored. The constant angle appearing in $v_2(t)$ results in a delay at carrier frequency. Since only the amplitudes of the sideband components of the FM wave are affected by the nonlinearity of the lines while the respective phases are shifted linearly with frequency, therefore the distortion in the line discriminator is caused only by the nonlinear variation in the amplitude of the sidebands. The equation of a frequency-modulated wave is

$$i(t) = I_0 \sin (\omega_c t + \beta \sin pt) \tag{13-29}$$

When this is applied to the circuit, each sideband is modified by the circuit response, and only the amplitude variation is contributing to the distortion. From Eqs. (13-27) and (13-28), we obtain

$$v_1(t) = Z_0 I_0 \sum_{\nu = -\infty}^{\infty} J_\nu(\beta) \sin (\omega_c + \nu p)t \cos\left(\frac{K\pi}{4} \frac{\omega_c + \nu p}{\omega_c}\right) \tag{13-30}$$

and

$$v_2(t) = Z_0 I_0 \sum_{\nu = -\infty}^{\infty} J_\nu(\beta) \sin (\omega_c + \nu p)t \sin\left(\frac{K\pi}{4} \frac{\omega_c + \nu p}{\omega_c}\right) \tag{13-31}$$

These expressions may be simplified by the following substitutions: $a = p/\omega_c$, $b = K\pi/4$, and normalizing with respect to $Z_0 I_0$. Therefore

$$v_1(t) = \sum_{\nu = -\infty}^{\infty} J_\nu(\beta) \sin (\omega_c + \nu p)t \cos b(1 + a\nu)$$

$$= \sin \omega_c t \sum_{\nu = -\infty}^{\infty} J_\nu(\beta) \cos \nu pt \cos b(1 + a\nu)$$

$$+ \cos \omega_c t \sum_{\nu = -\infty}^{\infty} J_\nu(\beta) \sin \nu pt \cos b(1 + a\nu)$$

$$= S_1 \sin \omega_c t + S_2 \cos \omega_c t \tag{13-32}$$

where

$$S_1 = \sum_{\nu = -\infty}^{\infty} J_\nu(\beta) \cos \nu pt \cos b(1 + a\nu) \tag{13-33}$$

and

$$S_2 = \sum_{\nu = -\infty}^{\infty} J_\nu(\beta) \sin \nu pt \cos b(1 + a\nu) \tag{13-34}$$

Therefore $\quad 2S_1 = \cos b \sum\limits_{\nu = -\infty}^{\infty} J_\nu(\beta)\,[\cos \nu(pt + ab) + \cos \nu(pt - ab)]$

$$-\sin b \sum\limits_{\nu = -\infty}^{\infty} J_\nu(\beta)[\sin \beta(pt + ab) - \sin \nu(pt - ab)]$$

$$(13\text{-}35)$$

or $\quad 2S_1 = \cos b\{\cos [\beta \sin (pt + ab)] + \cos [\beta \sin (pt - ab)]\}$

$$-\sin b\{\sin [\beta \sin (pt + ab)] - \sin [\beta \sin (pt - ab)]\}$$

$$(13\text{-}36)$$

Let $\quad \beta \sin (pt + ab) = \Phi_1$ $\qquad\qquad\qquad\qquad\qquad$ (13-37)

and $\quad \beta \sin (pt - ab) = \Phi_2$ $\qquad\qquad\qquad\qquad\qquad$ (13-38)

Therefore $\quad 2S_1 = (\cos b \cos \Phi_1 - \sin b \sin \Phi_1)$

$$+ (\cos b \cos \Phi_2 + \sin b \sin \Phi_2)$$

and $\qquad S_1 = \cos \left(b + \dfrac{\Phi_1 - \Phi_2}{2} \right) \cos \dfrac{\Phi_1 + \Phi_2}{2}$ $\qquad$ (13-39)

Similarly, it can be shown that

$$S_2 = \cos \left(b + \frac{\Phi_1 - \Phi_2}{2} \right) \sin \frac{\Phi_1 + \Phi_2}{2} \qquad (13\text{-}40)$$

From Eq. (13-32), we obtain

$$\left| v_1(t) \right| = \sqrt{S_1^2 + S_2^2} = \cos \left(b + \frac{\Phi_1 - \Phi_2}{2} \right) = \cos \bar\theta \qquad (13\text{-}41)$$

Likewise it can be shown that

$$\left| v_2(t) \right| = \sin \left(b + \frac{\Phi_1 - \Phi_2}{2} \right) = \sin \bar\theta \qquad (13\text{-}42)$$

where $\quad \bar\theta = b + \dfrac{\Phi_1 - \Phi_2}{2} = \dfrac{K\pi}{4} + \dfrac{\beta}{2}\,[\sin (pt + ab) - \sin (pt - ab)]$

or $\qquad \bar\theta = \beta \sin \left(\dfrac{K\pi}{4} \; \dfrac{p}{\omega_c} \right) \cos pt + \dfrac{K\pi}{4}$ $\qquad\qquad$ (13-43)

Finally, the differential rectified output, using Eq. (13-21), is

$$V = |v_2(t)| - |v_1(t)| = \sqrt{2}\ I_0 Z_0\ \sin\left[\beta \sin\left(\frac{K\pi}{4}\ \frac{p}{\omega_c}\right)\cos pt\right.$$

$$\left. + (K-1)\frac{\pi}{4}\right] \qquad\qquad (13\text{-}44)$$

Since K is odd, the term $(K-1)\pi/4$ will not affect the magnitude of V and may be omitted; therefore

$$V = \sqrt{2}\ I_0 Z_0\ \sin(x'\cos pt)$$

$$= 2\sqrt{2}\ I_0 Z_0 \sum_{n=0}^{\infty} (-1)^n J_{2n+1}(x')\ \sin(2n+1)pt \qquad\qquad (13\text{-}45)$$

where $\quad x' = \beta \sin\left(\frac{K\pi}{4}\ \frac{p}{\omega_c}\right)$ \qquad\qquad (13-46)

Note that for the quasi-stationary case, we have obtained

$$V = \sqrt{2}\ I_0 Z_0\ \sin(x\cos pt)$$

$$= 2\sqrt{2}\ I_0 Z_0 \sum_{n=0}^{\infty} (-1)^n J_{2n+1}(x)\ \sin(2n+1)pt$$

where $\quad x = \dfrac{K\pi}{4}\ \dfrac{\Delta\omega}{\omega_c}$

However, when the argument of the sine is small, x' reduces to x, since

$$x' = \beta \sin\left(\frac{K\pi}{4}\ \frac{p}{\omega_c}\right)$$

$$x' \doteq \beta\ \frac{K\pi}{4}\ \frac{p}{\omega_c} = \frac{\Delta\omega}{p}\ \frac{K\pi}{4}\ \frac{p}{\omega_c} = \frac{K\pi}{4}\ \frac{\Delta\omega}{\omega_c} = x$$

If we take again $f_c = 100$ Mc, $\Delta f = 10$ Mc, $p = 10$ Mc:

$$\beta = 1, \qquad K = 1,\ 5$$

The steady-state approach gives $D_3 = 0.025$ per cent for $K = 1$ and $D_3 = 0.62$ per cent for $K = 5$. Thus in this range of values, the two methods give nearly the same results. This is a consequence of the linear phase characteristic and the nearly linear amplitude characteristic.

4. *Distortion Analysis: Fourier Transform Method.* From Eqs. (13-16) and (13-17) and Fig. 13-3b, we obtain the transfer functions:

Open line: $$H_1(j\omega) = \frac{v_1(t)}{v(t)} = \cos\theta\, e^{-j\theta} = \frac{1 + e^{-2j\omega\, \ell/v}}{2}$$

$$(13\text{-}47)$$

Short-circuited line: $$H_2(j\omega) = \frac{v_2(t)}{v(t)} = j\sin\theta e^{-j\theta} = \frac{1 - e^{-2j\omega\, \ell/v}}{2}$$

$$(13\text{-}48)$$

where $\theta = \dfrac{2\pi\ell}{\lambda} = \dfrac{\omega\ell}{v}$

As before, let the input voltage be

$$v(t) = e^{j(\omega_c t + \beta\sin pt)} \qquad\qquad (13\text{-}49)$$

Using the convolution integral [Eq. (3-55)], the outputs are given by

$$v_1(t) = \int_{-\infty}^{\infty} h_1(t - \tau)v(t)\, d\tau \qquad\qquad (13\text{-}50)$$

$$v_2(t) = \int_{-\infty}^{\infty} h_2(t - \tau)v(t)\, d\tau \qquad\qquad (13\text{-}51)$$

where $h(t)$ is the response of the circuit to the unit impulse function $\delta_0(t)$.

From Eqs. (13-47) and (13-48), we obtain the following expressions for $h_1(t)$ and $h_2(t)$:

$$h_1(t) = \frac{1}{2\pi}\int_{-\infty}^{\infty} H_1(j\omega)e^{j\omega t}\, d\omega = \frac{1}{2}\left[\delta_0(t) + \delta_0\left(t - \frac{2\ell}{v}\right)\right] \quad (13\text{-}52)$$

$$h_2(t) = \frac{1}{2\pi}\int_{-\infty}^{\infty} H_2(j\omega)e^{j\omega t}\, d\omega = \frac{1}{2}\left[\delta_0(t) - \delta_0\left(t - \frac{2\ell}{v}\right)\right] \quad (13\text{-}53)$$

The output voltages are readily obtained from Eqs. (13-50) and (13-51) as follows:

$$v_1(t) = \tfrac{1}{2}\left(e^{j(\omega_c t + \beta \sin pt)} + e^{j[\omega_c(t - 2/v) + \beta \sin p(t - 2\ell/v)]}\right)$$

$$= \frac{v(t)}{2}\left(1 + e^{-j(\beta x + 2\omega_c \ell/v)}\right) = \frac{v(t)}{2}(1 + e^{-j\alpha}) \qquad (13\text{-}54)$$

$$v_2(t) = \frac{v(t)}{2}\left(1 - e^{-j(\beta x + 2\omega_c \ell/v)}\right) = \frac{v(t)}{2}(1 - e^{-j\alpha}) \qquad (13\text{-}55)$$

where $x = \sin pt - \sin p(t - 2\ell/v) = 2 \sin p(\ell/v) \cos p(t - \ell/v)$.

As before, $\ell = K\lambda_c/8$, where K is an odd integer; also $2\omega_c \ell/v = K\pi/2$ and $p\ell/v = (K\pi/4) p/\omega_c$.

From Eqs. (13-54) and (13-55), we readily derive the amplitudes of $v_1(t)$ and $v_2(t)$; thus

$$|v_1(t)| = \cos \frac{\alpha}{2} \qquad \text{and} \qquad |v_2(t)| = \sin \frac{\alpha}{2}$$

where $\quad \dfrac{\alpha}{2} = \beta \sin\left(\dfrac{K\pi}{4} \dfrac{p}{\omega_c}\right) \cos p\left(t - \dfrac{\ell}{v}\right) + \dfrac{K\pi}{4}$

Finally, the differential output is equal to

$$v_0 = |v_2(t)| - |v_1(t)| = \sqrt{2}\, I_0 Z_0 \sin\left[\beta \sin\left(\frac{K\pi}{4} \frac{p}{\omega_c}\right) \cos p\left(t - \frac{\ell}{v}\right)\right.$$

$$\left. + (K - 1)\frac{\pi}{4}\right] \qquad (13\text{-}56)$$

This result is identical to expression (13-44) except for the time delay ℓ/v in the message function. The difference results from the fact that we have neglected in Eq. (13-28) the linear variation of the phase in the line discriminator, which corresponds to a constant delay of all components of the frequency-modulated wave.

We note here that while the Fourier transform method is much shorter and mathematically more elegant, the Fourier spectrum approach affords a clearer physical picture of the changes which the sidebands are subjected to in the FM-to-AM conversion.

13.3   Cycle-counter Discriminator[4, 14-15]

It has been shown above that the tuned-circuit discriminator is basically a nonlinear device and consequently produces distortion components in the detection process. The line discriminator has been shown to have better linearity, wider bandwidth, and lower distortion. Another discriminator which is inherently a linear device is the cycle-counter discriminator, which can be made to operate linearly over a wide range with a minimum of critical circuit components and very low distortion. A characteristic of the cycle-counter-type discriminator is its low sensitivity, so that it must operate at a high signal level in order to give a high signal-to-noise ratio.

The incoming modulated signal is converted into a train of constant-amplitude pulses, and demodulation is effected by means of a "counter" circuit, which gives an output proportional to the repetition rate of the pulses. The basic circuit of the cycle counter is shown in Fig. 13-4. The incoming modulated signal is first heterodyned to produce an intermediate-frequency signal at about 200 to 1,000 kc/sec. After amplification, the signal is applied to several stages of limiting which produce a square wave of constant amplitude. The pulses are then applied to the diodes D_1 and D_2, which are connected in parallel but in opposite sense, so that the capacitor C is charged through D_2 and discharges through D_1.

The operation of the cycle counter over a complete cycle of the square wave is as follows. Consider first the quiescent condition with the limiter stage cutoff. The plate potential is that of the high voltage, and there is zero voltage across either diode or the diode load resistor R. As the grid of the limiter is driven positively, the tube conducts, and consequently the plate potential drops; diode D_1

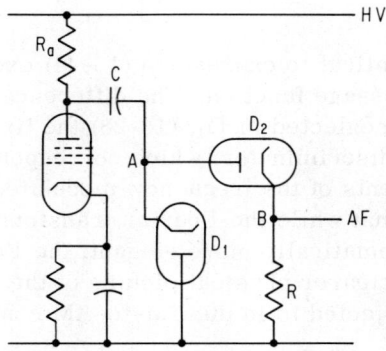

Fig. 13-4. Basic cycle-counter circuit.

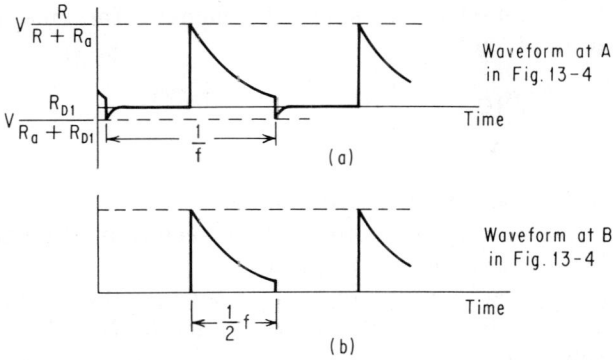

Fig. 13-5. Waveforms at points A and B of Fig. 13-4.

conducts while diode D_2 remains cut off. Because of the high ratio of
the resistance R_a to the resistance R_{D_1} of the diode D_1 when con-
ducting, the cathode of D_1 is not driven appreciably negative with respect to ground potential, and capacitor C discharges through R_a and
diode D_1 in series, until the cathode of D_1 returns to ground potential.
The discharging curve is exponential with a time constant $(R_a + R_{D_1})C$
sufficiently small for the potential of the D_1 cathode to approach
ground potential before the next part of the cycle, as shown in Fig.
13-5a. As the grid is driven negative beyond cutoff during the second half of the period of the square wave, the plate potential rises to
the high-voltage potential, and current flows through R_a, C, D_2, and
R in series; the voltage across R is shown in Fig. 13-5b.

The time constant of the combination is made small compared to
the time of one half-cycle of the frequency of the modulated wave
train, so that the capacitor is charged to the peak value of the square
wave and then completely discharged during the alternate half-cycles.
The output of the discriminator which appears across the resistor R
consists of unidirectional pulses of constant shape and amplitude.
Let V denote the amplitude of the voltage step at the plate of the
limiter which is applied to capacitor C; then if the diode forward resistance is negligible, the voltage across R at the beginning of the
charging period is $VR/(R + R_a)$. At the end of the period, this voltage has fallen to $V(R/R_1)e^{-1/2fCR'}$, where $R' = R + R_a$. By choosing
$R'C$ small compared to $1/2f$ we can calculate the mean value V_0 by
multiplying the peak output voltage by the ratio of the time constant
$(R + R_a)C$ to the period of one cycle. Thus

$$V_0 = \frac{VR}{R + R_a} \frac{(R + R_a)C}{1/f_c} = VRCf_c \qquad (13-57)$$

where f_c is the unmodulated carrier frequency. The rms output audio voltage is given by

$$V_{a(rms)} = \frac{VRC\Delta f}{\sqrt{2}}$$ (13-58)

where Δf is the peak frequency deviation.

The output of the discriminator is thus seen to be directly proportional to the instantaneous frequency.

The linearity, however, is not perfect because the capacitor C cannot charge and discharge completely in the half-cycle period as required. If the time constant is made very short to approach perfect linearity, the area of the pulses becomes smaller and the detected output decreases. In a practical circuit, the component values selected must be a compromise between the requirements of good linearity and sensitivity.

The degree of nonlinearity can be calculated from the exact expression for the area of the pulse. This is given by

$$\frac{VR}{R'} \int_0^{1/2f} e^{-1/CR't}\, dt = VCR(1 - e^{-1/2fCR'})$$

and the discriminator output is equal to this area multiplied by the signal frequency; i.e.,

$$V_{output} = VCRf(1 - e^{-1/2fCR'})$$ (13-59)

Thus the second term within the parentheses represents the departure from linearity.

A cycle-counter receiver of very low distortion operating at 1 Mc with modulating frequencies of about 150 kc is described in Ref. 15.

13.4 Microwave Linear Phase Discriminator[12, 16]

A discriminator which is commonly used for detection of frequency-modulated signals in the microwave region without heterodyning down to intermediate frequencies is the phase discriminator. In the phase discriminator, the frequency modulation is transformed linearly into amplitude modulation by causing interference between two frequency-modulated waves which originated from the same source but have traveled along different paths. This can be accomplished by using two magic T's combined with length of waveguide ℓ

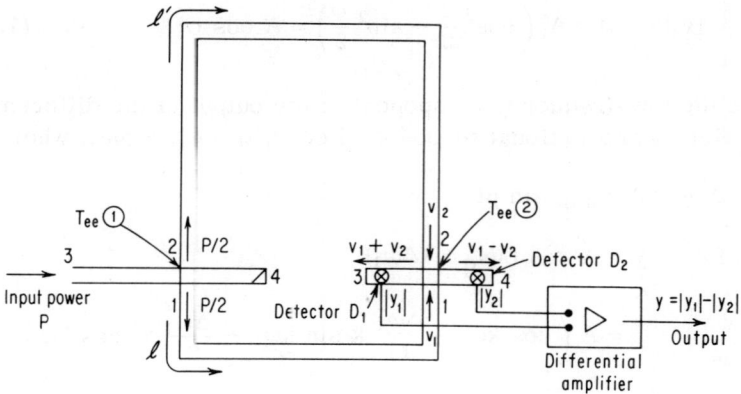

Fig. 13-6. Microwave phase discriminator. (From Fagot and Magne, [12] courtesy of Pergamon Press.)

and ℓ' where $\ell' \gg \ell$, as shown in Fig. 13-6. The properties of magic T's provide means for producing two interfering signals $(v_1 + v_2)$ and $(v_1 - v_2)$, which are applied to the rectifiers D_1 and D_2. With reference to Fig. 13-6, arms 1 and 2 are the colinear arms; arms 3 and 4 denote respectively the parallel inputs H and the series inputs E. Let ϕ denote the relative phase angle of the interfering signals,

$$\phi = 2\pi \frac{\ell' - \ell}{\lambda_g} \doteq k\omega_i(t) \tag{13-30}$$

so that ϕ can be assumed to be proportional to the instantaneous frequency $\omega_i(t)$. Assuming also square-law rectification, we obtain

$$y_1 = (v_1 + v_2)^2 = A_o^2 \{\cos \psi(t) + \cos [\psi(t) + \phi]\}^2$$

$$= 4A_o^2 \cos^2 \left[\psi(t) + \frac{\phi}{2}\right]^2 \cos^2 \frac{\phi}{2} \tag{13-61}$$

$$y_2 = (v_1 - v_2)^2 = A_o^2 \{\cos \psi(t) - \cos [\psi(t) + \phi]\}^2$$

$$= 4A_o^2 \sin^2 \left[\psi(t) + \frac{\phi}{2}\right] \sin^2 \frac{\phi}{2}$$

where $\psi(t) = \int_0^t \omega_i(t)\, dt$, the instantaneous phase angle of the FM signal. The rectified voltages $|y_1|$ and $|y_2|$ are applied to a differential amplifier whose output is

$$y = |y_1| - |y_2| = A_1^2 \left(\cos^2 \frac{\phi}{2} - \sin^2 \frac{\phi}{2} \right) = A \cos \phi \qquad (13\text{-}62)$$

Hence the low-frequency component of the output of the differential amplifier is proportional to $\cos \phi$. Let $\omega_i(t) = \omega_c + \Delta\omega_i$, where

$$\Delta\omega_i = \Delta\omega_{max} \sin pt$$

Therefore $y = A \cos (k\omega_c + k\Delta\omega_i)$

$$= A \left(\cos k\omega_c - \frac{\Delta\omega_i}{1!} k \sin k\omega_c - \frac{\Delta\omega_i^2}{2!} k^2 \cos k\omega_c \right.$$

$$\left. + \frac{\Delta\omega_i^3}{3!} k^3 \sin k\omega_c + \cdots \right) \qquad (13\text{-}63)$$

In order to obtain a sufficient conversion slope of volts per megacycle per second, the angle $\phi(\omega_i)$ is chosen so that

$$\phi_c = k\omega_c = (2n + 1) \frac{\pi}{2} \qquad (13\text{-}64)$$

Hence $k = (2n + 1) \dfrac{\pi}{2} \dfrac{1}{\omega_c}$ \qquad (13\text{-}65)

It is obvious from Eq. (13-63) that with this choice of k, the d-c term and the even harmonics are balanced out so that

$$y = A \left[(2n + 1) \frac{\pi}{2} \frac{\Delta\omega_i}{\omega_c} - \frac{(2n + 1)^3 \pi^3}{8 \cdot 3!} \frac{\Delta\omega_i^3}{\omega_c^3} + \cdots \right] \qquad (13\text{-}66)$$

The value of the third-harmonic distortion can be derived from Eq. (13-66):

$$D_3 = \frac{(2n + 1)^2 \pi^2}{24} \left(\frac{\Delta\omega_{max}}{\omega_c} \right)^2 \qquad (13\text{-}67)$$

13-5   Low-frequency Linear Discriminator[17-19]

In multichannel communication systems using the FM/FM method of frequency multiplexing, each of the information channels is made to modulate the frequency of a separate low-frequency subcarrier, lying in the range of 400 cps to 75 kc/sec, with large deviations amounting to 7.5 to 15 per cent of the subcarrier frequency. The modulated subcarriers are added together, and the resulting composite signal is used to frequency-modulate the main carrier frequency. At the receiving end, the composite modulating signal is first demodulated and then separated by bandpass channel filters into the original modulated subcarrier. Finally, the subcarriers are demodulated by low-frequency linear discriminators whose output is a true replica of the original information channels. Because of the low center frequency and large frequency deviation, it is difficult to achieve in practice a discriminator which is linear over the required bandwidth. The cycle-counter-type discriminator described previously offers a satisfactory solution to the design of a suitable discriminator for subcarrier signals. However, as we have noted above, the cycle-counter discriminator is characterized by its low sensitivity, and, consequently, it must operate at a high signal level in order to give a high signal-to-noise ratio in the output.

A simpler solution proposed by Tillman[18] and Stine[19] consists of passing the signal through two parallel-T, RC null networks with suitably staggered center frequencies, whose outputs are differentially rectified. These circuits have the merit of being without inductors but suffer from a lack of tunability on account of the inherently complicated structure of the twin-T network. The discriminator is simple in design, without inductors, and can readily be adjusted to provide stable, wideband, linear characteristics about the low center frequencies of the subcarriers.

References

1. Labin, E.: Sur un Point relatif à la détection des ondes modulées en fréquence, *L'Onde électrique*, vol. 28, pp. 60–69, February, 1948.
2. Zinn, M. K.: Transient Response of an FM Receiver, *BSTJ*, October, 1948.
3. Sturley, K. R.: The Phase Discriminator, *Wireless Engr. London*, February, 1944.
4. Johnstone, G. G.: Limiters and Discriminators for FM Receivers, *Wireless World*, January, 1957.
5. Guttinger, P.: Zur Theorie des Frequenz-Diskriminators, *Bull. Assoc. Suisse Electriciens*, no. 18, 1946.
6. Mayo, C. G., and J. W. Head: Foster-Seeley Discriminator, *Electron. Radio Engr.*, February, 1958.
7. Seeley, S. M., and J. Avins: The Ratio Detector, *RCA Rev.*, June, 1947.

8. Armstrong, E. H.: A Study of the Operating Characteristics of the Ratio-Detector and Its Place in Radio History, *Proc. Radio Am.*, vol. 25, no. 3, 1948.
9. Fancourt, B. A., and J. K. Skwirzynski: Design of a Simplified Linear Frequency Discriminator, *Marconi Rev.*, vol. 19, p. 61, 1956.
10. Panter, P. F., and W. Dite: Theory of Low Distortion Line Discriminators, Internal IT&T Federal Technical Laboratory Report, June, 1948.
11. Magne, P.: Le Discriminateur à lignes, *Ann. Radioelec.*, April, 1950.
12. Fagot, J., and P. Magne: "Frequency Modulation Theory," Pergamon Press, New York.
13. Clavier, A. G.: Application of Fourier Transform to Variable-frequency Circuit Analysis, *Proc. IRE*, November, 1949.
14. Scroggies, M. G.: Low Distortion FM Discriminator, *Wireless World*, April, 1956.
15. Vallarino, A. R., H. A. Snow, and C. Greenwald: Counter-circuit Multiplex Receiver, *Electronics*, July, 1953.
16. Familier, H.: "Discriminateur linéaire en hyperfréquences, *Ann. Radioelec.*, July, 1953.
17. Ganguly, U. S.: Linear Frequency Discriminator, *Electron. Technol.*, November, 1961.
18. Tillman, J. R.: Linear Frequency Discriminator, *Wireless Engr. London*, October, 1946, p. 281.
19. Stine, P. T.: Parallel-T Discriminator Design Technique, *Proc. Natl. Electron. Conf.*, vol. 9, p. 26, 1950.

14

SIGNAL-TO-NOISE IMPROVEMENT IN FM SYSTEMS

We alluded in Chap. 1 to the possibility of trading bandwidth for improved signal-to-noise ratio in certain modulation systems. One of the early and most widely used systems which takes advantage of this exchange is the wideband FM system proposed by E. H. Armstrong in 1936, in which he showed [1] in a qualitative manner that wideband FM has certain inherent noise- and interference-reducing properties not possessed by AM. In this application, the term wideband FM applies to that case in which the maximum frequency deviation of the modulated carrier is several times the highest frequency of the modulating signal. Such a system requires a wideband for the transmission of the modulated carrier with minimum distortion. Although there exist in practice other more efficient systems, as we shall see later on, frequency modulation represents a simple and convenient way of exchanging bandwidth for signal-to-noise ratio improvement.

Like the conventional envelope demodulator, phase and frequency demodulators exhibit a stronger-signal capture phenomenon, which results in a threshold that must be exceeded by the signal for proper demodulation. In this chapter, the concept of "FM threshold" will

427

be clarified with reference to the conventional FM demodulator. Techniques for threshold reduction, or as it is commonly called, "threshold extension," will be discussed in Chap. 16.

For the purpose of comparison of FM and AM, we shall assume the same carrier power and mean noise spectral power density at the input to each system and compare S/N ratios at the system outputs. The definition of output $(S/N)_o$ is taken to be the ratio of mean output signal power to mean output noise power, the signal power measured in the absence of noise, the noise power in the absence of signal (i.e., the carrier is unmodulated). This definition is valid for high $(S/N)_o$ ratio where it can be proved that the mean signal and noise powers may be assumed to add linearly, and signal power measured in the absence of noise does not differ substantially from that measured with noise present.

The assumption of linear addition of signal and noise powers is not valid in case of low-input carrier-to-noise ratio as the system is operating in the nonlinear region. It will be shown later that as the carrier-to-noise ratio drops below 0 db, signal suppression takes place.

In the case of high-input carrier-to-noise ratio, we shall first develop formulas for signal-to-noise improvement in a single FM communication channel and then extend the theory to multichannel FM systems. The advantages of the use of deemphasis will also be discussed.

The operation of the FM receiver in the case of low-input carrier-to-noise ratio will then be discussed, and several approaches to this very difficult problem will be outlined.

14.1 Signal-to-Noise Ratio in Exponential Modulation Systems [2-5]

Unlike linear modulation which was discussed in Chap. 6, we shall prove that in wideband FM systems there is an improvement in the output $(S/N)_o$ ratio.

The exponential-modulated signal of transmission bandwidth 2B cycles/sec is first passed through an ideal limiter, which removes all amplitude variations without affecting the phase variations of the signal. The limited output, after filtering, is applied to a discriminator whose output is directly proportional to the instantaneous phase or frequency of the signal. The output of the discriminator is band-limited by a low-pass filter of bandwidth f_m cycles/sec ($f_m < B$), where f_m is the maximum bandwidth of the information signal g(t) being transmitted. For a given carrier-to-noise receiver input $(C/N)_i$, the problem is to determine the signal-to-noise output

$(S/N)_O$ of the discriminator. The ratio of $(S/N)_O$ to $(C/N)_i$ is called signal-to-noise improvement ratio.

1. *Capture Effect (Threshold) with Random Fluctuation Noise.*
Before we derive the expression for the signal-to-noise improvement ratio in FM systems, we shall take a closer look at the threshold effect exhibited by FM systems. Let the exponential-modulated carrier at the output of the IF amplifier be given by

$$e_{exp}(t) = A_c \cos [\omega_c t + \psi(t)] \tag{14-1}$$

where $\psi(t)$ is proportional to $g(t)$ for phase modulation (PM) and to the integral of $g(t)$ for frequency modulation (FM). As in the case of linear modulation, we represent the total noise present within the receiver passband by the sinusoid

$$e_n(t) = V(t) \cos [\omega_c t + \phi(t)] \tag{14-2}$$

where the envelope $V(t)$ and phase $\phi(t)$ are slowly varying random time functions. For a narrow-band gaussian noise process, $V(t)$ is Rayleigh-distributed, and $\phi(t)$ is uniformly distributed over $0 \le \phi \le 2\pi$. The resultant of signal plus noise at the output of the IF amplifier is

$$e_{IF}(t) = A_c \cos [\omega_c t + \psi(t)] + V(t) \cos [\omega_c t + \phi(t)]$$

$$= R(t) \cos [\omega_c t + \theta(t)] \tag{14-3}$$

Using the vector diagram of Fig. 14-1a for $A_c \gg V(t)$ and the modified form of Fig. 14-1b for $V(t) \gg A_c$, the instantaneous phase of $e_{IF}(t)$ can be expressed in two equivalent forms, as

$$\omega_c t + \psi(t) + \tan^{-1} \frac{V(t) \sin [\phi(t) - \psi(t)]}{A_c + V(t) \cos [\psi(t) - \phi(t)]} = \omega_c t + \theta(t) \tag{14-4}$$

or as

$$\omega_c t + \phi(t) + \tan^{-1} \frac{A_c \sin [\psi(t) - \phi(t)]}{V(t) + A_c \cos [\psi(t) - \phi(t)]} = \omega_c t + \theta(t) \tag{14-5}$$

In Eq. (14-4), the instantaneous phase of the signal is taken as the reference, and the inverse-tangent term represents the phase disturbance caused by the noise. In Eq. (14-5), the instantaneous phase of the noise is used as a reference, and the disturbance term is caused by the signal.

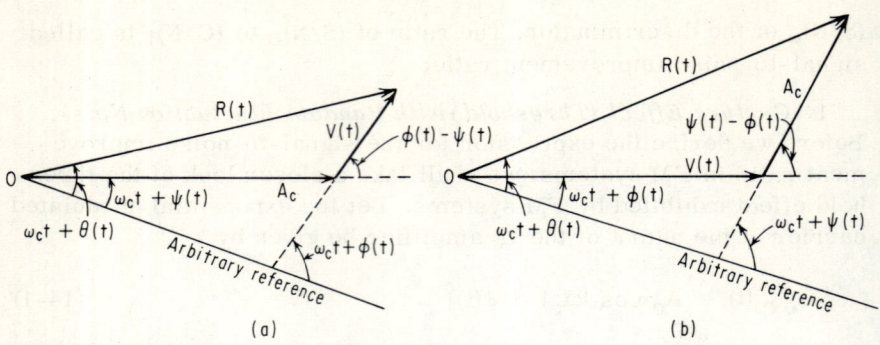

Fig. 14-1. Phasor diagram of resultant of modulated carrier and random noise: (a) High carrier-to-noise; (b) low carrier-to-noise.

In the first case, $A_c \gg V(t)$ most of the time, and Eq. (14-4) is closely approximated by

$$\omega_c t + \theta(t) \doteq \omega_c t + \psi(t) + \frac{V(t)}{A_c} \sin [\phi(t) - \psi(t)] \qquad (14\text{-}6)$$

In the second case, $V(t) \gg A_c$, and the instantaneous phase of $e_{IF}(t)$ is closely approximated by

$$\omega_c t + \theta(t) \doteq \omega_c t + \phi(t) + \frac{A_c}{V(t)} \sin [\psi(t) - \phi(t)] \qquad (14\text{-}7)$$

An examination of the last two equations reveals a very important clue to the threshold effect which is characteristic of wideband FM systems. When the signal is much stronger than the noise, a small random phase perturbation is introduced by the noise whose peak value is given by $V(t)/A_c$. However, when the noise is much stronger than the signal, the desired phase information $\psi(t)$ is buried in the noise, since it is the noise which controls the phase of the signal-plus-noise combination, and the various sidebands of the desired signal are no longer coherent in phase with the effective carrier and can no longer combine effectively to produce a large frequency deviation of the carrier.

The effect of the noise on the detected output signal, for large carrier-to-noise, can also be seen by reference to Fig. 14-1a. In this case, we note that variations in $V(t)$ and $\phi(t)$ produce small perturbations about $\psi(t)$. From the symmetry of the diagram and the uniform distribution of $\phi(t)$, we conclude that the ensemble average of $\theta(t)$ will in this case be just $\psi(t)$. However, as we shall show later on, the perturbations give rise to a nonzero mean-squared variation about the phase angle $\psi(t)$ and the information $\dot{\psi}(t)$, which is equivalent to noise at the output.

The effect of the noise when $V(t) \gg A_C$ can be seen by reference to Fig. 14-1b. In this case, the noise "captures" the signal, and the ensemble average of $\theta(t)$ becomes zero. This leads to the signal suppression characteristic which was mentioned above. Thus, the ensemble averages $\tilde{\theta}(t) = \tilde{\psi}(t)$ when $V(t) < A_C$, and $\tilde{\theta}(t) = 0$ when $V(t) > A_C$. Similarly, it can be shown that $\dot{\tilde{\theta}}(t) = \dot{\tilde{\psi}}(t) = kg(t)$ when $V(t) < A_C$, and $\dot{\tilde{\theta}}(t) = 0$ when $V(t) > A_C$. Since the noise envelope $V(t)$ is Rayleigh-distributed, the probability that $V(t) < A_C$ is equal to $1 - e^{-A_C^2/2\sigma^2} = 1 - e^{-(C/N)i}$. As we shall show later, this results in a signal-suppression effect which can be expressed by the following relation

$$\dot{\tilde{\psi}}(t) = kg(t)[1 - e^{-(C/N)i}] \qquad\qquad (14\text{-}8)$$

The capture phenomenon just described enables the signal to suppress the noise when $A_C \gg V(t)$, and the noise to suppress the signal when $V(t) \gg A_C$. When the signal and noise are of comparable strength, the output experiences a capture transition which is commonly referred to as the threshold.

2. *Performance of Conventional FM Receiver in the Presence of Random Fluctuation Noise.* The performance of a conventional FM receiver in the presence of random fluctuation noise is commonly judged on the basis of the variation of the $(S/N)_o$ power ratio as a function of $(C/N)_i$ measured at the input to the limiter. This relationship is usually presented graphically as in Fig. 14-2. The operation of a conventional FM receiver is characterized by the existence of thresholds. The first threshold, called the noise threshold, occurs when the carrier amplitude A_C is of such a magnitude as to bring about rapid quieting of the noise at the output. When A_C is increased a few decibels above this capture threshold which is denoted by A_{th}, another threshold is reached which is called the threshold of full FM improvement. For all values of A_C larger than the second threshold, the output $(S/N)_o$ average power ratio varies linearly with the input $(C/N)_i$.

The carrier level A_{th} of the first threshold can be expressed with reference to N_{rms}, the root mean square value of the noise in the IF.

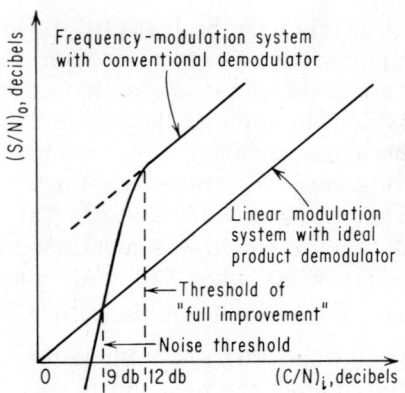

Fig. 14-2. Noise performance of conventional FM receiver.

$$N_{rms}^2 = \int_{B_{IF}} S'(f)\, df = N_0\, B_{IF}, \qquad \text{for white noise} \qquad (14\text{-}9)$$

where $S'(f)$ is the IF power spectral density in watts per cycle, and N_0 is a constant. As stated previously, in the capture transition region the signal and noise are of comparable strength; consequently the first threshold will occur when the carrier amplitude A_c is essentially equal to the peak value of $V(t)$. However, since the noise envelope $V(t)$ is Rayleigh-distributed, the term peak value of the noise envelope must be associated with the probability or percentage of time that $V(t)$ will exceed r times N_{rms}, which equals $e^{-r^2/2}$. For example, $V(t)$ may be expected to exceed $4N_{rms}$, 0.034 per cent of the time; and $3N_{rms}$, 1.111 per cent of the time. Based on experimental observations, we arbitrarily specify the peak value of the noise as the value which $V(t)$ may exceed for less than 1 per cent of the time, which is approximately equal to $3.7N_{rms}$. This results in $(C/N)_i = (3.7)^2/2 = 6.85$, or about 8.4 db. The value of 4 is often taken for simplicity, and it gives a $(C/N)_i = 4^2/2 = 8$ or 9 db, as shown in Fig. 14-2. As stated previously, rapid quieting of the noise occurs when A_c is increased slightly beyond A_{th}; and when A_c is a few decibels above A_{th}, the threshold of full improvement is reached as shown in Fig. 14-2.

In Chap. 16, we shall discuss techniques for reducing the threshold level in order to overcome the disadvantage of high-threshold requirement operation exhibited by conventional FM demodulators. The concept of threshold reduction is usually understood to mean an extension of the region in which the output $(S/N)_0$ is linearly related to the input $(C/N)_i$ below the so-called noise threshold without affecting the high $(S/N)_0$ performance.

3. *Signal-to-Noise Improvement for High Carrier-to-Noise Case.* In the following, we shall derive an expression for the improvement in the output signal-to-noise ratio $(S/N)_o$ of a conventional FM demodulator for the high carrier-to-noise case, namely, where the demodulator is operating in the linear region, as shown in Fig. 14-2. Following the same procedure as in the analysis of linear modulation discussed in Chap. 6, the following analysis will be carried out, using two methods. In the first, the noise is considered to consist of an infinite number of random-phased sinusoids, and in the second, the total noise within the IF passband is represented by the sinusoid of Eq. (14-2).

Signal-to-Noise Improvement by First Method.[2-4] As with linear modulation, we ignore the noise in the consideration of the message modulation of the carrier. For a phase discriminator, the input and output signals are given by

$$e_{exp}(t) = A_c \cos [\omega_c t + \Delta\Phi \, g(t) + \phi_c] \qquad (14\text{-}10)$$

$$e_{out}(t) = k_d \, \Delta\Phi \, g(t) \qquad (14\text{-}11)$$

and for a frequency discriminator,

$$e_{exp}(t) = A_c \cos [\omega_c t + \Delta\Omega \int_o^t g(u) \, du + \phi_c] \qquad (14\text{-}12)$$

$$e_{out}(t) = k_d \, \Delta\Omega \, g(t) \qquad (14\text{-}13)$$

where $\Delta\Phi$ and $\Delta\Omega$ are the peak phase and peak angular frequency deviation, respectively, provided $|g(t)|_{max} = 1$, and k_d is a constant of the discriminator.

Next we compute the mean square value of the output noise in the absence of the message modulation. The instantaneous phase deviation due to the noise contained within a narrow bandwidth Δf, as shown in Fig. 6-7, is given by Eq. (6-52), which is closely approximated by

$$\theta(t) \doteq \frac{A_n}{A_c} \sin [(\omega_n - \omega_c)t + \phi_n - \phi_c] \qquad (14\text{-}14)$$

where $A_n = \sqrt{2S'(f_n) \, \Delta f} \ll A_c$. The instantaneous angular frequency deviation caused by the incremental noise element is given by

$$\frac{d\theta(t)}{dt} \doteq (\omega_n - \omega_c) \frac{A_n}{A_c} \cos [(\omega_n - \omega_c)t + \phi_n - \phi_c] \qquad (14\text{-}15)$$

The output mean square value due to the discrete noise component is therefore given by

$$\overline{\theta^2(t)} = k_d^2 \frac{A_n^2}{2A_c^2}, \qquad \text{phase discriminator} \qquad (14\text{-}16a)$$

$$\overline{[\dot{\theta}(t)]^2} = k_d^2 (\omega_n - \omega_c)^2 \frac{A_n^2}{2A_c^2} \qquad \text{frequency discriminator} \quad (14\text{-}16b)$$

on the assumption that $f_n - f_c < f_m$. As in linear demodulation, the various noise components combine incoherently, and the mean square value of the total output noise power $(N)_o$ becomes

$$(N)_o = k_d^2 \frac{1}{A_c^2} \int_{-f_m}^{f_m} S'(f_c + f) \, df = \left(\frac{k_d}{A_c}\right)^2 2N_o f_m,$$

$$\text{phase discriminator} \qquad\qquad\qquad\qquad (14\text{-}17a)$$

$$(N)_o = k_d^2 \frac{1}{A_c^2} \int_{-f_m}^{f_m} \omega^2 S'(f_c + f) \, df = \left(\frac{k_d}{A_c}\right)^2 (2\pi)^2 \frac{2N_o}{3} f_m^3,$$

$$\text{frequency discriminator} \qquad\qquad\qquad\qquad (14\text{-}17b)$$

for white noise. From Eq. (14-15), we note that the noise components in the IF amplifier which are further away from the carrier produce a greater frequency deviation, and hence a larger voltage, at the output of the frequency discriminator. Thus, the amplitude spectrum of the detected noise voltage is directly proportional to f and is frequently referred to as being triangular in shape. However, it follows from Eq. (14-14) that the noise voltage at the output of a phase discriminator is uniformly distributed independent of f.

From Eqs. (14-17a) and (14-17b), it is seen that the two-sided spectral density of the output noise power $S_{No}(f)$ is

$$S_{No}(f) = \frac{k_d^2 S'(f_c + f)}{A_c^2}, \qquad \text{phase discriminator} \qquad (4\text{-}18a)$$

and

$$S_{No}(f) = \frac{k_d^2 \omega^2 S'(f_c + f)}{A_c^2}, \qquad \text{frequency discriminator} \qquad (4\text{-}18b)$$

Thus, in a phase discriminator, the output noise-power spectral density is flat, while in a frequency discriminator it is parabolic.

It should be noted here that because of the low-pass filter following the discriminator, only those input noise components within the range $(f_c \pm f_m)$ contribute to the output. In this calculation, we are neglecting the output noise due to the modulation of one noise component by another, on the assumption that for a large input carrier-to-noise ratio, the only significant noise components in the output are those due to the modulation of the carrier by the noise.

The output signal-to-noise power ratio $(S/N)_o$ will now be calculated as follows. For a phase discriminator, the output signal power, using Eq. (14-11), is given by

$$S_o = k_d^2 (\Delta\Phi)^2 \overline{g^2(t)} \tag{14-19}$$

and using Eq. (14-17a), we obtain the following expression for the $(S/N)_o$ of a phase discriminator.

$$\left(\frac{S}{N}\right)_o = \frac{(\Delta\Phi)^2 A_c^2 \overline{g^2(t)}}{\int_{-f_m}^{f_m} S'(f_c + f)\, df} = \frac{(\Delta\Phi)^2 A_c^2 \overline{g^2(t)}}{2N_o f_m}, \qquad \text{for white noise} \tag{14-20}$$

Similarly, from Eqs. (14-13) and (14-17b), we obtain for a frequency discriminator

$$\left(\frac{S}{N}\right)_o = \frac{(\Delta\Omega)^2 A_c^2 \overline{g^2(t)}}{\int_{-f_m}^{f_m} \omega^2 S'(f_c + f)\, df} = 3\left(\frac{\Delta F}{f_m}\right)^2 \frac{A_c^2 \overline{g^2(t)}}{2N_o f_m}, \qquad \text{for white noise} \tag{14-21}$$

where ΔF is the peak frequency deviation given by $\Delta\Omega = 2\pi\, \Delta F$. The expression $\dfrac{A_c^2/2}{2N_o f_m}$ is recognized as the input carrier-to-noise average power ratio of a reference AM receiver designed for the same class of messages as the FM receiver.

$$\left(\frac{C}{N}\right)_i = \frac{A_c^2/2}{2N_o f_m} \tag{14-22}$$

The ratio $(S/N)_o$ to $(C/N)_i$ is usually referred to as the signal-to-noise improvement ratio; for the special case where $g(t) = \cos(\omega_m t + \phi_m)$, $\overline{g^2(t)} = \frac{1}{2}$, the signal-to-noise improvement ratio is given by

$$\frac{(S/N)_o}{(C/N)_i} = (\Delta\Phi)^2, \qquad \text{phase discriminator} \tag{14-23}$$

$$\frac{(S/N)_o}{(C/N)_i} = 3 \left(\frac{\Delta F}{f_m}\right)^2 = 3\,\beta^2, \qquad \text{frequency discriminator} \quad (14\text{-}24)$$

or comparing FM with linear modulation (LM), we have

$$\frac{(S/N)_o, \ FM}{(S/N)_o, \ LM} = 3\,\beta^2 \qquad\qquad\qquad\qquad (14\text{-}25)$$

For large modulation index $(\beta \gg 1)$, the bandwidth approaches $2\,\Delta F$, and the signal-to-noise improvement ratio is large. As an example, in commercial FM broadcasting $\beta = 5$, and the output $(S/N)_o$ in FM is 75 times that of an equivalent AM system. But this requires increasing the transmitter bandwidth from $2f_m$ (AM case) to about $16f_m$ (FM case). As mentioned previously, FM provides a substantial improvement in $(S/N)_o$, but at the expense of bandwidth; this is characteristic of all noise improvement systems.

Signal-to-Noise Improvement by Second Method. [11,16] The same results will now be derived by representing the total noise present within the receiver IF in terms of the slowly varying envelope $V(t)$ and slowly varying phase $\phi(t)$ as in Eq. (14-2).

$$e_n(t) = V(t) \cos [\omega_c t + \phi(t)] = x_c(t) \cos \omega_c t - x_s(t) \sin \omega_c t$$
$$(14\text{-}26)$$

As stated above, the inphase and quadrature noise components $x_c(t)$ and $x_s(t)$ are slowly varying independent gaussian variables. The FM modulated signal as given by Eq. (14-1) combines with the noise voltage $e_n(t)$ to produce at the output of the IF amplifier a hybrid signal:

$$e_{IF}(t) = A_c \cos [\omega_c t + \psi(t)] + V(t) \cos [\omega_c t + \phi(t)]$$
$$= R(t) \cos [\omega_c t + \theta(t)] \qquad\qquad (14\text{-}27)$$

On the assumption that the resultant is amplitude-limited, the output of a balanced FM discriminator is proportional to $\theta(t)$, which contains the original signal plus the noise contribution.

From the phasor diagram, Fig. 14-1a or Eq. (14-4), it follows that the resultant phase angle $\theta(t)$ for $A_c \gg V(t)$ is given by

$$\theta(t) = \psi(t) + \tan^{-1} \frac{V(t) \sin [\phi(t) - \psi(t)]}{A_c + V(t) \cos [\phi(t) - \psi(t)]} \qquad (14\text{-}28)$$

Since $\phi(t)$ is uniformly distributed over the range 0 to 2π rad, the quantity $\phi(t) - \psi(t)$ will also be uniformly distributed over 2π rad. It follows therefore that the noise output, due to mean square perturbations of $\theta(t)$ about $\psi(t)$, is thus independent of the signal and depends only on the carrier and noise characteristics. As stated previously, in finding the noise output of the FM discriminator, we can therefore simplify our analysis by ignoring the modulation and considering only the unmodulated carrier plus narrow-band noise.

Thus for high carrier-to-noise ratio, we can write

$$e_{IF}(t) = [A_c + x_c(t)] \cos \omega_c t - x_s \sin \omega_c t = R(t) \cos [\omega_c t + \phi_n(t)]$$

(14-29)

where $$\phi_n(t) = \tan^{-1} \frac{x_s(t)}{A_c + x_c(t)}$$

(14-30)

This can also be derived directly from Eq. (14-27) by putting $\psi(t) = 0$ and noting from Fig. 14-1b that

$$x_c(t) = V(t) \cos \phi(t)$$

and $$x_s(t) = V(t) \sin \phi(t)$$

(14-31)

For large carrier-to-noise ratio, Eq. (14-30) can be simplified into the form

$$\phi_n(t) \doteq \frac{x_s(t)}{A_c}, \qquad A_c \gg \sigma$$

(14-32)

and $$\dot{\phi}_n(t) = -\frac{1}{A_c} \frac{dx_s(t)}{dt}, \qquad A_c \gg \sigma$$

(14-33)

where $\overline{x_c^2(t)} = \overline{x_s^2(t)} = \sigma^2$. Thus we have shown that the output noise for large carrier-to-noise ratio is proportional to the rate of change of the out-of-phase noise term $x_s(t)$.

In order to find the mean output noise power we must determine the spectral density of $\dot{\phi}_n(t)$ (for high carrier-to-noise). By the use of Eq. (4-232) we write

$$x_s(t) = \sum_{n=1}^{\infty} \sqrt{2S'(f_n)\,\Delta f} \,\sin\,[(\omega_n - \omega_c)t + \theta_n]$$

(14-34)

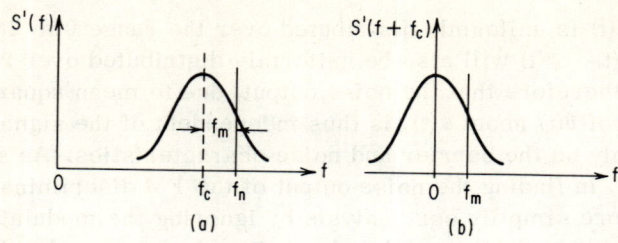

Fig. 14-3. Power spectral density of narrow-band noise: (a) spectral density of total noise; (b) spectral density of in-phase and out-of-phase components.

where $S'(f_n)$ denotes the IF noise spectral density (one-sided). As discussed in Chap. 4, $x_s(t)$ is a slowly varying time function since the frequency components $(\omega_n - \omega_c) \ll \omega_c$. This can be seen more clearly by defining a frequency $f_m = f_n - f_c$, as shown in Fig. 14-3. Using this notation we obtain

$$x_s(t) = \sum_m \sqrt{2S'(f_m + f_c)\,\Delta f}\;\;\sin(\omega_m t + \theta_m) \qquad (14\text{-}35)$$

The spectral density of $x_s(t)$ is therefore $S'(f + f_c)$ which is a two-sided spectral density as a result of the frequency translation about $f = 0$. A similar expression can readily be derived for $x_c(t)$.

The two-sided spectral density of $\dot{x}_s(t)$ is obtained directly from the equation

$$\dot{x}_s(t) = \sum_m \omega_m \sqrt{2S'(f_m + f_c)\,\Delta f}\;\;\cos(\omega_m t + \theta_m) \qquad (14\text{-}36)$$

Hence, $S_{\dot{x}_s}(f) = \omega^2 S'(f + f_c)$ \qquad\qquad\qquad\qquad\qquad (14-37)

and from Eq. (14-33) it follows that the two-sided spectral density of $\dot{\phi}_n t$ is given by

$$S_{\dot{\phi}_n}(f) = \frac{\omega^2 S'(f + f_c)}{A_c^2}, \qquad A_c \gg \sigma \qquad (14\text{-}38)$$

where $S'(f)$ is the one-sided noise spectral density of the IF noise $e_n(t)$.

As an example, we consider first a rectangular IF spectrum as shown in Fig. 14-4a where $S'(f) = N_0$, a constant. The two-sided spectral density of the detected output noise is therefore

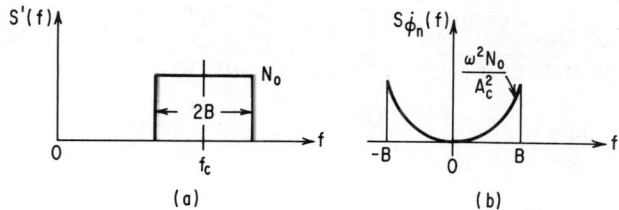

Fig. 14-4. FM noise spectrum input: rectangular IF spectrum: (a) spectrum of input IF noise; (b) spectrum of detected output noise.

$$S_{\dot{\phi}_n}(f) = \frac{\omega^2 N_0}{A_c^2}, \qquad -B < f < B \tag{14-39}$$

as shown in Fig. 14-4b. The last result can be expressed in terms of the input carrier-to-noise ratio $(C/N)_i$; where the total IF noise in a bandwidth $2B$ is $(N)_i = 2 N_0 B$ and input carrier $(C)_i = A_c^2/2$. Hence,

$$S_{\dot{\phi}_n}(f) = \frac{\pi^2}{B} \frac{f^2}{(C/N)_i}, \qquad -B < f < B \tag{14-40}$$

Assuming an ideal low-pass filter of cutoff frequency f_m at the FM detector's output, the total output noise power is

$$(N)_0 = k_d^2 \int_{-f_m}^{f_m} S_{\dot{\phi}_n}(f)\, df = k_d^2 \frac{2}{3} \frac{\pi^2}{B} \frac{f_m^3}{(C/N)_i} = \left(\frac{k_d}{A_c}\right)^2 (2\pi)^2 \frac{2N_0}{3} f_m^3 \tag{14-41}$$

which is identical with Eq. (14-17b).

For a gaussian IF spectrum with a 3-db bandwidth equal to $2B$, the one-sided spectral density is given by Eq. (4-171), namely:

$$S'(f) = \frac{Ne^{-(f - f_c)^2/2f_0^2}}{\sqrt{2\pi}\, f_0} \tag{14-42}$$

where the parameter $2f_0 = 1.7B$, and N equals the mean noise power at the IF output. The two-sided noise spectral density at the output of the FM detectors is then

$$S_{\dot{\phi}_n}(f) = \frac{\omega^2}{A_c^2} \frac{Ne^{-f^2/2f_0^2}}{\sqrt{2\pi}\,f_0} = \frac{2\pi^2}{\sqrt{2\pi}\,f_0} \cdot \frac{f^2 e^{-f^2/2f_0^2}}{(C/N)_i}$$

$$= \frac{0.94\,\pi^2}{B} \frac{f^2 e^{-f^2/1.4B}}{(C/N)_i} \tag{14-43}$$

It should be noted here that in the region of where $f \ll B$ the spectral distribution for the gaussian IF case is very nearly equal to that of the rectangular IF case, as illustrated in Fig. 14-5.

Assuming again an ideal low-pass filter at the detector output, the total output noise power for $f \ll B$ is

$$(N)_0 = k_d^2 \int_{-f_m}^{f_m} S_{\dot{\phi}_n}(f)\,df = k_d^2 \frac{2}{3}(0.94)\frac{\pi^2}{B}\frac{f_m^3}{(C/N)_i} \tag{14-44}$$

We shall now find the signal-to-noise improvement ratio for the high $(C/N)_i$ and rectangular IF spectrum. As before the $(S/N)_0$ is defined as the ratio of the mean squared output signal voltage (in the absence of noise) to the mean squared noise voltage at the output of the ideal FM detectors, in the absence of signal

$$S_0 = \frac{k_d^2\,(\Delta\Omega)^2}{2} \tag{14-45}$$

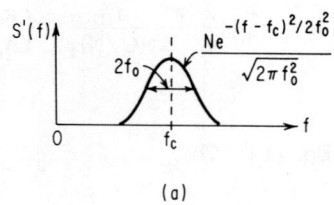

(a)

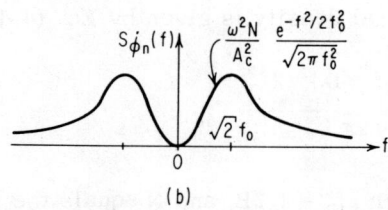

(b)

Fig. 14-5. FM noise spectrum input: gaussian IF spectrum: (a) spectrum of input IF noise; (b) spectrum of detected output noise.

Hence the output signal-to-noise ratio for high $(C/N)_i$ and rectangular IF spectrum is

$$\left(\frac{S}{N}\right)_o = 3\,\beta^2\left(\frac{B}{f_m}\right)\left(\frac{C}{N}\right)_i, \qquad \left(\frac{C}{N}\right)_i \gg 1. \qquad (14\text{-}46)$$

where $N_i = 2N_o B$, total IF output noise power. The last result is very often given in a different form when N_i is defined as the equivalent AM IF noise, namely $N_i = 2N_o f_m$, since only the IF noise power in the band $f_c \pm f_m$ contributes to the output noise power from the detector. Equation (14-46) reduces then to

$$\left(\frac{S}{N}\right)_o = 3\,\beta^2\left(\frac{C}{N}\right)_i \qquad (14\text{-}47)$$

which is identical with Eq. (14-24) where $N_i = 2N_o B$.
For large β, the transmission bandwidth $2B \doteq 2\Delta F$ and Eq. (14-46) reduces to

$$\left(\frac{S}{N}\right)_o = 3\left(\frac{B}{f_m}\right)^3\left(\frac{C}{N}\right)_i \qquad (14\text{-}43)$$

This result, which holds for $(C/N)_i \rightarrow \infty$, is derived by Rice.[16]

We shall return shortly to discuss the case of low carrier-to-noise ratio, but first we shall derive an expression for the signal-to-noise improvement in multiplex systems, which again is applicable only to high carrier-to-noise ratio.

4. *Signal-to-Noise Improvement in Multiplex Systems.* Equation (14-24) was derived for a single-channel FM system. A similar equation will presently be derived for the $(S/N)_o$ for a particular channel of a multiplex system. We consider the noise contribution due to a particular channel in the baseband $(0-f_m)$ occupying the frequency range from f_1 to f_2. From Eq. (14-18b), we calculate the output noise power due to this channel, as follows:

$$N_{o(f_1-f_2)} = \left(\frac{kd}{A_c}\right)^2 2\int_{f_1}^{f_2} f^2 S'(f_c + f)\,df \qquad (14\text{-}49)$$

For white noise this becomes

$$N_{o(f_1-f_2)} = \left(\frac{kd}{A_c}\right)^2 2N_o \int_{f_1}^{f_2} f^2\,df = \left(\frac{kd}{A_c}\right)^2 2N_o\,\frac{f_2^3 - f_1^3}{3}$$

Now $f_2^3 - f_1^3 = (f_2 - f_1)(f_1^2 + f_1 f_2 + f_2^2)$

Let the channel bandwidth $B_c = f_2 - f_1$ and $\overline{f} = (f_1 + f_2)/2$, the mid-frequency of the channel. In practice, the channel bandwidth $B_c \ll f_m$; therefore

$$\frac{f_2^3 - f_1^3}{3} \doteq \frac{3 B_c \overline{f}^2}{3} = B_c \overline{f}^2 \qquad (14\text{-}50)$$

The channel noise power is then

$$N_{0(f_1 - f_2)} = \left(\frac{k_d}{A_c}\right)^2 2 N_0 \overline{f}^2 B_c \qquad (14\text{-}51)$$

where B_c is the channel bandwidth in cycles, and $\overline{f}$ is the mid-channel frequency. Note the similarity between Eqs. (14-51) and (14-17b); it should be realized, however, that the noise contribution of a particular channel varies with the position of the channel in the baseband, being greater for the high-frequency channels. The problem of noise equalization in a multichannel system is therefore of great practical importance in the design of communication systems.

The output signal-to-noise in a particular channel is readily derived as follows. Let ΔF_m equal peak channel-frequency deviation, and $\overline{f}_n$ the mid-band channel frequency in the n'th channel. The input noise power to the FM demodulator equals $2 N_0 B$, and the carrier power is $A_c^2 / 2$. Therefore

$$\left(\frac{C}{N}\right)_i = \frac{A_c^2}{4 N_0 B} \qquad (14\text{-}52)$$

and from Eq. (14-51), we obtain

$$\left(\frac{S}{N}\right)_0 = k_d^2 \frac{\Delta F_m^2}{2} \frac{1}{2 N_0 (k_d / A_c)^2 \overline{f}_n^2 B_c}$$

or $$\left(\frac{S}{N}\right)_0 = \left(\frac{C}{N}\right)_i \left(\frac{B}{B_c}\right) \left(\frac{\Delta F_m}{\overline{f}_n}\right)^2 \qquad (14\text{-}53)$$

It is to be noted here that the IF bandwidth $2B$ does not affect the output signal-to-noise ratio as it may appear from Eq. (14-53); it has been put there for symmetry only, in order to balance out the factor B in $N_i = 2N_0B$. However, the IF bandwidth does play a very important role in setting the threshold level, since $N_i = 2N_0B$ and

$$A_{c,th} \doteq A_{n,th} = 4N_{rms} = 4\sqrt{2N_0B} \qquad (14\text{-}54)$$

It should be emphasized here that the above equations for $(S/N)_0$ hold provided $(C/N)_i$ is greater than a specified threshold level; thus the IF bandwidth determines the service range of the communication systems, and consequently, it should be designed as narrow as possible in keeping with the allowable distortion requirements.

The maximum operating range of any communication system is determined by the location at which the carrier falls below the improvement threshold. When this occurs, even in ordinary double-sideband amplitude modulation, there is a relatively sudden large rise of the noise level which effectively blankets the signal. This fact should be considered in the design of communication systems. If the $(C/N)_i$ ratio is much higher than the improvement threshold, the bandwidth of the IF prior to the limiter can be increased many-fold, and as long as the bandwidth is again narrowed to its optimum value by the audio (or video) amplifier, there will be no loss in $(S/N)_0$ ratio in the final output. However, in case of $(C/N)_i$ ratio near the improvement threshold, widening the predetection (IF) bandwidth to the extent that the $(C/N)_i$ ratio falls into the range of the improvement threshold will cause a rise in noise which cannot be eliminated by narrowing the bandwidth of the audio amplifier. This threshold level therefore determines the point at which the communication system fails, or determines the maximum operating range. It follows that any method or design which effectively lowers the threshold of the receiver will increase the operating range and enhance the reliability of the communication system. As stated previously, the problem of lowering the threshold or, as it is commonly referred to, threshold extension, will be discussed in Chap. 16.

14.2 Signal-to-Noise Improvement through De-emphasis[7-9]

We have seen in the last section that the noise-power spectrum at the output of the discriminator is emphasized at the higher frequencies and in fact is proportional to f^2 for large carrier-to-noise ratio. On the other end, in many communication systems, most of the energy of the modulating signal is found to be concentrated in the lower-frequency ranges. Since the smaller-amplitude high-frequency components of the signal produce on the average a much smaller frequency deviation, the assigned bandwidth may not be fully occupied by the FM signal, resulting in a poor $(S/N)_O$ ratio at the high-frequency end of the modulating signal.

In order to improve the $(S/N)_O$ ratio of the high-frequency end of the baseband, the modulating signal at the transmitting end is first passed through a network that emphasizes the higher signal frequencies but leaves the low signal components unaffected. This process, which is called pre-emphasis, tends to equalize the energy distribution throughout the frequency range of the modulating signal. At the output of the discriminator, the inverse process is carried out, or the higher-frequency components are de-emphasized in order to restore the original signal-power distribution. But in this de-emphasis process, the higher-frequency components of the noise are reduced, and thus the $(S/N)_O$ ratio is increased. Note that in this process of pre-emphasis and de-emphasis, we utilize the different statistical properties of signal and noise.

In practice, a simple RC circuit as shown in Fig. 14-6 is used to emphasize the high frequencies in many FM receivers, and a time constant of 75 μsec is commonly used for broadcast reception. The transfer function of Fig. 14-6 is given by

$$H(j\omega) = 1 + j\omega\tau \tag{14-55}$$

where the time constant $\tau = rC$, and $r \gg R$.

The amplitude response has two break frequencies given by $\omega_1 = 1/rC$ and $\omega_2 = 1/RC$; signals in the range between f_1 and f_2 are emphasized.

At the receiver, at the output of the discriminator, a de-emphasis network is used, as shown in Fig. 14-7, whose transfer characteristics are the inverse of those of the pre-emphasis circuit. Thus

$$H(j\omega) = \frac{1}{1 + j\omega\tau} \tag{14-56}$$

where $\omega_1 = 1/rC$; here again a time constant $\tau = rC$ of 75 μsec is used. Pre-emphasis and de-emphasis circuit responses, for time constants of 50, 75, and 100 μsec, are shown in Fig. 14-8.

In the following, we shall calculate the $(S/N)_O$ improvement due to the de-emphasis network for AM and FM.

1. *Amplitude Modulation with Pre-emphasis.* From Eq. (6-53), we obtain first an expression for the noise output without de-emphasis:

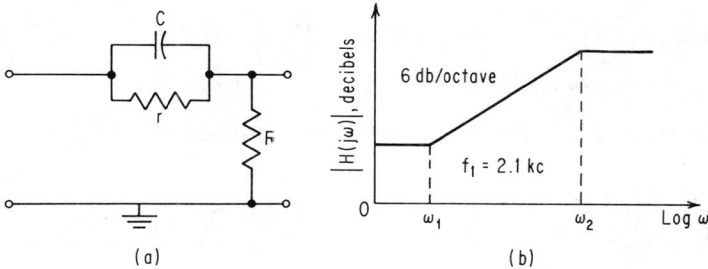

Fig. 14-6. Examples of pre-emphasis network: (a) Pre-emphasis network (r ≫ R, rC = 75 μsec); (b) asymptotic response ($\omega_1 = 1/rC$, $\omega_2 \doteq 1/RC$).

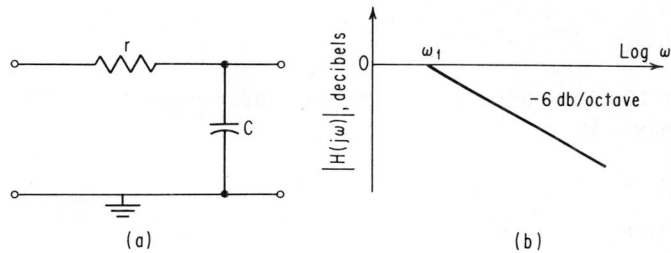

Fig. 14-7. Examples of de-emphasis network: (a) De-emphasis network (rC = 75 μsec); (b) asymptotic response ($f_1 = 2.1$ kc).

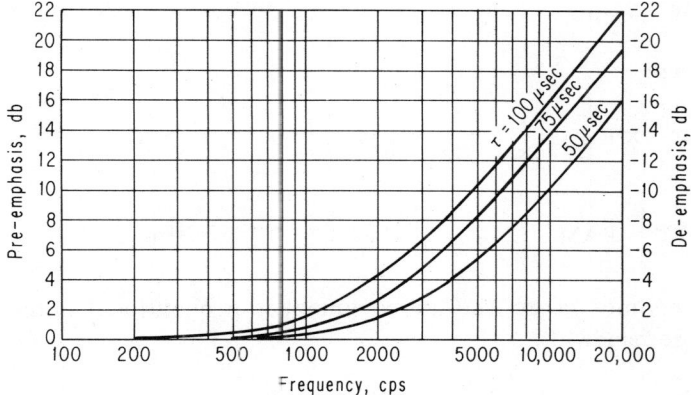

Fig. 14-8. Pre-emphasis and de-emphasis circuit response, for time constants of $\tau = 50$, 75, and 100 μsec. (From Tibbs and Johnstone,[9] courtesy of Chapman and Hall, Ltd.)

$$(N)_o = \left(\frac{kd}{A_c}\right)^2 \int_{-f_m}^{f_m} S'(f_c + f) \; df \; = \left(\frac{kd}{A_c}\right)^2 2N_o f_m \qquad (14\text{-}57)$$

With de-emphasis, the noise power is modified by $|H(j\omega)|^2$:

$$(N)_{o,d} = \left(\frac{kd}{A_c}\right)^2 N_o \int_{-f_m}^{f_m} \frac{df}{1 + 4\pi^2 \tau^2 f^2} = \left(\frac{kd}{A_c}\right)^2 \frac{N_o}{\pi\tau} \tan^{-1} 2\pi f_m \tau$$

$$(14\text{-}58)$$

The improvement in the $(S/N)_o$ due to de-emphasis is derived from the expression

$$\frac{(N)_o}{(N)_{o,d}} = \frac{2\pi f_m \tau}{\tan^{-1} 2\pi f_m \tau} = \rho_{AM} \qquad (14\text{-}59)$$

Thus for 100 per cent modulation, the improvement in $(S/N)_o$ due to de-emphasis is

$$\left(\frac{S}{N}\right)_{o,d} = \rho_{AM} \left(\frac{C}{N}\right)_i \qquad (14\text{-}60)$$

where ρ_{AM} is the improvement factor due to de-emphasis. It is of interest to evaluate ρ_{AM} for broadband AM systems. When $2\pi f_m \tau > 1$, we expand

$$\tan^{-1} 2\pi f_m \tau = \tan^{-1} x$$

as follows:

$$\tan^{-1} x = \frac{\pi}{2} - \frac{1}{x} + \frac{1}{3x^3} - \frac{1}{5x^5} + \cdots, \qquad x^2 > 1$$

Therefore $\quad \rho_{AM} = \dfrac{2\pi f_m \tau}{\pi/2 - 1/2\pi f_m \tau + \cdots} \doteq 4f_m \tau \qquad (14\text{-}61)$

As f_m becomes large, the improvement is proportional to the intelligence bandwidth.

    2. *Frequency Modulation with Pre-emphasis.* Here again, we start with the basic equation for the noise power at the output of the low-pass filter following the frequency discriminator. Using Eq. (14-18b), we have for white noise without pre-emphasis

$$(N)_o = \left(\frac{kd}{A_c}\right)^2 N_o \int_{-f_m}^{f_m} f^2 \, df = \left(\frac{kd}{A_c}\right)^2 \frac{2}{3} N_o f_m^3 \tag{14-62}$$

With pre-emphasis, the spectral distribution of the noise power is modified by $|H(j\omega)|^2$, and the total mean noise power at the output of an ideal low-pass filter of bandwidth f_m is given by

$$(N)_{o,d} = \left(\frac{kd}{A_c}\right)^2 N_o \int_{-f_m}^{f_m} \frac{f^2 \, df}{1 + 4\pi^2 \tau^2 f^2} = \left(\frac{kd}{A_c}\right)^2 2N_o \left(f_m - \frac{1}{2\pi\tau} \tan^{-1} 2\pi f_m \tau\right)$$

$$\tag{14-63}$$

Figure 14-9 illustrates the triangular distribution of the output noise voltage without de-emphasis and the effect of using de-emphasis. We note here that at high frequencies, the noise amplitude is constant, and the noise spectrum is similar to that of an AM receiver.

The improvement factor ρ_{FM} is given by

$$\rho_{FM} = \frac{(N)_o}{(N)_{o,d}} = \frac{(2\pi f_m \tau)^3}{3(2\pi f_m \tau - \tan^{-1} 2\pi f_m \tau)} \tag{14-64}$$

For narrow-band FM, $\qquad\qquad \rho_{FM} \rightarrow 1$

For wideband, f_m is large, and $\quad \rho_{FM} \rightarrow \dfrac{(2\pi f_m \tau)^2}{3}$

Thus for small f_m, the improvement factor due to de-emphasis reduces to unity. For large f_m the improvement factor is proportional to f_m^2 instead of f_m, as in the case of amplitude modulation.

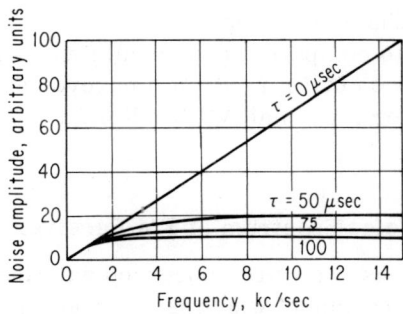

Fig. 14-9. The distribution of noise output voltage with frequency, with and without de-emphasis. (From Tibbs and Johnstone,[9] courtesy of Chapman and Hall, Ltd.)

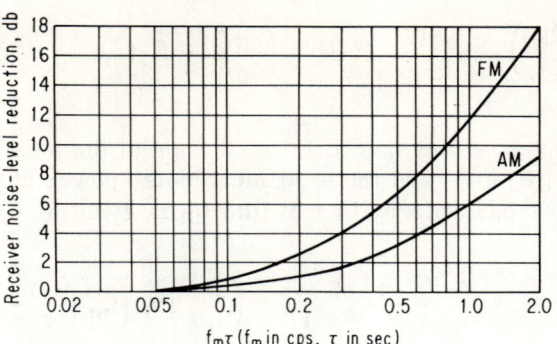

Fig. 14-10. The reduction in receiver noise output for FM and AM systems using pre-emphasis. (From Tibbs and Johnstone,[9] courtesy of Chapman and Hall, Ltd.)

This is shown in Fig. 14-10, where the reduction in receiver noise-power output for FM and AM systems due to pre-emphasis is plotted.

Finally, the mean S/N output power ratio for FM with de-emphasis is given by

$$\left(\frac{S}{N}\right)_{o,d} = \rho \, FM \left(\frac{S}{N}\right)_o \tag{14-65}$$

where

$$\left(\frac{S}{N}\right)_o = 3\beta^2 \left(\frac{C}{N}\right)_i \tag{14-66}$$

With $\tau = 75 \, \mu sec$ and $f_m = 15$ kc, $\rho = 20$, or the S/N improvement due to de-emphasis is 13 db. For $f_m = 21$ kc ($= 10 f_1$), the improvement due to de-emphasis is 16 db.

It is of interest to compare these results with the AM case; for $f_m = 15$ and 21 kc, we obtain in AM an improvement of only 6 and 8 db, respectively. For a normal AM band of $f_m = 5$ kc, the improvement is only 3 db.

14.3   The Output Noise Spectrum of FM Receivers [5,10-16]

In the past sections, we were concerned with the problem of the signal-to-noise improvement which is realized in wideband FM systems. We limited our discussion to high-input carrier-to-noise, namely, where the carrier is above the threshold level and the receiver is operating over the linear region of the $(S/N)_o$-$(C/N)_i$

characteristic curve. By restricting ourselves in this manner, it has been possible to simplify considerably the derivation of the results.

In this section, we consider the nonlinear region of the detection process, including the extreme case of the carrier being absent. The technique used in our analysis is a correlation function procedure in which the noise and signal spectrum at the discriminator output is found by first finding the correlation function at the output and then finding its inverse transform. Because of the complexity of the calculations, we shall start with the simpler case of zero carrier and then extend the method to low carrier-to-noise.

1. *Output Noise Spectrum of FM Receivers for Zero Carrier and IF Noise Input with Perfect Limiting.* [5,10] We consider the case of zero carrier, and we assume a perfect limiter ahead of the discriminator. Using Eq. (14-30), we obtain for $A_C = 0$ the expression

$$\phi_n(t) = \tan^{-1} \frac{x_s(t)}{x_c(t)} \tag{14-67}$$

and therefore the voltage at the balanced discriminator output is

$$\dot{\phi}_n(t) = \frac{d\phi_n(t)}{dt} = \frac{x_c(t)\dot{x}_s(t) - x_s(t)\dot{x}_c(t)}{x_c^2(t) + x_s^2(t)} \tag{14-68}$$

To find the spectrum of the noise output, we shall find first the correlation function of $\dot{\phi}(t)$, which is just the time average of $\dot{\phi}(t)\dot{\phi}(t+\tau)$.

$$R_n(\tau) = \overline{\dot{\phi}(t)\dot{\phi}(t+\tau)}. \tag{14-69}$$

To simplify the writing of equations, we use here a notation due to Rice, [11]

$$x_c(t) \equiv x_1, \quad x_c(t+\tau) \equiv x_2, \quad x_s(t) \equiv x_3, \quad x_s(t+\tau) \equiv x_4$$
$$\dot{x}_c(t) \equiv x_5, \quad \dot{x}_c(t+\tau) \equiv x_6, \quad \dot{x}_s(t) \equiv x_7, \quad \dot{x}_s(t+\tau) \equiv x_8 \tag{14-70}$$

so that

$$\dot{\phi}_n(t) = \frac{x_7 x_1 - x_3 x_5}{x_1^2 + x_3^2} \quad \text{and} \quad \dot{\phi}_n(t+\tau) = \frac{x_8 x_2 - x_4 x_6}{x_2^2 + x_4^2} \tag{14-71}$$

The correlation function is then

$$R_n(\tau) = \overline{\dot\phi_n(t)\dot\phi_n(t+\tau)} = \frac{x_7 x_1 - x_3 x_5}{x_1^2 + x_3^2} \frac{x_8 x_2 - x_4 x_6}{x_2^2 + x_4^2} \qquad (14\text{-}72)$$

where the variables $x_1, \ldots, x_8$ are gaussian-distributed. Assuming a stationary process, the time average may then be replaced by the ensemble average [Eq. (4-143)]:

$$R_n(\tau) = \int \cdots \int \dot\phi_n(t)\dot\phi_n(t+\tau) p(x_1, x_2, \cdots, x_8)\, dx_1, dx_2, \cdots, dx_8$$

$$(14\text{-}73)$$

where the probability density $p(x_1, x_2, \ldots, x_8)$ is an eight-dimensional gaussian distribution.

The direct evaluation of Eq. (14-73) is very tedious; a somewhat simpler method is to replace the eightfold probability distribution by its Fourier integral representation in terms of the characteristic function. Defining $F(z_1, z_2, \ldots, z_8)$ as the eight-dimensional Fourier transform of $\dot\phi_n(t)\dot\phi_n(t+\tau)$, and writing $C(z_1, z_2, \ldots, z_8)$ as the characteristic function of $p(x_1, x_2, \ldots, x_8)$, Eq. (14-73) can be replaced by

$$R_n(\tau) = \left(\frac{1}{2\pi}\right)^8 \int_{-\infty}^{\infty} \cdots \int_{-\infty}^{\infty} F(z_1, z_2, \ldots, z_8) C(z_1, z_2, \ldots, z_8)$$

$$\times\, dz_1, dz_2, \ldots dz_8 \qquad (14\text{-}74)$$

The integration of this expression is rather lengthy and is given in Ref. 10. The result is given in the following form:

$$R_n(\tau) = \frac{1}{2}\left(\frac{\ddot\rho}{\rho} - \frac{\dot\rho^2}{\rho^2}\right) \log(1 - \rho^2) \qquad (14\text{-}75)$$

where $\rho(\tau) = R(\tau)/N$, the normalized correlation function of either the inphase noise term $x_c(t)$ or the quadrature term $x_s(t)$, and N represents the IF noise power. The correlation function of the discriminator output is thus expressed in terms of the low-frequency equivalent correlation function of the IF output.

Although the output noise spectrum $S_n(f)$ is readily expressible by the Wiener-Khinchine theorem as the Fourier transform of $R_n(\tau)$, the actual integration can only be carried out numerically and depends on the IF noise spectrum. Rice [11] and Lawson and

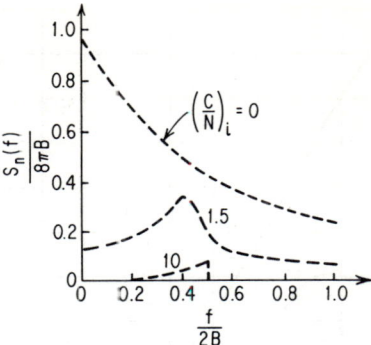

Fig. 14-11. Output FM noise-power spectrum (ideal limiting), rectangular IF spectrum. (After Lawson and Uhlenbeck, [10] courtesy of McGraw-Hill Book Company.)

Uhlenbeck [10] have plotted $S_n(f)$ vs. frequency f (normalized with respect to IF bandwidth) for the gaussian IF spectrum and for the rectangular spectrum, respectively. The curve for the rectangular IF spectrum is given in Fig. 14-11, and the corresponding curve for the gaussian IF spectrum in Fig. 14-12. Also included in Fig. 14-11 are analogous curves for input carrier-to-noise ratio $(C/N)_i$ = 1.5 and 10. Note the large increase in output noise as the carrier-to-noise ratio drops below 10. This can be seen from the following example. Consider the case where the audio bandwidth $f_m \ll B$, where $2B = $ IF bandwidth. The output noise power $(N)_o$ for the case of zero carrier can be obtained from the curve of Fig. 14-11, by taking the product of the zero-frequency ordinate and f_m.

$$(N)_o \doteq 8\pi B (0.955) f_m; \qquad C = 0, f_m \ll B \qquad (14\text{-}76)$$

The output noise power for $(C/N)_i \gg 1$ or high carrier-to-noise ratio is obtained from Eq. (14-17b), namely,

$$(N)_o = \frac{(\omega_m)^2 f_m}{6B} \left(\frac{N}{C}\right)_i, \qquad \left(\frac{C}{N}\right)_i \gg 1 \qquad (14\text{-}77)$$

The ratio of the output noise powers for the two cases of zero carrier to large carrier is

$$\frac{(N)_{o(C/N = 0)}}{(N)_{o(C/N \gg 1)}} = \frac{12}{\pi} (0.955) \left(\frac{B}{f_m}\right)^2 \left(\frac{C}{N}\right)_i \doteq 3\beta^2 \left(\frac{C}{N}\right)_i \qquad (14\text{-}78)$$

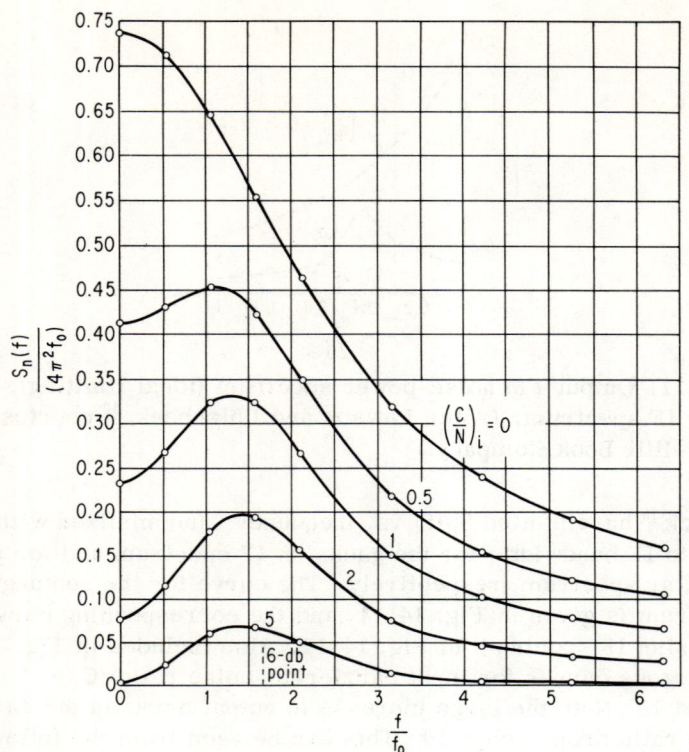

Fig. 14-12. Output FM noise-power spectrum (ideal limiting), gaussian IF spectrum.

with　$\beta = \dfrac{\Delta F}{f_m} \doteq \dfrac{B}{f_m}$

As an example, with $(C/N)_i = 10$, the ratio of the noise powers is $30\beta^2$. For $(C/N)_i \geq 10$, the noise power varies inversely with $(C/N)_i$. It is only for $(C/N)_i < 10$ db that the output noise power increases much more rapidly. In fact, the output noise power increases about 30 db when the carrier-to-noise ratio drops from 10 to 0 db.

The total output noise power may be found by integrating the output-noise spectrum curves over the assumed video bandwidth. This has been done by Stumpers [12] for both the rectangular and gaussian IF spectrum characteristics, assuming an ideal rectangular video filter. The resultant curves for the rectangular spectrum are

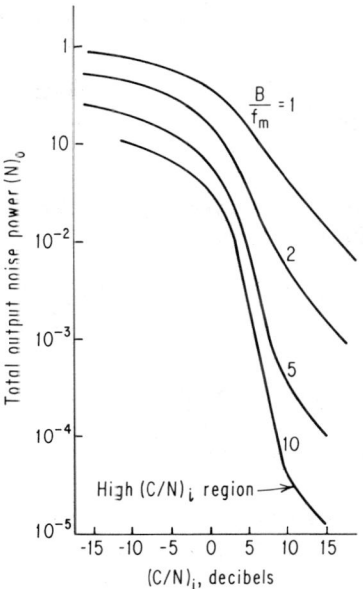

Fig. 14-13. Output FM noise power (ideal limiting), rectangular IF spectrum. (After F. L. H. M. Stumpers, [12] Proc. IRE.)

shown in Fig. 14-13. We note here the abrupt increase in output noise as the carrier-to-noise ratio drops below the threshold of 10 db. As an example, for $B/f_m = 10$, $(N)_o$ increases by about 30 db as $(C/N)_i$ decreases 10 db. For $(C/N)_{in}$ less than 0 db, the output noise levels off at an almost constant value.

2. *Output Noise Spectrum of FM Receivers for Carrier plus Noise Input: Carrier Unmodulated.* We shall now outline a method of analysis for the calculation of the output noise spectrum for unmodulated carrier plus noise. The calculations are much more cumbersome, and only the main steps will be indicated. For detailed analysis of an IF rectangular spectrum, reference may be made to Lawson and Uhlenbeck, [10] and a corresponding analysis for a gaussian IF spectrum is given by Rice. [11] Middleton [13] and Blachman [14,15] carried out extensive analyses, and plotted curves for modulated and unmodulated carriers with various degrees of limiting. Stumpers, [12] using a zero crossing procedure for obtaining the FM noise spectrum, has analyzed both cases of IF spectrum, namely, the flat and gaussian, and obtained identical results.

As for zero carrier, the spectral density $S_n(f)$ of the output noise power is derived by finding first the autocorrelation function of

$\dot{\phi}_n(t)$ of Eq. (14-30), and then obtaining its Fourier transform. In this case we have

$$\tan \phi_n(t) = \frac{x_S(t)}{A_c + x_c(t)} \tag{14-79}$$

The discriminator output is proportional to the instantaneous frequency of the resultant signal $\dot{\phi}_n(t)$, where

$$\dot{\phi}_n(t) = \frac{d\phi_n(t)}{dt} = \frac{[x_c(t) + A_c]\, \dot{x}_S(t) - x_S(t)\dot{x}_c(t)}{[x_c(t) + A_c]^2 + x_S^2(t)} \tag{14-80}$$

The autocorrelation of the instantaneous frequency is then

$$R_n(\tau) = \overline{\dot{\phi}_n(t)\dot{\phi}_n(t + \tau)} = \text{av.}\,[\dot{\phi}_n(t)\dot{\phi}_n(t + \tau)] \tag{14-81}$$

since a stationary process is assumed and the time average may be replaced by the ensemble average.

As before, we can simplify the equations by the following notation:

$$x_c(t) + A_c = x_1, \qquad x_c(t + \tau) + A_c = x_2, \qquad x_S(t) = x_3, \qquad x_S(t + \tau) = x_4$$

$$\dot{x}_c(t) = x_5, \qquad \dot{x}_c(t + \tau) = x_6, \qquad \dot{x}_S(t) = x_7, \qquad \dot{x}_S(t + \tau) = x_8 \tag{14-82}$$

and Eq. (14-81) can be written in the form

$$R_n(\tau) = \int_{-\infty}^{\infty} \cdots \int_{-\infty}^{\infty} \dot{\phi}_n(t)\dot{\phi}_n(t + \tau)p(x_1, x_2, \ldots, x_8)\, dx_1, dx_2, \ldots, dx_8 \tag{14-83}$$

where, as before, the probability density $p(x_1, x_2, \ldots, x_8)$ is an eight-dimensional gaussian distribution. According to Rice,[11] the algebra is greatly simplified by the eightfold probability distribution function as the Fourier transform of its characteristic function. Even with this simplification, the calculations are very elaborate and lengthy. Rice[11] has computed the power spectrum of $\dot{\phi}_n(t)$ for various ratios of $(C/N)_i$ for the case in which the carrier frequency is at the center of a narrow gaussian band of noise. The results are reproduced in Fig. 14-12, where f_0 is the frequency displacement from the center of the band at which the spectral density is down $\frac{1}{4}$ neper or 2.17 db. As stated above, corresponding results for a flat band of input noise have been given by Stumpers,[12] using an entirely different analytical approach, which is called the "zero crossing method"; the results are reproduced in Fig. 14-11.

3. *Decrease in Output Signal.*[5,16] It was indicated above on the basis of qualitative reasoning and the phasor diagram of Fig. 14-1b that when $V(t) > A_c$, the noise captures the signal and signal suppression takes place as the $(C/N)_i$ drops below 0 db. We have noted previously that as the $(C/N)_i$ drops from 10 to 0 db, the output signal remains nearly unchanged while the output noise increases by 30 db. However, as $(C/N)_i$ drops below 0 db, the $(S/N)_o$ begins to decrease while the output noise levels off to a fixed value. In the following, we shall outline the derivation of Eq. (14-8), which expresses the signal suppression effect.

From Eq. (14-28), we note that the instantaneous phase angle $\theta(t)$ of the resultant of a modulated sinusoidal carrier plus slowly varying gaussian noise consists of $\psi(t) = \int_o^t g(t)\,dt$ plus a noise term. Let

$$\theta(t) = \psi(t) + \phi_n(t) \tag{14-84}$$

so that the instantaneous noise output is $\dot\phi_n(t) = \dot\theta(t) - \dot\psi(t)$. The average signal output of the balanced discriminator equals the ensemble average of $\dot\theta(t)$, and the ensemble average noise output is $\tilde{\dot\phi_n}(t) = \tilde{\dot\theta}(t) - \tilde{\dot\psi}(t)$. Since the noise term $\dot\phi_n(t)$ is just the derivative with respect to time of the inverse tan term of Eq. (14-28), it is therefore a function of $V(t)$, $\dot V(t)$, $\phi(t)$, and $\dot\phi(t)$. The ensemble averaged noise is then given by

$$\tilde{\dot\phi_n}(t) = \iiiint \dot\phi_n(t)p(V,\ \dot V,\ \phi,\ \dot\phi)\ dV\ d\dot V\ d\phi\ d\dot\phi \tag{14-85}$$

The four-dimensional probability distribution $p(V,\ \dot V,\ \phi,\ \dot\phi)$ is found from the four-dimensional probability distribution $p[x_c(t),\ x_s(t),\ \dot x_c(t),\ \dot x_s(t)]$ by transforming to polar coordinates. Since in this analysis, the IF noise spectral density $S'(f)$ is assumed to be symmetrical about the center ω_c, the random variables $x_c(t)$, $x_s(t)$, $\dot x_c(t)$, and $\dot x_s(t)$ are independent gaussian variables, and $p[x_c(t), x_s(t), \dot x_c(t), \dot x_s(t)]$ is simply the product of four gaussian distributions, which will presently be evaluated.

First we integrate over V and $\dot\phi$ and obtain

$$\tilde{\dot{\phi}}_n(t) = \tilde{\dot{\theta}} - \dot{\psi} = -\frac{\dot{\psi}(t)}{2\pi} \int_0^\infty \int_{-\pi}^{\pi} \frac{(V/A_C) \cos(\phi - \psi) \, d\phi}{1 + (V/A_C)^2 + (2V/A_C) \cos(\phi - \psi)}$$

$$\times e^{-V^2/2\sigma^2} \cdot \frac{V \, dV}{\sigma^2}, \qquad N_i = \sigma^2 \qquad (14\text{-}86)$$

Next we integrate with respect to ϕ. As noted previously, when $V \gg A_C$, the noise captures the signal, and the signal output is reduced to zero. This is borne out by the integration with respect to ϕ.

For $\quad \dfrac{V}{A_C} > 1, \qquad$ integral $= 2\pi$

$\quad\quad \dfrac{V}{A_C} < 1, \qquad$ integral $= 0$

The resultant ensemble average of $\dot{\phi}_n(t)$ is given by

$$\tilde{\dot{\phi}}_n(t) = \tilde{\dot{\theta}} - \dot{\psi} = -\dot{\psi} \int_{A_C}^\infty \frac{V \, dV}{\sigma^2} e^{-V^2/2\sigma^2} \qquad (14\text{-}87)$$

and $\quad \tilde{\dot{\theta}} = \dot{\psi}(1 - e^{-(C/N)_i}) \qquad (14\text{-}88)$

with $C_i = A_C^2/2$. From Eq. (14-88), we note that the average amplitude of the discriminator output is reduced because of noise capture whenever the noise instantaneously exceeds the carrier. We also note that this signal suppression effect becomes significant when $(C/N)_i \leq 1$.

The average output signal power is proportional to the square of the frequency deviation, or

$$S_0 = \overline{(\tilde{\dot{\theta}})^2} = \overline{(\dot{\psi})^2} \left(1 - e^{-(C/N)_i}\right)^2 \qquad (14\text{-}89)$$

which for $(C/N)_i \ll 1$, reduces to

$$S_0 \doteq \overline{(\dot{\psi})^2} \left(\frac{C}{N}\right)_i^2 \qquad (14\text{-}90)$$

Since, as stated previously, the output noise power N_0 approaches a constant value for $(C/N)_i \ll 1$, we conclude therefore that in this case

$$\left(\frac{S}{N}\right)_o \propto \left(\frac{C}{N}\right)_i^2$$

We can summarize our conclusion with reference to Fig. 14-14. As $(C/N)_i$ drops below 10 db, there is a sharp increase in the output noise power. In fact, as $(C/N)_i$ drops from 10 to 0 db, the output noise power increases by about 30 db. As $(C/N)_i$ decreases below 0 db, the output noise power levels off at a large value, but the output signal is being suppressed. In the region where $(C/N)_i < 0$ db the output $(S/N)_o$ is proportional to $(C/N)_i^2$.

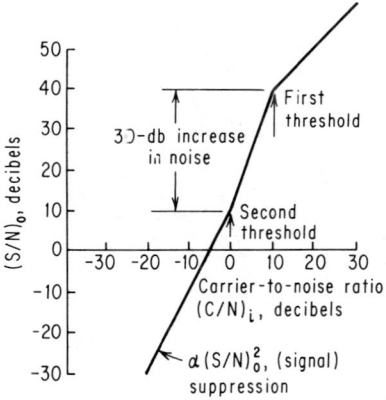

Fig. 14-14. Performance of conventional FM receiver in the non-linear region $(C/N)_i \ll 1$. (From M. Schwartz,[5] courtesy of Proc. Natl. Electronics Conf.)

A very interesting treatise, Noise in FM Receivers, is given by Rice.[16] In it he derives expressions for $(S/N)_o$ in the nonlinear region by calculating first the number of "clicks" which are heard in the FM detector output as the input carrier-to-noise ratio is reduced. The analysis is carried out for both rectangular and gaussian IF noise. Rice considers specifically the behavior of the FM detector in the region around the breaking point or the nonlinear region of the FM performance curve. According to Rice the output noise power $(N)_o$ is then approximately given by

$$(N)_o \doteq \frac{8\pi^2 B f_m}{\sqrt{3}} (1 - \text{erf }\sqrt{\rho}) + \frac{2\pi^2 f_m^3}{3\rho B} \qquad (14\text{-}91)$$

for the rectangular IF filter, where $\rho = (C/N)_i$. Similarly for the gaussian filter

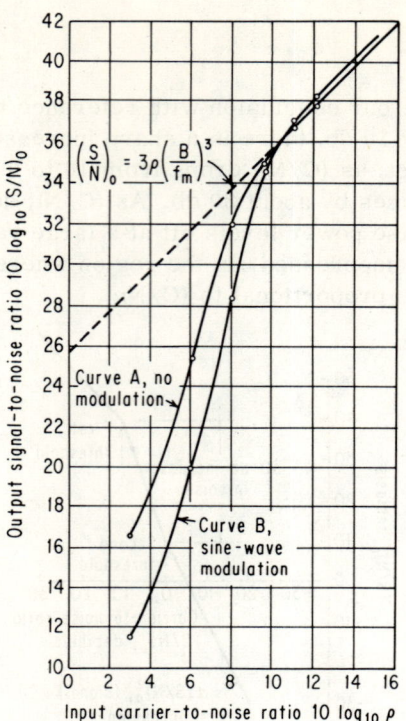

Fig. 14-15. Dependence of $(S/N)_o$ vs. $(C/N)_i$ in the nonlinear range 10 to 2 db. (From S. O. Rice, [16] Proc. of the Symposium on Time Series Analysis, Brown University, 1962.)

$$(N)_o \doteq 8\pi^2 f_o f_m \, (1 - \text{erf} \sqrt{\rho}) + \frac{4\pi^2 f_m^3}{3\rho f_o (2\pi)^{\frac{1}{2}}} \, (1 - \frac{3f_m^2}{10f_o^2} + \ldots) \tag{14-92}$$

Taking the output signal power to be $S_o = \Delta\Omega^2/2 = 2\pi^2 B^2$ in the rectangular case, and $S_o = 2\pi^2 B_n^2$, $B_n = \frac{1}{2}\sqrt{2\pi} \, f_o$ in the gaussian filter case, the respective $(S/N)_o$ ratios are

$$\left(\frac{S}{N}\right)_o \doteq \frac{3\rho (B/f_m)^3}{\rho \sqrt{3} \, (1 - \text{erf} \sqrt{\rho}) \, (2B/f_m)^2 + 1} \tag{14-93}$$

and

$$\left(\frac{S}{N}\right)_o \doteq \frac{3\rho (B_n/f_m)^3}{\rho (18/\pi)^{1/2} \, (1 - \text{erf} \sqrt{p}) \, (2B_n/f_m)^2 + [1 - 0.3 \, (f_m/f_o)^2 +]} \tag{14-94}$$

where $2B_n = \sqrt{2\pi}\ f_0$ is the equivalent rectangular bandwidth of the gaussian IF filter. As $\rho \to \infty$, the denominator of Eq. (14-93) tends to unity and the ratio $(S/N)_0 \to 3\rho (B/f_m)^3$ as in Eq. (14-48). A plot of Eq. (14-93) is given in curve A, Fig. 14-15, for a deviation ratio, $B/f_m = 5$. Of particular interest to our discussion is the nonlinear region of low $(C/N)_i$ from about 10 to 2 db, since in this region, the decrease in signal output due to the capture effect is small; the large decrease in $(S/N)_0$ is due to the large increase in the output noise power, as discussed above.

References

1. Armstrong, E. M.: A Method of Reducing Disturbances in Radio Signaling by a System of Frequency Modulation, *Proc. IRE*, May, 1936.
2. Baghdady, E. J. (ed.): "Lectures on Communication System Theory," McGraw-Hill Book Company, New York.
3. Baghdady, E. J.: FM Demodulators with Frequency Compressive Feedback, *Proc. Nat. Telemetering Conf.*, 1962.
4. Baghdady, E. J.: The Technique for Lowering the Noise Threshold of Conventional Frequency, Phase, and Envelope Demodulators, *IRE Trans. Commun. Systems*, September, 1961.
5. Schwartz, M.: Signal-to-Noise Effects and Threshold Effects in FM, *Proc. Nat. Electron. Conf.*, 1962.
6. Smith, J. E.: Theoretical Signal-to-Noise Ratios, *Electronics*, June, 1946.
7. Schwartz, M.: "Information Transmission, Modulation, and Noise," McGraw-Hill Book Company, New York.
8. Black, H. S.: "Modulation Theory," D. Van Nostrand Company. Inc., Princeton, N. J.
9. Tibbs, C. E., and G. G. Johnstone: "Frequency Modulation Engineering," John Wiley & Sons, Inc., New York.
10. Lawson, J. L., and G. E. Uhlenbeck (eds.): "Threshold Signals," chap. 13, McGraw-Hill Book Company, New York, 1950.
11. Rice, S. O.: Statistical Properties of a Sine Wave plus Random Noise, *BSTJ*, January, 1948.
12. Stumpers, F. L. H. M.: Theory of FM Noise, *Proc. IRE*, September, 1948.
13. Middleton, D.: On Theoretical Signal-to-Noise Ratios in FM Receivers: a Comparison with Amplitude Modulation, *J. Appl. Phys.*, April, 1949.

14. Blachman, N. M.: The Demodulation of an FM Carrier and Random Noise by a Limiter and Discriminator, *J. Appl. Phys.*, January, 1949.
15. Blachman, N. M.: The Demodulation of a Frequency-modulated Carrier and Random Noise by a Discriminator, *J. Appl. Phys.*, October, 1949.
16. Rice, S. O.: Noise in FM Receivers, *Proc. Symp. Time Ser. Anal.*, Brown University, June 11–14, 1962.

15

APPLICATION OF NEGATIVE FEEDBACK TO FREQUENCY-MODULATION SYSTEMS

In this chapter, methods will be discussed for improving the performance of frequency-modulation systems. In the broader aspects, these methods can be considered as an application of the principle of negative feedback to frequency modulation. First we shall discuss its application to a FM transmitter, where it will be shown that feedback reduces the distortion products which are generated in the modulation process, provided a limiter is used in the feedback path. We shall then discuss the application of negative feedback to FM receivers and show that a limiter is required before the discriminator to obtain constant output independent of rapid amplitude variations in the input signal. It will be shown that the use of inverse feedback has the advantage of reducing the required bandwidth of the IF amplifier without sacrificing the advantage of wideband frequency modulation. The method of negative feedback which is commonly referred to as "frequency following" or "frequency compression" will be shown to be applicable also to distance-measuring-equipment (DME) systems. The analysis will show that the frequency-following method may be used to reduce the inherent phase variation of DME systems where phase stability is of primary importance in the measurements

461

of range. The advantage of such a scheme to the $(S/N)_o$ improvement of a FM receiver, and the general problem of threshold extension, will be discussed in Chap. 16.

15.1 Application of Negative Feedback to FM Transmitters [1-3]

The problem of generation of FM signals with very low distortion components is of paramount importance in FM communication systems. We shall first develop a general feedback formula and then apply it to the problem of reducing distortion products generated in a frequency-modulation transmitter.

1. *General Feedback Formula.* Let the input–output characteristic of a nonlinear network be given by

$$e_o = \mu_1 e_i + \mu_2 e_i^2 + \mu_3 e_i^3 + \cdots \tag{15-1}$$

where $e_i(t)$ = effective input voltage
    $e_o(t)$ = output voltage
    μ = voltage gain of network
In Fig. 15-1, let $e(t)$ be the applied voltage. Then the effective input voltage is

$$e_i = e - \beta e_o \tag{15-2}$$

where β is the fraction of the output voltage fed back to the input.
 Substitution of (15-1) in (15-2) gives

$$e_i = e - \mu_1 \beta e_i - \mu_2 \beta e_i^2 - \mu_3 \beta e_i^3 - \cdots \tag{15-3}$$

or $e_i = \dfrac{e}{1 + \mu_1 \beta} - \dfrac{\mu_2 \beta e_i^2}{1 + \mu_1 \beta} - \dfrac{\mu_3 \beta e_i^3}{1 + \mu_1 \beta} - \cdots$ (15-4)

The problem is to find an expression for the output voltage $e_o(t)$ in terms of the applied voltage $e(t)$.

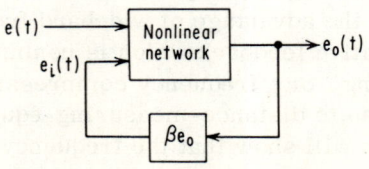

Fig. 15-1. Application of negative feedback.

Equation (15-4) may be written in the form

$$e_i(\xi) = a + \xi F(e_i) \tag{15-5}$$

where $a = \dfrac{e}{1 + \mu_1\beta}$

$$\xi = \dfrac{-\beta}{1 + \mu_1\beta} \tag{15-6}$$

and $F(e_i) = \mu_2 e_i^2 + \mu_3 e_i^3 + \cdots$

Equation (15-5) can be solved by the use of Lagrange's formula, which gives

$$f(e_i) = f(a) + \sum_{n=1}^{\infty} \frac{\xi^n}{n!} \frac{d^{n-1}}{da^{n-1}} [f'(a) F(a)^n] \tag{15-7}$$

where $f(e_i)$ is a function of e_i. Letting $f(e_i) = e_i$ and limiting ourselves to terms of the third degree, the effective input voltage e_i becomes

$$e_i = \frac{e}{1 + \mu_1\beta} - \frac{\mu_2\beta e^2}{(1 + \mu_1\beta)^2} - \left[\frac{\mu_3\beta}{(1 + \mu_1\beta)^4} - \frac{2\mu_2^2\beta^2}{(1 + \mu_1\beta)^5} \right] e^3 - \cdots \tag{15-8}$$

Since $e_i = e - \beta e_o$, we have from (15-8),

$$e_o = \frac{\mu_1 e}{1 + \mu_1\beta} + \frac{\mu_2 e^2}{(1 + \mu_1\beta)^3} + \left[\frac{\mu_3}{(1 + \mu_1\beta)^4} - \frac{2\mu_2^2\beta}{(1 + \mu_1\beta)^5} \right] e^3 + \cdots \tag{15-9}$$

Comparing the linear coefficients of Eqs. (15-1) and (15-9), we see that the gain has been reduced by the factor $1/(1 + \mu_1\beta)$. To restore the output to its level without feedback, the input voltage must be increased by the factor $(1 + \mu_1\beta)$. When this is done, Eq. (15-9) becomes

$$e_o = \mu_1 e + \frac{\mu_2 e^2}{1 + \mu_1\beta} + \left[\frac{\mu_3}{1 + \mu_1\beta} - \frac{2\mu_2^2\beta}{(1 + \mu_1\beta)^2} \right] e^3 + \cdots \tag{15-10}$$

This is the basic feedback equation, which will be used in the following analysis.

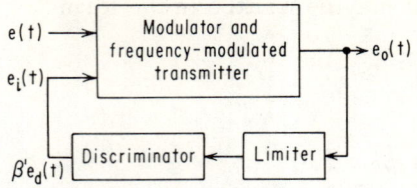

Fig. 15-2. Application of negative feedback to FM transmitter.

2. *Negative Feedback Applied to a Frequency-modulated Transmitter Limiter in Feedback Loop.* The results derived above will be applied to the analysis of a frequency-modulated transmitter with negative feedback, as shown in block diagram in Fig. 15-2, where a limiter is included in the feedback loop. We shall show that negative feedback can be used to linearize the modulation process.

The output of the transmitter, with $e_i(t)$ being the effective modulating voltage, is

$$e_0(t) = A(t)e^{j\int_0^t \omega_i(t)\,dt} \tag{15-11}$$

where $A(t)$, the amplitude of the modulated carrier, is given by

$$A(t) = A_0(1 + \lambda_1 e_i + \lambda_2 e_i^2 + \lambda_3 e_i^3) \tag{15-12}$$

and $\omega_i(t)$, the instantaneous angular frequency, is given by

$$\omega_i(t) = \omega_0(1 + \mu_1 e_i + \mu_2 e_i^2 + \mu_3 e_i^3) \tag{15-13}$$

Equations (15-12) and (15-13) express the nonlinearities which may arise in the modulation process. The limiter will be assumed to be ideal, so that the output of the discriminator is proportional to instantaneous frequency $\omega_i(t)$ only, and independent of the amplitude $A(t)$.

The output of the discriminator $e_d(t)$ is proportional to $\omega_i(t)$; hence

$$e_d(t) = k_d \omega_0 (\mu_1 e_i + \mu_2 e_i^2 + \mu_3 e_i^3) \tag{15-14}$$

where k_d is a proportionality factor. If a fraction β' of the output of the discriminator is fed back, then the effective modulating voltage of the transmitter is

$$e_i = e - \beta' e_d = e - \beta' k_d \omega_o (\mu_1 e_i + \mu_2 e_i^2 + \mu_3 e_i^3) \tag{15-15}$$

Letting $\beta = \beta' k_d \omega_o$, we have

$$e_i = \frac{e}{1 + \mu_1\beta} - \frac{\mu_2 \beta e_i^2}{1 + \mu_1\beta} - \frac{\mu_3 \beta e_i^3}{1 + \mu_1\beta}$$

which is of the same form as Eq. (15-4). Using Eq. (15-10), we find that the instantaneous frequency of the transmitter output is

$$\omega_i(t) = \omega_o \left\{ 1 + \mu_1 e + \frac{\mu_2 e^2}{1 + \mu_1\beta} + \left[\frac{\mu_3}{1 + \mu_1\beta} - \frac{2\mu_2^2 \beta}{(1 + \mu_1\beta)^2} \right] e^3 \right\} \tag{15-16}$$

Thus the output frequency can be made very nearly linear by increasing β. Remembering that the input voltage has been increased by a factor $(1 + \mu_1\beta)$, so that the feedback voltage from Eq. (15-16) is

$$\beta' e_d = \mu_1 \beta e + \frac{\mu_2 \beta e^2}{1 + \mu_1\beta} + \left[\frac{\mu_3 \beta}{1 + \mu_1\beta} - \frac{2\mu_2^2 \beta^2}{(1 + \mu_1\beta)^2} \right] e^3$$

the effective applied voltage is therefore

$$e_i = e - \frac{\mu_2 \beta e^2}{1 + \mu_1\beta} - \left[\frac{\mu_3 \beta}{1 + \mu_1\beta} - \frac{2\mu_2^2 \beta^2}{(1 + \mu_1\beta)^2} \right] e^3 \tag{15-17}$$

Equation (15-17) will now be used to evaluate the effect of negative feedback on the amplitude $A(t)$.

From Eq. (15-12), we have

$$A(t) = A_0 (1 + \lambda_1 e_i + \lambda_2 e_i^2 + \lambda_3 e_i^3) = A_0 \left\{ 1 + \lambda_1 e + \left(\lambda_2 - \lambda_1 \frac{\mu_2 \beta}{1 + \mu_1\beta} \right) e^2 \right.$$

$$\left. + \left[\lambda_3 - 2\lambda_2 \frac{\mu_2 \beta}{1 + \mu_1\beta} - \frac{\lambda_1 \mu_3 \beta}{1 + \mu_1\beta} + \frac{2\lambda_1 \mu_2^2 \beta^2}{(1 + \mu_1\beta)^2} \right] e^3 \right\} \tag{15-18}$$

For $\mu_1\beta \gg 1$, this reduces to

$$A(t) = A_0 \left\{ 1 + \lambda_1 e + \left(\lambda_2 - \lambda_1 \frac{\mu_2}{\mu_1} \right) e^2 \right.$$

$$\left. + \left[\lambda_3 - 2\lambda_2 \frac{\mu_2}{\mu_1} - \lambda_1 \frac{\mu_3}{\mu_1} + 2\lambda_1 \left(\frac{\mu_2}{\mu_1} \right)^2 \right] e^3 \right\} \tag{15-19}$$

It is seen that the linear term is unaffected by feedback. As $\mu_1 \gg \mu_2$ and $\mu_1 \gg \mu_3$, the higher terms are virtually unaffected by feedback. Thus, feedback of the type shown in Fig. 15-2 will linearize the frequency modulation but will not minimize the nonlinearity in amplitude modulation.

15.2   Application of Negative Feedback to FM Receivers [1-4]

We have seen earlier that the use of a large modulation index considerably increases the advantages of frequency over amplitude modulation but imposes severe bandwidth and linearity requirements on the intermediate-frequency amplifiers. As has been shown by Chaffee [1] and Carson,[2] the use of negative feedback allows a reduction in the bandwidth while preserving the advantages of a high modulation index. Earlier workers concluded that the output of the receiver is independent of the input amplitude when the negative feedback is great enough. However, we shall show that this conclusion is correct only if the amplitude varies at a sufficiently slow rate, but in the general case of a signal that is both amplitude- and frequency-modulated, a limiter is required to secure an output independent of the amplitude modulation of the input.

1. *Elementary Analysis of Frequency Following.* A simplified block diagram of a FM receiver with frequency following, or as it is sometimes called, frequency compressive feedback (FCF), is shown in Fig. 15-3. In this system, the output of the discriminator

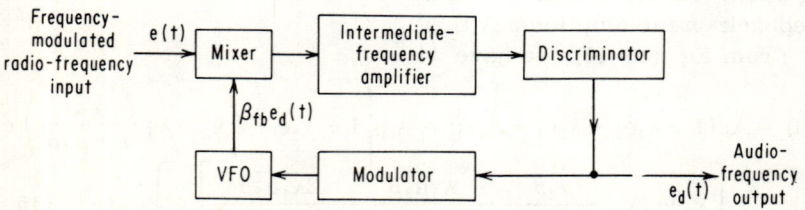

Fig. 15-3. Simplified block diagram of frequency-following receiver.

is used to modulate the frequency of the variable-frequency oscillator (VFO) in such a manner as to cause it to follow, but not coincide with, the instantaneous frequency of the incoming signal. The output signal from the mixer is frequency-modulated with an instantaneous frequency equal to the difference between the two frequencies of the input signals. Thus as the result of frequency following, the incoming wideband FM signal is processed into a narrow-band

signal whose frequency deviation is linearly compressed. In the
following simplified discussion, the time delays in the various
filters in the loop are neglected, and as the frequency of the VFO
varies, it is assumed that its amplitude remains constant. It is
also assumed for simplicity that the incoming signal is noiseless
described by

$$e(t) = A_c \cos[\omega_c t + \Delta\omega \int_0^t g(u)du] \tag{15-20}$$

where $\Delta\omega g(t)$ is the instantaneous-frequency deviation of the
incoming signal, and $g(t)$ is the message specification; the maximum
frequency deviation by the message being $\Delta\omega$ and $|g(t)|_{max} = 1$.

From Eq. (15-20), it follows that the instantaneous angular frequency of the receiver input signal is

$$\omega_i(t) = \omega_c + \Delta\omega g(t)$$

where ω_c is the carrier angular frequency.

If $e_d(t)$ is the output voltage of the discriminator, let $\beta_{fb}e_d(t)$
denote the change produced in the angular frequency of the local
oscillator as a result of feedback. The corresponding angular fre-
quency of the IF is $\omega_{IF} + \Delta\omega g(t) - \beta_{fb}e_d(t)$. Assuming a linear discriminator, the variable portion of the discriminator output is
given by

$$e_d(t) = k_d[\Delta\omega g(t) - \beta_{fb}e_d(t)] \tag{15-21}$$

where k_d = discriminator constant. Solving this relation of $e_d(t)$
gives

$$e_d(t) = \frac{k_d}{1 + k_d\beta_{fb}} \Delta\omega g(t) \doteq \frac{\Delta\omega}{\beta_{fb}}g(t) \tag{15-22}$$

provided $k_d\beta_{fb} \gg 1$. Thus, the effective index of modulation has been
reduced, and consequently the intermediate-frequency bandwidth
may be reduced to accept only one pair of sidebands of the frequency-
modulated wave. Since the output $e_d(t)$ is apparently independent of
the input signal amplitude, it might be concluded that a limiter is not
required in the loop. As is shown in the following section, this
statement is true provided the amplitude changes are slow.

2. *Simultaneous Frequency and Amplitude Modulation.*[3] We consider now a signal which is simultaneously modulated in amplitude
and frequency. The receiver input signal can be written in the exponential form,

$$e(t) = A(t)e^{j\left[\omega_c t + \Delta\omega \int_0^t g(\eta)\,d\eta\right]}$$

(15-23)

As a result of frequency following, the frequency deviation of the IF signal is reduced, so that

$$e_{IF}(t) = KA(t)e^{j\left(\omega_{IF}t + \int_0^t [\Delta\omega g(t) - \beta_{fb}e_d(t)]\,dt\right)} = KA(t)e^{j\int_0^t \Omega_1(t)\,dt}$$

(15-24)

where $\Omega_1(t)$ is the instantaneous angular frequency of the IF signal. Now we have shown in Chap. 9 that the response of a network of steady-state admittance $Y(j\omega)$ to an impressed signal

$$A(t)e^{\int_0^t \Omega_i(t)\,dt}$$

is of the form

$$e^{j\int_0^t \Omega_i(t)\,dt}\left[A(t)Y(j\Omega_i) + A'(t)\frac{dY(j\Omega_i)}{d(j\Omega_i)}\right]$$

(15-25)

For a general linear discriminator, we may write

$$Y(\omega) = a(\omega - \omega_{IF}) + b$$

Therefore $Y(j\omega_{IF}) = b$

or $Y(j\Omega_i) = a(\Omega_i - \omega_{IF}) + Y(j\omega_{IF}) = a[\Delta\omega g(t) - \beta_{fb}e_d(t)] + Y(j\omega_{IF})$

(15-26)

For a linear discriminator passing through the origin as in the case of an ordinary inductance used as a discriminator

$$b = Y(j\omega_{IF}) = a\omega_{IF}$$

and $Y(j\Omega_i) = a[\omega_{IF} + \Delta\omega g(t) - \beta_{fb}e_d(t)]$

In this special case, the discriminator acts as a pure differentiator of the applied signal, and its output becomes

$$e^{j\int_0^t \Omega_i(t)\,dt} \cdot \left[A(t)\ a\Omega_i(t) + A'(t)\frac{a}{j}\right]$$

which is proportional to

$$\frac{d}{dt}\left[A(t)e^{j\int_0^t \Omega_i(t)\,dt}\right]$$

However, the output of the general linear discriminator that does not pass through the origin may be expressed in the form

$$e^{j\int_0^t \Omega_i(t)\,dt} \cdot \left(A(t)\{a[\Delta\omega g(t) - \beta_{fb}e_d(t)] + Y(j\omega_{IF})\} - jaA'(t)\right)$$

$$= ae^{j\int_0^t \Omega_i(t)\,dt} \cdot \left\{KA(t)\left[1 + \frac{\Delta\omega g(t) - \beta_{fb}e_d(t)}{a}\right] - jA'(t)\right\}$$

where $K = Y(j\omega_{IF})/a$. Hence, the detected signal of the discriminator is

$$e_d(t) = k_d\left\{A'(t)^2 + K^2A(t)^2\left[1 + \frac{\Delta\omega g(t) - \beta_{fb}e_d(t)}{K}\right]^2\right\}^{\frac{1}{2}} \qquad (15-27)$$

For the sake of simplicity, the special differentiating discriminator will be discussed in the following. Here

$$K = \frac{Y(j\omega_{IF})}{a} = \omega_{IF}$$

and

$$e_d(t) = k_d\left\{A'(t)^2 + \omega_{IF}^2A(t)^2\left[1 + \frac{\Delta\omega g(t) - \beta_{fb}e_d(t)}{\omega_{IF}}\right]^2\right\}^{\frac{1}{2}} \qquad (15-28)$$

We note that for constant amplitude $A(t) = A_0$, $A'(t) = 0$, and Eq. (15-28) reduces to Eq. (15-21), as expected. Using the method of successive approximations, Eq. (15-28) can be shown[3] to reduce to

$$e_d(t) \doteq \frac{1}{\beta_{fb}}[\omega_{IF} - \Delta\omega g(t)]\left\{1 + \frac{1}{2}\left[A'(t)\frac{\beta_{fb}k_d}{\omega_{IF}}\right]^2\right\} \qquad (15-29)$$

which reduces to Eq. (15-22) for $A(t)$ = constant.

We conclude therefore that the output of a FM receiver with frequency following is independent of the amplitude of the input signal provided a limiter is used ahead of the discriminator.

3. *Reduction of Nonlinear Distortion in a Frequency-following Receiver.*[3,5] It has been shown above that the nonlinearities in a FM transmitter can be effectively reduced by the application of negative feedback. By making use of the same general feedback formulas as derived in Eqs. (15-9) and (15-10), we shall show that a similar reduction is effected in a frequency-following receiver. It should be noted, however, that only the nonlinearities in the forward path will be reduced by this method, while the distortion terms in the feedback path will be enhanced due to feedback. In the following, perfect phase subtraction is assumed in the mixer; the effect of imperfect subtraction as a result of loop delays may not only limit, but may also reverse the forward-path distortion reduction.

The input-output-modulation transfer characteristics of a frequency-following receiver (see Figs. 15-1 and 15-3) can be expressed as in Eq. (15-1) in the form

$$e_O = \mu_1 e_i + \mu_2 e_i^2 + \mu_3 e_i^3 + \cdots \tag{15-30}$$

where e_i is the effective deviation in the frequency-following receiver which is given by

$$e_i(t) = e(t) - \beta e_O(t) = \Delta\omega g(t) - \beta_{fb} e_d(t) \tag{15-31}$$

Here we assumed a distortionless feedback $\beta e_O(t)$. Using Eq. (15-9), we obtain directly an expression for the distortion terms in the detected output of the receiver.

$$e_O(t) = \frac{\mu_1 e(t)}{1 + \mu_1 \beta} + \frac{\mu_2 e^2(t)}{(1 + \mu_1 \beta)^3} + \left[\frac{\mu_3}{(1 + \mu_1 \beta)^4} - \frac{2\mu_2^2 \beta}{(1 + \mu_1 \beta)^5} \right] e^3(t) + \cdots \tag{15-32}$$

If we substitute the notation used above in Eq. (15-32), namely,

$$\mu_1 = k_d \qquad \beta = \beta_{fb} \qquad e(t) = \Delta\omega g(t) \qquad e_O(t) = e_d(t)$$

we obtain

$$e_d(t) = \frac{k_d \Delta\omega g(t)}{1 + k_d \beta_{fb}} + \frac{\mu_2 [\Delta\omega g(t)]^2}{(1 + k_d \beta_{fb})^3}$$

$$+ \left[\frac{\mu_3}{(1 + k_d \beta_{fb})^4} - \frac{2\mu_2^2 \beta_{fb}}{(1 + k_d \beta_{fb})^5} \right] [\Delta\omega g(t)]^3 + \cdots \tag{15-33}$$

We note that as the result of negative feedback, the distortion terms decrease rapidly as the total loop gain $k_d \beta_{fb}$ is increased. The reduction in the effective deviation of the output signal is also noted from the first term of the series, namely, $e_d(t) = k_d \Delta \omega g(t) / (1 + k_d \beta_{fb})$, as derived previously in Eq. (15-22).

Equation (15-33) was derived on the assumption that the feedback path is distortionless, equal to $\beta_{fb} e_d(t)$. In practice, the VFO may contribute nonlinearities in the modulation process so that the effective deviation is proportional to

$$e_i(t) \propto \left\{ \Delta \omega g(t) - \sum_{n=1} b_n [\beta_{fb} e_d(t)]^n \right\}, \qquad b_1 = 1 \qquad (15\text{-}34)$$

It can be shown[5] that the discriminator output signal is then given by

$$e_d(t) = \frac{k_d}{1 + k_d \beta_{fb}} \left\{ \Delta \omega g(t) - b_2 \frac{k_d^2 \beta_{fb}^2}{(1 + k_d \beta_{fb})^2} [\Delta \omega g(t)]^2 \right.$$
$$\left. + \left[2b_2^2 \frac{k_d^4 \beta_{fb}^4}{(1 + k_d \beta_{fb})^4} - b_3 \frac{k_d^3 \beta_{fb}^3}{(1 + k_d \beta_{fb})^3} \right] [\Delta \omega g(t)]^3 + \cdots \right\}$$

$$(15\text{-}35)$$

The distortion terms are seen to rise rapidly relative to $\Delta \omega g(t)$ as $k_d \beta_{fb} > 1$. Thus we have shown that by the use of frequency following, only the nonlinearities in the forward path are reduced while the distortion terms in the feedback path are enhanced.

The analysis of frequency following will be extended in Chap. 16 to encompass the problem of conditions for proper tracking and detection of FM signals in the presence of noise. This will finally lead to a discussion of threshold extension by means of frequency following. The remainder of this chapter will be devoted to the application of frequency following to DME systems.

15.3 Application of Frequency Following to DME Systems[6]

In a DME system, where range information is derived from phase measurements, phase stability of the electronic circuits is of great importance. In this section, the application of frequency following to DME systems is discussed, and analytical expressions applicable to the problem of phase stability of the system are derived.

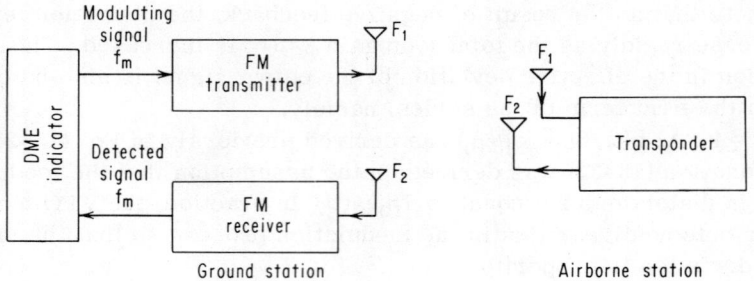

Fig. 15-4. Block diagram of a DME system.

A typical DME system shown in Fig. 15-4 is used to measure accurately and continuously the distance between a ground station and an airborne station.

The principle of the ranging system is as follows: A ranging video frequency f_m, generated in the DME indicator, is used to frequency-modulate the ground-station transmitter whose carrier frequency is F_1. In the transponder, the ranging signal f_m is first detected and then used to remodulate the carrier F_2, which is transmitted back to the ground receiver. The ranging signal is then detected and returned to the DME indicator where range information is derived by comparing the phase of the incoming signal and the reference signal. Since the range is determined by the round-trip phase shift of the modulation signal (excluding a fixed phase shift in the equipment which can be zeroed out), variations in the phase shift introduced by the ground equipment or transponder due to circuit variations may render the system useless. We shall show that by the use of frequency following, the inherent phase variation can be greatly reduced. In addition, frequency following causes the output frequency deviations to be approximately equal to the input deviations essentially independent of gain variations in the transponder. In a practical system, several modulation frequencies are used in order to obtain both high accuracy and large maximum range without ambiguity. The ranging signals which are generated in the DME indicator are added, and the sum is used to frequency-modulate the carrier in the ground station.

1. *System Analysis of Transponder.* In the following discussion, two methods of analysis are used. The first method uses the conventional steady-state approach applied to frequency modulation. The second method utilizes the transfer-function approach and yields results similar to those of the classical method with much

less effort. This method also provides a simple way of examining
the stability of the system.

Steady-state Approach. A functional block diagram of the frequency-following loop which is used in the transponder is shown in
Fig. 15-5.

The input signal at point 1 is frequency-modulated and is
assumed to be of the form

$$E_i \cos (\omega_i t + m_i \sin pt) \tag{15-36}$$

where E_i = input carrier amplitude
         ω_i = input carrier angular frequency
         m_i = input modulation index
         p = input modulation angular frequency
The output of the transponder at point 7 is given

$$E_o \cos [\omega_o t + m_o \sin (pt + \alpha_o) + \theta_o] \tag{15-37}$$

where E_o = output carrier amplitude
         ω_o = output carrier angular frequency
         m_o = output modulation index

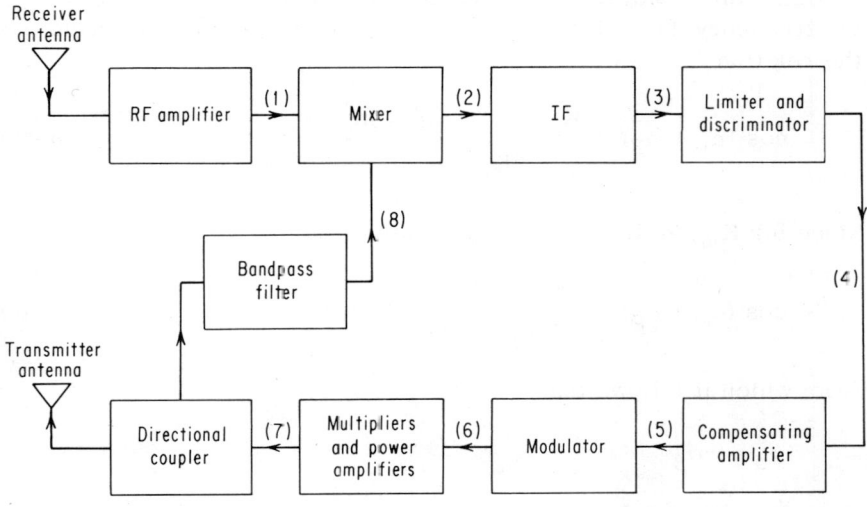

Fig. 15-5. Functional block diagram of frequency-following loop in
transponder.

α_0 = phase shift of the modulation signal in the output relative to input

θ_0 = phase shift of the output carrier relative to input

It should be noted that variations in the system parameters will result in variations in the phase angle α_0, which represent an error in the range measurement. In this analysis, it is assumed that the amplitude response of the IF bandpass circuits in the mixer is flat and the phase response is linear in the passband for all significant sidebands. We have proved previously in Chap. 7 that under these conditions, the phase shift introduced in the modulation signal of a frequency-modulated wave is equal to the difference in phase shift between the carrier frequency and the first-sideband frequency. It can readily be shown[6] that in a frequency-following system the output modulation index m_O is very closely equal to m_i, the input modulation index, provided the loop gain of the system $K_{\mu\beta}$ is large. This follows from the equation[6]

$$\frac{m_O}{m_i} = \sqrt{\frac{K_{\mu\beta}^2}{1 - 2K_{\mu\beta}\cos(\alpha_\mu + \alpha_\beta) + K_{\mu\beta}^2}} \tag{15-38}$$

where α_μ = phase contribution of forward path

and α_β = phase contribution of feedback path

Using the steady-state approach, it can also be shown that for the frequency-following transponder shown in Fig. 15-5, we have the relation

$$\cos(\alpha_0 + \alpha_\beta) = \frac{K_{\mu\beta}^2(m_O^2 + m_i^2) - m_O^2}{2K_{\mu\beta}^2 m_O^2 m_i} \tag{15-39}$$

Since for $K_{\mu\beta} \gg 1$, $m_O \doteq m_i$, we obtain

$$\cos(\alpha_0 + \alpha_\beta) \doteq \frac{2K_{\mu\beta}^2 - 1}{2K_{\mu\beta}^2} \doteq 1 \tag{15-40}$$

from which it follows that

$$\alpha_0 = {}^{-}\alpha_\beta$$

From Eq. (15-40). we conclude that any positive phase change in the feedback path appears as a negative change in range. In a practical case, α_β may be very small, and consequently α_0 is reduced as $K_{\mu\beta} \gg 1$. The physical significance of this result is that the absolute phase contribution α_0 of the transponder can be made small, provided the total loop gain $K_{\mu\beta} \gg 1$.

Transfer-function Approach. It will be of interest to derive the same results using the transfer-function approach. The same assumptions which were made previously are made here, namely, that the bandpass elements used in the transponder have a flat amplitude response and linear phase shift in the frequency region which includes all significant sidebands, so that a linear analysis does apply. This requirement is not overly stringent since these characteristics are essential for good inherent phase stability and low modulation distortion.

Figure 15-6 is a block diagram of the transponder showing the important operational parameters of the system. The following equations are derived from the block diagram by the use of Laplace transforms.

$$\frac{\Delta\omega_0(s)}{\Delta\omega_i(s)} = \frac{K_{FO}G_{FO}(s)}{K_T G_T(s) - 1} \tag{15-41}$$

where

$$\frac{\Delta\omega_0(s)}{\Delta\omega_e(s)} = K_{FO}G_{FO} \tag{15-42}$$

the transfer function of the forward loop, and

$$\frac{\Delta\omega_{BP}(s)}{\Delta\omega_e(s)} = K_{FO}G_{FO}(s) \cdot K_{BP}G_{BP}(s) \tag{15-43}$$

the transfer function of the total loop. The K's and G's are clearly indicated in the block diagram of Fig. 15-6 with the additional notation:

$\Delta\omega_i(s)$ = Laplace transform of frequency deviation of input carrier frequency

$\Delta\omega_{BP}(s)$ = Laplace transform of frequency deviation of output carrier frequency, after passing through bandpass filter

$\Delta\omega_e(s)$ = Laplace transform of frequency deviation of input intermediate frequency

The total loop gain at angular frequency ω is given by

$$K_T G_T(j\omega) = K_{\mu\beta}e^{j(\alpha_\mu + \alpha_\beta)} \tag{15-44}$$

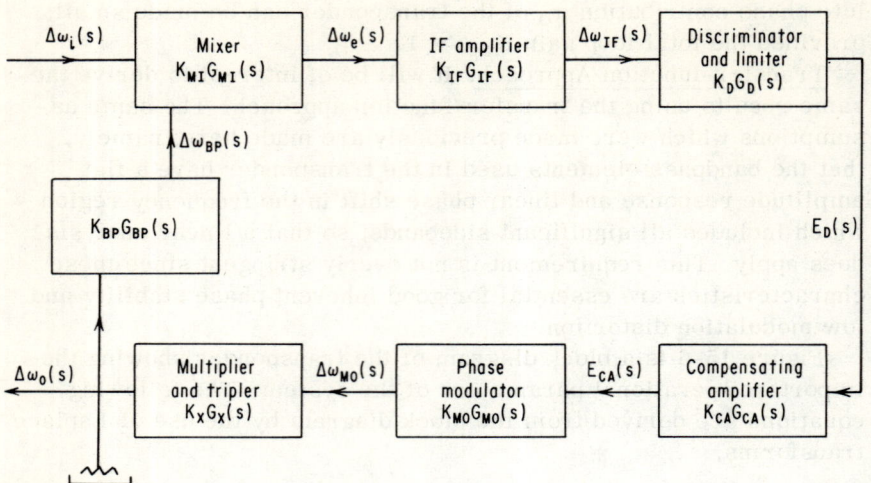

Fig. 15-6. Block diagram of transponder for transfer function analysis.

and similarly $\quad K_{FO}G_{FO}(j\omega) = K_\mu e^{j\alpha_\mu}$ $\qquad\qquad$ (15-45)

whereas before $K_{\mu\beta}$ = loop gain of system
$\qquad\qquad\alpha_\mu$ = phase contribution of forward path
$\qquad\qquad\alpha_\beta$ = phase contribution of feedback path

It is shown in Ref. 6 that the previous results which were derived using steady-state analysis can be derived from Eq. (15-41), which is equal to the transfer function of the transponder. Although the stability problem has not been discussed in our analysis, it should however be emphasized that in the design of the closed-loop system, proper attention should be given to the conventional stability criteria in order to achieve stable operation of the system.

References
1. Chaffee, J. A.: The Application of Negative Feedback to Frequency Modulating Systems, *BSTJ*, vol. 18, pp. 403-437, July, 1939.
2. Carson, J. R.: Frequency Modulation: Theory of the Feedback Receiving Circuit, *BSTJ*, July, 1939.
3. Panter, P. F., and W. Dite: Application of Negative Feedback to Frequency Modulating Systems, *Elec. Commun.*, vol. 26, pp. 173-178, June, 1949.

4. Bell, D. A.: Reduction of Bandwidth in F.M. Receivers, *Wireless Engr. London*, vol. 19, pp. 497-502, November, 1942.
5. Baghdady, E. J.: The Theory of FM Demodulation with Frequency-compressive Feedback, *IRE Trans. Commun. Systems*, vol. CS-10, September, 1962.
6. Panter, P. F., and R. C. Davis: Application of Frequency Following to DME Systems, 1st National Convention on Military Electronics, *IRE Natl. Conv. Record*, 1957.

16

THRESHOLD EXTENSION IN FREQUENCY-MODULATION RECEIVERS

It has been shown in Chap. 14 that conventional FM receivers are characterized by a threshold level, above which the output signal-to-noise ratio $(S/N)_O$ increases linearly with the received carrier level, and that below this level there is a very rapid deterioration of the $(S/N)_O$ as the received carrier level falls. This threshold level therefore determines the maximum operating range of the FM communication system. We have also noted that the threshold of a conventional FM receiver occurs when the carrier-to-noise ratio is approximately 12 db for wideband receivers. It follows therefore that any technique which will lower the threshold will enhance the system reliability, or for a given reliability specification, will either reduce the transmitter power requirements or extend the operating range. Since the threshold point depends on the $(C/N)_i$ ratio within the IF amplifier, reduction of noise at that point will lower the threshold. It thus becomes obvious that a reduction of the IF bandwidth will result in an improvement of threshold level. However, such a reduction cannot be realized without proper reduction in the frequency deviation at the transmitter; this follows from the fact that in a conventional FM system, the bandwidth required de-

478

pends on both the peak deviation and the highest modulating frequency, as shown in Chap. 7.

The concept of lowering the threshold, or as it is commonly called, threshold extension, is usually understood to mean an extension of the region relative to a conventional FM demodulator in which the output signal-to-noise ratio $(S/N)_o$ is linearly related to the input carrier-to-noise ratio $(C/N)_i$ without affecting the high $(S/N)_o$ performance.

Threshold extension techniques fall essentially into two major categories. One category follows the basic idea of frequency following or frequency feedback originated by Chaffee,[1] which was discussed in Chap. 15. The other major group of practical circuitry centers around what is known as a phase-locked detector, coherent detector, or phase oscillator. In this scheme, the instantaneous phase of the received FM signal is compared with the phase of the locally generated FM signal in a phase detector. The output of the phase detector is used to modulate the local oscillator and close the feedback loop. All threshold extension devices are essentially tracking filters, which can track only the slowly varying frequency of the modulated carrier and consequently respond to only a narrow band of noise centered about the instantaneous carrier frequency.

In this chapter, threshold extension techniques will be discussed, an analysis will be presented of the noise performance characteristics of the frequency-compressive feedback or frequency-lock demodulator and of a phase-locked loop FM demodulator, and a comparison will be made with the performance of a conventional FM demodulator. In making this comparison, the most important criteria for evaluation will be the input carrier-to-noise ratio $(C/N)_i$ required for a specified $(S/N)_o$. It is to be noted again that in all these systems, the threshold of improvement is extended without affecting the efficiency with which bandwidth is exchanged for signal-to-noise in conventional FM demodulators. However, this threshold reduction capability may be used in combination with an increase in the frequency deviation at the transmitter, to reflect some of the improvement directly into a higher-output signal-to-noise performance.

16.1 Threshold Extension Using Frequency Compression Feedback (FCF) [4-6,13]

As stated above, the frequency-feedback demodulator originated with Chaffee.[1] In the past few years, several papers have been published [2-6] to show that the application of frequency feedback could be used to decrease the threshold of signal-to-noise improvement in broadband systems. More recently, demodulators of this type were used in the projects Echo [7] and Telstar. [8-10]

The principle of frequency feedback or frequency following was discussed in Chap. 15. It is essentially a signal-tracking technique in which the frequency of a variable-frequency oscillator is made to follow the instantaneous frequency of the incoming wideband FM signal, resulting in a low modulation index or narrow-band signal. The IF bandwidth may then be reduced to twice the bandwidth of the message function without introducing distortion, with a consequent reduction of the noise power and hence of the noise threshold.

If the original receiver threshold is acceptable, it is possible by the use of frequency following to improve the $(S/N)_O$ without threshold deterioration, provided the IF bandwidth is unchanged and the frequency deviation at the transmitter is increased by a factor determined by the frequency-following loop gain. It has been shown by Chaffee [1] that feedback reduces both the signal level and the noise level at the output of the receiver, provided the disturbance is not too great. Thus when the modulation level is raised to offset the effect of feedback, an improvement in signal-to-noise ratio is realized without threshold deterioration.

It may be implied from the last statement that if the system were above the threshold prior to applying feedback, an unlimited $(S/N)_O$ ratio could theoretically be obtained by increasing the frequency deviation of the carrier and the amount of feedback simultaneously. However, following the discussion of Enloe, [4] it will be shown that there is an upper limit to the improvement in the $(S/N)_O$, due to the deterioration of the threshold as a function of the feedback action of the system.

In satellite communication systems, the output power of the satellite transmitter is in relatively short supply, and consequently the possibility of obtaining an overall system performance equivalent to an apparent increase in transmitter power is of great importance. This can be realized by applying feedback to the ground receiver and increasing the deviation at the satellite transmitter; thus the signal-to-noise ratio is improved, which is equivalent to an apparent increase in transmitter power.

A change in both the threshold point and $(S/N)_O$ may be realized by increasing the frequency deviation at the transmitter and reducing the IF bandwidth; the overall amount of both changes is determined by the gain of the receiver feedback loop. For instance, by applying 6 db feedback to the receiver, the frequency deviation at the transmitter may be doubled yielding a 6-db improvement in the $(S/N)_0$ ratio. Alternatively, the receiver IF bandwidth may be reduced to one-quarter of the original value, provided this is greater than two times the baseband, resulting in a threshold extension of 6 db. However, in order to realize both improvements, the doubled frequency deviation must be accommodated in one quarter of the original bandwidth; i.e., the incoming frequency deviation must be increased by a factor of 8, requiring 18 db receiver feedback.

In this example, the overall system performance has improved by 6 db; it means that while the output signal-to-noise ratio increased by 6 db, the threshold point decreased by 6 db.

A simple analysis of the frequency-feedback demodulator, considering a noiseless FM signal, has been given in Chap. 15. In the following, we shall present the theory of FCF demodulation of a FM signal-in-noise applicable to a demodulator operating above the threshold. No theory is at present available which completely describes the behavior of the FCF demodulator near or below the threshold. However, Enloe[4] has introduced a two-threshold concept which resulted in simple mathematical expressions for the calculation of threshold extension or threshold improvement. Following the analysis of Enloe[4] and others,[9-13] we shall show that Enloe's two-threshold concept can be used in formulating a generalized design procedure for designing FCF receivers for minimum power reception.

1. *Theory of FCF Demodulation of FM Signal-in-Noise, High Carrier-to-Noise Ratio.* As discussed in Chap. 14, when the carrier is much stronger than the noise, the system is linear and the mean square value of the output can be computed by ignoring the noise component. Similarly, the mean square value of the output noise is computed by ignoring the message modulation and considering the input to consist of an unmodulated carrier plus noise. We shall show now that the output $(S/N)_0$ ratio of a FCF demodulator is the same as that of a conventional FM demodulator receiving the same signal and noise-power density, provided the $(C/N)_i$ is sufficiently large.

A very interesting approach to the analysis of the operation of a frequency-feedback receiver on a FM signal can best be carried out by the use of a linear model of the FCF system, which is applicable to the case when the FCF loop gain is sufficient to reduce the peak phase deviation of the FM signal at the input of the loop bandpass filter to a value that is less than unity. Consider the block diagram of a frequency-feedback demodulator shown in Fig. 16-1; the principle of operation has already been discussed in Chap. 15. As shown

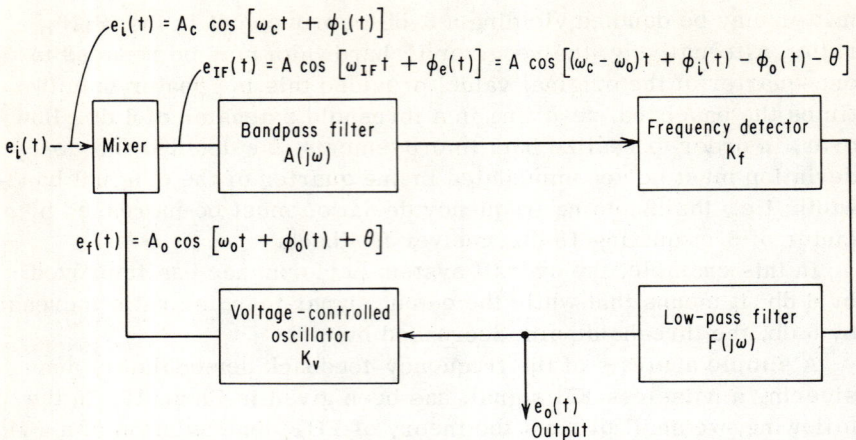

Fig. 16-1. Block diagram of a frequency-compressive feedback FM system.

in the figure, K_f and K_v are the gain constants of the frequency detector and the VCO, respectively. They relate radian frequency to voltage. Also, $A(j\omega)$ denotes the transfer function of the bandpass IF filter, and $F(j\omega)$ of the low-pass baseband filter in the feedback path. The linear model shown in Fig. 16-2 is the low-pass equivalent or the baseband analogue of the feedback demodulator. Mathematically, the receiver behaves as a linear negative-feedback system as far as the phase variations around the loop are concerned, as long as the modulation index of the modulated signal entering the bandpass IF filter is small compared to unity. The IF bandpass filter is represented by its low-pass equivalent $A_L(j\omega)$. The discriminator and VCO are considered ideal devices and therefore are

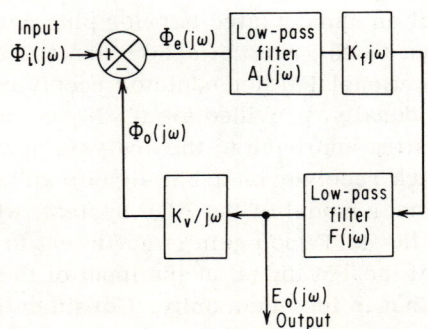

Fig. 16-2. Baseband analogue of FCF demodulator.

represented in the equivalent loop as a differentiator and integrator, respectively. The mixer is simply a phase subtractor. The low-pass filter $F(j\omega)$ is included to represent any low-pass filtering that is applied within the loop. The input to the equivalent system is the phase variation of the modulated signal plus gaussian noise. As stated above, the FCF demodulator behaves as a linear negative-feedback system as far as phase variations around the loop are concerned. The receiver closed-loop transfer function relates the phase of the VCO to that of the received signal. Let us define the closed-loop transfer function $H_c(j\omega)$ as the ratio $\Phi_0(j\omega)/\Phi_i(j\omega)$, i.e., the ratio of the phase variation of the VCO to the phase variation of the input signal to the mixer. From linear feedback theory, it follows that the closed-loop bandwidth is unavoidably larger than the open-loop bandwidth. As will be shown later on, this fact plays an important role in the determination of the threshold. The receiver closed-loop transfer function will be shown to play an important role in the optimum design procedure of FCF receivers using the two-threshold concept of Enloe. In order to gain insight into the mechanism of the threshold reduction property of the FCF system, we assume for a moment that the feedback loop in Fig. 16-1 is open and that the input to the mixer consists of an unmodulated carrier and a band of gaussian noise. The noise $n(t)$ can be separated into two components, one in phase with the carrier and the second in quadrature, so that

$$n(t) = n_c(t) \cos \omega_c t - n_s(t) \sin \omega_c t \qquad (16\text{-}1)$$

The input to the mixer is then given by

$$e_i(t) = \cos \omega_c t + n_c(t) \cos \omega_c t - n_s(t) \sin \omega_c t \qquad (16\text{-}2)$$

where $n_c(t)$ and $n_s(t)$ are random gaussian variables in phase quadrature with each other, normalized by the carrier amplitude. The input composite signal can be expressed in the well-known form of a hybrid wave, namely,

$$e_i(t) = \sqrt{[1 + n_c(t)]^2 + n_s^2(t)} \; \cos [\omega_c t + \phi_i(t)]$$
$$= V(t) \cos [\omega_c t + \phi_i(t)] \qquad (16\text{-}3)$$

where $\quad \tan \phi_i(t) = \dfrac{n_s(t)}{1 + n_c(t)} \qquad (16\text{-}4)$

Since in this discussion we consider the case of high carrier-to-noise ratio, the envelope $V(t)$ can therefore be approximated by

$$V(t) \doteq 1 + n_c(t) \qquad (16\text{-}5)$$

and the phase $\phi_i(t) \doteq n_s(t)$ $(16\text{-}6)$

These expressions show that the composite hybrid wave at the input to the discriminator consists of an envelope which is modulated by a small noise component which is in phase with the carrier and a phase angle modulated by the quadrature component of the noise.

Now let us apply feedback by closing the loop. The baseband noise modulates the VCO to produce a phase variation $\Phi_0(j\omega)$ which is assumed in this case to be small. Since the mixer acts as a phase subtractor, the phase modulation of the input to the IF is reduced by the phase variation of the VCO, which is equivalent to a reduction of the quadrature component $n_s(t)$ of the noise. Thus it is seen that for large carrier-to-noise ratio, the effect of the feedback loop is to reduce the phase deviation due to the quadrature noise in exactly the same manner as it operates on the phase deviation of the modulated carrier, without affecting the inphase noise. Signal and quadrature noise are reduced in the same proportion by feedback, with the result that the output $(S/N)_0$ ratio is independent of feedback and is equal to that of a conventional FM demodulator.

2. *Qualitative Discussion of Threshold.* We have just shown that the quadrature noise $n_s(t)$ reaching the discriminator is reduced by means of frequency feedback under conditions of high carrier-to-noise ratio. We arrived at this result by neglecting the inphase noise under open-loop conditions. However, as the carrier is reduced, the inphase component becomes significant and contributes to the phase variation of the composite wave, thus adding a significant amount of noise in addition to that derived from the quadrature noise. Now let us close the loop and examine the operation of the mixer as a "product modulator"; the mixer forms the product of the VCO wave with the receiver-input carrier and noise. Since the root-mean-square phase deviation of the VCO signal is small, its spectrum will consist primarily of a carrier and first-order sidebands which are in quadrature with the carrier. The product of the VCO carrier and incoming carrier results in the IF carrier, and the product of the VCO carrier and incoming noise components produces quadrature and inphase components in the IF. Finally, the product of the quadrature term (first-order sidebands) of the

VCO with the incoming carrier yields a second quadrature term in the IF but opposite in phase to the first quadrature term. Thus as a result of this cancellation, the quadrature noise in the IF, and consequently the noise in the baseband output, is reduced by feedback. In addition to the products enumerated so far, there are two more products which may be of significance under low carrier-to-noise condition, namely, the products of the quadrature term of the VCO signal with the incoming quadrature and inphase noise components. These will result in additional inphase and quadrature noise components in the IF, which will combine with the IF carrier to produce additional phase and envelope noise. The feedback threshold occurs when this additional phase noise becomes significant, as the root-mean-square phase deviation of the VCO wave caused by the detected noise is no longer small compared to unity. The theoretical results of Enloe,[4] which were confirmed experimentally, indicate that as the root-mean-square phase deviation of the VCO caused by the noise is greater than 1/3.11 rad., then the additional noise becomes noticeable and noise impulses appear in the baseband. The mean square phase $\overline{\phi_0^2(t)}$ of the wave generated by the VCO can be obtained by integrating the product of the input phase spectral power density and the square of the absolute value of the closed-loop transfer function as follows. The open-loop transfer function of the linear model of Fig. 16-2 is equal to

$$H_0(j\omega) = K_f K_V A_L(j\omega) F(j\omega) \tag{16-7}$$

and defining the closed-loop transfer function for the transmission from the signal input to the oscillator output

as $$H_c(j\omega) = \frac{\Phi_0(j\omega)}{\Phi_i(j\omega)} \tag{16-8}$$

we obtain

$$H_c(j\omega) = \frac{H_0(j\omega)}{1 + H_0(j\omega)} = \frac{K_V K_f A_L(j\omega) F(j\omega)}{1 + K_V K_f A_L(j\omega) F(j\omega)} \tag{16-9}$$

The feedback factor F, as usual, is related to the loop gain at the lowest frequencies, so that

$$F = 1 + H_0(0) = 1 + K_f K_V \tag{16-10}$$

and $$H_c(0) = \frac{F - 1}{F} \tag{16-11}$$

The two-sided closed-loop noise bandwidth B_C is obtained directly from the closed-loop transfer function, namely,

$$
B_C = \frac{\frac{1}{2\pi}\int_{-\infty}^{\infty}|H_C(j\omega)|^2 d\omega}{|H_C(0)|^2}
$$

$$
= \left(\frac{F}{F-1}\right)^2 \int_{-\infty}^{\infty}\left|\frac{K_V K_f A_L(j\omega)\,F(j\omega)}{1+K_V K_f A_L(j\omega)\,F(j\omega)}\right|^2 df \qquad (16\text{-}12)
$$

It follows from Eq. (16-9) that the mean square phase $\overline{\phi_0^2(t)}$ is given by

$$
\overline{\phi_0^2(t)} = N_0\int_{-\infty}^{\infty}|H_C(j\omega)|^2 df
$$

$$
= N_0\int_{-\infty}^{\infty}\left|\frac{K_V K_f A_L(j\omega)\,F(j\omega)}{1+K_V K_f A_L(j\omega)\,F(j\omega)}\right|^2 df = B_C N_0\left(\frac{F-1}{F}\right)^2
$$

$$
(16\text{-}13)
$$

where N_0 is the two-sided spectral power density of the quadrature noise $n_s(t)$. We recall from Chap. 4 that N_0 is also the one-sided spectral density of the IF noise $n(t)$, and that

$$
\overline{n^2(t)} = \overline{n_c^2(t)} = \overline{n_s^2(t)} = N_0 B_{IF} = \sigma^2 \qquad (16\text{-}14)
$$

The normalized input carrier-to-noise ratio to the mixer in a bandwidth equal to the closed-loop noise bandwidth can be expressed in terms of the mean-square phase of the VCO output signal $\overline{\phi_0^2(t)}$ and the feedback factor F as follows:

$$
\left(\frac{C}{N}\right)_i = \rho = \frac{1}{2B_C N_0} = \frac{1}{2\overline{\phi_0^2(t)}}\left(\frac{F-1}{F}\right)^2 \qquad (16\text{-}15)
$$

As stated above, Enloe has observed empirically that the threshold occurs when $\overline{\phi_0^2(t)} \geq (1/3.11)^2$ rad^2, and denoting under these conditions $\rho = \rho_T$, we obtain

$$
\rho_T = \frac{1}{2(1/3.11)^2}\left(\frac{F-1}{F}\right)^2 = 4.8\left(\frac{F-1}{F}\right)^2 \qquad (16\text{-}16)
$$

where $F = 1 + K_V K_f$, amount of feedback or frequency compression. For large amounts of feedback, this threshold is equal to 4.8 or 6.8 db.

Equation (16-16) plays a very important role in the understanding of the two-threshold concept. This equation tells us where the feedback threshold occurs in the FCF demodulator as a function of the feedback factor F and the input carrier-to-noise ratio in a bandwidth equal to the closed noise bandwidth of the system. Since in the development of this equation, we have assumed that the system was above threshold on an open-loop basis, it follows therefore that in order for a FCF demodulator to be above threshold, it must be above both the open-loop and the feedback threshold independently.

In an optimum-designed FCF demodulator, the open-loop threshold should coincide with the feedback or closed-loop threshold. This problem will be discussed more fully later on. It should be noted here, however, that for optimum performance of the FCF demodulator, full feedback should be maintained over the complete frequency range of the baseband, in order to fully compress the frequency deviation in the IF filter. Also, the closed-loop bandwidth should be as small as possible for a given amount of feedback, in order to minimize the feedback threshold. Finally, the bandwidth of the IF filter should be such that the open-loop and feedback thresholds occur simultaneously.

16.2 Transmitter Power Reduction Using FCF Demodulators [4,9-12]

In the last section, we established the existence of two separate thresholds in a FCF demodulator and discussed the factors which govern these thresholds. As pointed out by Enloe, the existence of a second threshold in a FCF demodulator tends to decrease substantially the actual power savings which might be expected from the use of frequency feedback. In this section, we shall formalize a design procedure which will maximize the power capability of the FCF system. In other words, the design procedure will result in a FM system using FCF receivers which will require a minimum amount of transmitted power to operate effectively, the prime constraints on the design criteria being the specification of a signal bandwidth and an acceptable output signal-to-noise power ratio $(S/N)_0$. The power capability of a FCF receiver will then be compared to that required in a conventional FM receiver yielding the same output $(S/N)_0$.

1. *FCF Receiver Design.*[11-12] As discussed previously, the key factor in the design of an optimum FCF receiver for minimum-power reception is the receiver closed-loop transfer function. The design procedure consists of adjusting the closed-loop transfer function to satisfy simultaneously the following conditions.

1. The two thresholds must be equal.

2. The common threshold power must be minimized.

3. The loop must satisfy the stability requirements.

For proper operation, the FCF receiver must also satisfy the following operating conditions:

1. The IF bandwidth should be wide enough to allow the passage of the compressed FM signal without appreciable distortion. However, to minimize open-loop threshold, it is necessary to use as narrow an IF as feasible. A reasonable compromise using Eq. (7-106) is a 3-db IF bandwidth given by

$$B_{IF} = 2(\Delta F + f_m) = 2 f_m (1 + \beta_F) \qquad (16\text{-}17)$$

where f_m is the highest modulating frequency and β_F is the effective reduced modulation index of the IF signal due to feedback, which has been shown in Chap. 15 to be equal to

$$\beta_F = \frac{\beta}{F} \qquad (16\text{-}18)$$

2. The carrier-to-noise input to the detector must exceed a threshold value called open-loop threshold. The exact value of the threshold is a function of the ratio between predetection and post-detection bandwidth. However, for simplicity, a constant value independent of the above ratio is often assumed. In the present design, an open-loop threshold ratio of 10 db will be assumed. The open-loop threshold can be found in terms of the closed-loop transfer function $H_c(j\omega)$ as follows. Assume a sinusoidally frequency-modulated carrier to be applied to the FCF receiver with a peak deviation of ΔF cps. From Fig. 16-2 we obtain

$$\Phi_e(j\omega) H_0(j\omega) = \Phi_i(j\omega) - \Phi_e(j\omega)$$

where $\Phi_e(j\omega) = \Phi_i(j\omega) - \Phi_0(j\omega)$

or $\Phi_e(j\omega) = \dfrac{\Phi_i(j\omega)}{1 + H_0(j\omega)}$

and using Eq. (16-9), this reduces to

$$\Phi_e(j\omega) = \Phi_i(j\omega)[1 - H_c(j\omega)] \qquad (16\text{-}19)$$

Hence, it follows that a frequency deviation of ΔF is reduced to

$$\Delta F_e = \Delta F \left| 1 - H_c(j\omega_m) \right| \text{cps} \qquad (16\text{-}20)$$

where $\omega_m = 2\pi f_m$ is the modulating frequency in radians per second. It follows therefore using Eq. (16-17) that the IF bandwidth B_{IF} is given by

$$B_{IF} = 2(\Delta F_e + f_m) = 2[\Delta F \left| 1 - H_c(j\omega_m) \right| + f_m] \qquad (16\text{-}21)$$

Assuming a single-tuned IF filter, the noise power in the IF may be approximated by $N_o \pi B_{IF}/2$, and consequently the minimum carrier power to satisfy the 10-db open-loop threshold is given by

$$C_{OL} = 10\pi N_o f_m [\beta \left| 1 - H_c(j\omega_m) \right| + 1] \qquad (16\text{-}22)$$

The mean square phase-noise variation of the VFO output must be small compared to unity in order to prevent interaction of input signal and feedback noise. For our purpose, a mean square value of 0.1 rad^2 is accepted as satisfactory. We can derive the minimum carrier power C_{CL} required to satisfy the closed-loop or feedback threshold in terms of the closed-loop transfer function $H_c(j\omega)$ from Eq. (16-13), namely,

$$\overline{\phi_o^2(t)} = \frac{N_o}{2C_{CL}} \int_o^\infty |H_c(j\omega)|^2 \, df \qquad (16\text{-}23)$$

Hence, $\qquad C_{CL} = 10 N_o \int_o^\infty |H_c(j\omega)|^2 \, df \qquad (16\text{-}24)$

on the assumption that $\phi_o^2(t)$ is limited to 0.1 rad^2.

Thus, we have related the two thresholds to the receiver closed-loop transfer function $H_c(j\omega)$. As pointed out above, an optimum receiver design for minimum power reception requires that the two thresholds be equal. This is accomplished by adjusting the parameters of $H_c(j\omega)$, namely the loop gain and filter time constants. Additional design requirements are to minimize the common threshold and to satisfy the loop stability and transient response requirements.

These two-threshold equations were first derived by Heitzman [12] using Enloe's two-threshold concept. Using the same concept, Gagliardi [11] subsequently derived slightly different results. In his design he assumes a 12-db threshold ratio, which results in

$$C_{OL} = 16 N_o B_{IF} \qquad (16\text{-}25)$$

where as before N_O is the power spectral density of B_{IF} and C_{OL} is the minimum carrier power to satisfy the 12-db open-loop threshold. Assuming Enloe's minimum mean square noise variation of $\overline{\phi_O^2(t)}$ = 0.1 rad^2, he then derives the minimum carrier power C_{CL} required to satisfy the closed-loop or feedback threshold in terms of the closed-loop noise bandwidth B_C as follows. From Eq. (16-15) we have

$$\rho = \frac{C}{N_O B_C} = \frac{1}{2\overline{\phi_O^2(t)}} \left(\frac{F-1}{F} \right)^2 \tag{16-26}$$

from which we derive for $\overline{\phi_O^2(t)}$ = 0.1 rad^2 and $F \gg 1$, the relation

$$C_{CL} = 5\,N_O B_C \tag{16-27}$$

The requirements of Eqs. (16-25) and (16-27) dictate that the received carrier power must exceed both C_{OL} and C_{CL} if the system is to operate above threshold. Since these threshold carrier powers are functions of B_{IF} and B_C respectively, they are practically independent and can be separately controlled. For minimum power-reception requirements, these thresholds should be made to coincide.

2. *Application of Threshold Equations to Specific Configurations.* We shall illustrate the application of the threshold equations to the design of a FCF receiver with an equivalent linearized loop consisting of specific low-pass filters.

With reference to Fig. 16-3a, we consider first the case of no filter in the detected output, i.e., $F(j\omega) = 1$. The low-pass equivalent of the IF filter is given by $A_L(s) = \dfrac{1}{1 + s/\pi B_{IF}}$ and the open-loop transfer function of the linear model is therefore given by

$$H_O(s) = \frac{K}{1 + s/\pi B_{IF}} \tag{16-28}$$

where $K = K_f K_v$ in Fig. 16-2. The closed loop transfer function $H_C(s)$ is

$$H_C(s) = \frac{H_O(s)}{1 + H_O(s)} = \frac{K\pi B_{IF}}{(K+1)\pi B_{IF} + s} \tag{16-29}$$

Using Eq. (16-24) we obtain

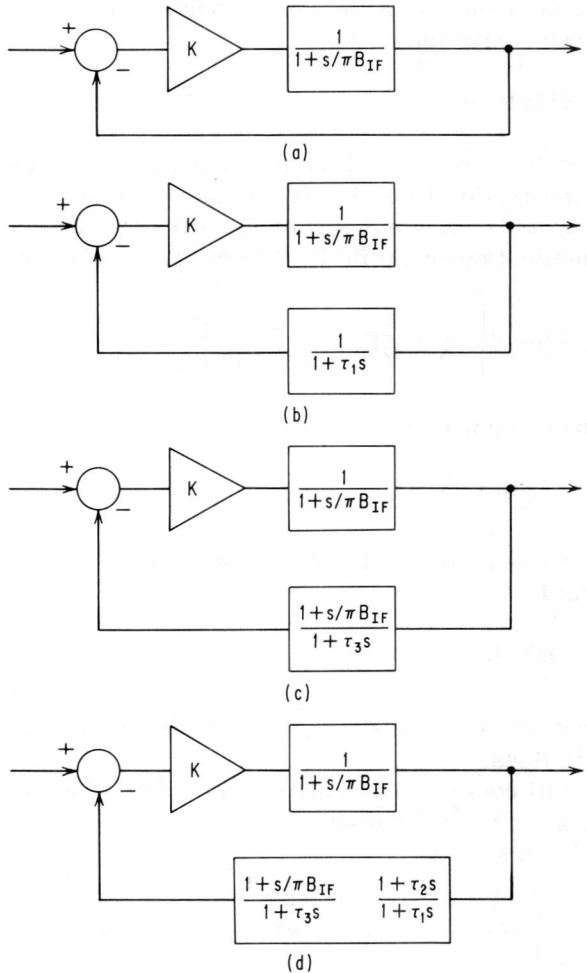

Fig. 16-3. Types of filters considered in the FCF loop: (a) No filter; (b) single RC section; (c) pole-compensating lag network; (d) tandem lag network. (From R. E. Heitzman,[12] IRE Trans.)

$$C_{CL} = \frac{10N_O}{2\pi} \int_O^\infty \frac{(K\pi B_{IF})^2 \, d\omega}{[(K + 1)\pi B_{IF}]^2 + \omega^2} = \frac{5\pi K^2 N_O B_{IF}}{2(K + 1)} \qquad (16-30)$$

For a 10-db open-loop threshold, the minimum carrier power is

$$C_{OL} = 10 \, N_O \pi \, (B_{IF}/2) = 5\pi \, N_O B_{IF} \qquad (16-31)$$

By adjusting the loop gain K the two thresholds can be equalized, from which we derive the result

K = 2.8 or 9 db

This optimum d-c loop gain which is independent of modulating frequency and modulation index is used to determine the threshold carrier power by use of Eqs. (16-22) or (16-24). This is minimized by using the smallest permissible B_{IF} which is specified by Eq. (16-21).

$$B_{IF} = 2f_m\left\{\beta\left[\frac{\pi^2 B_{IF}^2 + \omega_m^2}{(K+1)^2\pi^2 B_{IF}^2 + \omega_m^2}\right]^{\frac{1}{2}} + 1\right\} \tag{16-32}$$

which is closely approximated by

$$B_{IF} \doteq 2f_m(\beta/3.8 + 1) \tag{16-33}$$

Hence the minimum required carrier power to satisfy both threshold requirements is

$$C_M = 10\pi N_0 f_m[\beta/3.8 + 1] \tag{16-34}$$

We shall presently develop similar threshold formulas for other loop configurations.

Now we shall consider baseband filtering following the detector output using a simple RC low-pass filter, which is illustrated in Fig. 16-3b. In this model

$$A_L(s) = \frac{1}{1 + \tau_2 s}, \qquad \tau_2 = 1/\pi B_{IF} \tag{16-35}$$

$$F(s) = \frac{1}{1 + \tau_1 s}$$

Hence, $$H_0(s) = \frac{K}{(1 + \tau_1 s)(1 + \tau_2 s)} \tag{16-36}$$

and $$H_c(s) = \frac{K}{(K+1) + (\tau_1 + \tau_2)s + \tau_1\tau_2 s^2} \tag{16-37}$$

After some manipulation it is shown in Ref. 12 that the minimum carrier power to satisfy both threshold requirements is given by

$$C_M = 10\pi N_o f_m \left(\frac{\beta + 1 + \sqrt{2(\beta + 1)}}{1 + \sqrt{2(\beta + 1)}} \right) \qquad (16\text{-}38)$$

For the other two configurations illustrated in Figs. 16-3c and d, it is shown in Ref. 12 that the minimum carrier power for the configuration using a lag network (Fig. 16-3c) is given by

$$C_M = 10\pi N_o f_m \left(\frac{\beta + 1 + \sqrt{2(\beta + 1)}}{1 + \sqrt{2(\beta + 1)}} \right) \qquad (16\text{-}39)$$

which is identical with the threshold equation for the simple RC filter. Also, for the double-lag network (Fig. 16-3d) it is shown that

$$C_M = 10\pi N_o f_m (\beta^{1/3} + 1) \qquad (16\text{-}40)$$

The implication of the threshold expressions which were derived for the four FCF loop configurations can best be appreciated by reference to Fig. 16-4, where the output $(S/N)_o$ is plotted versus normalized carrier power $C/N_o f_m$ for various values of β. These lines are simply plots using the well-known relation

$$(S/N)_o = 3\beta^2 \left(\frac{C}{N} \right)_i = \frac{3}{2} \beta^2 \frac{C}{N_o f_m}$$

where $N_i = 2N_o f_m$.

When operating above threshold there is a linear relationship between $(S/N)_o$ and $(C/N)_i$. However, as the input carrier-to-noise ratio is decreased the operation becomes nonlinear as expected. The relative capabilities of FCF with different kinds of filtering are illustrated by the dashed lines connecting the threshold points which were determined by the use of the threshold expressions of Eqs. (16-34), and (16-38) to (16-40). A comparison of FCF demodulation with the other two demodulation methods, namely, the conventional discriminator and the phase-locked demodulator (which will be discussed in the next section), is made in Fig. 16-5. The advantage of using FCF demodulation over conventional FM is clearly indicated. For an interesting discussion on the dependence of the open loop threshold on the modulation index β see Refs. 4, 5, and 12.

Applications of the threshold equations to specific transfer functions $A_L(j\omega)$ and $F(j\omega)$ are also given by Gagliardi [11] and in many of the indicated references. From the published curves, one may conclude that the resultant power savings of optimized FCF receivers

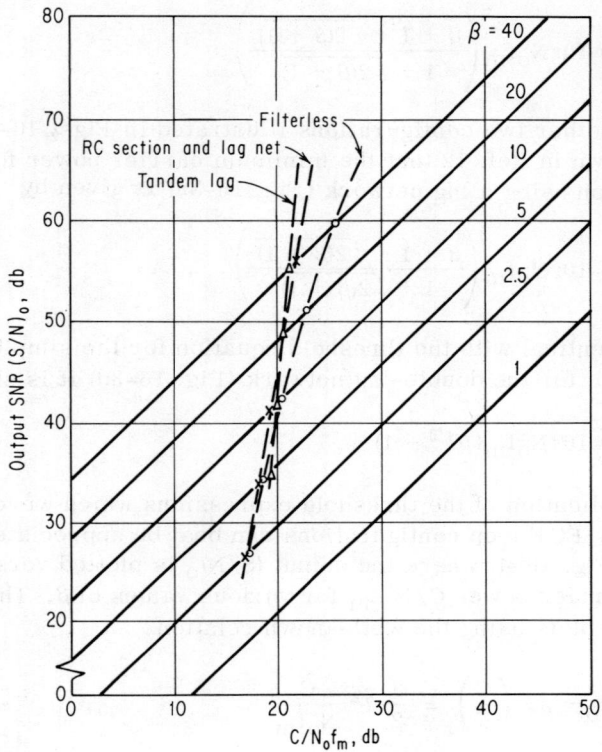

Fig. 16-4. FCF thresholds for different loop filters. (From R. E. Heitzman,[12] IRE Trans.)

over conventional receivers are approximately 6 db, relatively independent of the $(S/N)_0$ over most of the indicated range.

3. *Experimental Verification.* As mentioned above, a FCF demodulator was used in the Project Echo satellite communication experiment.[7] That experiment provided the first well-documented experimental evidence that the threshold of a FM receiver can indeed be improved by the use of negative feedback. While Project Echo is essentially a narrow-band system, Project Telstar[8,9] provided experimental evidence of threshold improvement in the case of a wideband FM system. Prior to Project Telstar, the design and performance of a broadband FCF demodulator were given by Ruthroff and Bodtman.[10] This type of demodulator was intended for use in intercontinental satellite communication systems suitable for either television or telephone service. In order to demonstrate conclusively that this demodulator has an improved threshold behavior for tele-

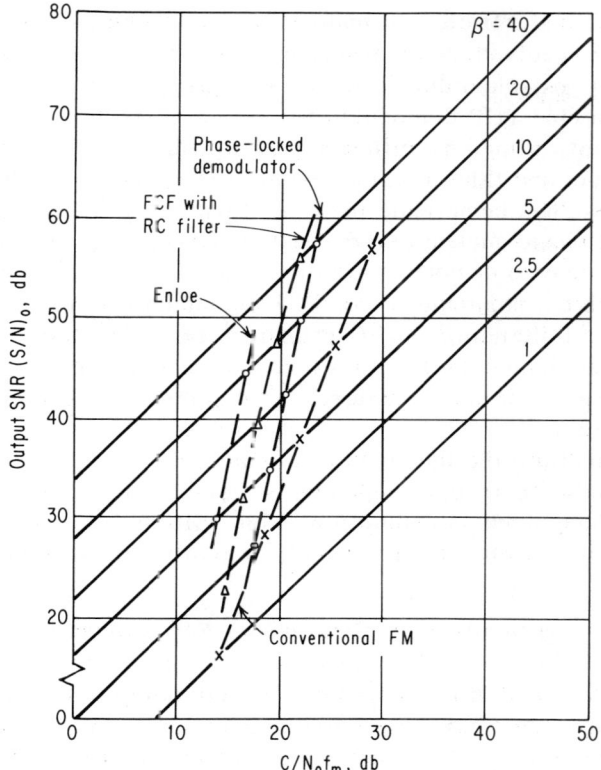

Fig. 16-5. Threshold comparison using three detection methods. (From R. E. Heitzman,[12] IRE Trans.)

vision relative to that of a conventional FM receiver, a comparison circuit was used which compared the performance of conventional FM with that of FCF. The threshold improvement which was realized in these experiments was about 5 to 6 db.

16.3 Threshold Extension Using Phase-locked Loop Demodulators[14-24]

The phase-locked loop demodulator provides a very attractive means of detecting a small signal in noise. This type of circuit is used increasingly by communication engineers in space-vehicle-to-earth data links, and whenever else the loss along the transmission path is very large or transmitter weight is at a premium. The phase-locked loop may even become the standard detector in the space vehicle of the future.

Phase-locked FM discriminators and other phase-locked techniques have received considerably more attention in the literature than frequency-locked discriminators, a term which is sometimes used to describe FCF demodulators. An analysis of phase-lock discriminators aimed specifically at the problem of demodulation of FM signals and threshold extension in the presence of poor carrier-to-noise has been made in several papers.[14-18] The threshold behavior of phase-locked discriminators designed for optimum reception has been discussed by Spilker[19] and Develet,[20,21] and a comparison with the performance of FCF demodulators has been made by Spilker,[19] Heitzman,[12] and Gagliardi.[11] Descriptions of several phase-locked demodulators for wideband frequency modulation have lately been given in the literature,[22,23] together with some experimental results.

In this section, the theory of phase-locked loop demodulators will be presented with special emphasis on establishing threshold criteria for phase-lock demodulation. This will be followed by a discussion of some experimental results which verify the theoretical analysis.

1. *Theory of Operation of Phase-lock FM Demodulator in the Absence of Noise.*[19-21,24] The phase-locked demodulator is essentially a coherent demodulator where the message function is recovered by operating upon the noise-corrupted receiver carrier with a replica of the carrier signal. The phase-locked demodulator uses a phase detector instead of a discriminator as the error-sensitive device for tracking the incoming frequency-modulated signal. As contrasted with FCF demodulation, the phase-locked loop does not cause a reduction in frequency deviation at any point in the receiver.

The operation of the phase-locked demodulator in case of high-input carrier-to-noise ratios is relatively simple to describe. As seen from the block diagram of Fig. 16-6, the system consists of a phase-detector or multiplier, a low-pass filter and a voltage-controlled oscillator (VCO). The phase detector or multiplier compares the instantaneous phase of the incoming frequency-modulated carrier with the phase of the locally generated replica of the modulated signal obtained from the voltage-controlled oscillator. The output of the phase detector is an error voltage proportional to the phase difference of the two signals, which, after passing through the low-pass filter, is fed back as a control voltage to the input of the VCO. In this manner, the VCO is locked in phase with the incoming signal. Since the instantaneous frequency of the VCO must be iden-

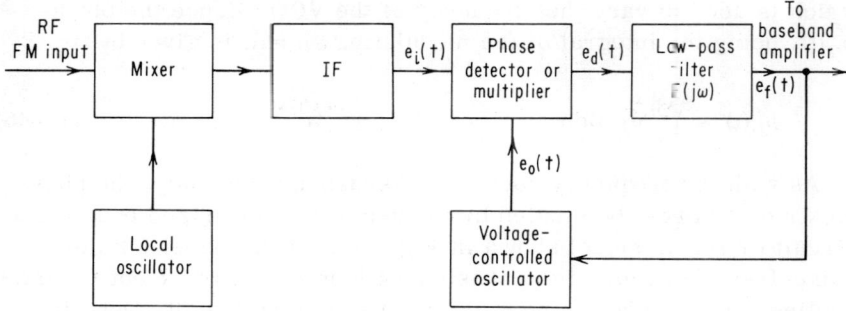

Fig. 16-6. Block diagram of a phase-locked FM demodulator.

tical with that of the incoming signal, the control voltage of the VCO represents the demodulated output signal.

It is useful to begin the analysis of the phase-lock loop by considering first the simple case of FM detection in the absence of noise. The received sinusoidal signal which is phase- or frequency-modulated, in the absence of noise, can be described by

$$e_i(t) = A_i \sin [\omega_0 t + \phi_i(t)] \qquad (16\text{-}41)$$

where $\qquad \phi_i(t) = \int_0^t s(u) du$

is the phase modulation produced by the message function $s(t)$. The output of the VCO can be expressed as

$$e_0(t) = A_0 \cos [\omega_0 t + \phi_0(t)] \qquad (16\text{-}42)$$

where $\phi_0(t)$ is the instantaneous phase of the locally generated replica signal. Referring to Fig. 16-6, the output of the multiplier is then

$$e_d(t) = e_i(t) e_0(t) = A_i A_0 \sin [\omega_0 t + \phi_i(t)] \cos [\omega_0 t + \phi_0(t)]$$

$$= \frac{A_i A_0}{2} \{\sin [\phi_i(t) - \phi_0(t)] + \sin [2\omega_0 t + \phi_i(t) + \phi_0(t)]\}$$

$$(16\text{-}43)$$

The higher frequency term is filtered out by the low-pass filter of the loop and the filtered output signal $e_f(t)$ is then

$$e_f(t) = \frac{A_i A_0}{2} \sin [\phi_i(t) - \phi_0(t)] \qquad (16\text{-}44)$$

which is used to vary the frequency of the VCO. Hence the phase $\phi_0(t)$, being the integral of the modulating signal, is given by

$$\phi_0(t) = \int_0^t e_f(u)du \qquad (16\text{-}45)$$

As with the frequency-lock loop discussed previously, the phase-lock loop can best be studied by the use of the linearized equivalent circuit or linear model shown in Fig. 16-7. This model for the noise-free operation of the phase-lock loop is linear except for the nonlinear sinusoidal element after the summing circuit. Denoting the phase error by $\phi_e(t) = \phi_i(t) - \phi_0(t)$, we can write as a first approximation $\sin \phi_e(t) \doteq \phi_e(t)$ for $\phi_e(t) \ll 1$ rad. Thus, in the region of the phase error $|\phi_e(t)| < 1$, the detector output can be approximated by

$$e_f(t) \doteq \frac{A_i A_o}{2} \phi_e(t) \qquad (16\text{-}46)$$

and the loop remains near the phase-lock condition, i.e., the phase of the VCO is closely tracking the phase of the incoming signal, and $\phi_0(t) \doteq \phi_i(t)$. It should be emphasized, however, that for the study of the pull-in and tracking capabilities of the loop when the initial phase error is large, the approximation is not valid and the non-linearity is essential.

2. *Performance of Phase-lock Demodulator in the Presence of Additive Noise.*[19,24] The linear model of Fig. 16-7 has been shown to be valid only in the absence of noise. In order to justify the analysis of the loop operation by means of the linear model when the input is noisy, we shall assume that for the linearized system the super-position principle holds. This assumption will be approximately

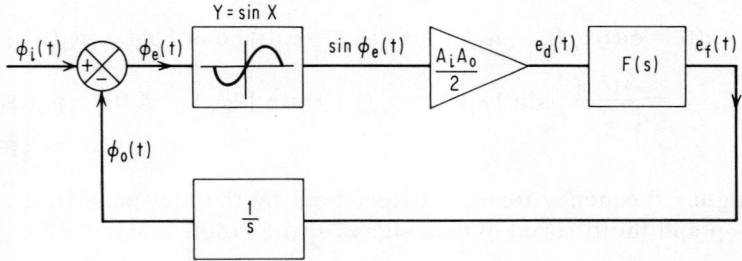

Fig. 16-7. Linearized model of phase-lock loop in the absence of noise.

valid as long as the total phase error, namely the error in the absence of the noise plus the phase jitter of the VCO output due to noise, will remain small compared with 1 rad most of the time.

Let us consider again the operation of the phase-lock loop but in the presence of noise without signal modulation. The input signal is

$$e_i(t) = A_i \sin \omega_0 t + n(t) \qquad (16\text{-}47)$$

where $n(t)$ represents additive band-limited noise.

$$n(t) = n_c(t) \cos \omega_0 t - n_s(t) \sin \omega_0 t \qquad (16\text{-}48)$$

The VCO output is

$$e_0(t) = A_0 \cos [\omega_0 t + \phi_n(t)] \doteq A_0 [\cos \omega_0 t - \phi_n(t) \sin \omega_0 t] \qquad (16\text{-}49)$$

where the phase jitter $\phi_n(t)$ is assumed small compared with 1 rad. The phase error signal is the product of these two signals

$$e_d(t) = e_i(t)e_0(t) = \frac{A_i A_0}{2} [\sin 2\omega_0 t - \phi_n(t)(1 - \cos 2\omega_0 t)]$$

$$+ A_0 n(t) [\cos \omega_0 t - \phi_n(t) \sin \omega_0 t] \qquad (16\text{-}50)$$

Since the loop filter will not pass the double frequency terms, Eq. (16-50) can be simplified in the form

$$e_d(t) = \frac{-A_i A_0}{2} \phi_n(t) + A_0 n(t) [\cos \omega_0 t - \phi_n(t) \sin \omega_0 t] \qquad (16\text{-}51)$$

on the assumption that $\phi_n(t) \sin \omega_0 t$ is small compared with $\cos \omega_0 t$; $e_d(t)$ can be further approximated by

$$e_d(t) \doteq \frac{A_i A_0}{2} \left[\frac{2}{A_i} n(t) \cos \omega_0 t - \phi_n(t) \right] \qquad (16\text{-}52)$$

The second term within the square bracket is simply the low-frequency VCO phase jitter due to noise or phase noise. In the first term, multiplication of white noise $n(t)$ by $(2/A_i) \cos \omega_0 t$ produces a signal whose autocorrelation function is a product of the autocorrelation functions of the two signals. This means that the

first term represents white noise of spectral density $N_O/(A_i^2/2)$.
Hence, we conclude from Eq. (16-52) that the total phase error due
to noise in the absence of signal is equal to $A_iA_O/2$ times the difference between white noise of spectral density $N_O/(A_i^2/2)$ and the VCO
phase jitter $\phi_n(t)$ due to the noise.

On the assumption that the superposition principle holds, the
transmission of the signal phase modulation and that of random
noise through the network can be considered to be independent and
consequently the linear model of the loop performance in the vicinity
of phase lock can be represented as shown in Fig. 16-8. In this
model $\Phi_i(s)$ is the Laplace transform of the input phase modulation
produced by the message function and $\Phi = N_O/(A_i^2/2)$ denotes the
noise spectral density normalized by the input carrier power $A_i^2/2$.
The Laplace transform of the total phase of the VCO is equal to
$\Phi_0(s) + \Phi_n(s)$.

The total phase error, which is the sum of the phase error due to
modulation and that due to noise, is given by $[\phi_i(t) - \phi_0(t) - \phi_n(t)]$.
The variance or mean square phase error is therefore given by

$$\overline{e_d^2(t)} = \lim_{T \to \infty} \frac{1}{2T} \int_{-T}^{T} [\phi_i(t) - \phi_0(t) - \phi_n(t)]^2 \, dt$$

$$= \lim_{T \to \infty} \frac{1}{2T} \int_{-T}^{T} \{[\phi_i(t) - \phi_0(t)]^2 + \phi_n^2(t)\} \, dt \qquad (16\text{-}53)$$

the cross term of $[\phi_i(t) - \phi_0(t)]$ and $\phi_n(t)$ being zero because the
message and noise functions are independent processes. Using
Parseval's theorem we can express the total mean square phase
error $\overline{e_d^2(t)}$ in terms of the transform

$$\overline{e_d^2(t)} = \frac{1}{2\pi j} \int_{-j\infty}^{j\infty} |\Phi_i(s) - \Phi_0(s)|^2 \, ds + \frac{1}{2\pi j} \int_{-j\infty}^{j\infty} |\Phi_n(s)|^2 \, ds$$

$$(16\text{-}54)$$

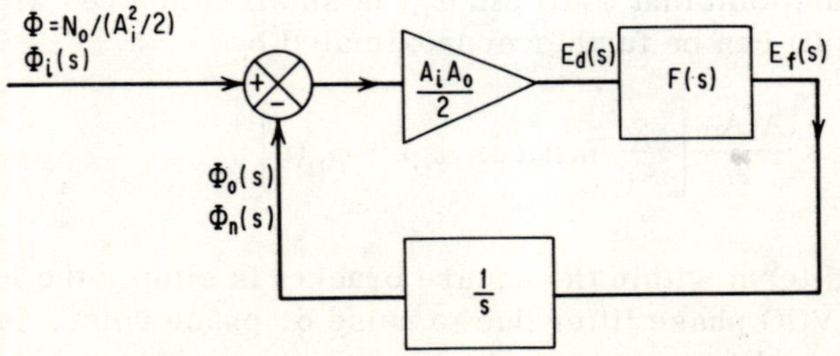

Fig. 16-8. Linearized model of phase-lock loop in the presence of
noise.

Thus, the total mean square phase error equals

$$\sigma_t^2 = \sigma_m^2 + \sigma_n^2 \tag{16-55}$$

where σ_m^2 denotes the mean square error or transient error due to modulation which is given by the first term, and σ_n^2 is the phase noise power or mean square noise interference which is given by the second term.

As in the previous analysis of FCF detection, we denote the closed-loop transfer function by $H_c(s) = \Phi_0(s)/\Phi_i(s)$; the total mean square phase error can then be expressed in terms of $H_c(s)$.

$$\sigma_t^2 = \frac{1}{2\pi j} \int_{-j\infty}^{j\infty} |\Phi_i(s)|^2 |1 - H_c(s)|^2 \, ds + \frac{1}{2\pi j} \int_{-j\infty}^{j\infty} \frac{N_0}{A_i^2} |H_c(s)|^2 \, ds \tag{16-56}$$

This result is based on the assumption that the linear analysis is a good approximation and consequently the principle of superposition can be applied.

3. *Threshold Effect.* [12] The linear approximation to the phase correction term of the actual phase-lock loop is limited to the range of phase error $|\phi_e(t)| < 1$ rad. If $\phi_e(t)$ should exceed 1 rad, the linear behavior no longer holds, the phase estimate becomes highly distorted and the loop can be unstable. In this analysis, we shall consider the phase-locked demodulator to reach its threshold when the $(C/N)_i$ decreases to the level where the probability of having $|\phi_e(t)| > 1$ rad exceeds a prescribed small probability. More specifically, the threshold of a phase-locked demodulator is defined to occur when its total rms phase error $\sigma_t = 0.5$ rad. Thus, its threshold definition may be written as

$$\sigma_t^2 = \sigma_m^2 + \sigma_n^2 = 0.25$$

where σ_m and σ_n are the rms values of the modulation-induced and noise-induced components, respectively.

To evaluate the threshold we consider a sinusoidal phase modulation of the form $\phi_i(t) = \dfrac{\Delta F}{f_a} \cos \omega_a t = \beta \cos \omega_a t$ where $f_a \le f_m$ is the maximum frequency of the modulating signal and ΔF is the peak frequency deviation. The mean square value of the modulation-induced phase error is therefore

$$\sigma_m^2 = \frac{\beta^2}{2} |1 - H_C(j\omega_a)|^2 \, , \quad \omega_a \leq \omega_m \qquad (16\text{-}57)$$

The phase noise error has a mean square value

$$\sigma_n^2 = \int_{-\infty}^{\infty} S_n(f) |H_C(j\omega_a)|^2 \, df \, , \quad S_n(f) = \frac{N_0}{2C} \qquad (16\text{-}58)$$

where $S(f)$ is the power spectral density of the input noise. Thus, the total rms phase error is

$$\sigma_t = (\sigma_m^2 + \sigma_n^2)^{1/2} \qquad (16\text{-}59)$$

In the most frequently encountered phase-locked demodulator, $F(s)$ takes the form

$$F(s) = \frac{1 + \tau_2 s}{1 + \tau_1 s} \qquad (16\text{-}60)$$

so that $\quad H_0(s) = \dfrac{KF(s)}{s + KF(s)}$

and $\quad H_C(s) = \dfrac{H_0(s)}{1 + H_0(s)} = \dfrac{K/\tau_1 + K(\tau_2/\tau_1)s}{K/\tau_1 + (1/\tau_1 + K\tau_2/\tau_1)s + s^2} \qquad (16\text{-}61)$

In practice $K\tau_2 \gg 1$, and $H_C(s)$ can be approximated by

$$H_C(s) = \frac{K/\tau_1 + K(\tau_2/\tau_1)s}{K/\tau_1 + K(\tau_2/\tau_1)s + s^2} = \frac{B_0 + 2\xi B_0 s}{B_0^2 + 2\xi B_0 s + s^2} \qquad (16\text{-}62)$$

Let ξ be arbitrarily set to $\xi = \sqrt{2}/2$; we have from Eq. (16-57)

$$\sigma_m^2 = \frac{\beta^2}{2} \left(\frac{\omega_m^2}{B_0^4 + \omega_m^4} \right) \doteq \frac{\beta^2 \omega_m^4}{2B_0^4} \qquad (16\text{-}63)$$

where $B_0 \gg \omega_m$. Similarly from Eq. (16-58) we derive

$$\sigma_n^2 = \frac{3N_0 B_0}{4\sqrt{2}\,C} \qquad (16\text{-}64)$$

The total mean square error σ_t^2 is

$$\sigma_t^2 = \frac{3N_0 B_0}{4\sqrt{2}\,C} + \frac{\beta^2 \omega_m^4}{2B_0^4} \qquad (16\text{-}65)$$

which can be minimized by setting $\dfrac{\partial \sigma_t^2}{\partial B_0} = 0$. Hence

$$B_0(\min) = \left[\frac{8\sqrt{2}\beta^2 C \omega_m^4}{3N_0}\right]^{1/5} \tag{16-63}$$

and $\quad \sigma_{t(\min)}^2 = 5\left[\dfrac{\beta^2 N_0^4 f_m^4}{2C^4}\right]^{1/5}$ $\qquad\qquad\qquad$ (16-67)

To find the minimum threshold carrier power C_M we set $\sigma_{t(\min)}^2 = 0.25$ so that

$$C_M = 35.5\beta^{1/2} N_0 f_m \tag{16-68}$$

This is plotted in Fig. 16-5 as the phase-locked threshold line.

In this figure a threshold comparison is made among the three demodulation methods, namely: FCF, phase-lock, and conventional demodulation. A similar comparison is given by Spilker in Ref. 19.

References

1. Chaffee, J. A.: "The Application of Negative Feedback to Frequency Modulating Systems," *BSTJ*, vol. 18, pp. 403-437, July, 1939.
2. Felix, M. O., and A. J. Buxton: The Performance of FM Scatter Systems Using Frequency Compression, *Proc. Natl. Electron. Conf.*, vol. 14, pp. 1029-1043, 1958.
3. Morito, M., and S. Ito: High Sensitivity Receiving System for Frequency Modulated Waves, *IRE Intern. Conv. Record*, 1960, part 5, pp. 228-237.
4. Enloe, L. H.: Decreasing the Threshold in FM by Frequency Feedback, *Proc. IRE*, vol. 50, pp. 18-30, January, 1962.
5. Enloe, L. H.: The Synthesis of Frequency Feedback Demodulators, *Proc. Natl. Electron. Conf.*, 1962, pp. 477-497.
6. Baghdady, E. J.: FM Demodulation with Frequency-compressive Feedback, *Proc. Natl. Telemetry Conf.*, vol. 2 (6-3), 1962; *IRE Trans. Commun. Systems*, September, 1962.
7. Ruthroff, C. L.: Project Echo: FM Demodulators with Negative Feedback, *BSTJ*, vol. 40, pp. 1149-1157, July, 1961.
8. Schill, J., and A. F. Perks: The Ground Station Transmitter and Receiver, *Bell Lab. Record*, special Telstar issue, April, 1963.
9. Giger, A. J., and J. A. Chaffee: The FM Demodulator with Negative Feedback, *BSTJ*, July, 1963.

10. Ruthroff, C. L., and W. F. Bodtmann: Design and Performance of a Broad-band FM Demodulator with Frequency Compression, *Proc. IRE*, December, 1962.

11. Gagliardi, R. M.: Transmitter Power Reduction with Frequency Tracking FM Receivers, *IEEE Trans. Space Electron. Telemetry*, March, 1963.

12. Heitzman, R. E.: A Study of the Threshold Power Requirements of FMFB Receivers, *IRE Trans. Space Electron. Telemetry*, December, 1962.

13. Wojnar, A.: An Analysis and Synthesis Procedure for Feedback FM Systems, *MIT Rept.* 415, Sept. 30, 1963.

14. Jaffe, R., and E. Rechtin: Design and Performance of Phase-locked Circuits Capable of Near Optimum Performance Over a Wide Range of Input Signal-to-Noise Ratios, *IRE Trans. Inform. Theory*, vol. IT-1, March, 1955.

15. Gilchriest, C. E.: Application of Phase-locked Loop to Telemetry as a Discriminator or Tracking Filter, *IRE Trans. Telemetry Remote Control*, vol. TRC-4, June, 1958.

16. Preston, G. W.: Basic Theory of Locked-oscillators in Tracking FM Signals, *IRE Trans. Space Electron. Telemetry*, vol. SET-5, March, 1959.

17. Margolis, S. G.: The Response of a Phase-locked Loop to a Sinusoid Plus Noise, *IRE Trans. Inform. Theory*, June, 1957.

18. Weaver, C. S.: A New Approach to the Linear Design and Analysis of Phase-locked Loops, *IRE Trans. Space Electron. Telemetry*, vol. SET-5, December, 1959.

19. Spilker, J. J., Jr.: Threshold Comparison of Phase-lock, Frequency-lock, and Maximum-likelihood Types of FM discriminators, *IRE WESCON Conv. Record*, 1961.

20. Develet, J. A., Jr.: An Analytic Approximation of Phase-lock Receiver Threshold, *IEEE Trans. Space Electron. Telemetry*, March, 1963.

21. Develet, J. A., Jr.: A Threshold Criterion for Phase-lock Demodulation, *Proc. IEEE*, February, 1963.

22. Ford, D. W.: A Phase-locked Detector for FM Multiplex Applications, *8th Natl. Commun. Symp. Record*, Utica, N. Y., October, 1962.

23. Booton, R. C., Jr.: Demodulation of Wideband Frequency Modulation Utilizing Phase-lock Technique, *Proc. Natl. Telemetering Conf.*, 1962.

24. Viterbi, J. V.: Phase-lock-loop Systems, Space Communications, Ch. 8, Ed. by A. V. Balakrishnan, McGraw-Hill Book Company, New York.

17

THE SAMPLING PRINCIPLE AND
INTRODUCTION TO PULSE MODULATION

The modulation systems which we have thus far discussed, namely, linear and exponential, can be classified as "CW" or "continuous-wave" communication. In this chapter, we shall discuss another method of transmitting intelligence in which, contrasted to CW communication, the carrier is transmitted in short bursts or pulses. In these systems, which can be classified as pulse-modulation systems, a series of discrete pulses rather than a continuously modulated carrier carries the message specification. As discussed in Chap. 1, pulse modulation exhibits many characteristics which make this modulation method particularly applicable to time-division multiplexing. Since with this method, the modulation channels are interleaved in time sequence and only one channel is transmitted at any one instant of time, nonlinearities in the transmission system do not introduce interchannel crosstalk products such as would be obtained in frequency-division multiplex systems. Another advantage of using pulse modulation is that improved signal-to-noise ratio can be obtained in exchange for increased bandwidth, as in pulse-time and pulse-code modulation.

Since pulses are characterized by the several parameters of amplitude, timing, duration, and frequency, a large number of modulation methods involving these characteristics are feasible. The modulated pulses which carry the information or message specification may be transmitted in various ways, or may undergo a second modulation process suitable to the transmission medium.

The simplest method of pulse modulation is pulse-amplitude modulation or PAM, where samples of the message function are taken at a rate exceeding twice the highest frequency necessary to represent the original signal, and the samples modulate in turn the amplitude of successive carrier pulses. Instead of modulating the amplitude, we may vary the pulse duration and obtain "pulse-duration modulation" PDM (or pulse-width modulation PWM). In this case, all the signal information is actually contained in the starting and stopping times of each pulse. A natural further development is "pulse-position modulation" or PPM, in which the positions of the pulses relative to regularly spaced reference instants are modulated by the message samples. Both modulation methods PDM and PPM may be classified as pulse-time modulation because in both the time of occurrence of some parameter of the carrier pulse is made to vary in accordance with the instantaneous value of the message sample. We have also pulse-frequency modulation (PFM), in which the number of pulses per unit time interval is proportional to the sample value. A more exotic modulation method is "pulse-code modulation" (PCM) in which the samples are "quantized," that is, replaced by the nearest one of a set of discrete numbers which can then be transmitted by coded pulse combinations. Another novel modulation method is "delta modulation" (DM). Delta (or unit increment modulation) is a form of digital transmission in which discrete increments relative to previous sample approximations are transmitted. We shall consider in the following chapters the important properties of these modulation systems from the point of view of spectral analysis and bandwidth, signal-to-noise improvement, and efficiency of information transmission. It will be shown that the outstanding characteristic common to all pulse-modulation systems (except PAM) is the noise-reducing properties.

Since in all pulse-modulation systems, the message function is sampled at regular intervals and the samples are used to modulate the carrier pulses, it behooves us to consider first the sampling principle in somewhat greater detail.

17.1   The Sampling Principle[1-9]

It has been mentioned previously that in pulse-modulation systems, the message function is sampled at a rate exceeding twice the highest

modulating frequency and the samples are used to modulate some
parameter of the carrier pulses. An immediate advantage derived
from transmitting only periodic samples of the message function and
not the complete signal is the conservation of time, whereby the time
saved may be used to transmit samples from other independent signal sources and thereby realize a time-division multiplex system.
We shall presently prove that under certain conditions, it is possible
to reconstruct the continuous-message signal from its periodic samples; this is the so-called sampling theorem in the time domain.

The sampling theorem specifies the least number of discrete
samples of an unknown function necessary for its complete and un-
ambiguous definition.[1] A restricted but widely used form of the
sampling theorem states: If a signal f(t) (which is a real function of
time) is sampled instantaneously at regular intervals and at a rate
slightly higher than twice the highest significant signal frequency,
then the samples contain all the information of the original signal.
Thus, if f_s denotes the sampling rate, and f_m is the highest significant frequency of the signal f(t), then

$$f_s > 2f_m \qquad\qquad\qquad\qquad (17\text{-}1)$$

This theorem is also frequently stated as follows: Any function of
time f(t) which is band-limited to B (or f_m) cycles/sec is completely described by its sample values every 1/2B sec, the samples
extending throughout the time domain. A corollary is that a channel
B cycles wide can be used to transmit 2B independent samples/sec.
A formal proof of this theorem will be given shortly; however, the
physical significance of it will be discussed first.[2]

There is theoretically no upper limit on f_s. However, as f_s is
increased, the available time between samples is decreased, and
fewer channels can be multiplexed. The lower limit on the sampling
frequency is highly significant. There is obviously a relation between the rate at which a signal varies and the number of pulses
needed to reproduce it exactly, for if the sampling rate is too low,
the signal may change radically between sampling times, resulting
in a loss of information. Equation (17-1) tells us that at least $2f_m$
uniformly spaced samples are needed every second in order eventually to reproduce the signal without distortion.

1. *The Sampling Theorem in the Time Domain.*[2-4] To prove the
sampling theorem, we assume that a signal f(t), band-limited to B
cycles/sec, has been sampled at intervals of 1/2B sec. We shall
first show that f(t) may be reconstructed from these samples and

shall then demonstrate that an ideal low-pass filter is called for in the reconstruction or demodulation process. The concept of "band-limited" function of time for which the Fourier transform vanishes outside a finite range of frequencies is a useful idealization in communication theory.

Let $F(j\omega)$ denote the Fourier transform of the band-limited function $f(t)$,

$$F(j\omega) = \int_{-\infty}^{\infty} f(t)e^{-j\omega t}\,dt \qquad (17\text{-}2)$$

or

$$f(t) = \frac{1}{2\pi}\int_{-\infty}^{\infty} F(j\omega)e^{j\omega t}\,d\omega \qquad (17\text{-}3)$$

and

$$F(j\omega) = 0, \qquad |\omega| > 2\pi B \qquad (17\text{-}4)$$

$F(j\omega)$ can be arbitrarily made periodic with a period of $4\pi B$, as shown in Fig. 17-1, and consequently it can be expanded in a Fourier series in the frequency variable ω of period $4\pi B$ to be used only within the interval $|\omega| \le 2\pi B$.

$$F(j\omega) = \sum_{n=-\infty}^{\infty} C_n e^{j(2\pi n/4\pi B)\omega} = \sum_{n=-\infty}^{\infty} C_n e^{jn\omega/2B}, \qquad |\omega| < 2\pi B$$

$$= 0, \qquad\qquad\qquad\qquad\qquad |\omega| > 2\pi B$$

$$(17\text{-}5)$$

where the Fourier coefficients C_n are given by

$$C_n = \frac{1}{4\pi B}\int_{-2\pi B}^{2\pi B} F(j\omega)e^{-jn\omega/2B}\,d\omega \qquad (17\text{-}6)$$

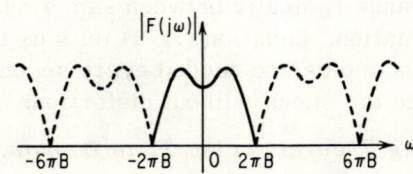

Fig. 17-1. The representation of $F(j\omega)$ as a periodic function.

This follows from Eqs. (2-21) and (2-22) by interchanging t and ω. Also, since $F(j\omega)$, the Fourier transform of $f(t)$, is assumed to be zero outside the band $\pm B$, it follows that

$$f(t) = \frac{1}{2\pi} \int_{-2\pi B}^{2\pi B} F(j\omega) e^{j\omega t} \, d\omega \qquad (17\text{-}7)$$

In particular, the specific values of the band-limited function $f(t)$ at the regular spaced instants $t = -n/2B$ are given by

$$f\left(-\frac{n}{2B}\right) = \frac{1}{2\pi} \int_{-2\pi B}^{2\pi B} F(j\omega) e^{-jn\omega/2B} \, d\omega = 2BC_n \qquad (17\text{-}8)$$

from Eq. (17-6).

From the last equation, we conclude that the Fourier coefficient C_n can be derived from the value of the function $f(t)$ at the corresponding sampling intervals $t = \ldots, -\dfrac{3}{2B}, -\dfrac{2}{2B}, -\dfrac{1}{2B}, 0, \dfrac{1}{2B}, \dfrac{2}{2B}, \dfrac{3}{2B}, \ldots$, and $F(j\omega)$ is thus uniquely determined from the values of the sampled ordinates as follows.

$$F(j\omega) = \sum_{n=-\infty}^{\infty} C_n e^{jn\omega/2B} = \sum_{n=-\infty}^{\infty} \frac{1}{2B} f\left(-\frac{n}{2B}\right) e^{jn\omega/2B} \qquad (17\text{-}9)$$

Knowing $F(j\omega)$, we find $f(t)$ for all possible times, namely,

$$f(t) = \frac{1}{2\pi} \int_{-2\pi B}^{2\pi B} F(j\omega) e^{j\omega t} \, d\omega$$

$$= \frac{1}{2\pi} \int_{-2\pi B}^{2\pi B} \frac{1}{2B} \left[\sum_{n=-\infty}^{\infty} f\left(-\frac{n}{2B}\right) e^{jn\omega/2B} \right] e^{j\omega t} \, d\omega$$

$$= \frac{1}{4\pi B} \int_{-2\pi B}^{2\pi B} \sum_{n=-\infty}^{\infty} f\left(-\frac{n}{2B}\right) e^{j\omega(t + n/2B)} \, d\omega \qquad (17\text{-}10)$$

We have just proved that a knowledge of $f(t)$ at sampling intervals $1/2B$ sec apart suffices to determine $f(t)$ at all times. This is equivalent to the statement that $f(t)$ may be reproduced completely from a knowledge of $f(t)$ at the periodic sampling intervals. If the order of integration and summation are now interchanged in Eq. (17-10), the resulting integral may be readily evaluated as follows:

$$f(t) = \sum_{n=-\infty}^{\infty} \frac{1}{4\pi B} \, f\left(-\frac{n}{2B}\right) \int_{-2\pi B}^{2\pi B} e^{j\omega(t+n/2B)} \, d\omega$$

$$= \sum_{n=-\infty}^{\infty} f\left(-\frac{n}{2B}\right) \frac{\sin 2\pi B(t + n/2B)}{2\pi B(t + n/2B)}$$

$$= \sum_{n=-\infty}^{\infty} f\left(\frac{n}{2B}\right) \frac{\sin 2\pi B(t - n/2B)}{2\pi B(t - n/2B)} \tag{17-11}$$

This result was derived for the sampling interval of $1/2B$ sec; however, it can readily be shown that if the samples are taken at sampling intervals $\alpha/2B$ sec apart where α is any fixed number satisfying the condition $0 < \alpha \le 1$, then Eq. (17-11) will be of the form

$$f(t) = \alpha \sum_{n=-\infty}^{\infty} f\left(\frac{n\alpha}{2B}\right) \frac{\sin 2\pi B(t - n\alpha/2B)}{2\pi B(t - n\alpha/2B)} \tag{17-12}$$

It follows from Eq. (17-11) that we can reconstruct $f(t)$ from the periodic samples by multiplying first each sample by a $(\sin x)/x$ weighting factor centered at the sample's time of occurrence and then summing the resultant terms. However, as we shall presently show, this process is equivalent to passing the samples through an ideal low-pass filter of a cutoff frequency of B cycles/sec.

Let the time function $f(t)$, whose spectrum is limited to B cycles/sec, be sampled periodically for a duration of τ sec (the sampling time) at intervals of $1/2B$ sec (Nyquist or sampling interval), as shown in Fig. 17-2. If $\tau \ll 1/2B$, $f(t)$ may be assumed very nearly constant during the sampling time, and the Fourier transform of the individual sample $f(n/2B)$ is given by

$$F_n(j\omega) = \int_{-\infty}^{\infty} f\left(\frac{n}{2B}\right) e^{-j\omega t} \, dt \doteq \tau f\left(\frac{n}{2B}\right) e^{-jn\omega/2B} \tag{17-13}$$

We recall from Eq. (2-114) that the Fourier transform of an impulse function displaced t_0 sec in time is given by

$$F(j\omega) = \int_{-\infty}^{\infty} \delta(t - t_0) e^{-j\omega t} \, dt = e^{-j\omega t_0} \tag{17-14}$$

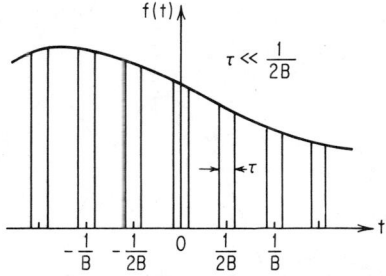

Fig. 17-2. The sampling process.

It follows therefore that $F_n(j\omega)$ is just the Fourier transform of an impulse function of strength $\tau f(n/2B)$ and located at $t = n/2B$ in time; by assuming the sample duration τ to be very small, we have effectively approximated the sample by an impulse of the same area $\tau f(n/2B)$.

Let this impulse be passed through an ideal low-pass filter of bandwidth B cycles/sec, which is assumed for simplicity to have zero phase shift and unity amplitude. The output response to the sample $f(n/2B)$ applied at the input is

$$g_n(t) = \frac{1}{2\pi} \int_{-2\pi B}^{2\pi B} F_n(j\omega)e^{j\omega t}\, d\omega = f\left(\frac{n}{2B}\right)\frac{\tau}{2\pi}$$

$$\times \int_{-2\pi B}^{2\pi B} e^{j\omega(t-n/2B)}\, d\omega$$

$$= 2B\tau f\left(\frac{n}{2B}\right)\frac{\sin 2\pi B(t - n/2B)}{2\pi B(t - n/2B)} \qquad (17\text{-}15)$$

This result can also be obtained by using Eq. (3-47), which gives the response of this idealized filter to an impulse $K\, \delta(t - t_0)$, $H(j\omega) = 1$, $|\omega| \le 2\pi B$.

$$h(t - t_0) = \frac{1}{2\pi} \int_{-\infty}^{\infty} H(j\omega)e^{j\omega(t - t_0)}\, d\omega = 2KB\frac{\sin 2\pi B(t - t_0)}{2\pi B(t - t_0)}$$

$$(17\text{-}16)$$

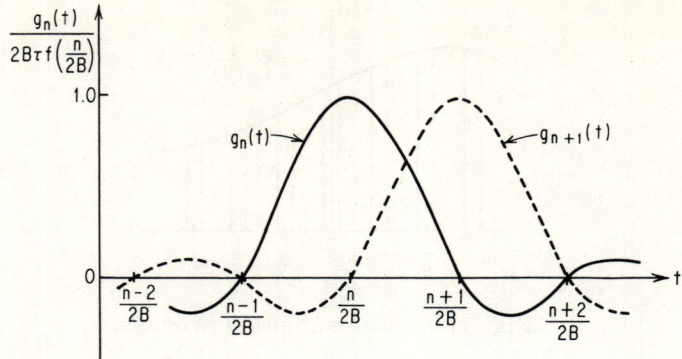

Fig. 17-3. Filter response to sampled inputs.

The output response $g_n(t)$ of the ideal low-pass filter to the individual sample $f(n/2B)$ can be identified with $h(t)$ by putting $K = \tau f(n/2B)$.

$$g_n(t) = 2B\tau \; f\left(\frac{n}{2B}\right) \; \frac{\sin 2\pi B(t - n/2B)}{2\pi B(t - n/2B)}$$

as in Eq. (17-15).

Figure 17-3 is a plot of $g_n(t)$; the output is a maximum at the given sampling point $t = n/2B$ and zero at all the other sampling points. The next sample occurring at $t = (n + 1)/2B$ likewise produces a maximum output at its sampling point and zero at all other sampling points.

The complete output of the ideal filter is just the superposition of the individual sample outputs, or

$$g(t) = \sum_{n=-\infty}^{\infty} g_n(t) = 2B\tau \sum_{n=-\infty}^{\infty} f\left(\frac{n}{2B}\right) \; \frac{\sin 2\pi B(t - n/2B)}{2\pi B(t - n/2B)}$$

$$= 2B\tau f(t) \qquad\qquad\qquad\qquad (17\text{-}17)$$

from Eq. (17-11). The output of the low-pass filter $g(t)$ is thus identically proportional to the original signal $f(t)$ at all instants of time, not only at the sampling points.

From the above discussion, it follows that Eq. (17-17) may also be interpreted as the response of the low-pass filter to the sampled input,

$$f_s(t) = \sum_{n=-\infty}^{\infty} f\left(\frac{n}{2B}\right) \; \delta\left(t - \frac{n}{2B}\right) \qquad\qquad (17\text{-}18)$$

so that $\quad g(t) = \int_{-\infty}^{\infty} f_S(x)h(t-x)\,dx$ $\qquad\qquad\qquad$ (17-19)

where $h(t)$ is the impulse response of the low-pass filter. This will be discussed at greater length in a subsequent paragraph.

It can be seen[4] that owing to the rapid attenuation of the sampling function $g_n(t)$, the contribution of any term in Eq. (17-17) will be appreciable only within a relatively small number of intervals in the neighborhood of the corresponding sampling point. For, consider an arbitrary function $f(t)$ whose spectrum is limited to the bandwidth B, and whose value is very small outside the range $T_1 < t < T_2$, as shown in Fig. 17-4. Let this function be sampled at points $1/2B$ apart so that the total number of sampling points in the range $T = T_2 - T_1$ equals 2TB. It follows therefore that the values of the function $f(t)$ at these 2TB sampling points contribute substantially to $f(t)$ of Eq. (17-17). However, because of the rapid attenuation of these sampling functions, it follows that the main contribution to the value of $f(t)$ in the range $T_1 < t < T_2$ is derived from the 2TB sampling values, even in the case where the function $f(t)$ is not small outside the range $T = T_2 - T_1$.

2. *The Sampling Theorem in the Frequency Domain.*[4] Corresponding to the sampling theorem which we have just derived for the time domain, we shall presently derive a sampling theorem for the frequency domain. Consider a function $f(t)$ which is zero outside the range $T_1 < t < T_2$, as shown in Fig. 17-4. The Fourier transform $F(j\omega)$ of this time-limited function $f(t)$ is given by

$$F(j\omega) = \int_{T_1}^{T_2} f(t)e^{-j\omega t}\,dt \qquad\qquad\qquad (17\text{-}20)$$

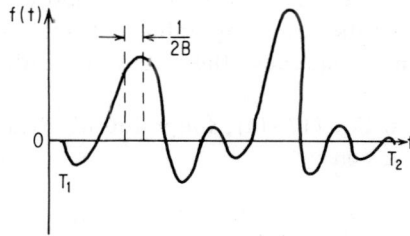

Fig. 17-4. A function $f(t)$, band-limited to B and time-limited to $T_1 < t < T_2$.

We shall show now that the function f(t) is completely determined by the values of the spectrum function $F(j\omega)$ at the angular-frequency sampling points given by

$$\omega_n = n\, \frac{2\pi}{T_2 - T_1} = n\omega_0 \tag{17-21}$$

where $\omega_0 = \dfrac{2\pi}{T_2 - T_1}$, $n = 0,\ \pm 1,\ \pm 2,\ \dots$ \hfill (17-22)

The proof of this statement which formulates the sampling theorem in the frequency domain is analogous to the previous proof in the time domain. The function f(t) can be expressed as a Fourier series in the time variable of period $(T_2 - T_1)$ in the interval $T_1 < t < T_2$; thus

$$f(t) = \sum_{n=-\infty}^{\infty} D_n e^{j2\pi nt/(T_2 - T_1)} \tag{17-23}$$

where $D_n = \dfrac{1}{T_2 - T_1} \displaystyle\int_{T_1}^{T_2} f(t) e^{-j2\pi nt/(T_2 - T_1)}\, dt$

$$= \frac{1}{T_2 - T_1}\, F\left(j\, \frac{2\pi n}{T_2 - T_1}\right) \tag{17-24}$$

From Eqs. (17-24) and (17-23), it follows that the function f(t) is completely determined from the sampling values of the spectrum $F(j\omega)$ at the frequency sampling points given by Eq. (17-21). From Eq. (17-20), it follows also that the spectrum $F(j\omega)$ is completely determined by these sampling values. The sampling theorem in the frequency domain may be stated as follows: The Fourier spectrum $F(j\omega)$ of a time-limited function f(t) which is zero everywhere except in the range $T_1 < t < T_2$ is completely determined for all values of ω by the series of its sampling values at points $2\pi/(T_2 - T_1)$ rad/sec apart in angular frequency, the series extending throughout the frequency domain.

Corresponding to Eq. (17-11), the value of $F(j\omega)$ in terms of its sampling points is given by

$$F(j\omega) = \sum_{n=-\infty}^{\infty} F\left(j\frac{2\pi n}{T}\right) \frac{\sin\,(\omega T/2 - \pi n)}{\omega T/2 - \pi n} \tag{17-25}$$

where $T_2 = T/2$ and $T_1 = -T/2$.

From Eqs. (17-23) and (17-24), f(t) may be expressed as a Fourier series in the form

$$f(t) = \sum_{n=-\infty}^{\infty} \frac{1}{T_2 - T_1} F\left(j \frac{2\pi n}{T_2 - T_1}\right) e^{j2\pi nt/(T_2 - T_1)} \qquad (17\text{-}26)$$

valid in the interval $T_1 < t < T_2$, which expresses f(t) in terms of its sampling values in the frequency domain.

3. *Application of the Convolution Theorem to the Process of Sampling.*[5-6]   A physical insight into the sampling process in terms of spectral analysis may be obtained by considering periodic sampling to be equivalent to multiplying the sampled function of time by a train of uniformly spaced infinitesimally narrow pulses of unit height. It has been shown in Chap. 2 that the spectrum of the product of the two signals is $1/2\pi$ times the convolution of their corresponding spectra. The sampling function $s_T(t)$ may be considered to consist of a periodic sequence of delta functions which has a spectrum consisting of lines at harmonics of the sampling frequency $f_s = 1/T$. If we now multiply a message function f(t) which has a spectrum $F(j\omega)$ by the sampling or scanning signal $s_T(t)$, the spectrum $F_s(j\omega)$ of the sampled message $f_s(t)$ is the convolution of the two spectra. Thus

$$f_s(t) = f(t) \ s_T(t) = f(t) \sum_{n=-\infty}^{\infty} \delta(t - nT) = \sum_{n=-\infty}^{\infty} f(nT) \ \delta(t - nT)$$

$$(17\text{-}27)$$

From Eq. (2-127), the Fourier transform of $s_T(t) = \sum_{n=-\infty}^{\infty} \delta(t - nT)$

is $S_T(j\omega) = \dfrac{2\pi}{T} \sum_{n=-\infty}^{\infty} \delta(\omega - n\omega_0)$. It follows therefore that the

Fourier transform $F_s(j\omega)$ of the sampled function $f_s(t)$ is given by

$$F_s(j\omega) = \frac{1}{2\pi} F(j\omega) * \frac{2\pi}{T} \sum_{n=-\infty}^{\infty} \delta(\omega - n\omega_0)$$

$$= \sum_{n=-\infty}^{\infty} \frac{1}{T} F[j(\omega - n\omega_0)] \qquad (17\text{-}28)$$

We conclude from the last equation that apart from the weighting factor $1/T$, $F_s(j\omega)$ consists of replicas of $F(j\omega)$ centered at $\omega = n\omega_0$, as shown in Fig. 17-5b. Interpolation—the recovery of the

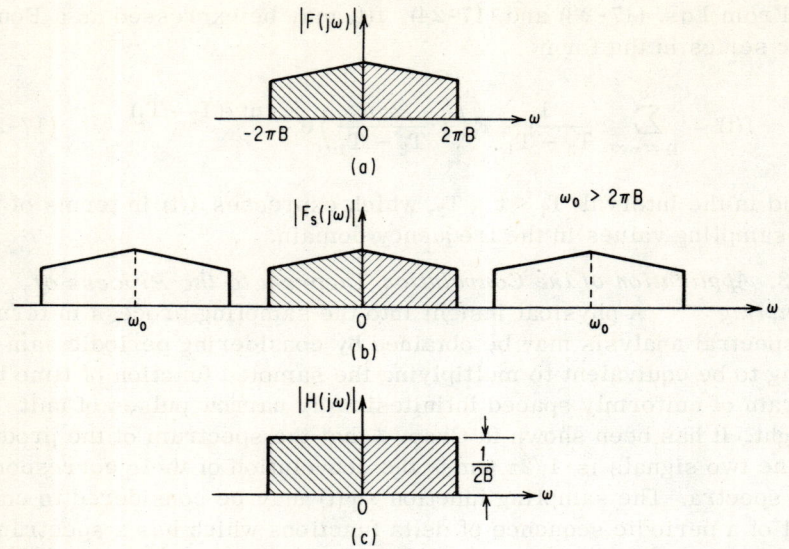

Fig. 17-5. First-order sampling of low-pass function.

original signal from its sample values — may be viewed in the frequency domain as a process of reconstructing the original spectrum by means of a spectral window. As shown in Fig. 17-5c, the original spectrum $F(j\omega)$ may be recovered by multiplying $F_S(j\omega)$ by the spectral window function $H(j\omega)$, the transfer function of the low-pass filter, so that by the use of Eq. (17-28), we obtain

$$F_S(j\omega) \, H(j\omega) = F(j\omega)$$

The corresponding operation in the time domain, of recovering the original signal, is the convolution of the sample pulses $f_S(t)$ by $h(t)$, the inverse transform of $H(j\omega)$, or the impulse response of the low-pass filter; thus

$$f(t) = h(t) * f_S(t) = h(t) * \sum_{n=-\infty}^{\infty} f(nT) \, \delta \, (t - nT)$$

$$= \sum_{n=-\infty}^{\infty} f(nT)h(t - nT) \tag{17-29}$$

where $h(t) = 2BT \dfrac{\sin 2\pi Bt}{2\pi Bt}$ \qquad $\tag{17-30}$

The original signal may be recovered provided $1/T \geq 2B$; the lowest permissible rate is $1/T = 2B$. Substituting this value in Eq. (17-30), Eq. (17-29) becomes

$$f(t) = \sum_{n=-\infty}^{\infty} f\left(\frac{n}{2B}\right) \frac{\sin 2\pi B(t - n/2B)}{2\pi B(t - n/2B)} \qquad (17\text{-}31)$$

in agreement with Eq. (17-11).

In this analysis, we have considered the sampling process as the result of sampling the function $f(t)$ by a train of uniformly spaced infinitesimally narrow pulses $s_T(t) = \sum_{n=-\infty}^{\infty} \delta(t - nT)$. In practice, the sampler remains closed for a finite interval of time, and consequently, the resulting "samples" are of finite duration τ. It follows therefore that, in practice, we may consider the sampling function to consist of a train of periodic pulses of unity amplitude, width τ and period T. The output pulses of the sampled signal will then correspond to a train of amplitude-modulated periodic pulses identical to that which is generated in a PAM system. When the modulated pulses follow the amplitude variation of the sampled time function during the scanning interval, we call this process exact scanning. In contrast with exact scanning, we may have square-topped scanning where the amplitude of the pulses is modulated by the message function $f(t)$ in such a manner that the value of $f(t)$ corresponding to (say) the center of the pulse is effective in the modulation process. In the following, we shall examine in greater detail the two processes of exact and square-topped scanning.

4. *Exact Scanning or Top Sampling.*[5]   As stated previously, in the process of exact scanning, the modulated pulses follow the sampled time function during the scanning interval. As illustrated in Fig. 17-6, exact scanning is equivalent to operating on the scanning pulses $p_T(t)$ by the modulating function $f(t)$; the result of the operation can be expressed by

$$f_S(t) = f(t)\, p_T(t) = f(t) \sum_{n=-\infty}^{\infty} s(t - nT) \qquad (17\text{-}32)$$

where $s(t - nT)$ represents a function of time which is defined as follows:

$$s(t - nT) = 1, \qquad nT - \frac{\tau}{2} < t < nT + \frac{\tau}{2}$$

$$= 0, \qquad \text{elsewhere}$$

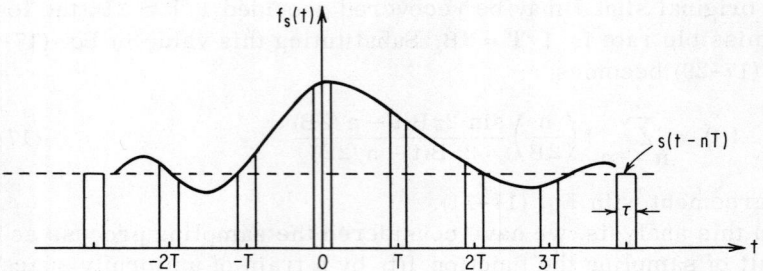

Fig. 17-6. Exact scanning or top sampling.

The scanning pulses can be represented according to Eq. (2-38) by the Fourier series expansion

$$p_T(t) = \frac{\tau}{T} \left[1 + 2 \sum_{n=1}^{\infty} \frac{\sin (n\omega_0 \tau/2)}{n\omega_0 \tau/2} \cos n\omega_0 t \right]$$

$$= \frac{\tau}{T} \sum_{n=-\infty}^{\infty} \frac{\sin (n\omega_0 \tau/2)}{n\omega_0 \tau/2} e^{jn\omega_0 t} \qquad (17\text{-}33)$$

where T is the period and $\omega_0 = 2\pi/T$ is the fundamental angular frequency. The corresponding frequency spectrum is derived by the use of Eq. (2-130), namely,

$$P(j\omega) = 2\pi \frac{\tau}{T} \left\{ \delta(\omega) + \sum_{n=1}^{\infty} \frac{\sin (n\omega_0 \tau/2)}{n\omega_0 \tau/2} [\delta(\omega - n\omega_0) + \delta(\omega + n\omega_0)] \right\}$$

$$(17\text{-}34)$$

The frequency spectrum resulting from the scanning operation is given by $F_S(j\omega)$, the Fourier transform of $f_S(t)$.

$$F_S(j\omega) = \int_{-\infty}^{\infty} f_S(t) e^{-j\omega t} \, dt = \int_{-\infty}^{\infty} [f(t) p_T(t)] e^{-j\omega t} \, dt \qquad (17\text{-}35)$$

Since in this equation, $F_S(j\omega)$ represents the spectrum of the product of two time functions, this is equal to $1/2\pi$ times the convolution of their corresponding spectra.

$$F_S(j\omega) = \frac{1}{2\pi} \int_{-\infty}^{\infty} F(j\omega_1) P[j(\omega - \omega_1)] \, d\omega_1 \qquad (17\text{-}36)$$

By the use of Eq. (17-34), this reduces to

$$F_S(j\omega) = \frac{\tau}{T} \int_{-\infty}^{\infty} F(j\omega_1) \left\{ \delta(\omega - \omega_1) + \sum_{n=1}^{\infty} \frac{\sin(n\omega_0\tau/2)}{n\omega_0\tau/2} \right.$$

$$\left. \times [\delta(\omega - \omega_1 - n\omega_0) + \delta(\omega - \omega_1 + n\omega_0)] \right\} \, d\omega_1 \qquad (17\text{-}37)$$

The first term under the integral equals $F(j\omega)$; that is to say, the original spectrum of the modulating signal is reproduced. Similarly, we obtain for the remaining terms the expression

$$\sum_{n=-\infty}^{\infty} \frac{\sin(n\omega_0\tau/2)}{n\omega_0\tau/2} \left\{ F[j(\omega - n\omega_0)] + F[j(\omega + n\omega_0)] \right\}$$

Finally, the resulting frequency spectrum $F_S(j\omega)$ is given by

$$F_S(j\omega) = \frac{\tau}{T} F(j\omega) + \frac{\tau}{T} \sum_{n=1}^{\infty} \frac{\sin(n\omega_0\tau/2)}{n\omega_0\tau/2} \left\{ F[j(\omega - n\omega_0)] \right.$$

$$\left. + F[j(\omega + n\omega_0)] \right\}$$

$$= \frac{\tau}{T} \sum_{n=-\infty}^{\infty} \frac{\sin(n\omega_0\tau/2)}{n\omega_0\tau/2} F[j(\omega - n\omega_0)] \qquad (17\text{-}38)$$

Comparing this result with that derived for the idealized case of sampling by a periodic sequence of delta functions [Eq. (17-28)], we note that the effect of sampling with pulses of finite duration is equivalent to multiplying the output spectrum $(1/T)F[j(\omega - n\omega_0)]$ by a constant scale factor $\tau \dfrac{\sin(n\omega_0\tau/2)}{n\omega\ \tau/2}$ for a given n. As an illustration of the physical significance of the result of Eq. (17-38), we consider a modulating signal whose frequency spectrum has a constant amplitude $|F(j\omega)| = A$, over the frequency band $\omega_1 \leq |\omega| \leq \omega_2$, as shown in Fig. 17-7. Let the modulated signal consist of a series of scanning pulses $p_T(t)$, as shown in Fig. 17-6. The output spectrum $F_S(j\omega)$ of the modulated pulses in the case of exact scanning is shown in Fig. 17-8. We note that the exact-scanning processes result in a spectrum consisting of sidebands about the harmonics of the scanning frequency $n\omega_0$ of constant amplitude A_n given by

$$A_n = \frac{\tau}{T} A \frac{\sin(n\omega_0\tau/2)}{n\omega_0\tau/2} \qquad (17\text{-}39)$$

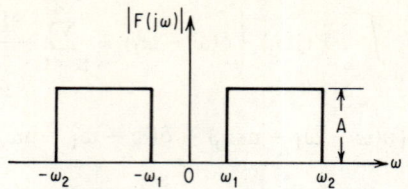

Fig. 17-7. Frequency spectrum of modulating signal ($\omega_1 \leq |\omega| \leq \omega_2$).

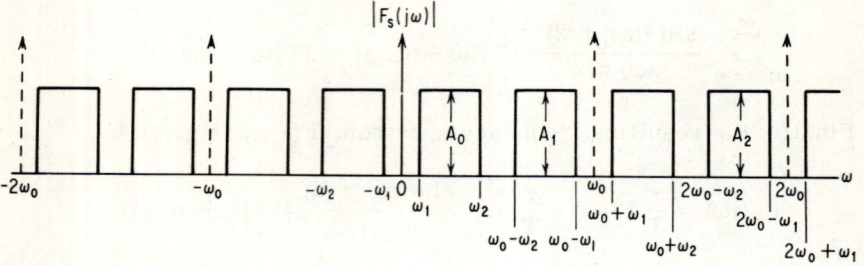

Fig. 17-8. Frequency spectrum of "exact scanning" process.

Thus, the original spectrum is reproduced without distortion, and consequently the original signal can be reconstructed by filtering. We also note that in order to be able to filter out the original modulating frequency band from the first lower sideband, a sufficient condition is that $\omega_0 > 2\omega_2$. As we shall show later on, this condition, while sufficient, is not always necessary. As seen from Fig. 17-9, provided $\omega_2 < 2\omega_1$, the lower sideband can be filtered out by choosing the scanning frequency such that $\omega_2 < \omega_0 < 2\omega_1$.

The above result is quite general and applies to the special case of a low-pass signal as well. In this case we have $\omega_1 = 0$, $\omega_2 = 2\pi B$, and $\omega_0 \geq 2\omega_2$. It follows therefore that in order to avoid overlapping of the spectra, we must choose the sampling period T to be $\leq 1/2B$; the smallest number of sampling points is obtained by taking $T = 1/2B$.

Before proceeding to the case of square-topped scanning, let us apply the above results to unipolar PAM of a cosine wave using top sampling. In this example, let

$$f(t) = 1 + m_a \cos \omega_m t \qquad (17\text{-}40)$$

so that $\quad F(j\omega) = 2\pi \, \delta(\omega) + m_a\pi \, [\delta(\omega - \omega_m) + \delta(\omega + \omega_m)] \qquad (17\text{-}41)$

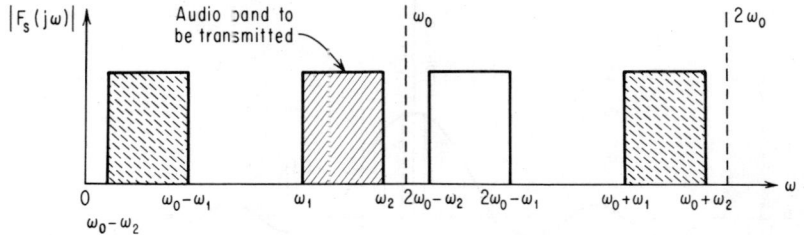

Fig. 17-9. Sideband distribution ($\omega_2 < \omega_0 < 2\omega_1$).

The modulation spectrum consists of two spectral lines at $\omega = \pm\omega_m$ and an impulse function at the origin. It follows therefore that

$$F[\,j(\omega - n\omega_0)] = 2\pi\,\delta(\omega - n\omega_0) + m_a\pi\,[\delta(\omega - \omega_m - n\omega_0)$$

$$+ \,\delta(\omega + \omega_m - n\omega_0)]$$

and by direct substitution in Eq. (17-38), we obtain for the spectrum of the PAM signal, the expression

$$F_s(j\omega) = \frac{\tau}{T}\sum_{n=-\infty}^{\infty}\frac{\sin(n\omega_0\tau/2)}{n\omega_0\tau/2}\{2\pi\,\delta(\omega - n\omega_0)$$

$$+ \,m_a\tau[\,\delta(\omega - \omega_m - n\omega_0) + \delta(\omega + \omega_m - n\omega_0)]\}$$

$$= P(j\omega) + m_a\frac{\tau}{T}\pi\sum_{n=-\infty}^{\infty}\frac{\sin(n\omega_0\tau/2)}{n\omega_0\tau/2}\,\delta[\omega - (n\omega_0 \pm \omega_m)]$$

$$(17\text{-}42)$$

where $P(j\omega)$ represents the spectrum of the scanning pulses [see Eq. (17-34)], and the remaining terms represent the sidebands about the harmonics of the scanning frequency ω_0.

5. *Square-topped Scanning.*[5,7]   As stated previously, in square-topped scanning, the modulated pulses do not follow the sampled time function $f(t)$ during the scanning interval, since the amplitude of the pulses is modulated by the message function $f(t)$ in such a manner that the value of $f(t)$ corresponding to the center (or any other fixed-reference position of the pulse) is effective in the modulation process. The result of such a process is shown in Fig. 17-10. The result of the operation can be expressed by

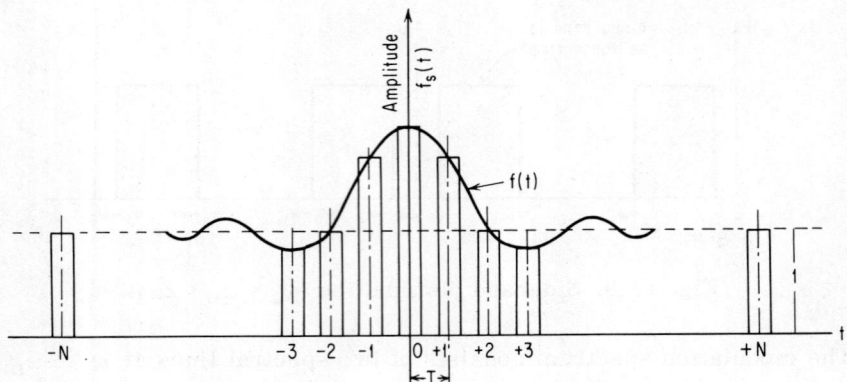

Fig. 17-10. Square-topped scanning.

$$f_S(t) = f(t)p_T(t) = \sum_{n=-\infty}^{\infty} f(nT)s(t - nT) \qquad (17\text{-}43)$$

This equation is analogous to Eq. (17-32) in the case of exact scanning; however, we may note that in the present case of square-topped scanning, only discrete values of $f(nT)$ are effective in the modulation process. The resulting frequency spectrum is then

$$F_S(j\omega) = \int_{-\infty}^{\infty} f_S(t)e^{-j\omega t}\, dt = \sum_{n=-\infty}^{\infty} \left[\int_{-\infty}^{\infty} f(nT)s(t - nT)e^{-j\omega t}\, dt \right] \qquad (17\text{-}44)$$

In order to evaluate Eq. (17-44), we note that since

$$f(t) = \frac{1}{2\pi} \int_{-\infty}^{\infty} F(j\omega_1)e^{j\omega_1 t}\, d\omega_1$$

it follows that

$$f(nT) = \frac{1}{2\pi} \int_{-\infty}^{\infty} F(j\omega_1)e^{j\omega_1 nT}\, d\omega_1$$

and Eq. (17-44) can be expressed in the following form:

$$F_S(j\omega) = \frac{1}{2\pi} \sum_{n=-\infty}^{\infty} \int_{-\infty}^{\infty} F(j\omega_1)e^{j(\omega_1 - \omega)nT}\, d\omega_1$$

$$\times \int_{-\infty}^{\infty} s(t - nT)e^{-j\omega(t - nT)} \, dt \qquad (17\text{-}45)$$

This can be simplified by noting that

$$\int_{-\infty}^{\infty} s(t - nT)e^{-j\omega(t - nT)} \, dt = S(j\omega) = \tau \, \frac{\sin (\omega\tau/2)}{\omega\tau/2}$$

where $S(j\omega)$ is the Fourier transform of the rectangular signal pulse

$$s(t) = 1, \qquad \frac{\tau}{2} < t < \frac{\tau}{2}$$

$$= 0, \qquad \text{elsewhere}$$

so that

$$F_s(j\omega) = \frac{1}{2\pi} \sum_{n=-\infty}^{\infty} S(j\omega) \int_{-\infty}^{\infty} F(j\omega_1)e^{j(\omega_1 - \omega)nT} \, d\omega_1 \qquad (17\text{-}46)$$

Equation (17-46) can be evaluated by the use of Eq. (2-130), namely,

$$\sum_{n=-\infty}^{\infty} e^{-jnT\omega} = \omega_0 \sum_{n=-\infty}^{\infty} \delta(\omega - n\omega_0) \qquad (17\text{-}47)$$

By direct substitution in Eq. (17-46), we obtain

$$F_s(j\omega) = \frac{1}{T} \sum_{n=-\infty}^{\infty} S(j\omega) \int_{-\infty}^{\infty} F(j\omega_1) \, \delta(\omega_1 - \omega - n\omega_0) \, d\omega_1$$

$$= \frac{1}{T} \sum_{n=-\infty}^{\infty} S(j\omega)F[j(\omega - n\omega_0)] = \frac{\tau}{T} \, \frac{\sin (\omega\tau/2)}{\omega\tau/2}$$

$$\times \sum_{n=-\infty}^{\infty} F[j(\omega - n\omega_0)] \qquad (17\text{-}48)$$

Comparing this result which is plotted in Fig. 17-11 with the analogous result given by Eq. (17-38) and plotted in Fig. 17-8, we note that in this case, the amplitude of the resulting frequency spectrum

$$A_n = \frac{\tau}{T} A \, \frac{\sin (\omega\tau/2)}{\omega\tau/2} \qquad (17\text{-}49)$$

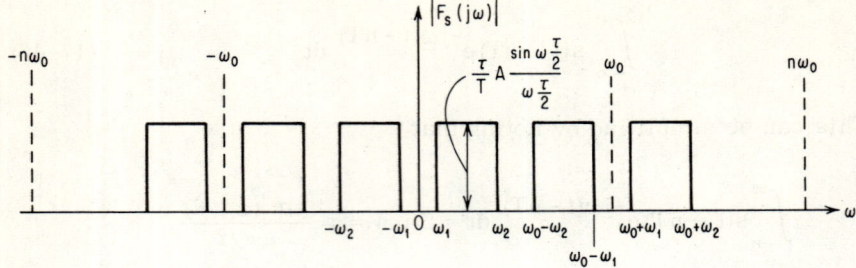

Fig. 17-11. Frequency spectrum of square-topped scanning process.

is not constant over the band (ω_1, ω_2), whereas in the case of exact scanning, the amplitude A_n is constant over the band (ω_1, ω_2) for a given n. We conclude therefore that square-topped sampling introduces frequency distortion over the frequency band of the modulating signal. This is sometimes referred to as the "aperture effect" and may be corrected by the use of an equalizer following the low-pass filter which is used in the reconstruction of the original signal.

6. *Sampling of a Bandpass Function* $(B_o,\ B_o + B)$.[6,8,9]   This problem of sampling a bandpass function has been considered previously; it is considered again in order to establish the minimum sampling rate necessary for reconstructing the original signal from its samples. We have shown above that the sampling process may be regarded in the time domain as a multiplication by a periodic sequence of delta time functions, its counterpart in the frequency domain being a convolution by a train of equispaced frequency delta functions. This led to the result of Eq. (17-28), namely,

$$F_S(j\omega) = \sum_{n=-\infty}^{\infty} \frac{1}{T} F[j(\omega - n\omega_0)]$$

where $F_S(j\omega)$ is the frequency spectrum of the sampled function, and $F(j\omega)$ is the frequency spectrum of the message function. We have also noted that, in practice, the sampling function will consist of a train of periodic pulses of finite width τ, resulting in samples of duration τ. However, as noted from Eqs. (17-38) and (17-48), except for a scale factor, the spectral distribution of the sampled function remains unaltered; consequently, we shall consider, for simplicity in the following analysis, idealized sampling by delta functions.

The spectrum of the bandpass function is assumed to occupy the

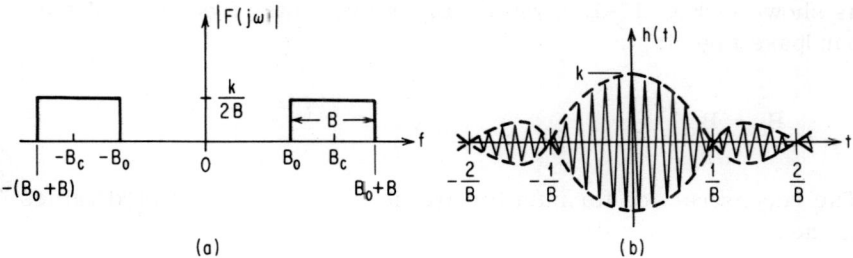

Fig. 17-12. Reconstruction of bandpass signal.

frequency range $B_0 \leq |f| \leq B_0 + B$ as shown in Fig. 17-12a. As before, only those values of the scanning period T which do not cause overlapping in the frequency range of the sampled function can be used. Making use again of the concept of "spectral window," we must, in order to recover the original signal from the sample values, pass the sampled signal

$$f_s(t) = \sum_{n = -\infty}^{\infty} f(nT)\ \delta(t - nT)$$

through a filter whose transfer function $H(j\omega)$ is given by

$$H(j\omega) = T, \qquad F(j\omega) \neq 0$$

$$= 0, \qquad \sum_{n \neq o} F\left[j\left(\omega - \frac{2\pi n}{T}\right)\right] \neq 0 \qquad (17\text{-}50)$$

$$\text{arbitrary otherwise}$$

If we take the simplest choice

$$H(j\omega) = T, \qquad 2\pi B_0 \leq |\omega| \leq 2\pi(B_0 + B) \qquad (17\text{-}51)$$

$$= 0, \qquad \text{otherwise}$$

then the impulse response $h(t)$ of the filter is given by

$$h(t) = \frac{T}{\pi t}\ [\sin 2\pi(B_0 + B)\ t - \sin 2\pi B_0 t]$$

$$= 2BT\ \frac{\sin \pi B t}{\pi B t}\ \cos 2\pi B_c t \qquad (17\text{-}52)$$

as shown in Fig. 17-12b, where B_C is the center frequency of the bandpass signal:

$$B_C = B_O + \frac{B}{2}$$

The reconstruction formula for $f(t)$ in terms of its sampled values is then

$$f(t) = \sum_{n=-\infty}^{\infty} f(nT)h(t-nT) = \sum_{n=-\infty}^{\infty} f(nT)\frac{T}{\pi t}[\sin 2\pi(B_O + B)$$

$$\times (t-nT) - \sin 2\pi B_O(t-nT)]$$

$$= 2BT \sum_{n=-\infty}^{\infty} f(nT)\frac{\sin \pi B(t-nT)}{\pi B(t-nT)}\cos 2\pi B_C(t-nT) \quad (17\text{-}53)$$

The permissible values of T are given by[8]

$$\frac{m}{2B_O} \leq T \leq \frac{m+1}{2(B_O + B)}, \qquad m = 0, 1, 2, \ldots \quad (17\text{-}54)$$

provided $B_O \neq 0$. There will be a largest m, call it M, which satisfies Eq. (17-54) and the relation $M \leq B_O/B$; the largest possible T will then be

$$T_{\max} = \frac{M+1}{2(B_O + B)} \leq \frac{1}{2B} \quad (17\text{-}55)$$

This expression agrees with the formula given by Feldman and Bennett[9] in terms of the minimum sampling rate f_S,

$$\frac{1}{T_{\max}} = f_S = 2B\left(1 + \frac{k}{M'}\right) \quad (17\text{-}56)$$

where $M' = M + 1$, largest integer not exceeding $(B_O + B)/B$,

and $k = (B_O + B)/B - M' = B_O/B - M$

The value of k in Eq. (17-56) varies between zero and unity. When the band is located between adjacent multiples of B, the value of k is zero, and it follows that $f_S = 2B$, no matter how high the frequency

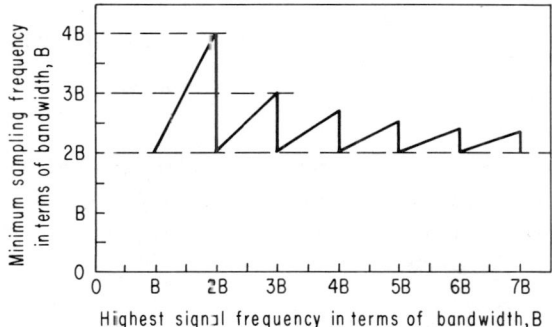

Fig. 17-13. Minimum sampling frequency for band of width B.

range of the signal may be. As k increases from zero to unity, the sampling increases from 2B to $2B(1 + 1/M)$. Regardless of band location, the theorem shows that the minimum permissible sampling rate necessary for signal reconstruction always lies between 2B and 4B samples/sec, as illustrated in Fig. 17-13. The highest sampling rate is required when $M' = 1$ and k approaches unity. This is the case of a signal band lying between $(B - \Delta f)$ and $(2B - \Delta f)$ with Δf small. The sampling rate needed is $2(2B - \Delta f)$, which approaches the value 4B as Δf approaches zero. Actually, when $\Delta f = 0$, there is a discontinuity, and $M' = 2$, $k = 0$, and $f_S = 2B$. The next maximum on the curve is 3B, which is approached when f_S is almost 3B. The sampling theorem as given by Eq. (17-56) is of considerable practical importance in the case of FDM where the baseband of a group of information channels may be in the form of bandpass function.

17.2 Higher-order Sampling[6, 8]

The sampling process can be generalized to encompass the use of several interlaced sampling trains in scanning the message function f(t); this process is usually referred to as higher-order sampling. We have seen that the output pulses of the sampled signal in case of square-topped scanning are given by

$$f_S(t) = \sum_{n = -\infty}^{\infty} f(nT)s(t - nT) \tag{17-57}$$

where s(t) specifies the shape of the pulse. The function $f_S(t)$ is called a first-order sampling of f(t). More generally, we define a p'th order sampling as a function

$$f_S(t) = \sum_{i=1}^{p} f_{S_i}(t) = \sum_{i=1}^{p} \left[\sum_{n=-\infty}^{\infty} f(nT_i + t_i) s_i(t - nT_i - t_i) \right]$$

$$(17\text{-}58)$$

where the i'th sampling $f_{S_i}(t)$ has a sampling period T_i and a time delay t_i. We shall presently show that if the functions $f(t)$ and $s_i(t)$ possess Fourier transforms, then the Fourier transform of $f_S(t)$, namely, $F_S(j\omega)$, is given by

$$F_S(j\omega) = \sum_{i=1}^{p} \frac{1}{T_i} S_i(j\omega) \left\{ \sum_{n=-\infty}^{\infty} F\left[j\left(\omega - \frac{2\pi n}{T_i}\right) \right] e^{-j(2\pi n/T_i)t_i} \right\}$$

$$(17\text{-}59)$$

Equation (17-59) represents the spectrum of a multiply-periodic, amplitude-modulated sequence of pulses.

Consider the function

$$f_{S_i}(t) = \sum_{n=-\infty}^{\infty} f(nT_i + t_i) s_i(t - nT_i - t_i) \qquad (17\text{-}60)$$

Using the integral property of the delta function, this can be rewritten in the following form

$$f_{S_i}(t) = \sum_{n=-\infty}^{\infty} \int_{-\infty}^{\infty} f(x) s_i(t - x) \, \delta(x - nT_i - t_i) \, dx \qquad (17\text{-}61)$$

Interchanging the order of summation and integration, we obtain

$$f_{S_i}(t) = \int_{-\infty}^{\infty} f(x) s_i(t - x) \left[\sum_{n=-\infty}^{\infty} \delta(x - nT_i - t_i) \right] dx \qquad (17\text{-}62)$$

and by the use of Eq. (2-128) it is seen that the expression in brackets in Eq. (17-62) represents a periodic function of period T_i given by

$$\sum_{n=-\infty}^{\infty} \delta(x - nT_i - t_i) = \sum_{n=-\infty}^{\infty} \frac{1}{T_i} e^{j(2\pi n/T_i)(x - t_i)} \qquad (17\text{-}63)$$

Therefore $\quad f_{S_i}(t) = \dfrac{1}{T_i} \displaystyle\int_{-\infty}^{\infty} f(x) s_i(t - x)$

$$\times \left(\sum_{n=-\infty}^{\infty} e^{j(2\pi n/T_i)(x-t_i)} \right) dx$$

$$= \frac{1}{T_i} \sum_{n=-\infty}^{\infty} \left[e^{-j(2\pi n/T_i)t_i} \int_{-\infty}^{\infty} s_i(t-x)f(x) \right.$$

$$\left. \times e^{j(2\pi n/T_i)x} dx \right] \tag{17-64}$$

Taking the Fourier transform of both sides of Eq. (17-64), we obtain, by the use of the convolution theorem and the shift theorem, the following result:

$$F_{s_i}(j\omega) = \frac{1}{T_i} \sum_{n=-\infty}^{\infty} \left\{ e^{-j(2\pi n/T_i)t_i} S_i(j\omega) F \left[j \left(\omega - \frac{2\pi n}{T_i} \right) \right] \right\} \tag{17-65}$$

so that $\quad F_s(j\omega) = \sum_{i=1}^{p} F_{s_i}(j\omega) \tag{17-66}$

which is identical to Eq. (17-59), which we set out to prove. We note here that in case of first-order sampling, Eq. (17-65) reduces to Eq. (17-48) as expected. In fact, Eq. (17-65) could have been derived directly from Eq. (17-48) by inspection, by the use of the shift theorem, in order to account for the relative delay t_i between the sampling trains.

As an illustration, we shall apply second-order sampling to the low-pass function f(t) which is band-limited to (0-B) cycles/sec. Let the two interlaced sampling trains be denoted by

$$s_{T_1}(t) = \sum_{n=-\infty}^{\infty} \delta \left(t - \frac{n}{B} \right) \tag{17-67}$$

and $\quad s_{T_2}(t) = \sum_{n=-\infty}^{\infty} \delta \left(t - \frac{n}{B} - t_o \right) \tag{17-68}$

where the sampling period of each sampling train is now twice as long as in first-order sampling, while the function f(t) is still being sampled at the rate of 2B samples/sec. The sampled functions are, using Eq. (17-60),

$$f_{s_1}(t) = \sum_{n=-\infty}^{\infty} f\left(\frac{n}{B} \right) \delta\left(t - \frac{n}{B} \right) \tag{17-69}$$

and $f_{S_2}(t) = \displaystyle\sum_{n=-\infty}^{\infty} f\left(\dfrac{n}{B} + t_0\right) \delta\left(t - \dfrac{n}{B} - t_0\right)$ (17-70)

The corresponding spectra are given by

$$F_{S_1}(j\omega) = \sum_{n=-\infty}^{\infty} BF[\, j(\omega - 2\pi nB)] \qquad\qquad (17\text{-}71)$$

and $F_{S_2}(j\omega) = \displaystyle\sum_{n=-\infty}^{\infty} Be^{-j2\pi nBt_0} F[\, j(\omega - 2\pi nB)]$ (17-72)

It can be shown[6] that interpolation functions or the inverse transforms of the transfer functions $H_1(j\omega)$ and $H_2(j\omega)$ are given by

$$h_1(t) = h_2(-t) = \frac{\cos\,(2\pi Bt - \pi Bt_0) - \cos \pi Bt_0}{2\pi Bt \sin \pi Bt_0} \qquad (17\text{-}73)$$

and the interpolation function $f(t)$ equals

$$f(t) = h_1(t) * f_{S_1}(t) + h_2(t) * f_{S_2}(t)$$

$$= \sum_{n=-\infty}^{\infty} \left[f\left(\frac{n}{B}\right)h_1\left(t - \frac{n}{B}\right) + f\left(\frac{n}{B} + t_0\right)h_1\left(-t + \frac{n}{B} + t_0\right) \right]$$

$$(17\text{-}74)$$

which reduces to the standard form of Eq. (17-31) for $t_0 = 1/2B$. For the application of second-order sampling to bandpass functions, the reader is referred to Linden[6] and Kohlenberg.[8]

17.3   Sampling of Analytic Signal[4,6]

The sampling principle can also be extended to a bandpass signal by the use of the analytic signal. Let the analytic signal $\psi(t)$ be represented by

$$\psi(t) = f(t) + j\hat{f}(t) \qquad\qquad (17\text{-}75)$$

where $\hat{f}(t)$ is the Hilbert transform of $f(t)$, and $f(t)$ represents the band-limited function in the range $(B_0,\ B_0 + B)$. Denote $\omega_1 = 2\pi B_0$ and $\omega_2 = 2\pi(B_0 + B)$. Let $F(j\omega) = R(\omega) + jX(\omega)$ represent the Fourier spectrum of $f(t)$, so that

$$f(t) = \frac{1}{2\pi} \int_{-\infty}^{\infty} F(j\omega) e^{j\omega t} \, d\omega = \frac{1}{\pi} \int_{\omega_1}^{\omega_2} [R(\omega) \cos \omega t - X(\omega)$$

$$\times \sin \omega t] \, d\omega \qquad (17\text{-}76)$$

It follows therefore from the last equation that

$$\hat{f}(t) = \frac{1}{\pi} \int_{\omega_1}^{\omega_2} [R(\omega) \sin \omega t + X(\omega) \cos \omega t] \, d\omega \qquad (17\text{-}77)$$

and $$\psi(t) = \frac{1}{\pi} \int_{\omega_1}^{\omega_2} [R(\omega) + jX(\omega)] [\cos \omega t + j \sin \omega t] \, d\omega$$

$$= \frac{1}{\pi} \int_{\omega_1}^{\omega_2} F(j\omega) e^{j\omega t} \, d\omega \qquad (17\text{-}73)$$

It should be noted here that this result is as expected from Eq. (5-56). Let us now expand $F(j\omega)$ in a Fourier series in the range ω_1 to ω_2, so that

$$F(j\omega) = \sum_{n=-\infty}^{\infty} \overline{C}_n \, e^{-j[2\pi n/(\omega_2 - \omega_1)]\omega} \qquad (17\text{-}79)$$

where $$\overline{C}_n = \frac{1}{\omega_2 - \omega_1} \int_{\omega_1}^{\omega_2} F(j\omega) e^{j[2\pi n/(\omega_2 - \omega_1)]\omega} \, d\omega \qquad (17\text{-}80)$$

By substituting $t = 2\pi n/(\omega_2 - \omega_1)$ into Eq. (17-78), we obtain

$$\psi\left(\frac{2\pi n}{\omega_2 - \omega_1}\right) = \frac{1}{\pi} \int_{\omega_1}^{\omega_2} F(j\omega) e^{j[2\pi n/(\omega_2 - \omega_1)]\omega} d\omega \qquad (17\text{-}81)$$

and combining this equation with Eq. (17-80), we obtain

$$\overline{C}_n = \frac{\pi}{\omega_2 - \omega_1} \, \psi\left(\frac{2\pi n}{\omega_2 - \omega_1}\right) \qquad (17\text{-}82)$$

From Eqs. (17-79), (17-82), and (17-78), we obtain

$$\psi(t) = \frac{1}{\pi} \int_{\omega_1}^{\omega_2} \sum_{n=-\infty}^{\infty} \frac{1}{\omega_2 - \omega_1} \, \psi\left(\frac{2\pi n}{\omega_2 - \omega_1}\right)$$

$$\times \int_{\omega_1}^{\omega_2} e^{-j[2\pi n/(\omega_2 - \omega_1)]\omega} e^{j\omega t} \, d\omega$$

$$= \frac{1}{\omega_2 - \omega_1} \sum_{n = -\infty}^{\infty} \psi\left(\frac{2\pi n}{\omega_2 - \omega_1}\right)$$

$$\times \int_{\omega_1}^{\omega_2} e^{j\omega[t - 2\pi n/(\omega_2 - \omega_1)]} \, d\omega$$

$$= \frac{1}{\omega_2 - \omega_1} \sum_{n = -\infty}^{\infty} \psi\left(\frac{2\pi n}{\omega_2 - \omega_1}\right) \left[\frac{e^{j\omega[t - 2\pi n/(\omega_2 - \omega_1)]}}{j[t - 2\pi n/(\omega_2 - \omega_1)]}\right]_{\omega_1}^{\omega_2}$$

$$(17\text{-}83)$$

Since $f(t)$ is the real part of $\psi(t)$, we derive $f(t)$ from the following expression by taking the real part of the right-hand side:

$$\psi(t) = f(t) + j\hat{f}(t) = \frac{1}{\omega_2 - \omega_1} \sum_{n = -\infty}^{\infty} \left[f\frac{2\pi n}{\omega_2 - \omega_1} + j\hat{f}\left(\frac{2\pi n}{\omega_2 - \omega_1}\right)\right]$$

$$\times \left[\frac{e^{j\omega[t - 2\pi n/(\omega_2 - \omega_1)]}}{j[t - 2\pi n/(\omega_2 - \omega_1)]}\right]_{\omega_1}^{\omega_2} \qquad (17\text{-}84)$$

Therefore

$$f(t) = \frac{2}{\omega_2 - \omega_1} \sum_{n = -\infty}^{\infty} \left\{f\left(\frac{2\pi n}{\omega_2 - \omega_1}\right) \frac{\sin\left[(\omega_2 - \omega_1)/2\right][t - 2\pi n/(\omega_2 - \omega_1)]}{t - 2\pi n/(\omega_2 - \omega_1)}\right.$$

$$\times \cos\frac{\omega_2 + \omega_1}{2}\left(t - \frac{2\pi n}{\omega_2 - \omega_1}\right) + \hat{f}\left(\frac{2\pi n}{\omega_2 - \omega_1}\right)$$

$$\times \frac{\sin\left[(\omega_2 - \omega_1)/2\right][t - 2\pi n/(\omega_2 - \omega_1)]}{t - 2\pi n/(\omega_2 - \omega_1)}$$

$$\left.\times \sin\left[\frac{\omega_2 + \omega_1}{2}\left(t - \frac{2\pi n}{\omega_2 - \omega_1}\right)\right]\right\} \qquad (17\text{-}85)$$

or $$f(t) = \frac{2}{\omega_2 - \omega_1} \sum_{n = -\infty}^{\infty} \sqrt{f_n^2 + \hat{f}_n^2}$$

$$\times \frac{\sin\left[(\omega_2 - \omega_1)/2\right]\left[t - 2\pi n/(\omega_2 - \omega_1)\right]}{t - 2\pi n/(\omega_2 - \omega_1)}$$

$$\times \cos\left(\frac{\omega_2 + \omega_1}{2}\right)\left(t - \frac{2\pi n}{\omega_2 - \omega_1} - \phi_n\right) \qquad (17\text{-}86)$$

where $\quad f_n = f\left(\frac{2\pi n}{\omega_2 - \omega_1}\right)$

$$\hat{f}_n = \hat{f}\left(\frac{2\pi n}{\omega_2 - \omega_1}\right)$$

and $\qquad \phi_n = \tan^{-1}\dfrac{\hat{f}_n}{f_n}$

It follows from Eq. (17-86) that a knowledge of f(t) and of $\hat{f}$(t) at all the sample points t, given by

$$t = \frac{2\pi n}{\omega_2 - \omega_1} = \frac{n}{f_2 - f_1} = \frac{n}{B} \qquad (17\text{-}87)$$

completely determines f(t) everywhere. It should be noted here that the sample points determined by Eq. (17-87) are twice as far apart as in the low-frequency case. However, since the values of both f(t) and $\hat{f}$(t) must be known at these sample points, the total number of separate information items is still 2TB, as before.

A result identical to Eq. (17-85) is derived by Linden,[6] using the convolution method outlined above. He derives the following interpolation functions:

$$h(t) = \frac{\sin \pi Bt}{\pi Bt} \cos 2\pi \left(B_0 + \frac{B}{2}\right)t \qquad (17\text{-}88)$$

$$\hat{h}(t) = -\frac{\sin \pi Bt}{\pi Bt} \sin 2\pi \left(B_0 + \frac{B}{2}\right)t \qquad (17\text{-}89)$$

where h(t) and $\hat{h}$(t) are the impulse response function of the spectral window functions $H(j\omega)$ and $\hat{H}(j\omega)$, respectively. Then using these relations, the interpolation formula for f(t) is given by

$$f(t) = \sum_{n=-\infty}^{\infty}\left[f\left(\frac{n}{B}\right)h\left(t - \frac{n}{B}\right) + \hat{f}\left(\frac{n}{B}\right)\hat{h}\left(t - \frac{n}{B}\right)\right] \qquad (17\text{-}90)$$

It can readily be shown that these equations are identical to the results derived previously.

17.4   The Spectrum of Modulated Pulses[11-13]

We have already noted that in a pulse-modulation system, the message information is transmitted in the form of discrete pulses. With PAM, PDM, and PPM systems, one pulse normally specifies each sample of the message signal, whereas in PCM systems, each sample of the message signal is specified by a group of pulses. In these cases, the minimum number of pulses is 2B per sec per channel.

We have also pointed out that characteristic of pulse modulation is the comparative ease with which channels may be multiplexed by time division as distinguished from frequency division. Multiplexing by time division simplifies instrumentation since relatively simple synchronous switches or gating circuits replace the modulators, demodulators, carrier generators, and bandpass filters ordinarily employed in a frequency-division system.

In practice, a complete communication system may utilize both pulse-modulation and CW modulation techniques. In general, the output of the pulse-modulation system will ordinarily be used to amplitude- or frequency-modulate in turn a sine-wave carrier for radio communication.

Pulse-position-modulation and pulse-duration-modulation systems require much greater bandwidths, however, than do the CW systems for they depend on the accurate location of the pulse edges. The increased bandwidth in PPM will be shown later to lead to signal-to-noise improvement, just as wideband FM provides signal-to-noise improvement. There is no signal-to-noise improvement in PAM, just as AM provides no improvement.

In order to evaluate pulse-modulated multichannel communication systems, the frequency spectra of the modulated pulse trains must be known.

One of the early methods of analysis of the spectrum of modulated pulses is the pseudo-static method.[10] This involves setting up the formula for the spectrum of a train of unmodulated pulses and then modulating any of the required parameters such as the duration, phase, etc., of the repetitive frequency of the pulse.

If there is no rational relation between the modulating frequency and the pulse-repetition frequency, a Fourier series approach is evidently inapplicable since the waveform will then never be periodic. On the other hand, if such a relation does exist, a Fourier

series method can be applied directly, but then the spectrum will be
degenerate, for some sidebands of harmonics of the pulse-repetition
frequency will coincide. It would seem reasonable then to use a
Fourier integral method; the difficulty there lies in specifying in an
explicit form the moments at which the edges occur.

This difficulty is overcome by the method of staircase analysis[11],
where the spectrum of an infinite train of pulses is derived from the
sum of the spectra of an ascending and falling staircase. The spectrum of the modulated pulse train is then obtained by modulating the
time scales of the staircases.

In the following analysis we shall use a modified staircase method
of analysis for the derivation of the spectra of modulated pulses and
compare them with the spectra of amplitude- and angular-modulated
CW systems.

1. *General Consideration of Time Modulation.* In order to obtain a
clear understanding of the relationship between the modulating wave
and the modulated pulse train, the picture of the modulation process
as shown in Fig. 17-14 may be assumed.

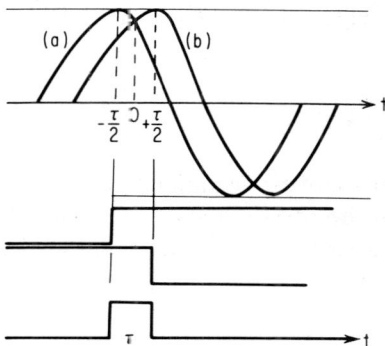

Fig. 17-14. Modulation process (modified).

Here a and b are two cosine waves of angular frequency ω_r displaced relative to each other by an amount τ, the width of the unmodulated pulse. The positive and negative steps which give rise to the
pulse train are assumed to occur at the peaks of the waveforms a
and b, respectively. Without any modulation, the time of occurrence
of the positive and negative steps is given by

$$\omega_r\left(t + \frac{\tau}{2}\right) = 2n\pi \qquad (17\text{-}91)$$

and $\quad \omega_r \left(t - \dfrac{\tau}{2} \right) = 2n\pi$ $\hfill$ (17-92)

which corresponds to the time when the values of the phase of the sine wave are separated by 2π.

For pulse-frequency modulation, the two waveforms a and b are regarded as being frequency-modulated, and the moments at which the positive peaks of the waveforms recur can be said to be frequency-modulated.

The expressions for the unmodulated case are

(a) $\quad f(t) = A \cos \omega_r \left(t + \dfrac{\tau}{2} \right)$ $\hfill$ (17-93)

(b) $\quad f(t) = A \cos \omega_r \left(t - \dfrac{\tau}{2} \right)$ $\hfill$ (17-94)

and for the modulated case

(a) $\quad f(t) = A \cos \left[\omega_r \left(t + \dfrac{\tau}{2} \right) + \beta \sin (\omega_m t + \phi) \right]$ $\hfill$ (17-95)

(b) $\quad f(t) = A \cos \left[\omega_r \left(t - \dfrac{\tau}{2} \right) + \beta \sin (\omega_m t + \phi) \right]$ $\hfill$ (17-96)

where ω_m is the modulating frequency, $\beta = \Delta\omega/\omega_m$ is the modulation index, and ϕ is the phase angle of the modulating wave.

With modulation, the time of occurrence of the positive and negative steps is given by

$$\omega_r \left(t + \frac{\tau}{2} \right) + \beta \sin (\omega_m t + \phi) = 2n\pi \qquad (17\text{-}97)$$

$$\omega_r \left(t - \frac{\tau}{2} \right) + \beta \sin (\omega_m t + \phi) = 2n\pi \qquad (17\text{-}98)$$

Pulses whose moments of occurrence satisfy these equations are said to be frequency-modulated. Pulse-frequency modulation is defined in this manner because of the connection with carrier-frequency modulation, which is well known.

Examination of Eqs. (17-97) and (17-98) shows that the moment at which the n'th pulse or edge occurs is determined by the value of the modulating voltage at the instant when the edge occurs. This fact is of fundamental importance. In general, these equations are transcendental, which prevents the use of the Fourier integral method.

It will be convenient to define phase-modulated pulses in an analogous manner. The two cosine waves, a and b, are phase-modulated when the modulation index β is constant independent of the modulating frequency.

2. *Pulse-frequency Modulation.* The expression for an infinite train of pulses which we derived in Chap. 2 [Eq. (2-38)] can be expressed in the following form:

$$F(t) = \frac{A}{2\pi j} \sum_{k=-\infty}^{\infty} \frac{1}{k} \left(e^{jk\omega_r(\tau/2)} - e^{-jk\omega_r(\tau/2)} \right) e^{jk\omega_r t}$$

(17-99)

where A is the amplitude of the pulses, and ω_r is the pulse-repetition frequency. This expression clearly shows the individual effects of the leading and the trailing edges, each of which can be modulated in any desired manner. The spectrum of frequency-modulated pulses will now be obtained. We have noted earlier that the steps occur when the phase of the modulated fundamental frequency is a multiple of 2π, resulting in frequency-modulated steps. There are really two distinct types of frequency-modulated pulse waveforms. In the first type, the displacement of the waveform b of Fig. 17-14 from its unmodulated position at any instant of time t is determined by the value of the modulating signal at that instant. In the second type, this displacement is determined by the value of the modulating signal at time $(t - \tau)$. It is clear that in the first type there is bound to be a variation in the width of the pulse along the modulation cycle as well as a positional displacement. Considering the first type for which Eq. (17-97) and (17-98) apply, it is seen from these equations that modulation can be taken into account by substituting for $\omega_r\tau/2$ in the expressions for the leading and trailing edge in Eq. (17-99) the expressions

$$\frac{\omega_r \tau}{2} + \beta \sin (\omega_m t + \phi) \quad \text{and} \quad \frac{\omega_r \tau}{2} - \beta \sin (\omega_m t + \phi)$$

Then the expression for the modulation pulse train becomes

$$F(t) = \frac{A}{2\pi j} \sum_{k=-\infty}^{\infty} \frac{1}{k} \left(e^{j[k\omega_r\tau/2 + k\beta \sin(\omega_m t + \phi)]} \right.$$
$$\left. - e^{-j[k\omega_r\tau/2 - k\beta \sin(\omega_m t + \phi)]} \right) e^{jk\omega_r t}$$

(17-100)

Using the formula $e^{jz \sin \theta} = \sum\limits_{n=-\infty}^{\infty} J_n(z)e^{jn\theta}$, where $J_n(z)$ is the Bessel function of the first kind and of order n, the following expression will be obtained:

$$F(t) = \frac{A\omega_r\tau}{2\pi} + \frac{A\omega_r\tau}{\pi} \sum_{k=1}^{\infty} \frac{\sin (k\omega_r\tau/2)}{k\omega_r\tau/2} \left(J_0(k\beta) \cos k\omega_r t \right.$$

$$+ \sum_{n=1}^{\infty} J_n(k\beta)\{\cos [(k\omega_r + n\omega_m) t + n\phi]$$

$$\left. + (-1)^n \cos [(k\omega_r - n\omega_m)t - n\phi]\} \right) \tag{17-101}$$

This expression may be compared with that for the spectrum of a frequency-modulated continuous wave given by

$$F(t) = AJ_0(\beta) \cos \omega_r t + A \sum_{n=1}^{\infty} J_n(\beta)\{\cos [(\omega_r + n\omega_m)t + n\phi]$$

$$+ (-1)^n \cos [(\omega_r - n\omega_m)t - n\phi]\} \tag{17-102}$$

Comparison of Eqs. (17-101) and (17-102) leads to the following conclusions:

1. With this type of pulse-frequency modulation, there will be no sideband accompanying the zero or d-c component of the pulse spectrum. Hence a low-pass filter used for the demodulation will not yield any modulating frequency component.

2. The k'th harmonic of the pulse-repetition frequency is frequency-modulated, the modulation index being $k\beta$.

3. As the order of the pulse-repetition frequency increases, the amplitude of the carrier and sidebands will diminish, as shown by the factor $(\sin x)/x$, where $x = k\omega_r\tau/2$. It is clear that pulses frequency-modulated in this manner can be demodulated by means of a bandpass filter which will extract one of the repetition frequency harmonics along with its sidebands and by applying it to a frequency-discriminator circuit. It is worth noting that there will be no harmonic distortion accompanying this method of demodulation.

The other type of pulse-frequency modulation will now be considered. In this type the displacement of the waveform b of Fig. 17-14 from its unmodulated position at any instant of time t will depend on the value of the modulating voltage at $(t - \tau)$. Hence the equations which will determine the position of the leading and trailing edges of the pulses are:

$$\omega_r\left(t + \frac{\tau}{2}\right) + \beta \sin(\omega_m t + \phi) = 2n\pi \qquad (17\text{-}103)$$

$$\omega_r\left(t - \frac{\tau}{2}\right) + \beta \sin(\omega_m \overline{t - \tau} + \phi) = 2n\pi \qquad (17\text{-}104)$$

On comparing these with Eqs. (17-91) and (17-92), it is found that modulation can be taken into account by the substitution of $\omega_r \tau/2 + \beta \sin(\omega_m t + \phi)$ and $\omega_r \tau/2 - \beta \sin(\omega_m \overline{t - \tau} + \phi)$ for $\omega_r \tau/2$ in the expressions for the leading and trailing edges, respectively, in Eq. (17-99). Then the expressions for the modulation pulse train become

$$F(t) = \frac{A}{2\pi j} \sum_{k=1}^{\infty} \frac{1}{k} \left(e^{j[k\omega_r \tau/2 + k\beta \sin(\omega_m t + \phi)]} \right.$$

$$\left. - e^{-j[k\omega_r \tau/2 - k\beta \sin(\omega_m \overline{t - \tau} + \phi)]} \right) e^{jk\omega_r t} \qquad (17\text{-}105)$$

After simplification, this becomes

$$F(t) = \frac{A\omega_r \tau}{2\pi} + A(\Delta\omega/2\pi)\tau \, \frac{\sin(\omega_m \tau/2)}{\omega_m \tau/2} \cos\left(\omega_m t + \phi - \frac{\omega_m \tau}{2}\right)$$

$$+ \frac{A\omega_r \tau}{\pi} \sum_{k=1}^{\infty} \left(J_0(k\beta) \, \frac{\sin(k\omega_r \tau/2)}{k\omega_r \tau/2} \cos k\omega_r t \right.$$

$$+ \sum_{n=1}^{\infty} J_n(k\beta) \left\{ \frac{\sin \overline{(k\omega_r + n\omega_m)\tau/2}}{k\omega_r \tau/2} \right.$$

$$\times \cos\left[k\omega_r + n\omega_m)t + n\phi - n\omega_m \frac{\tau}{2}\right] + (-1)^n$$

$$\times \frac{\sin \overline{(k\omega_r - n\omega_m)\tau/2}}{k\omega_r \tau/2} \cos\left[(k\omega_r - n\omega_m)t - n\phi \right.$$

$$\left.\left.\left. + n\omega_m \frac{\tau}{2}\right]\right\}\right) \qquad (17\text{-}106)$$

It will be interesting to compare this result with Eqs. (17-101) and (17-102), which are the expressions for the frequency-modulated pulses of the first type and the frequency-modulated continuous wave, respectively. This comparison leads to the following conclusions.

1. The zero or the d-c component of the pulse spectrum has a sideband of the modulating frequency of amplitude $\dfrac{A\Delta\omega\tau}{2\pi} \dfrac{\sin(\omega_m\tau/2)}{\omega_m\tau/2}$. Modulation can therefore be recovered by means of a low-pass filter, and there will be no harmonic distortion as there are no harmonic terms in the zero-order pulse spectrum. The only form of distortion which can occur comes from the lower sidebands of harmonics of the pulse-repetition frequency which penetrate into the demodulating filter passband.

2. If the sidebands of the k'th pulse-repetition frequency harmonic are considered, it is seen that the upper and lower sidebands of the same order are not equal in amplitude, whereas in pulse-frequency modulation of the first type they are equal, as seen from Eq. (17-101). It may therefore be concluded that if the signal is demodulated by selecting any of the pulse-repetition frequency harmonics and sidebands and applying it to a frequency discriminator, there is bound to be a certain measure of distortion.

3. *Pulse-phase Modulation.* The expression for a train of modulated pulses can be obtained by phase-modulating the two waveforms a and b of Fig. 17-14. Here again there are two distinct types of phase-modulated pulses. In one of them the phase waveform b at any instant of time is determined by the value of the modulating voltage at that instant. Therefore the displacement of the leading and trailing edges are not equal, and hence there will be a variation in the width of the pulses along the modulation cycle, as well as a positional displacement. Considering this type of modulation first, the expression for the modulated waveforms of a and b of Fig. 17-14 are

$$\text{(a)} \quad f(t) = A \cos \omega_r \left[\left(t + \frac{\tau}{2} \right) + \tau_d \sin (\omega_m t + \phi) \right] \tag{17-107}$$

$$\text{(b)} \quad f(t) = A \cos \omega_r \left[\left(t - \frac{\tau}{2} \right) + \tau_d' \sin (\omega_m t + \phi) \right] \tag{17-108}$$

where $\omega_r\tau_d$ and $\omega_r\tau_d'$ are the phase deviations of the waveform a and b, respectively. The two deviations are assumed as being different to make the treatment applicable to the case of pulse-width modulation. By comparing these equations with Eqs. (17-95) and (17-96), it is seen that modulation can be taken into account by substituting $\dfrac{\omega_r\tau}{2} + \omega_r\tau_d \sin(\omega_m t + \phi)$ and $\dfrac{\omega_r\tau}{2} - \omega_r\tau_d' \sin(\omega_m t + \phi)$

for $\omega_r \tau / 2$ in the expressions for the leading and trailing edges, respectively, in Eq. (17-99). The spectrum is therefore given by

$$F(t) = \frac{A}{2\pi j} \sum_{k=-\infty}^{\infty} \frac{1}{k} \left(e^{j[k\omega_r \tau/2 + k\omega_r \tau_d \sin(\omega_m t + \phi)]} \right.$$

$$\left. - e^{-j[k\omega_r \tau/2 - k\omega_r \tau_d' \sin(\omega_m t + \phi)]} \right) e^{jk\omega_r t} \qquad (17\text{-}109)$$

For pulse-phase modulation, the two deviations are equal, i.e., $\tau_d = \tau_d'$, and the above equation can be simplified into

$$F(t) = \frac{A\omega_r \tau}{2\pi} + \frac{A\omega_r \tau}{\pi} \sum_{k=1}^{\infty} \frac{\sin(k\omega_r \tau/2)}{k\omega_r \tau/2} \left(J_0(k\omega_r \tau_d) \cos k\omega_r t \right.$$

$$+ \sum_{n=1}^{\infty} J_n(k\omega_r \tau_d) \{ \cos [(k\omega_r + n\omega_m)t + n\phi]$$

$$\left. + (-1)^n \cos [(k\omega_r - n\omega_m)t - n\phi] \} \right) \qquad (17\text{-}110)$$

This expression is similar to Eq. (17-101), with the difference that $k\beta$ is replaced by $k\omega_r \tau_d$. Note that the expression $\omega_r \tau_d$ is the phase deviation of waveform a, which is equal to $\beta (= \Delta\phi)$ when β is constant. Evidently each of the pulse-repetition frequency harmonics is phase-modulated, the maximum deviation being $k\omega_r \tau_d$. The amplitudes of the pulse-repetition frequency harmonics and their sidebands decrease with k according to the term $\frac{\sin(k\omega_r \tau/2)}{k\omega_r \tau/2}$. Also there is no sideband accompanying the zero or the d-c component of the pulse spectrum, and hence modulation cannot be recovered by means of a low-pass filter.

The spectrum of the other type of pulse-phase modulation can be derived directly from Eq. (17-106) by substituting $\omega_r \tau_d$ for β and setting $\tau_d = \tau_d'$.

$$F(t) = \frac{A\omega_r \tau}{2\pi} + \frac{A\omega_r \omega_m \tau \tau_d}{2\pi} \frac{\sin(\omega_m \tau/2)}{\omega_m \tau/2}$$

$$\times \cos \left(\omega_m t + \phi - \frac{\omega_m \tau}{2} \right) + \frac{A\omega_r \tau}{\pi}$$

$$\times \sum_{k=1}^{\infty} \left(J_0(k\omega_r \tau_d) \frac{\sin(k\omega_r \tau/2)}{k\omega_r \tau/2} \cos k\omega_r t \right.$$

$$+ \sum_{n=1}^{\infty} J_n(k\omega_r \tau_d) \left\{ \frac{\sin(k\omega_r + n\omega_m)\tau/2}{k\omega_r \tau/2} \right.$$

$$\times \cos\left[(k\omega_r + n\omega_m)t + n\phi - n\omega_m \frac{\tau}{2}\right]$$

$$+ (-1)^n \frac{\sin(k\omega_r - n\omega_m)\tau/2}{k\omega_r \tau/2}$$

$$\left. \times \cos\left[(k\omega_r - n\omega_m)t - n\phi + n\omega_m \frac{\tau}{2}\right]\right\}\right) \qquad (17\text{-}111)$$

This equation is very similar to Eq. (17-106); the same conclusions which were arrived at for the case described by Eq. (17-106) are applicable to this equation.

4. *Pulse-width Modulation. Symmetrical Double-edge Modulation.* The spectrum for width-modulated pulses can be obtained by considering the spectrum of phase-modulated pulses given by Eq. (17-109). If the trailing edge, instead of being displaced in the same direction as the leading edge, is displaced in the opposite direction, then pulse-width modulation will be produced. Considering the case of symmetrical double-edge modulation, the equation becomes

$$F(t) = \frac{A}{2\pi j} \sum_{n,k=-\infty}^{\infty} \frac{1}{k}\left[J_n(k\omega_r \tau_d)e^{jk\omega_r \tau/2} - J_n(-k\omega_r \tau_d)\right.$$

$$\left. \times e^{-jk\omega_r \tau/2}\right] e^{j[(k\omega_r + n\omega_m)t + n\phi]} \qquad (17\text{-}112)$$

which simplifies into

$$F(t) = \frac{A\omega_r \tau}{2\pi} + \frac{A\omega_r m\tau}{2\pi}\sin(\omega_m t + \phi) + \frac{A\omega_r \tau}{2\pi}$$

$$\times \sum_{k=1}^{\infty}\left[2J_0\left(k\omega_r m \frac{\tau}{2}\right)\frac{\sin(k\omega_r \tau/2)}{k\omega_r \tau/2}\cos k\omega_r t\right.$$

$$+ 4J_1\left(k\omega_r m \frac{\tau}{2}\right)\frac{\cos(k\omega_r \tau/2)}{k\omega_r \tau/2}\cos(k\omega_r t)\sin(\omega_m t + \phi)$$

$$+ 4J_2\left(k\omega_r m \frac{\tau}{2}\right)\frac{\sin(k\omega_r \tau/2)}{k\omega_r \tau/2}\cos(k\omega_r t)\cos(2\omega_m t + 2\phi)$$

$$\left. + \cdots\right] \qquad (17\text{-}113)$$

In the above equation, the term τ_d has been replaced by $\frac{1}{2}m\tau$, where m is the modulation index. Thus if m = 1, then the maximum and minimum values of the pulse width will be 2τ or 0, which corresponds to the case of 100 per cent modulation. From the above equation, it is seen that the zero order of the pulse spectrum has a modulating-frequency term of amplitude $A\omega_r m\tau/2\pi$ and no harmonics. Therefore modulation can be recovered by means of a low-pass filter, and distortion will be only due to the lower sidebands of the pulse-repetition frequency harmonics.

Single-edge Width Modulation. If only the leading edge is being modulated, the spectrum for this case can be obtained by setting $\tau'_d = 0$ in Eq. (17-109). The resulting expression on simplification becomes

$$F(t) = \frac{A\omega_r\tau}{2\pi} + \frac{A\omega_r m\tau}{2\pi} \sin(\omega_m t + \phi) + \frac{A\omega_r\tau}{2\pi}$$

$$\times \sum_{k=1}^{\infty} \left[\frac{\sin k\omega_r(t - \tau/2)}{k\omega_r\tau/2} + \frac{J_0(k\omega_r m\tau)}{k\omega_r\tau/2} \sin k\omega_r\left(t + \frac{\tau}{2}\right) \right.$$

$$+ \frac{2J_1(k\omega_r m\tau)}{k\omega_r\tau/2} \cos k\omega_r\left(t + \frac{\tau}{2}\right) \sin(\omega_m t + \phi)$$

$$\left. + \frac{2J_2(k\omega_r m\tau)}{k\omega_r\tau/2} \sin k\omega_r\left(t + \frac{\tau}{2}\right) \cos(2\omega_m t + 2\phi) + \cdots \right]$$

$$(17\text{-}114)$$

In this case also the modulated signal can be extracted by means of a low-pass filter, and there will be distortion due only to the lower sidebands of the pulse-repetition frequency harmonics.

5. *Pulse-amplitude Modulation.*[11]   As in the process of scanning, we distinguish between two types of pulse-amplitude modulation. One type produces pulses whose amplitude is everywhere modulated by the signal voltage existing at the corresponding instant. This is called top modulation (exact scanning). Such pulses may be regarded as vertical slices of the modulation waveform. The other gives flat-topped pulses, the amplitude of each pulse being determined by the value of the modulating voltage at a single instant, say that at which the leading edge of the pulse occurs.

Top Modulation (or Exact Scanning). It was stated above that the pseudo-static method gives erroneous results when applied to the spectral analysis of modulated pulses. In top modulation, however, the pseudo-static method may be applied. For if all components in a train of unmodulated rectangular pulses are simultaneously modulated in amplitude, it will result in top modulation. This follows from the fact that at each instant, if all components are altered in the same ratio, then the sum will change in the same way. Consequently, if a train of rectangular pulses is top-modulated by a single tone of the form $\cos(\omega_m t + \phi)$ with modulation depth m_a, then from Eqs. (17-32) and (17-33) we obtain

$$
F(t) = [1 + m_a \cos(\omega_m t + \phi)]
$$

$$
\times \frac{A\omega_r \tau}{2\pi} \left[1 + 2 \sum_{k=1}^{\infty} \frac{\sin(k\omega_r \tau/2)}{k\omega_r \tau/2} \cos k\omega_r t \right]
$$

$$
= \frac{A\omega_r \tau}{2\pi} \left(1 + m_a \cos(\omega_m t + \phi) + 2 \sum_{k=1}^{\infty} \frac{\sin(k\omega_r \tau/2)}{k\omega_r \tau/2} \right.
$$

$$
\times \left\{ \cos k\omega_r t + \frac{m_a}{2} \cos[(k\omega_r + \omega_m)t + \phi] \right.
$$

$$
\left. \left. + \frac{m_a}{2} \cos[(k\omega_r - \omega_m)t - \phi] \right\} \right)
\qquad (17\text{-}115)
$$

The output contains a d-c term, a low-frequency term corresponding to the modulating signal of amplitude $A\omega_r \tau m_a/2\pi$, and a set of harmonics of the fundamental repetition frequency of the pulse train, each with a pair of sidebands. Demodulation by a low-pass filter may be rendered distortionless simply by using a repetition frequency greater than twice the bandwidth of the modulating signal. In the same conditions, demodulation may also be achieved by using a bandpass filter containing one repetition frequency harmonic with its sidebands and applying the output to an envelope detector.

The problem of top modulation or exact scanning has already been discussed in Sec. 17.1 by the use of Fourier transform. The frequency spectrum of a top-modulated pulse train by a general modulating function $f(t)$ is given by Eq. (17-38), and the special case when $f(t) = 1 + m_a \cos \omega_m t$ is considered in Eq. (17-42). It should be noted here that Eq. (17-42) represents the Fourier spectrum of the modulated pulse train of Eq. (17-115). The advantage of using the more general Fourier transform approach is obvious since the modulating signal is not confined to a simple sinusoid.

Flat-topped Modulation (or Square-topped Scanning). A general expression for square-topped scanning was derived in Eq. (17-48). The expression for a train of flat-topped amplitude-modulated pulses by a cosine wave can be obtained from the general result by taking its inverse Fourier transform and putting $f(t) = 1 + m_a \cos \omega_m t$. However, it will be instructive to derive the result directly from Eq. (17-115). The waveform of a train of very short pulses of duration $\delta\tau$, top-modulated by a function of the form $\cos(\omega_m t + \phi)$, is

$$\delta F(t) = \frac{A\omega_r \delta\tau}{2\pi} \left(1 + m_a \cos(\omega_m t + \phi) + 2 \sum_{k=1}^{\infty} \frac{\sin(k\omega_r \tau/2)}{k\omega_r \tau/2}\right.$$

$$\left\{\cos k\omega_r t + \frac{m_a}{2} \cos[(k\omega_r + \omega_m)t + \phi]\right.$$

$$\left.\left. + \frac{m_a}{2} \cos[(k\omega_r - \omega_m)t - \phi]\right\}\right) \tag{17-116}$$

But top-modulated and flat-top-modulated pulses must be represented by the same expression when the pulse duration $\delta\tau$ is very short. To obtain the formula for a train of flat-top-modulation pulses, it is only necessary to add together a large number of identical trains of amplitude-modulated elementary impulses, each successive train delayed by an additional interval.

The entire collection of elementary impulses, delayed by amounts ranging from zero to $t_0(=\tau)$, will then add to determine the complete pulse. The desired sequence of flat-top pulses is obtained by summing the contributions of all of the trains of elementary impulses over the range $-t_0/2$ to $+t_0/2$.

$$F(t) = \int_{-t_0/2}^{t_0/2} \delta F(t-\tau)\, d\tau$$

$$= \frac{A\omega_r \tau}{2\pi}\left\{1 + m_a \frac{\sin(\omega_m \tau/2)}{\omega_m \tau/2} \cos(\omega_m t + \phi)\right.$$

$$+ 2\sum_{k=1}^{\infty} \frac{\sin(k\omega_r \tau/2)}{k\omega_r \tau/2} \cos k\omega_r t$$

$$+ \frac{m_a}{2} \frac{\sin (k\omega_r + \omega_m)\tau/2}{(k\omega_r + \omega_m)\tau/2} \cos [(k\omega_r + \omega_m)t + \phi]$$

$$+ \frac{m_a}{2} \frac{\sin (k\omega_r - \omega_m)\tau/2}{(k\omega_r - \omega_m)\tau/2} \cos [(k\omega_r - \omega_m)t - \phi] \Bigg\}$$

$$(17\text{-}117)$$

Comparing the results of Eqs. (17-115) and (17-117), we conclude that except for an amplitude variation, the distributions of the spectral components in both cases are identical. We recall that the same conclusion was reached in the general case in Sec. 17.1 where the problem of exact and square-topped scanning was discussed. Once more a low-pass filter would provide distortionless demodulation, but a bandpass filter and an envelope detector would give some distortion since the two sidebands about any repetition frequency harmonic are unequal. Notice however that the output of the low-pass filter is not constant but varies as a $(\sin x)/x$ function, where x is proportional to the modulating frequency. This variable factor of the modulation signal $\dfrac{\sin (\omega_m \tau/2)}{\omega_m \tau/2}$ illustrates the basic theorem for aperture effect in sampling. Equation (17-117) was derived for flat-top modulation, where the height of the pulse is determined by the modulating signal at the instant of the center of the pulse. For the case where the voltage affecting the pulse is that existing at its leading edge, the signal must be of the form $\sin [\omega_m(t - \tau/2) + \phi]$.

Demodulation Methods.   Almost all types of amplitude-modulated pulses are easily demodulated by a low-pass filter, or network, having an appropriately shaped loss-frequency characteristic. The process amounts to selection of the modulating wave from the modulated pulses. Network characteristics are determined from the spectrum of the pulses. As noted earlier, from Eq. (17-115), the case of top modulation, the gain characteristic of the filter should be flat in the passband for distortionless demodulation.

Above the passband, the filter must introduce enough loss for adequate suppression of other spectrum components. It must suppress the lower sidebands about the sampling frequency ω_r, which fall close to the modulating frequency ω_m.

The flat-top pulse train as given by Eq. (17-95) is likewise readily demodulated, but, as previously mentioned, a simple equalizing network is sometimes needed in addition to the low-pass filter. The equalizer must compensate for the aperture effect, namely,

$$\frac{\sin (\omega_m \tau/2)}{\omega_m \tau/2}.$$

References

1. Black, H. S.: "Modulation Theory," D. Van Nostrand Company, Inc., Princeton, N. J.
2. Schwartz, M.: "Information Transmission, Modulation, and Noise," McGraw-Hill Book Company, New York.
3. Oliver, B. M., J. R. Pierce, and C. E. Shannon: Philosophy of PCM, *Proc. IRE*, November, 1948, p. 1324.
4. Goldman, S.: "Information Theory," Prentice-Hall, Inc., Englewood Cliffs, N. J.
5. Clavier, A. G., P. F. Panter, and D. D. Grieg: Distortion in a Pulse Count Modulation System, *AIEE Tech. Paper* 47-152, April, 1947.
6. Linden, D. A.: A Discussion of Sampling Theorems, *Proc. IRE*, July, 1959, pp. 1219-1226.
7. Hönicke, H.: Untersuchungen an Pulsmodulationsverfahren, Teil II, *Nachrichtentechnik*, vol. 8, no. 11, 1958.
8. Kohlenberg, A.: Exact Interpolation of Band-limited Functions, *J. Appl. Phys.*, vol. 24, no. 12, pp. 1432-1436, December, 1953.
9. Feldman, C. B., and W. R. Bennett: Bandwidth and Transmission Performance, *BSTJ*, vol. 28, pp. 490-595, July, 1949.
10. Roberts, F. F., and J. C. Simmonds: Multichannel Communication Systems, *Wireless Engr. London*, vol. 22, pp. 538-549, November, 1945; pp. 576-580, December, 1945.
11. Fitch, E.: The Spectrum of Modulated Pulses, *J.IEE London*, vol. 94, part 3A, pp. 556-564, 1947.
12. Rao, V. N.: Some Aspects of the Spectrum of Modulated Pulses, *Jour. Indian Inst. Sci.*, vol. 35, sec. B., pp. 125-136, July, 1953.
13. Moss, S. H.: Frequency Analysis of Modulated Pulses, *Phil. Mag.* and *J. Sci.*, ser. 7, vol. 39, pp. 663-691, September, 1948.

18

TIME-DIVISION MULTIPLEX SYSTEMS (TDM)

We have noted earlier that pulse modulation exhibits many characteristics which make this mode of modulation particularly suitable for time-division multiplexing. Multiplexing by time division
as contrasted with frequency division offers many advantages in
system instrumentation; this applies to processes of modulation as
well as detection.

As discussed in Chap. 1, in FDM systems the individual channels
are used to modulate different carrier frequencies called subcarriers; the modulated subcarriers form a baseband which is used
to modulate the final carrier to be transmitted. At the receiver, the
baseband is first detected, and then the individual modulated subcarriers are separated by means of bandpass filters. If the frequency separation between adjacent channels is small, the filter
specifications must be very stringent, and because each channel
occupies a different frequency band, it requires a different filter
design. In TDM systems, since the channels are connected to the
transmission path sequentially and consequently are separated in
time only, identical circuits are used for the channels, consisting of
relatively simple synchronous switches or gating circuits. The only
filters in the detection process are the low-pass filters which are
identical for each channel and usually are not required to meet very

stringent specifications. This simplified circuitry is to be contrasted with the modulators, demodulators, carrier generators, and
bandpass filters ordinarily used in a FDM system.

An additional advantage of TDM systems is that circuit design
required to meet the low crosstalk requirements in a multiplex system is less stringent in this system compared to a FDM system.
In a multichannel carrier system, nonlinearities in the transmitter
amplifiers or in the receiver IF amplifier will introduce crosstalk
in the different channels. The phase and amplitude linearity requirements of the amplifiers used must therefore be more stringent
than those required for a single channel. However, nonlinear distortion does not introduce interchannel crosstalk in a TDM system
because the signals of the different channels are allotted separate
time slots and are not applied to the system at the same time.
Consequently, the linearity requirements of a TDM system are no
more stringent than those for a single channel. However, as we
shall show later on, in order to confine each sample of the individual
channels to its time slot, the transmission system must have sufficient bandwidth and linear transfer characteristics. Insufficient
bandwidth and/or phase and attenuation distortion will cause the
pulse waveforms to spread in time into neighboring time slots,
which will result in interchannel crosstalk.

This chapter is devoted to the principles of time-division multiplex systems. First we shall discuss the features which are common to TDM systems in general, such as methods of generation,
synchronization, etc. Then we shall analyze in more detail the particular characteristics of PAM, PDM, and PPM systems. In each
system, methods of modulation and demodulation will be discussed,
and particular consideration will be given to the problem of bandwidth, signal-to-noise improvement, crosstalk and other salient
features.

18.1   Fundamentals of Pulse-modulation Systems[1,2]

The elementary TDM system shown in Fig. 18-1 consists of a
transmitting path, at each end of which is connected a rotating
switch; the switches rotate in synchronism, so that the sending and
receiving apparatus of each channel are connected together during
their allotted time interval and disconnected throughout the remainder of each revolution of the switch. The form of signal which
is transmitted by the TDM system shown in Fig. 18-2 consists of
interlaced trains of pulses of identical repetition frequency, each
pulse train being modulated by the sampled value of the signal of
one channel, as shown in Fig. 18-2b.

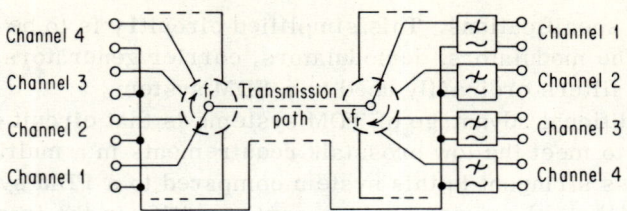

Fig. 18-1. Elementary TDM system.

In a PAM system, the periodic pulse trains consist of amplitude-modulated pulses corresponding to the amplitudes of the sampled signal. As discussed in Chap. 17, the modulation process may be considered as a scanning or sampling operation where the modulated pulses either follow the sampled time function during the scanning interval, a process which is called exact scanning or top modulation, or the amplitude of the scanning pulse is determined by the modulating signal at the instant of the center of the pulse or at some other arbitrary instant. The latter type of modulation is called flat-topped modulation or square-topped scanning. In both cases, the modulation process results in a series of pulses proportional in magnitude to the sampled values. We have also noted in the last chapter that the frequency spectrum of a train of amplitude-modulated pulses consists of components of the modulating signal of the channel together with the scanning frequency and its harmonics, and upper and lower sidebands about the scanning frequency and each harmonic. Besides amplitude modulation, several other methods are available for modulating pulse trains. In PDM, shown in Fig. 18-2c, pulses of constant amplitude are made to vary in duration, proportional to the amplitude of the sampled signal. In pulse-duration modulation or pulse-width modulation (PWM), the trailing edge of the pulse occurs at a fixed time in the cycle, and the width of the pulse is modulated by varying the time at which the leading edge of the pulse occurs. Since all the information contained in the modulating signal is conveyed by the leading edge of the pulse, transmitter power can be saved by using much shorter pulses of equal amplitude and duration, whose times of occurrence are advanced or retarded by the modulating signal, as shown in Fig. 18-2d. This method of modulation is generally known as pulse-position modulation (PPM).

In pulse-frequency modulation (PFM), the repetition frequency of a train of identical pulses is made proportional to the amplitude of the sampled signal. However, pulse-frequency modulation cannot be used in TDM systems because interlacing of the pulse trains of

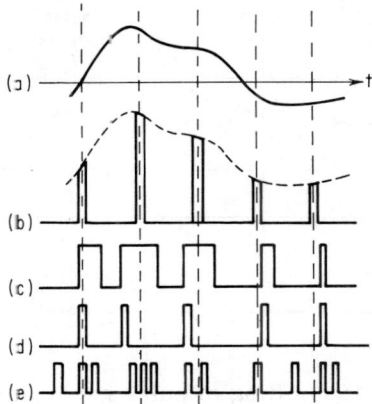

Fig. 18-2. Signal waveforms in TDM systems: (a) Modulating signal waveform; (b) pulse-amplitude modulation; (c) pulse-length modulation; (d) pulse-position modulation; (e) pulse-code modulation.

the different channels can only be maintained if they all have the same repetition frequency. It is often used in FDM systems where the resulting baseband is used to modulate the repetition frequency of a pulsed carrier of constant amplitude.

It should be pointed out that our discussion will be limited to the analysis of the video pulses of the time-division multiplex signal without regard to the specific method of radio-frequency transmission. Any of the well-known methods of modulation of the RF carrier such as amplitude-, frequency-, or phase-shift keying may be utilized for the transmission of the TDM video signal. This will be discussed in some detail in Chap. 23. As mentioned above, time-division multiplex systems such as PPM, which use pulses of constant amplitude, are usually employed instead of PAM because they offer considerably improved signal-to-noise ratios. In Chap. 21, we shall show that the greatest immunity from noise is obtained with systems such as binary PCM in which the receiver has only to detect the presence and absence of each pulse; however, a greater bandwidth than for simpler systems is required because of the greater number of pulses per second which must be transmitted for each channel.

1. *Pulse Generation and Synchronization.* There are several available methods for the generation of the pulse trains which are modulated by the individual channels. These pulse trains which are accurately interleaved in time throughout the pulse-repetition cycle are used to operate the gates of the channels comprising the TDM

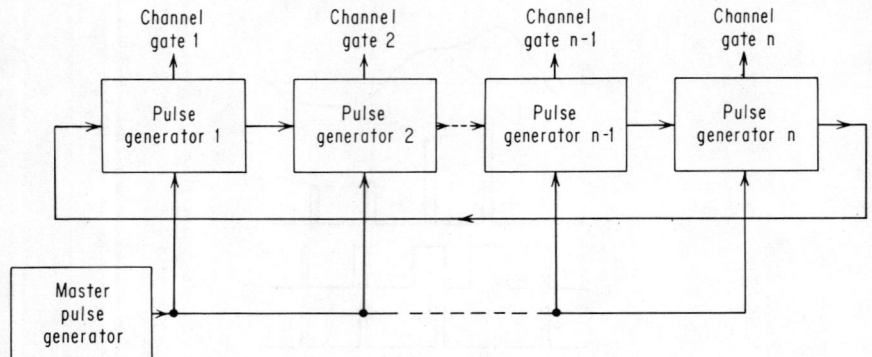

Fig. 18-3. Ring of pulse generators.

system. A typical method of generating the pulses is shown in Fig. 18-3 where a set of pulse generators such as flip-flop circuits or blocking oscillators are connected to form a ring. In order to make it possible to generate pulses at precise regular intervals without imposing excessive tolerance requirements on the components used, each pulse generator is arranged to operate only when it receives simultaneous triggering pulses from the preceding generator and from a master pulse source. The master pulse generator is crystal-controlled, and its repetition frequency equals the channel PRF multiplied by the number of channels.

Another method of generating the pulses required for the channel gates is by means of a delay-line distributor, as shown in Fig. 18-4. As before, the master pulse generator is crystal-controlled with a PRF equal to the channel PRF multiplied by the number of channels. The pulse generator which drives the delay line is locked to the master pulse generator and is triggered by the output pulses from the far end of the delay line whose time delay is made equal to the channel pulse-repetition period. Since the channel gates are connected to the delay line at equal intervals along its length, the gates are operated in sequence at the required PRF of the channel.

In order to separate the multiplex channels at the receiving end, and ensure that a given send channel remains connected to the same receive channel, some method of synchronizing is required in a TDM system. This is usually accomplished by having one channel in the system allotted to a distinctive signal, a marker pulse, such as a specially long pulse or a combination of pulses. These marker pulses, when received and separated from the message-bearing pulses, can be used to control the synchronizing circuit at the receiving terminal whose function is to open and close channel gates in proper sequence.

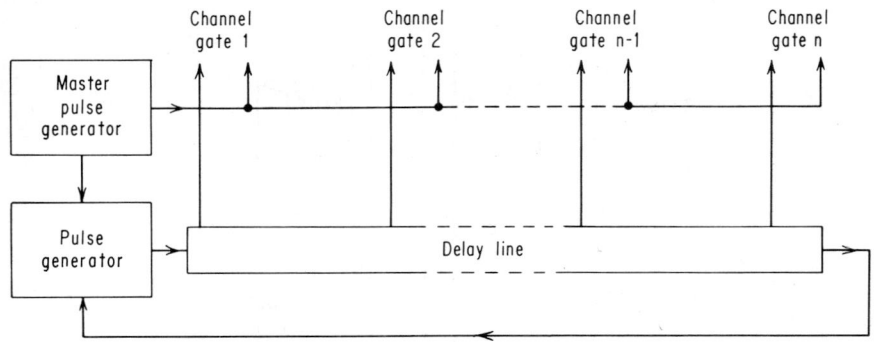

Fig. 18-4. Delay-line pulse distributor.

18.2   Pulse-amplitude-modulation Systems [1-7]

Pulse-amplitude modulation was originally chosen by many workers in the field of TDM because of its simplicity and ease of instrumentation. It was subsequently superseded by alternative pulse-time-modulation systems, which offer improvement in signal-to-noise ratio. However, pulse-amplitude modulation is still being used as an intermediate step in many systems which use finally another type of modulation such as PCM, where the message signals are pulse-amplitude-modulated before quantization.

PAM can be instrumented by two distinct methods. The first produces a variation of the amplitude of a pulse sequence about a fixed nonzero value, or pedestal, and constitutes ordinary double-sideband amplitude modulation. The pulses are of positive polarity as shown in Fig. 18-5a, which corresponds to top modulation, and Fig. 18-5b for flat-topped modulation. In the second method, the pedestal is zero and the magnitude of the modulated pulses may be positive or negative with reference to the zero pedestal, as shown in Fig. 18-5c and d. The output signal consists of double-polarity modulated pulses and resembles double-sideband suppressed-carrier modulation. Physically, this may be accomplished by periodically closing a normally open switch in series with the continuous-data signal.

1. *Spectra of Amplitude-modulated Pulses.* [2,3]   The spectra involved in the different modulation processes of PAM are summarized below.

Unit Sampling Function.   The unit sampling function shown in Fig. 18-5e consists of a train of unmodulated periodic pulses of unit amplitude. From Eq. (17-33),

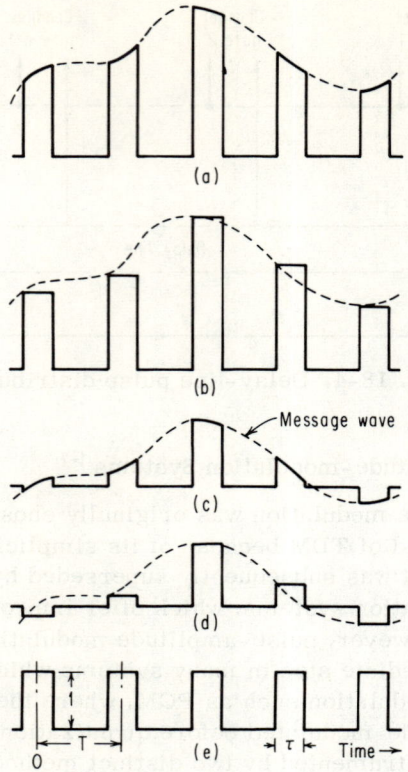

Fig. 18-5. Various shapes of amplitude-modulated pulses: (a) Single-polarity pulses; (b) single-polarity flat-top pulses; (c) double-polarity pulses; (d) double-polarity flat-top pulses; (e) unit sampling function. (From H. S. Black,[2] courtesy of D. Van Nostrand Company, Inc.)

$$p_T(t) = \frac{\tau}{T} \sum_{n=-\infty}^{\infty} \frac{\sin(n\pi\tau/T)}{n\pi\tau/T} e^{jn\omega_0 t} \tag{18-1}$$

where $\omega_0 = 2\pi f_0 = 2\pi/T$ is the fundamental angular frequency of the pulse train, τ is the duration of the pulse, and τ/T the duty cycle.

Double-polarity Amplitude-modulated Pulses: Exact Scanning. The double-polarity pulses shown in Fig. 18-5c are obtained by multiplying the message signal $f(t)$ by the unit sampling function $p_T(t)$. Let $f(t) = A \cos(\omega_m t + \phi)$. The output $f_{S_1}(t)$ is given by

$$f_{S_1}(t) = f(t) \ p_T(t)$$

or

$$f_{S_1}(t) = \frac{\tau}{T} A \cos(\omega_m t + \phi) + \frac{\tau}{T} A \sum_{n=1}^{\infty} \frac{\sin(n\pi\tau/T)}{n\pi\tau/T}$$

$$\times \cos[(n\omega_o \pm \omega_m)t \pm \phi] \tag{18-2}$$

In the general case, the message function $f(t)$ is band-limited, and its spectrum is $F(j\omega)$. The output spectrum is, from Eq. (17-38),

$$F_{S_1}(j\omega) = \frac{\tau}{T} F(j\omega) + \frac{\tau}{T} \sum_{n=1}^{\infty} \frac{\sin(n\pi\tau/T)}{n\pi\tau/T} \left\{ F[j(\omega - n\omega_o)] \right.$$

$$+ F[j(\omega + n\omega_o)] \Big\} = \frac{\tau}{T} \sum_{n=-\infty}^{\infty} \frac{\sin(n\omega_o\tau/2)}{n\omega_o\tau/2} F[j(\omega - n\omega_o)] \tag{18-3}$$

The spectrum of the double-polarity amplitude-modulated pulses consists of the original modulation spectrum and an infinite number of upper and lower sidebands around ω_o and its harmonics.

Single-polarity Amplitude-modulated Pulses: Exact Scanning. These are shown in Fig. 18-5a. In this case

$$f_{S_2}(t) = [1 + m_a \cos(\omega_m t + \phi)] \cdot \frac{\tau}{T} \sum_{n=-\infty}^{\infty} \frac{\sin(n\pi\tau/T)}{n\pi\tau/T} e^{jn\omega_o t} \tag{18-4}$$

or $\quad f_{S_2}(t) = \frac{\tau}{T} \sum_{n=-\infty}^{\infty} \frac{\sin(n\pi\tau/T)}{n\pi\tau/T} e^{jn\omega_o t} + f_{S_1}(t) \tag{18-5}$

In the general case, $\quad F_{S_2}(j\omega) = P(j\omega) + F_{S_1}(j\omega) \tag{18-6}$

where $P(j\omega)$ is the Fourier transform of $p_T(t)$.

Double-polarity Flat-topped Modulation. As stated in Chap. 17, flat-topped modulation, as shown in Fig. 18-5d, may be considered to consist of a large number of elementary impulses, delayed by amounts ranging from zero to τ and added together to form the complete pulse. The spectrum of the resulting amplitude-modulated pulses of duration $d\tau$, using Eq. (18-2), is

$$f_{s_\delta}(t) = A \frac{d\tau}{T} \sum_{n=-\infty}^{\infty} \cos [(n\omega_0 + \omega_m)t + \phi] \qquad (18\text{-}7)$$

and the contribution of all the trains of short pulses over the range $(0\text{-}\tau)$ is

$$f_{s_3}(t) = \frac{A}{T} \sum_{n=-\infty}^{\infty} \int_0^\tau \cos [(n\omega_s + \omega_m)(t - \tau)] \, d\tau$$

$$= \frac{\tau}{T} A \sum_{n=-\infty}^{\infty} \frac{\sin [\pi(\tau/T)(n\omega_0 + \omega_m)/\omega_0]}{[\pi (\tau/T)(n\omega_0 + \omega_m)/\omega_0]}$$

$$\times \cos \left[(n\omega_0 + \omega_m)\left(t - \frac{\tau}{2}\right) + \phi\right] \qquad (18\text{-}8)$$

As shown in Eq. (17-48) in the general case, the amplitude of the sidebands is modified by a factor $\dfrac{\sin (\omega\tau/2)}{\omega\tau/2}$ so that the output spectrum is given by

$$F_{s_3}(j\omega) = \frac{\tau}{T} \frac{\sin (\omega\tau/2)}{\omega\tau/2} \sum_{n=-\infty}^{\infty} F [j(\omega - n\omega_0)] \qquad (18\text{-}9)$$

which illustrates the aperture effect of sampling discussed in the last chapter.

Single-polarity Flat-topped Modulation. Single-polarity flat-topped pulses are shown in Fig. 18-5b. They can be derived from double-polarity modulation by adding a constant to the modulating signal as in exact scanning. The resulting spectrum is

$$F_{s_4}(j\omega) = P(j\omega) + F_{s_3}(j\omega) \qquad (18\text{-}10)$$

2. *Demodulation Methods.* [1,2] The term ''demodulation'' when applied to TDM systems means the separation of the pulse train corresponding to one channel from all the other pulse trains and, in addition, the detection of the message function from this separated

modulated pulse train. With amplitude and width modulation, as noted in the last chapter, the message function can be extracted from the modulated pulse train simply by means of a low-pass filter. This can be seen, for example, from the low-frequency term of Eq. (18-2), which applies to double-polarity amplitude-modulated pulses. The low-pass filter should be flat in the passband for distortionless demodulation. Above the passband, the filter cutoff characteristics should be such as to attenuate the lower sideband of the scanning frequency. Since the spectrum of the single-polarity amplitude-modulated pulses differs from the double-polarity pulses only in the additional scanning frequency and its harmonics, as seen from Eq. (18-5), the design characteristics of the low-pass demodulation filters are identical to those discussed previously. Only in flat-topped modulation is an equalizing network sometimes needed in addition to the low-pass filter, to compensate for the aperture effect. The amount of equalization will depend on the duration of the scanning pulses, and in many practical applications it is rather small.

   3. *Interchannel Crosstalk.* [5-7]   Interchannel crosstalk arises in PAM time-multiplexed transmission systems when each sample is not confined to its assigned time slot. Insufficient bandwidth and nonlinearities of the transfer character of the common transmission path will cause each sample pulse of the PAM time-multiplexed signal to tail off into neighboring time slots, rather than be confined to its assigned time slot. The amount of the resulting interchannel crosstalk is dependent on the upper and lower cutoff frequencies of the transmission path.

   Attenuation and phase distortion at high frequencies will prevent a pulse of one channel from decaying to zero before the time at which the gate of the next channel is opened to receive its pulse, as shown in Fig. 18-6. Distortion at low frequencies causes modulation to be developed on the d-c level present between the pulses, and thus causes crosstalk from each channel into all the other channels of the system, as shown in Fig. 18-7. It is therefore necessary for the crosstalk caused by low-frequency distortion to be attenuated more than that caused by high-frequency distortion, which is usually only appreciable between adjacent channels. The low-frequency cutoff of the common transmission path must therefore be considerably lower than the lowest modulating frequency employed.

   Interchannel crosstalk in TDM systems is contrasted to that in a FDM system in the following manner. In the latter, the translated message functions which form the baseband signal are transmitted simultaneously over the common transmission medium. In such systems, crosstalk arises as a result of the harmonic and inter-

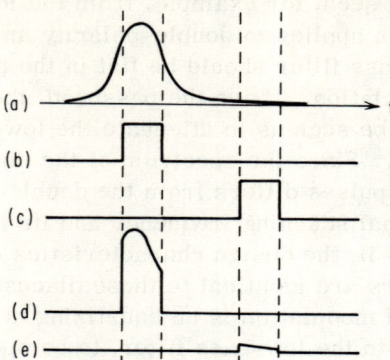

Fig. 18-6. Crosstalk caused by bandwidth limitation at high frequencies: (a) Signal pulse of disturbing channel; (b) gating pulse of disturbing channel; (c) gating pulse of disturbed channel; (d) signal received by disturbing channel; (e) crosstalk received by disturbed channel.

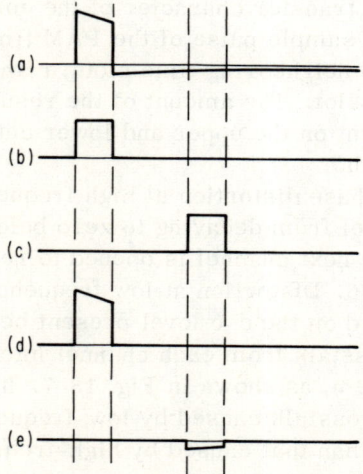

Fig. 18-7. Crosstalk caused by bandwidth limitation at low frequencies: (a) signal pulse of disturbing channel; (b) gating pulse of disturbing channel; (c) gating pulse of disturbed channel; (d) signal received by disturbing channel; (e) crosstalk received by disturbed channel.

modulation products which are generated because of insufficient bandwidth or nonlinear phase and amplitude characteristics of the common transmission path. In a TDM system, the message functions are amplitude-sampled in sequence, and the resulting AM pulses, each in a separate time slot, are transmitted over a common medium. Transmission of such a signal through a common medium of limited bandwidth causes the pulse waveforms to spread in time so that the received pulses of the successive channels interfere with each other.

Crosstalk Due to Nonlinearities of Transfer Characteristics of Transmission Path. Bennett[5] considered the case of crosstalk introduced in an AM-TDM system due to bandwidth limitation of an ideal filter in the common transmission path. His results are of practical interest where bandwidth economy is to be considered in the design of the TDM system. In the following, we shall briefly review the results of Bennett, and then we shall consider the effect of nonlinearities of the transfer characteristics of the transmission path on the interchannel interference or crosstalk.

Consider the unmodulated periodic pulse train shown in Fig. 18-8a which, according to Eq. (18-1), can be represented by the Fourier series,

$$p_T(t) = \frac{\tau}{T}\left[1 + 2\sum_{n=1}^{\infty} \frac{\sin(n\pi\tau/T)}{n\pi\tau/T}\cos n(\omega_0 t - \pi\frac{\tau}{T})\right] \qquad (18\text{-}11)$$

where the origin $t = 0$ has been shifted to coincide with the leading edge of the pulse. The equation of the amplitude-modulated pulse train by a signal of angular frequency ω_m is obtained from Eq. (18-4), namely,

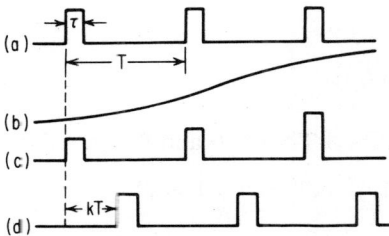

Fig. 18-8. Waveforms in TDM system: (a) modulating pulse; (b) modulated signal; (c) amplitude-modulated pulse; (d) demodulating pulse.

$$f_s(t) = \frac{\tau}{T}\left(1 + m_a \cos \omega_m t + 2 \sum_{n=1}^{\infty} \frac{\sin (n\pi\tau/T)}{n\pi\tau/T} \cos n(\omega_0 t - \pi \frac{\tau}{T})\right.$$

$$+ m_a \sum_{n=1}^{\infty} \frac{\sin (n\pi\tau/T)}{n\pi\tau/T} \left\{\cos \left[(n\omega_0 + \omega_m)t - n\pi \frac{\tau}{T}\right]\right.$$

$$\left.\left.+ \cos \left[(n\omega_0 - \omega_m)t - n\pi \frac{\tau}{T}\right]\right\}\right) \tag{18-12}$$

where m_a is the depth of modulation.

Let the train of modulated pulses be applied to a network whose transfer function is given by $A(\omega)e^{-j\theta(\omega)}$; the output voltage $g_s(t)$ is then

$$g_s(t) = \frac{\tau}{T}\left\{A_0 + m_a A(\omega_m) \cos [\omega_m t + \theta(\omega_m)] + 2 \sum_{n=1}^{\infty} \frac{\sin (n\pi\tau/T)}{n\pi\tau/T}\right.$$

$$\times \cos \left[n\omega_0 t - n\pi \frac{\tau}{T} + \theta(n\omega_0)\right] + m_a \sum_{n=1}^{\infty} A(n\omega_0 + \omega_m)$$

$$\times \frac{\sin (n\pi\tau/T)}{n\pi\tau/T} \cos \left[(n\omega_0 + \omega_m)t - n\pi \frac{\tau}{T} + \theta(n\omega_0 + \omega_m)\right]$$

$$+ m_a \sum_{n=1}^{\infty} A(n\omega_0 - \omega_m) \frac{\sin (n\pi\tau/T)}{n\pi\tau/T} \cos \left[(n\omega_0 - \omega_m)t\right.$$

$$\left.\left.- \frac{n\pi\tau}{T} + \theta(n\omega_0 - \omega_m)\right]\right\} \tag{18-13}$$

The output signal is demodulated or "gated" for a period τ beginning at time $t_0 = kT$ after the start of each modulated pulse, as shown in Fig. 18-8d. The gating process can be represented operationally by the relation

$$g_g(t) = g_s(t) f_g(t) \tag{18-14}$$

where $f_g(t) = 0$, $kT > t > 0$ and $T > t > (kT + \tau)$

$$= 1,   (kT + \tau) > t > kT \tag{18-15}$$

Thus $$f_g(t) = \frac{\tau}{T}\left[1 + 2 \sum_{q=1}^{\infty} \frac{\sin (q\pi\tau/T)}{q\pi\tau/T} \cos \left(q\omega_0 t - q\pi \frac{\tau}{T} - 2q\pi k\right)\right]$$

$$\tag{18-16}$$

Demodulation is accomplished by passing the gated signal $g_g(t)$ through a low-pass filter of cutoff frequency equal to one-half the scanning frequency; since $\omega_c \geq 2\omega_m$, the detected output will contain only a d-c term in addition to terms in ω_m. Neglecting the d-c term, we have

$$g_d(t) = m_a\left(\frac{\tau}{T}\right)^2 \cos \omega_m t \left(A(\omega_m) \cos \theta(\omega_m) + \sum_{n=1}^{\infty} \left[\frac{\sin (n\pi\tau/T)}{n\pi\tau/T}\right]^2\right.$$

$$\times \{A(n\omega_0 + \omega_m) \cos [2\pi nk + \theta(n\omega_0 + \omega_m)] + A(n\omega_0 - \omega_m)$$

$$\left.\times \cos [2\pi nk + \theta(n\omega_0 - \omega_m)]\}\right)$$

$$- m_a \left(\frac{\tau}{T}\right)^2 \sin \omega_m t \left(A(\omega_m) \sin \theta(\omega_m) + \sum_{n=1}^{\infty} \left[\frac{\sin (n\pi\tau/T)}{n\pi\tau/T}\right]^2\right.$$

$$\times \{A(n\omega_0 + \omega_m) \sin [2\pi nk + \theta(n\omega_0 + \omega_m)] - A(n\omega_0 - \omega_m)$$

$$\left.\times \sin [2\pi nk + \theta(n\omega_0 - \omega_m)]\}\right) \qquad (18\text{-}17)$$

By limiting the bandwidth to $\omega \leq M\omega_0 + \omega_m$, where M is an integer, and putting $A(\omega) = 1$ in the passband, Bennett calculated the crosstalk to be

$$\frac{1 + 2 \sum_{n=1}^{M} \left[\frac{\sin (n\pi\tau/T)}{n\pi\tau/T}\right]^2 \cos 2\pi nk}{1 + 2 \sum_{n=1}^{M} \left[\frac{\sin (n\pi\tau/T)}{n\pi\tau/T}\right]^2} \qquad (18\text{-}18)$$

Bennett deduced from Eq. (18-18) that the crosstalk attenuation is poor unless M is chosen to be large.

Crosstalk Due to Insufficient High-frequency and Low-frequency Bandwidth.[7] The contribution of insufficient bandwidth and non-linearities of the transfer function of the common transmission medium to interchannel crosstalk can conveniently be evaluated by considering separately the two cases of insufficient high- and low-frequency bandwidth. The analysis will be carried out by the use of the simple RC transmission-medium models shown in Fig. 18-9, which will be applied as an illustration to a four-channel PAM time-multiplexed signal shown in Fig. 18-10.

As noted earlier, insufficient high-frequency transmission contributes to crosstalk by increasing the decay time of each pulse. This is illustrated in Fig. 18-11 where the pulse in channel 1, when transmitted by the low-pass transmission model, results in an output pattern shown in Fig. 18-11b. We note here that as a result of deficient high-frequency response, the signal tails off into time slots reserved for channels 2, 3, and 4. However, the major crosstalk occurs from the interfering time slot to the next, while very little crosstalk appears in the remaining time slots.

The contribution of insufficient low-frequency transmission to crosstalk is shown in Fig. 18-12 where the "overshoot tail" due to the high-pass filter model constitutes crosstalk since it also carries information from channel 1 into adjacent time slots. It is significant to note that this overshoot tail does not decay appreciably in the range of several time slots, and consequently, the crosstalk contributions extend many time slots beyond the interfering one. We shall presently derive approximate analytical expressions for the crosstalk resulting from low-pass and high-pass transmission models.

Crosstalk Due to Upper Cutoff Frequency. First we shall calculate the crosstalk between adjacent time slots. This can be done with reference to Fig. 18-11b by noting that the crosstalk voltage V_1 at time t sec after the end of time slot 1 is given by

$$V_1 = Ve^{-t/R_1C_1} \tag{18-19}$$

where R_1C_1 is the time constant of the low-pass transmission model. The crosstalk voltage in channel 2 is found by taking $t = \tau_g + \Delta$, where τ_g denotes the guard interval between channels, and Δ is a time interval within channel 2 which will depend on the particular mechanism by which crosstalk is detected in slot 2. Hence

$$V_1 = Ve^{-(\tau_g + \Delta)/R_1C_1} \tag{18-20}$$

In the extreme case, when the detection is governed mainly by the amplitude near the beginning of slot 2, the value of $\Delta = 0$ should be used. Generally the crosstalk ratio is given by

$$CTR = \frac{V_1}{V} = e^{-(\tau_g + \Delta)/R_1C_1} \tag{18-21}$$

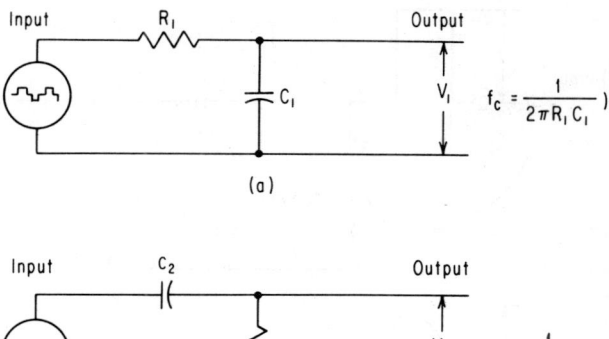

Fig. 18-9. Transmission-medium models: (a) Low-pass transmission model; (b) high-pass transmission model. (From H. M. Straube,[7] IRE Trans.)

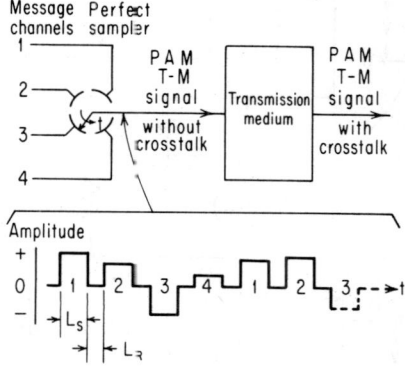

Fig. 18-10. Four-channel PAM time-multiplexed signal. (From H. M. Straube,[7] IRE Trans.)

or the crosstalk attenuation in decibels is

$$(CT)_{db} = 8.686 \frac{\tau_g + \Delta}{R_1 C_1} \qquad (18\text{-}22)$$

As stated previously, the crosstalk in subsequent time slots due to insufficient high-frequency response will be attenuated rapidly. This

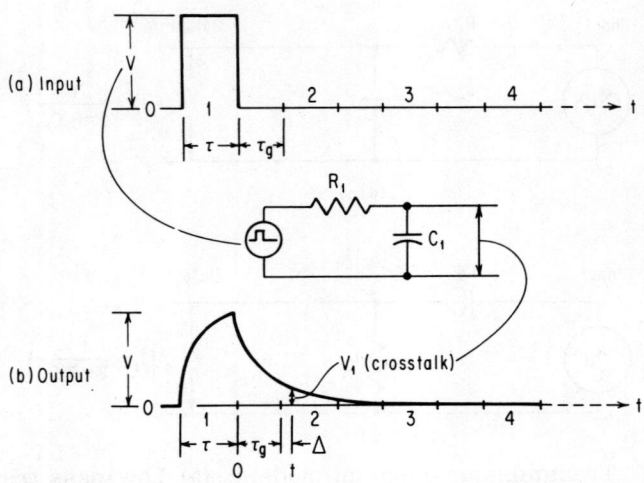

Fig. 18-11. Crosstalk due to insufficient high-frequency transmission. (From H. M. Straube,[7] IRE Trans.)

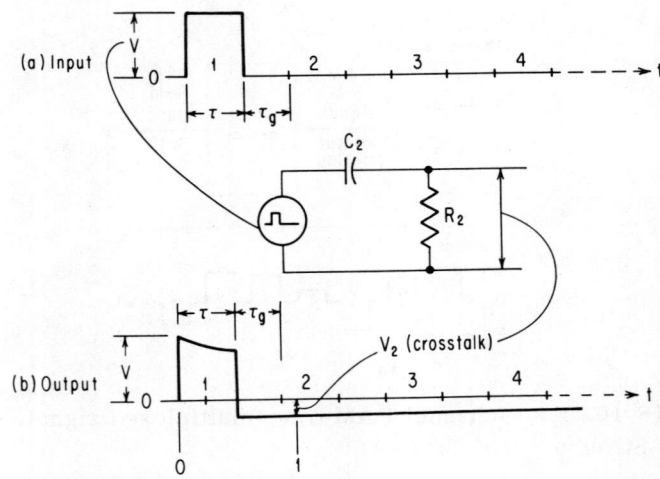

Fig. 18-12. Crosstalk due to insufficient low-frequency transmission. (From H. M. Straube,[7] IRE Trans.)

can be demonstrated simply by finding the crosstalk voltage in channel 1. Since in practice $\tau_g \doteq \tau$, we can take $t = 3\tau_g$ in Eq. (18-22), and assuming $\Delta = 0$, we obtain for channel 3,

$$(CT)_{db} = 8.686 \frac{3\tau_g}{R_1 C_1} \qquad (18\text{-}23)$$

For a typical adjacent-channel crosstalk of 60 db, the crosstalk in channel 3 due to channel 1 will be 120 db less than the crosstalk in channel 2.

Effect of Lower Cutoff Frequency. As before, we shall calculate first the crosstalk between adjacent time slots. With reference to Fig. 18-12b, the crosstalk voltage V_2 is given by

$$V_2 = Ve^{-t/R_2C_2} + (-V)e^{-(t-\tau)/R_2C_2} \tag{18-24}$$

where the first term of the rms represents a voltage step $+V$ applied at $t = 0$, and the second term a step $-V$ applied at $t = \tau_g$.
Equation (18-24) can be rewritten in the form

$$V_2 = V\left(1 - e^{\tau/R_2C_2}\right)e^{-t/R_2C_2}, \qquad t > \tau \tag{18-25}$$

The crosstalk in channel 2 can be evaluated by taking $t = \tau + \tau_g$, whence

$$V_2 = V\left(1 - e^{\tau/R_2C_2}\right)e^{-(\tau+\tau_g)/R_2C_2} \tag{18-26}$$

Considering the fact that the overshoot tail is decaying very slowly it is obvious that the total interference in channel 2 due to a message in channel 1 will include, in addition to the contribution given by Eq. (12-26), also the contributions arising from previous samples in channel 1. The calculations of the general case of crosstalk in one channel due to all previous samples in another channel are straightforward and are given in Ref. 7.

4. *Signal-to-Noise Ratio.*[8] Let P denote the average power of the unmodulated RF pulse trains; when modulation is introduced, the power is increased by $\frac{1}{2}m_a^2P$, where m_a is the modulation index, just as in CW amplitude modulation, due to the energy in the sidebands. Furthermore, let B denote the bandwidth of the RF signal; then it follows that the ratio of the unmodulated input carrier power to noise power is

$$\left(\frac{C}{N}\right)_i = \frac{P}{N_0 B} \tag{18-27}$$

where N_O = noise-power density in watts per cycle, and the input signal-to-noise power ratio is

$$\left(\frac{S}{N}\right)_i = \frac{\frac{1}{2} m_a^2 P}{N_O B} \tag{18-28}$$

Assuming a perfect video detector, the output of the first detector contains unidirectional video pulses of mean power P in the unmodulated state. The pulses when modulated are given by

$$v(t) = \sqrt{P}\,\frac{\tau}{T}\,(1 + m_a \sin \omega_m t)\left[1 + 2\sum_{n=1}^{\infty} \frac{\sin (n\pi\tau/T)}{n\pi\tau/T}\cos n\omega t\right] \tag{18-29}$$

The power in the d-c component is $P\,\tau/T$ which is equal to the power of the RF carrier; the peak pulse power is $\dfrac{P}{\tau/T}$, both in the RF and video circuits. With constant signal power, the signal-to-noise ratio at the output of the video detector is

$$\left(\frac{S}{N}\right)_{o(v)} = \frac{\frac{1}{2} m_a^2 P}{N_O B/2} \tag{18-30}$$

This follows from the fact that the noise in the sidebands is not coherent and adds powerwise, while the sidebands of the signal add voltagewise. The final detection is assumed to be effected by filtering out the low-frequency message signal from the video-frequency spectrum. The low-frequency signal amplitude is obtained from Eq. (18-29) and is equal to $m_a \sqrt{P\,\dfrac{\tau}{T}}$; consequently the low-frequency signal power is equal to $(m_a^2 \tau/2T)P$.

Let the bandwidth of the low-pass filter be f_m, where f_m is the top frequency of the message function. The signal-to-noise ratio in the output is

$$\left(\frac{S}{N}\right)_o = \frac{\frac{1}{2} m_a^2(\tau/T)P}{N_O f_m} \tag{18-31}$$

By blocking the receiver between pulses to eliminate the noise in the interpulse period, which can be done, for example, by means of bottom limiting, the noise power is reduced in the ratio τ/T, and the signal-to-noise power ratio at the output of the low-pass filter is

$$\left(\frac{S}{N}\right)_{o(LP)} = \frac{\frac{1}{2} m_a^2 (\tau/T) P}{N_0 f_m \, \tau/T} = \frac{\frac{1}{2} m_a^2 P}{N_0 f_m} \qquad (18\text{-}32)$$

We have thus obtained the same result as for the conventional CW carrier-amplitude-modulation system. Only a part of the actual signal power is used, however, when the low-frequency band is filtered out of the video-frequency spectrum. Each harmonic in the pulse spectrum has a pair of sidebands associated with it, and they contain signal intelligence power. Any such sideband may be separated out with a bandpass filter, and the message function may then be detected with a normal amplitude-modulation detector. It can be shown [8] that in the process of detecting the intelligence function outside the low-frequency region, there is a deterioration of the signal-to-noise power ratio.

In practice, pulse-amplitude modulation provides a poorer signal-to-noise ratio than conventional amplitude modulation. This is due to the fact that the receiver is unblocked for rather longer than the pulse-duration time, owing to the sloping sides of the pulse.

The advantage of PAM lies not in the signal-to-noise improvement but rather in the feature of time-division multiplexing. We shall see that other pulse-modulation schemes do provide improved signal-to-noise ratio.

18.3   Pulse-duration and Pulse-position modulation [2,9,10]

The improvement in signal-to-noise ratio which is obtained by the use of time-modulated pulses of constant amplitude instead of amplitude-modulated pulses led to the development of systems using pulse-duration and pulse-position modulation.

Pulse-width or pulse-duration modulation is sometimes referred to in the literature as pulse-length modulation. In PDM systems, the trailing or leading edge of the pulse, and sometimes both, may be made to vary as a function of the sampled value of the modulating signal. Pulse-duration modulation has an advantage analogous to that of frequency modulation. Noise and interference can be reduced at the expense of increasing bandwidth, provided the noise level is below the improvement threshold. In PDM, the part of the signal power that carries no information to the receiver is wasted. When the useless part is subtracted from PDM, we have PPM.

Pulse-position modulation is a particular form of pulse-time modulation in which the position of a standard-width pulse is varied as a function of the sampled value of a modulating wave. Pulse-position modulation is essentially the same as PDM except that the variable edge is now replaced by a short pulse. In practice, this can be accomplished by differentiating the waveform of PDM, thus conserving power over that required for PDM. It will be shown that the modulation on a PPM wave may be recovered by passing the wave through a low-pass filter as in the PAM system.

1. *Modulation Methods.* As mentioned previously, the modulating signal may vary the time of occurrence of the leading edge, the trailing edge, or both edges of the pulse. This is illustrated in Fig. 18-13 where the solid lines indicate the durations of the unmodulated pulses and dotted lines show the two extremes for maximum modulation. "Guard interval" refers to the minimum interval between the trailing edge of one pulse and the leading edge of the next when the pulses are fully modulated.

As discussed in Chap. 17, the sampling associated with pulse modulation may be either uniform or natural. Uniform sampling may be defined as a process of sampling where the variation in the parameter of the pulse is proportional to the modulating signal at uniformly spaced sampling times. Natural sampling may be defined as a process of sampling where the time of sampling coincides with the time of appearance of the position-modulated pulse. The two methods of sampling are illustrated in Fig. 18-14 as applied to pulse-duration modulation.

In the process of natural sampling (Fig. 18-14a), the pulse duration τ_n corresponds to the modulation values $M(t_n)$, and conse-

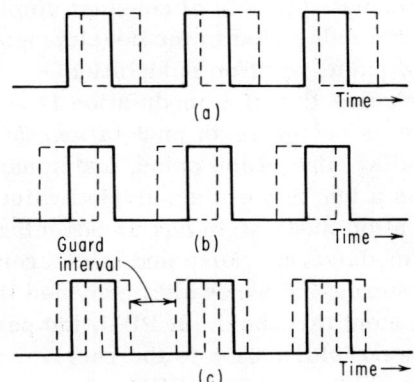

Fig. 18-13. Pulse-duration modulation (PDM): (a) Trailing edge modulated, leading edge fixed; (b) leading edge modulated, trailing edge fixed; (c) both edges modulated.

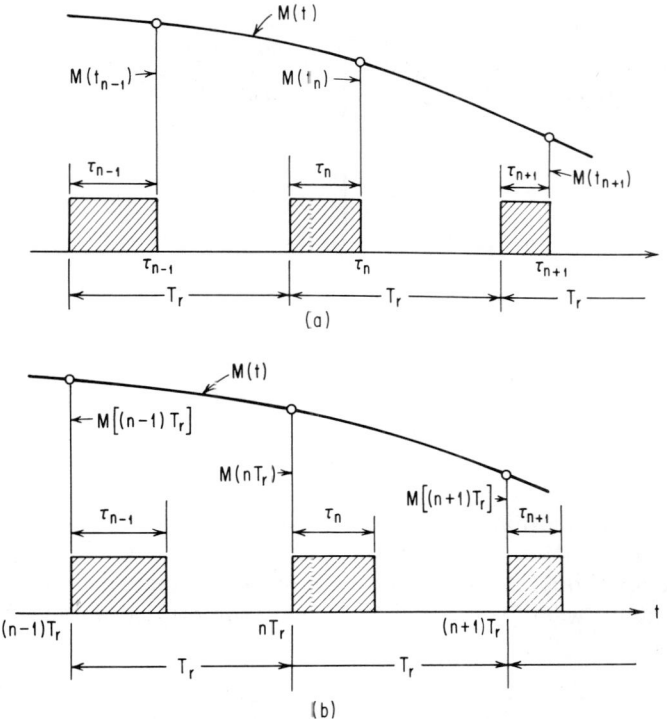

Fig. 18-14. PDM using natural and uniform sampling: (a) Natural sampling; (b) uniform sampling.

quently the sampling intervals t_n are not equal but depend on the modulation level. In the process of uniform sampling, the width of the pulses is proportional to the modulation values $M(t_n)$, which are sampled at equal time intervals $t_n = nT_r$ and are independent of the modulation process.

There are many schemes for producing duration-modulated pulses. In the usual method, the modulating signal voltage is added to a sawtooth-voltage waveform, as shown in Fig. 18-15a, and the composite signal is applied to a tube which is biased beyond cutoff. The tube begins to conduct when the voltage applied to its grid is about half the peak voltage of the sawtooth. When a modulating voltage is added to the sawtooth, the time at which the tube conducts is advanced by a positive signal and retarded by a negative one. The tube is cut off by the trailing edge of the sawtooth, so that its anode current consists of a train of pulses whose trailing edges occur at fixed instants and whose leading edges are modulated as shown in

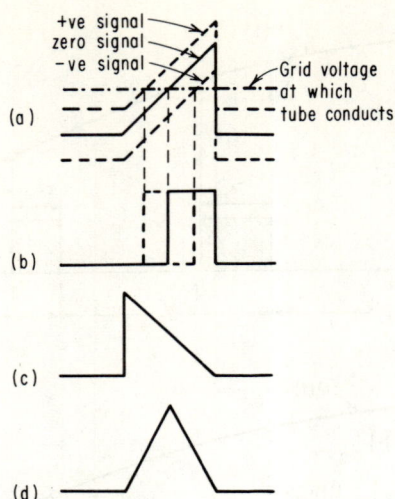

Fig. 18-15. Generation of PDM using sawtooth wave: (a) Sawtooth wave plus modulating signal; (b) resulting length-modulated pulse (leading-edge modulation); (c) waveform for producing trailing-edge modulation; (d) waveform for producing symmetrical pulse-length modulation.

Fig. 18-15b. If, however, the shape of the sawtooth is as shown in Fig. 18-15c, the leading edges of the pulse will occur at fixed instants and the trailing edges will be modulated. If a symmetrical triangular waveform as shown in Fig. 18-15d is used, the time of the center of the pulse is fixed and both leading and trailing edges are modulated. Another method of generating PDM pulses is shown in Fig. 18-16 where the trailing edge alone is modulated. In Fig. 18-16a, the derived AM pulses of maximum duration are added to the sawtooth signal, and the combination is applied to the input of a slicer. This process produces a duration-modulated pulse train where the pulse durations are proportional to the modulating signal level at uniformly spaced sampling times.

In Fig. 18-16b, the modulating signal varies during the scanning process, and the duration of a duration-modulated pulse is proportional to the magnitude of the modulating signal at the trailing edge of the pulse. Thus, the time of sampling varies and coincides in time with each trailing edge.

When this method is used to produce pulses with leading edges modulated, the shape of the sawtooth wave is reversed. Pulses with both edges modulated can be produced with sweep waves in the form of isosceles triangles.

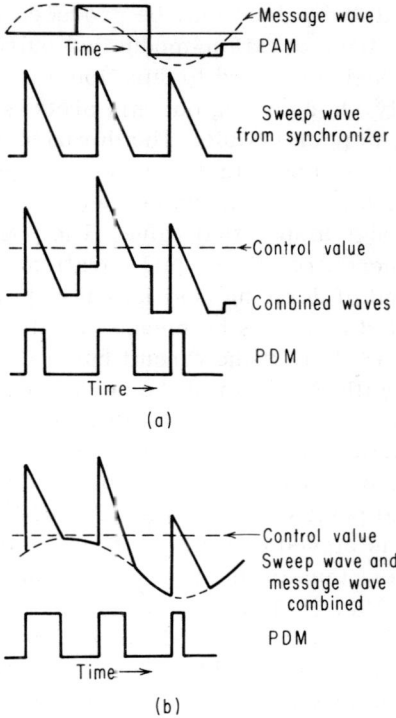

Fig. 18-16. Generation of PDM - trailing edge only: (a) Uniform sampling; (b) natural sampling. (From H. S. Black,[2] courtesy of D. Van Nostrand Company, Inc.)

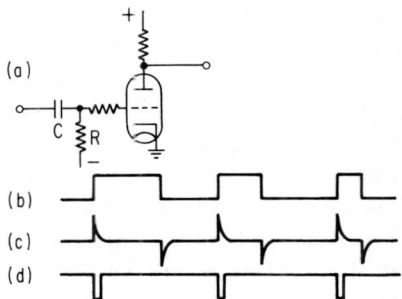

Fig. 18-17. Derivation of PPM from PDM by differentiation: (a) Circuit; (b) length-modulated input pulses; (c) voltage across R; (d) position-modulated output pulses.

Position-modulated pulses may be produced in many ways. One is to derive them from duration-modulated pulses. The modulating wave is sampled and converted to duration-modulated pulses, as explained previously. In carrying out this process, either uniform or natural sampling is permissible. The duration-modulated pulses are then applied to a circuit that generates a rectangular pulse of short duration each time the modulated edge of a duration-modulated pulse passes through a specified value. The position-modulated pulses thus produced are all of equal duration.

Another method of deriving position-modulated pulses from duration-modulated pulses is by means of a differentiating circuit, as shown in Fig. 18-17a. If the circuit has a time constant which is small compared with the width of the duration-modulated pulses of the input, then the output will consist of short spikes of opposite polarity at the leading and trailing edges of each pulse, as shown in Fig. 18-17c. The polarity of the duration-modulated input pulses is chosen so that the positive spike coincides with their modulated edges. The tube is biased beyond cutoff and only conducts during the positive spikes, thus producing position-modulated output pulses, as shown in Fig. 18-17d.

In natural sampling, the PPM signal is produced by making the displacement of successive pulses from their unmodulated positions proportional to the magnitudes of the modulating wave at the instants of sending the pulses. In uniform sampling, the modulating wave is assumed to be instantaneously sampled at regular intervals T_r, and the pulses are subsequently position-modulated. To carry out this operation, it is necessary to delay the sampling information by an interval greater than the maximum excursion of a pulse from its unmodulated time of occurrence.

2. *Demodulation Methods.* The most common method for deriving the modulating signal is to use a low-pass filter. For example, consider a single-edge pulse with modulation with natural sampling. From Eq. (17-114) it is seen that the modulating signal may be recovered without harmonic distortion; however, when uniform sampling is used, the output of the low-pass filter will contain not only the modulating signal but also its harmonics.

Another method of detection of the wanted modulating signal from a duration-modulated pulse train is to convert the PDM signal to PAM. Uniform sampling and conversion to PAM followed by a filter can provide exact recovery of the wanted signal. Conversion to PAM may be accomplished by means of a holding circuit. With this arrangement, a capacitor is charged to a voltage proportional to the duration of the duration-modulated pulse, and the voltage is then held at a virtually constant value while a new pulse proportional in magnitude to the voltage across the capacitor is generated. The amplitudes of the AM pulses are then precise samples of the original modulating signal. With uniform sampling, the modulating signal may be recovered in undistorted form by feeding the amplitude-modulated pulses to the input of an appropriate low-pass filter and equalizer.

With natural sampling, the amplitudes of the derived amplitude-modulated pulses are not proportional to samples taken at equally spaced instants. Instead, they are proportional to samples taken at times coinciding with the positions of the modulated pulse edge. With the above-mentioned method of detection, this results in distortion of the modulating wave at the receiving terminal.

The distortion terms of the detected signal may be calculated as follows. With reference to Fig. 18-18, let the linear sawtooth voltage be given by $e_a = k_1(t_0 - t)$, and the modulating signal by $e_m = k_2 \sin \omega_m t$. The width of the modulated pulse equals $(t_m - t_0)$, where t_m is determined by

$$(e_a + e_m)_{t=t_m} = 0 \qquad (18\text{-}33)$$

Therefore $\quad k_2 \sin \omega_m t_m + k_1(t_0 - t_m) = 0$

or $\quad \omega_m t_0 = \omega_m t_m - \omega_m \dfrac{k_2}{k_1} \sin \omega_m t_m \qquad (18\text{-}34)$

It follows therefore that the time shift $(t_m - t_0)$ of the pulse edge is proportional to the instantaneous modulating signal at the instant t_m at which the pulse edge actually occurs, which is natural sampling. The problem is to solve for t_m in terms of t_0. It is known from Bessel's solution that if

$$\psi = \phi - e \sin \phi \qquad (18\text{-}35)$$

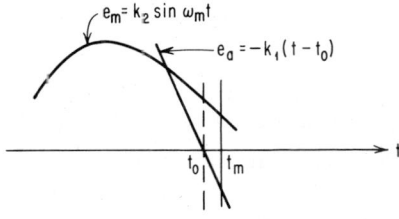

Fig. 18-18. Detection of PDM with natural sampling.

then $\quad \phi = \psi + 2 \sum_{n=1}^{\infty} \dfrac{J_n(ne)}{n} \sin n\psi$ $\hspace{2cm}$ (18-36)

or $\quad \omega_m t_m = \omega_m t_o + 2 \sum_{n=1}^{\infty} \dfrac{J_n(n\omega_m \, k_2/k_1)}{n} \sin n\omega_m t_o$ $\hspace{1cm}$ (18-37)

Therefore $\quad t_m = t_o + \dfrac{2}{\omega_m} \sum_{n=1}^{\infty} \dfrac{J_n(n\omega_m \, k_2/k_1)}{n} \sin n\omega_m t_o$ $\hspace{0.5cm}$ (18-38)

Equation (18-38) can be used to calculate the distortion terms, for example, the second-harmonic distortion $= J_2(2x_0)/2J_1(x_0)$, where

$\quad x_o = \omega_m \, (k_2|k_1), \; \omega_m = 2\pi f_m$

$\quad k_1 = $ slope of sawtooth in volts per second

$\quad k_2 = $ peak audio amplitude in volts

$\quad k_2 = k_1\tau_0$, where τ_0 is the unmodulated pulse duration

and $\quad x_o = \omega_m \, k_2/k_1 = \omega_m \tau_0$

Let $\rho = 2\pi\tau_0/T_r$ radians. Hence

$$x_o = \frac{2\pi f_m \rho}{2\pi f_r} = \rho \, \frac{f_m}{f_r}$$

The following example will illustrate the application of the distortion formula. Let $\rho = 1$, $f_m = 0.5$ Mc/sec, $f_r = 5.0$ Mc/sec. Hence $x_o = 0.1$.

Second-harmonic distortion $\quad \dfrac{J_2(0.2)}{2J_1(0.1)} = 5\%$

Third-harmonic distortion $\quad \dfrac{J_3(0.3)}{3 \times J_1(0.1)} = 0.4\%$

The demodulation of PPM pulses can be effected by passing the PPM signal through a low-pass filter followed by an appropriate equalizer, provided the modulation index is made moderately small so that a delayed copy of the original signal with little distortion is obtained. This is true for either uniform or natural sampling; however, there is less distortion with uniform sampling, as indicated by Eq. (17-40).

A common method of demodulating position-modulated pulses is to convert them first to duration-modulated pulses. This may be done by means of a double-stability circuit that is thrown to its ON position by the received position-modulated pulses. Either the leading or trailing edge may be made to control the operation. The duration-modulated pulses may then be demodulated by any of the methods described previously.

18.4   Crosstalk in PDM and PPM Systems [10,11]

As stated above, crosstalk occurs between the channels of a time-division-multiplex communication system if the transfer characteristics of the transmission path vary sufficiently with frequency to cause the pulse waveforms to spread in time and thus produce at the allotted time of one channel a disturbing voltage due to another channel. The effect of the interfering signal in a PAM system was examined in a previous section. In this section, we shall extend the analysis to systems using pulse-duration and pulse-position modulation.

1. *Effect of a Disturbing Voltage and Operation of Slicing Circuit.*   Since in both PDM and PPM systems, the modulated pulses are of constant amplitude but of variable duration or time of occurrence, it is possible to remove to a considerable degree the effect of the interfering signal without distorting the modulating signal. This is accomplished by a technique called slicing, whereby both the lower part of the pulse together with the interfering signal in the interpulse period and the upper part of the pulse together with the disturbing signal superimposed on the top of the pulse are removed by means of a trigger circuit. It should be noted here that this technique of slicing is applicable only to time-modulated pulses and cannot be used in PAM where the interfering signal present on the tops of the pulses is inseparable from the modulation signal. It will be shown in the next section that in the application of a slicing circuit to time-modulated pulses, the only residual noise which cannot be removed is that which is superimposed on the rise and fall of each pulse. Thus by the use of slicing techniques, it is possible to achieve an improvement in signal-to-noise ratio in PDM and PPM systems compared to PAM. For the same reason, it is possible to obtain greater interchannel crosstalk attenuation in time-modulated TDM systems than in PAM systems.

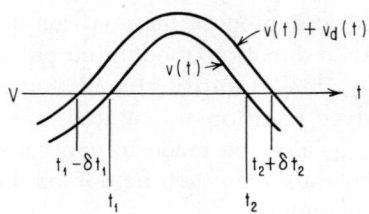

Fig. 18-19. Pulse in the presence of a disturbing voltage.

The effect of the disturbing signal $v_d(t)$ on the pulse $v(t)$ is shown in Fig. 18-19. Let the leading edge of the pulse cut the slicing level V at time t_1, in the absence of disturbance. As seen from the illustration, the presence of a small disturbing voltage $v_d(t)$ will cause the leading edge of the pulse to cut the slicing level at time $(t_1 - \delta t_1)$ given by

$$\delta t_1 = \left(\frac{dt}{dv}\right)_{t_1} v_d(t_1) \tag{18-39}$$

where $(dv/dt)_{t_1}$ denotes the slope of the leading edge of the pulse at time t_1. Similarly, the effect of the disturbing voltage on the time t_2 of the intersection of the trailing edge with the slicing level is given by

$$\delta t_2 = \left(\frac{dt}{dv}\right)_{t_2} v_d(t_2) \tag{18-40}$$

If the disturbing voltage v_d and the displacement time of the pulse t_m are both varying sinusoidally with time, then

$$v_d(t_1) = V_1 \cos (\omega_d t + \psi) \tag{18-41}$$

$$v_d(t_2) = V_2 \cos (\omega_d t + \psi) \tag{18-42}$$

$$t_m = T_m \cos (\omega_m t + \phi) \tag{18-43}$$

In a PPM system, the demodulated output voltage is proportional to the displacement of the leading edge of the pulse from the slicer. It follows therefore from Eqs. (18-39), (18-41), and (18-43) that the crosstalk ratio of the amplitudes of the disturbing and signal voltages after demodulation is

$$\text{CTR} = \frac{\delta t_1}{T_m} = \frac{V_1}{T_m} \left(\frac{dt}{dv}\right)_{t_1} \tag{18-44}$$

In a PDM system, the width of the pulse is usually varied by modulating the time of occurrence of the leading edge of the pulse while keeping the time of occurrence of the trailing edge fixed. If the leading edge of the pulse is displaced by the modulation signal by t_m, the duration of the output pulse from the slicing circuit equals $(t_2 - t_1 - t_m + \mathcal{E}t_1 - \delta t_2)$. It follows from Eqs. (18-39) to (18-44) that the crosstalk ratio in a PDM system is given by

$$\text{CTR} = \frac{V_2}{T_m}\left(\frac{dt}{dv}\right)_{t_2} - \frac{V_1}{T_m}\left(\frac{dt}{dv}\right)_{t_1} \tag{18-45}$$

Let the rise time of the pulse T_R and the fall time T_F be defined as

$$T_R = V_p\left(\frac{dt}{dv}\right)_{t_1} \tag{18-46}$$

$$T_F = -V_p\left(\frac{dt}{dv}\right)_{t_2} \tag{18-47}$$

where V_p is the peak amplitude of the pulse. Hence for PPM, the crosstalk ratio becomes

$$\text{CTR} = \frac{V_1}{V_p}\frac{T_R}{T_m} \tag{18-48}$$

and for PDM,

$$\text{CTR} = \frac{V_2 T_F + V_1 T_R}{V_p T_m} \tag{18-49}$$

2. *Crosstalk Caused by Distortion at Low Frequencies.*   It has been shown earlier that the spectrum of a width- or position-modulated pulse train comprises a d-c component, a component of the angular frequency ω_m of the modulating signal, and harmonics of the pulse-repetition frequency $m\omega_r$, together with upper and lower sidebands about each harmonic $(m\omega_r \pm n\omega_m)$. It follows therefore that in order to effect demodulation without introducing large distortion components, the pulse-repetition frequency ω_r must be chosen high enough so that the lower-sideband components of the PRF which overlap with the modulating frequencies will be of negligible amplitude.

As in the case of PAM, let the transfer function of the network of the common transmission path be given by $A(\omega)e^{-j\theta(\omega)}$, and let the component of the modulating signal be denoted by

$$v_m(t) = V_m \cos \omega_m t \tag{18-50}$$

As a result of the distorting network, the low-frequency component of the output signal will be distorted and given by

$$v_o(t) = V_m [A(\omega_m) \cos \theta(\omega_m) \cos \omega_m t + A(\omega_m) \sin \theta(\omega_m) \sin \omega_m t]$$

$$= v_m(t) + v_d(t) \tag{18-51}$$

where the distortion component $v_d(t)$ is given by

$$v_d(t) = V_d \cos (\omega_m t + \psi) = V_m \{[A(\omega_m) \cos \theta(\omega_m) - 1] \cos \omega_m t$$

$$+ A(\omega_m) \sin \theta(\omega_m) \sin \omega_m t\} \tag{18-52}$$

Thus, each channel in the system produces a disturbing voltage at the allotted time of each other channel which is independent of the time separation between the pulses of the disturbing and disturbed channels.

As an illustration, consider the RC coupling circuit shown in Fig. 18-20a,

$$A(\omega) = \frac{\omega}{\sqrt{\omega^2 + \omega_o^2}}$$

$$\theta(\omega) = \tan^{-1} \frac{\omega_o}{\omega}$$

where $\omega_o = \dfrac{1}{R_g C_c}$

If $\omega \gg \omega_o$, then $A(\omega) \doteq 1$ and $\theta(\omega) \doteq \omega_o/\omega$; hence, using Eq. (18-52), the amplitude of the disturbing signal is given by

$$V_d = \frac{\omega_o}{\omega_m} V_m \tag{18-53}$$

For an amplifier with n RC coupling circuits with equal time constants

$$A(\omega) \doteq 1, \qquad \theta(\omega) \doteq n \frac{\omega_o}{\omega}$$

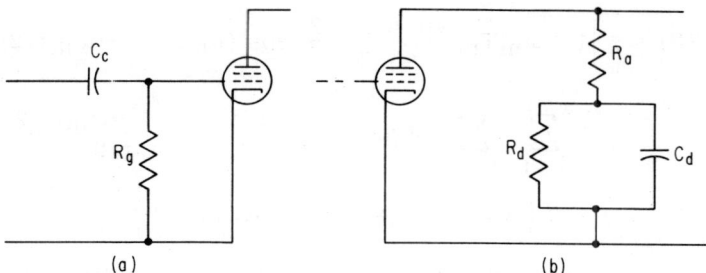

Fig. 18-20. Circuits causing distortion at low frequencies: (a) Coupling circuit; (b) decoupling circuit.

Therefore $V_d = n \dfrac{\omega_0}{\omega_m} V_m$ (18-54)

For the RC decoupling circuit shown in Fig. 18-20b, we obtain

$$A(\omega) \cos \theta(\omega) = \frac{\omega^2 + \omega_1 \omega_2 + \omega_2^2}{\omega^2 + \omega_2^2}$$

$$A(\omega) \sin \theta(\omega) = - \frac{\omega \omega_1}{\omega^2 + \omega_2^2}$$

where $\omega_1 = \dfrac{1}{R_a C_d}$ and $\omega_2 = \dfrac{1}{R_d C_d}$

If $\omega \gg \omega_1$ and ω_2, then $A(\omega) \doteq 1$, and $\theta(\omega) \doteq -\omega_1/\omega$. Hence

$$V_d = \frac{\omega_1}{\omega_m} V_m \tag{18-55}$$

For n stages with RC decoupling circuits with equal time constants, we obtain

$$V_d = n \frac{\omega_1}{\omega_m} V_m \tag{18-56}$$

In the following, we shall apply the results which were just derived to pulse-position and pulse-duration modulation.

Pulse-position Modulation.   The spectrum of a train of rectangular pulses, of unity amplitude-position modulated by a sinusoidal signal, can be derived from Eq. (17-106) in the form

$$F(t) = S \left\{ 1 + \omega_m T_m \frac{\sin (\omega_m \tau/2)}{\omega_m \tau/2} \cos (\omega_m t + \phi - \omega_m \tau/2) \right.$$

$$+ 2 \sum_{k=1}^{\infty} \sum_{n=-\infty}^{\infty} J_n(k\omega_r T_m) \frac{\sin (k\omega_r + n\omega_m)\tau/2}{k\omega_r \tau/2}$$

$$\left. \times \cos [(k\omega_r + n\omega_m) + n\phi - n\omega_m \tau/2] \right\} \qquad (18\text{-}57)$$

where the displacement of each pulse from the mean time of occurrence is given by Eq. (18-43), the duration of each pulse is τ, and S
is the duty ratio. It follows therefore from Eq. (18-57) that the
modulating-frequency component of the modulated pulse train is

$$v_m(t) = 2 \frac{T_m}{\tau} S \sin \left(\omega_m \frac{\tau}{2}\right) \cos \left(\omega_m t + \phi - \omega_m \frac{\tau}{2}\right) \qquad (18\text{-}58)$$

Since $\omega_m \tau \ll 1$, we have

$$v_m(t) \doteq \omega_m S T_m \cos \left(\omega_m t + \phi - \omega_m \frac{\tau}{2}\right) \qquad (18\text{-}59)$$

and for $\omega_0 \ll \omega_m$, the disturbing voltage caused by the RC coupling
circuit of Fig. 18-20a is

$$V_d = \frac{\omega_0}{\omega_m} V_m = \omega_0 S T_m \qquad (18\text{-}60)$$

If the pulses were perfectly rectangular, their rise time would be
zero, resulting in zero crosstalk in the output from the slicing circuit. In practice, limited high-frequency bandwidth causes the
pulses to have finite rise times, and the crosstalk ratio, using Eq.
(18-48), is

$$CTR = \omega_0 S T_R \qquad (18\text{-}61)$$

Similarly, for the RC decoupling circuit, the crosstalk ratio is

$$CTR = \omega_1 S T_R \qquad (18\text{-}62)$$

The crosstalk ratio is independent of the frequency of the modulating
signal and of the time separation between the pulses of the disturbed
and disturbing channels. If, however, ω_0 or ω_1 is not small compared with ω_r, the pulse shape is distorted by the network and the
crosstalk is no longer independent of time.

Pulse-duration Modulation. In this case, we make use of Eq. (17-114) from which we derive, for the spectrum of a duration-modulated pulse train of unity amplitude, the expression

$$F(t) = S \left\{ 1 + \sin(\omega_m t + \phi) + \sum_{k=1}^{\infty} \frac{\sin k\omega_r(t - \tau/2)}{k\omega_r \tau/2} \right.$$

$$\left. + \sum_{k=1}^{\infty} \sum_{n=-\infty}^{\infty} \frac{J_n(k\omega_r \tau)}{k\omega_r \tau} \sin\left[(k\omega_r + n\omega_m)t + k\omega\frac{\tau}{2} + n\phi \right] \right\} \tag{18-63}$$

In this expression it is assumed that the leading edge alone is modulated by a waveform $\sin(\omega_m t + \phi)$, and the depth of modulation is unity ($T_m = \tau$). The modulating-frequency component is

$$v_m(t) = S \sin(\omega_m t + \phi) \tag{18-64}$$

From Eq. (18-53), the disturbing voltage caused by the RC coupling circuit is

$$V_d = \frac{\omega_o}{\omega_m} V_m = \frac{\omega_o}{\omega_m} S \tag{18-65}$$

Since $T_m = \tau$ and $V_p = 1$, we obtain for $\omega_o \ll \omega_m$, by the use of Eq. (18-49), the crosstalk ratio

$$CTR = \frac{\omega_o}{\omega_m} S \frac{T_F + T_R}{\tau} \tag{18-66}$$

Similarly for the RC decoupling circuit of Fig. 18-20b, the crosstalk ratio is

$$CTR = \frac{\omega_1}{\omega_m} S \frac{T_F + T_R}{\tau} \tag{18-67}$$

In the case of PDM, we note that the crosstalk ratio varies inversely with the modulating frequency and is independent of the time of separation between the pulses of the disturbed and disturbing channels. As in PPM, if ω_o or ω_1 is not small compared with ω_r, the pulse is distorted, and the crosstalk is no longer independent of time.

3. *Crosstalk Caused by Distortion at High Frequencies.* The adjacent-channel crosstalk ratio for PPM and for PDM is derived in Ref. 10, for an n-stage RC amplifier whose stages have equal time constants. Only the results will be given below.

Pulse-position Modulation. Figure 18-21 shows two adjacent pulses in a PPM system displaced from their mean positions by the times t_{m_1} and t_{m_2}, respectively. The time interval between the mean positions of the two pulses is t_c. It is shown in Ref. (10) that the crosstalk ratio is given by

$$\text{CTR} = \alpha T_R \frac{U}{V_p} \tag{18-68}$$

where $\alpha = 1/R_a C_s$, R_a equals the plate resistance of the RC-coupled amplifier stage, and C_s is the total plate-to-ground capacitance. T_R is the rise time of the output pulse from the amplifier given by Eq. (18-46), and U is given by

$$U = e^{-\alpha(t_c + t_1 - \tau)} \sum_{r=0}^{n-1} \frac{\alpha^r (t_c + t_1 - \tau)^r}{r!} \tag{18-69}$$

where t_1 is the time taken for the response of the amplifier to reach the slicing level when a rectangular pulse of unit height is applied. The optimum slicing level at the amplifier output is shown in Fig. 18-22, and the peak output voltage V_p is plotted in Fig. 18-23. For minimum crosstalk, the slicing level should be adjusted at the point where the slope of the leading edge is maximum, which is given by

$$\alpha t_1 = n - 1 \tag{18-70}$$

and T_R is given in Fig. 18-24.

For a single-stage amplifier $t_1 = 0$, and Eq. (18-68) reduces to

$$\text{CTR} = \frac{\alpha T_R}{V_p} e^{\alpha(\tau - t_c)} \tag{18-71}$$

Pulse-duration Modulation. Two adjacent channel pulses of a PDM system are shown in Fig. 18-25. The crosstalk ratio is shown to be given by the expression

$$\text{CTR} = \alpha T_R \frac{W}{V_p} \tag{18-72}$$

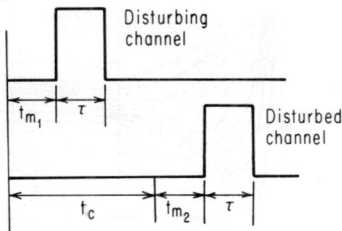

Fig. 18-21. Adjacent channel pulses of a PPM system.

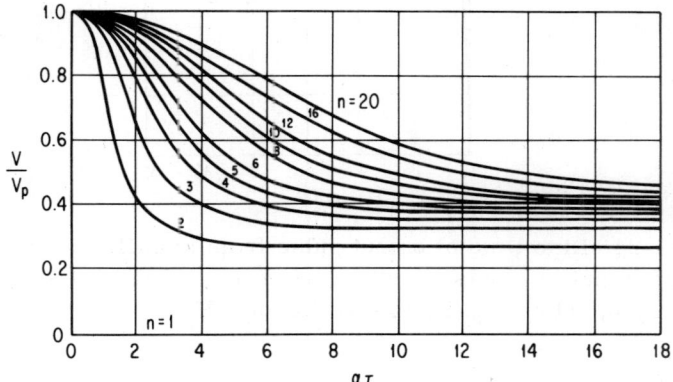

Fig. 18-22. Optimum slicing level at output of amplifier: V = slicing voltage; V_p = peak voltage of pulse; n = number of stages;
τ = pulse width. $\alpha = 1/C_s R_a$. (From J. E. Flood, [10] Proc. IEE.)

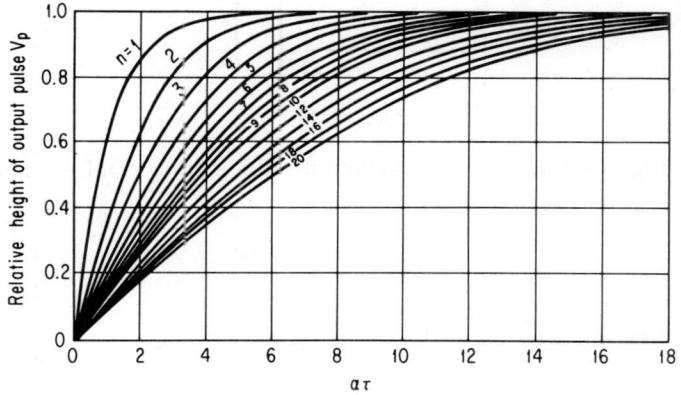

Fig. 18-23. Peak height of output pulse from amplifier: n = Number
of stages; τ = pulse width. $\alpha = 1/C_s R_a$. (From J. E. Flood, [10]
Proc. IEE.)

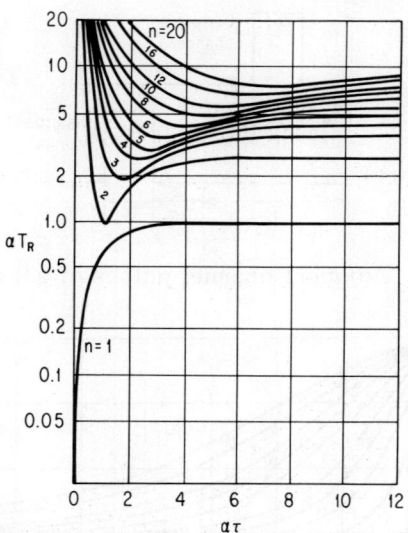

Fig. 18-24. Rise time of output pulse from amplifier: T_R = Rise time at output; τ = pulse width at input; n = number of stages; $\alpha = 1/C_S R_a$. (From J. E. Flood,[10] Proc. IEE.)

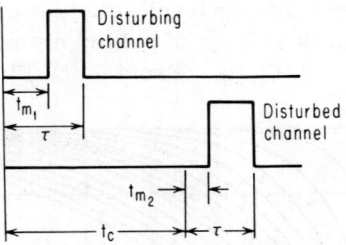

Fig. 18-25. Adjacent-channel pulses in a PDM system.

where
$$W = e^{-\alpha(t_c + t_1)} \sum_{r=o}^{n-1} \frac{\alpha^r (t_c + t_1)^r}{r!} \qquad (18\text{-}73)$$

As before, for minimum crosstalk, the slicing level should be that at which the slope of the leading edge is maximum; t_1 is then given by Eq. (18-70) and T_R by Fig. 18-24.

18.5   Signal-to-Noise Improvement Ratio [2,11-12]

In PDM, the noise manifests itself as jitter in the leading and trailing edges of the recovered pulses. Since the positions of the leading and trailing edges define pulse duration, it is evident that the receiver cannot distinguish between signal modulation and noise. Slopes of the leading and trailing edges influence noise reduction. The absolute value of the slope is proportional to the bandwidth. Since both leading and trailing edges of the pulses are affected by noise, the noise contributions from the two edges will combine to an extent depending upon their correlation. With resistance noise, the correlation is negligible; consequently, the S/N ratio is 3 db poorer than when only one edge is affected.

A PPM system can be considered the pulse-modulation analogue of a phase-modulation or frequency-modulation system. We have noted that in a PPM system, information is transmitted by displacement of the positions of the train of pulses constituting the carrier. Since the pulses are maintained at fixed width and amplitude, the positional information is also carried by the location of the leading edge of a pulse or by the point of zero crossing of the leading edge. Thus, it is obvious that PPM systems are affected by noise in the same manner as PDM systems.

In order to evaluate the effect of noise on the location of the edge of the pulse, consider a train of unmodulated trapezoidal pulses as shown in Fig. 18-26. A_c represents the fixed amplitude of these pulses, and t_0 represents the maximum modulation displacement in any one direction away from the unmodulated position of the pulse. The position of the pulse will vary sinusoidally in time about the unmodulated pulse location given by $\cos \omega_m t$, where ω_m is the frequency of the modulating signal. The rms displacement of the pulse is therefore $t_0/\sqrt{2}$ sec, and the output of the demodulator proportional to t_0 is given by $k t_0 \cos \omega_m t$.

If noise is superimposed on the pulse, the pulse amplitude and the point of zero crossing will both be perturbed, as shown in Fig. 18-27. An amplitude limiter can be used to eliminate the pulse-amplitude variation due to the noise, just as in the FM system; however, the uncertainty in the pulse position due to noise fluctuations remains and gives rise to a noise voltage at the output of the demodulator.

Let Δt sec be the rms positional error due to the rms noise voltage σ. For a given mean noise power, the noise effect may be minimized by using pulses with sharply defined leading edges; but this requires either decreasing the rise time, correspondingly

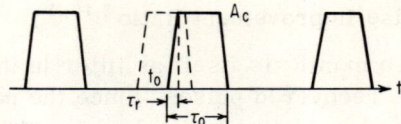

Fig. 18-26. Pulse-position modulation of trapezoidal pulses.

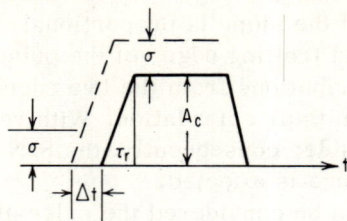

Fig. 18-27. Variation in pulse position due to noise or interference.

widening the transmission bandwidth, or increasing the pulse amplitude A_c. This is analogous to the FM case where an exchange of
input signal-to-noise for bandwidth is possible; here decreasing the
rise time (or increasing the bandwidth) enables a smaller amplitude
pulse to be used for a given rms positional error.

A simple expression for output signal-to-noise ratio may be
derived as follows:[12] The rms output signal voltage is $kt_0/\sqrt{2}$, and
the rms noise voltage due to the rms positional displacement Δt
will similarly be $k \Delta t$ at the output of the demodulator. The S/N
power ratio at the demodulator output is then

$$\left(\frac{S}{N}\right)_o = \left(\frac{t_0/\sqrt{2}}{\Delta t}\right)^2 \tag{18-74}$$

From Fig. 18-27, we have $\dfrac{\Delta t}{\tau_r} = \dfrac{\sigma}{A_c}$ (18-75)

valid for $A_c \gg \sigma$, and in the absence of a signal. Substituting for
Δt in Eq. (18-74), the output $(S/N)_o$ is given by

$$\left(\frac{S}{N}\right)_o = \frac{1}{2}\left(\frac{t_0}{\tau_r}\right)^2 \left(\frac{A_c}{\sigma}\right)^2 \tag{18-76}$$

But $(A_c/\sigma)^2$ is the ratio of peak pulse power to mean noise power
and is analogous to the input carrier-to-noise ratio in FM. Therefore

$$\left(\frac{C}{N}\right)_i = \left(\frac{A_c}{\sigma}\right)^2$$

and $\quad \left(\frac{S}{N}\right)_o = \frac{1}{2}\left(\frac{t_o}{\tau_r}\right)^2 \left(\frac{C}{N}\right)_i$ (18-77)

Equation (18-77) indicates the improvement in $(S/N)_o$ possible in a PPM system by decreasing the pulse rise τ_r or correspondingly widening the transmission bandwidth. If the channel has reasonably sharp passband characteristics, then $B \doteq 1/\tau_r$, where B is the video bandwidth necessary to pass the trapezoidal pulses chosen. The output $(S/N)_o$ ratio becomes

$$\left(\frac{S}{N_o}\right) = \frac{1}{2} t_o^2 B^2 \left(\frac{C}{N}\right)_i$$ (18-78)

If $B = 1/2\tau_r$ is used instead,

$$\left(\frac{S}{N}\right)_o = 2t_o^2 B^2 \left(\frac{C}{N}\right)_i$$ (18-79)

As in the case of FM, the $(S/N)_o$ ratio cannot be improved indefinitely because the noise power introduced at the receiver increases with bandwidth and eventually becomes comparable to the signal and "takes over" the system. A threshold level thus also exists just as in the FM case. This threshold level is usually taken as $A_c/\sigma = 2$, or $(C/N)_i = 4$ (6 db).

It is instructive at this time to compare further the expressions for the output power C/N in FM and PPM. [12]

FM: $\quad \left(\frac{S}{N}\right)_o = 3\beta^2 \left(\frac{C}{N}\right)_i$ (18-80)

PPM: $\quad \left(\frac{S}{N}\right)_o = \frac{1}{2}\left(\frac{t_o}{\tau_r}\right)^2 \left(\frac{C}{N}\right)_i$ (18-81)

The modulation index β is proportional to the frequency deviation Δf, which in turn is one-half the IF bandwidth for large β. The bandwidth required in a PPM system is proportional to $1/\tau_r$, so that for both FM and PPM, the improvement in the $(S/N)_o$ power ratio is proportional to the square of the bandwidth, and the improvement in the S/N voltage ratio, $(S/N)_{o(v.r.)}$, varies linearly with the bandwidth. Thus, in the FM case, we have $\beta = \Delta f/f_m = 2B/2f_m$, where $\beta \gg 1$, and in the PPM case we have $1/\tau_r = B$. B cycles/sec is then the frequency bandwidth required for PPM transmission, and 2B cycles/sec for FM transmission. For a single-channel PPM system, the minimum spacing between pulses is the Nyquist sampling interval $1/2f_m$, with f_m the maximum-frequency component of the signal to be transmitted. For this single-channel system, the maximum modulation displacement t_o is just one-half the sampling interval, assuming $t_o > \tau_o$, the pulse width, and assuming no guard time between pulses. Then

$$t_o \doteq \frac{1}{4f_m} \qquad (18\text{-}82)$$

In terms of bandwidth, the rms $(S/N)_{o(v.r.)}$ voltage ratios at the system output become

FM: $$\left(\frac{S}{N}\right)_{o(v.r.)} = \sqrt{3}\,\beta \left(\frac{C}{N}\right)_{i(v.r.)} = \frac{2\sqrt{3}\ B}{2f_m} \left(\frac{C}{N}\right)_{i(v.r.)} \qquad (18\text{-}83)$$

Single-channel PPM: $$\left(\frac{S}{N}\right)_{o(v.r.)} = \frac{t_o/\sqrt{2}}{\tau_r} \left(\frac{C}{N}\right)_{i(v.r.)}$$

$$= \frac{B}{4\sqrt{2}\ f_m} \left(\frac{C}{N}\right)_{i(v.r.)} \qquad (18\text{-}84)$$

In both cases, then, the output-voltage S/N is linearly proportional to the ratio of transmission bandwidth to signal bandwidth f_m. The larger the ratio of B/f_m, the greater the S/N improvement, provided that the input C/N always exceeds a specified threshold level.

Both FM and PPM systems are examples of uncoded, or analogue-type, signal transmission systems where the rms noise improvement is linearly proportional to B/f_m. Information transmission is also possible by so-called coded, or digital-type, transmission systems. Pulse-code modulation is an example of such a system, as we shall see in Chap. 20. For these coded systems, bandwidth widening also improves the output signal-to-noise ratio; but the improvement will be shown to be exponential with bandwidth, rather than linear. Such systems are thus inherently capable of better transmission efficiency than the uncoded types, as represented by FM, PPM, and other examples.

References

1. Flood, J. E.: Time Division Multiplex Systems, *Electron. Eng.*, January, February, March, April, 1953.
2. Black, H. S.: "Modulation Theory," D. Van Nostrand Company, Inc., Princeton, N. J.
3. Fitch, E.: The Spectrum of Modulated Pulses, *J. IEE London*, vol. 94, part 3A, pp. 556–564, 1947.
4. Roberts, F. F., and J. C. Simmonds: Multichannel Communication Systems, *Wireless Engr. London*, vol. 22, pp. 538–549, November, 1945; pp. 576–589, December, 1945.
5. Bennett, W. R.: Time Division Multiplex Systems, *BSTJ*, vol. 20, p. 199, 1941.
6. Flood, J. E., and J. R. Tillman: Crosstalk in Amplitude-modulated Time-division-Multiplex Systems, *Proc. IEE London*, vol. 98, part 3, p. 279, 1951.
7. Straube, H. M.: Dependence of Crosstalk on Upper and Lower Cut-off Frequencies in PAM Time-multiplexed Transmission Paths, *IRE Trans. Commun. Systems*, September, 1962.
8. Haard, B.: Signal-to-Noise Ratios in Pulse Modulation Systems, *Ericsson Technics*, no. 47, 1948.
9. Sanchez, M., and F. Popert: Über die Berechnung der Spektren Modulierter Impulsfolgen, *Arch. Elek. Übertragung*, October, 1955, pp. 441–452.
10. Flood, J. E.: Crosstalk in TDM Communication Systems Using Pulse-position and Pulse-length Modulation, *Proc. IEE London*, vol. 99, part 4, p. 64, 1952.
11. Kretzmer, E. R.: Interference Characteristics in PTM, *Proc. IRE*, vol. 38, p. 252, 1950.
12. Schwartz, M.: "Information Transmission, Modulation, and Noise," McGraw-Hill Book Company, New York.

19

INTRODUCTION TO INFORMATION THEORY— WITH SPECIAL APPLICATION TO PCM

Information theory provides a fundamental insight into the nature of communication systems. It provides a means of assessing quantitatively the "information content" of a message and can be used to show to what extent existing modulation methods and communication systems fall short of theoretically attainable performance.

The term "information theory" as used in the current technical literature applies to a variety of disciplines which are comprised in the statistical theory of communication. In the strict-sense definition, information theory is the study of the three concepts: information content of message or entropy, channel capacity, and coding. These are essentially the three key concepts which Shannon[1] used to describe the specific mathematical model of communication systems in his pioneering paper in 1948. In the extended sense, information theory includes also the study of filtering and prediction, signal detection, noise theory, information processing, and many other aspects of modulation theory which are sometimes referred to as statistical communication theory. The purpose of this chapter is to cover only the elementary principles and basic concepts of information theory, with special emphasis on the application to PCM. We shall limit our discussion to the original concepts of Shannon's

theory with the view of developing the proper tools for evaluating
the performance of the various transmission systems which are
treated in this book. In particular, the theory will provide us with a
definition and measure of the quantity of information in a message,
as well as a definition and measure of the capacity of a communication channel used to transmit information. These definitions and
measures, which are common to all modulation systems, will provide us with means of measuring their relative efficiency and establishing criteria for making comparisons between the various communication systems.

Figure 19-1 represents the generalized communication system
which will be examined from the point of view of information theory.
It consists essentially of five elements: [2]

1. An information source, which selects one message from a set
of possible messages to be transmitted to the receiver. The message
may consist of a finite sequence of discrete symbols such as letters
or numbers as in telegraphy or teletype, or a continuous function of
time such as speech or music.

2. A transmitter, which operates on the message in order to
produce a signal suitable for direct transmission over the transmitting medium or channel. This operation on the message differs
for various modulation systems. In voice communication, it amounts
simply to a CW modulation process of a carrier, whereas in telegraphy we have an encoding operation which produces a sequence of
dots, dashes, and spaces corresponding to the letters of the message.
This encoding process may assume a more complex form as in
multiplex PCM telephony, where the different continuous time functions of the voice channels are compressed, sampled, quantized, and
encoded in a group of interleaved pulses.

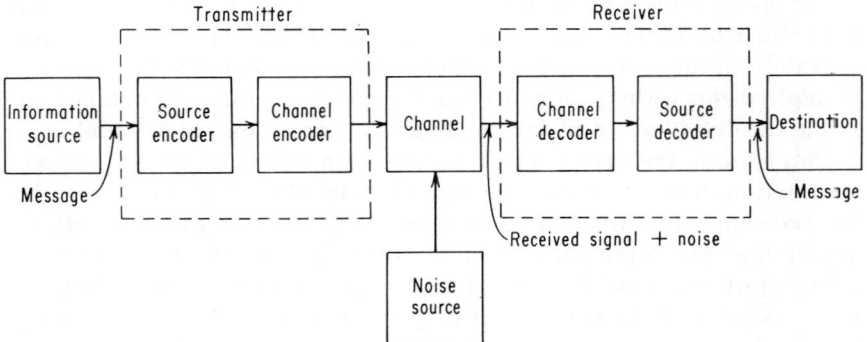

Fig. 19-1. Model of generalized communication system.

3. The channel, which is merely the medium used to transmit the signal from the transmitter to the receiver. During transmission, or at the receiving terminal, the signal may be perturbed by noise or distortion. This is represented in the generalized communication system by the noise source. Although the distortion is deterministic in nature, the noise can be treated only in a statistical manner.

4. The receiver, which operates on the received signal in order to retrieve or decode the original message as accurately as possible before delivery to the destination. Just as the transmitter is introduced to match the channel to the information source, it is the function of the receiver to match the channel to the destination.

5. The destination, which is the person or instrument for whom the message is intended.

19.1   Measure of Information [1-5]

Fundamental to the concept of measure of information is the notion that the more uncertain we are about the composition of the message, the greater will be the amount of information we acquire when reception has completely removed the uncertainty as to what message was sent. Indeed, if the message were known at the receiver beforehand, there would be little point in transmitting it since no new information would be acquired at the receiver. At the receiver, only the class of messages is known, from which a selection is made at the input by the information source, together with the relative probabilities of selection of messages from that class. It is this lack of detailed knowledge about the message which is intimately tied in with the information content of the message.

In addition to the lack of knowledge of the message, the behavior of the noise source in the channel can be described only statistically, in probability terms, and consequently introduces an additional uncertainty in the composition of the message. These are the fundamental reasons why probability theory plays such an important role in the description and understanding of communication systems.

Intrinsic to the strict-sense information theory of Shannon is a restriction to discrete or telegraphlike signals. Continuous signals can be reduced to discrete values by the process of sampling and quantizing. The information source consists of a discrete set or alphabet of symbols. In general, any message to be transmitted consists of symbols selected at discrete moments from the ensemble or alphabet of symbols. Thus, the information source may be considered as a stationary random process whose sample sequences are strings of symbols out of the available alphabet. Let n denote the number of symbols used in a message, and s the number of symbols in the alphabet of symbols; then the possible number of different messages or events $x_1, x_2, \ldots, x_k, \ldots, x_M$ of length n will be given by

$$M = s^n \qquad (19\text{-}1)$$

If the information source selects a message purely at random from the total number of M equiprobable messages, each message will have an equal probability of being selected, which is given by $P(x_k) = 1/M$. A quantitative measure of the information content of the message selected may be derived as follows.[4] Following Nyquist[6] and Hartley,[7] it is convenient to define the information content of a message as some function $F[P(x)]$ of its probability $P(x)$ of being selected from all possible messages. Intuitively, since we have assumed in this case that all the possible messages are equally probable, the information content must obviously be proportional to the number of messages, and $F[P(x)]$ must therefore possess this additive property. As an example, we consider the information content of r messages. Since the probability of selecting r messages is $[P(x)]^r$, therefore the information content of r messages must be given by

$$F[P^r(x)] \equiv rF[P(x)] \qquad (19\text{-}2)$$

Differentiating with respect to P, we obtain

$$F'[P^r(x)]rP^{r-1}(x) = rF'[P(x)]$$

or $\quad P^r(x)\, F'[P^r(x)] = P(x)\, F'[P(x)] = k_1, \quad$ a constant

or $\quad F'[P(x)] = \dfrac{k_1}{P(x)}$

The solution of this differential equation is

$$F[P(x)] = k_1 \log P(x) + k_2 \qquad (19\text{-}3)$$

Thus, we have shown that a function which satisfies the identity given by Eq. (19-2) is the logarithm. In order to make the information content positive, we define the information content of the message as

$$I(x) = -k \log P(x) = k \log M = kn \log s \qquad (19\text{-}4)$$

This result was introduced by Hartley in 1927 in a paper delivered to the International Congress of Telegraphy and Telephony. [7]

The simplest message would consist of one symbol (n = 1) selected from two equally probable symbols (s = 2) and the information content would therefore be proportional to log 2, the proportionality factor being dependent on the base of logarithm used. If this is to represent one unit of information, the logarithm must be taken to base 2, and the unit is then called a "binary digit," usually referred to as a "bit." It follows therefore that whenever a choice is made between two a priori equally likely alternatives, one bit of information is transmitted by the choice. Equation (19-4) is then modified into the form

$$I(x) = \log_2 M = n \log_2 s \qquad \text{bits/message} \qquad (19\text{-}5)$$

where M represents the total possible number of different messages, each consisting of n symbols selected from an available alphabet of s symbols. As an example, consider the message AAABBC selected from a set of messages resulting from all possible combinations of the symbols A, B, C, taken six at a time; the information content of this message is

Information = $6 \log_2 3 = 9.5$ bits

1. *Measure of Information for Independent Selections.* [4,5] In the previous discussion, we have assumed that all the possible M messages of the source have an equal probability of being selected. We shall now extend the definition of measure of information to the general case, where the messages are generated by independent selections from a discrete source, and to each symbol in the message there is associated its own probability of occurrence. The information source is then considered in the general case as a discrete source of an alphabet of symbols. Statistically, the source is completely characterized by the probability of occurrence $P(x_1)$, $P(x_2)$, ..., $P(x_S)$ of the available alphabet symbols of the information source $x_1, x_2, ..., x_S$. It follows that a very long message of N symbols will contain $N_1 = P(x_1)N$ symbols of type x_1, $N_2 = P(x_2)N$ symbols of type x_2, etc., so that the probability of occurrence of the extremely long message is

$$P(N) = [P(x_1)]^{N_1} \cdot [P(x_2)]^{N_2} \cdots [P(x_S)]^{N_S}$$

and its information content according to our definition is

$$I(N) = -\log_2 P(N) = -N \sum_{k=1}^{s} P(x_k) \log_2 P(x_k) \qquad (19\text{-}6)$$

Hence the average information content per symbol in the message is given by

$$I(x) = -\sum_{k=1}^{s} P(x_k) \log_2 P(x_k) \equiv H(x) \qquad \text{bits/symbol} \qquad (19\text{-}7)$$

This expression is very basic, and it was used by Shannon as the starting point in his original presentation of the theory; since it is identical in form to the expression for "entropy" in statistical mechanics, the information content I(x) is often referred to as the entropy of the system and is denoted by H(x); thus H(x) and I(x) stand for the same quantity.

Returning to the special case considered previously where the s symbols were considered equally probable, we have $P(x) = 1/s$, and the information content per symbol is $I(x) = \log_2 s$. The information content of a message consisting of n symbols is therefore $I(n) = I(x) = n \log_2 s$, in agreement with Eq. (19-5).

We shall show later on that the information content per symbol is maximum when all the s symbols have equal probability of being selected; the same applies to M equiprobable messages. As an illustration, consider the simple case when $s = 2$, and hence $P(x_1) + P(x_2) = 1$. The average information content per symbol using Eq. (19-7) is given by

$$I(x) = -\{P(x_1) \log_2 P(x_1) + [1 - P(x_1)] \log_2 [1 - P(x_1)]\} \qquad (19\text{-}8)$$

The behavior of I(x) as a function of $P(x_1)$ is shown in Fig. 19-2; the maximum value of I(x) occurs for $P(x_1) = \frac{1}{2} = P(x_2)$ and is zero for $P(x_1) = 0$ and $P(x_1) = 1$. The maximum value is

$$I(x)_{max} = -[\tfrac{1}{2}\log_2 \tfrac{1}{2} + \tfrac{1}{2}\log_2 \tfrac{1}{2}] = 1 \text{ bit/symbol}$$

The same result can be derived by considering the message to consist of one symbol ($n = 1$) selected from two equally probable symbols ($s = 2$). The information content per message, using Eq. (19-5), is equal to $I(x) = n \log_2 s = \log_2 2 = 1$ bit.

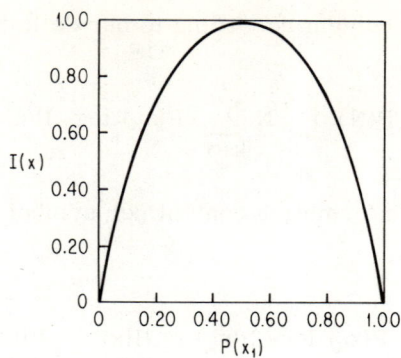

Fig. 19-2. Information of a binary alphabet as a function of the probability of one of the two symbols.

2. *Measure of Information of Continuous Messages.* As stated earlier, messages comprising continuous waveforms can be converted to discrete messages through the process of sampling and quantization. We have shown by the use of the sampling theorem that a band-limited signal can be uniquely specified by means of 2BT samples in T sec where B cycles/sec is the highest-frequency component of the sampled time function. Thus to encode a continuous waveform of highest frequency B cycles/sec, samples are first taken at intervals 2B apart and the samples are then quantized into discrete amplitude; the discrete levels may be finally encoded by the use of symbols selected from a given alphabet of symbols as discussed above. It will be shown in the next chapter that the process of quantization introduces distortion in the signal, which decreases with increasing number of quantizing levels. This effect, which is usually referred to as quantizing noise, will be discussed more fully in the next chapter.

It would be natural to extend the definition of entropy to continuous messages as a limiting case of the entropy of the discrete case. The information content per sample of a continuous message can be obtained by replacing the discrete probabilities $P(x_k)$ in Eq. (19-7) by a continuous probability density function $p(x)$ and by replacing the summation by integration over the whole range of possible amplitudes; thus

$$H(x) \equiv I(x) = - \int_X p(x) \log_2 p(x)\ dx \qquad (19-9)$$

The information content of the complete message waveform of duration T sec and bandwidth B cycles/sec is obtained by noting that

2BT samples are required for its specification. It follows that the information content of the complete waveform is

$$I(2BT) = 2BT \left[- \int_X p(x) \log_2 p(x) \, dx \right] \tag{19-10}$$

based on the assumption that all the samples are statistically independent. The information content transmitted per second or the rate of transmitting information is

$$I = -2B \int_X p(x) \log_2 p(x) \, dx \tag{19-11}$$

The information content per sample will obviously depend on the probability density function $p(x)$ and will therefore vary for different types of signals. By imposing some restrictions upon the statistical behavior of the function, it is possible to determine the form of $p(x)$ which will make the information content a maximum. As an example, let us consider a signal whose mean power P is fixed. For maximum information content, we have to maximize the integral

$$I(x) = - \int_{-\infty}^{\infty} p(x) \log_2 p(x) \, dx$$

subject to the conditions

$$\int_{-\infty}^{\infty} x^2 p(x) \, dx = P, \qquad \text{a given constant} \tag{19-12}$$

and $\quad \int_{-\infty}^{\infty} p(x) \, dx = 1 \tag{19-13}$

Using the method of Lagrange multipliers, it can be shown [4,8,9] that under these conditions a gaussian distribution of amplitude will make the information content per sample a maximum, namely,

$$p(x) = \frac{1}{\sqrt{2\pi P}} e^{-x^2/2P} \tag{19-14}$$

The maximum information content per sample is then given by

$$\frac{I(x)_{\max}}{\log_2 e} = - \frac{1}{\sqrt{2\pi P}} \int_{-\infty}^{\infty} e^{-x^2/2P} \log_e \left(\frac{1}{\sqrt{2\pi P}} e^{-x^2/2P} \right) dx$$

$$= -\frac{1}{\sqrt{2\pi P}} \int_{-\infty}^{\infty} e^{-x^2/2P} \left(\log_e \frac{1}{\sqrt{2\pi P}} - \frac{x^2}{2P} \right) dx$$

where the logarithms were converted to the base e for ease of calculation. But

$$\int_{-\infty}^{\infty} e^{-x^2/2P}\, dx = \sqrt{2\pi P}$$

and $\quad \displaystyle\int_{-\infty}^{\infty} \frac{x^2}{2P}\, e^{-x^2/2P}\, dx = \tfrac{1}{2}\sqrt{2\pi P}$

Therefore $\quad \dfrac{I(x)_{max}}{\log_2 e} = -\left[\log_e \dfrac{1}{2\pi P} - \dfrac{1}{2} \right] = \dfrac{1}{2} \log_e 2\pi e P$

Hence $\quad I(x)_{max} = \tfrac{1}{2} \log_2 e \, \log_e 2\pi e P = \tfrac{1}{2} \log_2 2\pi e P \qquad (19\text{-}15)$

which is the entropy of a gaussian distribution.

Examples of other types of constraint on the waveform and the resulting probability distribution are given in Table 19-1.

19.2   The Capacity of a Communication Channel [2,8-10]

Perhaps the most important measure in the comparison of communication systems is the rate of transmission of information which can be attained. It is implied in this statement that this transmission of information be either free from errors or that it remain within some specified limit of error frequency. In the following, we shall consider first the discrete noiseless channel, and then we shall extend the analysis to the discrete channel with noise.

1. *The Discrete Noiseless Channel.*   No communication system is ever noiseless; there is always some noise present in the system. However, if the noise level is low compared with the signal level so that it does not significantly alter the message as it passes along the channel, then the system can be regarded as a noiseless system.

As stated above, in any communication system, the transmitter is so chosen as to match the information source to the channel. The signals which are selected by the source are first encoded by the transmitter and then transmitted along the channel. By the capacity of the channel we mean the number of different signals that can be transmitted over it in a given interval of time. Let M(T) denote the

Table 19.1. Types of Constraint on Waveform and Resulting Probability Distributions

Type and limitation	Probability distribution	Information content
Symmetrical signal distribution: Mean power..................	$p(x) = \dfrac{1}{\sqrt{2\pi P}}\, e^{-x^2/2P}$	$\frac{1}{2}\log_2 2\pi e P$
Peak power limitation $\hat{P}$ at sampling points	$\mu(x) - \dfrac{1}{\sqrt{2\hat{P}}}$	$\frac{1}{2}\log_2 4\hat{P}$
Peak power limitation in P throughout time		$\frac{1}{2}\log_2 \dfrac{4P}{e^2} > I(x) > \frac{1}{2}\log_2 4P$
Unidirectional signal distribution: Positive average A_0...............	$p(x) = \dfrac{1}{A_0}\, e^{-x/A_0}$	$\log_2 e A_0$
Average power P................	$p(x) = \sqrt{\dfrac{2}{\pi P}}\, e^{-x^2/2P}$	$\frac{1}{2}\log_2 \frac{1}{2}\pi e P$

number of different signals which satisfy the following three properties:

1. Each signal which is selected by the source is transmitted by
the transmitter.

2. Each signal is compatible with the physical characteristics of
the channel.

3. Each signal is of duration T time units.

It has been shown in the last section that if each of these M(T)
signals were equally likely, then there would be $\log_2 M(T)$ bits/sig-
nal of duration T time units, or the information rate would be

$$C(T) = \frac{\log_2 M(T)}{T} \quad \text{bits/(signal)(unit time)} \quad (19\text{-}16)$$

We have also stated that in the case when each signal is equally
likely, then the information transmitted is maximum. Since there
can be only one signal on the channel at a time, it is reasonable to
suppose that C(T) is the maximum number of bits per unit time that
may be handled by the channel, provided, of course, it is possible to
distinguish reliably the M(T) different signal functions of duration T
on the channel. It must be emphasized here that the channel must
be considered in relation to the information source and the transmitter. For in practical systems these are constrained to generate
symbols chosen from a finite set and having a particular duration.
Since the stored energy in the system can be dissipated only at a
finite rate in any physical network, there is a minimum allowable
time between message symbols. This minimum time is the time
necessary for the energy stored by one symbol to be sufficiently
dissipated so that the next symbol may be resolved. This interrela-
tion of source and channel was recognized by Shannon in his definition of channel capacity for the noiseless case. The channel capacity
for the noiseless case is defined by Shannon [2] as

$$C = \lim_{T \to \infty} C(T) = \lim_{T \to \infty} \frac{\log_2 M(T)}{T} \quad (19\text{-}17)$$

For the noiseless case, it is assumed that there need be no error
in the decoding of the original message at the receiver.

2. *The Capacity of a Channel in the Presence of White Thermal
Noise.* The inherent fluctuation of the system due to ''noise'' limits
the number of distinguishable amplitude levels into which the signal
may be subdivided. Let us assume that the noise in the system is

white thermal noise band-limited to bandwidth B cycles/sec, and that it is added to the transmitted signal to produce the received signal. The perturbation of the signal amplitude in the presence of noise will result in a minimum detectable signal-amplitude change. The question to be answered is how many different signals can be distinguished at the receiving point, in spite of the perturbation due to noise. An approximate estimate can be obtained from the plausible assumption that signal-amplitude changes can be distinguished only if they are at least comparable to the rms noise level. Indeed, in order to relate the number of distinguishable amplitude levels to the signal-to-noise ratio of the system, let us assume that a signal-voltage change is distinguishable if it is equal to the rms noise voltage N_V, and assuming a maximum signal voltage of S_V volts, the number of distinguishable signal levels will then be equal to S_V/N_V. Including 0 volts as an additional possible signal level, the total number of distinguishable signal levels (or symbols) in the presence of noise is equal to $1 + S_V/N_V$. Or in terms of carrier-to-noise ratio of the system, this is approximately equal to $\sqrt{1 + P/N}$, where P is the carrier power, and N is the average noise power. Since in time T there are 2BT independent signal amplitudes, the total number of reasonably distinct signals or messages which can be sent in time T, using Eq. (19-1), is

$$M = \left(\sqrt{1 + \frac{P}{N}}\right)^{2BT} = \left(1 + \frac{P}{N}\right)^{BT} \tag{19-18}$$

The maximum number of bits that can be sent in time T, or the information content of the message, equals $\log_2 M$; the rate of transmission is therefore given by

$$\frac{\log_2 M}{T} = B \log_2 \left(1 + \frac{P}{N}\right) \quad \text{bits/sec} \tag{19-19}$$

We have just shown that if it is possible to distinguish reliably M different signals of duration T on a channel, then we can say that the channel can transmit $\log_2 M$ bits in time T, or the rate of transmission is then $(\log_2 M)/T$. A theorem due to Shannon states that by sufficiently complicated encoding systems, it is possible to transmit binary digits at a rate given by Eq. (19-19) with as small a frequency of errors as desired. Furthermore, it is not possible by any encoding method to send at a higher rate and have an arbitrarily low frequency of errors. It follows therefore that the system capacity can be defined as the maximum rate of transmitting information

$$C = B \log_2 \left(1 + \frac{P}{N}\right) \quad \text{bits/sec} \qquad (19\text{-}20)$$

Equation (19-20) shows that the rate $B \log_2 (1 + P/N)$ measures in a sharply defined way the capacity of the channel for transmitting information. Since the quantity $TB \log_2 (1 + P/N)$ gives, for large T, the number of bits that can be transmitted in time T, it can be considered as an exchange relation between the different parameters. For example, we can increase the capacity of the system by increasing either the bandwidth B or the P/N ratio. In fact, the individual parameters T, B, P, and N can be varied without changing the amount of information we can transmit, provided $TB \log (1 + P/N)$ is held constant. It should be pointed out, however, that these parameters cannot be varied at will without regard to the modulation system employed. In amplitude modulation, for example, the channel bandwidth should be equal to the message bandwidth. For a given message, a greater channel bandwidth does not improve the transmission and does not permit the use of a lower carrier-to-noise ratio. A channel bandwidth narrower than the message bandwidth will inevitably result in loss of part of the information, and this loss cannot be compensated for by increasing the carrier-to-noise ratio.

The channel capacity as a function of bandwidth will now be examined as follows. As B is increased, the noise power N in the band will increase proportionally. Let N_0 denote noise power per cycle; hence $N = N_0 B$, and the channel capacity C is then given by

$$C = B \log_2 \left(1 + \frac{P}{N_0 B}\right) \text{ bits/sec} \qquad (19\text{-}21)$$

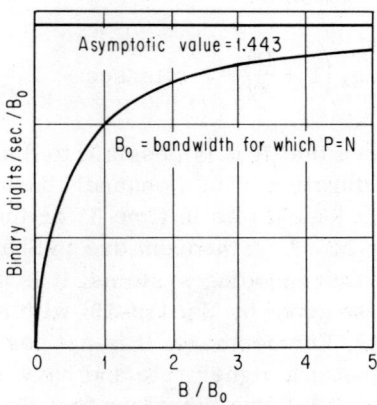

Fig. 19-3. Channel capacity as a function of bandwidth.

Let B_0 denote the bandwidth for which the noise power is equal to the carrier power; i.e., $B_0 = P/N_0$. Equation (19-21) can be written in the form

$$\frac{C}{B_0} = \frac{B}{B_0} \log_2 \left(1 + \frac{B_0}{B}\right) \tag{19-22}$$

which is plotted in Fig. 19-3 as a function of B/B_0. As we increase B, the capacity C increases rapidly until the total noise power is about equal to the carrier power; after this, the increase is slow, and it approaches an asymptotic value of $\log_2 e$ for $B \gg B_0$.

19.3   Approximation to an Ideal System [3,9,10]

A system capable of transmitting without errors at the rate C [Eq. (19-20)] is called an ideal system. Such a system cannot be achieved with any finite encoding process but can be approximated as closely as desired. An approximately ideal system is characterized by the following properties: (1) The rate of transmission of binary digits approaches $C = B \log_2 (1 + P/N)$. (2) The frequency of error approaches zero. (3) The statistical properties of the transmitted signal approach those of white noise. (4) The threshold effect is very sharp; i.e., the frequency error increases very rapidly

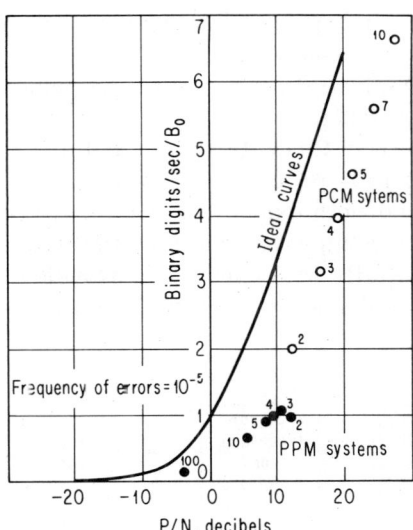

Fig. 19-4. Comparison of PCM and PPM with ideal performance. (From C. E. Shannon,[2] Proc. IRE.)

as the noise exceeds the value for which the system was designed. In the next chapter on PCM, it will be shown that a practical PCM system approaches the idealized system. This is illustrated in Fig. 19-4, where the circles represent PCM system of the binary, ternary, etc., types, using positive and negative pulses and adjusted to give one error in about 10^5 binary digits.

1. *Coded Transmissions.* The trading of bandwidth for signal-to-noise ratio has been shown earlier to be characteristic of all wideband systems such as FM and PPM. These systems, however, are much less efficient than PCM and certainly do not approach the performance of the idealized system postulated by Shannon. If we examine the various methods of modulation, we find that these fall naturally into two classes of uncoded and coded systems. In the uncoded systems, one symbol in the message space is transformed into one symbol in the signal space. For example, in AM each possible amplitude of the message to be transmitted results in a particular amplitude of the RF signal, while in PTM each possible amplitude of the message function results in a particular pulse position. In the coded systems, of which PCM is the best-known example, each message symbol or amplitude is transformed into a number of signal symbols. It has been shown by Tuller, [12] in a general way, that only with coded systems is it possible to achieve full exchange of bandwidth for signal-to-noise ratio, while uncoded systems like PTM are inherently incapable of achieving efficient exchange of bandwidth and signal-to-noise ratio.

As an illustration, we shall compare the performance of a PCM system with FM or PTM.

It has been shown in Chap. 18 that in FM and PTM systems the signal-to-noise ratio is proportional to the bandwidth and hence proportional to bandwidth expansion. We shall presently show that in a PCM system the improvement resulting from bandwidth widening varies exponentially with bandwidth. Consider the wideband information-transmission system shown in Fig. 19-5. Let us assume

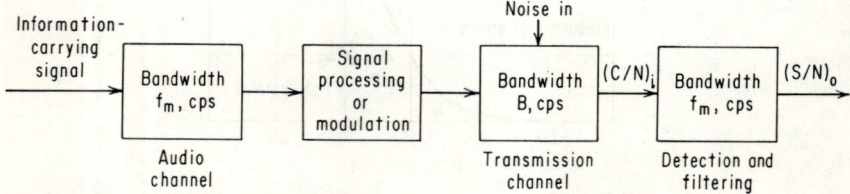

Fig. 19-5. Wideband information transmission system.

a sequence of discrete voltage levels to be transmitted through the
system in such a manner that the capacity of the system in bits per
second is everywhere the same; i.e., the information per second
emerging at the system output is the same as that going in. Initially
the sequence of voltage levels requires f_m cycles/sec bandwidth for
transmission. However, the encoded message requires B cycles/sec
for transmission with an rms noise voltage of $\sqrt{N}$ volts introduced
⋯⋯⋯ ⋯⋯⋯ity at the receiver is

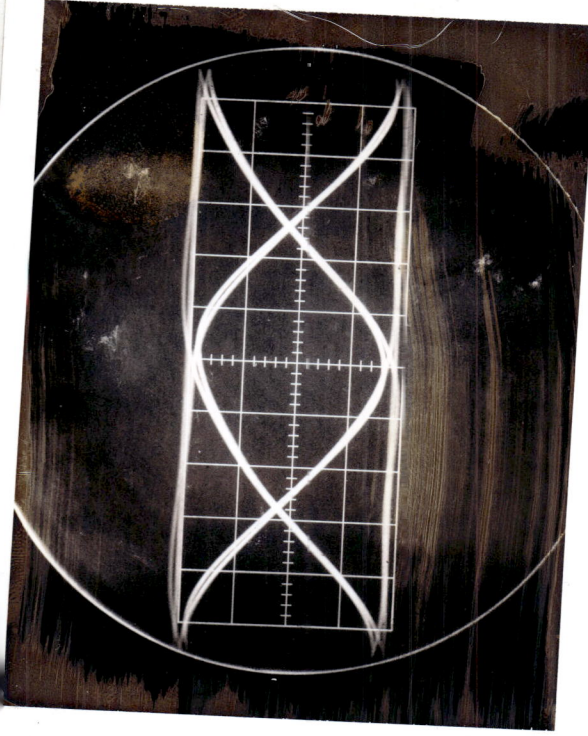

$$(19\text{-}23)$$

s must be encoded
d through a filter
ts in an output signal–
a constant informa-
nformation transmission

$$(19\text{-}24)$$

apacity throughout the
for system capacity to

$$1 + \left(\frac{S}{N}\right)_o = \left[1 + \left(\frac{C}{N}\right)_i\right] \qquad (19\text{-}25)$$

or for $(S/N)_o \gg 1$ and $(C/N)_i \gg 1$, we obtain the important result
that

$$\left(\frac{S}{N}\right)_o \doteq \left(\frac{C}{N}\right)_i^{B/f_m} \qquad (19\text{-}26)$$

where B/f_m is the ratio of channel bandwidth to message bandwidth.
We have just shown that for $B > f_m$, the output $(S/N)_o$ increases

exponentially with the bandwidth expansion. Thus, a hypothetical coded system of the type described in Fig. 19-5 is theoretically capable of producing a much greater S/N improvement as compared with FM or PPM systems.

19.4   Reception or Decoding when Noise is Present[4,5,13-16]

In a practical communication system, the transmitted signal is subject to noise from various sources. The presence of noise will introduce an uncertainty in the received signal; i.e., the effect of the noise upon the transmitted signal is to make it impossible for the observer at the receiver to identify with certainty which message or symbol was transmitted from the source. Thus, the ''a posteriori probability'' for a transmitted symbol x_k will not in general be unity but $P(x_k|y_i)$ say, which denotes the conditional probability that the transmitted symbol was x_k on receiving the signal y_i.

The effect of noise on the transmitted messages can best be described by the use of the generalized model of a noisy communication channel shown in Fig. 19-6 and the probability relations developed in Chap. 4 [Eq. (4-10)], namely

$$P(x_k,y_i) = P(x_k)P(y_i|x_k) = P(y_i)P(x_k|y_i) \tag{19-27}$$

In this model, the set X represents the source which consists of n possible discrete events or states denoted by $x_1, x_2, \ldots, x_k, \ldots,$ x_{n-1}, x_n. The set Y represents the possible results of observations performed at the destination, which are denoted by $y_1, y_2, \ldots,$ $y_i, \ldots, y_{m-1}, y_m$, where m is in general different from n. Let $P(x_k)$ represent the a priori probability that the source is in state x_k, $P(y_i|x_k)$ the conditional probability or the likelihood that y_i will be received, given that x_k is transmitted, and $P(x_k|y_i)$ the a posteriori probability that x_k was transmitted, given that y_i is received. Equation (19-27) states that the probability $P(x_k,y_i)$ of the joint event of message x_k being transmitted and y_i received is equal to the conditional probability $P(y_i|x_k)$ weighted by the probability $P(x_k)$ of x_k being transmitted. Similarly, $P(x_k,y_i)$ equals $P(x_k|y_i)$ weighted by the probability $P(y_i)$ of receiving y_i. With reference to Fig. 19-6, it is obvious that the a priori probability of an event at the source is equal to the weighted sum over all conditional probabilities, namely,

$$P(x_k) = \sum_Y P(x_k,y) = \sum_Y P(y)P(x_k|y) \tag{19-28}$$

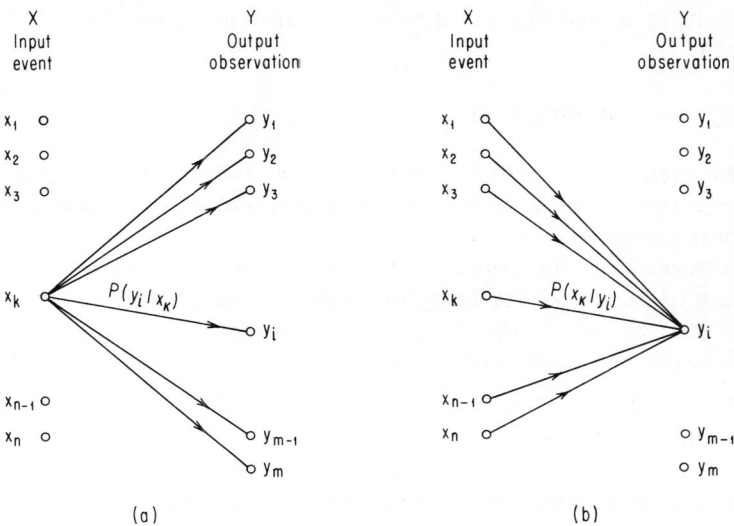

| X | Y | X | Y |
| Input event | Output observation | Input event | Output observation |

(a)   (b)

Fig. 19-6. Generalized model of a noisy communication channel.

Also $\quad P(y_i) = \sum_{X} P(x,y_i) = \sum_{X} P(x)P(y_i|x)$ $\qquad$ (19-29)

which means that $P(y_i)$ is the weighted sum over all conditional probabilities of set X.

1. *A Measure of Information.* As stated previously the effect of noise is to introduce an uncertainty in the received signal. When a message y_i is received, the information provided by y_i about x_k consists of changing the probability of x_k from the a priori value $P(x_k)$ to the a posteriori value $P(x_k|y_i)$. The mutual information associated with the states x_k and y_i is defined by

$$I(x_k;y_i) = \log_2 \frac{P(x_k|y_i)}{P(x_k)} \qquad \text{bits} \qquad (19\text{-}30)$$

The mutual information $I(x_k;y_i)$ has the property of being symmetrical with respect to x_k and y_i. Using Eq. (19-27), we obtain

$$I(x_k;y_i) = \log \frac{P(x_k|y_i)\,P(y_i)}{P(x_k)\,P(y_i)} = \log \frac{P(x_k,y_i)}{P(x_k)\,P(y_i)} = I(y_i;x_k) \qquad (19\text{-}31)$$

This means that the information provided by y_i about x_k is equal to the information provided by x_k about y_i. When the two events of transmitting x_k and receiving y_i are statistically independent, that is, when

$$P(x_k, y_i) = P(x_k) P(y_i) \tag{19-32}$$

then the measure of mutual information, or the amount of information received in the message, equals zero, which corresponds to a very noisy channel.

However, when the transmission is over a "noiseless" channel, the state of the source is completely determined from the observation of the state of the destination. Consequently, $P(x_k | y_i)$ is then equal to one, and the amount of information received in the message is

$$I(x_k) = - \log P(x_k) \tag{19-33}$$

This is the so-called self-information associated with the state x_k. Similarly, the quantity

$$I(y_i) = - \log P(y_i) \tag{19-34}$$

is a measure of the self-information of the state y_i. This is in agreement with the definition of information content of a message set forth in Equation (19-4).

In general, the mutual information $I(x_k; y_i)$ between x_k and y_i is less than $I(x_k)$, the self-information of x_k, unless an observation of y_i completely specifies the state of the source x_k, that is to say, when $P(x_k | y_i) = 1$. This follows from the inequalities

$$P(x_k | y_i) \leq 1 \text{ and } P(y_i | x_k) \leq 1 \tag{19-35}$$

and the relations

$$I(x_k; y_i) = \log \frac{P(x_k | y_i)}{P(x_k)} = \log \frac{P(y_i | x_k)}{P(y_i)} \tag{19-36}$$

Hence $I(x_k; y_i) \leq \log \dfrac{1}{P(x_k)} \equiv I(x_k)$

$$\leq \log \frac{1}{P(y_i)} \equiv I(y_i) \tag{19-37}$$

We conclude therefore that the self-information of x_k is the maximum amount of information that can possibly be provided about x_k.

From Eq. (19-36) we derive the relations

$$I(x_k;y_i) = I(x_k) - I(x_k|y_i) = I(y_i) - I(y_i|x_k) \qquad (19\text{-}38)$$

where $\quad I(x|y) = -\log P(x|y) \qquad\qquad\qquad\qquad\qquad (19\text{-}39)$

is the conditional self-information. Equation (19-38) can be interpreted to mean that the information provided by y_i about x_k is equal
to the difference between the amounts of information required to
specify x_k, before and after the destination has been observed to be
in state y_i.

The mutual information between x_k and y_i can also be expressed
in the following symmetrical form by making use of Eqs. (19-27)
and (19-38):

$$I(x_k;y_i) = I(x_k) + I(y_i) - I(x_k, y_i) \qquad (19\text{-}40)$$

where $\quad I(x_k, y_i) \equiv -\log P(x_k,y_i)$

is the self-information of the joint event of symbol x_k being transmitted and y_i received.

2. *Communication Entropy.* We have shown that the self-information of a message is given by the logarithm of the probability of
its occurrence. The average value of self-information per transmitted message or symbol, averaged over the ensemble X, is called
the entropy and is given by

$$I(X) = -\sum_X P(x_k) \log P(x_k) = H(X) \qquad (19\text{-}41)$$

The quantity $H(X)$ may be considered as a measure of the uncertainty existing about the message before its reception. In the special
case of M equiprobable messages in the X space we have $P(x)$
$= 1/M$, and the information per message is given by

$$I(x) = \log M \qquad (19\text{-}42)$$

We shall now prove that the entropy $H(X)$ is maximum when all
M messages in the X space are equally probable; i.e., we shall
prove that

$$H(X) \leq \log M \qquad (19\text{-}43)$$

Consider the expression

$$H(X) - \log M = \sum_X P(x) \log \frac{1}{P(x)} - \sum_X P(x) \log M$$

$$= \sum_X P(x) \log \frac{1}{MP(x)} \qquad (19\text{-}44)$$

Making use of the inequality $\ln w \leq w - 1$ which follows from the fact that the curve $u = \ln w$ is tangent to the straight line $u = w - 1$ at the point $w = 1$, we find

$$H(X) - \log M \leq \sum_X \left[\frac{1}{M} - P(x) \right] \log e = 0 \qquad (19\text{-}45)$$

The equality sign holds only when $P(x) = 1/M$, which corresponds to all messages being equiprobable. We conclude therefore that for any given alphabet of symbols, the average information per symbol is maximum when all symbols have equal probability of being selected.

In a manner similar to the definition of entropy or average value of self-information, we define the conditional entropy or average value of the conditional self-information as

$$I(X|Y) = - \sum_Y \sum_X P(x,y) \log P(x|y) = H(X|Y) \qquad (19\text{-}46)$$

A better insight into the meaning of $H(X|Y)$ can be obtained by expressing it in the form

$$H(X|Y) = - \sum_Y P(y_i) \sum_X P(x_k|y_i) \log P(x_k|y_i) \qquad (19\text{-}47)$$

and noting that the expression $-P(x_k|y_i) \log P(x_k|y_i)$ represents the uncertainty that the observed message y_i originated from message x_k. Similarly, we note that the expression

$$H(X|y_i) = - \sum_X P(x_k|y_i) \log P(x_k|y_i) \qquad (19\text{-}48)$$

represents the average uncertainty in the received information y_i summed over all possible transmitted states x_k. This is called the entropy of equivocation per received symbol y_i. When averaged over all states y_i, it becomes $H(X|Y)$, the per symbol equivocation or conditional entropy.

3. *Average Mutual Information.* The mutual information $I(x_k;y_i)$ between x_k and y_i has been defined as

$$I(x_k;y_i) = \log \frac{P(x_k|y_i)}{P(x_k)} = \log \frac{P(x_k,y_i)}{P(x_k)\,P(y_i)}$$

The average information is obtained by weighting the mutual information $I(x_k;y_i)$ with the probability $P(x_k,y_i)$ of the joint occurrence of x_k being transmitted and y_i received. Consequently, the average value of the mutual information over the product ensemble XY is given by

$$I(X;Y) = \sum_X \sum_Y P(x,y)\,I(x;y) = \sum_X \sum_Y P(x,y)\,\log\frac{P(x|y)}{P(x)} \qquad (19\text{-}49)$$

The average mutual information $I(X;Y)$ can be expressed in terms of the entropies that pertain to the product ensemble XY as follows:

$$I(X;Y) = \sum_X \sum_Y P(x,y)\,\log P(x|y) - \sum_X \sum_Y P(x,y)\,\log P(x)$$

$$= \sum_X \sum_Y P(y)\,P(x|y)\,\log P(x|y) - \sum_X P(x)\,\log P(x)$$

$$= H(X) - H(X|Y) \qquad (19\text{-}50)$$

where, as discussed previously, the entropy $H(X)$ represents the average value of self-information transmitted per message, averaged over the ensemble X, and its conditional entropy or equivocation $H(X|Y)$ is the average uncertainty that the observed state y_i originated with message x_k, averaged over the product ensemble XY.

Equation (19-50) is of particular interest to the problem of transmission of information through a noisy communication channel. It states that the average amount of information $I(X;Y)$ received about the transmitted message is equal to $H(X)$, the average amount of information per transmitted message at the source or the entropy of the source, minus the conditional entropy or equivocation $H(X|Y)$, which equals the average amount lost per message because of the noise in the channel.

In a similar manner it can be shown that

$$I(X;Y) = H(Y) - H(Y|X) \qquad (19\text{-}51)$$

where

$$H(Y) = -\sum_Y P(y_i) \log P(y_i) \qquad (19\text{-}52)$$

is the average received information per symbol and

$$H(Y|X) = -\sum_X P(x_k) \sum_Y P(y_i|x_k) \log P(y_i|x_k) \qquad (19\text{-}53)$$

is the average uncertainty that the transmitted symbol x_k is received as y_i, averaged over all x_k and y_i in the product ensemble XY. The average uncertainty in the received information $H(Y|X)$ is called the error entropy. It follows therefore from Eq. (19-51) that the average amount of information per received symbol is equal to the entropy $H(Y)$ of the received message minus the average uncertainty about the received message or error entropy $H(Y|X)$ resulting from the noise in the channel. The relations between the various information averages are shown graphically in Fig. 19-7.

We shall illustrate the application of these relations to a PCM channel. Consider a PCM channel of eight discrete amplitude levels, as shown in Fig. 19-8. Let the probability of error in transmitting a binary pulse be equal to p; hence the conditional probability $P(y_i|x_k)$ of receiving level 5 if level 5 was sent is q^3, where $q = 1 - p$; the conditional probabilities for other levels are as shown in the figure. For simplicity, let all levels be equiprobable so that $P(x_k) = \frac{1}{8}$; also in this case $P(y_i|x_k) = P(x_k|y_i)$.

The equivocation if y_i is received is

$$H(X|y_i) = \sum_X -P(x_k|y_i) \log_2 P(x_k|y_i)$$

$$= p^3 \log \frac{1}{p^3} + 3p^2q \log \frac{1}{p^2q} + 3pq^2 \log \frac{1}{pq^2} + p \log \frac{1}{q^3}$$

$$= 3(p \log \frac{1}{p} + q \log \frac{1}{q})$$

In general, for n pulses/code group,

$$H(X|y_i) = n(p \log \frac{1}{p} + q \log \frac{1}{q})$$

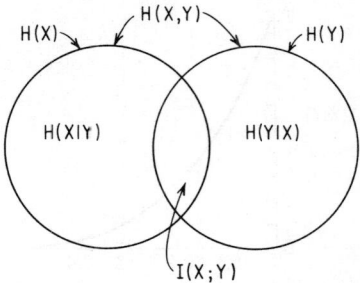

Fig. 19-7. Schematic presentation of different entropies associated with a simple communication system.

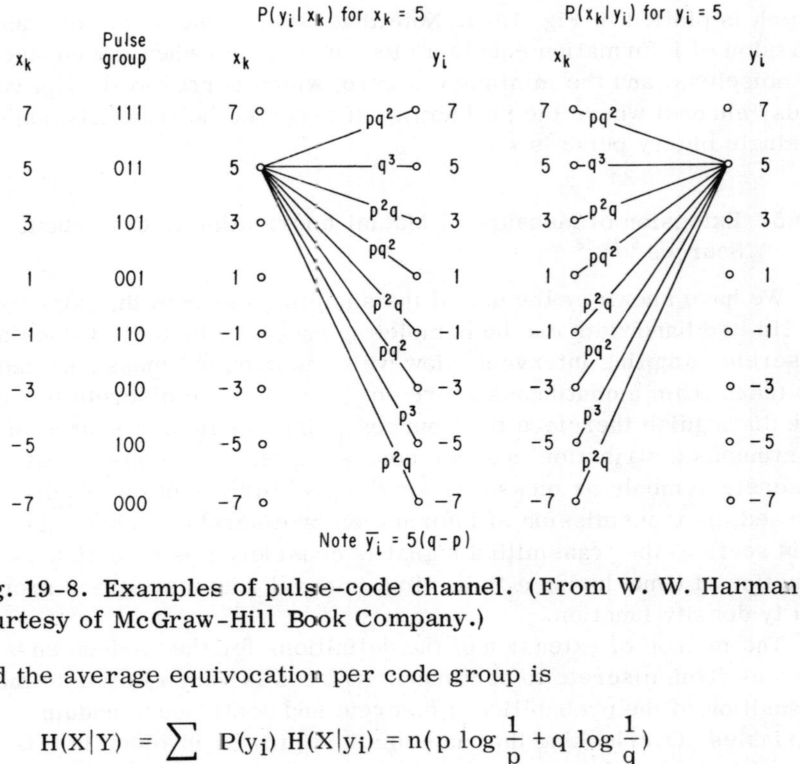

x_k	Pulse group	$P(y_i \mid x_k)$ for $x_k = 5$		$P(x_k \mid y_i)$ for $y_i = 5$	
		x_k	y_i	x_k	y_i
7	111	7 ○	pq^2 ○ 7	7 ○ pq^2	○ 7
5	011	5 ○	q^3 ○ 5	5 ○ q^3	○ 5
3	101	3 ○	p^2q ○ 3	3 ○ p^2q	○ 3
1	001	1 ○	pq^2 ○ 1	1 ○ pq^2	○ 1
-1	110	-1 ○	p^2q ○ -1	-1 ○ p^2q	○ -1
-3	010	-3 ○	pq^2 ○ -3	-3 ○ pq^2	○ -3
-5	100	-5 ○	p^3 ○ -5	-5 ○ p^3	○ -5
-7	000	-7 ○	p^2q ○ -7	-7 ○ p^2q	○ -7

Note $\bar{y}_i = 5(q - p)$

Fig. 19-8. Examples of pulse-code channel. (From W. W. Harman,[15] courtesy of McGraw-Hill Book Company.)

and the average equivocation per code group is

$$H(X|Y) = \sum_Y P(y_i)\, H(X|y_i) = n\left(p \log \frac{1}{p} + q \log \frac{1}{q}\right)$$

The rate of transmission is thus

$$I(X;Y) = H(X) - H(X|Y) = n(1 + p \log_2 p + q \log_2 q) \qquad \text{bits/pulse group}$$

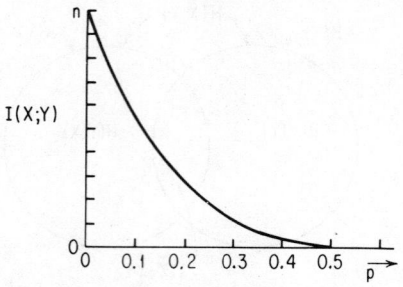

Fig. 19-9. Rate of transmission in pulse-code channel when all levels x_k are equiprobable.

which is plotted in Fig. 19-9. Note that the maximum rate of transmission of information equals n bits/pulse group when the channel is noiseless, and the minimum is zero, which corresponds to a very noisy channel where the probability of error in the transmission of a single binary pulse is $\frac{1}{2}$.

19.5 Extension of Measure of Mutual Information to Continuous Sources [5,15,16]

We have shown by the use of the sampling theorem that signals of limited bandwidth can be completely specified by their values at discrete sampling intervals. However, the sampled messages can be taken from a continuous source or from a set of discrete levels. We distinguish therefore two sources of information: a source of continuous distribution, and one consisting of a finite number of discrete symbols or messages. In the preceding section, we discussed the transmission of information by discrete symbols. In this section, the transmitted signal is considered as a continuous function of time during a finite time interval with a specified probability density function.

The method of extension of the definitions for the various entropies from discrete to continuous sources is very similar to the transition of the probability of discrete and continuous random variables. Overlooking the mathematical pitfalls involved in this transition, we define the average amount of information which is obtained about the state x from observation of state y as

$$I(x,y) = \int_{-\infty}^{\infty} \int_{-\infty}^{\infty} p(x,y) \log \frac{p(x,y)}{p(x)\ p(y)}\ dx\ dy \qquad (19\text{-}54)$$

and as before, it can be shown that

$$I(x;y) = H(x) - H(x|y) = H(y) - H(y|x)$$
$$= H(x) + H(y) - H(x,y) \qquad (19\text{-}55)$$

where

$$H(x) = -\int_{-\infty}^{\infty} p(x) \log p(x) \, dx, \qquad \text{transmitted entropy}$$
$$(19\text{-}56)$$

$$H(y) = -\int_{-\infty}^{\infty} p(y) \log p(y) \, dy, \qquad \text{received entropy}$$
$$(19\text{-}57)$$

$$H(x|y) = -\iint_{-\infty}^{\infty} p(x,y) \log \frac{p(x,y)}{p(y)} \, dx \, dy, \qquad \text{equivocation}$$
$$(19\text{-}58)$$

$$H(y|x) = -\iint_{-\infty}^{\infty} p(x,y) \log \frac{p(x,y)}{p(x)} \, dx \, dy, \qquad \text{error entropy}$$
$$(19\text{-}59)$$

$$\text{and} \quad H(x,y) = -\iint_{-\infty}^{\infty} p(x,y) \log p(x,y) \, dx \, dy \qquad (19\text{-}60)$$

It should be pointed out, however, that owing to some mathematical difficulties the concept of self-information can no longer be associated with the entropies $H(x)$, $H(y)$, and the conditional entropies. We shall refer to $H(x)$ and $H(y)$ as the entropy functions of the transmitted and received signal, respectively, or the average information per sample.

1. *Capacity of Continuous Channel Perturbed by White Gaussian Noise.* As an example, let us consider a continuous source transmitting information over a noisy channel, the noise source to consist of white gaussian noise, that is, noise with uniform spectral density and gaussian amplitude distribution. The signal mean power is assumed to be limited to P watts, and the average noise power within the bandwidth B of the channel equals N watts. We shall now proceed to determine the rate at which information is received by the receiver.

Let $x(t)$ be the transmitted signal, $n(t)$ the noise in the channel, and $y(t)$ the received signal. Suppose the noise is simply added to the signal so that

$$y(t) = x(t) + n(t) \tag{19-61}$$

and

$$\overline{y^2(t)} = \overline{x^2(t)} + \overline{n^2(t)} = P + N \tag{19-62}$$

on the assumption that the received signal $x(t)$ and the noise $n(t)$ are independent.

Furthermore,

$$H(x,y) = H(x,n) = H(x) + H(n)$$

Hence $I(x;y) = H(y) - H(n)$

The entropy $H(y)$ of the received signal $y(t)$ will be maximum if its amplitude distribution is gaussian, and since the noise source is assumed gaussian, the information source $x(t)$ must also have a gaussian distribution.

By analogy with Eq. (19-15), we obtain

$$H(y) = \tfrac{1}{2} \log_2 2\pi e(P + N) \tag{19-63}$$

The average transmitted information in bits per sample is, however, less than $H(y)$, due to the average error $H(y|x)$ introduced by the noise; this error is the relative uncertainty about the received signal $y(t)$ when a given $x(t)$ is sent. Since the noise distribution is assumed gaussian, we have

$$H(n) = H(y|x) = \int_{-\infty}^{\infty} \int_{-\infty}^{\infty} p(x,y) \log_2 \frac{1}{p(y|x)} \, dx \, dy = \tfrac{1}{2} \log_2 2\pi eN \tag{19-64}$$

Thus the average transmitted information in bits per sample is

$$I(x;y) = H(y) - H(y|x) = \tfrac{1}{2} \log_2 2\pi e(P + N) - \tfrac{1}{2} \log_2 2\pi eN$$

$$= \tfrac{1}{2} \log_2 \frac{P + N}{N} \qquad \text{bits/sample} \tag{19-65}$$

In time T sec, there are 2BT samples so that the total gain of information in time T is

$$I(T)_{max} = BT \log_2 \frac{P + N}{N} \tag{19-66}$$

Therefore the maximum information rate R_{max}, or the capacity of the system C, is given by

$$C = R_{max} = B \log_2 \left(1 + \frac{P}{N}\right) \quad \text{bits/sec} \qquad (19\text{-}67)$$

which is identical to Eq. (19-20). This is Shannon's fundamental channel-capacity equation for band-limited time function—one of the most important results of information theory, and as stated previously, this rate cannot be exceeded by any encoding method without increasing the desired small frequency of errors.

We shall return to this fundamental equation in Chap. 24 in connection with our study of comparison of modulation systems. Since band-limited signals are commonly used in communication systems, it is obvious why the above equation for the channel capacity for band-limited time functions plays a very important role in evaluating the optimum performance of communication systems.

References
1. Shannon, C. E.: A Mathematical Theory of Communication, *BSTJ*, vol. 27, pp. 379, 623, 1948.
2. Shannon, C. E.: Communication in the Presence of Noise, *Proc. IRE*, vol. 37, p. 10, 1949.
3. McMillan, B., and D. Slepian: Information Theory, *Proc. IRE*, vol. 50, pp. 1151–1157, May, 1962.
4. Holroyed, P., and G. P. Jones: Information Theory, *Electron. Technol.*, February, 1961.
5. Fano, R. M.: "Transmission of Information," John Wiley & Sons, Inc., New York.
6. Nyquist, H.: Certain Factors Affecting Telegraph Speed, *BSTJ*, vol. 3, p. 324, April, 1924.
7. Hartley, R. V. L.: The Transmission of Information, *BSTJ*, vol. 3, p. 535, July, 1928.
8. Brillouin, L.: "Science and Information Theory," Academic Press, Inc., New York, 1956.
9. Goldman, S.: "Information Theory," Prentice-Hall, Inc., Englewood Cliffs, N. J.
10. Leifer, M., and W. F. Schreiber: Communication Theory, *Advan. Electron.*, vol. 3, Academic Press, Inc., New York, 1951.
11. Schwartz, M.: "Information Transmission, Modulation, and Noise," McGraw-Hill Book Company, New York.
12. Tuller, W. G.: Theoretical Limits on the Rate of Transmission of Information, *Proc. IRE*, vol. 37, p. 468, 1949.

13. Hancock, J. C.: "An Introduction to the Principles of Commun-
    ication Theory," McGraw-Hill Book Company, New York.
14. Cherry, C.: "On Human Communication," Science Editions,
    Inc., New York, 1961.
15. Harman, W. W.: "Principles of the Statistical Theory of Com-
    munication," McGraw-Hill Book Company, New York.
16. Reza, M. F.: "An Introduction to Information Theory,"
    McGraw-Hill Book Company, New York.

20

PRINCIPLES OF PULSE-CODE MODULATION (PCM)

It has been shown earlier that pulse modulation exhibits many important characteristics which make this method of modulation particularly applicable to communication transmission systems. One of the fundamental properties is the ease of transmitting many intelligence signals through multiplexing by means of time division. We have also noted that just as in the case of CW exponential modulation, pulse-modulation systems employing pulse-time modulation are also characterized by the trade-off between increased bandwidth and signal-to-noise improvement.

Modern communication systems consist, in general, of a very large number of repeater sections or radio links. In most transmission systems, the noise and distortion from the individual links cumulate, and, consequently, the accumulation of noise sets a limit to the distance a signal can be transmitted or to the number of links in the system. It follows therefore that for a given quality of overall transmission, the requirements on each link become more severe as the length of the circuit increases. For example, if 100 links are to be used in tandem, the noise power added per link can only be 1/100 of the total permissible output system noise. A modulation system where the transmission requirements for the individual

619

links are almost independent of the total length of the system is represented by pulse-code modulation (PCM).

Pulse-code-modulation systems were first proposed by A. H. Reeves [1] of the ITT Corporation in a French patent specification in 1939 and an American specification in 1942. In PCM, the modulating-signal waveform is sampled at regular intervals, as in conventional methods of pulse modulation, but the samples are first quantized and then transmitted over the system by means of groups of pulses, which uniquely represent the values of the samples in some code. The code groups are then transmitted, either as a time sequence of pulses by time division over the same channel, or by frequency division over separate channels. If a pulse, although considerably distorted by noise, is recognized as such at the receiving end of the link, it can be reshaped and retransmitted like new over the next link. Reshaping or pulse regeneration can take place at any or all repeater points, as required, with the effect that the code groups are reconstructed without appreciable error at the receiving end of the circuit. The only condition is that the noise within each link remain below a comparatively large threshold level, at which discrimination between mark and space becomes impossible. At the receiver, the regenerated code groups are decoded to form a series of impulses proportional to the original quantized samples from which the signal wave is recovered by means of a low-pass filter.

The process of transforming the analogue samples into digital form is called quantizing. It inherently introduces an initial error in the amplitude of the samples, giving rise to quantization noise. As we shall see later, the quantization noise depends on the number of discrete amplitude levels used in the quantizing process and can be minimized by nonuniform spacing of levels.

It is the purpose of this chapter to cover the basic principles of pulse-code modulation. Consideration will be given particularly to the problems of quantization and the resulting quantization noise, encoding and decoding, transmission requirements and performance of a PCM system. It will be shown that this system, as an example of a coded wideband system, is much more efficient than the uncoded systems, but is still about 9 db less efficient than Shannon's ideal binary transmission system.

20.1   Basic Operations in PCM System [2-4]

A PCM system embodies several basic processes, such as sampling, quantizing, pulse shaping, and decoding. These operations will be introduced, and the part each plays in PCM will be explained in this section.

1. *Sampling.*   The first basic operation in PCM is sampling, which is based on the sampling theorem discussed earlier in Chap. 17. The input signal $F(t)$ which is band-limited to the frequency band $(0-f_m)$ cycles/sec is sampled at a constant frequency $f_0 = 1/T_0$, where $f_0 \geq 2f_m$. The sampled function consists of short pulses $F_1, F_2, \ldots, F_n$, of equal duration τ at equal intervals T_0 apart, as shown in Fig. 20-1. According to the sampling theorem, the band limited function $F(t)$ can be reconstructed from these samples provided $f_0 \geq 2f_m$, or

$$T_0 = \frac{1}{2f_m} \tag{20-1}$$

in the limiting case. Hence, in order to transmit a band-limited signal of duration T and maximum frequency f_m, it suffices to send a finite set of $2f_m T$ independent amplitude samples obtained by sampling the instantaneous amplitude of the signal at a regular rate of $2f_m$ samples/sec. We recall that the original signal function $F(t)$ can be reconstructed from the discrete samples F_n by passing the samples through a low-pass filter of cutoff frequency f_m and a suitable amplifier.

2. *Quantization.*   A continuous signal, such as speech, has a continuous range of amplitudes, and, therefore, its samples have a continuous amplitude range. However, it is impossible and also not necessary to transmit the exact amplitudes of the samples for the following reasons. As discussed above, when the amplitude of a sample is transmitted as an amplitude of a pulse or as the time position of a pulse, it is subject to disturbance due to noise and crosstalk in the system, which will cause errors in the amplitude of the recovered sample. The effect of noise in the system can, however, be minimized by representing the samples by a finite number of discrete allowed levels. If the sampled amplitude of the

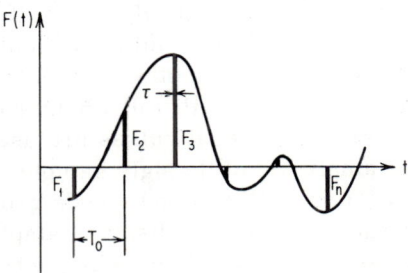

Fig. 20-1. Sampled time function.

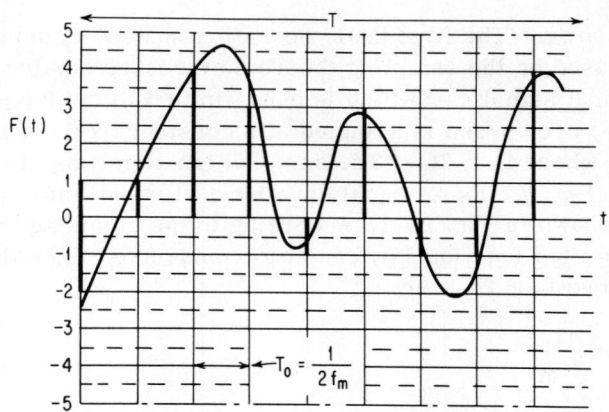

Fig. 20-2. Quantization with equal steps.

signal falls between two levels, either the lower or upper level is transmitted, depending upon the level which is closest, as shown in Fig. 20-2.

Representing the signal by certain discrete allowed levels only is called quantizing. It inherently introduces an initial error in the amplitude of the samples, giving rise to quantization noise. But once the signal is in a quantized state, it can be relayed many times without further loss in quality, provided only that the added noise in the signal received at each repeater is not too great to prevent recognition of the particular level each given signal is intended to represent.

The dependence of the quantization noise on the number of levels will be discussed later; at that time, both cases of uniform and non-uniform level quantization will be analyzed.

3. *Encoding.* A quantized sample could be sent as a single pulse which would have certain possible discrete amplitudes, the resultant system being a modified form of PAM. However, if many allowed sample amplitudes are required, it would be difficult to make circuits to distinguish these one from another. On the other hand, it is very easy to make a circuit which will tell whether or not a pulse is present. Suppose then that several pulses are used as a code group to describe the amplitude of a single sample. For example, each pulse can be ON (1) or OFF (0), and a code group of n ON-OFF pulses (binary code) can represent 2^n discrete amplitudes. A binary code is just one special case of possible coding schemes; in ternary PCM, three pulse levels, which might be $+1$, -1, and 0, are used to represent 3^n amplitudes. In general, in an s-ary PCM system, any

one quantized signal sample is coded into a group of n pulses each with s possible levels. The number of quantized amplitude levels the code group can express including zero level is given by

$$M = s^n \qquad\qquad (20\text{-}2)$$

We note here that the number M can be identified with the possible number of different messages of the information source discussed in Chapter 19, each message consisting of n symbols using an alphabet of s symbols.

For the purpose of encoding, one characterizes the M discrete amplitude levels by integers, the state numbers $0, 1, 2, \ldots, (M-1)$. The state numbers are then considered to be one-digital numbers of a number system of base M. The process of encoding consists merely in encoding these numbers into another number system of base s.

Equation (20-2) expresses the possibility of transforming the primary signal F(t) with M states into the secondary signal G(t) with only s states, at the expense that one needs for each sample of F(t) a group of

$$n = \frac{\log M}{\log s} \qquad\qquad (20\text{-}3)$$

pulses for G(t). These n pulses must be transmitted during the original sampling interval $T_0 = 1/2f_m$ allotted to the quantized sample. In other words, the pulse rate of an n-digit s-ary PCM system is $2nf_m$ pulses/sec, instead of the original rate of $2f_m$ samples/sec used in the transmission of the primary signal F(t). Reduction of the number of states has thus become possible at the expense of an expanded channel-frequency band.

According to the choice of s, F(t) can be transformed into a large variety of signal functions G(t). As we shall show later, the minimum value s = 2, or a binary number system, offers the greatest advantages in reducing the effect of noise. The secondary signal function G(t) consists then only of two discrete amplitude levels: ON and OFF. With the binary number system, all numbers are expressed by two symbols, 0 and 1, which are called "binary symbols."

If a stands either for 0 or 1, the binary notation with n digits

$$a_1\, a_2, \ldots, a_n$$

represents the number

$$a_1 \, 2^0 + a_2 2^1 + a_3 \, 2^2 + \cdots + a_n \, 2^{n-1}$$

As an example, the binary notation 101 represents the (absolute) number $1 + 0 + 4 = 5$. (Note that the binary notations are written backwards.)

In the ternary number system, a stands for the pulse amplitude 0, 1, 2, and the code group of n digits will represent the number

$$a_1 \, 3^0 + a_2 3^1 + a_3 3^2 + \cdots + a_n 3^{n-1}$$

Table 20-1 shows how the 64 numbers from 0 to 63 are represented in the binary, quaternary, and octonary notation. Thus a total of 64 sample levels can be transmitted either by 6 two-valued pulses, 3 four-valued pulses, or 2 eight-valued pulses. That is to say, a

Table 20-1. Encoding into Binary, Quaternary, and
Octonary Numbers

Decimal no.	Binary no.	Quaternary no.	Octonary no.
0	000000	000	00
1	000001	001	01
2	000010	002	02
3	000011	003	03
4	000100	010	04
5	000101	011	05
6	000110	012	06
7	000111	013	07
8	001000	020	10
9	001001	021	11
10	001010	022	12
11	001011	023	13
12	001100	030	14
..	......	...	..
..	......	...	..
62	111110	332	76
63	111111	333	77

6-digit binary PCM is equivalent to 3-digit quarternary PCM and 2-digit octonary PCM in terms of signal levels per pulse group or "word."

4. *Decoding.* To decode a code group, a pulse must be generated which is the linear sum of all the pulses in the group, each multiplied by its place value (s^0, s, s^2, s^3, ...) in the code. Perhaps the simplest way which has been used involves sending the code group in reverse order, i.e., the "units" pulse first, and the pulse with the highest place value last. The pulses are then stored as charge on a capacitor-resistor combination with a time constant such that the charge decreases by a factor $1/s$ between pulses. After the last pulse, the charge (voltage) is sampled.

At the output terminals of the decoding circuit, we obtain a series of pulses, spaced at regular time intervals $T_0 = 1/2f_m$ sec, of equal short duration τ and of $M = 2^n$ discrete amplitude levels 0, 1, 2, ..., $(M - 1)$. Therefore, if we send these pulses through a low-pass filter of bandwidth f_m and amplifier with a linear gain T_0/τ, we obtain, according to the sampling theorem, the continuous signal function F(t) at the output terminals of the circuit.

As already mentioned, the recovered signal $F_0(t)$ will differ somewhat from the original signal F(t) which was produced by the source, due to the quantization of F(t). These random differences, which give rise to quantization noise, can be kept well within tolerable limits by using a large number of quantization steps in the encoding process. In tolerating this small amount of noise, which is independent of the length of the circuit, we get rid of the channel noise, as long as this noise remains below a comparatively large threshold value, at which discrimination between mark and space becomes impossible. In an ON-OFF binary PCM system with the threshold of pulse detection set at half the peak signal pulse, it is necessary only for the pulse height to exceed twice the highest value of noise and interference to avoid error. A bipolar or plus-and-minus PCM signal with recognition threshold at zero can withstand disturbances up to the pulse height itself. It will be shown later that the ratio of peak-to-peak signal excursion to peak-to-peak noise is the same for the two systems, but the bipolar system has the advantage of operating with a lower value of average signal to average noise power.

20.2   Transmission Requirements for PCM [2,3,5]

In this section we shall consider the transmission requirements of the channel which is to carry the encoded PCM signal, such as bandwidth, threshold power, and average power.

1. *Bandwidth.*   Let B denote the channel bandwidth of the PCM system. As noted earlier, the pulse rate of an n-digit s-ary PCM system is $2nf_m$ pulses/sec, or the time interval between the pulses is $1/2nf_m$ sec. We shall presently show that it is possible to send up to 2B independent pulses/sec over the channel, and hence $2B = 2nf_m$ or $B = nf_m$.

Let the pulses occur (or not occur) at the time t = 0, τ_0, $2\tau_0$, ..., $m\tau_0$, where $\tau_0 = 1/2B$, and let each pulse as received be of the form

$$V = V_0 \frac{\sin [\pi/\tau_0 (t - m\tau_0)]}{\pi/\tau_0 (t - m\tau_0)} \tag{20-4}$$

which is illustrated in Fig. 20-3. Since the channel has a bandwidth of B cycles/sec, it follows that the pulse given by Eq. (20-4) contains no frequencies higher than B; in fact, it is identical to the impulse response of an ideal low-pass filter of cutoff frequency B. As seen from Fig. 20-3, the pulse centered at time $m\tau_0$ will be zero at $t = k\tau_0$ where $k \neq m$, and consequently pulses occurring at intervals τ_0 apart will not interfere with each other. Hence the pulse rate in the channel is $1/\tau_0 = 2B$, which requires a bandwidth of

$$B = nf_m \tag{20-5}$$

The pulses may be transmitted in time sequence over one channel or by frequency division; in each case, the bandwidth required for

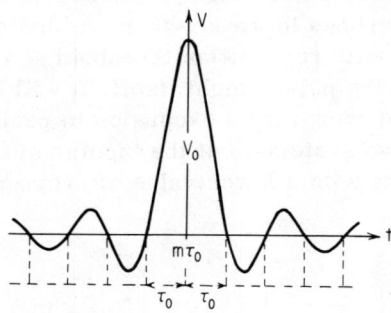

Fig. 20-3. Output pulse at time $m\tau_0$.

PCM is n times as great as that required for direct transmission of the signal, where n is the number of pulses per code group.

2. *Threshold Power.*   The reliability of detecting the presence or absence of a pulse is a function of the signal-to-noise ratio. If the pulse power is too low compared to the noise, the output of the detector will occasionally be mistaken as a pulse when there is none. In the case of gaussian noise, the probability of exceeding any value no matter how great is not zero, so that, in theory, a threshold detector has a finite though small probability of error, no matter how much pulse power is used. However, the probability of error decreases very rapidly as the ratio of the peak pulse to rms noise is increased. It will be shown in Chap. 21 that the probability of error in the presence of white gaussian noise is given by

$$p = \tfrac{1}{2}\left(1 - \operatorname{erf} \frac{V_O}{2\sigma}\right) \tag{20-6}$$

where V_O = peak signal pulse
  σ = rms noise amplitude
and erf x is the well-known error function

$$\operatorname{erf} x = \frac{1}{\sqrt{2\pi}} \int_{-X}^{X} e^{-u^2/2}\, du$$

As the ratio of peak pulse to rms noise is increased, the error rate can be made negligible with only a nominal increase in pulse power, provided that the signal-to-noise ratio P/N is large enough to make the signal at all intelligible. The rapid decrease of the probability of error as the signal-to-noise ratio is increased can be seen from Table 20-2. We conclude, therefore, that there exists in PCM a fairly definite threshold, at about 20 db, above which the interference is negligible. Comparing this figure of 20 db with the signal-to-noise ratio of 60 to 70 db required for high-quality AM transmission of speech, it will be seen that PCM requires much less signal power, even though the noise power has been increased by the n-fold increase in channel bandwidth.

3. *Average Signal Power.*   To calculate the average signal power, we assume that the quantized signal $F_0(t)$ of M discrete levels has been encoded into a group of n pulses each with s possible levels. The spacing between these levels must be chosen so that the decoder at the receiver will be able to distinguish between ad-

Table 20-2. Relationship of Probability of
Error to Signal-to-Noise Ratio

P/N, db	Probability of error	Error frequency: one error every
13.3	10^{-2}	10^{-3} sec
17.4	10^{-4}	10^{-1} sec
19.6	10^{-6}	10 sec
21.0	10^{-8}	20 min
22.0	10^{-10}	1 day
23.0	10^{-12}	3 months

jacent levels with as little error as desired. It is obvious that if the spacing is chosen to be $K\sigma$ (Fig. 20-4) (K = constant), σ being the rms noise voltage at the input to the decoder, then as K increases, the chance of an error decreases. The average signal power may now be calculated as follows. Let V_O volts equal the total voltage swing at the decoder.

For the binary code group s = 2, there are two possibilities, as seen in Fig. 20-5a and b.

1. With ON-OFF signals used, the ON signal is a pulse of V_O volts amplitude, and the OFF signal is zero. The decoder converts the binary pulse group back to the original M levels by determining whether a pulse is present or not. This it does by checking to see whether or not the voltage (signal plus noise) in a given interval exceeds $V_O/2$ volts. If the pulses are assumed equally likely to be ON or OFF, then the average signal power will be

$$P = \frac{V_O^2}{2} \tag{20-7}$$

2. In a bipolar binary system, plus and minus pulses are sent instead of amplitude $V_O/2$ and $- V_O/2$, respectively. The peak-to-peak voltage swing is still V_O volts, but in this case the decoder must judge the polarity of the pulse present in any given interval. This it does by determining whether or not the pulse exceeds the threshold, which is set at 0 volts. Again assuming either polarity

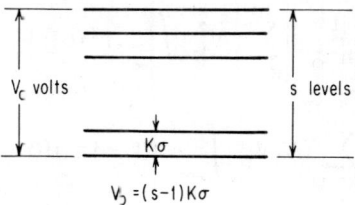

Fig. 20-4. Spacing of levels of code pulse.

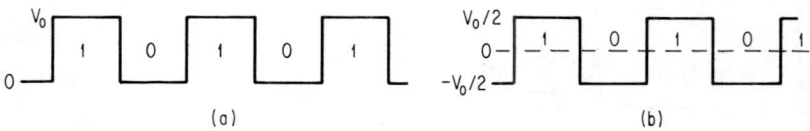

Fig. 20-5. Binary code pulses at the decoder: (a) ON-OFF pulses;
(b) plus-minus or bipolar pulses.

of pulse equally likely, the peak power is $V_0/4$ and is the same as
the average power:

$$P = \frac{V_0^2}{4} \qquad (20\text{-}8)$$

Less power is thus required in a bipolar binary system.

The previous analysis was based on the simplified assumption
that the detected pulses are of constant amplitude. In practice,
their amplitude will not be constant but will be of the form

$$f(t) = \frac{\sin (\pi t/\tau_0)}{\pi t/\tau_0} \qquad (20\text{-}9)$$

occurring at the regular rate of $1/\tau_0 = 2B$. The signal wave consisting of the series of pulses can be expressed by

$$v(t) = \sum_{k=1}^{m} V_k f(t - k\tau_0) \qquad (20\text{-}10)$$

where V_k = peak amplitude of pulse occurring at the time $t = k\tau_0$.
The average power P or the mean square amplitude of the signal
is given by

$$P = \overline{v^2(t)} = \lim_{m \to \infty} \frac{1}{m\tau_0} \int_{-\infty}^{\infty} v^2(t) \, dt$$

$$= \lim_{m \to \infty} \frac{1}{m\tau_0} \left[\sum_{k=1}^{m} V_k^2 \int_{-\infty}^{\infty} f^2(t - k\tau_0) \, dt \right.$$

$$\left. + \sum_{j=1}^{m} \sum_{k=1}^{m} V_j V_k \int_{-\infty}^{\infty} f(t - j\tau_0)f(t - k\tau_0) \, dt \right]_{j \neq k}$$

For the assumed pulse shape, the first integral is equal to τ_0, while the second integral is equal to zero. Thus

$$P = \lim_{m \to \infty} \frac{1}{m} \sum_{k=1}^{m} V_k^2 \qquad (20\text{-}11)$$

which is simply the mean square value of the individual pulse peak amplitudes. This may also be written as the second moment of the probability density function of the pulse amplitude, namely,

$$P = \int_{-\infty}^{\infty} V^2 p(V) \, dV \qquad (20\text{-}12)$$

where $p(V) \, dV$ = probability that the pulse amplitude lies between V and $(V + dV)$.

Now we can proceed to calculate the average power for the code with s possible discrete pulse levels, each spaced $K\sigma$ apart, ranging from 0 to $(s - 1)K\sigma$, as shown in Fig. 20-4. Each pulse has then an amplitude $mK\sigma$ where m is an integer. Assuming that the amplitudes of the pulses are uniformly distributed, the probability of amplitude $mK\sigma$ is $p(m) = 1/s$. The average signal power is given by

$$P = K^2\sigma^2 \sum_{m=0}^{s-1} p(m)m^2 = K^2\sigma^2 \frac{1}{s} \sum_{m=0}^{s-1} m^2$$

$$= K^2\sigma^2 \frac{(s - 1)(2s - 1)}{6} \qquad (20\text{-}13)$$

The average signal power will be least if the amplitude range of the pulses is symmetrical about zero, ranging from $-K\sigma(s - 1)/2$ to $+K\sigma(s - 1)/2$. The average signal power P, assuming all levels to be equally likely, is then

$$P = K^2\sigma^2 \left[\frac{(s - 1)(2s - 1)}{6} - \frac{(s - 1)^2}{4} \right]$$

$$= K^2 \sigma^2 \frac{s^2 - 1}{12} \qquad\qquad (20\text{-}14)$$

This may also be written as

$$P = \frac{V_o^2}{12} \frac{s + 1}{s - 1} \qquad\qquad (20\text{-}15)$$

where $V_o = (s - 1)K\sigma$ is the total amplitude range, as shown in Fig. 20-4.

It will be noticed from Eq. (20-14) that the required signal power P increase is approximately proportional to the square of the possible amplitude levels of the pulses in the pulse group.

20.3   Channel Capacity of a PCM System [2,3]

We have seen that PCM requires more bandwidth and less power than is required with direct transmission of the signal itself. We have, in a sense, exchanged bandwidth for power. A good measure of the bandwidth efficiency is the information capacity of the system as compared with the theoretical limit for a channel of the same bandwidth and power. According to the concepts developed in the last chapter, the information capacity of a system may be thought of as the number of independent symbols or discrete amplitude levels which can be transmitted without error in unit time. We have seen in Chap. 19 that the simplest, most elementary character is a binary digit, and it is convenient to express the information capacity C as the equivalent number of binary digits per second which the channel can handle. By the use of Eq. (19-20), an ideal system has the capacity

$$C = B \log_2 \left(1 + \frac{P}{N}\right) \quad \text{bits/sec} \qquad\qquad (20\text{-}16)$$

where B = channel bandwidth
  P = average signal power
  N = noise power

In order to compare the ideal system capacity with that of a PCM system, assume that the original signal was sampled at the minimum Nyquist rate. For a band-limited signal of f_m cycles/sec bandwidth, this corresponds to $2f_m$ samples/sec. By quantizing each sample to M discrete voltage levels, and assuming all levels equally likely, each sample carries $\log_2 M$ bits of information. The rate of information transmission or system capacity is

$$C = 2f_m \log_2 M \quad \text{bits/sec} \tag{20-17}$$

The encoding process does not change the rate of transmission of information but only converts the M quantized levels to a code group of n pulses of s levels each, such that

$$M = s^n \tag{20-18}$$

The capacity can therefore be written as

$$C = 2nf_m \log_2 s = nf_m \log_2 s^2 = B \log_2 s^2 \quad \text{bits/sec} \tag{20-19}$$

Alternatively, we can derive the same expression by considering that n pulses are sent for each previous pulse sample. The pulse rate at the output of the encoder is thus $2nf_m$ pulses/sec. But with s equally likely levels in each pulse, the average information content of each pulse is $\log_2 s$ bits, or

$$C = 2nf_m \log_2 s = B \log_2 s^2 \quad \text{bits/sec}$$

as before. The PCM system is thus a coded wideband system in which the transmission bandwidth is purposely widened from f_m to B cycles/sec to increase the effective S/N ratio at the system output.

In order to compare the PCM system capacity as given in Eq. (20-19) with the ideal system capacity of Eq. (20-16), we make use of Eq. (20-14), from which we derive the expression

$$s^2 = 1 + \frac{12P}{(K\sigma)^2} \tag{20-20}$$

and by direct substitution in Eq. (20-19), we obtain the expression

$$C = B \log_2 \left(1 + \frac{12P}{K^2 N}\right) \quad \text{bits/sec} \tag{20-21}$$

Comparing this result with Eq. (20-16) for the ideal system capacity, we see that PCM requires $K^2/12$ times the power theoretically required to realize a given channel capacity for a given bandwidth.

The PCM expression involves the parameter K which determines the separation between levels desired. As K increases, the system error rate decreases. From Table 20-2, we note that the transmission of binary digits with an average error rate of 1 digit in 10^3 digits requires a peak signal-to-rms-noise voltage about 10. Since for the binary pulses $V_O = K\sigma$, it follows that K must have the value 10 for an average error rate of 10^6. For this value of K, the PCM system transmitting binary digits requires seven times as much power (8.5 db) as the ideal system for the same channel capacity.

The PCM system, as an example of a coded wideband system, is much more efficient than the uncoded systems, but is still 8.5 db less efficient than Shannon's ideal binary transmission system.

20.4 Quantization Distortion or Noise in a PCM System I: Uniform Spacing of Levels [3,6-10]

As discussed earlier, the process of quantizing a continuous signal into discrete steps gives rise to a random error variation which may be considered as a form of distortion or quantization noise. We have also shown that the quantization process of converting the sample value of the signal in a discrete set of amplitude levels permits digital encoding which results essentially in noise-free transmission in the medium. However, the deliberate error imparted to the signal by quantization or the "rounding off" process is the major source of "noise" in the PCM output signal. As will be shown, this type of noise can be reduced to an acceptable minimum by choosing the quantum step or level separation fine enough. In order to determine the number of discrete steps required to transmit specific signals, we shall present in the following, several methods which will lead to establishing relations between distortion and step size for several types of signals.

The quantizing error (noise) may also be reduced by varying the size of quantizing steps, without adding to their number, so as to provide smaller steps for weaker signals. This means that for a given number of quantized levels or steps, coarser quantization is applied near the peak of large signals, but the larger absolute errors are tolerable here because they are small compared to the larger signal amplitudes. This type of nonuniform quantization or nonuniform spacing between levels will be the subject of the subsequent section, where it will be shown that this can be achieved directly by nonlinear encoding, or by first compressing the input signal and then applying uniform quantization. The choice of quantization or coding function will be shown to be governed by the statistics of the message signal.

1. *Quantizing Error as a Function of Step Size.* The quantizing error may conveniently be expressed in terms of the total mean square error voltage between the exact and quantized samples of the signal. In evaluating the mean square error voltage, we consider the pulsed samples of a complex signal such as speech distributed within all the steps assigned to the signals peak-to-peak voltage range with a probability density governed by statistics of the signal. Furthermore, it is assumed that the probability density is effectively constant within each step, although it is expected to vary from step to step. This assumption is justified provided the steps are sufficiently small and therefore numerous.

With reference to Fig. 20-6, we consider a pulse sample of amplitude v falling in the quantizing interval Δv_k. If the mid-step voltage v_k is assigned to all pulse samples falling in the quantizing interval Δv_k, the absolute value of error in any of these pulse samples will be limited to <u>values</u> between zero and $\Delta v_k/2$. The mean square error voltage $\overline{(v - v_k)^2}$ associated with the quantization of the pulse samples assigned to the k'th voltage level v_k is adopted in our analysis as the significant measure of the error introduced by quantization. From Fig. 20-6, we have

$$v_{k-\frac{1}{2}} = v_k - \frac{\Delta v_k}{2} \leq v \leq v_k + \frac{\Delta v_k}{2} = v_{k+\frac{1}{2}} \qquad (20\text{-}22)$$

and $\quad \overline{(v - v_k)^2} = \int_{v_{k-\frac{1}{2}}}^{v_{k+\frac{1}{2}}} (v - v_k)^2 p(v) \, dv \qquad (20\text{-}23)$

where $(v - v_k)$ is the voltage error imparted to the sample by quantization, and p(v) is the probability density of the signal. By locating v_k at the center of the voltage range assigned to this level, we minimized $\overline{(v - v_k)^2}$ since we shall assume that $p(v) = p(v_k)$, a constant value within the range of this step. It follows that

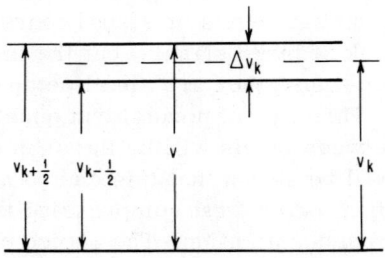

Fig. 20-6. Quantization step.

$$\overline{(v - v_k)^2} = \frac{(\Delta v_k)^3}{12} p(v_k) \tag{20-24}$$

The quantized noise power N_q is equal to the sum of the mean square errors introduced at each level; hence

$$N_q = \sum_k \overline{(v - v_k)^2} = \frac{1}{12} \sum_k p(v_k)(\Delta v_k)^3$$

$$= \frac{1}{12} \sum_k (\Delta v_k)^2 p(v_k) \Delta v_k \tag{20-25}$$

The last expression can be approximated by

$$N_q \doteq \frac{1}{12} \sum_k (\Delta v_k)^2 p_k(v) \tag{20-26}$$

where the discrete probability appropriate to the k'th step is given by

$$p_k(v) = \int_{v_{k-\frac{1}{2}}}^{v_k + \frac{1}{2}} p(v) \, dv \doteq p(v_k) \Delta v_k \tag{20-27}$$

Hence $\quad N_q \doteq \frac{1}{12} \text{ av. } (\Delta v)^2 = \frac{\overline{(\Delta v)^2}}{12} \tag{20-28}$

We have just shown that the total mean square quantizing error voltage is equal to $\frac{1}{12}$ the weighted average of the square of the size of the voltage steps. We shall return to this result in connection with the analysis of nonuniform spacing of levels. In our present discussion of uniform spacing of levels $\Delta v_k = \Delta v = $ constant, and Eq. (20-26) reduces to

$$N_q = \frac{(\Delta v)^2}{12} \tag{20-29}$$

By the use of Eq. (20-29), it is possible to estimate the magnitude of the quantizing noise power with respect to the power of the un-quantized samples. Let S_q denote the signal power recovered from the quantized samples, and let M denote the total number of discrete levels assigned to the message signal. Assuming that the amplitudes of the samples are uniformly distributed, the average signal power can be derived directly from Eq. (20-14) by replacing $K\sigma$ with Δv and s with M; hence

$$S_q = \frac{M^2 - 1}{12} (\Delta v)^2 \tag{20-30}$$

From Eqs. (20-29) and (20-30) it follows that the power of the unquantized samples is equal to

$$S = S_q + N_q = \frac{M^2}{12} (\Delta v)^2 \tag{20-30a}$$

This expression can be derived directly from the relation

$$S = \int_{-\infty}^{\infty} v^2 \, p(v) \, dv = \int_{-\frac{M}{2}(\Delta v)}^{\frac{M}{2}(\Delta v)} v^2 \, p(v) \, dv$$

On the assumption that the samples are uniformly distributed, we obtain

$$S = \frac{1}{M(\Delta v)} \int_{-\frac{M}{2}(\Delta v)}^{\frac{M}{2}(\Delta v)} v^2 \, dv = \frac{1}{M(\Delta v)} \frac{v^3}{3} \Bigg]_{-\frac{M}{2}(\Delta v)}^{\frac{M}{2}(\Delta v)}$$

$$= \frac{M^2}{12} (\Delta v)^2$$

The ratio of the signal power to the quantizing noise power gives a measure of fidelity of the PCM system:

$$\frac{S_q}{N_q} = M^2 - 1 \doteq M^2, \qquad M \gg 1 \tag{20-31}$$

or expressed in decibels,

$$D = 20 \log_{10} M \qquad db \tag{20-32}$$

Table 20-3 shows how the signal-to-noise ratio increases with the number of amplitude levels. Since quantizing noise is uniformly distributed throughout the signal band, it sounds like thermal noise of equal mean power.

The number of quantum levels M is dictated by the desired trans-

Table 20-3. Quantization
Signal-to-Noise Ratio

M	4	8	16	32	64	128
D	12	18	24	30	36	42

mission fidelity. It has been established experimentally that 8 or
16 levels are just sufficient to obtain good intelligibility of speech,
but that quantization noise can easily be heard in the background.
For commercial use, a minimum of 32 levels is considered; however, for toll quality, 128 levels are used with a theoretical signal-
to-noise ratio of 42 db. This will require a binary code group of 7
pulses, or seven times the bandwidth of the original quantized signal.

2. *Distortion in PCM Applied to a Sinusoidal Signal.* The
following analysis is carried out in order to provide a clearer insight into the mechanism of quantization. The process of quantization can be considered to result from the application of the message
signal to a "staircase transducer" whose instantaneous output vs.
input characteristic is shown in Fig. 20-7. As an example, the
corresponding output when the input is a sinusoid is illustrated in
Fig. 20-8. Because of the nonlinearities introduced in the output
signal as a result of the nonlinear staircase characteristic, it is to
be expected that harmonic distortion would be produced, dependent
on the number of amplitude levels used in the quantizing process.

It has been shown in Chap. 17 that when an infinite train of
square-topped pulses is used in PAM to scan a sine wave, the output spectrum does not contain any harmonics of the modulating frequency. In order to calculate the distortion introduced in the system
when PCM is applied to a sine wave, we can replace the signal with
a step function consisting of 2N steps of equal height, as shown in

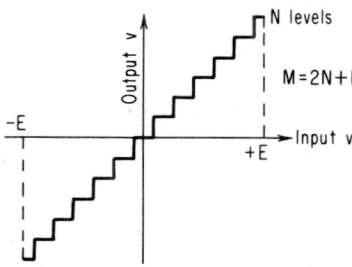

Fig. 20-7. Transfer function characteristic in PCM (uniform quantization).

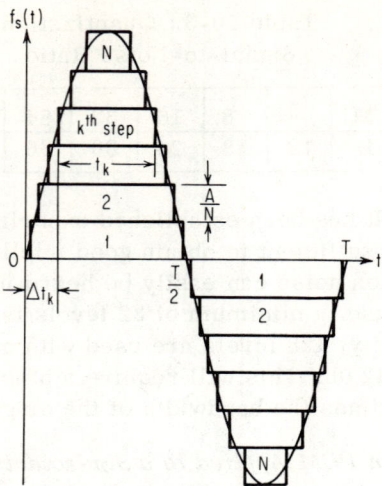

Fig. 20-8. Step representation of a sinusoid.

Fig. 20-8. In this representation, the system will recognize amplitudes to the nearest half level. Obviously, applying PAM to such a step function is equivalent to applying PCM to a sine wave, for the modulated pulses carrying the intelligence will be identical for the two cases, and thus the harmonic distortion of such a step function must equal the distortion introduced in the process of PCM.

In order to calculate the harmonic distortion of the step function, we consider a sine wave $f(t) = A \sin \omega t$ to be replaced by a step function $f_s(t)$, as shown in Fig. 20-8. This step function may be considered as a superposition of N pairs of rectangular steps per cycle of the same period T, equal to that of the fundamental sine wave, of the same height A/N, and of different width t_k such that

$$t_k = \frac{T}{2} - \frac{2}{\omega} \sin^{-1} \frac{2k-1}{2N} \qquad (20\text{-}33)$$

The k'th rectangular step pair may be represented by the Fourier series

$$f_k(t) = \sum_{n=1}^{\infty} B_{n,k} \sin n\omega t$$

where

$$B_{n,k} = \frac{A}{N} \frac{\omega}{\pi} \int_{\Delta t_k}^{T/2-\Delta t_k} \sin n\omega t \, dt = \frac{4A}{\pi n N} \cos n\omega \Delta t_k, \qquad n \text{ odd}$$

but $\quad \Delta t_k = \frac{1}{\omega} \sin^{-1} \frac{2k-1}{2N}$

Hence

$$f_k(t) = \sum_{\nu=0}^{\infty} \frac{4A}{\pi N(2\nu+1)} \cos\left[(2\nu+1) \sin^{-1} \frac{2k-1}{2N}\right] \sin(2\nu+1)\omega t$$

$$(20\text{-}34)$$

For N pair of rectangular pulses, we obtain

$$f_s(t) = \sum_{k=1}^{N} f_k(t) = \sum_{k=1}^{N} \sum_{\nu=0}^{\infty} \frac{4A}{\pi N(2\nu+1)} \cos\left[(2\nu+1) \sin^{-1} \frac{2k-1}{2N}\right]$$

$$\times \sin(2\nu+1)\omega t \qquad (20\text{-}35)$$

The harmonic distortion of $f_s(t)$ is defined by

$$D_h = \frac{\sqrt{\sum_{\nu=1}^{\infty} (A_{2\nu+1})^2}}{A_1} \qquad (20\text{-}36)$$

where $A_{2\nu+1}$ equals the amplitude of the $(2\nu+1)$'th harmonic of the step function $f_s(t)$, namely,

$$A_{2\nu+1} = \frac{4A}{\pi N(2\nu+1)} \sum_{k=1}^{N} \cos\left[(2\nu+1) \sin^{-1} \frac{2k-1}{N}\right] \qquad (20\text{-}37)$$

From the theory of harmonic analysis, we can write the relation

$$\sum_{\nu=0}^{\infty} (A_{2\nu+1})^2 = \frac{2}{T} \int_0^T [f_s(t)]^2 \, dt \qquad (20\text{-}38)$$

so that Eq. (20–36) reduces to

$$D_h = \frac{\sqrt{\frac{2}{T} \int_0^T [f_S(t)]^2\, dt - A_1^2}}{A_1} = \frac{\sqrt{\overline{A^2} - A_1^2}}{A_1} \tag{20–39}$$

where

$$\overline{A^2} = \frac{2}{T} \int_0^T [f_S(t)]^2\, dt = \frac{2A^2}{N^2} \left[N^2 - \frac{2}{\pi} \sum_{k=1}^N (2k-1)\, \sin^{-1} \frac{2k-1}{2N} \right] \tag{20–40}$$

and

$$A_1 = \frac{4A}{\pi N} \sum_{k=1}^N \cos\left(\sin^{-1} \frac{2k-1}{2N} \right) \tag{20–41}$$

The harmonic distortion for various values of N has been calculated by means of Eq. (20–39) and is shown in Table 20–4. From Table 20–4, we note that for N large, the harmonic distortion

$$D_h \doteq \frac{1}{\sqrt{6}\,N} \tag{20–42}$$

a result which is to be expected from Eq. (20–29); for, in the case of a sine wave, the signal power S is equal to

$$S = \frac{(N.\Delta v)^2}{2} = \frac{N^2 (\Delta v)^2}{2}$$

Table 20–4. Per Cent Distortion in a PCM System for Different Levels

N number of levels per half-cycle	A_1 amplitude of first harmonic in % of A	D_h % of distortion	$\frac{1}{\sqrt{6}\,N} \times 100\%$
5	100.97	7.55	8.16
10	100.34	3.80	4.09
20	100.12	2.00	2.04
50	100.03	0.81	0.816

and the quantizing noise power as given by Eq. (20–29) is

$$N_q = \frac{(\Delta v)^2}{12}$$

Hence the harmonic distortion can be expressed in the form

$$D_h = \sqrt{\frac{N_q}{S}} = \frac{1}{\sqrt{6}\,N}$$

where $2N \doteq M$, the total number of discrete levels in the PCM system. Equation (20–42) can also be expressed as a function of the total number of levels M, namely,

$$D_h = \frac{\sqrt{2}}{\sqrt{3M}} \tag{20–43}$$

20.5   Quantization Distortion or Noise in a PCM System II.[8,9,11]
Nonuniform Spacing of Levels

We have shown in the last section that the process of quantization in a PCM system introduces distortion. Thus far our analysis has been limited to the case in which the quantized steps are equal. In actual systems designed for transmission of speech, it is found advantageous to taper the steps in such a way that finer divisions are available for weak signals. In this section, it will be shown that quantization noise can be minimized by a proper level distribution, which is a function of the probability density of the signal. In practice, nonuniform quantization is realized by compression, followed by uniform quantization. The most common form of compression is the so-called logarithmic one, where the levels are crowded near the origin and spaced farther apart near the peaks. A complementary device, the expandor, employs a characteristic inverse to that of the compressor to restore the proper quantized distribution of pulse amplitudes after transmission and decoding. Taken together, the compressor and expandor constitute a compandor. It will be shown that with logarithmic compression, the distortion is largely independent of the statistical properties of the signal.

1. *Distortion for Nonuniform Spacing of Levels.*   Consider a quantized signal v(t), as shown in Fig. 20–9. Let the levels be symmetrically disposed about zero level, but otherwise placed in an arbitrary manner in an interval $(-V, V)$ so that $v_k = -v_{-k}$ and

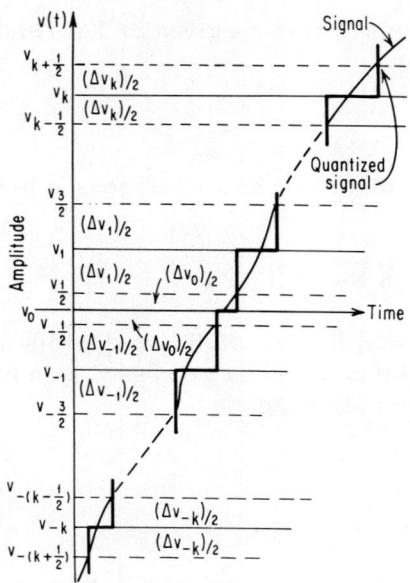

Fig. 20-9. Nonuniform quantization.

$v_0 = 0$. As before, we adopt the measure of error to be equal to the mean square error, which has been shown according to Eq. (20-23) to be equal to

$$\overline{(v - v_k)^2} = \int_{v_k - \frac{1}{2}}^{v_k + \frac{1}{2}} (v - v_k)^2 p(v)\, dv = \sigma_k^2 \tag{20-44}$$

where σ_k is the root mean square error in the k'th level. We shall consider now the problem of minimizing the error of the quantized signal by properly spacing the levels.

First we shall derive a relation between the various v's in Eq. (20-44) so that σ_k in the k'th level is a minimum. We shall show that by locating v_k at the center of the voltage range assigned to this level, the error of the k'th level is minimized. Suppose that the levels are so close that p(v) may be considered as nearly constant over the region of integration and equal to $p(v_{av})$ where

$$v_{av} = \frac{v_{k+\frac{1}{2}} - v_{k-\frac{1}{2}}}{2}$$

It follows from Eq. (20-44) that σ_k^2 is given by

$$\sigma_k^2 = \frac{p(v_{av})}{3} \left[(v_{k+\frac{1}{2}} - v_k)^3 + (v_k - v_{k-\frac{1}{2}})^3 \right] \qquad (20\text{-}45)$$

Differentiating σ_k^2 with respect to v_k gives

$$\frac{d(\sigma_k^2)}{dv_k} = p(v_{av}) \left[-(v_{k+\frac{1}{2}} - v_k)^2 + (v_k - v_{k-\frac{1}{2}})^2 \right] = 0$$

or $\qquad v_k = \dfrac{v_{k+\frac{1}{2}} + v_{k-\frac{1}{2}}}{2} = v_{av} \qquad (20\text{-}46)$

Thus, the condition for minimizing σ_k is to locate v_k halfway between $v_{k+\frac{1}{2}}$ and $v_{k-\frac{1}{2}}$, a condition which was assumed earlier in a heuristic manner in deriving Eq. (20-24). Hence

$$v_{k+\frac{1}{2}} = v_k + \frac{\Delta v_k}{2}$$

$$v_{k-\frac{1}{2}} = v_k - \frac{\Delta v_k}{2}$$

and substituting these values in Eq. (20-45), we obtain

$$\sigma_k^2 = \frac{(\Delta v_k)^3}{12} p(v_k) \qquad (20\text{-}47)$$

The total mean square error voltage is found by summing over all levels, giving

$$N_q = \frac{1}{12} \sum_{-N}^{N} p(v_k)(\Delta v_k)^3 \qquad (20\text{-}48)$$

Now it will be proved that the total mean square error N_q is a minimum when σ_k^2 is constant, independent of the k'th level. By the definition of an integral, we may write

$$\sum_{-N}^{N} p(v_k) \, \Delta v_k = \int_{-V}^{V} [p(v)]^{\frac{1}{3}} \, dv = K, \qquad \text{a constant} \qquad (20\text{-}49)$$

since the integral is a function of its limits only. Let $\mu_k = [p(v_k)]^{\frac{1}{3}} \Delta v_k$; then Eqs. (20-48) and (20-49) reduce to

$$N_q = \frac{1}{12} \sum_{-N}^{N} \mu_k^3 \qquad (20\text{-}50)$$

and

$$K = \sum_{-N}^{N} \mu_k \qquad (20\text{-}51)$$

The problem has now been reduced to minimizing the sum of cubes subject to the condition that the sum of the variables is constant.

From Lagrange's method of undetermined multipliers, it follows that Eq. (20-50) is a minimum when

$$\mu_{-N} = \mu_{-N+1} = \cdots\cdots = \mu_N = \frac{K}{2N + 1} \qquad (20\text{-}52)$$

From this it follows that

$$[p(v_k)]^{\frac{1}{3}} \Delta v_k = \frac{K}{2N + 1}$$

or

$$\sigma_k^2 = \frac{1}{12} \frac{K^3}{(2N + 1)^3} \qquad (20\text{-}53)$$

and the total minimum error power is

$$(N_q)_{\min} = \frac{1}{12} \frac{K^3}{(2N + 1)^2} = \frac{1}{12(2N + 1)^2} \left\{ \int_{-V}^{V} [p(v)]^{\frac{1}{3}} \, dv \right\}^3 \qquad (20\text{-}54)$$

Since $p(v)$ is an even function, and letting $M = 2N + 1$ be the total number of steps, we obtain

$$(N_q)_{\min} = \frac{2}{3M^2} \left\{ \int_{o}^{V} [p(v)]^{\frac{1}{3}} \, dv \right\}^3 \qquad (20\text{-}55)$$

The ratio of the mean square error voltage to the mean square signal is

$$D_m^2 = \frac{(N_q)\min}{S} = \frac{2}{3M^2} \frac{\left[\int_0^V p^{\frac{1}{3}}(v)\,dv\right]^3}{\int_0^V v^2 p(v)\,dv} \tag{20-56}$$

Equation (20-56) gives the minimum distortion resulting with optimum level spacing.

2. *Nonuniform Quantization Through Uniform Quantization of a Compressed Signal.* Nonuniform quantization is logically equivalent to uniform quantization of a "compressed" version of the original input signal. By inserting a nonlinear transducer of the "compressing" type ahead of the quantizer, the weak pulse amplitudes are amplified while the high level pulse amplitudes are compressed; thus a weak signal is effectively allocated a larger number of levels when the compressed signal is quantized uniformly.

The instantaneous compressor is essentially a nonlinear pulse amplifier which modifies the distribution of pulse amplitudes of the PAM signal of the sampler output. The general compression characteristic of such a device is illustrated in Fig. 20-10, where the amplification factor u/v varies from a large value for small inputs to unity gain for the largest input signal V. Figure 20-10 also illustrates how uniform quantization of the compressor output is equivalent to tapered quantizing of the input signal.

An approximate method of obtaining optimum level spacing is as follows. With reference to Fig. 20-9, the k'th level can be written as

$$
\begin{aligned}
v_k &= \frac{\Delta v_0}{2} + \Delta v_1 + \cdots + \Delta v_{k-1} + \frac{\Delta v_k}{2} \\
&= \frac{K}{M}\left[\frac{1}{2p^{\frac{1}{3}}(v_0)} + \frac{1}{p^{\frac{1}{3}}(v_1)} + \cdots + \frac{1}{p^{\frac{1}{3}}(v_{k-1})} + \frac{1}{2p^{\frac{1}{3}}(v_k)}\right] \\
&= \frac{K}{2V}\left[\frac{1}{2p^{\frac{1}{3}}(v_0)} + \frac{1}{p^{\frac{1}{3}}(v_1)} + \cdots + \frac{1}{p^{\frac{1}{3}}(v_{k-1})} + \frac{1}{2p^{\frac{1}{3}}(v_k)}\right]\frac{2V}{M}
\end{aligned}
\tag{20-57}
$$

This series may be approximated by the integral

$$v = A \int_0^u \frac{1}{p^{\frac{1}{3}}(u)}\,du \tag{20-58}$$

where the variable on the right has been changed to u to avoid con-
fusion. A is a constant of proportionality so chosen that when
u = V, v = V.

$$v = V \; \frac{\displaystyle\int_o^u \frac{1}{p^{\frac{1}{3}}(u)}\, du}{\displaystyle\int_o^V \frac{1}{p^{\frac{1}{3}}(u)}\, du} \qquad (20\text{-}59)$$

By letting u vary from 0 to V, v will describe a curve as shown in
Fig. 20-10. As u takes on the values of

$$u = 0, \; u_1 = \frac{2V}{M}, \; \ldots, \; u_k = \frac{2kV}{M}, \; \ldots, \; u_N = \frac{2NV}{M}$$

we obtain the points $v_o = 0, v_1, v_2, \ldots, v_k, \ldots, v_N$. This is illustrated in Fig. 20-10, where it is shown that uniform spacing along
the u axis is transformed into a nonuniform spacing along the v
axis. The compressed output u is then subjected to uniform quantization.

In summary, nonuniform spacing of levels may be realized by
passing the signal through a "compressor" with a given characteristic and applying uniform quantization to its output. A complimen-
tary device, the expandor, whose characteristic is the inverse of
that of the compressor, is incorporated in the receiver. The function of the compressor is to restore the proper distribution of pulse

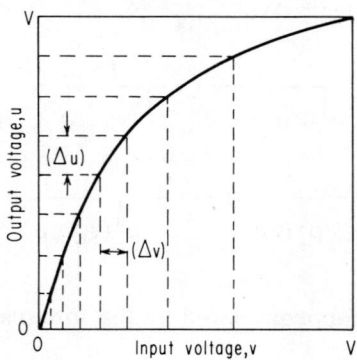

Fig. 20-10. Compression characteristic of "compressor."

amplitudes after transmission and decoding. The overall combined
characteristics of the "compressor" and "expandor" which constitute a "compandor" are linear. The sole purpose of the compandor in the PCM system is to minimize the quantizing noise of
the signal by converting uniform to effectively nonuniform quantization.

3. *Logarithmic Compression.* A compression curve that is
relatively easy to obtain and has been used in practice is the so-
called logarithmic compression curve.

$$u = k \log \left(1 + \frac{\mu v}{V}\right)$$
(20-60)

where v = input voltage

          u = output voltage

          μ = compression parameter

          k = an undetermined constant

To find k, let u = V when v = V, so that the maximum values of
the input and compressed waves are equal. This gives

$$u = \frac{V \log (1 + \mu v/V)}{\log (1 + \mu)}, \qquad 0 \le v \le V$$
(20-61)

and $$u = \frac{-V \log (1 - \mu v/V)}{\log (1 + \mu)}, \qquad -V \le v \le 0$$

The parameter μ controls the degree of compression and may be
chosen so that large changes in the input produce relatively small
changes in the output. Typical compression characteristics, corresponding to various choices of the compression parameter μ, are
shown in Fig. 20-11. An expanded picture of small-amplitude behavior is shown in the logarithmic replot of Fig. 20-12, where it
is seen that for $\mu = 1,000$, a 60-db change in the input will introduce only a 20-db change in the output. When μ is large, the levels
are crowded about zero; $\mu = 0$ corresponds to uniform spacing.

The significance of the parameter μ can also be appreciated by
evaluating the ratio of the largest to the smallest step size which
can be found, as follows. Differentiating Eq. (20-61) yields

$$\frac{du}{V} = \frac{\mu}{\log (1 + \mu)} \frac{1}{1 + \mu v/V} \frac{dv}{V}$$
(20-62)

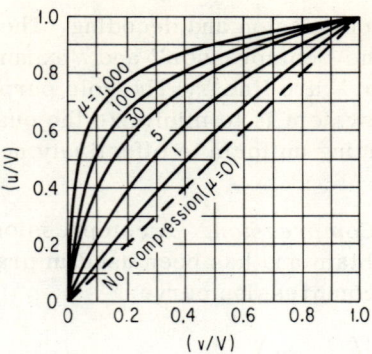

Fig. 20-11. Logarithmic compression characteristics.

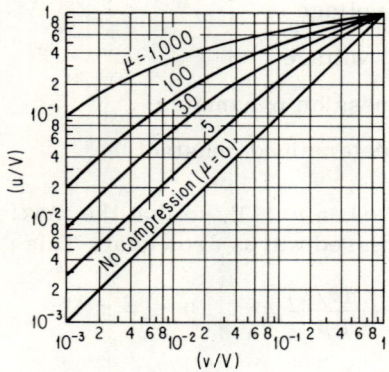

Fig. 20-12. Logarithmic replot of Fig. 20-11.

As previously noted, the compressed signal is divided into M uniformly spaced levels; if we designate the uniform voltage step size by Δu, then

$$\Delta u = \frac{2V}{M} \qquad\qquad (20\text{-}63)$$

For M sufficiently large, we can approximate the differentials dv and du by the step sizes Δv and Δu, so that Eq. (20-62) when combined with (20-63) yields

$$\Delta v = \alpha(V + \mu v), \qquad 0 \le v \le V \qquad\qquad (20\text{-}64)$$

and $\quad \Delta v = \alpha(V - \mu v), \qquad -V \le v \le 0$

where $\quad \alpha = \dfrac{2 \log (1 + \mu)}{\mu M} \qquad\qquad\qquad\qquad$ (20-65)

From Eq. (20-64), we find that the ratio of the largest to the small-est step is approximately given by

$$\frac{(\Delta v)_{v = V}}{(\Delta v)_{v = o}} = 1 + \mu \qquad\qquad\qquad\qquad (20\text{-}66)$$

It is interesting to note that the ratio below is also equal to $(1 + \mu)$.

$$\frac{(du/dv)_{v = o}}{(du/dv)_{v = V}} = 1 + \mu \qquad\qquad\qquad\qquad (20\text{-}67)$$

This ratio, when expressed in decibels, is often referred to as the compression of the system.

4. *Calculation of Quantizing Error with Logarithmic Compression.* The quantizing error or noise power is given approximately by Eq. (20-48), namely,

$$N_q = \frac{1}{12} \sum_{-N}^{N} p(v_k)(\Delta v_k)^3$$

$$= \frac{\alpha^2}{12} \sum_{-N}^{N} (V + \mu v_k)^2 p(v_k) \Delta v_k$$

$$= \frac{\alpha^2}{12} \sum_{-N}^{N} [V^2 p(v_k) + 2V \mu v_k p(v_k) + \mu^2 v_k^2 p(v_k)] \Delta v_k$$

$$\qquad\qquad\qquad\qquad\qquad\qquad\qquad\qquad (20\text{-}68)$$

Since $p(v_k)$ is an even function and $v_k = -v_{-k}$, the second term in the summation vanishes. Thus

$$N_q = \frac{\alpha^2}{12} \sum_{-N}^{N} [V^2 p(v_k) + \mu^2 v_k^2 p(v_k)] \Delta v_k$$

$$\doteq \frac{\alpha^2 V^2}{12} \int_{-V}^{V} p(v) \, dv + \frac{\alpha^2 \mu^2}{12} \int_{-V}^{V} v^2 p(v) dv$$

$$= \frac{\alpha^2}{12} \, (V^2 + \mu^2 S) \tag{20-69}$$

where S equals the average signal power, and V is the peak value of
the signal. Hence the distortion is given by

$$D = \left(\frac{N_q}{S}\right)^{\frac{1}{2}} = \frac{\alpha}{2\sqrt{3}} \left(\frac{V^2}{S} + \mu^2\right)^{\frac{1}{2}}$$

$$= \frac{\log \, (1 + \mu) \, [1 + (\rho/\mu)^2]^{\frac{1}{2}}}{\sqrt{3} \, M} \tag{20-70}$$

where
$$\rho = \frac{V}{\sqrt{S}} \tag{20-71}$$

is the ratio of peak to root-mean-square value of signal. For a
given ρ, the distortion is a function of μ only and will be a minimum
for optimum μ, as shown in Fig. 20-13. When μ is large compared
to ρ, the distortion D reduces to

$$D = \frac{\log \, (1 + \mu)}{\sqrt{3} \, M} \tag{20-72}$$

It follows therefore that when μ is large, the distortion is largely
independent of the signal.

The distortion for uniform quantization may be derived from Eq.
(20-70) by letting $\mu \rightarrow 0$. Thus

$$D_0 = \frac{\rho}{\sqrt{3} \, M} \tag{20-73}$$

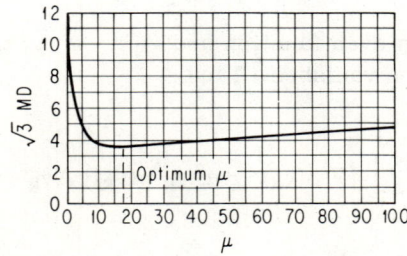

Fig. 20-13. Distortion plotted vs. the parameter, μ.

For a sine wave $\rho = \sqrt{2}$, and hence

$$D_0 = \frac{\sqrt{2}}{\sqrt{3} \, M} \qquad\qquad\qquad (20\text{-}74)$$

as in Eq. (20-43). Thus, when a signal is uniformly quantized, the distortion depends on the ratio of the peak to the root-mean square value of the signal. This is illustrated in curve 1 of Fig. 20-14. Curve 2 gives the minimum distortion for optimum logarithmic compression when the compression parameter μ is chosen according to the relation

$$\rho = \left[\frac{\mu^2}{(1+\mu)/\mu \, \log \, (1+\mu) - 1} \right]^{\frac{1}{2}} \qquad\qquad (20\text{-}75)$$

which is plotted in curve 3.

It is shown in Ref. 8 that for small values of ρ there is little advantage in using optimum spacing of levels. This is illustrated in Fig. 20-15, where it is seen that the chief advantage of using optimum spacing occurs when the crest factor of the signal is high.

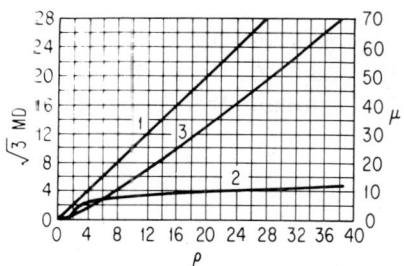

Fig. 20-14. Distortion characteristics as a function of ρ.

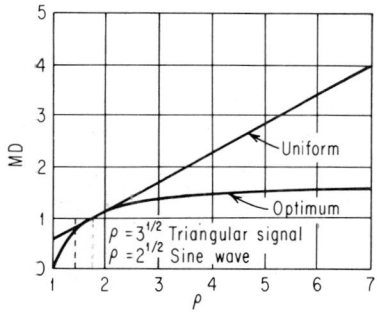

Fig. 20-15. Distortion plotted vs. crest-factor ρ.

References
1. Reeves, A. H.: Electrical Signaling System, International Standard Electric Corp., U. S. Patent 2,272,070, Feb. 3, 1942.
2. Oliver, B. M., J. R. Pierce, and C. E. Shannon: The Philosophy of PCM, *Proc. IRE*, November, 1948, pp. 1324-1331.
3. Mayer, H. F.: Principles of Pulse Code Modulation, *Advan. Electron.*, vol. 3, Academic Press Inc., New York, 1951.
4. Flood, J. E.: Time Division Multiplex Systems: part 4, Pulse-code Modulation, *Electron. Eng.*, April, 1953.
5. Schwartz, M.: "Information Transmission, Modulation, and Noise," McGraw-Hill Book Company, New York.
6. Clavier, A. G., P. F. Panter, and D. D. Grieg: PCM Distortion Analysis, *Trans. AIEE*, vol. 66, pp. 1110-1122, November, 1947.
7. Bennett, W. R.: "Spectra of Quantized Signals," *BSTJ*, vol. 27, pp. 946-972, 1948.
8. Panter, P. F., and W. Dite: Quantization Distortion in PCM with Non-uniform Spacing of Levels, *Proc. IRE*, January, 1951.
9. Smith, B.: Instantaneous Companding of Quantized Signals, *BSTJ*, vol. 36, pp. 653-709, 1957.
10. Groenewout, H. W. F.: Non-linear Distortion in Pulse Code Modulation Systems, in W. Jackson (ed.), "Communication Theory," Academic Press Inc., New York.
11. Shennum, R. H., and J. R. Gray: Performance Limitations of a Practical PCM Terminal, *BSTJ*, vol. 41, pp. 143-171, January, 1962.

21

OUTPUT SIGNAL-TO-NOISE IMPROVEMENT RATIO IN PCM SYSTEMS

The deliberate quantization error or noise imparted to the signal at the transmitting terminal of a PCM system has been discussed at some length in the previous chapter. The error or noise produced by "quantization" is the major source of signal impairment and originates only at the transmitting end of the system and nowhere else. We have shown that for a given bandwidth this type of noise can be minimized by nonuniform quantization, which can be achieved by first compressing the input signal and then applying uniform quantization.

The other type of noise which characterizes a PCM system originates primarily at the receiving end of the system. This is a "false pulse noise," caused by incorrect interpretation of the intended amplitude of a pulse by the receiver or by any repeater which is caused by noise spikes breaking through the threshold or trigger level. However, we shall show that this noise decreases so rapidly as the signal power is increased above threshold that in any practical system it would be made negligible by design. Aside from causing errors occasionally, fluctuation noise will have no other effect on the output signal. This is to be contrasted with the modulation systems previously considered—AM, FM, PAM, PPM—in

653

which the noise affected the output signal continuously. For a given quality of overall transmission, the longer the system, the more severe are the requirements on each link. For example, if K links are to be used in tandem, the noise power added per link can only be 1/K of that which would be permissible in a single link. (This does not hold for tandem links in troposcatter systems, owing to the statistical properties of the noise contributions in each link.) This is the basic reason for transmitting coded signals; the problem becomes one of recognizing the presence or absence of a pulse (binary code) or of the amplitude level of a pulse (s > 2). So long as the noise does not cause a 0 symbol to be mistaken for a 1 symbol, for example, the effect of fluctuation noise can be completely removed. With the signal power great enough, mistakes can be made to occur rather infrequently.

In the simple binary code, each digit is equally liable to error, owing to noise on the transmission path; however, the amounts of noise which these errors cause at the output of the decoder are unequal. Thus, in a 5-digit code, an error in the most significant digit results in an error in the amplitude of the output signal 16 times as great as that caused by an error in the least significant digit. The signal-to-noise ratio of the system should be improved by making the most significant digits less liable to error than those of less significance, so that they contribute equally to the output noise. In the following paragraphs, possible methods of equalizing the noise contribution will be investigated.

Because the signal in a PCM system can be regenerated as often as necessary, the effects of amplitude and phase and nonlinear distortions in one link, if not too great, produce no effect whatever on the regenerated input signal to the next link. The transmission requirements for a PCM link are almost independent of the total length of the system.

In this chapter we shall consider the effect of fluctuation noise on the output signal-to-noise ratio and the use of noise-reducing codes to minimize the false-pulse noise resulting from the noise spikes which exceed the threshold level. The problem of pulse regeneration and its effect on the signal-to-noise output of a PCM system with many repeating stations will also be considered. It will be shown that a small increase in input carrier-to-noise ratio is sufficient to overcome the cumulative effect of noise along a chain of many repeaters, provided use is made of pulse regeneration in the repeaters.

21.1   Generation of False-pulse Noise and the Effect on Output Signal-to-Noise Ratio [1-4]

In order to evaluate the effect of channel noise on the decoded signal at the receiver, we shall consider a binary PCM system where the quantized samples of the message function are trans-mitted by a code group of n binary ON-OFF pulses, giving in all 2^n discrete amplitude levels including zero level. The contributions which the pulses make to the decoded output voltage are

$$2^0 A, \; 2^1 A, \; 2^2 A, \; \ldots, \; 2^{n-1} A$$

where A is the height of a quantizing step or level. Since with white noise no sharp threshold noise power exists, there is always a finite probability that the noise will exceed the amplitude of the code pul-ses and thus produce a fault in the code group. If the amplitude of the code pulses is taken as V_c, the following simplifying assump-tions are made as to the effect of noise bursts.

1. A negative noise burst of any shape or duration, and of ampli-tude greater than $V_0/2$ when combined with a code pulse, will cause complete obliteration of the code pulse.

2. A positive noise burst of any shape or duration, and of ampli-tude greater than $V_0/2$, occurring in the absence of a code pulse, will cause a spurious code pulse to appear.

1. *Transition Probabilities.*   The effect of the false pulses in-troduced in the code group on the decoded signal can be described by transition probabilities, as discussed in Chapter 19 and illustra-ted in Fig. 21-1. At the input of the encoder, the primary quantized signal to be transmitted has M states, represented by M inputs: 0, 1, ..., (M − 1). The secondary signal, after encoding, has only two states 0 and 1; a group of n = $\log_2 M$ ON-OFF pulses being transmitted for each particular state of the primary signal. With

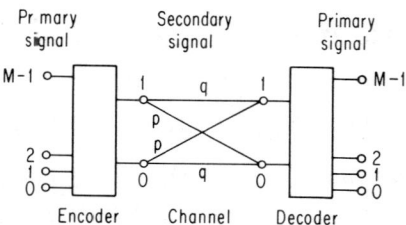

Fig. 21-1. Graphical representation of PCM transmission in the presence of noise.

ideal transmission, these groups are decoded into the original states without any errors.

Due to channel noise, there is a certain probability p that we send out state 1 but receive state 0 and vice versa. Consequently, the probability that no transmission fault occurs is q, and p + q = 1. In the limiting cases: with no noise, p = 0, q = 1; and with infinite noise, we have $p = q = \frac{1}{2}$. Assuming that the transition probabilities p and q of the secondary signal are known, the problem is to calculate the error due to noise introduced in the received primary signal during the transmission between the input terminals of the encoder and the output terminals of the decoder.

Consider the decoding of the secondary signal which consists of a series of ON-OFF pulses, a fraction p of the pulses being wrong, the fraction q = 1 − p of them being the right ones, and the faulty pulses being distributed in a random manner within the sequence. The decoded primary sample at the output of the decoder will be in error whenever a fault occurs in the pulse group. As an example, if we transmit with a 3-digit code, and state 6 = 011 is transmitted, the probability of receiving state 3 = 110 will be pqp = p^2q. Table 21-1 shows the probabilities that any one of the eight possible states will be received with a 3-digit code if nothing but a sequence of state 6 is sent. It should be noted that without noise p = 0 and q = 1, we receive only state 6 as expected, and with infinite noise $p = q = \frac{1}{2}$, all probabilities are equal ($= \frac{1}{8}$). In this case, we receive a random sequence of all possible states without any relation to the transmitted state.

2. *Calculation of Noise Power.* To calculate the mean square error or noise power introduced in the decoded signal, we proceed as follows. Let x_k denote the discrete signal amplitude corresponding to state k at the transmitter, measured in arbitrary units, and correspondingly, we have the same set of discrete amplitude y_i at

Table 21-1. Transition Probabilities

Transmission sequence (state 6)	011	011	011	011	011	011	011	011
Received sequence	000	100	010	110	001	101	011	111
Received state	0	1	2	3	4	5	6	7
Probability	qpp	ppp	qqp	pqp	qpq	ppq	qqq	pqq

the receiver. This means that no matter what happens on the channel, no other than these discrete amplitudes y_i can be produced by the decoder. The transition probability or the likelihood $P(y_i|x_k)$ is the probability that the signal amplitude y_i will be received if we send the signal amplitude x_k. From the definition of transition probability (see Fig. 19-6), the following general relations hold:

$$\sum_{\text{all } k} P(x_k|y_i) = \sum_{\text{all } i} P(y_i|x_k) = 1 \qquad (21\text{-}1)$$

We shall now determine the expected value or the statistical average of received state y_i, when the transmitted state x_k is given, on the assumption that there are n independent chances for error in the transmission of the n digit binary code. Representing x_k and y_i as binary numbers (in arbitrary units) we write

$$x_k = \sum_{r=0}^{n-1} \alpha_r 2^r$$

and
$$y_i = \sum_{r=0}^{n-1} \beta_r 2^r$$

where　$\alpha_r = 0$ or 1 and $\beta_r = 0$ or 1.

Hence　$(x_k - y_i) = \sum_{r=0}^{n-1} (\alpha_r - \beta_r) 2^r$

The expected value of $(x_k - y_i)$ is

$$E\left[(x_k - y_i)\right] = E\left[\sum_{r=0}^{n-1} (\alpha_r - \beta_r) 2^r\right] = \sum_{r=0}^{n-1} 2^r E\left[(\alpha_r - \beta_r)\right] \qquad (21\text{-}2)$$

For correct transmission of the rth digit $(\alpha_r - \beta_r) = 0$. However, if an error occurs in the rth digit $(\alpha_r - \beta_r) = +1$ or -1 depending on whether α_r is 1 or 0. The probability that the quantity $(\alpha_r - \beta_r)$ equals $+1$ or -1 clearly depends on the choice of α_r. We shall now calculate the conditional expectation of the quantity $(\alpha_r - \beta_r)$ for the two possible values of α_r.

1. $\alpha_r = 1$. The conditional probability $P(\alpha_r = 1 | (\alpha_r - \beta_r) = 0) = q$ and $P(\alpha_r = 1 | (\alpha_r - \beta_r) = 1) = p$. Hence, the conditional expectation of $(\alpha_r - \beta_r)$ given $\alpha_r = 1$ is

$$E[(\alpha_r - \beta_r)] = q.0 + p.1 = p$$

2. $\alpha_r = 0$; $P(\alpha_r = 0 | (\alpha_r - \beta_r) = 0) = q$ and $P(\alpha_r = 0 | (\alpha_r - \beta_r) = -1) = p$. Hence, given $\alpha_r = 0$,

$$E[(\alpha_r - \beta_r)] = q.0 + p(-1) = -p$$

The conditional expectations of $(\alpha_r - \beta_r)$ for both values of α_r can be uniquely expressed by

$$E[(\alpha_r - \beta_r)] = p(2\alpha_r - 1) \tag{21-3}$$

By direct substitution of Eq. (21-3) in Eq. (21-2), we obtain an expression for the statistical average of $(x_k - y_i)$

$$E[(x_k - y_i)] = \sum_{r=o}^{n-1} 2^r p(2\alpha_r - 1) = 2p \sum_{r=o}^{n-1} \alpha_r 2^r - p \sum_{r=o}^{n-1} 2^r$$

$$= 2px_k - p(2^n - 1) = 2px_k - p(M-1) \tag{21-4}$$

Since state x_k is given, it follows that the statistical average of the received amplitude y_i is given by

$$E[y_i | x_k] = x_k(1 - 2p) + p(M - 1) \tag{21-5}$$

Let us refer the amplitude of both transmitted and received states to an origin selected in the middle of the range of possible amplitudes and proceed with the calculation of the statistical average of the received amplitude y_i relative to the new origin. The transmitted and received amplitudes relative to the middle of the range are given by

$$x_k' = \left(k - \frac{M-1}{2}\right) \Delta v \tag{21-6}$$

$$y_i' = \left(i - \frac{M-1}{2}\right) \Delta v \tag{21-7}$$

where Δv denotes the quantizing step.

The expected value of y_i' given x_k is

$$E[y_i'] = E\left[\left(i - \frac{M-1}{2}\right)\right]\Delta v = E[i]\,\Delta v - \frac{(M-1)\,\Delta v}{2}$$

$$= \left[k(1-2p) + p(M-1) - \frac{M-1}{2}\right]\Delta v$$

$$= \left(k - \frac{M-1}{2}\right)(1-2p)\,\Delta v$$

$$= \left(k - \frac{M-1}{2}\right)\Delta v\,(q-p) = x_k'\,(q-p) \qquad (21\text{-}8)$$

In the notation of transitional probabilities Eq. (21-8) can be written as

$$\sum_{\text{all } i} y_i\,P(y_i|x_k) = (q-p)\,x_k \qquad (21\text{-}9)$$

where it is understood that the amplitudes x_k and y_i refer to the new origin. Eq. (21-9) states that the statistical average of the received amplitude is proportional to the transmitted signal amplitude. The received signal y_i can be considered to be equivalent to receiving the signal amplitude x_k together with a noise amplitude equal to $y_i - x_k$. Generally, one can as well assume that the received signal amplitude was mx_k and the noise amplitude was $y_i - mx_k$, where m is a constant, independent of the amplitude, to be determined in such a way that the noise power at the output becomes a minimum.

The mean square error between the fixed output signal amplitude mx_k and the possible output amplitudes y_i, taking into account the transitional probabilities, is given by

$$N_k = \sum_{\text{all } i} (y_i - mx_k)^2\,P(y_i|x_k) \qquad (21\text{-}10)$$

The mean square error N_k represents the noise power at the output terminals of the decoder when one sends the signal amplitude x_k into the encoder. This can be minimized by solving the equation

$$\frac{\partial N_k}{\partial m} = 2x_k \sum_{\text{all } i} (y_i - mx_k)\,P(y_i|x_k) = 0 \qquad (21\text{-}11)$$

and making use of Eq. (21-9). The factor m for minimum noise power, or maximum signal power, is then found to be given by

$$m = q - p$$

The mean square error or noise power is therefore given by

$$N_k = \sum_{\text{all } i} P(y_i|x_k) \, y_i^2 - 2 \, (q - p) \, x_k \sum_{\text{all } i} P(y_i|x_k) \, y_i$$

$$+ (q - p)^2 \, x_k^2 \sum_{\text{all } i} P(y_i|x_k) = \sum_{\text{all } i} P(y_i|x_k) \, y_i^2 - (q - p)^2 \, x_k^2$$

$$\tag{21-12}$$

The first term of the right-hand side of Eq. (21-12) represents the decoder output power if one sends a fixed signal power x_k^2; the second term of the right-hand side represents the received signal power. Thus, the difference between the decoder output power and signal power equals the noise power N_k.

3. *Output Signal-to-Noise Ratio.* The effect of channel noise on the entire transmitted signal will now be evaluated. If $P(x_k)$ is the probability of occurrence of sample amplitude x_k then the quantized signal power at the transmitting end will be

$$S_q = \sum_{\text{all } k} x_k^2 \, P(x_k) \tag{21-13}$$

and the total power at the receiving end is

$$P_o = \sum_{\text{all } i} y_i^2 \, P(y_i) \tag{21-14}$$

but using Eq. (19-29) we write

$$P(y_i) = \sum_{\text{all } k} P(x_k) \, P(y_i|x_k) \tag{21-15}$$

Hence, $$P_o = \sum_{\text{all } k} \sum_{\text{all } i} y_i^2 \, P(y_i|x_k) \, P(x_k) \tag{21-16}$$

The decoder output power is partly signal and partly noise. The received signal power is

$$S_o = (q - p)^2 \sum_{\text{all } k} x_k^2 \, P(x_k) = (q - p)^2 \, S_q \qquad (21\text{-}17)$$

and the received noise power is therefore given by

$$N_o = \sum_{\text{all } k} N_k \, P(x_k) = P_o - S_o \qquad (21\text{-}18)$$

From Eq. (21-16), we note that the total output power P_o depends on the statistical structure of the primary signal and on the transition probabilities. We also note that $P_o \rightarrow S_q$ in case $p \rightarrow 0$. Furthermore, if the statistical structure of the signal at the transmitter is such that all possible amplitudes x_k are uniformly distributed, then all possible amplitudes y_i at the receiver occur also with equal probabilities, or $P_o = S_q$. In this case, we have

$$S_o = (q - p)^2 S_q = (1 - 4pq) S_q \qquad (21\text{-}19)$$

since $p + q = 1$, and

$$N_o = S_q - S_o = 4pqS_q \qquad (21\text{-}20)$$

The output signal-to-noise ratio is therefore

$$\left(\frac{S}{N}\right)_o = \frac{1}{4pq} - 1 \qquad (21\text{-}21)$$

The output signal-to-noise ratio drops from infinity with a noiseless channel ($p = 0$, $q = 1$) to zero in the case of an infinitely large channel noise ($p = q = \frac{1}{2}$).

4. *Signal-to-Noise Ratio of K Links in Tandem.* We shall consider now a PCM system to consist of K links in tandem, and we shall calculate the $(S/N)_o$ at the end of the K'th link. In such a system, the transmission faults occur within each link and cumulate over the links. As an example we consider the case of two links; some pulses which were right on the first link will get wrong on the second link, but we must also recognize that some pulses which were wrong on the first link will become right again on the second link.

Consequently, the probability that a pulse is received right after transmission over two links is $q_2 = q^2 + p^2$, and the probability that it will be wrong is $p_2 = pq + qp$. Therefore, for two links, we obtain

$$q_2 - p_2 = (q - p)^2 \tag{21-22}$$

or, in general, with transmission over K links

$$q_k - p_k = (q - p)^K \tag{21-23}$$

The received signal power after transmission over K channels will therefore be, according to Eq. (21-19),

$$S_O = (q - p)^{2K} S_q \tag{21-24}$$

and the accompanying noise power will be

$$N_O = S_q - S_O = [1 - (q - p)^{2K}] S_q \tag{21-25}$$

Finally, the resulting output signal-to-noise ratio is therefore

$$\left(\frac{S}{N}\right)_{o,K} = \frac{(q - p)^{2K}}{1 - (q - p)^{2K}} \tag{21-26}$$

We have just derived an expression for the signal-to-noise output in a PCM system in terms of the probability of false pulses in the code group due to channel noise. In the following section, we shall derive a relation between error probabilities and the carrier-to-noise in the channel. This will lead to an expression relating the output signal-to-noise ratio $(S/N)_o$ to the input carrier-to-noise ratio $(C/N)_i$ of the channel.

21.2 Output Signal-to-Noise Ratio in a PCM System vs. Input Carrier-to-Noise Ratio [1-2, 5-9]

We have shown in Chap. 20 in the discussion of transmission requirements for PCM that the channel bandwidth $B = nf_m$, where n equals the number of digits in the code group, and f_m is the bandwidth of the primary message signal. Using the notation of the previous chapter, the noise power in bandwidth B is $N = \sigma^2$, where σ = rms noise amplitude. As the noise is restricted to the channel bandwidth B, it can be described by discrete noise samples $1/2B$ sec apart which coincide with the code pulses. The effect of the channel noise on the ccde pulses will obviously depend on the statistical properties of the noise; various types of noise with equal noise power, but different statistical properties, will give rise to different error probabilities.

1. *Error Probability and Channel Carrier-to-Noise Ratio.*[1,6] In order to establish a relation between the probability of false pulses in the code group or error probability and carrier-to-noise ratio in the channel, we shall start with the statistical properties of white noise. As discussed in Chap. 4, the probability density function of white noise is gaussian or normal, so that the probability of finding a noise sample in the amplitude range v and (v + dv) is given by

$$p(v) \, dv = \frac{1}{\sqrt{2\pi}\,\sigma} e^{-v^2/2\sigma^2} \, dv \qquad (21\text{-}27)$$

where σ is determined by the mean noise power N in the channel bandwidth B; i.e., $\sigma^2 = N$. As discussed above, any positive noise burst which exceeds $V_0/2$ and occurs in the absence of a code pulse will cause a spurious pulse to be registered by the decoder. However, any positive noise bursts which occur during a code pulse will not introduce false code pulses. Assuming that the presence and absence of a code pulse is equally likely, then, on the average, half the positive noise bursts exceeding $V_0/2$ will result in errors. Similar reasoning leads to the conclusion that half the negative noise bursts of amplitude exceeding $V_0/2$ will cause errors by canceling the code pulses which are present. This intuitive analysis applies to both unipolar and bipolar code pulses, the only difference being in the setting of the threshold level which is at $V_0/2$ for ON-OFF pulses and 0 for bipolar pulses.

Consider first the case of transmitting a 0, so that no pulse is present at the time of decoding. The probability of error in this case is just one-half times the probability that the noise voltage v exceeds $V_0/2$; using Eq. (4-83), we have

$$P\left(v > \frac{V_0}{2}\right) = \frac{1}{\sqrt{2\pi}\,\sigma} \int_{V_0/2}^{\infty} e^{-v^2/2\sigma^2} \, dv \qquad (21\text{-}28)$$

Now let a 1 be transmitted by the encoder; in this case a negative noise burst of amplitude $|v| > V_0/2$ will cancel the code pulse, and the decoder will register a 0. The probability of $-v < -V_0/2$ is given by

$$P\left(-v < -\frac{V_0}{2}\right) = \frac{1}{\sqrt{2\pi}\,\sigma} \int_{-\infty}^{-V_0/2} e^{-v^2/2\sigma^2}\, dv \qquad (21\text{-}29)$$

The probability of error of mistaking a binary 1 for a 0 and vice versa is therefore equal to

$$p = \frac{1}{2}\frac{1}{\sqrt{2\pi}\,\sigma} \int_{-\infty}^{-V_0/2} e^{-v^2/2\sigma^2}\, dv + \frac{1}{2}\frac{1}{\sqrt{2\pi}\,\sigma} \int_{V_0/2}^{\infty} e^{-v^2/2\sigma^2}\, dv$$
$$(21\text{-}30)$$

Because of the symmetry of the probability density function of Eq. (21-27) which is plotted in Fig. 21-2, the probability of error p is given by

$$p = \frac{1}{\sqrt{2\pi}\,\sigma} \int_{V_0/2}^{\infty} e^{-v^2/2\sigma^2}\, dv = \frac{1}{2}\left(1 - \operatorname{erf}\frac{V_0}{2\sigma}\right) \qquad (21\text{-}31)$$

where erf x is the well-known error function integral,

$$\operatorname{erf} x \equiv \frac{1}{\sqrt{2\pi}} \int_{-x}^{x} e^{-y^2/2}\, dy \qquad (21\text{-}32)$$

From these equations, we readily derive the following relations:

$$q = 1 - p = \tfrac{1}{2}\,(1 + \operatorname{erf} x)$$

and $q - p = \operatorname{erf} x$

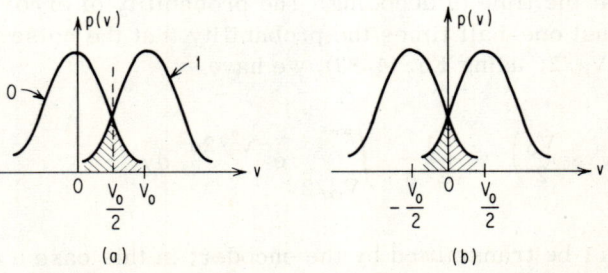

(a)   (b)

Fig. 21-2. Probability density function of signal plus white gaussian noise: (a) ON-OFF case; (b) bipolarity case.

Using Eq. (21-26), we can relate the output signal-to-noise ratio of K links in tandem to the input carrier-to-noise ratio $(C/N)_i$.

$$\left(\frac{S}{N}\right)_{o,K} = \frac{(\text{erf } x)^{2K}}{1 - (\text{erf } x)^{2K}} \qquad (21\text{-}33)$$

where $x = V_0/2\sigma$. For high-input carrier-to-noise ratio, $x \gg 1$, and erf x can be approximated by

$$\text{erf } x \doteq 1 - \sqrt{\frac{2}{\pi}} \; \frac{e^{-x^2/2}}{x} \qquad (21\text{-}34)$$

In this case, we obtain for a single link $(K = 1)$, the approximate result

$$\left(\frac{S}{N}\right)_{o} \doteq \sqrt{\frac{\pi}{8}} \; x \; e^{x^2/2} = \sqrt{\frac{\pi}{8}} \left(\frac{V_0}{2\sigma}\right) e^{\frac{1}{2}(V_0/2\sigma)^2} \qquad (21\text{-}35)$$

Using Eqs. (20-7) and (20-8), we can now express the $(S/N)_o$ in terms of $(C/N)_i$. For unipolar or ON-OFF binary system we have $C_i = V_0^2/2$, and hence

$$\left(\frac{S}{N}\right)_{o} = \tfrac{1}{4} \sqrt{\pi} \; \sqrt{(C/N)_i} \; e^{\frac{1}{4}(C/N)_i} \qquad (21\text{-}36)$$

For bipolar system we have $C_i = V_0^2/4$ and

$$\left(\frac{S}{N}\right)_{o} = \sqrt{\frac{\pi}{8}} \; \sqrt{\left(\frac{C}{N}\right)_i} \; e^{\frac{1}{2}(C/N)_i} \qquad (21\text{-}37)$$

The rapid improvement in the output signal-to-noise ratio from very low to very high values is illustrated graphically in Fig. 21-3 for K = 1, 10, and 100 links in tandem. This is especially true for 10 links or even more so for 100 links. With 100 links, for example, a small increase of the channel carrier-to-noise ratio from 10 to 14 db improves the output signal-to-noise ratio from about 4 to 40 db.

2. *Signal-to-Noise Improvement in PCM: an Alternate Approach.* [2,5]   As before, we consider a binary code group of n pulses:

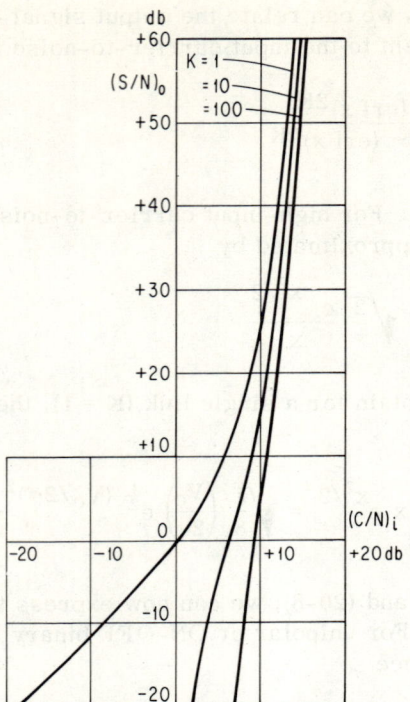

Fig. 21-3. Output signal-to-noise ratio for PCM. (From H. F. Mayer,[1] courtesy of Academic Press, Inc.)

the contributions which the pulses make to the decoded output voltage are

$$2^0A, \; 2^1A, \; 2^2A, \; \ldots, \; 2^{r-1}A, \; \ldots, \; 2^{n-1}A$$

If the probability of an error in the r'th pulse of the group is p_r, the mean square noise voltage produced at the output is

$$N_0 = \overline{V_n^2} = A^2 \sum_{r=1}^{n} p_r 2^{2(r-1)} \tag{21-38}$$

As we shall show below, the probability of error p_r may be expressed as a function of the average number $\mu(v)$ of positive noise bursts per second of amplitude $v > V_0/2$, where V_0 equals the amplitude of the code pulses.

To calculate $\mu(v)$, the noise is assumed as before, to be distributed normally about the mean value, and, as discussed previously,
one-half of both the positive and negative noise bursts of amplitude
$|v| > V_0/2$ will result in errors. Consequently, the average rate at
which errors occur due to both positive and negative bursts of noise
is $\mu(v)$. If the circuit includes an ideal low-pass filter with a cutoff
frequency f_0, it has been shown by Rice [7] that the probable number
of positive noise bursts per second of amplitude greater than v is
given by the expression

$$\mu(v) = \frac{f_0}{\sqrt{3}} e^{-v^2/2\sigma^2}$$

(21-39)

In the following analysis of deriving the functional relationship
between p_r and $\mu(v)$, the noise bursts are considered as being independent of each other; furthermore, the number of noise bursts in
a given time interval is considered to be independent of what has
occurred in any other interval. By assuming the noise bursts to
have a Poisson distribution, the probability of exactly k noise
bursts occurring in an interval T, as given by Eq. (4-99), is

$$p(k) = \frac{[\mu(v)T]^k}{k!} e^{-\mu(v)T}$$

(21-40)

In order to evaluate the effect of the channel noise on the pulses
of the code group, and consequently on the decoded signal at the receiver, we shall assume that a noise burst will cause the insertion
or cancellation of only one code pulse except for very rare cases
which may be neglected. This assumption is justified, provided the
cutoff frequency f_0 of the filter is chosen as low as the duration of
the code pulse will allow and the input carrier-to-noise ratio is
sufficiently large. Of course, when the noise power is comparable
to the signal power, this is no longer true, but the system would
then become inoperative. Let the interval T in Eq. (21-40) correspond to the interval τ_r of one digit pulse of the code group, and
since we have assumed that only one error can occur (k = 1), and if
the errors are relatively infrequent, $\mu(v)\tau \ll 1$. Equation (21-40)
then becomes

$$p_r \doteq \mu(v_r)\tau_r = \frac{f_0}{\sqrt{3}} \tau_r e^{-v_r^2/2\sigma^2}$$

(21-41)

By direct substitution for p_r in Eq. (21-38), the mean square noise voltage produced at the output is

$$N_O = \overline{V_n^2} = \frac{A^2}{\sqrt{3}} \sum_{r=1}^{n=1} f_O \tau_r \, 2^{2(r-1)} e^{-v_r^2/2\sigma^2} \qquad (21\text{-}42)$$

This is the output noise voltage for the general case of a binary code group in which the pulses are not equally liable to error.

The output signal power is determined by the amplitude and form factor of the quantized signal. If the modulating signal is a sine wave, the greatest possible amplitude is $(A/2)(2^n - 1)$. The mean square signal output voltage or signal power is then

$$S_O = \overline{V_s^2} = A^2 \frac{(2^n - 1)^2}{8} \qquad (21\text{-}43)$$

In the simple binary code, all the pulses are of equal height V_O and duration τ_O and equally liable to error with probability p. Equation (21-42) reduces to

$$N_O = \overline{V_n^2} = \frac{A^2 f_O \tau_O}{\sqrt{3}} \, e^{-(V_O/2)^2/2\sigma^2} \sum_{r=1}^{n} 2^2 \,(r-1)$$

$$= (4^n - 1) \frac{f_O \tau_O A^2}{\sqrt{3}} \, e^{-(V_O/2)^2/2\sigma^2} = \frac{1}{3} A^2 p (2^{2n} - 1) \qquad (21\text{-}44)$$

where $\quad p = \dfrac{f_O \tau_O}{\sqrt{3}} \, e^{-(V_O/2)^2/2\sigma^2}$

The output signal-to-noise ratio is

$$\left(\frac{S}{N}\right)_O = \frac{3\sqrt{3}\,(2^n - 1)}{8 f_O \tau_O (2^n + 1)} \, e^{\frac{1}{2}(V_O/2\sigma)^2} = \frac{3\,(2^n - 1)}{8\,(2^n + 1)} \frac{1}{p} \qquad (21\text{-}45)$$

It is to be noted here that the product $f_O \tau_O$ is a significant parameter of a pulse system which determines the shape of the pulse. An acceptable minimum value for this quantity is 0.5. Consequently, Eq. (21-45) can be approximated

by $\quad \left(\dfrac{S}{N}\right)_O \doteq 1.3 \, e^{\frac{1}{4}(C/N)_i}, \qquad$ for ON-OFF binary system $\quad (21\text{-}46)$

and $\quad \left(\dfrac{S}{N}\right)_o \doteq 1.3\ e^{\frac{1}{2}\,(C/N)_i}$, $\qquad$ for bipolar binary system $\qquad$ (21-47)

where both $(C/N)_i$ and $(S/N)_o$ are expressed as power ratios; expressing $(S/N)_o$ in decibels, we obtain for the bipolar case the relation

$$\left(\dfrac{S}{N}\right)_{o\,(db)} = 2.2\ \left(\dfrac{C}{N}\right)_{i\ (power\ ratio)} \qquad\qquad (21\text{-}43)$$

It should be noted here that for high $(C/N)_i$ this method produces a result very similar to Eq. (21-37).

Thus, the output signal-to-noise ratio expressed in decibels is proportional to the input carrier-to-noise ratio expressed as a power ratio. This result is independent of n, the number of code pulses, provided the number is large enough: that is, more than 3 or 4. The results of Eqs. (21-47) and (21-48) are shown graphically in Figs. 21-4 and 21-5.

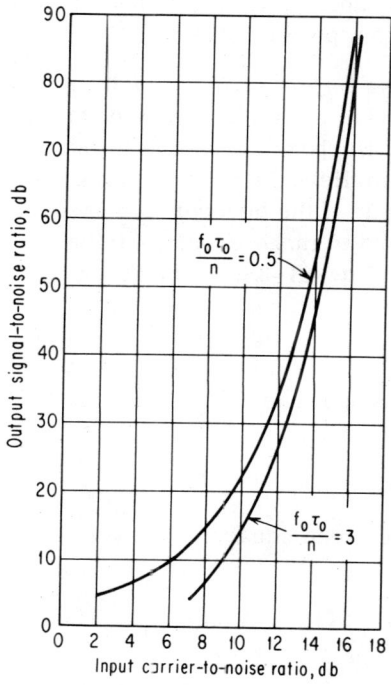

Fig. 21-4. Signal-to-noise improvement in a PCM system $(S/N)_o$ in decibels vs. $(C/N)_i$ in decibels.

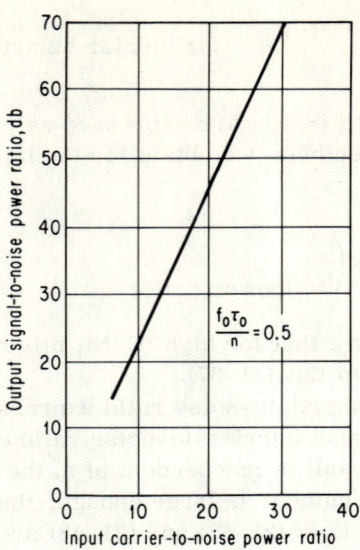

Fig. 21-5. Signal-to-noise improvement in a PCM system $(S/N)_o$ in decibels versus $(C/N)_i$ power ratio.

3. *Signal-to-Noise Improvement with Weighted Code Pulses.* [5,8,9] In the previous analysis a simple binary code was assumed where each digit is equally liable to error due to false noise bursts. However, as discussed earlier, the noise powers which these errors introduce at the output of the decoder are unequal. If errors in each of the code pulses are to make equal contributions to the output noise power, then in Eq. (21-38) we must put

$$p_r = p_1 2^{2(1-r)} \qquad\qquad (21\text{-}49)$$

so that Eq. (21-38) reduces to

$$\overline{V_n^2} = np_1 A^2$$

and consequently the mean square noise voltage produced at the output is

$$N_{o1} = \overline{V_n^2} = \frac{nf_1 \tau_1 A^2}{\sqrt{3}} e^{-(V_1/2)^2/2\sigma^2} \qquad\qquad (21\text{-}50)$$

where V_1 is the amplitude of the first code pulse.

For a sine-wave-modulating signal, the output signal-to-noise ratio is

$$\left(\frac{S}{N}\right)_0 = \frac{\sqrt{3}\,(2^n - 1)}{8nf_1\tau_1}\,e^{(V_1/2)^2/2\sigma^2} \tag{21-51}$$

Thus as a result of the contribution of equal amounts of noise power resulting from the errors in the weighted code pulses, the output noise power has been changed in the ratio

$$\frac{N_{0_1}}{N_0} = \frac{3nf_1\tau_1}{(4^n - 1)f_0\tau_0}\,e^{[(V_0/2)^2/2\sigma^2 - (V_1/2)^2/2\sigma^2]} \tag{21-52}$$

The parameters $f_1\tau_1$ and $f_0\tau_0$ determine the shape of the pulses with an acceptable minimum of 0.5; for $f_1\tau_1 = f_0\tau_0$, the improvement in output noise power is

$$\Delta = 10\,\log_{10}\frac{4^n - 1}{3n} + 10\left[\frac{(V_1/2)^2}{2\sigma^2} - \frac{(V_0/2)^2}{2\sigma^2}\right]\log_{10}e \qquad \text{db} \tag{21-53}$$

Several methods have been proposed [5] for making the digits of a binary code group contribute equally to the output noise, which will result in a better signal-to-noise ratio than that provided by simple binary PCM. For example, this can be accomplished by varying the pulse height [8] or duration [5] from digit to digit. It is shown that by these methods an improvement in output signal-to-noise ratio is obtained over a wide range of input carrier-to-noise ratios. However, the proposed systems are much more complicated than a simple binary system, and in view of the already excellent noise performance of a simple binary PCM, the additional complexity is too high a price to pay for the gain in the signal-noise improvement.

21.3 Effect of Noise on PCM Systems with Repeating Stations [2,3]

The advantage of pulse regeneration in a PCM repeater system can best be illustrated by comparing the signal-to-noise output of a regenerative repeater system with a nonregenerative system.

Let the relay system in Fig. 21-6 consist of K identical linear amplifiers and K identical paths, the gain of an amplifier being just sufficient to overcome the path attenuation. In the nonregenerative system, the various noise sources of rms value σ may then be replaced by an equivalent noise source of rms value of $\sqrt{K}\,\sigma$ at the input, and Eq. (21-48) gives

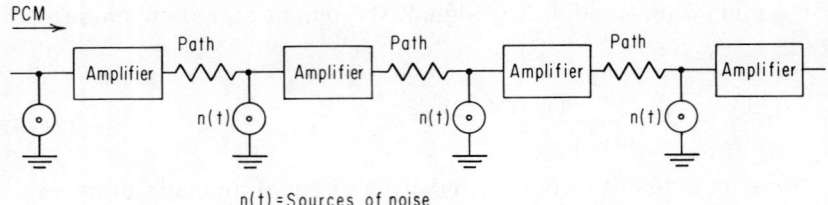

n(t) = Sources of noise

Fig. 21-6. Relay system.

$$\left(\frac{S}{N}\right)_{o\,(db)} \doteq \frac{2.2}{K}\left(\frac{C}{N}\right)_{i\,(power\ ratio)} \tag{21-54}$$

In practice, the results obtained may be somewhat worse, owing to the progressive deterioration of the pulse shape.

In the regenerative type of relay system, the repeaters regenerate the code pulses so that the output consists of idealized PCM pulses. The gain, again, is just sufficient to overcome the path attenuation. As noise enters in the form of wrong code pulses, each noise source contributes n noise bursts, and consequently μ wrong codes independently of the others. If we neglect the possibility of some noise bursts canceling the effect of others, Eq. (21-44) for the output noise power becomes

$$N_O = \frac{K\,A^2\,f_O\,\tau_O\,(4^n - 1)}{3\sqrt{3}}\,e^{-(V_O/2)^2/2\sigma^2} \tag{21-55}$$

Using this value, Eq. (21-48) becomes

$$\left(\frac{S}{N}\right)_{o\,(db)} \doteq 2.2\left(\frac{C}{N}\right)_{i\,(p.r.)} - 10\log K \tag{21-56}$$

From this we see that the curve of Fig. 21-5 is moved parallel to itself, whereas in the nonregenerative case considered, its slope is changed.

To show the effect more clearly, the input carrier-to-noise ratio for an assumed output signal-to-noise ratio of 60 db vs. the number of repeaters is plotted in Fig. 21-7. It is seen that, for the linear nonregenerative repeaters, the input level rises considerably with the number of repeaters, and requires a 17-db increase for 50 repeaters. On the other hand, for regenerative repeaters a small increase of 1 db takes care of the cumulative effect of noise. This represents a considerable economy in total power installed, since each repeater power must necessarily be increased.

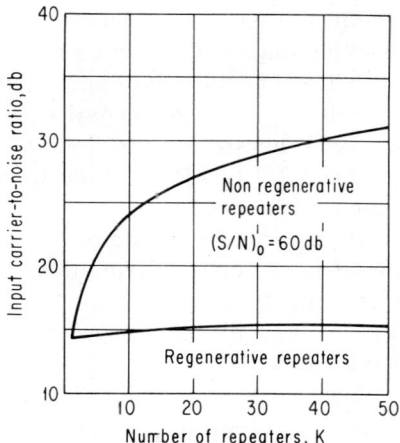

Fig. 21-7. Input carrier-to-noise ratio vs. number of repeaters for constant-output signal-to-noise ratio.

21.4 Rate of Transmission [1,ε,10]

Up to now we have discussed the effect of noise on the signal-to-noise output of a PCM system. In this section, we shall discuss the loss of information during transmission over a noisy channel.

We have shown in Chap. 20 [Eq. (20-19)] that the maximum information rate or capacity of a PCM system is given by

$$C = 2B \log_2 s \qquad \text{bits/sec}$$

That is, if we transform a signal of bandwidth B_1 and number of states s_1 into another signal of bandwidth B_2 and number of states s_2, the rate of information remains constant with all possible code transformations:

$$C = 2B_1 \log_2 s_1 = 2B_2 \log_2 s_2 = \cdots = 2B \log_2 s \qquad (21\text{-}57)$$

It follows therefore that the maximum rate of a noiseless binary PCM system equals 2 bits/(sec) (cycle bandwidth). When noise is added to the system, there will arise equivocation, which has the effect of reducing the channel capacity. The question is: How many bits are lost during the transmission over a noisy channel?

Let us consider again the primary quantized signal of M states to be transmitted before encoding. The encoder, channel, and decoder establish a link between the input and output states. With ideal transmission, without channel noise, any input state is connected only with the corresponding output state, but with none of the others. As shown in Fig. 19-6, with channel noise, any input state is connected with all output states by a set of transition probabilities $P(y_i|x_k)$; that is, if we send the signal amplitude x_k into the encoder, there is a finite probability $P(y_i|x_k)$ that we will receive the amplitude y_i at the output of the decoder.

Consider a PCM system with $M = 2^n$ states. As discussed in Chap. 19, let the probability of error in transmitting a binary pulse be equal to p and the probability of transmitting the pulse correctly be equal to q, so that $p + q = 1$. We have shown that the average equivocation per code group is

$$H(X|Y) = -n(p \log_2 p + q \log_2 q) \qquad \text{bits} \qquad (21\text{-}58)$$

and that

$$I(X; Y) = H(X) - H(X|Y) = H(Y) - H(Y|X) \qquad (21\text{-}59)$$

where

$$H(X) = - \sum_{\text{all } k} P(x_k) \log P(x_k) \qquad (21\text{-}60)$$

and

$$H(Y) = - \sum_{\text{all } i} P(y_i) \log P(y_i) \qquad (21\text{-}61)$$

It has also been shown in Chap. 19 that the entropy is maximum if all states occur with equal probabilities; namely, $P(x_k) = P(y_i) = 1/M$. Hence

$$H(X) = H(Y) = -\sum_{i}^{M} \frac{1}{M} \log \frac{1}{M} = n \qquad (21\text{-}62)$$

The decoder output can be considered to consist of two sources: the information source which produces $H(Y)$ bits of information per sample, and the noise source which works against the information source and which produces an average of $H(Y|X)$ bits/sample. The effective rate at which information is transmitted is therefore given by

$$R = I(X; Y) = H(Y) - H(Y|X)$$

or

$$R = n(1 + p \log_2 p + q \log_2 q) \qquad \text{bits/sample} \qquad (21\text{-}63)$$

If f_m denotes the primary-signal bandwidth, and consequently nf_m is the channel bandwidth, we can send a maximum of $2f_m$ samples/ sec into the encoder. The maximum rate of transmission or the channel capacity of the noisy system is

$$C = 2nf_m(1 + p \log_2 p + q \log_2 q) \qquad \text{bits/sec} \qquad (21\text{-}64)$$

where the values of p must be determined as a function of $(C/N)_i$.

In the absence of noise, $p = 0$ and $q = 1$, we transmit 2 bits/(sec) (cycle) which, as we have seen previously, is the characteristic maximum for binary PCM. With infinite noise $p = q = \frac{1}{2}$, we obtain $C = 0$, and no information is transmitted at all.

Since the rate in bits per second per cycle is a function of p and q only, there exists a unique relation between rate of information and output signal-to-noise ratio, given by Eq. (21-21). This relation is plotted in Fig. 21-8. For high signal-to-noise ratios, the rate approaches the limit of 2 bits/(sec)(cycle) of bandwidth, and approaches zero as the noise becomes infinitely large.

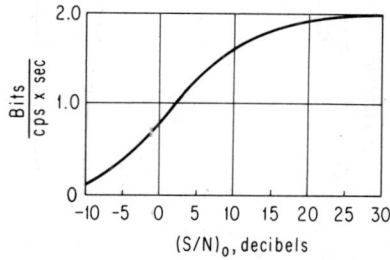

Fig. 21-8. Rate of information and output signal-to-noise ratio.

The value of p and q in Eq. (51-64) depends primarily on the channel-input carrier-to-noise ratio, and in case of white noise the probability of error p is given by Eq. (21-31). Let the input carrier-to-noise voltage ratio be denoted by

$$r_i = \frac{V_o}{\sigma}, \qquad \text{peak signal to rms noise}$$

We can readily show that Eq. (21-31) reduces to

$$p = 0.5 - \frac{1}{\sqrt{2\pi}} \int_o^{r_i/2} e^{-t^2/2} \, dt \qquad (21\text{-}65)$$

This function is plotted in Fig. 21-9. Corresponding to a given r_i, the value of p is obtained from the curve, and when substituted in Eq. (21-64), the channel capacity can be calculated.

We can also derive a relation between the output signal-to-noise ratio as a function of input carrier-to-noise ratio, with n the number of pulses per code group as parameter. Since after demodulation the bandwidth is reduced to f_m, retaining the same capacity, hence

$$C = f_m \log_2 \left[1 + \left(\frac{S}{N} \right)_o \right] = 2nf_m \left[1 + p \log_2 p + q \log_2 q \right]$$

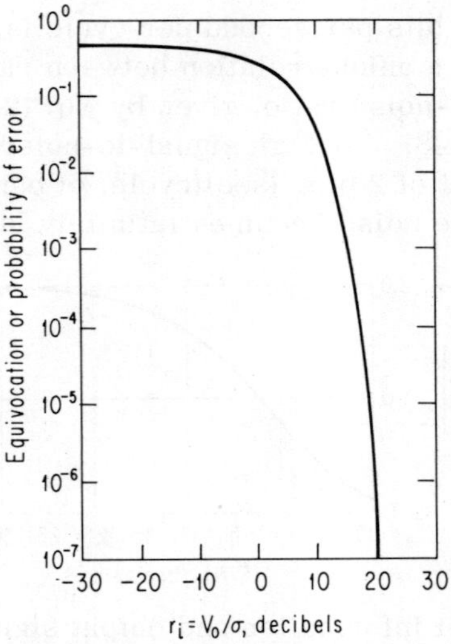

Fig. 21-9. Probability of error in any single pulse as a function of input signal-to-noise voltage ratio.

Hence

$$\left(\frac{S}{N} \right)_o = 2^{2n(1 + p \log_2 p + q \log_2 q)} - 1 \tag{21-66}$$

This result combined with Eq. 21-65 is plotted in Fig. 21-10.

The rate of transmission as a function of $(C/N)_i$, with K the number of links in tandem, is shown plotted in Fig. 21-11. Curve C is the channel capacity given by the well-known Eq. (20-16), namely,

$$C = B \log_2 \left[1 + \left(\frac{C}{N} \right)_i \right] \qquad \text{bits/sec}$$

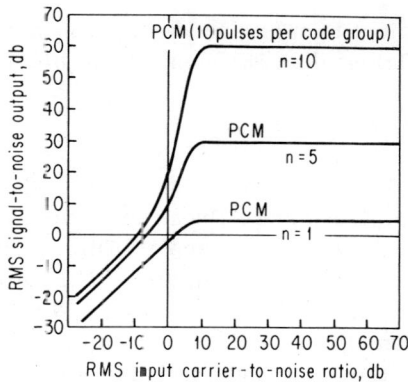

Fig. 21-10. $(S/N)_o$ vs. $(C/N)_i$ for binary PCM. (From R. M. Page,[3] IRE Conv. Record.)

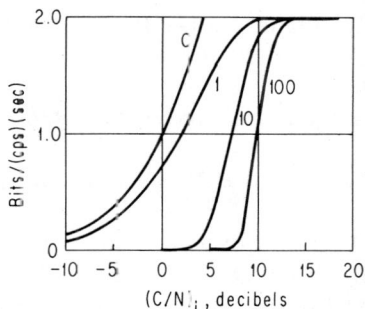

Fig. 21-11. Rate of transmission and channel carrier-to-noise ratio. (From H. F. Mayer,[1] courtesy of Academic Press, Inc.)

where $B = nf_m$ is the channel bandwidth. This represents the maximum rate which can be obtained by ideal encoding. The three curves designated with 1, 10, 100 refer to PCM circuits consisting of 1, 10, and 100 links in tandem. It is of interest to note that with an input carrier-to-noise rate of 13 to 15 db, one approaches the maximum rate of 2 bits/(sec)(cycle) in all three cases, practically independent of the number of links. We also note from curve C that in order to transmit 2 bits/(sec)(cycle), we need with ideal encoding a carrier-to-noise ratio of 5 db, while with PCM and white noise, we need a carrier-to-noise ratio of 13 db, an increase of about 8 db. This is in agreement with the result derived previously in Chap. 20, where it has been shown that while PCM as an example of a coded system is much more efficient than an uncoded one, nevertheless, it is still about 9 db less efficient than Shannon's ideal binary transmission system.

References

1. Mayer, H. F.: Principles of Pulse Code Modulation, *Advan. Electron.*, vol. 3, Academic Press Inc., New York, 1951.
2. Clavier, A. G., P. F. Panter, and W. Dite: Signal-to-Noise Ratio Improvement in a PCM System, *Proc. IRE*, April, 1949.
3. Yates-Fish, N. L., and Fitch, E.: Signal/Noise Ratio in Pulse Code Modulation, *Proc. IEE*, vol. 102, pt. B, March, 1955.
4. Viterbi, A. J.: Lower Bounds on Maximum Signal-to-Noise Ratios for Digital Communication Over the Gaussian Channel, *IEEE Trans. Commun. Systems*, March, 1964.
5. Flood, J. E.: Noise-reducing Codes for PCM, *IEE Monograph* 291R, February, 1958.
6. Griffith, J. W. R.: Signal/Noise Ratio in Pulse Code Modulation Systems—Use of the "Ideal Observer" Criterion, *J. Brit. IRE*, March, 1959.
7. Rice, S. O.: Mathematical Analysis of Random Noise, *BSTJ*, July, 1944.
8. Bedrosian, E.: Weighted PCM, *IRE Trans. Inform. Theory*, March, 1958.
9. Purton, R. F.: A Survey of Telephone Speech-Signal Statistics and their Significance in the Choice of a PCM Compounding Law, *IEE Paper* No. 3773 E, January, 1962.
10. Page, R. M.: Comparison of Modulation Methods, *IRE Conv. Record*, 1953.

22

DELTA MODULATION (DM)

It has been shown in the last chapter that PCM is the most efficient among the existing communication systems. However, the implementation of such a system is quite complex; the circuitry used in modulation and demodulation is complicated and expensive. In this chapter, we shall discuss a more novel coded modulation system known as delta modulation (DM), which is almost as efficient as PCM, requires wider bandwidth than PCM, but has much simpler circuitry. Delta modulation is a differential PCM or a 1-digit code PCM which was invented in the ITT French Laboratories.[1] Whereas in the PCM system, an n-digit binary code is used to transmit information, in delta modulation, a code comprising only one digit is used. The transmitted pulses carry the message information corresponding to the derivative of the amplitude of the message function, and at the receiving end these pulses are integrated to obtain the original waveform. One of the significant differences between PCM and DM is the relative simplicity and low cost of the coding equipment in the latter. In PCM, the analogue message signal is first converted into PAM samples, which are encoded in groups of pulses. In the delta process, the encoding into the binary form is done in a single operation, without going through the PAM and the binary coding processes.

In this chapter we shall cover the basic principles of delta modulation, with particular emphasis on its inherent quantizing noise and signal-to-noise performance. Several variations of delta modulation will be introduced, such as delta-sigma modulation (Δ-ΣM) and high-information delta modulation (HIDM), which were designed to overcome some of the deficiencies of ordinary delta modulation. Finally, a comparison between DM and PCM systems will be made from the point of view of information theory and system performance.

22.1   Theory of Delta Modulation [2-5]

In a DM system, instead of the absolute signal amplitude being transmitted at each sampling, only the changes in signal amplitude from sampling instant to sampling instant are transmitted. The principle of operation of delta modulation can be described with reference to Fig. 22-1. It is essentially a quantized feedback system consisting of a pulse generator, a pulse modulator, an integrator, and a difference circuit or comparator. In the encoder, the pulses from the pulse generator are treated by the modulator, which delivers positive pulses if the sign of the difference signal $\epsilon(t)$ is positive and otherwise delivers negative pulses. As shown in Fig. 22-2, the transmitted pulse train $e_2(t)$ of positive and negative pulses at the output of the encoder can be assumed to be generated at a constant clock rate. The difference circuit or comparator at the transmitter decides on the basis of feedback whether the output

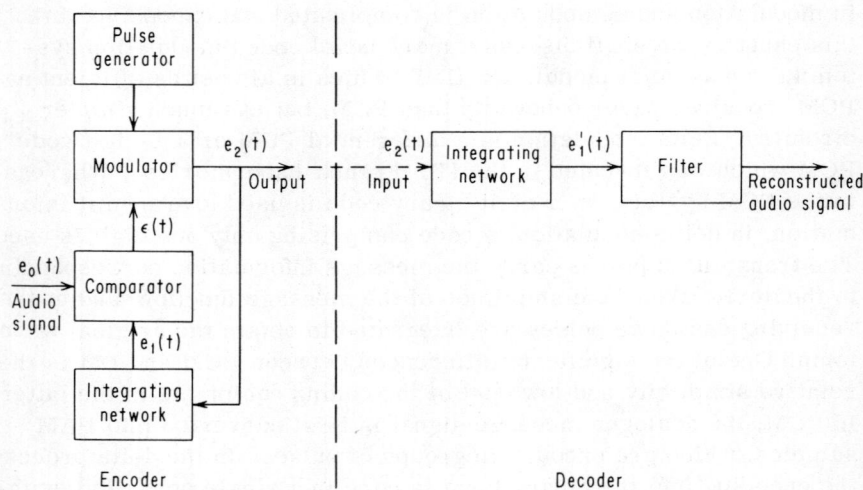

Fig. 22-1. Basic circuit for delta modulation.

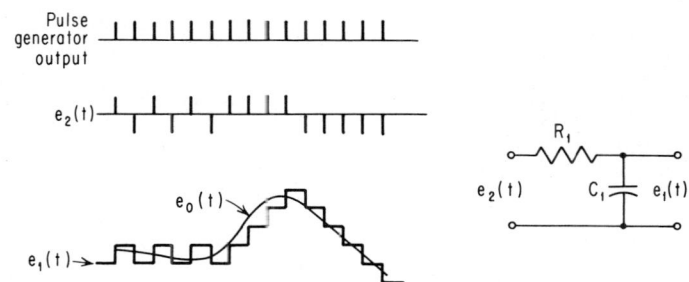

Fig. 22-2. Delta modulation waveforms using single integration.

pulse should be positive or negative. The difference signal from the comparator is derived in the following manner. The delta-modulated output pulse train $e_2(t)$ is synthesized by passing it through the integrating network of the feedback loop, and the resulting waveform $e_1(t)$, which consists of a series of unit steps up and down, is then compared with the original message signal $e_0(t)$ by the comparator. The output of the comparator $\epsilon(t) = e_0(t) - e_1(t)$ decides what the polarity of the output pulse should be in order to correct for the difference between the two voltages. The feedback system tends to reduce the difference, so that the synthesized signal $e_1(t)$ in the form of a step wave follows the message signal $e_0(t)$. Sampling of the voltage $\epsilon(t)$ is done at a constant rate by pulses from the pulse generator.

In practice, the negative pulses of the output signal $e_2(t)$ may be omitted in the transmission path without affecting the signal-to-noise ratio at the receiving end. For the addition of a periodic series of positive pulses to the original pulse series of $e_2(t)$ has only the effect of raising the d-c level by doubling the amplitude of the positive pulses and deleting the negative ones, as all other frequencies are cut off by the low-pass filter. In the following discussion, however, we shall consider for reasons of simplicity the approximation signal as being generated by narrow positive and negative pulses.

In the decoder, the delta-modulated pulse train $e_2(t)$ is again integrated into the voltage $e_1'(t)$ which consists of the original message function plus noise components due to sampling. These are eliminated by a low-pass filter, so that the reconstructed signal of the final output is a close replica of the original modulating signal $e_0(t)$, as shown in Fig. 22-2. The difference between the original and reconstructed signals gives rise to a "quantizing noise," which can be decreased by increasing the pulse frequency of the pulse generator. In contrast to the quantization principle of

PCM, the information is quantized in a 1-unit code, and the "sampling frequency" is made equal to the pulse frequency. This results in very rough quantizing, which is compensated by the fact that the signal samples are taken as often as indicated by the pulse interval and thus n times as often as for PCM at the same pulse frequency, where n equals the number of pulses in the PCM code group.

As seen from Fig. 22-2, the message signal $e_0(t)$ is approximated by a step curve $e_1(t)$, which is constructed in such a way that at each sampling time a unit step upward or downward is made, depending on the synthesized stepped signal at the encoder being lower or higher than the message signal.

A better approximation of the signal can be obtained by using, instead of a 1-unit code, an n-unit code for the construction of the quantized signal, in such a way that at each sampling time a choice of 2^n possibilities is available, consisting of the steps $(2^n - 1)\sigma$, $(2^n - 3)\sigma, \ldots, 3\sigma, \sigma, -\sigma, -3\sigma, \ldots, -(2^n - 3)\sigma, -(2^n - 1)\sigma$, where σ denotes the height of a unit step. In this case not only the polarity of the difference between the signal and the approximation must be determined, but also the magnitude, which can be represented by 2^n possible choices. For n = 2, for instance, there are four possible choices, namely, a step of $+3\sigma$, $+\sigma$, $-\sigma$, -3σ, as shown in Fig. 22-3a; the coded approximation signal is shown in Fig. 22-3b.

1. *Single Integration.* The simplest system of DM is obtained by using an integrating network in the feedback path, as shown in Fig.

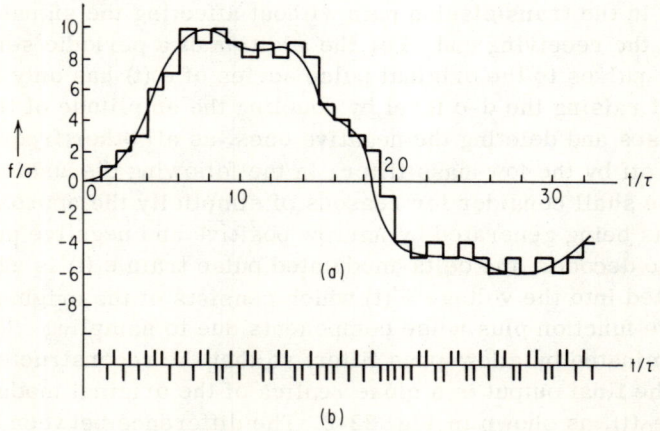

Fig. 22-3. Delta modulation waveforms using single integration and 2-unit code: (a) Approximated step curve; (b) coded approximated signal that is transmitted.

22-1. This network has a large time constant, and the response to
an impulse is practically a unit step function. The input to the integrator $e_2(t)$ is a delta-modulated bipolar pulse train. The output
signal $e_1(t)$ is then built up from positive and negative pulses in the
form of a step curve oscillating around the information signal $e_0(t)$,
as shown in Fig. 22-2. By applying the same bipolar pulse train to
another integrating network at the decoder, the same approximate
replica of the message signal is obtained at the receiving end,
which is then smoothed further by passing it through a low-pass
filter. As mentioned above, this method of quantization is coarse,
resulting in large quantizing noise which is audible as a type of
noise approximately correlated to the audio signal. At a pulse frequency of 40 kc/sec, the intelligibility of the speech is good, but the
quantizing noise has an effect on the speech which is called by
deJager [2] "sandiness." This effect is diminished by the better
approximation of the signal by the step curve when higher pulse
frequencies are used.

In the absence of an input signal (zero level), the approximation
signal $e_1(t)$ consists of an alternating pulse series, as shown in
Fig. 22-4. Since only high frequencies are present in the signal, the
output signal of the low-pass filter at the receiving end is zero.
However, since small signal levels of the information signal $e_0(t)$
cannot affect this sequence of pulses, there will be, in this case, a
noticeable threshold effect.

The problem of overloading will be considered next. In a DM
system, the information contained in the transmitted pulses is
mainly correlated to changes of input signal and not to its amplitude.
However, since the synthesized wave can change only one level per
clock pulse, DM has no fixed maximum amplitude but overloads
when the slope of the signal is too large. The largest slope the
system can reproduce is one changing by one level or step every
pulse interval, so that the maximum signal power depends on the
type of signal. If the magnitude of one quantum step is σ volts, and
the time between sampling instants is $T_S = 1/f_S$ sec, then the maximum rate of change of amplitude which it can register is σ/T_S or
σf_S volts/sec. For a sine wave of angular frequency $\omega = 2\pi f$, the
maximum slope is $A\omega$, where A is the peak amplitude. It follows

$e_1(t)$

Fig. 22-4. The approximation signal $e_1(t)$ in the absence of an input
signal.

therefore that for a sine wave of frequency f applied to the integrator, the maximum amplitude which can be transmitted is

$$A_{max} = \frac{f_s \sigma}{2 \pi f} \qquad (22\text{-}1)$$

Thus both the maximum amplitude and the number of distinguishable levels decrease with increase of frequency to be transmitted. This limitation is minimized in the case of speech where the higher frequencies contain less energy than the lower ones, so that in this respect an integrating network is well adapted to the transmission of speech.

It has been observed experimentally that the DM system can transmit a speech signal without overloading if the amplitude of the signal does not exceed the maximum sine-wave amplitude that can be transmitted at a frequency of 800 cps. We shall now prove that this reference frequency $\overline{f} = \overline{\omega}/2\pi$ is related to the speech spectrum as given by Eq. (22-2),

$$\overline{\omega}^2 = \frac{\int_0^{\omega_m} A^2(\omega) \omega^2 \, d\omega}{\int_0^{\omega_m} A^2(\omega) d\omega} \qquad (22\text{-}2)$$

where $A(\omega)$ represents the amplitude–frequency response of the speech signal, and ω_m corresponds to the highest frequency to be transmitted. To prove this functional relationship, we consider a transmission system where the amplitude $A(\omega)$ of the input signal is limited to V_1 as shown in Fig. 22-5, and another transmission system where the slope of the input signal is limited to S_1 as shown in Fig. 22-6a. For the behavior of the second system, we consider the system shown in Fig. 22-6b, where the input signal is first applied to a differentiating network $\psi(\omega)$, and then the output is limited in amplitude to a voltage V_2. In both systems, the phase of the different frequencies of the input signal may be considered as being random. Let E_1 and E_2 denote the rms voltage at the input of the systems, as

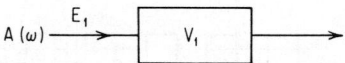

Fig. 22-5. A transmission system in which the amplitude is limited to the voltage V_1.

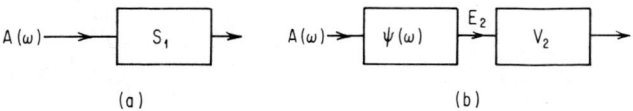

(a)   (b)

Fig. 22-6. (a) A transmission system in which the slope of the input signal is limited to S_1. (b) Equivalent circuit.

shown in Figs. 22-5 and 22-6, respectively; then both systems will give rise to distortion during an equal number of time intervals if

$$\frac{E_1}{V_1} = \frac{E_2}{V_2} \tag{22-3}$$

and $\quad E_1^2 = \int_{-\omega_m}^{\omega_m} A^2(\omega)\,d\omega = 2\int_0^{\omega_m} A^2(\omega)\,d\omega \tag{22-4}$

If $\psi(\omega)$ is the frequency characteristic of the differentiating network, then the spectrum of the signal at the input of V_2 is $A(\omega)\psi(\omega)$, so that

$$E_2^2 = 2\int_0^{\omega_m} |A(\omega)\psi(\omega)|^2\,d\omega \tag{22-5}$$

Now consider the case of applying a sine wave of frequency $\overline{\omega}$ and of voltage $\overline{V}$ such that limiting just occurs in both systems of Figs. 22-5 and 22-6. Let $\overline{V_1}$ and $\overline{V_2}$ denote the rms voltages at the input of the limiters V_1 and V_2, so that $\overline{V_1} = \overline{V}$ and $\overline{V_2} = \overline{V}\,\psi(\omega)$. Limiting of signals in both systems occurs at the same time when

$$\frac{\overline{V_1}}{V_1} = \frac{\overline{V_2}}{V_2} \tag{22-6}$$

Let the output of the differentiating network be given by $\psi(\omega) = kj\omega$; then by the use of the previous relations, the reference frequency $\overline{\omega}$ is readily shown to be given by

$$\overline{\omega}^2 = \frac{\int_0^{\omega_m} A^2(\omega)\omega^2\,d\omega}{\int_0^{\omega_m} A^2(\omega)\,d\omega}$$

Thus if both systems can transmit a sine wave of frequency $\overline{\omega}$ with the same amplitude, they can also transmit a signal of the defined frequency spectrum with the same amplitude, and if overloading

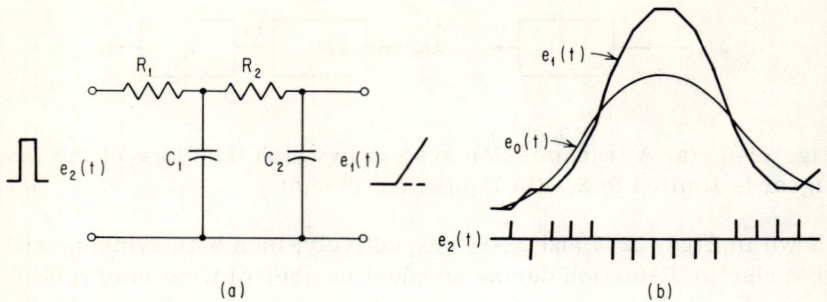

Fig. 22-7. Delta modulation with double integration.

occurs, it will take place for an equal number of time intervals in both systems.

2. *Double Integration*. We have just noted that single integration in the feedback loop and at the receiving end results in a coarse approximation of the original signal $e_O(t)$, where the quantizing noise is very large and the threshold effect is high.

A better approximation of the information signal $e_O(t)$ is obtained by the use of a double-integrator circuit, as shown in Fig. 22-7a. The time constants R_1C_1 and R_2C_2 are large, so that the response due to a pulse is a unit step at C_1 and a voltage of constant slope at C_2. The effect of this circuit is to change the slope of e_1 with every pulse input to e_2. The resulting waveform, shown in Fig. 22-7b, is seen to be a much smoother curve than for the case of single integration where every received pulse had the effect of increasing or decreasing the level of the received signal by a unit step. When the signal is smoothed by a low-pass filter, it will approximate the original message function much closer than in single integration.

A disadvantage of a circuit using double integration is that it may not recognize changes of slope in the message signal $e_O(t)$ soon enough, since the comparator only recognizes differences in amplitude. To correct for this defect, a modified double integrator can be used in the encoder, which makes use of some information about the future, by using the principle of prediction. As shown in Fig. 22-8a, the modification consists of a prediction time constant τ made arbitrarily equal to the pulse interval. With this prediction, the response of the circuit is a step followed by a voltage of constant slope. By means of this step, the output $e_1(t)$ knows in advance what the voltage at capacitor C_2 will be rising to. This is equivalent to extrapolating. The advantage of extrapolation can be seen by comparing in Fig. 22-8b the two approximating curves $e_1(t)$ for the same signal $e_O(t)$.

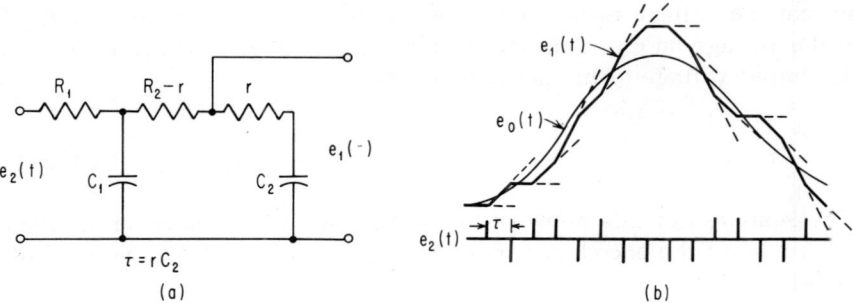

Fig. 22-8. Double integrator with prediction.

At the decoder, a conventional double-integrator circuit is used with time constants R_1C_1 and R_2C_2.

The mathematical analysis of the principle of prediction will now be given. Essentially, the principle of prediction amounts to extrapolation of the approximating curve to the dotted points in Fig. 22-8b, and comparing these values with the values of the original function. Let the output signal at the second capacitor of the double-integrating network at the time $(t + \tau)$ be denoted by e_2'; the problem is to make this signal equal to the input signal e_0 at time t. This is accomplished by the network of Fig. 22-8a which generates the extrapolated value of e_2' so that

$$e_1(t) = e_2'(t + \tau) \tag{22-7}$$

Placing this network in the feedback loop of the encoder, a decision is made by the comparator whether the slope of the approximation signal should be increased or decreased by one unit, i.e., whether the following pulse should be positive or negative. The extrapolated voltage is given by

$$e_2'(t + \tau) = e_2'(t) + \tau \frac{de_2'(t)}{dt} \tag{22-8}$$

where $e_2'(t)$ denotes the output signal on the second capacitor of the double integrator. If I denotes the current through the resistor R_2, then

$$\frac{de_2'(t)}{dt} = \frac{I}{C_2} \tag{22-9}$$

and $$e_1(t) = e_2'(t) + \frac{\tau I}{C_2} \tag{22-10}$$

so that the voltage $e_1(t)$ is found by adding a correction term $\tau I/C_2$ to the voltage on C_2. This means that $e_1(t)$ in Fig. 22-8a is equal to the output voltage of the network where

$$\tau = rC_2 \tag{22-11}$$

The response of this network to an impulse, when placed in the feedback loop of the encoder, consists of a step followed by a voltage of constant slope.

22.2 Signal-to-Noise Ratio in DM [2,3]

It has been mentioned previously that the quantization noise can be decreased by increasing the pulse frequency. Generally speaking, the quality of transmission is improved in two ways by this increase: first, the number of steps can be increased so that signals of greater amplitude can be transmitted, and, second, a corresponding change in the frequency spectrum of the noise will cause a further reduction of the quantizing noise in the low-frequency band. As these two effects are different for the two systems of single and double integration, we shall therefore discuss them separately.

1. *The Quantizing Noise in the System of Single Integration.* In the single-integration system, the noise is produced by the difference signal $\epsilon(t) = e_1(t) - e_0(t)$, which at the receiving end is smoothed by means of a low-pass filter. The difference signal, which is shown in Fig. 22-9, is usually nonperiodic and has therefore a continuous-frequency spectrum which can be shown to be uniform for frequencies which are small compared to the sampling frequency f_s. The power of the quantizing noise received after the low-pass filter with cutoff frequency $f_m = \omega_m/2\pi$ will thus be proportional to f_m. Similarly, it can be reasoned that the unfiltered noise power is inversely proportional to the sampling frequency. For if the pulse frequency is doubled, using the same value σ for the height of one step, the new difference signal ϵ will have the same total power in a frequency band twice as large as before; consequently, the noise power in the band $(0 - f_m)$ will be reduced by a

Fig. 22-9. The characteristic form of the difference signal $\epsilon(t)$ in the system of single integration.

factor of 2. In fact, it can be shown[3] that the quantized noise power N_O is approximately given by

$$N_O = \frac{2}{3}\left(\frac{f_m}{f_s}\right)\sigma^2 \tag{22-12}$$

The signal power in the calculation of signal-to-noise ratio is taken as the power of the sinusoidal tone, which is just below the overload point. In signals which have approximately uniform energy distribution, the single-tone frequency should be the top frequency, since in this case the point just below overload is determined by the highest frequency. In signals with a nonuniform energy distribution, some other than the top frequency may be used; as mentioned earlier, the reference frequency of a speech signal is 800 cps. Assuming a sinusoidal signal $A \sin \omega t$, the maximum amplitude that can be transmitted with single integration without overloading is, according to Eq. (22-1),

$$A = \frac{f_s\sigma}{2\pi f}$$

and, consequently, the average signal power is

$$S_O = \frac{A^2}{2} = \frac{f_s^2\sigma^2}{4\pi^2 f^2} \tag{22-13}$$

so that the signal-to-noise ratio for single integration is

$$\left(\frac{S}{N}\right)_O = \frac{3}{2}r^3\left(\frac{f_m}{\pi f}\right)^2 = C_1\frac{f_s^3}{f_m f^2} \tag{22-14}$$

where $r = f_s/2f_m$ = bandwidth expansion factor. This number r represents the relationship between the sampling frequency f_s which determines the transmission bandwidth and the bandwidth of the message signal.

2. *The Quantizing Noise in the System of Double Integration.* The signal-to-noise ratio may be correspondingly obtained for the double integration[3]

$$\left(\frac{S}{N}\right)_O = \frac{3}{2}r^5\left(\frac{f_m}{\pi f}\right)^4 = C_2\frac{f_s^5}{f_m^3 f^4} \tag{22-15}$$

The improvement in signal-to-noise ratio varies with f_s^3 for the system with single integration, whereas it varies with f_s^5 for double integration.

3. *Experimental Results.* Experimental verification of the formulas for $(S/N)_O$ were obtained by de Jager[2] using a reference signal of 800 cps and a frequency band from 200 to 3,800 cps. His published results indicate a fair agreement with the theoretical values, especially for the higher pulse frequencies. The quantizing noise in the system of double integration is not only smaller in amount but also differs in character from that of single integration, especially in the reproduction of low-level signals. The noise produced with single integration has more or less the character of a line spectrum with many sharp interference tones, whereas that in the double system is more evenly smoothed in a continuous spectrum. According to de Jager, a very good reproduction of speech is possible with a pulse frequency of 100 kc/sec. In this case, a bandwidth of 50 kc/(sec)(channel) is necessary in the transmission path, which is of the same order as usually applied in PCM. In order to compare DM with PCM, we make use of Eq. (20-32) for the signal-to-noise ratio in decibels, using a code of n digits, or $M = 2^n$ levels, namely,

$$\left(\frac{S}{N}\right)_O = 20 \log_{10} M \qquad db \qquad (22\text{-}16)$$

which gives n = 8 for $(S/N)_O$ = 50 db. In this case, the bandwidth needed for transmission with DM is about 50 per cent greater than by using PCM. For taking a sampling frequency of 8 kc/sec, as is usual for the transmission of speech, the 8-unit code would require a pulse frequency of 64 kc/sec.

Several experimental delta-modulation systems have recently been described[6-8] with very interesting experimental results. One is illustrated in Figs. 22-10 and 22-11 for voice application. We note from Fig. 22-10 that at a sampling rate of 40,000 pps, delta modulation is equal in performance to a 5-bit PCM system.

22.3 Delta-Sigma Modulation (D-ΣM) [9-11]

We have noted in the introduction that delta modulation is characterized by simpler circuitry as compared with PCM. However, in spite of the simplified circuitry, the use of DM has been limited to the transmission of such signals as speech, which does not contain a d-c component and has less energy in higher frequencies. Delta

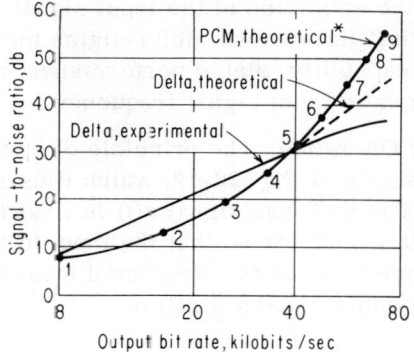

Fig. 22-10. Signal-to-noise ratio for delta modulation and PCM.
*Numbers indicate PCM coding digits. (From Lender and Kozuch,[7]
Globecom. Conv. Record.)

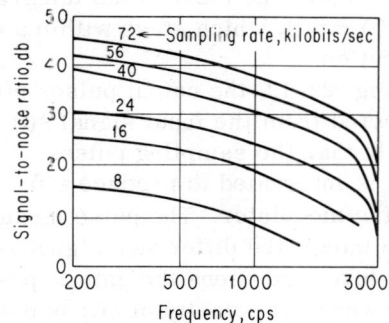

Fig. 22-11. Signal-to-noise ratio vs. frequency. (From Lender and
Kozuch,[7] Globecom. Conv. Record.)

modulation is incapable of transmitting d-c signals, its dynamic
range and signal-to-noise ratio are inversely proportional to the
signal frequency, and the integration at the receiving end causes an
accumulative error in the demodulated signal when the system is
subject to transmission disturbances such as noise.

The modified DM system called delta-sigma modulation, which
will presently be described, has been designed to meet the requirements for digital transmission of video signals and the like, which
are characterized by more uniform spectra with d-c components,
through adverse transmitting conditions. To compensate for the
inevitable differentiation of the input signal, the modified DM system has a signal integration process added at the input of the original delta modulator, so that the output pulses carry information

corresponding to the amplitude of the input signal. As we shall show in the following discussions, delta-sigma modulation offers d-c transmission capability, stable performance, and independence of signal-to-noise ratio from signal frequency.

1. *Principle of Operation.* The principle of operation of the D-ΣM system is shown in Fig. 22-12, which illustrates a modified delta modulator. The message signal s(t) is first integrated and then applied to the delta modulator so that the input to the pulse modulator ϵ(t) is the difference between the integrated message signal $\int$ s(t) dt and the integrated output pulses $\int$ p(t) dt

$$\epsilon(t) = \int s(t)\ dt - \int p(t)\ dt \qquad (22\text{--}17)$$

Since the configuration of Fig. 22-12 requires an integrator with a very large dynamic range, the two integrators can be combined as shown in Fig. 22-13, where the input to the integrator is the difference signal Δ(t) = s(t) $-$ p(t), which stays within a certain limit for proper system operation.

In the D-ΣM of Fig. 22-13, the output pulses p(t) are fed back to the input and subtracted from the input signal s(t), which varies sufficiently more slowly than the sampling pulses. The difference signal Δ(t) = s(t) $-$ p(t) is integrated to produce ϵ(t) = $\int$ Δ(t) dt, which is applied to the pulse modulator. The pulse modulator compares the amplitude of the integrated difference signal ϵ(t) with a predetermined reference level and opens the gate to pass a pulse from the pulse generator when the polarity of ϵ(t) is positive, i.e., when

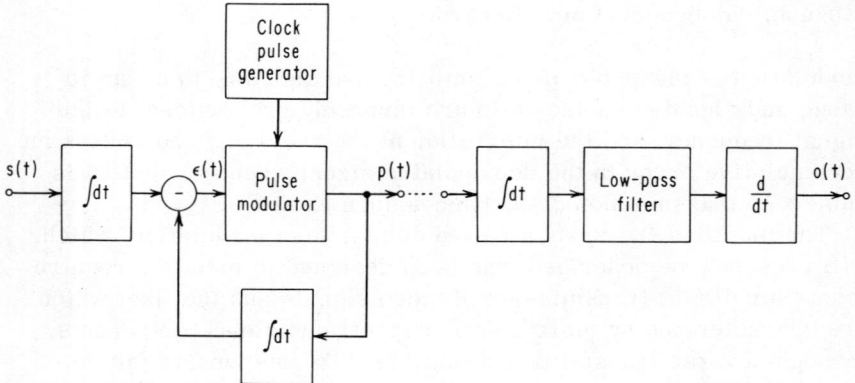

Fig. 22-12. Block diagram of a modified delta modulator. (From Inosi and Yasuda,[11] Proc. IEEE.)

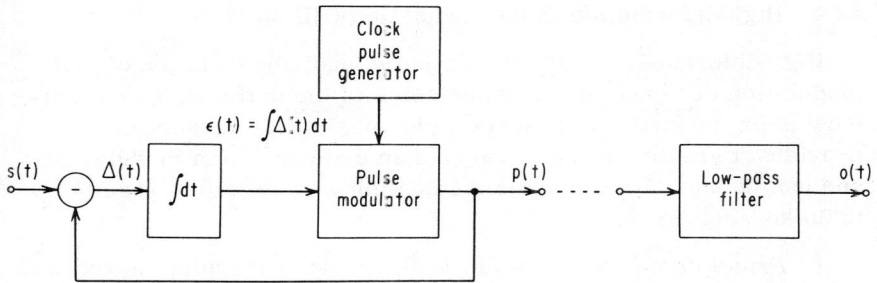

Fig. 22-13. Block diagram of the delta-sigma system. (From Inosi and Yasuda,[11] Proc. IEEE.)

$\epsilon(t)$ is larger than the reference level, and closes the gate to inhibit the pulses when $\epsilon(t)$ is negative. Thus through this negative-feedback procedure, the integrated difference signal is always kept in the vicinity of the reference level of the pulse modulator, provided that the input signal is not too large. It follows therefore that the output pulses appear more frequently as the amplitude of the input signal becomes large. In other words, the output pulses carry the information corresponding to the input-signal amplitude.

At the receiving end, no integration of the pulses is required, and demodulation is carried out by passing the pulses through a low-pass filter. Since no integration process is involved in the demodulation, no accumulative error due to transmission disturbances results in the demodulated signal.

2. *Signal-to-Noise Characteristics.* Theoretical calculations given by Inosi and others [10,11] show that the signal-to-quantizing noise ratio is given by the equation

$$\left(\frac{S}{N}\right)_o = \rho^2 \frac{9}{16\pi^2} \left(\frac{f_s}{f_m}\right)^3 \qquad (22\text{-}18)$$

where ρ is the ratio of the signal amplitude to the maximum amplitude that does not overload the modulator. Comparing this result with Eq. (22-14) for delta modulation, we note that in both systems the $(S/N)_o$ is proportional to the pulse frequency cubed, while, in contrast to the delta-modulation system, the signal-to-noise output in a Δ-ΣM system is independent of the signal frequency.

22.4 High-information Delta Modulation (HIDM) [12]

High-information delta modulation is another variation of delta modulation designed to overcome some of the deficiencies of ordinary delta modulation while retaining most of its advantages. It provides a greater dynamic range than ordinary delta modulation and can be used for voice communication with excellent results up to 20 kc/sec.

1. *Principle of Operation.* In ordinary delta modulation, there is a unit amplitude-level change per pulse. This suggests that with a sequence of n pulses, all of the same polarity, the level change could vary with proper instrumentation as 2^n, or exponentially. This is accomplished in HIDM where the changes in level vary exponentially with time.

The essential difference between HIDM and ordinary delta modulation is in the manner of counting amplitude levels. In HIDM, counting is performed in binary steps, resulting in an exponential variation of 2^n of the amplitude levels due to a sequence of pulses of one polarity such as 1, 2, 4, 8, 16, 64, etc.

Should the increment be too large, the pulse direction reverses, reducing in magnitude by a factor of 2. This is illustrated in Fig. 22-14 where a step function starting at $\frac{1}{2}$ unit and rising to $39\frac{1}{2}$ units might be approximated as shown in the time plot as follows: 1, 2, 4, 8, 16, 32, 64, 48, 32, 40, 36, 38, 40, 39, 40, etc. Thus in less than 7 pulse periods, the magnitude of $39\frac{1}{2}$ levels has been exceeded. As shown in Fig. 22-14, ordinary DM would require 40 pulse periods to cover this magnitude.

The method of instrumenting HIDM is similar to that for DM, as shown in Fig. 22-15, the only difference being in the demodulator.

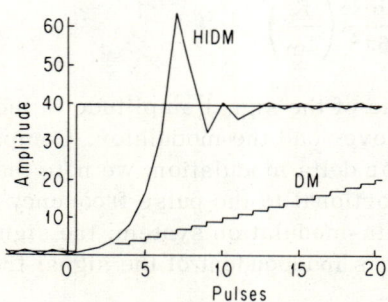

Fig. 22-14. HIDM response to step function. (From Winkler,[12] IRE Conv. Record.)

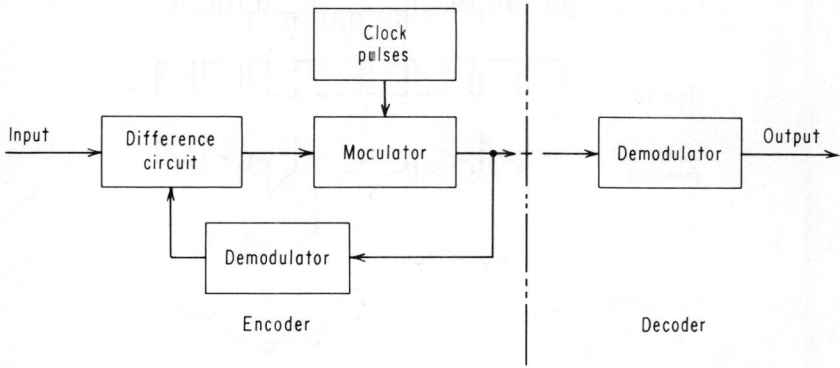

Fig. 22-15. Block diagram of HIDM system. (From Winkler,[12] IRE Conv. Record.)

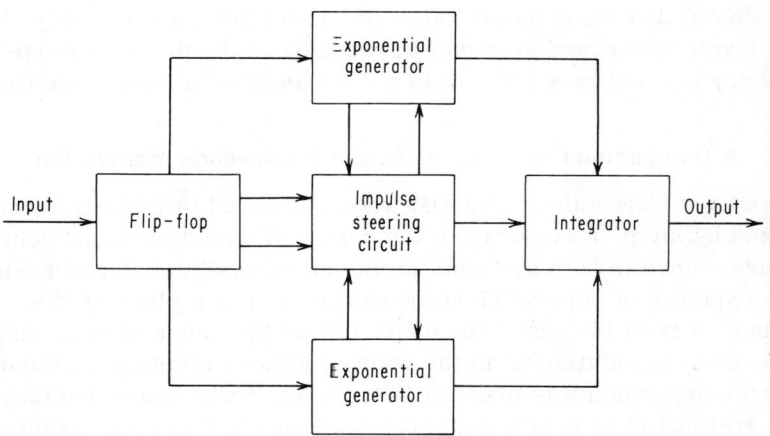

Fig. 22-16. Block diagram of HIDM demodulator. (From Winkler,[12] IRE Conv. Record.)

A block diagram of the demodulator is shown in Fig. 22-16. The flip-flop takes a state depending on the polarity of the pulses. The exponential generators are used alternately, one generating positive waveforms, the other negative waveforms. The impulse device delivers the proper initial impulse to the exponential generators. The output of the exponential generators consists of a series of exponential ramp signals, as shown in Fig. 22-17, which are summed in an integrator to provide the desired demodulator output.

It has been demonstrated experimentally that HIDM has many characteristics which make it very suitable for voice transmission.

Pulse train

Flip-flop

Exponential
generator

Demodulator
output

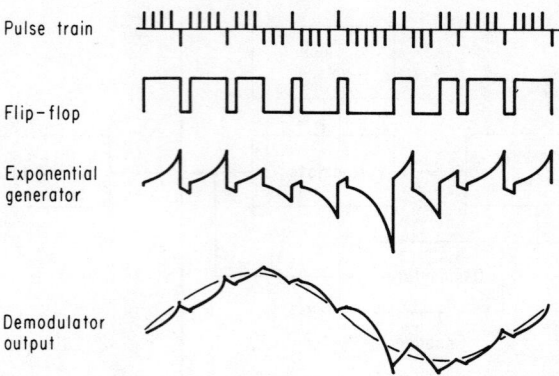

Fig. 22-17. Waveforms in HIDM demodulator. (From Winkler,[12] IRE Conv. Record.)

It is shown that the dynamic range of HIDM with slope limiting of 11.1 levels/pulse period is much greater than the dynamic range of ordinary DM and is equivalent to a 7-bit PCM with linear quantizing.

22.5 A Comparison between Delta and Pulse-code Modulation [13]

We have shown above that DM when compared to PCM is characterized by simpler circuitry in the design of modulators and demodulators. Since in both systems the information signal is represented by a sequence of pulses before transmission takes place on the channel, it is of interest to compare the performance of delta and pulse-code modulation from the point of view of information theory. Such a comparison was made by Zetterberg,[13] who concluded that both systems have nearly the same capacity for transmitting information for an equal number of amplitude levels, provided the number is larger than 10 levels. Zetterberg has also shown that for moderately high equal-output signal-to-noise ratios, DM needs a much larger bandwidth than PCM.

1. *Method of Analysis.* Following Shannon, the information source is considered by Zetterberg to consist of M discrete quantized signals which are delivered to the encoder. The time interval between the amplitude values for DM is equal to τ sec, and for PCM it is equal to $n\tau$, where n denotes the number of pulses in the PCM code group, and the number of quantizing levels $M = 2^n$. Thus at the same pulse frequency $f_s = 1/\tau$, the quantized signal samples are taken in a DM system n times as often as for PCM. For DM, the value of the function changes one positive or negative step of the height σ at

every pulse, while for PCM, any sample can follow a given sample among the $M = 2^n$ possible samples. In PCM, it is assumed that in the sampling process the probability of occurrence $P(x_i)$ of the discrete signal amplitude x_i corresponding to the state i (i = 1, 2 ..., M) is independent of the previous state. However, because of the greater sampling frequency in DM, the assumption of independence in this case is no more valid and consequently we must introduce the concept of transition probabilities p_{ij} to reach from state i to state j in one trial. The transition probabilities fulfill the following conditions:

$$p_{ij} \qquad = 0; \qquad j = i; \ |j - i| \geq 2$$
$$\qquad\qquad\qquad\qquad i = 2, \ldots, M - 1$$

$$p_{M, M-1} \quad = 1$$
$$p_{1,2} \qquad = 1$$
$$p_{i,i-1} + p_{i,i+1} = 1; \qquad i = 2, \ldots, M - 1 \qquad\qquad\qquad (22\text{-}19)$$

The source of information was considered by Zetterberg to be a set of functions represented by finite Markov chains. A Markov chain is a stochastic process in which a future state is determined only by the previous state and the transition probabilities remain fixed. For speech signals, or any signal likely to be transmitted, there will be a much stronger interdependence between the various states. But according to Shannon, a stronger interdependence between the values of the sample functions gives a smaller entropy for the process, and consequently a lower rate of information. It follows therefore that the use of Markov chains will give only an upper bound on the channel capacity.

2. *Rate of Information and Channel Capacity.* Confining himself to discrete signals which are generated by a Markov process, Zetterberg assumed that the signals can take only M possible values, thus imposing a restriction on the possible signals for the DM system.

With PCM every signal value is considered statistically independent which gives the highest rate of information. The channel capacity is shown to be given by

$$C = \frac{1}{n\tau} \log M = \frac{1}{\tau} \log_2 2 = f_s \qquad \text{bits/sec} \qquad\qquad (22\text{-}20)$$

For DM, the channel capacity is shown to be given by

$$C = \frac{1}{\tau} \log_2 \left(2 \cos \frac{\pi}{M + 1} \right) \qquad \text{bits/sec} \qquad\qquad (22\text{-}21)$$

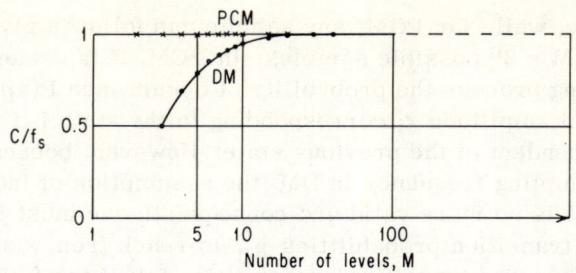

Fig. 22-18. Channel capacity for PCM and DM at same pulse frequency. (From Zetterberg,[13] Ericsson Technics.)

A comparison of the channel capacity for PCM and DM at the
same pulse frequency is plotted in Fig. 22-18. It can be seen that as
the number of levels M increases to 10 or greater, the channel capacity of a DM system approaches very closely that of a PCM system.
Since in a practical application, the number of levels is likely to be
greater than 10, DM can be seen to compare favorably on the basis
of channel capacity.

3. *Amplitude Distribution in Signals and Signal Power.* Calculation of the channel capacity provides us with the knowledge of the
statistical structure of the signals which best suit the systems.
With PCM, all levels are equally probable, but when using DM, the
levels near the mean are most probable. This means that less signal power is required with delta modulation. For PCM, the modulation signal power according to Eq. (20-30) is given by

$$S_P = \frac{M^2 - 1}{12} (\sigma)^2 \tag{22-22}$$

The corresponding expression for DM is given by [13]

$$S_D = \left[\frac{(M + 3)(M - 1)}{12} - \frac{1}{2} \cot^2 \frac{\pi}{M + 1} \right] (\sigma)^2 \tag{22-23}$$

which for large values of M reduces to

$$S_D \doteq \left[(M^2 + 2M) \left(1 - \frac{6}{\pi^2} \right) - 1 \right] \frac{(\sigma)^2}{12} \tag{22-24}$$

For large values of M, the ratio between the powers S_D/S_P approaches the constant $(1 - 6/\pi^2)$ which corresponds to about 60 per
cent less signal power for delta modulation than for PCM. This
ratio is shown plotted in Fig. 22-19.

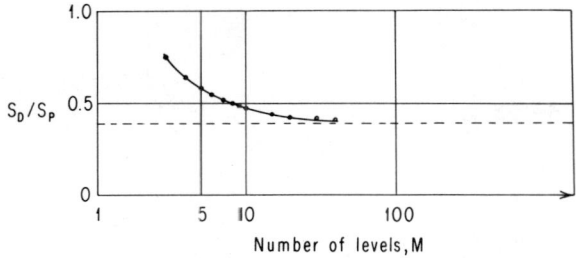

Fig. 22-19. Ratio of signal power for DM and PCM. (From Zetter-berg,[13] Ericsson Technics.)

References

1. Deloraine, E. M., S. Van Mierlo, and B. Derjavitch: French Patent 932,140, Aug. 10, 1946; U.S. Patent 2,629,857, Feb. 24, 1953.
2. de Jager, F.: Deltamodulation: a Method of PCM Transmission Using 1-unit Code, *Philips Res. Rept.* 7, pp. 442–466, 1952.
3. Lender, A.: Delta Modulation Study, *Internal IT&T Fed. Lab. Rept.* 1565.
4. Van de Weg, H.: Quantizing Noise of a Single Integration Delta Modulation System with an N-digit Code, *Philips Res. Rept.* 8, pp. 367–385, 1953.
5. Schouten, J. F., F. de Jager, and J. A. Greefkes: Delta Modulation: a New Modulation System for Telecommunication, *Philips Tech. Rev.*, vol. 13, pp. 237–245, March, 1952.
6. Lender, A., and M. Kozuch: Single Bit Delta Modulation Systems, *Electronics*, Nov. 17, 1961, pp. 125–129.
7. Lender, A., and M. Kozuch: An Experimental Delta Modulator, *Globecom Conv. Record*, Chicago, May, 1961.
8. Arter, F. W.: Simple Delta Modulation System, *Proc. IRE Australia*, September, 1962.
9. Nakamaru, Y., and H. Kalenko: Delta Modulation Encoder, NEC Research and Development, October, 1960.
10. Inosi, H., Y. Yasuda, and J. Murakami: A Telemetering System by Code Modulation: Δ-Σ Modulation," *IRE Trans. Space Electron. Telemetry*, pp. 204–209, September, 1962.
11. Inosi, H., and Y. Yasuda: A Unity Bit Coding Method by Negative Feedback, *Proc. IEEE*, November, 1963, pp. 1524–1535.
12. Winkler, M. R.: High Information Delta Modulation, *IRE Conv. Record*, 1963.
13. Zetterberg, L. H.: A Comparison between Delta and Pulse Code Modulation, *Ericsson Technics*, vol. 11, no. 1, pp. 95–154, 1955.

23

DIGITAL DATA MODULATION SYSTEMS

This chapter deals with the problem of RF carrier modulation where the information or modulation signal consists of discrete messages, that is, messages that can assume only a finite number of different specified forms. The set of discrete messages may, for example, consist of the 26 letters of the English alphabet, or it may be the "mark" and "space" in binary pulse-code modulation. In general, in a digital communication system the information source consists of a finite number of discrete messages which are coded into a sequence of waveforms or symbols; each waveform is selected from a finite alphabet of signal waveforms. Thus the problem of transmitting information is reduced to the problem of transmitting a sequence of waveforms, each one selected from a specified and finite set. This is in contrast to the problem of transmitting analogue information where the resulting set of waveforms is infinite.

As discussed in Chap. 19, a generalized communication channel can be represented diagrammatically as shown in Fig. 23-1. As with other forms of communication, the basic problem in digital communication is to transmit the information from the information source to the destination or information sink. The digital information to be transmitted is presented to the modulator, whose function is to trans-

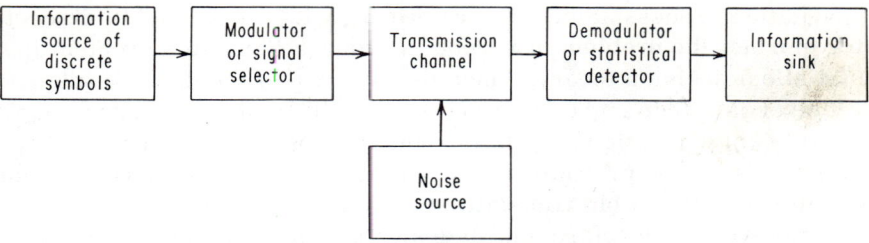

Fig. 23-1. Idealized model of data transmission system.

late the digital waveform into a signal wave which is suitable for
transmission over the transmission medium. As in analogue modulation, the information-carrying digital waveforms are used to modulate a sinusoidal carrier in order to place the relatively low-frequency energy of the video signals into a higher-frequency band.
The transmission channel or the medium coupling the transmitter
to the receiver is assumed to add stationary-white-zero mean
gaussian noise to the transmitted signal, but otherwise is assumed
to be distortion-free. It is further assumed that synchronous detection is used at the receiver; i.e., the receiver is time-synchronized
with the transmitter. In our analysis we shall distinguish between
coherent detection where the receiver is phase-locked to the transmitter and noncoherent detection where the receiver is not phase-
synchronized with the transmitter. However, time synchronization
will be assumed in all the digital modulation systems under discussion. For efficient transmission of the information from the information source to the information sink the modulation system must
meet the following requirements:[1]

1. The most fundamental requirement placed on the modulation
system is to represent sequences of digits as waveforms acceptable
to the channel. For example, in order to transmit a sequence of
binary symbols, a sequence of two-state signals must be produced
which will pass recognizably through the channel. This means that
the distortion introduced by the channel due to amplitude and phase
nonlinearities should not be excessive to the degree of making the
recognition of the two-state signal impossible.

2. The waveforms generated by the modulation process should
not be greatly affected by the presence of noise signals in the channel. The most common way of minimizing the effect of noise on the
transmitted signals is to generate the different states of the signal
by means of phase or frequency modulation of the carrier. The
noise impairment of the signal can be further minimized by making
use of the different statistics of the signal and noise during the de-

modulation process. Since we are dealing with a finite set of discrete waveforms, the problem of reception reduces itself to a problem in statistical decision theory. Since the decision of the discrete-signal receiver is either right or wrong, the criterion of performance of a digital communication system is ordinarily based on the probability of error, i.e., the probability of choosing an incorrect message from a finite set of possible transmitted messages. The basic problem at the receiver is therefore how to design an optimized receiver which will make as few errors as possible.

3. The rate of sending information should be as high as possible. This can be achieved by having the discrete modulated signals occupy a short average time consistent with the available bandwidth and maximum allowable error rate.

As stated above, for optimum performance it is necessary to minimize intersymbol interference due to channel nonlinearities or bandwidth restriction. In the following presentation we shall limit our analysis to the case of distortionless medium and consider only the effect of white gaussian noise at the receiver input. The error probability in digital transmission due to white gaussian noise will be shown to depend on the modulation and detection method. However, before entering into the calculation of probability of error for the various digital transmission systems, we shall briefly review the most common methods of digital modulation.

23.1 Review of Digital Modulation Methods [1-6]

In this section we shall briefly review the most common modulation methods of producing discrete carrier-modulated signals. For the sake of simplicity and without loss of generality, we shall primarily be concerned with binary or two-state modulation of the carrier. However, we shall have occasion to generalize the results to multistate modulation systems.

In binary modulation systems the information to be transmitted is assumed to be coded into binary form using two signals. The two elementary signals, called "mark" and "space," "one" or "zero," are of equal and finite duration and occur with equal probability. The encoding of these signals may represent many kinds of messages. The two elementary signals of the binary coded message are generated by modulating a sinusoidal carrier in amplitude, frequency, or phase in a time sequence of two mutually exclusive states. Reception of the message depends upon the successful determination of the state of the modulated carrier during each interval of the sequence. Transmission of digital information by binary rather than by multi-

level pulses offers significant advantages in system design. It simplifies the implementation of regenerative repeaters and the terminal
equipment, and as we shall see later, it is more efficient from the
point of view of system performance as a function of receiver
carrier-to-noise ratio.

In the following we shall analyze and compare the performance of
three basic data-transmission systems, namely: amplitude-shift
keying (ASK), frequency-shift keying (FSK), and phase-shift keying
(PSK).

1. *Amplitude-shift Keying (ASK)*. The simplest two-state carrier
modulation uses double-sideband amplitude modulation in which a
sinusoidal carrier is pulsed, so that one of the binary states is
represented by the presence of the carrier while the other state is
represented by its absence. ON-OFF modulation schemes were his-
torically the first used for transmission of binary information. Such
systems behave in the presence of noise in a manner similar to a
simple d-c modulation system but require twice as much bandwidth.
It is relatively simple to instrument by using envelope detection, and
therefore it was extensively used in early carrier telegraph systems.

2. *Frequency-shift Keying (FSK)*. In this system the two binary
states are represented by two different frequencies. Frequency-
shift keying is basically equivalent to amplitude-shift keying using
two different carriers, the one being keyed ON while the other is
OFF. Demodulation is usually accomplished by using two frequency-
tuned sections, one tuned to each of the two-bit frequencies. Digital
transmission using FSK is less sensitive to noise than ASK but requires slightly more bandwidth.

3. *Phase-shift Keying (PSK)*. The most promising modulation
system, which is gaining in popularity, is phase-shift keying or
double-sideband suppressed-carrier modulation. In such systems
great immunity from noise can be achieved, together with insensi-
tivity to level variations. In a two-phase system, one phase of the
carrier frequency is used to represent one binary state, and another
phase which is usually 180° apart is used for the second state. The
two phases are generally detected by multiplying the information signal with a reference signal at the receiver, which is of the same frequency as the incoming carrier and known phase with respect to it
(coherent detection). The reference signal is usually arranged to be
exactly in phase with one of the binary signals and 180° out of phase
with the other. It will be shown later that in this demodulation
scheme, the largest output is obtained when the signal is of the same

frequency as the reference and of the same phase, or 180° out of phase with that reference, which results in an increased signal-to-noise improvement.

An improved phase-modulation system uses the delayed preceding bit as a phase reference and is usually referred to as a relative phase system. In this system a binary 1 is represented by sending a signal of the same phase as that of the previous signal sent, while a binary 0 is represented by a signal of a phase opposite to that of the previous signal transmitted. At the receiver, each received signal is first stored for one bit period and then compared in phase with the next signal element. In this stored reference system there is a loss of about 1 db in signal-to-noise improvement, due to effect of the noise on the reference signal. Both phase-modulation systems perform very well in the presence of noise. With typical instrumentation, the phase systems will operate, with the same error rate, with a carrier-to-noise ratio of 1 to 3 db less than an FM system, and 5 to 7 less than an AM system.

4. *Compound Modulation.* In addition to these modulation schemes, a number of more complex modulation systems are in use. A typical compound modulation scheme is exemplified by the Kineplex system which uses two double-sideband channels with quadrature carriers (or four-phase modulation). Recently, several investigations have been carried out analyzing the performance of combined amplitude- and phase-modulated systems.

In the following we shall briefly analyze the performance of the various digital data-transmission systems under no-fading conditions in the presence of white additive gaussian noise, from the aspect of probability of error or error rate as a function of input carrier-to-noise ratio, system bandwidth, and threshold. In the discussion of error rates, we shall not be concerned with error-correction codes which make it possible to detect or even correct errors which occur as a result of the additive noise in the transmission channel. Our discussion will be limited only to the analysis of the probability of errors due to the channel noise, without the benefit of introducing redundancy into the coding scheme.

The average error or probability of error is defined as the fraction of the total number of intervals for which an incorrect decision is made. The problem will be to calculate the probability of error of amplitude-, frequency-, and phase-shift keying in terms of receiver-input carrier-to-noise ratio. For each type of modulation it will be assumed that the ideal detecting device uses a reference threshold upon which to base its decision concerning the state of the modulated

carrier. As an example, for amplitude modulation the threshold is
set at a fraction of the amplitude of the carrier. The analysis will
be carried out for coherent and noncoherent detection of binary data
systems. The performance of multilevel digital modulation systems
will also be discussed.

23.2   Amplitude-shift Keying (ASK) [2,3,7,8]

In a binary amplitude-shift-keying system the two states of
"mark" and "space" of the modulated carrier are represented by
ON-OFF carrier pulses, respectively. At the receiver the pulse-
modulated carrier representing a sequence of "marks" and
"spaces" or "ones" and "zeros" is disturbed by noise, resulting
in errors made by the receiver in determining the actual symbols
transmitted. The problem is to calculate the probability of error as
a function of receiver-input carrier-to-noise ratio.

In ASK, two methods of detection are used: synchronous detection
and envelope detection. A synchronous detector is a device that
measures amplitudes only during small time intervals which are
synchronized with the peaks of the carrier wave. This method of
detection requires the availability of a local oscillator in phase
coherence with the incoming modulated carrier. The use of envelope detection in a binary ASK transmission system obviates the
requirement of carrier phase coherence and results in a system
which is easier to implement.

1. *Coherent Detection of ASK.* In this modulation system the information carried by the transmitted waveform is contained in the
amplitude of the carrier. The set of message waveforms of a m-
level ASK system can be described by

$$s_i(t) = \sqrt{\frac{2E_i}{T_0}} \cos \omega_0 t, \qquad 0 \le t \le T_0$$

$$0, \qquad\qquad \text{elsewhere; } i = 1, 2, \ldots, m \qquad (23\text{-}1)$$

where E_i is the energy content of $s_i(t)$, T_0 is the duration of the
waveform $s_i(t)$, and $\omega_0 = 2\pi n_0/T_0$ for some fixed integer n_0. In a
binary ASK system which will be analyzed presently, we have only
two levels or states of the modulated carrier, which are represented
by ON-OFF carrier pulses, respectively.

At the receiver, the modulated carrier and noise are assumed to
be band-limited. The disturbing noise can therefore be represented
in terms of the inphase and quadrature components with reference to

the phase of the carrier input. Using Eq. (4-231), the gaussian band-limited noise is given by

$$n(t) = x_c(t) \cos \omega_0 t - x_s(t) \sin \omega_0 t = V(t) \cos [\omega_0 t + \phi(t)]$$

$$(23-2)$$

where $\quad \overline{n^2(t)} = \dfrac{\overline{x_c^2(t)}}{2} + \dfrac{\overline{x_s^2(t)}}{2} = \dfrac{\overline{V^2(t)}}{2} = \sigma^2 = N \qquad (23-3)$

The two components, $x_c(t)$ and $x_s(t)$, are uncorrelated, and each is a gaussian-distributed random variable so that

$$\overline{n^2(t)} = \overline{x_c^2(t)} = \overline{x_s^2(t)} = N \qquad (23-4)$$

where N represents the average noise power of the band-limited noise process, and the probability density function of $x_c(t)$ and $x_s(t)$ is given by

$$p(x_c) = \frac{1}{\sqrt{2\pi}\ \sigma} e^{-x_c^2/2\sigma^2} \qquad (23-5)$$

and $\quad p(x_s) = \dfrac{1}{\sqrt{2\pi}\ \sigma} e^{-x_s^2/2\sigma^2}$

The noise envelope V(t) has been shown to have a Rayleigh distribution, namely,

$$p(V) = \frac{V}{\sigma^2} e^{-V^2/2\sigma^2} \qquad (23-6)$$

Since coherent detection is used, only the inphase component of the noise, $x_c(t)$ say, is effective as interference. As discussed previously, the ideal detecting device uses a reference threshold to base its decision concerning the state of the receiver carrier. For ASK, the threshold is set at a fraction kA_c of the amplitude A_c of the carrier, which may be optimized for minimum probability of error.

We shall calculate now the probability of error, or the fraction of the total number of intervals for which an incorrect decision is made by the receiver due to the inphase noise component $x_c(t)$ at the output of the coherent detector or product demodulator. An incorrect decision will be made by the receiver whenever $x_c(t)$ is in phase with the vector A_c during an interval of carrier omission and out of phase with A_c during carrier transmission, provided $x_c(t)$ is greater than the decision threshold kA_c. Consider first the case of transmitting a zero. The probability that the amplitude of $x_c(t)$ is greater than the decision threshold kA_c, using Eq. (4-83), is given by

$$P(x_c \geq kA_c) = 1 - \Phi\left(\frac{kA_c}{\sigma}\right) = \frac{1}{2}\left(1 - \text{erf } \frac{kA_c}{\sigma}\right) \qquad (23\text{-}7)$$

The probability that $x_c(t)$ will be in phase with the vector A_c during the carrier omission as opposed to being out of phase is $\frac{1}{2}$. Consequently, the probability of error of mistaking a binary 0 for a 1 is equal to

$$P_{x_c, kA_c} = \frac{1}{4}\left(1 - \text{erf } \frac{kA_c}{\sigma}\right) \qquad (23\text{-}8)$$

Similarly the probability of error during carrier transmission of mistaking a 1 for a 0 is given by

$$P_{x_c, A_c - kA_c} = \frac{1}{4}\left(1 - \text{erf } \frac{A_c - kA_c}{\sigma}\right) \qquad (23\text{-}9)$$

The average probability of error P_e or the fraction of the total number of incorrect decisions made by the receiver is therefore equal to

$$P_e = P_{x_c, kA_c} + P_{x_c, A_c - kA_c} = \frac{1}{2} - \frac{1}{4}\text{ erf } \frac{kA_c}{\sigma} - \frac{1}{4}\text{ erf }(1-k)\frac{A_c}{\sigma}$$
$$(23\text{-}10)$$

which is at a minimum when $k = \frac{1}{2}$. For this optimum threshold the probabilities of error of the two carrier states are equal, and

$$P_e = \frac{1}{2}\left(1 - \text{erf } \frac{A_c}{2\sigma}\right) \qquad (23\text{-}11)$$

For high-input carrier-to-noise ratio $A_c \gg \sigma$ and

$$\text{erf } t \doteq 1 - \sqrt{\frac{2}{\pi}} \frac{e^{-t^2/2}}{t}, \qquad t = \frac{A_c}{2\sigma} \qquad (23\text{-}12)$$

So that for large $(C/N)_i$, we obtain

$$P_e = \frac{1}{2}\sqrt{\frac{2}{\pi}} e^{-A_c^2/8\sigma^2} \frac{2\sigma}{A_c} = \frac{1}{\sqrt{\pi(C/N)_i}} e^{-\frac{1}{4}(C/N)_i} \qquad (23\text{-}13)$$

where $\quad \left(\dfrac{C}{N}\right)_i = \dfrac{A_c^2}{2\sigma^2}$

2. *Envelope Detection of ASK.* As stated previously, envelope detection is simpler to implement than coherent detection. Signal phase coherence is not required in the detection process, and a simple envelope detector is used following the IF amplifiers. In order to calculate the probability of error we shall consider first the probability distribution of the envelope of narrow-band noise with and without a sinusoidal carrier corresponding to the state of "mark" and "space" of the received signal. We have seen before that the noise envelope V(t) is Rayleigh-distributed, so that

$$p(V) = \frac{V}{\sigma^2}\, e^{-V^2/2\sigma^2} \tag{23-14}$$

The probability density of carrier plus noise is given by Eq. (4-259), namely,

$$q(R) = \frac{R}{\sigma^2}\, e^{-(R^2+A_c^2)/2\sigma^2} \cdot I_0\left(\frac{RA_c}{\sigma^2}\right), \qquad R(t) \geq 0 \tag{23-15}$$

where R(t) is the envelope of carrier plus narrow-band noise.

As before, we assume that the 0's and 1's are equiprobable of being transmitted. At the receiver, two kinds of error can occur: a "space" is transmitted and a "mark" received; and a "mark" is transmitted and a "space" received. Here again, the decision at the detector output is based on a threshold kA_c; if the detector output voltage is greater than some fixed threshold kA_c, the signal is judged to be a "mark"; otherwise, it is called a "space." The probability of error of sending "space" and receiving "mark" is equal to the product of the probability of sending "space" times the probability that the noise envelope V(t) exceeds kA_c; hence

$$P_{e,s} = \frac{1}{2} \int_{kA_c}^{\infty} p(V)\, dV = \frac{1}{2} \int_{kA_c}^{\infty} \frac{V}{\sigma^2} e^{-V^2/2\sigma^2}\, dV$$

$$= \frac{1}{2} e^{-(kA_c)^2/2\sigma^2} \tag{23-16}$$

Similarly, the probability of sending "mark" and receiving "space" is equal to the probability of sending "mark" multiplied by the probability that the noise envelope $V(t)$ superimposed on the carrier wave produces a resultant envelope $R(t) \leq kA_C$.

$$P_{e,m} = \frac{1}{2} \int_0^{kA_C} q(R) \, dR = \frac{1}{2} \int_0^{kA_C} \frac{I_0(A_C R/\sigma^2)}{\sigma^2} \, R e^{-(R^2+A_C^2)/2\sigma^2} \, dR$$

$$(23\text{-}17)$$

By making use of the indefinite integral

$$\int z^n I_{n-1}(az) \, dz = \frac{z^n I_n(az)}{a} \qquad\qquad (23\text{-}18)$$

the integral of Eq. (23-17) can be evaluated by successive integration by parts in terms of the sum

$$P_{e,m} = \frac{1}{2} e^{-(1+k^2)A_C^2/2\sigma^2} \cdot \sum_{n=1}^{\infty} (k)^n I_n\left(\frac{A_C^2}{\sigma^2}\right) \qquad (23\text{-}19)$$

The two types of error, namely, $P_{e,s}$ and $P_{e,m}$, are not in general equiprobable, but can be made so for a given carrier-to-noise ratio by a proper choice of the decision threshold kA_C which will render the two hatched areas of Fig. 23-2 equal. It should be noted here that in this case the threshold level is set at more than half the peak carrier amplitude for making the two types of error equal. This choice of kA_C will not, however, result in an overall minimum probability of error. It can be shown that choosing kA_C at the intersection between the two curves, as shown in Fig. 23-2, will result in a minimum overall probability of error, provided the a priori probabilities of sending "mark" or "space" are equal.

Determination of the proper value of kA_C for making the probabilities of error of both types equal requires the solution of a transcendental equation. The results are plotted in Fig. 23-3, where it is seen that the decision factor k approaches the value 0.5 for high carrier-to-noise ratio. Included is also a plot of P_e vs. input

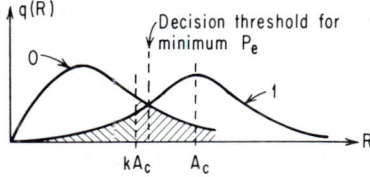

Fig. 23-2. Decision threshold for envelope detection of binary ASK.

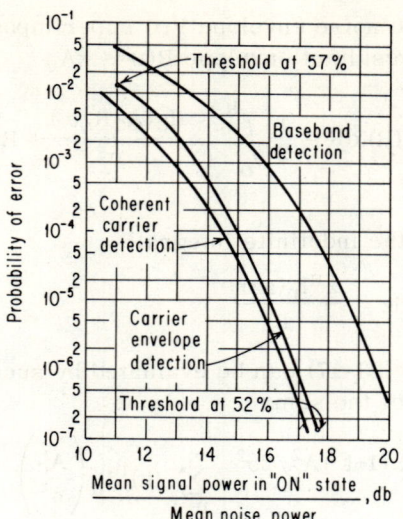

Fig. 23-3. Probability of error for binary ASK due to gaussian noise. (After Bennett,[7] Proc. IRE.)

carrier-to-noise ratio for coherent ASK for a fixed k = 0.5. It can be shown that for high carrier-to-noise ratio, both detection methods give essentially the same results.

23.3   Frequency-shift Keying (FSK) [2,8-16]

In a multiple-coded frequency-shift-keying system, digital information is transmitted by using as a code the sequential transmission of carrier pulses of constant amplitude and several different frequencies. A typical set of signal waveforms is described by

$$s_i(t) = \sqrt{\frac{2E}{T_o}} \cos \omega_i t, \qquad 0 \le t \le T_o$$

$$= 0, \qquad\qquad \text{elsewhere} \qquad\qquad (23\text{-}20)$$

where E is the energy content of $s_i(t)$, constant for all signals, and

$$\omega_i = 2\pi \frac{n_o + i}{T_o}, \qquad \text{for some fixed integer } n_o; \; i = 1, 2, \ldots, m$$

$$(23\text{-}21)$$

In the FSK noncoherent case the i'th signal is of the form

$$s_i(t) = \sqrt{\frac{2E}{T_o}} \cos (\omega_i t + \alpha) \qquad\qquad (23\text{-}22)$$

where the unknown angle α is assumed to be a random variable uniformly distributed over a 2π interval.

In a binary system only two carriers are used, with the code block being any desired length, and the two angular frequencies ω_1 and ω_2 are used to designate "mark" and "space" or 1 and 0, respectively.

As an example of multiple-coded frequency-shift keying we may consider the transmission of the quantized samples of an analogue information signal by using as a code the sequential transmission of s carriers of fixed frequency in a code block of n signals where $M = s^n$, M being the number of levels in the quantized samples. On the other hand, the quantized samples can be used to directly frequency-modulate a carrier; the modulated carrier will then be identical to signal waveforms of a multiple FSK system.

As stated previously, there are two basic methods of detection: a coherent or phase-locked system using coherent detection in the demodulation process, and a noncoherent system using envelope detection. Although noncoherent detection is normally used in FSK systems, the analysis of coherent FSK is included for comparison purposes. A block diagram illustrating both methods of detection

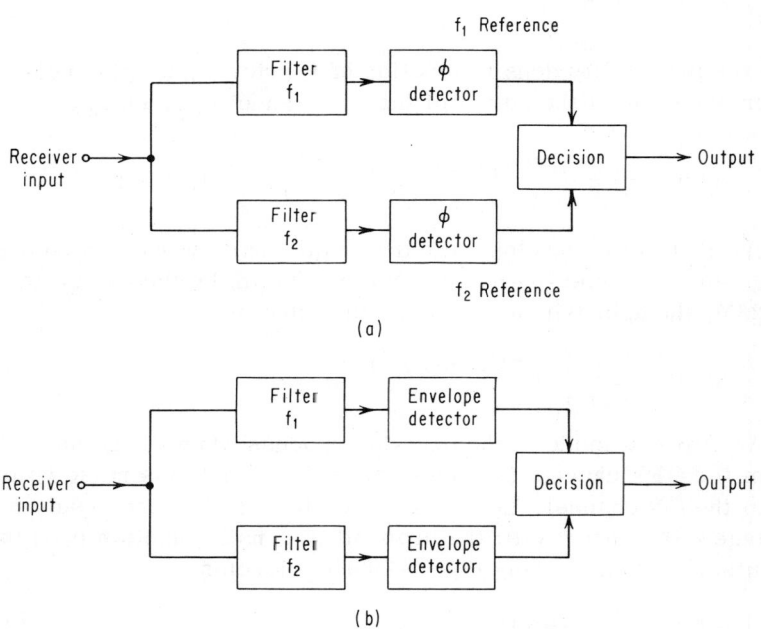

Fig. 23-4. Detection methods in FSK systems: (a) Coherent FSK system; (b) noncoherent FSK system.

is shown in Fig. 23-4. Sampling normally takes place at the end of the signal interval T_0, and the decision as to whether a "mark" or "space" was transmitted is made on the basis of the highest amplitude of the detector outputs.

1. *Noncoherent Detection.* A noncoherent FSK system uses envelope detection, and the decision as to whether a "mark" or a "space" was transmitted is made on the basis of which detector output has the highest amplitude at the sampling time. It should be noted here that the decision as to which message to accept is always the same, irrespective of signal level; i.e., the channel having the larger envelope is more likely to contain the signal. This is contrasted with envelope detection in a binary ASK (ON-OFF) system, where a threshold must be set, above which the signal is assumed present, and below which only noise is assumed at the detector output.

An expression for the error probability for the noncoherent FSK system will presently be derived. The probability density function of the noise voltage at the output of the envelope detector in the OFF channel has a Rayleigh distribution which, as in the ASK case, is given by

$$p(V) = \frac{V}{\sigma^2} e^{-V^2/2\sigma^2}$$

and the probability density function of carrier signal plus noise after detection at the output of the ON channel is given by

$$q(R) = \frac{R}{\sigma^2} e^{-(R^2+A_c^2)/2\sigma^2} \cdot I_0\left(\frac{RA_c}{\sigma^2}\right), \qquad R(t) \geq 0$$

where $R(t)$ is the envelope of carrier plus narrow-band noise. For a large-input carrier-to-noise ratio, we obtain, by the use of Eq. (4-265), the approximate gaussian distribution

$$q(R) \doteq \frac{1}{\sqrt{2\pi}\,\sigma} e^{-(R-A_c)^2/2\sigma^2} \tag{23-23}$$

An error is made in the decision process when the noise voltage from the OFF channel is greater than the signal plus noise voltage from the ON channel. Let the decision be based on the resultant voltage $z(t) = R(t) - V(t)$; the probability density function $p_1(z)$ of the resultant voltage z using Eq. (4-108) is therefore

$$p_1(z) = \int_{-\infty}^{\infty} q(R)p(R-z)\,dR \tag{23-24}$$

which is the probability density function for all $R(t)$ when $V(t)$ is in the range $R(t) - z(t)$. By direct substitution of $q(R)$ and $p(R - z)$ into the integrand of Eq. (23-24) it can be shown that the probability density function of the resultant voltage $z(t)$ reduced to

$$p_1(z) = \left(\frac{A_c - z}{2\sqrt{2}\,\sigma^2}\right) e^{-(A_c-z)^2/4\sigma^2} \tag{23-25}$$

Since an error is made in the decision process whenever $z < 0$, the probability of error can be determined by integrating the probability density function over all $z < 0$. The probability of error for noncoherent FSK can be shown to be given by

$$P_e(z < 0) = \tfrac{1}{2} e^{-A_c^2/4\sigma^2} = \tfrac{1}{2} e^{-\frac{1}{2}(C/N)_i} \tag{23-26}$$

This result is very often given in terms of the energy of the signal during the interval T_0 and the noise-power density N_0. Since

$$C_i = A_c^2/2 \quad\text{and}\quad N_- = BN_0 \doteq \frac{N_0}{T_0}$$

hence
$$\left(\frac{C}{N}\right)_i = \frac{E}{T_0}\frac{T_0}{N_0} = \frac{E}{N_0}$$

and
$$P_e = \tfrac{1}{2} e^{-E/2N_0} \tag{23-27}$$

2. *Coherent Detection.* As before, the probability density function of the signal plus noise voltage at the output of the phase detector for a large carrier-to-noise input is approximated by a gaussian distribution as in Eq. (23-23). For a coherent system the output noise voltage of the OFF channel has a probability density function which is also a gaussian distribution:

$$p(x) = \frac{1}{\sqrt{2\pi}\,\sigma} e^{-x^2/2\sigma^2} \tag{23-28}$$

Following the same method as in the noncoherent case, the probability density function of the resultant voltage $z(t) = R(t) - x(t)$ is given by the expression

$$p_1(z) = \frac{1}{2\sqrt{\pi}\,\sigma} e^{-(A_c-z)^2/4\sigma^2} \tag{23-29}$$

The probability of error is again determined by integrating the probability density function over all z < 0. Finally it can be shown that for coherent FSK the probability of error is given by

$$P_e = \frac{1}{2}\left(1 - \mathrm{erf}\,\frac{A_c}{2\sigma}\right)$$

$$= \frac{1}{2}\left(1 - \mathrm{erf}\sqrt{\frac{E}{2N_0}}\right) \qquad (23\text{-}30)$$

These results will be derived in a subsequent section, using the statistical decision theory. A plot of error probability as a function of input signal-to-noise ratio is shown in Fig. 23-5 for both coherent and noncoherent FSK.

3. *Probability of Error in Multiple FSK Systems.*[8,11] The derivation of the expressions for the probability of error in multiple FSK systems is quite involved and will not be given here. In the following we shall discuss both coherent and noncoherent detection.

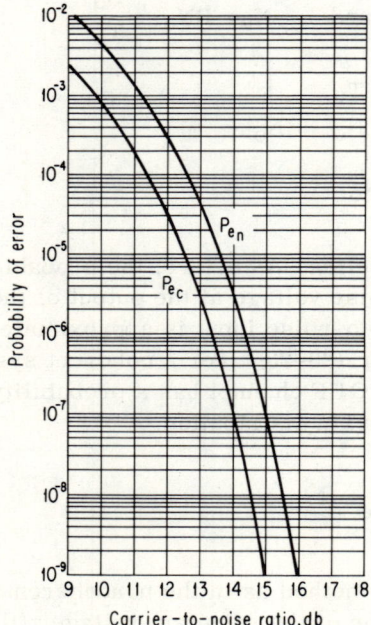

Fig. 23-5. Probability of error in binary FSK system: P_{e_n} = probability of error for a noncoherent FSK system; P_{e_c} = probability of error for a coherent FSK system.

The error rate in multiple FSK systems with coherent detection is given by[11]

$$P_e = \frac{1}{\sqrt{2\pi}} \int_{-\infty}^{\infty} e^{-\frac{1}{2}(x - A_c/\sigma)^2} \left[1 - \left(\frac{1}{\sqrt{2\pi}} \int_{-\infty}^{x} e^{-u^2/2} \, du \right)^{m-1} \right] dx$$

$$(23\text{-}31)$$

where m is the number of keying frequencies. The integral appearing in Eq. (23-31) does not appear to be available in terms of standard functions for m > 2. The values of P_e are calculated by an electronic computer as a function of m and $(C/N)_i$ and are shown in Fig. 23-6.

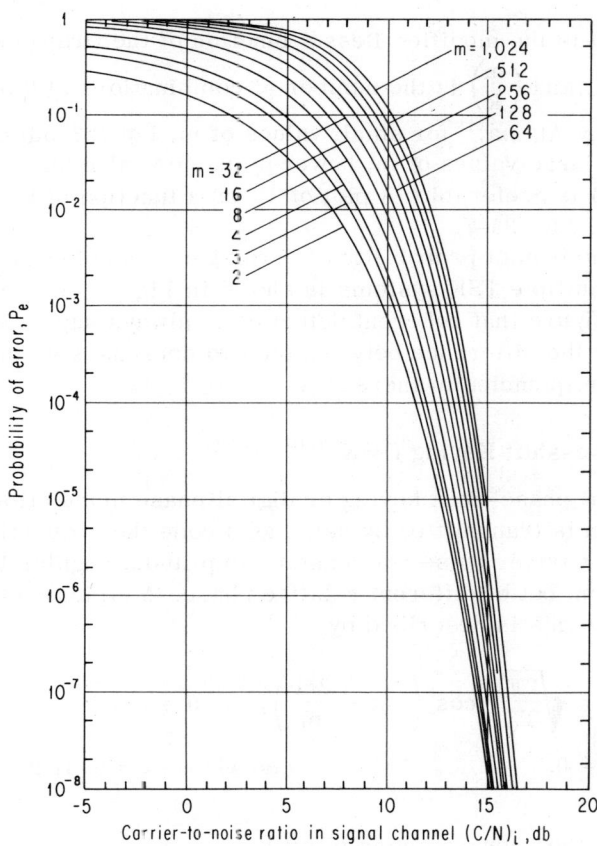

Fig. 23-6. Probability of error in multiple coherent FSK system. (From Akima,[11] Natl. Bur. Standards.)

The error rate in multiple FSK systems with noncoherent detectors is given by [11]

$$P_e = \int_0^\infty x\, e^{-(x^2 + A_c^2/\sigma^2)/2} \cdot I_0\left(\frac{xA_c}{\sigma}\right) \left[1 - \left(1 - e^{-x^2/2}\right)^{m-1}\right] dx$$

(23-32)

or by its equivalent expression

$$P_e = \frac{1}{m}\left[\sum_{k=2}^m (-1)^k \binom{m}{k} e^{-(k-1)(C/N)_i/k}\right]$$

(23-33)

where $I_0(z)$ is the modified Bessel function of the first kind and of zero order, and $\binom{m}{k}$ is the number of combinations of k out of m. As stated by Akima,[11] for small values of m, Eq. (23-33) can be used. For large values of m, however, numerical integration of Eq. (23-32) is preferable. A plot of P_e as a function of m and $(C/N)_i$ is shown in Fig. 23-7.

A comparison of probability of error between coherent and noncoherent multiple FSK systems is shown in Fig. 23-8. It is clear from this figure that coherent detection is always superior to incoherent, but the difference between the two decreases as the number of keying frequencies m increases.

23.4   Phase-shift Keying (PSK) [8,13-15,17-19]

In digital phase-shift keying or digital phase modulation, digital information is transmitted by using as a code the sequential transmission of carrier pulses of constant amplitude, angular frequency, and duration, but of different relative phase. A typical set of message waveforms is described by

$$s_i(t) = \sqrt{\frac{2E}{T_0}} \, \cos\left(\omega_0 t + \frac{2\pi i}{m}\right), \qquad 0 \le t \le T_0$$

$$= 0, \qquad\qquad \text{elsewhere; } i = 0, 1, 2, \ldots, (m-1)$$

(23-34)

where E is the energy content of $s_i(t)$.

$$\omega_0 = \frac{2\pi n_0}{T_0}, \qquad \text{for some fixed integer } n_0$$

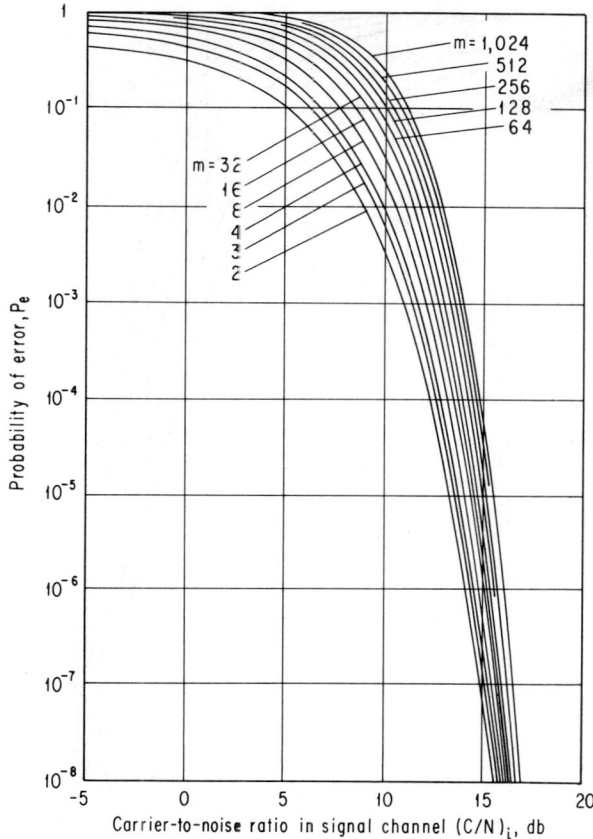

Fig. 23-7. Probability of error in multiple incoherent FSK system. (From Akima,[11] Natl. Bur. Standards.)

and m represents the number of possible transmitted signals or waveforms. The detection problem at the receiver is to determine the phase $\theta = 2\pi i/m$ under conditions of no fading and with additive gaussian noise

$$n(t) = x_c(t) \cos \omega_0 t - x_s(t) \sin \omega_0 t \qquad (23\text{-}35)$$

where $\qquad \overline{n^2(t)} = \dfrac{\overline{x_c^2(t)}}{2} + \dfrac{\overline{x_s^2(t)}}{2} = \sigma^2 = N$

An error in transmission will occur whenever the noise causes the received signal to be displaced in phase by more than $\pm \pi/m$ rad.

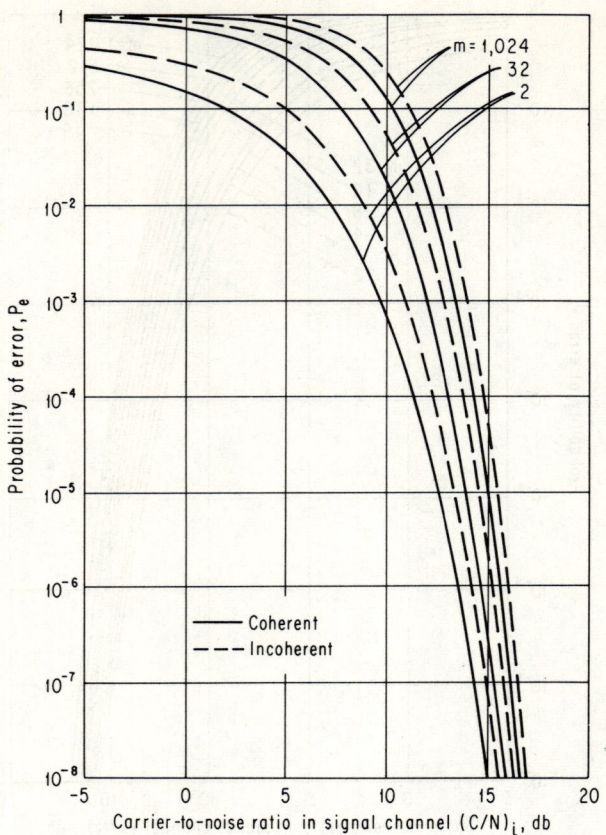

Fig. 23-8. A comparison of probability of error between coherent and noncoherent multiple FSK systems. (From Akima,[11] Natl. Bur. Standards.)

As stated above, the detection process may be carried out by using either coherent or noncoherent detection. In coherent detection, a phase reference is provided in the receiver, which permits the receiver to be phase-synchronized with the transmitter. In many practical systems, however, obtaining and maintaining a coherent phase reference in the receiver is not feasible. In such instances, phase comparison of successive samples is utilized in the detection process, and the information is conveyed by the phase transitions between carrier pulses rather than by the absolute phases of the pulses.

The basic difference between the two methods of detection is that

in the coherent case the received signal is being compared with a
clear reference. In the noncoherent case, however, two noisy signals
are being compared with each other. Thus, we might intuitively ex-
pect in this case twice as much noise as in the coherent case or a
3-db degradation in performance. As we shall show later, this is
approximately true in case of high $(C/N)_i$ and m > 2.

In this section we shall analyze the performance of digital phase
modulation under conditions of no fading in the presence of white
gaussian noise. We shall derive an expression for the error rate as
a function of input carrier-to-noise ratio with m, the number of discrete phases, as parameter. It will be shown that multiphase modulation provides an efficient trade of bandwidth for input carrier-to-
noise ratio in comparison with multilevel amplitude modulation.

1. *Coherent Phase Detection.* In this phase-modulation system a
coherent phase reference is available in the receiver to facilitate
the signal processing. The analysis can be carried out by describing
the set of m possible transmitted signals by m equally spaced
phasors in the complex plane, as shown in Fig. 23-9, for m = 8. As
seen from the figure, the addition of the noise components to the
zero phase signal introduces a distortion both in amplitude and
phase, with a possible transition of the decision threshold. The
probability of an error P_e due to noise is the probability that the
phase of signal plus noise will be distorted outside the sector
$-\pi/m < \theta < \pi/m$, where zero is the undistorted or true phase. The
probability of error P_e can be derived as a function of carrier-to-
noise ratio from the probability density function $q(\theta)$ of the phase of
a carrier plus gaussian noise given by Eq. (4-267), namely,

$$P_e = 1 - \int_{-\pi/m}^{\pi/m} q(\theta) \, d\theta \qquad (23\text{-}36)$$

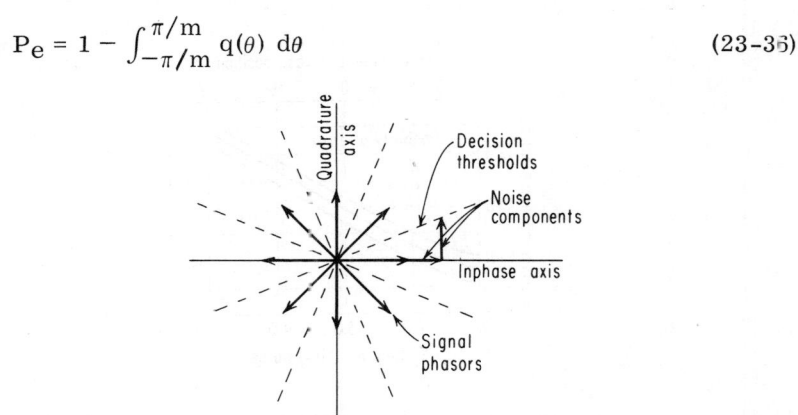

Fig. 23-9. Phasor representation of digital multiphase signals.

where

$$q(\theta) = \frac{1}{2\pi} e^{-(C/N)_i}$$

$$\times \left[1 + \sqrt{4\pi \left(\frac{C}{N}\right)_i} \cos \theta \, e^{(C/N)_i \cos^2 \theta} \cdot \Phi(\sqrt{2\,(C/N)_i} \, \cos \theta) \right]$$

$$(23\text{-}37)$$

and $\Phi(x)$ is the probability integral defined by

$$\Phi(x) = \frac{1}{\sqrt{2\pi}} \int_{-\infty}^{x} e^{-t^2/2} \, dt \qquad (23\text{-}38)$$

The probability of error P_e as a function of input carrier-to-noise ratio and the number of phase positions can be obtained by numerical and graphical integration. The required $(C/N)_i$ for a specified probability of error vs. the number of phase positions m is shown plotted in Fig. 23-10.

An analytic approximation to these curves can be derived as follows. For high $(C/N)_i$ or large x, the probability integral $\Phi(x)$ may be approximated by the asymptotic expansion

$$\Phi(x) \doteq 1 - \frac{e^{-x^2/2}}{\sqrt{2\pi} \, x} \qquad (23\text{-}39)$$

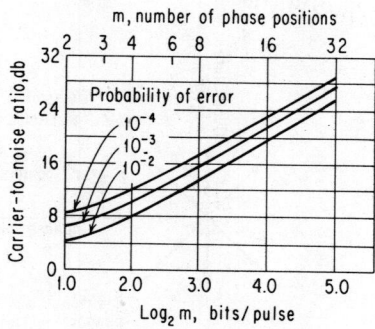

Fig. 23-10. Input carrier-to-noise ratio vs. number of phase positions. (From Cahn,[17] IRE Trans. Commun. Systems.)

and the corresponding probability density of phase for a sine wave in gaussian noise may be approximated by

$$q(\theta) \doteq \sqrt{\frac{(C/N)_i}{\pi}} \, \cos\theta \cdot e^{-(C/N)_i \sin^2\theta}, \qquad |\theta| < \pi/2 \qquad (23\text{-}40)$$

where θ denotes the phase displacement from the noise-free carrier. Hence,

$$P_e = 1 - \int_{-\pi/m}^{\pi/m} q(\theta)\, d\theta \doteq 2\left\{1 - \Phi\left[\sqrt{2\left(\frac{C}{N}\right)_i}\, \sin\frac{\pi}{m}\right]\right\}$$

$$\doteq \frac{e^{-(C/N)_i \sin^2(\pi/m)}}{\sqrt{\pi(C/N)_i}\,\sin(\pi/m)} \qquad (23\text{-}41)$$

Equation (23-41) represents a "normalized" or universal curve providing an approximation to P_e for any number of phase positions m. This approximation is exact in the limit as m $\rightarrow \infty$ and is plotted in Fig. 23-11. The other curves for small values of m are exact, obtained by numerical integration; the approximation (m = ∞) is extremely good for m > 3.

For binary coherent PSK it can be shown [13] that

$$P_e = \frac{1}{2}\left[1 - \text{erf}\left(\frac{E}{N_o}\right)^{\frac{1}{2}}\right] \qquad (23\text{-}42)$$

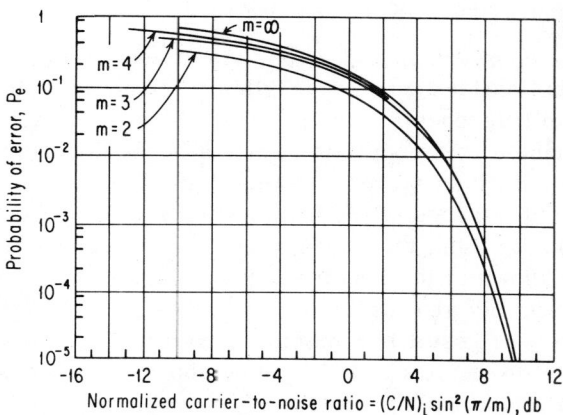

Fig. 23-11. Probability of error for PSK with coherent phase detection. (From Cahn,[17] IRE Trans. Commun. Systems.)

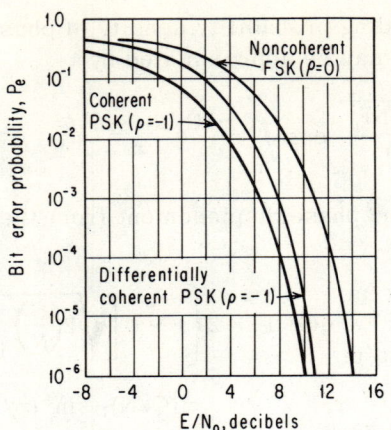

Fig. 23-12. Error rates for binary phase modulation. (From Law-ton,[13] Proc. 2d Natl Conf. Military Electron.)

where E is the average signal energy per received symbol, and N_0 is the noise-power density which in the ideal case when the channel bandwidth $B \doteq 1/T_0$ reduces to

$$P_e = \frac{1}{2}\left[1 - \mathrm{erf}\left(\frac{C}{N}\right)_i^{\frac{1}{2}}\right] \qquad (23\text{-}43)$$

This is shown plotted in Fig. 23-12 for coherent PSK where for comparison noncoherent FSK is also included.

2. *Phase-comparison Detection.* As stated earlier in many practical applications it is not feasible to obtain and maintain a coherent phase reference in the receiver. A practical variation of the coherent phase-shift-modulation system is the differentially coherent or phase-comparison system. In this system information is conveyed by the phase difference of the transmitted signal between adjacent signal elements; thus, the absolute phase of the received signals need not be known at the receiver. It is obvious that because the reference phase is also perturbed by the noise, this type of detection process will result in a higher error rate for any specified input carrier-to-noise ratio. It can be shown that in case of multiphase modulation for which the angular deviations are small, its degradation introduced by phase-comparison detection is essentially 3 db over coherent detection. This can be demonstrated by proving

that the probability density of the phase is approximately gaussian for small angular deviations as follows.

For high $(C/N)_i$ we obtain from Eq. (23-40) the approximate expression for the probability density

$$q(\theta) \doteq \sqrt{\frac{(C/N)_i}{\pi}} \cos \theta \, e^{-(C/N)_i \sin^2 \theta}$$

$$\doteq \sqrt{\frac{(C/N)_i}{\pi}} \, e^{-(C/N)_i \theta^2}, \qquad |\theta| < \frac{\pi}{2}$$

$$\doteq 0, \qquad\qquad\qquad |\theta| > \frac{\pi}{2} \qquad\qquad (23\text{-}44)$$

which is a gaussian distribution with zero mean and variance

$$\mu^2 = \frac{1}{2(C/N)_i}$$

or $\quad q(\theta) \doteq \dfrac{1}{\sqrt{2\pi}\,\mu} \, e^{-\theta^2/2\mu^2} \qquad\qquad\qquad (23\text{-}45)$

It follows therefore that for phase comparison of successive samples in case of high $(C/N)_i$ and small angular deviations (large m), the difference phase has approximately a gaussian distribution with twice the variance of the distribution for the phase of a single sample, which is equivalent to a degradation of 3 db. For a binary system (m = 2), it is shown [13] that for high $(C/N)_i$ the degradation approaches zero. It can be shown [19] that the carrier-to-noise ratios required for equal probabilities of error with the two phase-detection methods are asymptotically related by

$$\frac{(C/N)_i \text{ (phase comparison)}}{(C/N)_i \text{ (coherent)}} = \frac{\sin^2 (\pi/m)}{2 \sin^2 (\pi/2m)} \qquad (23\text{-}46)$$

This is shown plotted in Fig. 23-13, where it is seen that the degradation approaches 3 db for a large number of phases.

Cahn [19] obtained a universal curve for probability of error with phase-comparison detection in the form

$$P_e \doteq 2 \left[1 - \Phi\left(2\sqrt{\left(\frac{C}{N}\right)_i} \, \sin \frac{\pi}{2m} \right) \right] \doteq \frac{e^{-2(C/N)_i \sin^2 (\pi/2m)}}{\sqrt{\pi 2(C/N)_i}\, \sin (\pi/2m)} \qquad (23\text{-}47)$$

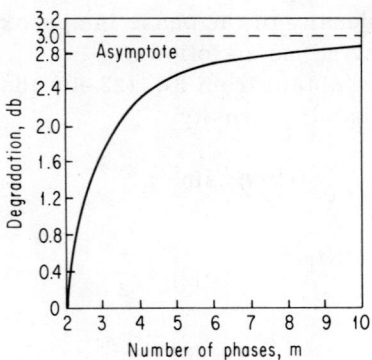

Fig. 23–13. Degradation of phase-comparison detection compared to coherent phase detection. (From Cahn,[19] IRE Trans. Commun. Systems.)

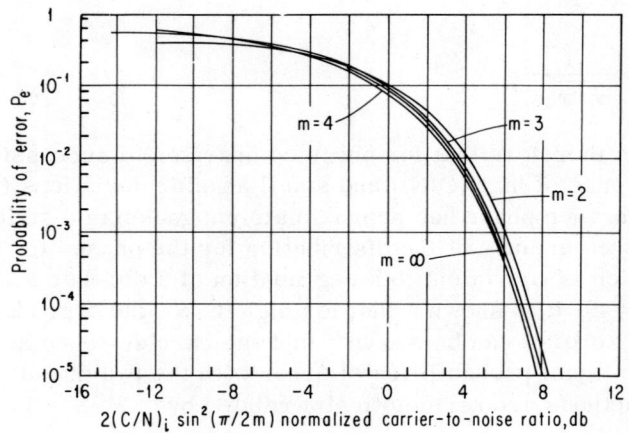

Fig. 23–14. Probability of error for PSK systems using phase-comparison detection. (From Cahn,[19] IRE Trans. Commun. Systems.)

which is shown plotted in Fig. 23–14 for $m = \infty$. For comparison, exact curves obtained by numerical integration for small values of m are also included. It is thus seen from these curves that Eq. (23–47) is an excellent approximation for all values of m.

An exact expression for the probability of error for binary or phase-reversal PSK, using phase-comparison detection,[8,13] is derived in the form

$$P_e = \tfrac{1}{2} e^{-E/N_0} = \tfrac{1}{2} e^{-(C/N)_i} \qquad\qquad (23\text{-}48)$$

This is plotted in Fig. 23-12, where it is seen that the error probability in this case is larger than in the coherent system. For large
E/N_0, the difference between the two systems is small. At E/N_0 below 3 db, there is a rapid degradation of the differentially coherent
PSK system.

We conclude therefore that for high $(C/N)_i$, phase-comparison
detection yields about a 3-db degradation over coherent detection,
except in the binary case where the degradation approaches zero
with high-input carrier-to-noise ratio.

23.5 Combined Phase-and-Amplitude-Shift Keying [19-21]

It has been shown by Shannon that a gaussian amplitude distribution provides maximum transmission capacity when the signals are
average-power-limited. However, since multiphase modulation produces an essentially constant amplitude signal, it has been suggested
that a combination of phase-shift- and amplitude-shift-keying modulation would produce a more efficient signal.

In this section we shall investigate the performance of combined
digital phase- and digital amplitude-modulated systems. For the
high carrier-to-noise case, two types of combined modulation schemes
will be considered. In the type I system the number of phase positions used is the same, regardless of the amplitude level. In the
type II system, which is a more efficient one, the number of phase
positions available is increased with increasing amplitude level.
Approximate expressions for the probability of error and channel
capacity for the combined modulation schemes will be derived and
compared with the corresponding results for a purely digital phase-
modulated system. It will be shown that from a channel-capacity
point of view the combined digital modulation systems make more
efficient use of the channel for carrier-to-noise ratios greater than
11 db.

1. *Combined Multiamplitude and Multiphase Digital Modulation.*
In this combined modulation system, the phase and amplitude of the
transmitted carrier are used to transmit the digital information.
During each carrier pulse length of T_0 sec, the phase and amplitude
of the transmitted carrier signal assume values chosen from a discrete set of possible phases and amplitudes. Each combination of a
particular amplitude level and phase position represents a transmitted symbol; a combination of n amplitude levels and m phase

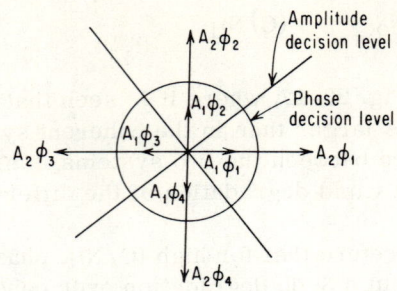

Fig. 23-15. Decision levels in type I system.

positions will allow transmission of digital data having an alphabet
size s = mn. As an example we consider a system illustrated in
Fig. 23-15 with two possible amplitude levels, A_1 and A_2, and four
possible phase positions, $\phi_1 - \phi_4$, a total of eight possible selections
in the decision diagram. At the receiver, the phase and amplitude of
an incoming pulse are detected, and the decision as to which symbol
or alphabet letter has been sent is made on the basis of the amplitude and phase measurements. In this system the number of phase
positions is constant, independent of the amplitude level; such a
system will be designated a type I system.

In the type I system just described, the probability of an error in
phase decreases with increasing amplitude levels. In order to produce a more efficient system where the probability of phase error
is constant, it is required to increase the number of phase positions
available with increasing amplitude level. Thus, for a given peak
power, the number of available symbols is increased. This type of
system, which is designated as the type II system, is illustrated in
Fig. 23-16, where the number of phase positions is increased from
3 on the first amplitude level to 9 on the second level, for a total of
12 transmission symbols.

2. *Mathematical Analysis of Combined Modulation System.* Only
coherent detection using a synchronized local reference will be considered. First we consider a type I system. For large $(C/N)_i$ the
probability of a phase error at the k'th amplitude level A_k is given
by the asymptotic approximation of Eq. (23-41), namely,

$$P_{e\phi} = \frac{e^{-(A_k^2/2\sigma^2)\,\sin^2(\pi/m)}}{\sqrt{\pi}\,(A_k/\sqrt{2}\,\sigma)\,\sin(\pi/m)} \qquad (23\text{-}49)$$

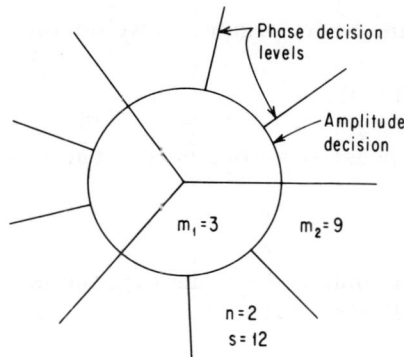

Fig. 23-16. Decision levels in type II system.

where σ is the rms input noise voltage and m is the number of phase positions, which is constant, independent of the amplitude level A_k. The spacing between amplitude levels may be obtained by consideration of the probability distribution of the detected envelope, which is essentially gaussian for large-input signal-to-noise ratios. Using this approach, Cahn [19] has shown that for a type I system the amplitude levels are given by the relation

$$A_k = A_1 \left[1 + 2(k - 1) \sin \frac{\pi}{m} \right] \qquad (23-50)$$

Under these conditions Cahn has shown that the probability of an amplitude error at each amplitude level is equal to the probability of a phase error on the first amplitude level. The placing of A_1 is determined by the number of phase positions m and the desired probability of error, as seen from Eq. (23-49) for k = 1.

For a type II system the probability of phase error $P_{e\phi}$ is maintained constant by increasing the number of phase positions, and is equal to the probability of an amplitude error. It follows from Eq. (23-49) that for constant phase error, the argument

$$\frac{A_k}{\sqrt{2}\sigma} \sin \frac{\pi}{m_k} = \frac{A_1}{\sqrt{2}\sigma} \sin \frac{\pi}{m_1} = \text{constant}$$

Since the amplitude levels are not changed in the type II system, Eq. (23-50) may be used to relate A_k to A_1; hence

$$\sin \frac{\pi}{m_1} = \left[1 + 2(k - 1) \sin \frac{\pi}{m_1} \right] \sin \frac{\pi}{m_k} \qquad (23-51)$$

approximating the sine by its argument, we obtain

$$m_k \doteq m_1 + 2\pi(k - 1) \qquad (23\text{-}52)$$

Since the number of phase positions must be an integer, we obtain

$$m_{k+1} = m_k + 6 \qquad (23\text{-}53)$$

The alphabet size s is obtained by summing the number of phase positions on all amplitude levels

$$s = \sum_{k=1}^{n} m_k \qquad (23\text{-}54)$$

where n is the number of amplitude levels. Using Eq. (23-53), we obtain for the alphabet size the expression

$$s = 3n(n - 1) + m_1 n$$

Hancock and Lucky [20] have shown that the minimum probability of error that can be attained for a given $(C/N)_i$, alphabet size, and the system parameters A_1, m_1, and n is given by

$$P_{e(min)} = \frac{2e^{-(C/N)_i/(8/9s - 4/3)}}{\sqrt{\pi} \, \dfrac{(C/N)_i}{8/9s - 4/3}} \qquad (23\text{-}55)$$

This is shown plotted in Fig. 23-17. Note that the carrier-to-noise ratio improvement obtained by the addition of amplitude modulation to phase modulation increases approximately 3 db every time the alphabet size is doubled.

Hancock and Lucky have evaluated the channel capacities of the combined modulation schemes and compared the results with the capacity of a simple digital phase-modulation system. Their findings are shown plotted in Fig. 23-18, where a plot of the ideal channel capacity given by

$$C = B \log \left[1 + \left(\frac{C}{N} \right)_i \right]$$

is also included for comparison. It is seen that the combined digital

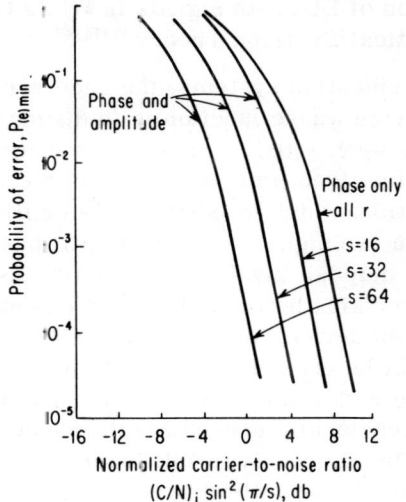

Fig. 23-17. Probability of error vs. $(C/N)_i$. (From Hancock and Lucky,[20] IRE Trans. Commun. Systems.)

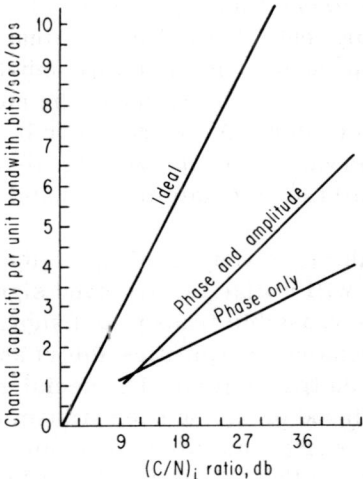

Fig. 23-18. Channel capacity vs. $(S/N)_i$. (From Hancock and Lucky, [20] IRE Trans. Commun. Systems.)

phase and amplitude system makes much more efficient use of the channel for high-input carrier-to-noise ratios. The reason for it lies in the fact that it is more efficient to use larger alphabet sizes at high power and carry more bits of information per sample.

23.6   The Detection of Discrete Signals in White Gaussian Noise, Using Statistical Decision Theory [13,15,22-25]

In digital communication systems, the receiver may be considered a decision device whose function is to decide which one of the elementary signals $s_i(t)$, $s_1(t)$, . . . , $s_m(t)$ was transmitted during each interval T_0 of a given time sequence. In the simplest case of binary communication, which we shall consider first, the choice is to be made between two signals, $s_1(t)$ and $s_2(t)$, based on a decision rule which may or may not involve a threshold. As an illustration let us consider a decision between the two elementary signals by means of a threshold device. In the presence of noise alone the threshold should not be exceeded. Since the noise can have theoretically any amplitude with a finite probability, it follows therefore that noise alone will occasionally exceed the threshold, and consequently the decision made by the receiver will inherently have a certain probability of error. In the following we shall discuss the problem of optimum detection or "guessing" of signals in the presence of additive white gaussian noise with minimum probability of error.

1. *Statistical Decision Theory*. The two elementary signals, $s_1(t)$ and $s_2(t)$, which represent a 0 and 1 of the binary coded message are assumed to be equiprobable with no intersymbol dependence. Under these assumptions, as shown by Shannon, the message will have the highest information content. At the receiver it is known that at a particular time interval, one of the two signals has been transmitted; the problem is to calculate the probability of error per decision.

Let $y(t)$ denote the received signal in the interval $0 \le t \le T_0$. The ideal receiver will utilize the received signal $y(t)$ and the knowledge of the two possible transmitted signals $s_1(t)$ and $s_2(t)$ to compute the a posteriori probabilities $P(0|y)$ and $P(1|y)$ that the symbols 0 or 1 have been transmitted. The decision as to which symbol has a greater likelihood of having been transmitted is then made on the basis of the greater of the two a posteriori probabilities. Now, Woodward [23] has shown that the probabilities $P(0|y)$, $P(1|y)$ that signals $s_1(t)$ and $s_2(t)$ were sent when signal $y(t)$ has been received are given by

$$P(0|y) = kP(0)e^{-\frac{1}{N_0}\int_0^{T_0}[y(t)-s_1(t)]^2dt}$$

(23-56)

$$\text{and} \quad P(1|y) = kP(1)e^{-\dfrac{1}{N_0} \int_0^{T_0} [y(t)-s_2(t)]^2 \, dt} \tag{23-57}$$

where k is a normalizing constant adjusted so that

$$P(0|y) + P(1|y) = 1 \tag{23-58}$$

$P(0)$ and $P(1)$ are the a priori probabilities of signal 0 $[s_1(t)]$ and 1 $[s_2(t)]$ being sent, and N_0 is the noise-power density per cycle per second of bandwidth (one-sided spectrum). Since we have assumed the signals $s_1(t)$ and $s_2(t)$ to be equiprobable, $P(0) = P(1) = \frac{1}{2}$. Let us assume that $s_1(t)$ was sent and that the received signal $y(t) = s_1(t) + n(t)$, where $n(t)$ is the additive gaussian noise. It follows from Eqs. (23-56) and (23-57) that the decision is, according to the inequality,

$$\int_0^{T_0} [y(t) - s_1(t)]^2 \, dt \gtrless \int_0^{T_0} [y(t) - s_2(t)]^2 \, dt \tag{23-59}$$

where > indicates a decision in favor of $s_2(t)$ (an error) and < indicates a decision in favor of $s_1(t)$ (no error). It should be noted here that the decision according to the inequality of Eq. (23-59) corresponds to the well-known mean square difference criterion. If we assume that $s_1(t)$ and $s_2(t)$ have equal energies during the interval T_0, then

$$\int_0^{T_0} s_1^2(t) \, dt = \int_0^{T_0} s_2(t) \, dt = E \tag{23-60}$$

and the decision equation reduces to

$$\int_0^{T_0} y(t)s_1(t) \, dt \underset{s_2(t)}{\overset{s_1(t)}{\gtrless}} \int_0^{T_0} y(t)s_2(t) \, dt \tag{23-61}$$

The evaluation of the integrals in Eq. (23-61) can be interpreted either as a process of correlation detection, i.e., of selecting that waveform whose cross-correlation with $y(t)$ is the largest, or as a process of matched-filter detection, where the received signal $y(t)$ is passed through filters which are matched to the signals $s_1(t)$ and $s_2(t)$, and the filter output which is largest at $t = T_0$ is selected. Mathematically, correlation detection and matched-filter detection are identical. The process of correlation detection is illustrated in

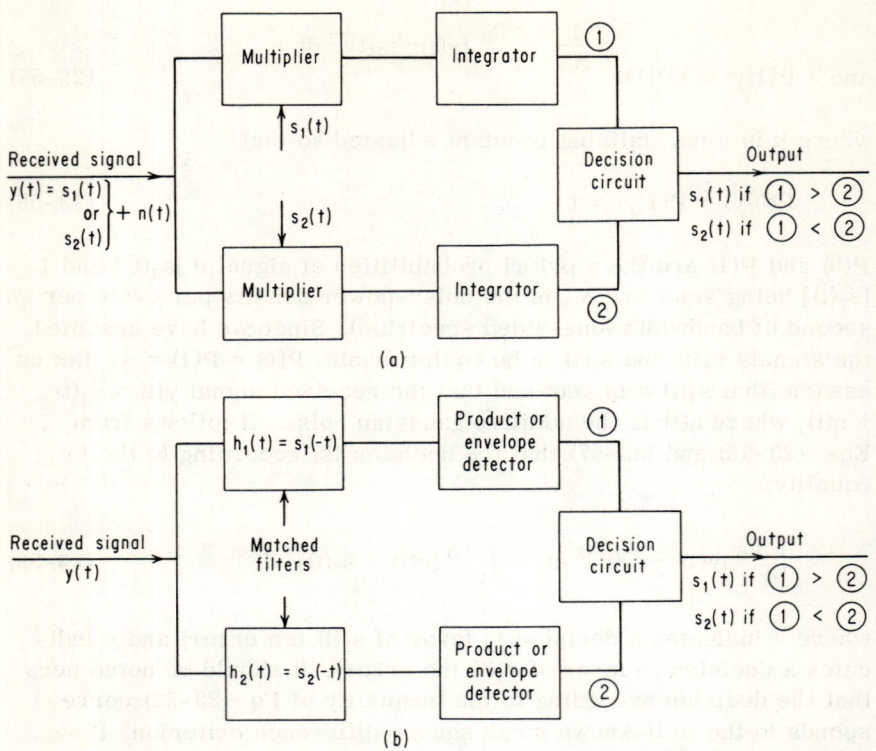

Fig. 23-19. Optimum signal detectors: (a) Correlation detection;
(b) matched-filter detection.

Fig. 23-19a, where the received signal y(t) is first multiplied by s(t)
and then the product integrated for the duration of the interval T_0.
This process requires a convenient means of generating the signals
$s_1(t)$ and $s_2(t)$ at the receiver plus a precise knowledge of the timing
of the received symbol waveform. In the process of matched-filter
detection, the signal y(t) is applied to a linear filter whose impulse
response is h(t). By the use of the convolution integral, the filter
output y(t) is given by

$$g(t) = \int_{-\infty}^{\infty} y(\tau)h(t - \tau) \, d\tau \qquad (23\text{-}62)$$

and by choosing a filter whose impulse response h(t) = s(−t), the
filter output is then

$$g(t) = \int_{-\infty}^{\infty} y(\tau)s(\tau - t) \, d\tau \qquad (23\text{-}63)$$

At $t = 0$ this integral is identical with that given by Eq. (23-61), which is required in making the decision as to which signal was most probably transmitted. In this process of matched-filter detection, which is illustrated in Fig. 23-19b, it is not necessary to know in advance exactly when the signal occurs, since the filter computes the required integral continuously in time. However, this method has the disadvantage that the matched filters may be difficult to realize in practice and can only be approximated by crystal or mechanical filters.

2. *Error Probabilities in Coherent Binary Systems*. If $s_1(t)$ and $s_2(t)$ are equally likely to be transmitted, then the probability of error can be shown [13] to be

$$P_e = \frac{1}{2}\left[1 - \text{erf}\sqrt{\frac{E(1-\rho)}{2N_0}}\right] \qquad (23\text{-}64)$$

where E is the average signal energy per received symbol,

$$\rho = \frac{1}{E}\int_0^{T_0} s_1(t)s_2(t)\,dt \qquad (23\text{-}65)$$

is the finite time correlation coefficient of the two waveforms $s_1(t)$ and $s_2(t)$ which ranges over $-1 \leq \rho \leq 1$, and N_0 is the noise-power spectral density (one-sided spectrum). Equation (23-64) shows that the error probability increases as the noise-power density increases, and decreases as the signal energy increases, as expected. For a fixed E and N_0, P_e is a minimum when $\rho = -1$, which requires that

$$s_1(t) = -s_2(t)$$

This corresponds to binary coherent PSK, where the phase of $s_1(t)$ differs from that of $s_2(t)$ by 180°. The probability of error is then

$$P_e = \frac{1}{2}\left[1 - \text{erf}\sqrt{\left(\frac{E}{N}\right)_0}\right] \qquad (23\text{-}63)$$

given earlier in Eq. (23-42).

As $\rho \rightarrow +1$, or as $s_1(t) \rightarrow s_2(t)$, the error probability approaches $1/2$ as expected, since both waveforms are then identical and consequently do not carry any information. The case $\rho = 0$ applies to two orthogonal functions such as in ON-OFF and FSK systems. For these systems the probability of error is

$$P_e = \frac{1}{2}\left(1 - \text{erf}\sqrt{\frac{E}{2N_0}}\right) \qquad (23\text{-}67)$$

as in Eqs. (23-11) and (23-30). From the last two equations it is seen that for the same error probability P_e the orthogonal systems suffer a 3-db disadvantage in average transmitted energy over the optimum phase-reversal system.

3. *Error Probabilities in Differentially Coherent PSK System.* As discussed above, the probability of error of this system is higher than in the ideal coherent PSK system. Lawton[13] derived for this system the expression

$$P_e = \frac{1}{2}e^{-E/N_0} \qquad (23\text{-}68)$$

It is of interest to compare the relative power levels needed in both coherent and differentially coherent PSK systems for the same probability of error. In order to achieve the same error probability with the two systems let

$$\left(\frac{E}{N_0}\right)_{\text{d. coherent}} = k\left(\frac{E}{N_0}\right)_{\text{coherent}}$$

or $\quad \frac{1}{2}\left(1 - \text{erf}\sqrt{\frac{E}{N_0}}\right) = \frac{1}{2}e^{-kE/N_0}$

For large E/N_0 we obtain

$$\left(1 - \text{erf}\sqrt{\frac{E}{N_0}}\right) = \frac{2}{\sqrt{\pi}}\frac{e^{-E/N_0}}{2\sqrt{E/N_0}}$$

so that $\quad \dfrac{2}{\sqrt{\pi}}\dfrac{e^{-E/N_0}}{2\sqrt{E/N_0}} = e^{-kE/N_0}$

from which we derive

$$-\frac{E}{N_0} = \frac{1}{2}\ln\left(\frac{\pi E}{N_0}\right) - \frac{kE}{N_0}$$

hence $\quad k = \dfrac{\frac{1}{2}\ln(\pi E/N_0)}{E/N_0} + 1 \qquad (23\text{-}69)$

We conclude therefore that the ratio of powers required to achieve the same error probability with the two systems approaches unity for large E/N_0. This is seen from Fig. 23-12 where for high $(C/N)_i$ the performance of the two systems becomes identical.

So far we have considered decision problems between two alternative symbols. Extension of the theory to m symbols of equal energy leads to integrals which have not been tabulated, and consequently no simple formula exists for the probability of error of multiple digital systems.

References

1. Wier, J. M.: Digital Data Communication Techniques, *Proc. IRE*, January, 1961, pp. 196–209.
2. Sundae, E. D.: Ideal Binary Pulse Transmission by AM and FM, *BSTJ*, vol. 38, November, 1959.
3. Montgomery, G. G.: A Comparison of Amplitude and Angle Modulation for Narrow-band Communication of Binary-coded Messages in Fluctuation Noise, *Proc. IRE*, February, 1954, pp. 447–454.
4. Critchlow, D. L., R. H. Dennard, and E. Hopner: A Vestigial Sideband Phase-reversal Data Transmission System, *IBM J.*, January, 1964.
5. Hopner, E.: An Experimental Modulation–Demodulation Scheme for High-speed Data Transmission, *IBM J.*, January, 1959.
6. Clark, A. P.: Considerations in the Choice of the Optimum Data Transmission Systems for Use Over Telephone Circuits, *J. Brit. IRE*, May, 1962.
7. Bennett, W. R.: Methods of Solving Noise Problems, *Proc. IRE*, May, 1956.
8. Arthurs, E., and H. Dym: On the Optimum Detection of Digital Signals in the Presence of White Gaussian Noise, *IRE Trans. Commun. Systems*, December, 1962, pp. 336–372.
9. Feldman, J. R., and J. N. Faraone: An Analysis of Frequency Shift Keying Systems, *Proc. Natl. Electron. Conf.*, vol. 17, pp. 530–542, 1961.
10. Akima, H.: Error Rate in a Multiple-frequency-shift System, *Proc. IEE London*, vol. 3, no. 3, March, 1964.
11. Akima, H.: The Error Rates in Multiple FSK Systems, *Natl. Bur. Standards Tech. Note* 167, March, 1963.
12. Pierce, J. N.: Theoretical Diversity Improvement in Frequency-shift Keying, *Proc. IRE*, May, 1958.
13. Lawton, J. G.: Comparison of Binary Data Transmission Systems, *Proc. 2d Natl. Conf. Military Electron.*, 1958, pp. 54–61.

14. Glenn, A. B.: Performance Analysis of Data Link Systems, *IRE Trans. Commun. Systems*, vol. CS-7, no. 1, pp. 14-24, May, 1959.

15. Glenn, A. B.: Comparison of PSK vs. FSK and PSK-AM vs. FSK-AM Binary Coded Transmission Systems, *IRE Trans. Commun. Systems*, vol. CS-8, no. 2, pp. 87-100, June, 1960.

16. Helstrom, C. W.: The Comparison of Digital Communication Systems, *IRE Trans. Commun. Systems*, vol. CS-8, no. 3, pp. 141-150, September, 1960.

17. Cahn, C. R.: Performance of Digital Phase-modulation Communication Systems, *IRE Trans. Commun. Systems*, vol. CS-7, no. 1, pp. 3-6, May, 1959.

18. Bussgang, J. J., and M. Leiter: Error Rate Approximation for Differential Phase Shift Keying, 9th National Communication Symposium, October, 1963.

19. Cahn, C. R.: Combined Digital Phase and Amplitude Modulation Communication Systems, *IRE Trans. Commun. Systems*, vol. CS-8, no. 3, pp. 150-155.

20. Hancock, J. C., and R. W. Lucky: Performance of Combined Amplitude and Phase-modulated Communication Systems, *IRE Trans. Commun. Systems*, vol. CS-8, no. 4, pp. 232-237, December, 1960.

21. Campopiano, C. N., and B. G. Glazer: A Coherent Digital Amplitude and Phase Modulation Scheme, *IRE Trans. Commun. Systems*, vol. CS-10, pp. 90-95, March, 1962.

22. Lerner, R. M.: Modulation and Signal Selection for Digital Data Systems, *AIEE Trans. Commun. Electron.*, January, 1962.

23. Woodward, P. M.: "Probability and Information Theory," Pergamon Press, New York.

24. Helstrom, C. W.: The Resolution of Signals in White Gaussian Noise, *Proc. IRE*, September, 1955.

25. Reiger, S.: Error Probabilities of Binary Data Transmission Systems in the Presence of Random Noise, *IRE Conv. Record*, 1953, part 8, pp. 72-79.

24

COMPARATIVE ANALYSIS OF MODULATION SYSTEMS

The process of modulation has been introduced in Chap. 1 as a transformation or mapping process, mapping from the message space into a signal space; similarly, the purpose of the demodulation process is to retransform the transmitted signal into the original message. We have considered in the past two classes of message waveforms, namely: messages which are continuous functions of time, and messages in the form of a sequence of digital data. In the first class, the modulation process amounts to varying one or more of the carrier parameters as a function of the instantaneous value of the message, whereas in the second class the carrier parameters are varied in a digital manner.

In this chapter we shall compare the various modulation systems from several points of view. We shall see that in making these comparisons, different figures of merit are required for analogue and digital systems. In the analogue modulation systems which we have discussed previously, the signal-to-noise power ratio was used as a criterion for system performance. This figure of merit stems from the ability of a listener to interpret a message as being a function of signal-to-noise ratio. In digital systems, however, a more meaningful figure of merit is the probability of error and/or

information rate. In the following sections, the relative performance of analogue systems will be made on the basis of signal-to-
noise power ratio, and of digital systems on the basis of probability
of error. For digital systems, we shall introduce also the concept
of minimum required energy per transmitted bit of information,
which will provide us with an absolute standard for making these
comparisons. This concept is becoming increasingly important in
space communication where power is at a premium, and consequently minimization of the energy required per bit transmitted in
the presence of a given noise spectral density is of primary importance.

24.1   Comparison of Analogue Modulation Systems [1-3]

The theoretical information capacity of a communication system
as formulated by Shannon, namely,

$$C = B \log_2 \left[1 + \left(\frac{S}{N} \right)_i \right] \quad \text{bits/sec} \tag{24-1}$$

was used in Chap. 19 to derive an expression for the optimum trade-
off between output signal-to-noise ratio and bandwidth. It was shown
there that for high-input signal-to-noise ratio $(S/N)_i$, the output
signal-to-noise ratio $(S/N)_o$ increases exponentially with the bandwidth expansion, so that

$$\left(\frac{S}{N} \right)_o \doteq \left(\frac{S}{N} \right)_i^{B/f_m} \tag{24-2}$$

where B/f_m is the ratio of channel bandwidth to message bandwidth.
It will be instructive to compare the trade-offs for the various
analogue modulation systems with the optimum as predicted by
Eq. (24-2). This is shown in Table 24-1, where the actual trade-
offs of the analogue systems which were discussed in the earlier
chapters are compared with the theoretical optimum.

1. *Relative Comparison of Analogue Modulation Systems.*   In the
systems under discussion, the additive interference in the channel
is assumed to be white gaussian noise of spectral density (one-sided)
N_0 watts/(cycle)/(sec), so that the total noise power at the input to
the demodulator $N_i = N_0 B$. The expression $(S/N)_i$ represents the
ratio of signal power to total noise power at the input to the demodu-

Table 24-1. Characteristics of Analogue Systems

Mode of operation	B vs. f_m	$\left(\dfrac{S}{N}\right)_o$ vs. $\left(\dfrac{S}{N}\right)_i$ (actual) $N_i = N_o B$	$\left(\dfrac{S}{N}\right)_o$ vs. $\left(\dfrac{S}{N}\right)_i$ (actual) $N_i = N_o f_m$	$\left(\dfrac{S}{N}\right)_o$ vs. $\left(\dfrac{S}{N}\right)_i$ (ideal) $N_i = N_o B$
AM-DSB-SC coherent detection	$B = 2f_m$	$\left(\dfrac{S}{N}\right)_o = 2\left(\dfrac{S}{N}\right)_i$	$\left(\dfrac{S}{N}\right)_o = \dfrac{S_i}{N_o f_m}$	$\left(\dfrac{S}{N}\right)_o = \left(\dfrac{S}{N}\right)_i^2$
AM-SSB coherent detection	$B = f_m$	$\left(\dfrac{S}{N}\right)_o = \left(\dfrac{S}{N}\right)_i$	$\left(\dfrac{S}{N}\right)_o = \dfrac{S_i}{N_o f_m}$	$\left(\dfrac{S}{N}\right)_o = \left(\dfrac{S}{N}\right)_i$
AM-DSB linear envelope detection	$B = 2f_m$	$\left(\dfrac{S}{N}\right)_o = 0.916\left(\dfrac{S}{N}\right)_i^2$ $\left(\dfrac{S}{N}\right)_i < 1$ — $\left(\dfrac{S}{N}\right)_o = 2\left(\dfrac{S}{N}\right)_i$ $\left(\dfrac{S}{N}\right)_i > 10$	$\left(\dfrac{S}{N}\right)_o = 0.229\left(\dfrac{S_i}{N_o f_m}\right)^2$ $\left(\dfrac{S}{N}\right)_i < 1$ — $\left(\dfrac{S}{N}\right)_o = \dfrac{S_i}{N_o f_m}$ $\left(\dfrac{S}{N}\right)_i > 10$	$\left(\dfrac{S}{N}\right)_o = \left(\dfrac{S}{N}\right)_i$
WBFM	$B = 2m_f f_m$ $m_f > 10$	$\left(\dfrac{S}{N}\right)_o = 3m_f^3\left(\dfrac{S}{N}\right)_i$ $\left(\dfrac{S}{N}\right)_i > 10$	$\left(\dfrac{S}{N}\right)_o = \dfrac{3}{2}m_f^2\left(\dfrac{S_i}{N_o f_m}\right)$	$\left(\dfrac{S}{N}\right)_o = \left(\dfrac{S}{N}\right)_i^{2m_f}$

lator. It should be noted here that in the case of AM-DSB, the term S_i does not include the carrier power.

The following brief remarks will serve to highlight the results of Table 24-1.

1. The apparent improvement of 3 db of DSB-SC over SSB has already been discussed in Chap. 6. It should be noted, however, that the input-noise terms N_i are different for the two systems, the N_i for DSB-SC being double that for SSB. The relative rating for the systems is clearly illustrated in the fourth column, where the input noise power N_i is taken to be identical for all systems, namely $N_i = N_o f_m$. Using this approach, this column, which is more representative of the relative performance of the systems shown, indicates that both DSB-SC and SSB systems offer identical performance.

2. In the fourth column it is also shown that using the approach of expressing the input noise in terms of $N_i = N_o f_m$ only, the two systems AM-DSB-SC and AM-DSB are identical in their performance. However, as shown in Chap. 6, if we include the carrier power in the total signal power, the DSB-SC shows an advantage of a factor of 3 over the conventional AM-DSB system.

3. The problem of linear envelope detection for both high- and low-input signal-to-noise ratio was discussed in Chap. 6. In case of high-input signal-to-noise ratio it was shown that an envelope detector is as efficient as the coherent product demodulator. However, in case of low $(S/N)_i$, Eq. (6-83) is applicable and is presented in Table 24-1.

4. The results for FM systems are approximate and are valid only for wideband FM systems operating above threshold. The FM improvement formula listed in the table can be derived from the conventional form as follows. As shown in Chap. 7, for large modulation index m_f, the IF bandwidth required for such a system is $B \doteq 2\Delta F$, where ΔF equals the maximum frequency deviation of the modulated carrier. Using the FM improvement formula of Eq. (14-46), we obtain

$$\left(\frac{S}{N}\right)_o = 3m_f^2\left(\frac{B}{2f_m}\right)\left(\frac{S}{N}\right)_i \doteq 3m_f^3\left(\frac{S}{N}\right)_i \qquad (24-3)$$

where $N_i = N_o B$, the total IF noise power and $m_f = \Delta F/f_m$.

2. *Absolute Comparison between Analogue Modulation Systems.* As stated above, the theoretical information capacity of a communication system as formulated by Shannon can be used to make a com-

parison between modulation systems against an absolute standard. Let C_c denote the information capacity of the channel, and C_s that of the message bandwidth at the demodulator output. Assuming an ideal system of modulation and demodulation, the information rates are the same at both input and output points of the receiver, and consequently we have the relations

$$C_c = B \log_2 \left[1 + \left(\frac{S}{N} \right)_i \right] \quad \text{bits/sec} \qquad (24\text{-}4)$$

$$\text{and} \quad C_s = f_m \log_2 \left[1 + \left(\frac{S}{N} \right)_o \right] \quad \text{bits/sec} \qquad (24\text{-}5)$$

$$\text{where} \quad C_c = C_s \qquad (24\text{-}6)$$

$$\text{Hence,} \quad \left(\frac{S}{N} \right)_o = \left[1 + \left(\frac{S}{N} \right)_i \right]^{B/f_m} - 1 \doteq \left(\frac{S}{N} \right)_i^{B/f_m} \qquad (24\text{-}7)$$

for $(S/N)_i \gg 1$, which is shown in the last column of Table 24-1, for the various modulation systems. Equation (24-7) can be put in the more convenient form

$$\left(\frac{S}{N} \right)_o = \left(1 + \frac{S_i}{N_o f_m} \frac{f_m}{B} \right)^{B/f_m} - 1 \qquad (24\text{-}8)$$

from which we obtain an upper bound for $(S/N)_o$ vs. $(S_i/N_o f_m)$ with parameter B/f_m. The results of Table 24-1 are plotted in Fig. 24-1, together with the upper bounds given by Eq. (24-8). Note the equivalence of the AM systems as previously discussed, as well as the limiting performance for (B/f_m) of unity, coinciding with the AM-SSB result. We also note from the figure that for small bandwidth expansions, FM is nearly ideal, but for bandwidth expansions exceeding 10:1, FM is less attractive.

24.2 Efficiency of Communication Systems [4,5]

In the last section we have assumed an ideal system of modulation and demodulation with no information loss in the receiver. With reference to Eq. (24-4) we note that the information capacity of a channel may be expressed in terms of two parameters only, namely, the channel bandwidth B and the input signal-to-noise ratio $(S/N)_i$. Similarly, as expressed by Eq. (24-5), the information capacity of the message bandwidth is expressed in terms of the information

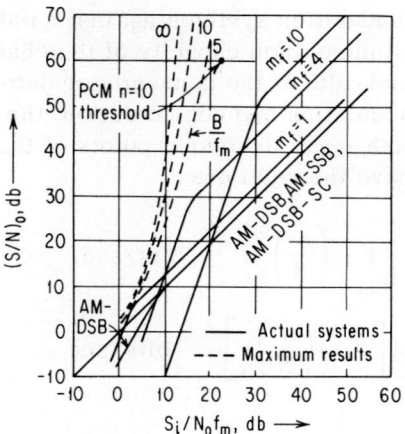

Fig. 24-1. Performance of analogue communication systems. (From Hancock,[2] Proc. Natl. Electron. Conf.)

bandwidth f_m and output signal-to-noise ratio $(S/N)_o$. The information bandwidth is dictated by the system requirements; consequently, we are left with the problem of maximizing the second system parameter, namely, the output signal-to-noise ratio $(S/N)_o$, by selecting the optimum modulation method.

The above idealization, however, does not apply to most practical systems since the information rates at the input and output points of the receiver are not equal, and, in general, C_S is less than C_C. But, by raising the power of the distant transmitter, the receiver signal is increased from S_i to S_i', and the resulting new information capacity C_S' of the message channel may thus be made equal to C_C.

As stated earlier, the efficiency of a communication system may be defined in a number of ways; both the methods of modulation and the type of message processing used determine the system efficiency. One method of comparing actual and ideal systems is to express each in terms of the information rates per cycle per second of RF bandwidth occupied. This is shown in Fig. 24-2, where C/B ratios are plotted for an ideal system and for actual systems vs. receiver-input signal-to-noise ratio $(S/N)_i$ in RF bandwidth B. It should be noted here that wideband FM performs very poorly compared to PCM.

Another method of assessing modulation-system efficiency is by means of a factor β_0 defined by

$$\beta_0 = \frac{\text{transmitter power required in an actual system}}{\text{transmitter power required in an ideal system}} = \frac{S_i}{S_i'}$$

$$(24-9)$$

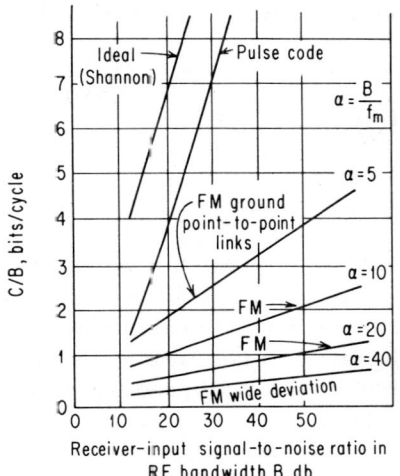

Fig. 24-2. Information capacities of several communication systems. (From Wright,[5] courtesy of Elsevier Publishing Co.)

Rewriting Eq. (24-7) for the actual or nonideal case, we obtain

$$\left(\frac{S'}{N}\right)_o = \left[1 + \left(\frac{S}{N}\right)_i\right]^\alpha - 1 \tag{24-10}$$

where $\alpha = B/f_m$ and S'_o corresponds to the increased input signal S'_i required for an equivalent nonideal system. Since for a given modulation system the relationship between the predetector and postdetector signal-to-noise ratios in the receiver is known, it is seen from Eqs. (24-9) and (24-10) that, in general, β_o is a function of $(S/N)_i$, the input signal-to-noise ratio, as well as being a function of the modulation-demodulation process.

In the following we shall compare the efficiencies of the various modulation systems in terms of their β_o values.

1. *Efficiency of Single-sideband (SSB) Modulation System.* In this system the channel bandwidth B equals the information bandwidth f_m, so that

$$\alpha = \frac{B}{f_m} = 1$$

Also, $\left(\dfrac{S'}{N}\right)_o = k_d \left(\dfrac{S'}{N}\right)_i$

where k_d is the detector efficiency. From Eqs. (24-9) and (24-10) we obtain the result

$$\frac{k_d}{\beta_o}\left(\frac{S}{N}\right)_i = \left[1 + \left(\frac{S}{N}\right)_i\right] - 1 = \left(\frac{S}{N}\right)_i \tag{24-11}$$

or $\quad \beta_o = k_d$

Thus, we have shown that in this simple modulation system the efficiency as defined by β_o is equal to that of the detector in the receiver. Since the factor k_d may be made to approach unity, even for low values of input signal-to-noise ratio and coherent detection, a SSB system will be regarded for all practical purposes as having an efficiency $\beta_o = 1$, as shown in (a) Fig. 24-3.

It should be noted here that in practice the $(S/N)_o$ requirement may be impractical to meet in a SSB system, and consequently more complex modulation methods must be resorted to, which have the property of exchanging bandwidth for $(S/N)_o$. In these wideband systems, however, the communication efficiencies are smaller than that of the SSB case, as shown in Fig. 24-3.

2. *Efficiency of Frequency-modulation (FM) Systems.* In this case the output-input signal-to-noise relation is given by the well-known FM improvement equation.

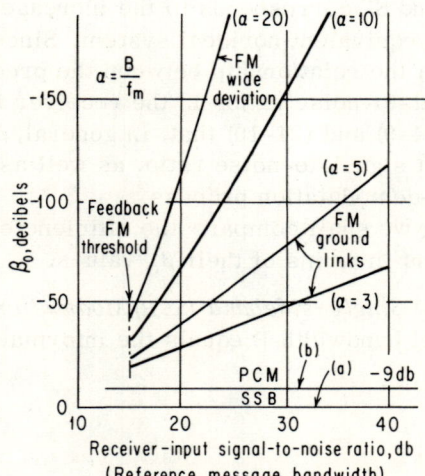

Fig. 24-3. Efficiencies of various modulation systems. (From Wright and Jolliffe,[4] J. Brit. IRE.)

$$\left(\frac{S}{N}\right)_o = 3m_f^2 \frac{S_i}{2N_o f_m} = \frac{3}{2}\left(\frac{S}{N}\right)_i \alpha \left(\frac{\Delta F}{f_m}\right)^2 \tag{24-12}$$

where $N_i = N_o B$. In a FM system the minimum bandwidth B as given by Eq. (7-106) is

$$B = 2(\Delta F + f_m)$$

or $\quad \dfrac{\Delta F}{f_m} = \dfrac{B}{2f_m} - 1 = \dfrac{\alpha}{2} - 1$ \hfill (24-13)

By substitution in Eq. (24-12) we have

$$\left(\frac{S'}{N}\right)_o = \frac{3}{2}\left(\frac{S'}{N}\right)_i \alpha \left(\frac{\alpha}{2} - 1\right)^2 \tag{24-14}$$

and using Eqs. (24-9) and (24-10), we obtain

$$\frac{3\alpha}{2}\left(\frac{\alpha}{2} - 1\right)^2 \frac{1}{\beta_o}\left(\frac{S}{N}\right)_i = \left(1 + \frac{S_i}{N_i}\right)^\alpha - 1$$

Hence, $\quad \beta_o = \dfrac{(3\alpha/2)(\alpha/2 - 1)^2 (S/N)_i}{[1 + (S/N)_i]^\alpha - 1}$

$$\tag{24-15}$$

where $(S/N)_i$ is the input signal-to-noise ratio referred to an RF bandwidth of $2f_m$.

The efficiency, as might be expected, varies with α, the factor which governs the exchange of bandwidth for output signal-to-noise ratio. As seen from Fig. 24-3, the larger the value of α, i.e., the greater the $(S/N)_o$ improvement, the worse the efficiency β_o becomes. This is clearly seen in case of wideband FM systems in which a typical factor of $\beta_o = 20$ is illustrated. For an input signal-to-noise ratio of 20 db, we note that the β_o value is -115 db, or the equivalent FM system would require a transmitter power level of 115 db above that of the transmitter in an ideal system, working at the same information rate within the same radio-frequency bandwidth.

3. *Efficiency of a Pulse-code-modulation System* (PCM). The problem of comparing the PCM system capacity with the ideal system capacity was discussed in Chap. 20 where it was shown that $(S'/S)_i \doteq 8$ for an average error rate of 10^6, or

$$\beta_0 \equiv \left(\frac{S'}{S}\right)_i \doteq 8$$

which represents about 9 db loss in input signal power in comparison with the ideal system for the same channel capacity, as shown in (b) Fig. 24-3.

24.3 Absolute Comparison between Digital Modulation Systems [2,6-8]

As stated in the introduction to this chapter, in case of digital systems, a convenient absolute standard for comparing the different modulation systems is obtained through the concept of signal power required per bit per second in the presence of a given noise spectral density. A figure of merit for comparing modulation systems based on this concept leads to the emphasis on the minimization of the total transmitted energy required to transmit a given amount of data, which is of increasing importance in satellite communication systems. To this end, we return to Shannon's formulation of channel capacity,

$$C_c = B \log_2 \left[1 + \left(\frac{S}{N}\right)_i \right] = B \log_2 \left(1 + \frac{S_i}{N_0 B} \right) \qquad (24\text{-}16)$$

or

$$\frac{C_c}{B} = \log_2 \left(1 + \beta \, \frac{C_c}{B} \right) \qquad (24\text{-}17)$$

where

$$\beta = \frac{S_i}{N_0 C_c} \qquad (24\text{-}18)$$

The parameter β provides a useful criterion for comparing the efficiency of digital modulation systems; it can be interpreted as the minimum input signal power required per bit of information per second in the presence of a given uniform gaussian noise-power spectral density of N_0 watts/cycle for a given error rate. The parameter β can also be written in the form

$$\beta = \frac{E_{min}}{N_0} \qquad (24\text{-}19)$$

where E_{min} is the minimum received signal energy required per bit of information for a given error rate. Equation (24-17) can be written in the form

$$\beta = \frac{B}{C_c} \, (2^{C_c/B} - 1) = \gamma (2^{1/\gamma} - 1) \qquad (24\text{-}20)$$

where $\gamma = B/C_c$, channel bandwidth/(bit)(sec). From Eq. (24-20) it follows that the lower bound on β can be obtained by letting $\gamma \to \infty$, which corresponds to systems of unlimited bandwidth. Thus,

$$\lim_{\gamma \to \infty} \beta = \log_e 2 = 0.693 \qquad (24-21)$$

In order to facilitate the calculations of communication efficiency of the various systems, the parameter β can be written in the form

$$\beta = \frac{S_i}{N_0 C_c} = \frac{S_i}{N_0 B} \frac{B}{C_c} = \left(\frac{S}{N}\right)_i \gamma \qquad (24-22)$$

It follows therefore that for any given digital system, β may be computed from the minimum-input signal-to-noise ratio required for a given error rate, and the ratio of bandwidth to information rate.

As a typical application of the β factor to a modulation system we consider a PCM system where successive 0's and 1's are transmitted by means of phase modulation of a sinusoidal carrier. The detector output is $+V_0/2$ when a 0 is sent under noise-free conditions, $-V_0/2$ for a 1, and is an odd function of signal plus noise between these limits when the maximum-likelihood threshold is set at 0. Assuming band-limited gaussian noise in the output, the probability of error P_e of mistaking a binary 1 for a 0 and vice versa is given by Eq. (21-31):

$$P_e = \frac{1}{2}\left(1 - \text{erf} \frac{V_0}{2\sigma}\right) = \frac{1}{\sqrt{2\pi}} \int_{-\infty}^{-z} e^{-t^2/2} \, dt \qquad (24-23)$$

where $\quad \text{erf } z = \frac{1}{\sqrt{2\pi}} \int_{-z}^{z} e^{-t^2/2} \, dt$

and $z = V_0/2\sigma$ = output signal-to-noise voltage ratio (rms).

The receiver input signal-to-noise power ratio required to produce a given output signal-to-noise power ratio $(S/N)_0 = z^2$ varies with the modulation and detection system used. In this case which is equivalent to suppressed-carrier double-sideband modulation, $B = 2f_m$, and using coherent detection, we have $(S/N)_i = \frac{1}{2}(S/N)_0 = z^2/2$. From the sampling theorem it follows that the sampling frequency $f_s \geq 2f_m$, and consequently the information rate or channel capacity is equal to

$$C_c = 2 \frac{B}{2} = B$$

Hence, assuming that the error probability is small, we obtain

$$\gamma = \frac{B}{C_c} = 1$$

$$\beta = \left(\frac{S}{N} \right)_i \gamma = \frac{z^2}{2} \gamma = \frac{z^2}{2} \qquad\qquad (24\text{-}24)$$

For example, if $P_e = 10^{-6}$, $z = 4.753$ and $\beta = 11.30$.

It is of interest to note here that for PCM–DSB–SC coherent detection the β factor is independent of the channel bandwidth B. However, for a noncoherent PCM/AM system β is significantly larger. Sanders[6] has shown that the transmitted power required to communicate at a given transmission rate is about 3 db greater than that for a coherent PCM system, using the same system parameters.

Figure 24-4 illustrates the β factor for several digital modulation systems. We observe here that the PSK system is the most efficient and has a factor ranging from 5 to 10 in the region of low error rates. If we recall that the lower bound on β is 0.693, it follows that even the PSK system falls short of the optimum system by 10 to 13 db. Digital systems can be made more efficient by increasing the number of quantizing levels. Sanders[6] has considered the β factors as a function of M, the number of quantizing levels, assuming orthogonal signals for the transmission of the quantized samples. His results are plotted in Fig. 24-5 for an error rate of 10^{-3} and 10^{-6}; as M $\rightarrow \infty$, the curves will approach the lower bound of β.

As stated in the introduction, the purpose of this chapter was to establish meaningful criteria for system performance of analogue and digital communication systems. The selection of the most suitable modulation system for space communication, where power is at a premium, is of particular interest to the system engineer. This is evidenced by the numerous investigators working on this problem in recent years and the frequent publications describing their results. Our analysis is by no means complete; only fundamental concepts were introduced which might be considered as introductory to the general problem of communication efficiency.

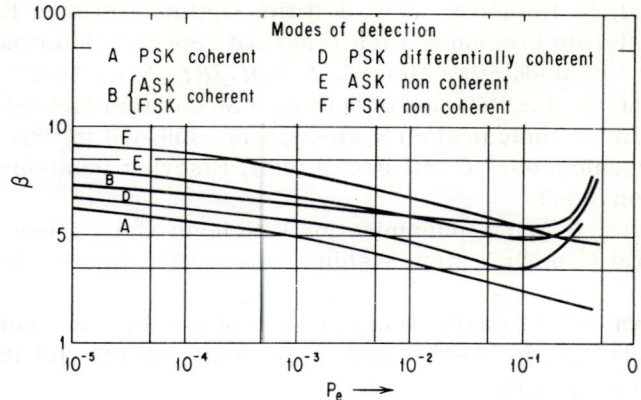

Fig. 24-4. Comparison of binary systems on a β factor basis. Modes of detection: A, PSK coherent; B, ASK coherent and FSK coherent; D, PSK differentially coherent; E, ASK noncoherent; F, FSK noncoherent. (From Hancock,[2] Proc. Natl. Electron. Conf.)

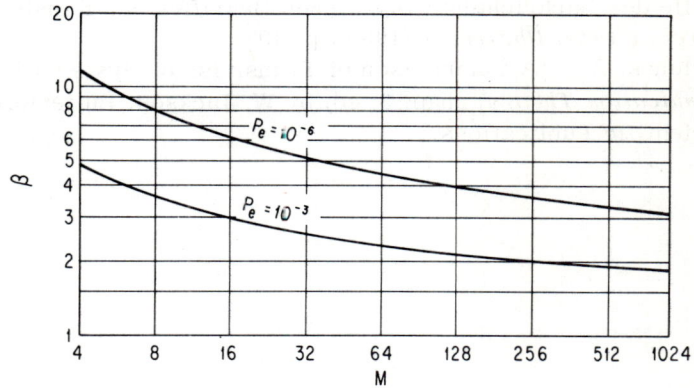

Fig. 24-5. Plot of β factor as a function of the number of quantizing levels L. (From Hancock,[2] Proc. Natl. Electron. Conf.)

References

1. Hancock, J. C.: "An Introduction to the Principles of Communication Theory," McGraw-Hill Book Company, New York.
2. Hancock, J. C.: On Comparing the Modulation Systems, *Proc. Natl. Electron. Conf.*, 1962.
3. Page, R. M.: Comparison of Modulation Methods, *IRE Conv. Record*, 1953.

4. Wright, N. L., and S. A. W. Jolliffe: Optimum System Engineering for Satellite Communication Links with Special Reference to the Choice of Modulation Method, *J. Brit. IRE*, May, 1962.

5. Wright, W. L.: "Choice of Optimum Modulation Method in Active Satellite Communication Systems," pp. 409–446 in "Space Radio Communication," G. M. Brown (ed.), Elsevier Publishing Co., London, 1962.

6. Sanders, R. W.: Communication Efficiency Comparison of Several Communication Systems, *Proc. IRE*, vol. 48, April, 1960.

7. Viterbi, A.: Classification of Coherent Synchronous Sampled Data Telemetry System, *JPL Tech. Rept.* 32–123, Jet Propulsion Laboratory, Pasadena, Calif.

8. Dupraz, J.: Comparaison théorique de trois systèmes digitaux pour télémesures spatials, *L'Onde électrique*, March, 1963.

9. Stewart, J. A., and Huber, E. A.: Comparison of Modulation Methods for Multiple-access Synchronous Satellite Communication Systems, *Proc. IEE*, vol. III, no. 3, March, 1964.

10. Piloty, R.: Über Die Beurteilung der Modulationssysteme mit Hilfe des Nachrichtentheoretischen Begriffes der Kanalkapagität, *Arch. elektr. Übertr.* 4 (1950), p. 493.

11. Jelonek, Z.: "A Comparison of Transmission Systems," *Communication Theory*, chap. 3, ed. by W. Jackson, Butterworth Scientific Publications.

INDEX

AM (see Amplitude modulation)
AM reception, 212–218
  comparison with DSB, 215
  signal-to-noise ratio, for high $(S/N)_i$, 213
    for low $(S/N)_i$, 216
Amplitude, instantaneous, 238
Amplitude characteristic, network from
  transfer function, 64
Amplitude detector, linear, 228
  square-law, 233
Amplitude frequency distortion, 523
Amplitude modulation (AM), of carrier, 176
  definition of, 175
  digital data systems, 705–710
    (See also Amplitude-shift keying)
  modulation factor, 176
  modulation index, 176
  of nonperiodic signal, 177
  of periodic signal, 176
  piecewise linear modulator for, 189
  power of, average, 216
    peak, 216
  pulse (see Pulse-amplitude modulation)
  sidebands in, 177–178
  S/N improvement through de-emphasis,
    444–446
  S/N ratio, comparison with FM, 436
    using envelope detection, 226, 227
    using product demodulation, 224
  square-law modulator for, 188
  typical spectra, 177–178
Amplitude modulation detector, envelope, 212
  phasor diagram of, 213
Amplitude modulators, 187–192
  piecewise-linear, 189
    switching function of, 191
  square-law, 188
Amplitude-shift keying (ASK), 705–710
  coherent detection of, 705
  envelope detection of, 708
Amplitude spectrum, of AM signals, 176 177
  of complex wave, 11
  of FM signals, 249, 252, 254–260
  from Fourier coefficients, 17
  of sinusoidal wave, 11
Analog modulation systems, comparison of,
  738–741
Analytic signal, concept of, 193–200
  relation to Hilbert transform, 199
  sampling of, 202, 530
Angle modulation (see Exponential modulation; Frequency modulation)
Angular velocity, 238
Aperiodic signals, correlation functions of,
  58–61
Aperture effect, 524, 546
Armstrong's indirect FM modulator, 383
  distortion in, 385
  double-channel, 390
Asymmetrical sidebands, 261
Asymptotic error rate for phase detection,
  721
Asymptotic expansion, Stumper's, 310
Autocorrelation function, aperiodic signal, 59
  from energy density function, 59
  properties of, 60
  input-output relationship for linear sys-
    tems, 152
  of narrow-band noise process, 143

Autocorrelation function, of periodic time
  function, 28, 30
  from power spectrum, 28
  properties of, 28
  of random binary telegraph signal, 149
  of random function, 135
    with aperiodic component, 146
  of signal plus noise, 147
  of white gaussian noise, 145
Average power, in AM, 216
  in exponentially modulated signal, 250
  in multitone combined amplitude and fre-
    quency modulation, 266
  in SSB, 184
  and spectral density, 138
Averages, ensemble, 133
  as measure of correlation, 135
  statistical, 109–114
  time, 133

Balanced modulator, in Armstrong's phase-
    shift modulator, 383
  in DSB, 172
  in SSB, 181, 194, 195
  in vestigial-sideband system, 179, 180
Band-limited white noise, 139, 159–163
Bandpass signal, sampling of, 524
Bandwidth, exchange of S/N and, 4, 427
  in FM, 270
  improvement of S/N with, in FM systems,
    433–441
  in frequency-division multiplex system,
    441
  in wideband systems, 605, 738
  in multitone FM, 269
  in PCM systems, 626
  in PTM systems, 588
  and system capacity, 692
Baseband signals, 174, 441
Beat wave, 98
Bessel functions, of first kind, 74, 250
  Fourier series expansion of, 248
  plot of, 253
  zero order, asymptotic series of, 167
    and imaginary argument, 166
Binary coding, 622, 625
Binary digit, 594
Binary error probability, in ASK, 705–710
  in baseband detection, 710
  in coherent detection, 707
  in envelope detection, 708
Binary modulation, 702
Binary notation, 624
Bipolar pulse transmission, 628
Bipolar signal, 629
Bit (unit of information) defined, 594

Capacity (see Channel capacity)
Capture effect in FM, 351, 429
Capture ratio, 374
Carrier, in CW modulation, amplitude
    modulation, 12, 175, 245
  in exponential modulation, 245
Carrier delay, 99
Carrier keying, 703
Carson and Fry expansion, 274, 304, 305, 308
Cauchy principal value, 88
CCIR, 197
Central-limit theorem, 127

Central-limit theorem, examples of, 127, 130
Central moments, 111
Channel capacity, 598
  approximation to ideal system, 603
  defined, 600
  of DM system, 693
  as function of bandwidth, 602
  of noiseless channel, 598
  of noisy channel, 600, 607, 615
  of PCM system, 631
    comparison with ideal system, 632
Characteristic functions, 124–131
  of gaussian distributions, 128
  of joint probability density function, 124
  of probability density function, 124
  of random phase distributions, 129, 130
  of uniform distributions, 126
Code, binary, 622, 625
  ternary, 622
Coded transmission, 604
Coherent detection, 206
  in ASK systems, 705
  in FSK systems, 713
  in LM systems, 221
  in PSK systems, 719
Communication systems, efficiency of, 741–746
    FM, 744
    ideal, 741
    PCM, 745
    SSB, 743
  model of, 591, 701
    wideband, ideal, 604
Compandors, 647
Comparison of modulation systems, analog, 738–741
  digital, 746–749
Compatible single-sideband (CSSB), 196, 203
Compound-modulation systems, 6, 704
Conditional probability, 104
Confluent hypergeometric function, 230
Continuous spectrum, 31
Continuous-wave (CW) modulation, 3
Convolution integral, 79, 80
  application to FM distortion, 302
  and characteristic function, 125
  derivation of, 79
  examples of, 80, 81
  and transform of product, 37
Convolution theorem, 37
Correlation, measure of, 136
Correlation detection, 731
Correlation functions, of aperiodic signals, 50–61
  normalized, 159
  of periodic signals, 26–31
  of random signals, 135–152
    at output of linear system, 152–155
    telegraph, 149–152
  of signal plus noise, 147
Cosine time function, spectrum of, 52
Cross-correlation function, of aperiodic signals, 60
  of linearly transformed random functions, 155
  of periodic signals, 26
Cross-power density spectrum, periodic signals, 29
  random signals, 148
Cross-spectral density, spectrum of, 60

CW modulation, 3, 505
Cycle-counter discriminator, 420

Data transmission system, idealized model, 701
Decay function, exponential, 46
Decision levels in PSK system, 726, 727
Decision theory, statistical, 730
  error probabilities, coherent binary, 733
    differentially binary, 734
Decision threshold, 628, 704, 707, 709
Decoding, in PCM, 625
  in presence of noise, 606–614
    measure of information, 607
    self-information, 608
      conditional, 609
De-emphasis network, 445
Delay, carrier, 99
  envelope, 97
  group, 99
  phase, 97
Delay distortion, 97
Delay time of linear filter, 68
Delta function, (see Impulse function)
Delta modulation (DM), 5, 679–699
  comparison with PCM, 6, 696–699
    analysis, 696
    channel capacity, 697
    signal power, 698
  delta-sigma modulation, 690–693
    principle of operation, 692
    signal-to-noise ratio, 693
  high-information DM, 694–696
    principle of operation, 694
    signal-to-noise ratio, 688–690
    experimental results, 690
    quantization noise, 688–689
  theory of, 680–688
    basic circuit, 680
    double integration, 686
    single integration, 682
Demodulation (see Detection)
Density function, energy, 40
  probability (see Probability density functions)
  spectral (see Spectral density)
Detection of FM signals, 405–426
  conversion of FM to AM, 406–410
    balanced discriminator, 409
    off-resonance, 319
  cycle-counter discriminator, 420–422
    distortion of, 422
    waveforms of, 421
  line discriminator, 410–420
    distortion, by Fourier transform, 418
      quasi-stationary approach, 413
      steady-state approach, 414
    general theory of, 410
  microwave phase discriminator, 422–424
    distortion, 424
Detection of linear modulation, 206-235
  AM signals, 212–218
  DSB signals, 207–212
  envelope, 212
  SSB signals, 218–219
  synchronous, 221
Deterministic signals, 101
Deviation, frequency, 246, 269
  phase, 244
Digital data systems, 700–736

Digital data systems, combined phase- and
     amplitude-shift keying, 725–729
  comparison of performance, 746–749
  frequency-shift keying, 703, 710–716
  phase-shift keying, 703, 716–725
Digital errors, probability of, 733
Digital information, 701
Digital message representation, binary, 702
  in PCM, 622
Diode reactance modulator, 391
Direct FM, 391, 402
Dirichlet's conditions, 16
Distortion analysis using paired echoes, 72–76
Distortion of FM signals, (see Frequency-
     modulation distortion)
Distortionless transmission, 63–71
  of bandpass filter, 69
  ideal characteristics for, 71
  of low-pass filter, 66–69
Distribution function, 102–130
  of continuous random variable, 105
  cumulative, 119
  discrete, 103
  examples of, 117–124
  gaussian (see Gaussian distribution
     functions)
  joint, 106
  Poisson, 121–124
  probability density, 103, 105, 106
  Rayleigh, 120
  of sum of random variables, 124
  uniform, 117
Double-sideband baseband signal, spectrum
     of, 174
Double-sideband modulation systems, com-
     parison with AM, 215
  detection, 221
  phase and frequency synchronization of
     LO, 211
  reception, 207–212
  signal-to-noise improvement, 209
  spectral analysis, 208
  synchronization, 211
  transmission, 173–175
  waveforms, 175
Double-sideband suppressed carrier (DSB-
     SC), 173
Duality theorem, 36

Echo distortion, 289, 370–373
  resulting in phase distortion, 371
  (See also Paired echoes)
Encoding in PCM, 622
Energy density function (energy spectral
     density), 40
Ensemble, average, 135
  defined, 131
Ensemble method, 131, 132
Entropy, communication, 609
  conditional, 611
  defined, 595
  error, 615
  of gaussian distribution, 598
  received, 615
  transmitted, 615
Envelope, distortion by nonlinear phase
     characteristic, 98
  effect of large carrier component of sine
     wave plus noise, 167

Envelope, distortion by nonlinear phase
     characteristic, of symmetrical narrow-
     band gaussian noise, 163
Envelope delay, 97–99
Envelope detection, of AM, 224
  compared with product detection, 226
  of on-off carrier pulses, 708
Envelope detector, 212
  AC power output, 233
  mean values of output, 229
  noise of output, 230, 232
  $(S/N)_O$, 231
Envelope distribution function, 166, 229
Equivalent-noise bandwidth, defined, 140
  of single-tuned circuit, 140
  of staggered-tuned circuit, 140
Equivocation, 611, 615
  per code group, 613
Ergodic process, 134
Error, probability of, in coherent binary
     systems, 733
  in differentially coherent PSK systems,
     734
  in PCM, 627, 663
Error function, 120, 627
Expectation (see Statistical averages)
Exponential decay function, 46
  transform of, 47
Exponential modulation, 236–272
  application to complex signal, 240
  definitions, 237–240
  fundamental considerations, 242–245
  multitone combined amplitude and fre-
     quency modulation, 263–266
    multitone, 265
    two-tone, 264
  multitone frequency modulation, 254–259
    bandwidth, 260
    by periodic complex signal, 255
    by signals not harmonically related, 254
    by square wave, 257
  simultaneous amplitude, and angular
     modulation, 259, 263
    and frequency, 260
    and phase, 262
  single-tone, 243, 245–254
    frequency modulation, 243, 251, 252
    phase modulation, 243, 245–250
    spectral distribution of AM/FM, 266–
       268
  zero crossings, 239

Feedback, negative (see Negative feedback)
Feedback factor, 485
Filters, bandpass ideal, 69–71
  equivalent low-pass, 82
  impulse response, 84
  with nonlinear amplitude character-
     istic, 72
  with nonlinear phase characteristic, 74
    combined nonlinear amplitude and
       phase, 75
  response to modulated input, 84–87
  gaussian, 144, 145
  linear, and convolution integral, 79
  low-pass, ideal, 66–69
    response to, rectangular pulse, 67
       unit impulse, 78
  matched, 732

FM (*see* Frequency modulation)
Foster-Seeley discriminator, 317
Fourier integral (*see* Fourier transform)
Fourier series, 15–25
  amplitude spectrum, 17
  of envelope, 353
  exponential form, 17
  of instantaneous frequency, 354
  of periodic pulses, 22
    frequency spectrum, 23
  phase characteristic, 17
  of pulsed RF wave, 24
    frequency spectrum, 25
  real form, 15
  and sampling theorem, 508
  of sawtooth wave, 21
  of square wave, 19
  of triangular wave, 20
Fourier transform, 31–48
  amplitude spectrum, 31
  of autocorrelation function, 28
  condition for existence, 31
  conjugate, 32
  and continuous spectrum, 31–34
  of cosine time function, 52
  of even and odd functions, 33
  of exponential decay function, 46
  of gaussian distribution function, 47
  of impulse function, 49
  inverse, 31
  pair, 32
  of periodic function, 53
  phase spectrum, 31
  properties, 34–40
    change of sign of variable, 34
    differentiation, 38
    energy integrals, Parseval's theorem, 40
    integration, 38
    interchange and functions, duality
      theorem, 36
    linear addition, superposition theorem,
      34
    multiplication, 39
    transform of product, 36
    translation of variable, shift theorem, 34
  of pulse train, 54
  of pulsed RF signal, 43
  of rectangular frequency pulse, 42
  of rectangular signal pulse, 41
  of signum function, 56
  of sine time function, 52
  of step function, 44
  symmetrical form of, 32
Frequency characteristic of linear system,
  64
Frequency compressive feedback, 481
Frequency detection (*see* Detection of FM
  signals)
Frequency deviation, 246
  and FM bandwidth, 269
Frequency discriminators (*see* Detection of
  FM signals)
Frequency division multiplexing (FDM) 2,
  441, 548
Frequency following (*see* Negative feedback)
Frequency modulation (FM), analysis by AM
  synthesis, 244, 247
  bandwidth consideration in, 270

Frequency modulation (FM), capture effect,
  429
  conversion to AM, 317
  de-emphasis and pre-emphasis networks,
    445
  detection of (*see* Detection of FM signals)
  generation of, 381–404
    direct, 391–402
      crystal oscillator (FMQ), 399
      diode reactance modulator, 391
      klystron tube modulator, 403
      reactance tube modulator, 395
      saturable reactor modulator, 392
      by varying C and L, 393
    indirect, Armstrong's double-channel
      modulator, 390
      Armstrong's phase-shift modulator,
        383, 385
      serrasoid modulator, 388
      wideband modulators, 402–403
  interference in (*see* Frequency-modulation,
    reception)
  modulation index, 242
  multitone (*see* Exponential modulation)
  narrow-band, 244
    compared to AM, 245
    generation by balanced modulation, 384
  noise in conventional receiver, 431
  noise reduction, 453
    de-emphasis networks for, 443
  reactance-tube modulator, 395
  sidebands in, 249
  $(S/N)_O$ improvement ratio, 433, 436
    in multiplex systems, 441
  spectrum using, sinusoidal modulation,
    249
    square-wave modulation, 257
  threshold effects, 429–432
  transient response (*see* Transient
    response)
Frequency-modulation distortion, asymptotic
  method, 301–324
  Carson and Fry expansion, 304
    Fourier integral, 308
  convolution integral approach, 302
  quasi-stationary approach, 312
  single-tuned circuit, 321
  Van der Pol's approximation, 323
  Van der Pol-Stumper's expansion, 310
  Fourier method, 273–298
    derivation of, 276–280
    effect of linear phase shift, 281
    exact solution, 292–298
    first-order distortion, extended, 286
      by method of Medhurst, 282
    small sinusoidal variation of transmis-
      sion characteristics, 288–292
Frequency-modulation receiver, decrease
  in signal output, 455
  design for interference rejection, 373–379
    cascaded narrow-band limiters, 376
    wideband approach, 374
Frequency-modulation reception, interference in, 350–380
    common and adjacent-channel, 357–363
    both signals modulated, 361
    both signals unmodulated, 360
    general problem, 357
    two unmodulated carriers, 352–357

Frequency-modulation reception, two un-
    modulated carriers, Fourier series
    analysis, 353, 354
  multipath transmission, 363–370
    general distortion formula, 364
    phase distortion in feeder lines, 371
    phase error, 368
Frequency modulators (*see* Frequency mod-
    ulation, generation of)
Frequency pulse, 42
  transform of, 42
Frequency-shift keying, 703, 710–716
  coherent detection, 713
  noncoherent detection, 712
  probability of error, 714
Frequency spectrum, of AM signals, 177, 178
  and complex Fourier coefficients, 17
  of complex wave, 11
  continuous, 31
  discrete, 15
  of FM signals (*see* Frequency modulation)
  of periodic functions, 54
  of periodic pulse train, 23
  of pulsed RF wave, 25
  of sinusoidal waves, 11

Gaussian distribution functions, 47, 118–120
  characteristic function of, 128
  Fourier transform of, 47
  frequency spectrum of, 48
  of noise, 162
  for one variable, 118
  for two variables, 158
Gaussian filter, bandpass, 145
  low-pass, 144
Gaussian noise, 116
  narrow-band, 143
  at output of square-law device, 116
  resolution into in-phase and quadrature
    components, 161
  white, 162
Gaussian power spectrum, 144
  autocorrelation function of, 145
Gaussian random process, 157–170
  of narrow-band noise, 159–163
  of sine wave plus narrow-band noise, 164–
    170
Gaussian random variables, distributions of,
    127
Group delay, 99
Group velocity, 98

Half-power bandwidth, 140
Harmonic distortion (*see* Frequency-modu-
    lation distortion)
Harmonics in Fourier series, 15
Hilbert transforms, 87–92
  applied to SSB, 202–204
  properties of, 200–202
Hybrid-modulated wave, in CSSB, 197
  in FM, 277, 302

Ideal filters (*see* Filters, bandpass ideal)
Ideal system, approximation to, 603–606
IF, noise spectrum of, 222
Impulse function, defined, 48–53
  examples of, 50–53
  integral definition for, 49

Impulse function, and network response,
    77
  periodic, 50
  in power spectral analysis, 146
  in sampling theory, 510
  spectrum of, 50, 51, 52
  transform of, 49
  and unit step, 57
Impulse response, of linear systems, 77–81
  of low-pass filter, 78
  and network transfer function, 77
  of RC network, 81
  resolution into even and odd components, 89
Independent events in statistics, 104
Indirect FM, 382–391
Information, bit as unit of, 594
  capacity of communication channel, 598–
    603
    discrete noiseless channel, 598
    discrete noisy channel, 600
    as function of bandwidth, 602
  measure of, 592–598
    defined, 594
    continuous messages, 596
    continuous sources, 614–617
    independent selections, 594
  mutual, average, 611
  rate of transmission, 600–613
Instantaneous amplitude, 238
Instantaneous frequency, 239, 240
  application to complex signal, 240
  using density of zero crossings, 239
Instantaneous phase, 238
Interference in FM (*see* Frequency-
  modulation reception)
Interpolation, 515
  filter, 512
ITT corporation, 620, 679

Klystron modulator, 403

Lagrange multipliers, 597, 644
Lagrange's formula, 463
Likelihood (*see* Probability, conditional)
Limiter, amplitude, 3
  in negative feedback systems, 464, 466
Line discriminator, 410
Line spectra, 53–55
Linear amplitude detector (*see* Envelope
  detector)
Linear modulation (LM), 171–173
  AM transmission, 175–178
    nonperiodic, 177
    periodic, 176
  comparison of systems, 219–221
  detection (*see* Detection of linear modu-
    lation)
  signal-to-noise ratio, 221
  SSB transmission, 180–187
    average-to-peak power ratio, 184–187
    square-wave signal, 182
  vestigial-sideband transmission, 178–180
    equivalent of, 180
Linear phase shift, and distortionless trans-
    mission, 68–71
  effect on FM signal, 281
  and signal time delay, 68
"Linear" rectifier in modulator circuit,
    189
Linear system, defined, 63

Linear system, differential equation of, 64
  distortionless signal transmission, 68
  equivalent noise bandwidth, 140
  FM signal transmission, asymptotic
      method, 301–324
    Fourier method, 273, 298
    quasi-stationary method, 275
    quasi-stationary response, 312
    response to AM-FM signal, 313–317
    transient response, 336
  impulse response of, 77
  input-output autocorrelation relationship,
      152
  response to random functions, 152–157
  system response of, 63–65
  transfer function of, 64
  low-pass filter, 66
Linear time-varying system as modulator,
    187
LM (see Linear modulation)
Logarithmic compression, 647
Low-pass filter (see Filters, low-pass)
Low-pass signal, sample of, 516

Mapping message to signal space, 1
Marconi FMQ transmitter, 401
Markov process, 697
Matched filters, 732
Maximum likelihood, 730
Mean, 110
Mean-square-error criterion, 731
Median, 119, 120
Messages, continuous, 2
    measure of information, 596
  discrete, 592
    measure of information, 594
Modulation, comparative analysis of
    systems, 737–750
  compound, 6, 704
  continuous-wave, 3
  defined, 1, 2
  evolution of, 3, 4
  methods of, 1–6
  (See also Amplitude modulation; Exponen-
      tial modulation; Frequency modulation;
      Linear modulation; other specific
      methods and systems)
Modulation index, of AM, 176
  of FM, 242, 251
    and bandwidth, 252
  of PM, 242, 245
Modulator, amplitude (see Amplitude
    modulators)
  balanced (see Balanced modulator)
  frequency (see Frequency modulation,
      generation of)
Moments, central, 111
  of product of random variables, 113
  of two random variables, 112
  (See also Statistical averages)
Multiplex systems, frequency-division
    (FDM), 2, 441, 548
  time-division (TDM) (see Time-division
      multiplexing)
Multiplication process, 171
Multiplication theorem in Fourier transform,
    39

Narrow-band FM (see Frequency modula-
    tion)

Narrow-band gaussian process (see
    Gaussian random process)
Negative feedback, in DME systems, 471–
    476
  system analysis of transponder, 472–476
    steady-state approach, 473
    transfer-function approach, 475
  in FM systems, 461–477
    FM receivers, 466–471
      elementary analysis, 466
      reduction in nonlinear distortion, 469
      simultaneous amplitude and frequency
          modulation, 467
    FM transmitters, 462–466
      general formula, 462
      with limiter, 464
Noise, autocorrelation of, 142
  equivalent-noise bandwidth, 140
  gaussian distribution of, 162
    (See also Gaussian noise)
  mathematical representation of, 157–170
    resolution into in-phase and quadrature
        components, 161
  power spectral density of narrow-band,
      438
  quantization (see Quantization noise)
  sine-wave representation of, 160
  spectrum of, FM receiver output, 446–459
    gaussian, IF, 440
    rectangular IF, 439
    unmodulated carrier, 453
    zero carrier, 449
  white (see White noise)
Noise-equivalent bandwidth, closed-loop,
    486
  of network, 139, 140
Noise-improvement systems, coded, 604
  uncoded, FM and PPM, 587
Nonlinear distortion in FM (see Frequency-
    modulation distortion)
Normal distribution (see Gaussian distri-
    bution functions)
Nyquist (samples) interval, 510

Optimum signal detectors, 732

Paired echoes, 72–76
  amplitude distortion, 72–73
  combined amplitude and phase, 75–76
  phase distortion, 74–75
Parseval's theorem, in Fourier series, 18
  in Fourier transforms, 40
Periodic functions, autocorrelation of, 28
  cross-correlation of, 26–27
  cross-powers spectrum of, 28
Periodic pulse train, Fourier series of, 22
  frequency spectrum of, 23
Phase-amplitude-shift keying, 725–729
Phase delay, 97–99
Phase deviation, 244
Phase discriminator, 422
Phase-locked FM demodulator, 496
Phase-modulated signals, transient response
    of linear networks, 343–348
Phase modulation (see Exponential modulation)
Phase-shift keying, 703, 716–725
  coherent phase detection, 719
  phase comparison detection, 722
Phasor diagram, of AM detector, 213, 217
  of carrier and noise, 225

Phasor diagram, of digital multiphase
    signals, 719
  of FM transients, 329
Piecewise-linear modulator, 189
Poisson distribution function, 121-124
Power spectrum, from autocorrelation func-
    tion, 28
  of narrow-band noise, 438
  of periodic functions, 28, 30
  of random process, 137
  spectral density of, 28
Pre-emphasis, in AM, 444
  in FM, 446
  network, 445
Probability, a posteriori, 606
  a priori, 606
  conditional, 104, 606
  discrete, defined, 104
    and message information content, 594
  joint, 104
    of two independent random variables, 105
  transition, in PCM, 655
Probability density functions, definition of,
    105
  of envelope of narrow-band noise plus
    sinusoid, 167
  examples of, 117-124
  joint, of two random functions, 162
  of phase of sine wave plus noise, 168
  transformation of, 114-117
Probability distribution function, 103, 106
Probability integral, 120, 168
Product detection, of AM, 223
  of DSB, 222
Product modulator (see Balanced modulator)
Pseudo-Fourier method in distortion
    analysis, 274
Pseudo-static method, 534
Pulse, double-polarity, 554, 629
  single-polarity, 554, 629
Pulse-amplitude modulation (PAM), 4, 543,
    553-567
  demodulation methods, 546, 556
  flat-topped modulation, 545
  interchannel cross-talk, 557-565
  signal-to-noise ratio, 565
  spectra, 553-556
  top modulation, 544
Pulse-code modulation (PCM), basic opera-
    tions, 620-625
  channel capacity, 631-633
  error probability in, 628, 663
  quantization noise, nonuniform levels, 641
    -651
    of compressed signals, 645
    formula, 645
    logarithmic compression, 647
  uniform levels, 633-641
    as function of step size, 634
    of sinusoidal signal, 637
  S/N improvement, 653-678
    effect of false pulse noise, 655-662
      calculation of noise power, 656
      $(S/N)_O$ in tandem links, 660
      transition probabilities, 655
    as function of $(C/N)_i$, 662-671
      error probabilities, 663, 665
      weighted code pulses, 670
    rate of transmission, 673-677
    repeater system, 671-672
  threshold effect, 625

Pulse-code modulation (PCM), trans-
    mission requirements, 626-631
  bandwidth, 626
  signal power, 627
  threshold power, 627
Pulse-duration modulation (PDM), 4, 506,
    567
  cross-talk, 576, 581
  demodulation methods, 572
  distortion, 574
  generation of, 570
  modulation methods, 568
  S/N improvement, 585
Pulse-frequency modulation (PFM), 506, 536
Pulse-phase modulation, 540
Pulse-position modulation (PPM), 4, 388,
    506, 567
  cross-talk, 579, 582
  derivation from PDM, 571
  S/N improvement, 586
  compared with FM, 587, 588
Pulse signal, rectangular, 40
  spectrum of, 41
Pulse-time modulation (PTM), 4, 535
Pulse-width modulation (PWM), 4, 506, 542,
    567
  double-edge, 542
  single-edge, 543

Quantization noise, in delta modulation, 688,
    689
  in pulse-code modulation
    (see Pulse-code modulation)
Quantization process, 5
Quasi-stationary approximation in FM, 275
Quasi-stationary response, 312

Random-fluctuation noise (see Noise)
Random process, 131-152
  autocorrelation of, 135-137
    and spectral density, 137-138
    with periodic component, 146
  cross-correlation of linear transformation,
    155-157
  gaussian (see Gaussian random process)
  power spectral density of, 137-139
  sample function of, 132
  stationary, 133
  statistical average of, 133
Random signal, theory of, 101-170
Random variables, continuous, 105-109
  average, product of, 113
  joint probability density function, 106-108
  probability density function, 105
  statistical average of, 109
  statistical independence of, 108
  discrete, 103-105
  conditional, probabilities, 104
  joint probability of, 104
  probability distribution function of, 103
  statistical average of, 109, 110
  statistical independence of, 105
  typical examples, 117-124
Rayleigh distribution function, 120
Reactance tube modulator, 395
RF signal pulse, 43
  spectrum of, 44

Sample function of random process, 132
Sample point, 102

Sample space, 102
Sampling, natural, 569
  uniform, 569
Sampling frequency, in DM, 682
  in PCM, 621
Sampling function, 515, 553
Sampling interval, 507
Sampling principle, 506–534
  of analytic signal, 530
  of bandpass function, 524
    minimum sampling frequency, 527
    reconstruction formula, 526
  first-order, 515
  higher-order, 527
  square-topped scanning, 521
    spectrum of, 524
  theorem, in frequency domain, 513
    in time domain, 507
  top sampling (exact scanning), 517
    spectrum of, 520
Sampling rate, 507
Saturable-reactor frequency modulator, 392
Serrasoid FM modulator, 388
Shannon's theorem, 602, 617
Shift theorem, 34
Sidebands, in AM, 177
  in FM, 249
Signal-to-noise ratio, in AM, 213, 224, 226,
    227, 232
  in delta-sigma modulation, 693
  in DM, 689
  in DSB, 215, 223
  in FDM, 442
  in FM, 427–460
  in PAM, 566
  in PCM, 665, 671
  in PPM, 587
Signal-suppression effect, 431, 455
Signal transmission, distortion analysis,
    72–76
  distortionless, 68–71
    of bandpass filter, 70–71
    of low-pass filter, 69
Signum function, 56
(Sin x)/x function, 23, 67
Sine integral, 68
Sine time function, spectrum of, 53
Single-sideband modulation (SSB), detection
    of, 218
  $(S/N)_O$, 218
  equivalent of, 203
  generation of, 192–196
    frequency-discrimination method, 192
    phase-shift method, 192–194
    third method, 194–196
  receiving system, 180
  spectrum of, 181
    using Hilbert transform, 202
    using square wave, 182
  transmission system, 180
    compatible transmission (CSSB), 196–
      197
    compatibility problem, 206
Singularity functions, 53–58
  signum function, 56
  step function, 56
    approximation of, 58
Sinusoid, autocorrelation function of, 30
  power spectrum of, 30
  spectrum presentation of, 11, 52

Sinusoidal modulation, amplitude, 12
  frequency, 251
  phase, 248
Spectral density, narrow-band white gaussian
    process, 161
  nonperiodic function, 40
  normalized, 159
  periodic function, 28
  random functions, 137
    with periodic component, 146
  random telegraph signal, 149
  related, to autocorrelation function, 141
    to Fourier transform, 40
  two-sided, 139
Spectrum, amplitude (see Amplitude
    spectrum)
  of complex wave, 11
  of continuous and nonperiodic functions,
    31–34
  of discrete and periodic functions, 15–25
  of single-tone AM signal, 13
Square-law detection, 233
Square-law modulator, 188
Square wave, autocorrelation function, 30
  frequency modulator, 257
  power spectrum of, 30
  signal, 19
SSB (see Single-sideband modulation)
Staircase analysis, 535
Standard deviation, 111
Stationary phase, 92–96
  application to, location of signal, 94
    spectral distribution of FM signal, 95
Statistical averages, 109–114
  of continuous random variable, 110
  of discrete random variable, 109
  of joint probability distribution, 111
  of product of random variables, 113
  of sum of random variables, 112
Statistical decision theory (see Decision
    theory)
Statistical independence, of continuous
    random variables, 108
  of discrete random variables, 104
  of signal and noise, 147
Statistical regularity, 102
Step function, 44, 56, 57
  approximation of, 58
  spectrum of, 44, 57
Stochastic process (see Random process)
Subcarriers, 2
Superposition of linear networks, 63
Superposition theorem in Fourier transform,
    34
Suppressed-carrier system (see Double-
    sideband suppressed carrier)
Switching function, 191
Synchronization, in DSB, 211
  in pulse-modulation systems, 551
Synchronous detection (see Coherent
    detection)
System capacity (see Channel capacity)
System function (see Transfer function)

Telegraph signal, random binary, 150
  autocorrelation function of, 151
  spectral density of, 151
Threshold, closed-loop, 488, 489
  open-loop, 488, 489
Threshold effects, in FM systems, 431

Threshold effects, in PCM systems, 627
  in PPM systems, 587
Threshold extension, 478–504
  concept of, 478
  transmitter power reduction, 487–495
    FCF receiver design, 487
      application to specific configuration,
        490
  using frequency-compressive feedback
    (FCF), 479–487
    carrier-to-noise input, 486
    theory of, 481
    threshold effects, 484
  using phase-lock loop demodulators, 495
    –503
    noise absent, 496
    noise present, 498
    threshold effects, 501
Time average, 135
Time delay of linear filter, 68
Time-division multiplexing (TDM), 2, 548–
  589
  cross-talk in, 575–585
    from high-frequency distortion, 582
    from low-frequency distortion, 577
  pulse-amplitude-modulation systems, 553
    –567
  pulse duration and pulse position, 567–575
    demodulation methods, 572
    modulation methods, 568
  pulse generation and synchronization, 551
    –553
  pulse-modulation systems, fundamentals,
    549–553
  signal-to-noise improvement, 585–588
  signal waveforms, 551
Time function, periodic, 8
  random, 9
  transient, 9
Time modulation, 535
Transfer function, amplitude characteristic
  of, 64
  of closed loop, 483
  derivation from differential equation, 65
  Fourier transform of, 77
  of open loop, 485
  phase characteristic, 64
  relationship between frequency character-
    istics of, 87
Transform properties, Fourier (see Fourier
  transform, properties)

Transient response, in FM, 325–348
  of linear networks, to FM signals, 336–343
    to PM signals, 343–348
  theoretical model of, 326
    experimental results, 324
Transmission rate, 600, 614
Transponder system, analysis of, 472
Travis discriminator, 317, 409
Triangular wave, 20

Unit impulse function (see Impulse function)
Unit step function (see Step function)

Van der Pol's equation, 323
Van der Pol-Stomper's expansion, 304, 305,
  310
Variance, 111
  of random process, 137
  of sum of random variables, 113
Vector diagrams, amplitude-modulated
  carrier, 12, 248
  fixed reference, 13
  general RF signal, 13
  in multipath interference, 365
  rotating reference, 14
    of modulated carrier, 14, 15
    of unmodulated carriers, 352
Vectors, addition of, 241
  conjugate, 10
  rotating, 9
Vestigial-sideband modulation, 178–180
  derivation from SSB, 179
  transmission system, 180
  equivalent of, 179, 180

Weighting function, 63
White noise, 139, 140
  autocorrelation function, of narrow-band
    noise process, 143
    at output of low-pass filter, 142
  band-limited, 139, 140, 143, 159–163
  defined, 139
  gaussian, 144–145
  spectral density, two-sided, 139
Wideband PM, 248
Wiener-Khinchine theorem, 141–142
  applied, to band-limited white noise, 142
    to gaussian noise spectrum, 144

Zero crossings, density of, 239